PE CIVIL
STUDY GUIDE
17TH EDITION

MICHAEL R. LINDEBURG, PE

PPI2PASS.COM
A **KAPLAN** COMPANY

Report Errors for This Book

PPI is grateful to every reader who notifies us of a possible error. Your feedback allows us to improve the quality and accuracy of our products. Report errata at **ppi2pass.com**.

Digital Book Notice

All digital content, regardless of delivery method, is protected by U.S. copyright laws. Access to digital content is limited to the original user/assignee and is non-transferable. PPI may, at its option, revoke access or pursue damages if a user violates copyright law or PPI's end-user license agreement.

PE CIVIL STUDY GUIDE
Seventeenth Edition

Current release of this edition: 2

Release History

date	edition number	revision number	update
Aug 2022	17	1	New edition.
Dec 2022	17	2	Minor corrections.

© 2022 Kaplan, Inc. All rights reserved.

All content is copyrighted by Kaplan, Inc. No part, either text or image, may be used for any purpose other than personal use. Reproduction, modification, storage in a retrieval system or retransmission, in any form or by any means, electronic, mechanical, or otherwise, for reasons other than personal use, without prior written permission from the publisher is strictly prohibited. For written permission, contact permissions@ppi2pass.com.

Printed in the United States of America.

PPI
ppi2pass.com

ISBN: 978-1-59126-877-2

Topics

Topic I: Breadth
Topic II: Construction
Topic III: Geotechnical
Topic IV: Structural
Topic V: Transportation
Topic VI: Water Resources and Environmental

Table of Contents

Preface vii

Acknowledgments ix

How to Use This Book xi

Codes and References Used to Prepare This Book xiii

Topic I: Breadth

Project Planning 1-1
Means and Methods 2-1
Soil Mechanics 3-1
Structural Mechanics 4-1
Hydraulics and Hydrology 5-1
Geometrics 6-1
Materials 7-1
Site Development 8-1
Engineering Economics 9-1

Topic II: Construction

Earthwork Construction and Layout 10-1
Estimating Quantities and Costs 11-1
Construction Operations and Methods 12-1
Scheduling 13-1
Material Quality Control and Production 14-1
Temporary Structures 15-1
Health and Safety 16-1

Topic III: Geotechnical

Site Characterization 17-1
Soil Mechanics, Laboratory Testing, and Analysis 18-1
Field Materials Testing, Methods, and Safety 19-1
Earthquake Engineering and Dynamic Loads 20-1
Earth Structures 21-1
Groundwater and Seepage 22-1
Problematic Soil and Rock Conditions 23-1
Earth Retaining Structures (ASD or LRFD) 24-1
Shallow Foundations (ASD or LRFD) 25-1
Deep Foundations (ASD or LRFD) 26-1

Topic IV: Structural

Analysis of Structures: Loads and Load Applications 27-1
Analysis of Structures: Forces and Load Effects 28-1
Design and Details of Structures: Concrete 29-1
Design and Details of Structures: Steel 30-1
Design and Details of Structures: Timber 31-1
Design and Details of Structures: Masonry 32-1
Codes, Standards, and Guidance Documents 33-1
Temporary Structures and Other Topics 34-1

Topic V: Transportation

Traffic Engineering 35-1
Horizontal Design 36-1
Vertical Design 37-1
Intersection Geometry 38-1
Roadside and Cross-Section Design 39-1
Signal Design 40-1
Traffic Control Design 41-1
Geotechnical and Pavement 42-1
Drainage 43-1
Alternatives Analysis 44-1

Topic VI: Water Resources and Environmental

Analysis and Design 45-1
Hydraulics—Closed Conduit 46-1
Hydraulics—Open Channel 47-1
Hydrology 48-1
Groundwater and Wells 49-1
Wastewater Collection and Treatment 50-1
Water Quality 51-1
Drinking Water Distribution and Treatment 52-1
Engineering Economics Analysis 53-1

Index I-1

Preface

In 2022, the National Council of Examiners for Engineering and Surveying (NCEES) changed the Principles and Practice of Engineering (PE) Civil licensing exams from a pencil-and-paper format to a computer-based format.

The new format for the exam is not in itself a big change. Instead of using a pencil to fill in circles on an answer sheet, you now record your answers by clicking on a computer screen. The kinds of questions you are asked on the exam and the knowledge areas that they cover have not changed much.

NCEES introduced another change at the same time, however, and if you are preparing to take the exam, this change has enormous consequences for you. It used to be that, for the pencil-and-paper exam, you would bring your own reference materials to the exam with you. That meant that you could use the *PE Civil Reference Manual* while preparing for the exam and get to know it well—highlight things in the text, write notes in the margins, even glue tabs to its pages to make it easy for you to find the most important equations, tables, and graphs quickly—and then on the big day, you could bring it right into the exam room with you along with your other references.

For the computer-based exam, though, you must leave your books at home—all of them. Instead, you are given onscreen access to a searchable electronic copy of the *NCEES PE Civil Reference Handbook* (*NCEES Handbook*). You can—and you should—download a PDF copy to use and annotate while you are studying, but you can't have those notes with you during the exam itself. An unmarked, onscreen copy of the *NCEES Handbook*, along with onscreen copies of selected codes and standards, will be the only references available to you.

This drastically changes how you must study for the exam. It is no longer enough to know how to solve the problems on the exam efficiently. Now you must know how to solve them efficiently using only the equations and data you can find in one specific source, the *NCEES Handbook*.

But while the *NCEES Handbook* contains hundreds of pages of equations, tables, and figures, it doesn't say much about how to use them. There's still a need for a book like PPI's *PE Civil Reference Manual* to help you with your review.

That is the reason for this new book. The *PE Civil Study Guide* shows the connections between the *NCEES Handbook*, the *PE Civil Reference Manual*, and PPI's depth manuals for the PE Civil disciplines. In addition to clear and concise explanations of the most important equations and terms in the *NCEES Handbook*, the *PE Civil Study Guide* contains lists of key concepts to know, important sections from codes and standards to review, exam tips to help you solve problems efficiently, and pointers to longer, in-depth coverage of exam topics in other PPI books. All this is organized according to the exam's own specifications, making it easy to find what you need.

Future editions of this book will be very much shaped by what you and others want to see in it. PPI is braced for the influx of comments and suggestions from those readers who (1) want more topics or (2) want more detail in existing topics. We appreciate suggestions for improvement, additional questions, and recommendations for expansion so that new editions will better meet the needs of future examinees. You can reach us through the Contact Us and Errata pages on our website at **ppi2pass.com**.

We wish you all the best as you study for your exam. With the *PE Civil Study Guide* as your road map and the *PE Civil Reference Manual* and *NCEES Handbook* as your primary sources, we are confident you will have a study experience that is challenging but targeted to ensure success.

Acknowledgments

The *PE Civil Study Guide* required the development of a completely new kind of book. Accordingly, it was a team effort to design and publish this book. PPI would like to thank the following subject matter experts for their contributions to the *PE Civil Study Guide*. Their extensive efforts and technical guidance were absolutely instrumental in the creation of this book.

Breadth: Clarissa Hageman, PE; Diana Galofaro, PE; Jon Kuchem, PE; John Connor Styles, PE

Construction: McKenzie Lehman, PE; Matt McBurnie, PE

Geotechnical: Bill Simpson, PE

Structural: Diana Galofaro, PE; Erika Weber, PE, SE; Kaylea Michaelis, PE; Yi-Chin Chen, PE

Transportation: Fatih Adam, PE; Liam Wallace, PE

Water and Environmental: Clarissa Hageman, PE; Michael Goldrich, PE

A special thanks to Erika Weber, PE, SE, for her guidance on organizing and developing the structural depth chapters in this book.

Many thanks to the following staff at PPI, a Kaplan Company, who assisted throughout the publication process.

Editorial: Bilal Baqai, Grace Wong, Michael Wordelman, Scott Marley, Susan Bedell, Tyler Hayes

Art, Cover, and Design: Tom Bergstrom, Sam Webster

Production and Project Management: Beth Christmas, Crystal Clifton, Damon Larson, Jeri Jump, Kimberly Burton-Weisman, Richard Iriye, Sean Woznicki, Stan Info Solutions

Content and Product: Anna Howland, Joe Konczynski, Megan Synnestvedt, Meghan Finley, Nicole Evans, Scott Rutherford

A current list of known errata for this book is maintained at the PPI website (**ppi2pass.com**), where you can also let PPI know of any errors you find. PPI greatly appreciates the time you take to help keep this book accurate and up to date.

Suggestions for improvement are appreciated. PPI hopes this book serves you well. Expect future updates as NCEES refines the computer-based exam.

How to Use This Book

Congratulations on your decision to earn your PE license! You have chosen a path that will open many doors to exciting and interesting opportunities.

The path to PE licensure is one of the most difficult and rewarding experiences in the life of an engineer. The adventure you've embarked on is an intense and rigorous one. It will be daunting; at times it will even feel impossible. However, once successful, you will stand out from your peers.

GOALS

The *PE Civil Study Guide* (and the range of PPI publications that support it) is designed to prepare you for success on the exam. It is a bridge between the *NCEES PE Civil Reference Handbook* (*NCEES Handbook*), the *PE Civil Reference Manual* (CERM), and the civil depth reference manuals. Using the *PE Civil Study Guide* and the in-depth information in CERM, you will gain a comprehensive understanding of the equations and concepts most likely to be encountered on the exam.

The primary focus of this book is the knowledge contained in the *NCEES Handbook*, which is the only reference you will have access to during the exam. However, past experience—and NCEES's own exam specifications—show that there are a multitude of key concepts required for success on the exam that are not included in the *NCEES Handbook*. With this in mind, the *PE Civil Study Guide* has been written with three main goals.

- **To ensure that you study exactly what you need to study—no more and no less.** In addition to providing detailed explanations of *NCEES Handbook* equations and content, the *PE Civil Study Guide* also covers key concepts not included in the *NCEES Handbook*.

- **To guide you in using the *NCEES Handbook* with maximum efficiency.** Because this is the only reference guide available to you during the exam (other than the NCEES-approved design standards), it is crucial that you are able to reference it with ease and confidence.

- **To show you where to look for more vital information.** In most cases, this is the *PE Civil Reference Manual*. In some cases, this is one of PPI's civil depth reference manuals (e.g., the *Structural Depth Reference Manual*).

BOOK STRUCTURE

The chapters of the *PE Civil Study Guide* align with the knowledge areas in the NCEES examination specifications for each of the five civil engineering disciplines, with a few notable exceptions. Go to NCEES.org and download a PDF of the exam specifications so you can reference it while you study.

The eight NCEES knowledge areas for the breadth exam specifications are covered in the first eight chapters of this book. Engineering economics is covered in the ninth chapter. These first nine chapters make up the breadth section of this book.

The depth sections follow the breadth section in alphabetical order by discipline (construction, geotechnical, structural, transportation, and water resources and environmental). With the exception of the structural section of this book, the chapters for each depth discipline align with a knowledge area in the NCEES exam specifications. For structural, we chose to cover knowledge areas 10.A (Materials and Material Properties) and 10.B (Component Design and Detailing) in four chapters rather than two, with each chapter focusing on one of the four materials in structural engineering (concrete, steel, timber, and masonry).

CHAPTER STRUCTURE

Each chapter is organized as follows.

- exam information about the knowledge area
- subtopics within the knowledge area

The organization of subtopics within each chapter follows the order of subtopics given by NCEES for the related knowledge area. The following diagram illustrates the subtopics for the second knowledge area (Chapter 2 in this book).

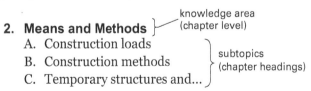

SUBTOPIC STRUCTURE

At the beginning of each subtopic, we have listed the key concepts you need to know to successfully answer exam questions relevant to the subtopic. When applicable, we have also included some or all of the following lists.

- sections in the *NCEES Handbook* where you can find equations, figures, tables, and concepts relevant to the subtopic
- sections in CERM where you can find information relevant to the subtopic (if only the chapter is listed, refer to all sections in that chapter)
- sections in the PPI depth manual where you can find information relevant to the subtopic (if only the chapter is listed, refer to all sections in that chapter)
- sections from NCEES-approved codes and standards relevant to the subtopic (NCEES-approved codes and standards are taken from the design standards published with each of the five civil exam specifications)

A collection of entries discussing relevant equations, figures, tables, and concepts follows the lists.

ENTRY STRUCTURE AND CONTENT

Each subtopic contains a collection of entries discussing relevant equations, figures, tables, and concepts you will need to know for the exam. An entry includes some or all of the following components.

- relevant *NCEES Handbook* equations
- relevant CERM and depth manual equations, figures, and tables
- citations for the *NCEES Handbook*, CERM, depth manual, and NCEES-approved codes and standards, when applicable
- useful commentary

Equations in blue are directly from the *NCEES Handbook*. For each subtopic, such equations are covered in the order they appear in the *NCEES Handbook*. For equations that are identical or nearly identical in presentation in both the *NCEES Handbook* and CERM or the related PPI depth reference manual, you will see a single blue equation with a reference to its location in each book. Some *NCEES Handbook* equations, however, differ substantially in presentation from the same equation in CERM or the related depth manual. For example, the *NCEES Handbook* equation may use different variables or subscripts, or it may be a rearranged version of the same equation in CERM or the related depth manual. In these instances, both the blue *NCEES Handbook* equation and the CERM or depth manual equation will be shown.

Citations in blue refer to relevant *NCEES Handbook* sections, tables, and figures. Relevant equations and section numbers from CERM and the depth manual are also cited, so you can easily find more in-depth information. You will also find citations to NCEES-approved codes and standards when applicable. Citations appear just below the entry titles and within the running text of the entries.

Perhaps the most valuable component of this book is the useful commentary included in each of the entries. Features of this useful commentary include

- explanations of terms used in equations
- explanations of underlying assumptions
- important equation derivations
- simplified versions of equations you can apply to the exam
- units used for certain variables
- useful information on commonly confused concepts
- valuable explanations of how concepts align between the *NCEES Handbook* and CERM or the related depth manual

In addition to the features listed, some entries conclude with helpful exam tips that will be priceless to you on exam day.

Codes and References Used to Prepare This Book

BREADTH

PE Civil Reference Manual, Michael R. Lindeburg, 16th ed., 2021, PPI, **ppi2pass.com**.

CONSTRUCTION

Code of Federal Regulations, Title 29: Labor, July 2020, U.S. Department of Labor, Washington, DC.

- Part 1903: Inspections, Citations, and Proposed Penalties
- Part 1904: Recording and Reporting Occupational Injuries and Illnesses
- Part 1926: Safety and Health Regulations for Construction

Construction Depth Reference Manual for the Civil PE Exam, Thomas M. Korman, 2nd ed., 2022, PPI, **ppi2pass.com**.

Design and Control of Concrete Mixtures, 17th ed., 2021, Portland Cement Association, Skokie, IL, www.cement.org.

Design Loads on Structures During Construction (ASCE/SEI 37), 2nd ed., 2015, American Society of Civil Engineers, Reston, VA, www.asce.org.

Formwork for Concrete (ACI SP-4), 8th ed., 2014, American Concrete Institute, Farmington Hills, MI, www.concrete.org.

Guide to Formwork for Concrete (ACI 347R), 2014, American Concrete Institute, Farmington Hills, MI, www.concrete.org.

Manual on Uniform Traffic Control Devices for Streets and Highways—Part 6 Temporary Traffic Control, 2009, including Revisions 1 and 2 dated May 2012, U.S. Department of Transportation, Federal Highway Administration, Washington, DC, www.fhwa.dot.gov.

PE Civil Reference Manual, Michael R. Lindeburg, 16th ed., 2021, PPI, **ppi2pass.com**.

Standard Practice for Bracing Masonry Walls Under Construction, 2012, Council for Masonry Wall Bracing, Mason Contractors Association of America, Lombard, IL, www.masoncontractors.org.

Steel Construction Manual, 14th ed., 2011, American Institute of Steel Construction, Chicago, IL, www.aisc.org.

GEOTECHNICAL

Code of Federal Regulations, Title 29, Part 1926, Safety and Health Regulations for Construction, U.S. Department of Labor, Washington, DC, July 2020.

- Subpart E, Personal Protective and Life Saving Equipment, Secs. 95–107
- Subpart M, Fall Protection, Secs. 500–503 with Apps. A–E
- Subpart P, Excavations, Secs. 650–652 with Apps. A–F
- Subpart CC, Cranes and Derricks in Construction, Secs. 1400–1442 with Apps. A–C

Design and Construction of Driven Pile Foundations: Geotechnical Engineering Circular No. 12 (FHWA NHI-16-009), Vols. I and II, 2016, U.S. Department of Transportation, Federal Highway Administration, Washington, DC, www.fhwa.dot.gov.

Drilled Shafts: Construction Procedures and Design Methods: Geotechnical Engineering Circular No. 10 (FHWA NHI-18-024), 2018, U.S. Department of Transportation, Federal Highway Administration, Washington, DC, www.fhwa.dot.gov.

Foundations & Earth Structures, Design Manual 7.02 (NAVFAC DM-7.02), 1986, U.S. Army Corps of Engineers, Naval Facilities Engineering Command, Alexandria, VA.

Geotechnical Aspects of Pavements (FHWA NHI-05-037), 2006, U.S. Department of Transportation, Federal Highway Administration, Washington, DC, www.fhwa.dot.gov.

Geotechnical Site Characterization: Geotechnical Engineering Circular No. 5 (FHWA NHI-16-072), 2017, U.S. Department of Transportation, Federal Highway Administration, Washington, DC, www.fhwa.dot.gov.

Guide to Design of Slabs-on-Ground (ACI 360R), 2010, American Concrete Institute, Farmington Hills, MI, www.concrete.org.

LRFD Seismic Analysis and Design of Transportation Geotechnical Features and Structural Foundations Reference Manual: Geotechnical Engineering Circular No. 3 (FHWA NHI-11-032), 2011, U.S. Department of Transportation, Federal Highway Administration, Washington, DC, www.fhwa.dot.gov.

Minimum Design Loads for Buildings and Other Structures (ASCE/SEI 7), 3rd printing, 2010, American Society of Civil Engineers, Reston, VA, www.asce.org.

PE Civil Reference Manual, Michael R. Lindeburg, 16th ed., 2021, PPI, **ppi2pass.com**.

Soils and Foundations Reference Manual (FHWA NHI-06-088, FHWA NHI-06-089), Vols. I and II, 2006, U.S. Department of Transportation, Federal Highway Administration, Washington, DC, www.fhwa.dot.gov.

Unified Facilities Criteria (UFC): Dewatering and Groundwater Control (UFC 3-220-05), 2004, U.S. Army Corps of Engineers, Naval Facilities Engineering Command, Air Force Civil Engineer Center, Washington DC.

Unified Facilities Criteria (UFC): Geotechnical Engineering (UFC 3-220-01), 2012, U.S. Army Corps of Engineers, Naval Facilities Engineering Command, Air Force Civil Engineer Center, Washington DC.

Unified Facilities Criteria (UFC): Soil Mechanics (UFC 3-220-10N), 2005, U.S. Army Corps of Engineers, Naval Facilities Engineering Command, Air Force Civil Engineer Center, Washington DC.

USACE Engineering and Design: Slope Stability (EM 1110-2-1902), 2003, U.S. Army Corps of Engineers, Washington, DC, www.publications.usace.army.mil.

STRUCTURAL

AASHTO LRFD Bridge Design Specifications, 7th ed., 2014, with 2016 interim revisions, American Association of State Highway & Transportation Officials, Washington, DC, www.transportation.org.

Building Code Requirements and Specification for Masonry Structures and companion commentaries (TMS 402/602), 2013, The Masonry Society, Longmont, CO, www.masonrysociety.org.

Building Code Requirements for Structural Concrete and *Commentary* (ACI 318 and ACI 318R), 2014, American Concrete Institute, Farmington Hills, MI, www.concrete.org.

Code of Federal Regulations, Title 29, U.S. Department of Labor, Washington, DC, July 2020.

- Part 1910, Occupational Safety and Health Standards
  - Subpart D, Walking-Working Surfaces, Secs. 28–30
  - Subpart F, Powered Platforms, Manlifts, and Vehicle-Mounted Work Platforms, Secs. 66–68, with Apps. A–D to Sec. 66
  - Subpart I, Personal Fall Protection Systems, Sec. 140

- Part 1926, Safety and Health Regulations for Construction
  - Subpart E, Personal Protective and Life Saving Equipment, Sec. 104
  - Subpart L, Scaffolding Specifications, App. A
  - Subpart M, Fall Protection, Secs. 500–503, with Apps. B–D
  - Subpart Q, Concrete and Masonry Construction, Secs. 703–706, with App. A
  - Subpart R, Steel Erection, Secs. 752, 754–758

International Building Code (IBC), 2015 edition (without supplements), International Code Council, Falls Church, VA, www.iccsafe.org.

Minimum Design Loads for Buildings and Other Structures (ASCE/SEI 7), 2010, American Society of Civil Engineers, Reston, VA, www.asce.org.

National Design Specification for Wood Construction with Commentary and *National Design Specification Supplement, Design Values for Wood Construction* (NDS), 2015, American Wood Council, Leesburg, VA, www.awc.org.

PCI Design Handbook: Precast and Prestressed Concrete, 7th ed., 2010, Precast/Prestressed Concrete Institute, Chicago, IL, www.pci.org.

PE Civil Reference Manual, Michael R. Lindeburg, 16th ed., 2021, PPI, **ppi2pass.com**.

Special Design Provisions for Wind and Seismic with Commentary, 2015, American Wood Council, Leesburg, VA, www.awc.org.

Steel Construction Manual, 14th ed., 2011, American Institute of Steel Construction, Chicago, IL, www.aisc.org.

Structural Depth Reference Manual for the PE Civil Exam, Alan Williams, 5th ed., 2022, PPI, **ppi2pass.com**.

TRANSPORTATION

Guide for Design of Pavement Structures, 4th ed., 1993 with 1998 supplement, American Association of State Highway & Transportation Officials, Washington, DC, www.transportation.org.

Guide for the Planning, Design, and Operation of Pedestrian Facilities, 1st ed., 2004, American Association of State Highway & Transportation Officials, Washington, DC, www.transportation.org.

Highway Capacity Manual (Vols. 1–4), 6th ed., 2016, Transportation Research Board, National Research Council, Washington, DC, www.mytrb.org.

Highway Safety Manual, 1st ed., 2010, with 2014 Supplement (including September 2010, February 2012, and March 2016 errata), American Association of State Highway & Transportation Officials, Washington, DC, www.transportation.org.

Hydraulic Design of Highway Culverts (FHWA HIF-12-026), Hydraulic Design Series No. 5, 3rd ed., April 2012, U.S. Department of Transportation, Federal Highway Administration, Washington, DC, www.fhwa.dot.gov.

Manual on Uniform Traffic Control Devices for Streets and Highways, 2009, including Revisions 1 and 2 dated May 2012, U.S. Department of Transportation, Federal Highway Administration, Washington, DC, www.mutcd.fhwa.dot.gov.

Mechanistic-Empirical Pavement Design Guide: A Manual of Practice, 2nd ed., August 2015, American Association of State Highway & Transportation Officials, Washington, DC, www.transportation.org.

PE Civil Reference Manual, Michael R. Lindeburg, 16th ed., 2021, PPI, **ppi2pass.com**.

A Policy on Geometric Design of Highways and Streets (the "Green Book"), 7th ed., 2018 (including October 2019 errata), American Association of State Highway & Transportation Officials, Washington, DC, www.transportation.org.

Roadside Design Guide, 4th ed., 2011 (including February 2012 and July 2015 errata), American Association of State Highway & Transportation Officials, Washington, DC, www.transportation.org.

Transportation Depth Reference Manual for the PE Civil Exam, Norman R. Voigt, 3rd ed., 2020, PPI, **ppi2pass.com**.

WATER RESOURCES AND ENVIRONMENTAL

PE Civil Reference Manual, Michael R. Lindeburg, 16th ed., 2021, PPI, **ppi2pass.com**.

Recommended Standards for Wastewater Facilities, 2014, Great Lakes—Upper Mississippi River Board of State and Provincial Public Health and Environmental Managers.

Recommended Standards for Water Works, 2018, Great Lakes—Upper Mississippi River Board of State and Provincial Public Health and Environmental Managers.

Water Resources and Environmental Depth Reference Manual for the Civil PE Exam, Jonathan A. Brant and Gerald J. Kauffman, 2020, PPI, **ppi2pass.com**.

Topic I: Breadth

Chapter

1. Project Planning
2. Means and Methods
3. Soil Mechanics
4. Structural Mechanics
5. Hydraulics and Hydrology
6. Geometrics
7. Materials
8. Site Development
9. Engineering Economics

Project Planning

Content in blue refers to the *NCEES Handbook*.

A. Quantity Takeoff Methods 1-1
B. Cost Estimating 1-2
C. Project Schedules 1-3
D. Activity Identification and Sequencing 1-4

The knowledge area of Project Planning makes up between four and six questions out of the 80 questions on the PE Civil exam. The organization of this chapter follows the order of subtopics given by NCEES for this knowledge area. Each subtopic is covered in the following sections.

A. QUANTITY TAKEOFF METHODS

Key concepts: These key concepts are important for answering exam questions in subtopic 1.A, Quantity Takeoff Methods.

- bills of quantity (BOQs)
- quantity and cost estimates
- takeoff reports

***NCEES Handbook*:** To prepare for this subtopic, familiarize yourself with these sections in the *Handbook*.

- Quantity Takeoff Methods
- Cost Estimate Classification Matrix for Building and General Construction Industries
- Cost Indexes
- Time-Cost Trade-Off

***PE Civil Reference Manual* (CERM):** Study these sections in CERM that either relate directly to this subtopic or provide background information.

- Section 86.2: Budgeting
- Section 87.62: Economic Order Quantity

The following figures, tables, and concepts are relevant for subtopic 1.A, Quantity Takeoff Methods.

Takeoff Report

Handbook: Quantity Takeoff Methods

CERM: Sec. 86.2

Estimators compile and analyze data on all the factors that can influence costs, such as materials, labor, location, project duration, and special machinery. The quantity takeoff process can be done manually using a printout, a red pen, or a clicker, but it can also be done with a digitizer. The user can take measurements from paper bid documents or with an integrated takeoff viewer program that interprets electronic bid documents.

See CERM Table 86.1 for a sample takeoff report.

Quantity Surveys

Handbook: Quantity Takeoff Methods

CERM: Sec. 86.2

During the takeoff process, the estimator must make decisions concerning equipment needs, the sequence of operations, the size of the crew required, and physical constraints at the site. Allowances for wasted materials, inclement weather, shipping delays, and other factors that may increase costs also must be incorporated in the estimate.

CERM and the *Handbook* do not include quantity estimation methods for materials and equipment.

Cost Estimating

Handbook: Cost Estimate Classification Matrix for Building and General Construction Industries

Handbook table Cost Estimate Classification Matrix for Building and General Construction Industries provides a generic and acceptable classification system that can be used as a guideline to compare and assess a firm's standards with generally accepted cost practices. See Sec. 1.B, Cost Estimating, for more information on using the cost estimate classification matrix.

Methods for calculating earthwork volumes include

- average end area method
- prismoidal formula
- borrow pit grid method
- coordinate method
- trapezoidal method
- Simpson's rule

See Chap. 2 in this book for detailed information on the calculations used in each of these methods.

Bills of Quantity (BOQs)

Handbook: Time-Cost Trade-Off

CERM: Sec. 87.62

CERM Fig. 87.18 shows the simplest economic order quantity (EOQ) model for minimizing inventory costs per unit time.

Figure 87.18 Inventory with Instantaneous Reorder

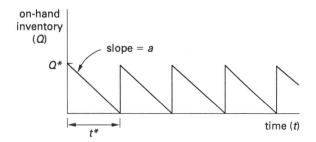

The slope, a, is the constant depletion rate (items/unit time), Q is the order quantity (original quantity on hand), and t^* is the time at depletion.

The model is based on several assumptions: reordering is instantaneous, shortages are not allowed, demand for the inventory item is deterministic (i.e., not random), demand is constant with respect to time, and an order is placed when the inventory is zero.

Time-cost trade-off is a strategy for optimizing project time and cost. The strategy identifies either least-cost solutions to add resources, which shortens the duration of critical path activities, or least-time solutions to reallocate resources over an extended duration of critical path activities, which reduces total project cost.

B. COST ESTIMATING

Key concepts: These key concepts are important for answering exam questions in subtopic 1.B, Cost Estimating.

- budgeting
- cost estimating
- cost index

NCEES Handbook: To prepare for this subtopic, familiarize yourself with these sections in the *Handbook*.

- Cost Estimate Classification Matrix for Building and General Construction Industries
- Cost Indexes

PE Civil Reference Manual **(CERM):** Study these sections in CERM that either relate directly to this subtopic or provide background information.

- Section 86.2: Budgeting

The following equations and tables are relevant for subtopic 1.B, Cost Estimating. Engineering economics is also relevant to this subtopic. As part of your preparation, be sure to review Chap. 9 in this book.

Cost Estimate Classification

Handbook: Cost Estimate Classification Matrix for Building and General Construction Industries

CERM: Sec. 86.2

Handbook table Cost Estimate Classification Matrix for Building and General Construction Industries provides a generic and acceptable classification system that can be used as a guideline to compare and assess a firm's standards with generally accepted cost practices. The matrix breaks down cost estimation classes based on the maturity of the project, end usage, type of estimating applicable to each class, and expected accuracy of each class.

Class 1 estimating is applicable to pre-bid, award, and change order phases of project development and involves detailed takeoffs with unit costs.

Class 2 estimating is typically more detailed and applicable to bidding phases of project development. It involves detailed unit cost input and detailed takeoffs.

Class 3 estimating is typically applicable to design development phases of project development and involves budget authorizations and project feasibility studies with semi-detailed unit costs and line items.

Class 4 estimating is typically applicable to schematic design phases of project development and involves parametric studies and assembly-driven models.

Class 5 estimating is typically applicable during early project phases and typically involves area factoring, parametric models, and judgment.

Cost Index

Handbook: Cost Indexes

CERM: Sec. 86.2

$$\text{current \$} = (\text{cost in year } M)\left(\frac{\text{current index}}{\text{index in year } M}\right)$$

A cost index is an indicator of the average cost movement over time of goods and services in the construction industry. The *Handbook* equation determines the current cost for an item of equipment in year M given the historical purchase cost and the current index value. Tables of index values are available from government and industry sources. Since the *Handbook* does not include a table of index values, expect that any index values needed on the exam will be provided in the problems themselves.

C. PROJECT SCHEDULES

Key concepts: These key concepts are important for answering exam questions in subtopic 1.C, Project Schedules.

- activity-on-arc networks
- activity-on-node networks
- critical path method (CPM)
- Gantt charts
- resource scheduling and leveling
- time-cost trade-off

NCEES Handbook: To prepare for this subtopic, familiarize yourself with these sections in the *Handbook*.

- CPM Precedence Relationships
- Resource Scheduling and Leveling
- Time-Cost Trade-Off

PE Civil Reference Manual (**CERM**): Study these sections in CERM that either relate directly to this subtopic or provide background information.

- Section 86.1: Project Management
- Section 86.3: Scheduling
- Section 86.4: Resource Leveling
- Section 86.5: Activity-on-Node Networks
- Section 86.6: Solving a CPM Problem
- Section 86.7: Activity-on-Arc Networks

The following equations, figures, and concepts are relevant for subtopic 1.C, Project Schedules.

Critical Path Method Precedence Relationships

Handbook: CPM Precedence Relationships

CERM: Sec. 86.5

$$\text{total float} = LS - ES \text{ or } LF - EF$$
$$\text{free float} = \text{earliest ES of successors} - EF$$

In these equations,

EF	early finish	days
ES	early start	days
LF	late finish	days
LS	late start	days

Note that the *Handbook* equation for total float is equivalent to CERM Eq. 86.1.

Critical path method (CPM) is the most common method used by project managers on projects of every size. Unlike many other methods of project management, most of which are stochastic, CPM is deterministic. A deterministic activity is one in which the duration of the project is defined, while a stochastic activity has a duration based on the probability of events happening.

With *activity-on-node networks* such as the one shown in CERM Fig. 86.4, each activity (task) is traditionally represented by a node (junction). Each activity can be thought of as a continuum of work, each with its own implicit "start" and "finish" event.

Figure 86.4 Activity-on-Node Network

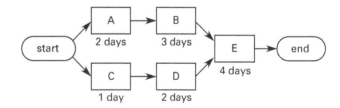

With *activity-on-arc networks* (also known as *activity-on-branch networks*) such as the one shown in CERM Fig. 86.6, the continuum of work occupies the arcs, while the nodes represent instantaneous starting and ending events. The arcs have durations, while the nodes do not.

Figure 86.6 Activity-on-Arc Network with Predecessors

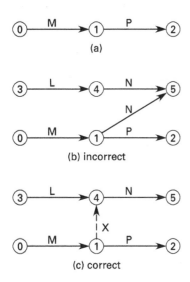

Resource Leveling

Handbook: Resource Scheduling and Leveling

CERM: Sec. 86.4

Resource scheduling is the process of tabulating project resource demands over time. Many variables can affect construction time. Most can be controlled, but others, such as weather, are beyond control. Material delivery times and labor availability have a large impact on construction schedules.

Resource leveling is used to address overallocation (i.e., situations that demand more resources than are available). Two common methods are used to level resources.

- Tasks can be delayed until resources become available.
- Tasks can be split so that the parts are completed when planned, and the remainders can be completed when resources become available.

Time-Cost Trade-Off

Handbook: Time-Cost Trade-Off

CERM: Sec. 86.3

The essence of the time-cost trade-off is that the cost increases as the project time is decreased and vice versa. The project cost is the sum of the direct and indirect costs. Direct costs are connected to a specific construction activity. Indirect costs cannot be directly attributed to a construction activity. Instead, they are costs that help maintain the functioning of a company.

The concept that indirect costs tend to be steady over the duration of the project is very important in planning fast-track projects. The project manager is responsible for assembling the schedule and project cost, and has some control over the critical path of a project.

The *crash time* of an activity is the least amount of time in which a project can be completed. Shortening the time to complete an activity results in higher costs (e.g., requiring workers to work overtime requires higher pay). However, it may help reduce overall project costs.

Gantt Charts

CERM: Sec. 86.3

A Gantt chart allows a project manager to identify the activities in a project and the duration of these activities. A project manager will often use computer software to build a Gantt chart for a project. These programs can then be used to forecast and manage the resources (equipment, money, time, etc.) on a given project. They can also be useful to coordinate resources between several simultaneous projects.

One advantage of a Gantt chart is that it is simple to make and understand. A disadvantage is that it cannot show all the sequences and dependencies of activities.

Project management software is often used to coordinate the interdependency of various activities in a project. It is important to have a plan and sequence laid out to avoid confusion, keep a project on track, and predict pinch points in the schedule or with labor/materials. See CERM Fig. 86.3 for an example of a Gantt chart.

Exam tip: Problems on the exam that involve Gantt charts will require students to understand the general concepts of a Gantt chart.

D. ACTIVITY IDENTIFICATION AND SEQUENCING

Key concepts: These key concepts are important for answering exam questions in subtopic 1.D, Activity Identification and Sequencing.

- design sequencing
- earned-value analysis
- fast tracking
- lead and lag relationships

NCEES Handbook: To prepare for this subtopic, familiarize yourself with these sections in the *Handbook*.

- Lead and Lag Relationships
- Earned-Value Analysis

- Earned-Value Analysis: Variances
- Earned-Value Analysis: Indices
- Earned-Value Analysis: Forecasting

PE Civil Reference Manual (**CERM**): Study these sections in CERM that either relate directly to this subtopic or provide background information.

- Section 86.3: Scheduling
- Section 86.12: Earned Value Method

The following equations, figures, tables, and concepts are relevant for subtopic 1.D, Activity Identification and Sequencing.

Lead and Lag Relationships

Handbook: Lead and Lag Relationships

In lead and lag relationships, activities A and B have a finish-to-start relationship with lead or lag time. The figure in *Handbook* section Lead and Lag Relationships shows how the scheduling works for these relationships.

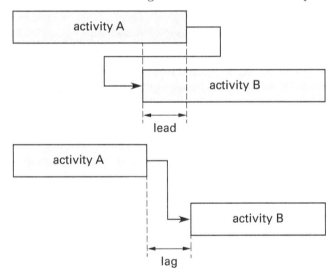

Earned-Value Analysis: Variances

Handbook: Earned-Value Analysis; Earned-Value Analysis: Variances

CERM: Sec. 86.12

CERM Eq. 86.6 can be used to determine the *cost variance*, CV.

$$\text{CV} = \text{BCWP} - \text{ACWP} \qquad 86.6$$

CV is the difference between the planned and actual costs of the work completed.

CERM Eq. 86.7 can be used to determine the *schedule variance*, SV.

$$\text{SV} = \text{BCWP} - \text{BCWS} \qquad 86.7$$

SV is the difference between the value of work accomplished for a given period and the value of the work planned.

In these equations,

ACWP	actual cost of work performed	$
BCWP	budgeted cost of work performed	$
BCWS	budgeted cost of work scheduled	$

BCWS is a spending plan for the project as a function of schedule and performance. ACWP is the actual spending as a function of time or performance. It is the cumulative actual expenditures on the project viewed at regular intervals within the project duration. BCWP is the actual earned value based on the technical accomplishment. BCWP is the cumulative budgeted value of the work actually completed.

Earned-Value Analysis: Indices

Handbook: Earned-Value Analysis; Earned-Value Analysis: Indices

CERM: Sec. 86.12

CERM Eq. 86.8 can be used to determine the *cost performance index*, CPI.

$$\text{CPI} = \frac{\text{BCWP}}{\text{ACWP}} \qquad 86.8$$

CPI is a cost efficiency factor representing the relationship between the actual cost expended and the earned value. A CPI of 1.0 or more suggests efficiency.

CERM Eq. 86.9 can be used to determine the *schedule performance index*, SPI.

$$\text{SPI} = \frac{\text{BCWP}}{\text{BCWS}} \qquad 86.9$$

SPI is a measure of schedule effectiveness, as determined from the earned value and the initial planned schedule. An SPI of 1.0 or more suggests that work is ahead of schedule.

In these equations,

ACWP	actual cost of work performed	$
BCWP	budgeted cost of work performed	$
BCWS	budgeted cost of work scheduled	$
CPI	cost performance index	–

Earned-Value Analysis: Forecasting

Handbook: Earned-Value Analysis; Earned-Value Analysis: Forecasting

CERM: Sec. 86.12

CERM Eq. 86.10 can be used to determine the *estimate to complete*, ETC.

$$ETC = BAC - BCWP \quad\quad 86.10$$

ETC is a calculated value representing the cost of work required to complete remaining project tasks.

CERM Eq. 86.11 can be used to determine the *estimate at completion*, EAC.

$$EAC = ACWP + ETC \quad\quad 86.11$$

EAC is a calculated value that represents the projected total final cost of work when completed.

In these equations,

ACWP	actual cost of work performed	$
BAC	budget at completion	$
BCWP	budgeted cost of work performed	$

Budget at completion, BAC, also known as the *performance measurement baseline*, is the sum total of the time-phased budget. It equals the original project estimate.

The project completion percentage can be calculated as

$$\text{project completion (\%)} = \frac{\text{actual units complete}}{\text{total units budgeted}} \cdot 100$$

Design Sequencing Phases

CERM: Sec. 86.3

A project may combine the various phases of a project, or the phases may be informal based on the scope of a project. For example, a property owner wants to build a small garden shed that must be engineered per the local building department specifications. The owner contacts a contractor who has an engineer on staff for this type of project. The engineer may perform the schematic, design development, and construction document phases in a single meeting with the owner.

Contractors should consider construction sequencing when bidding on a project. After procuring a contract for a job, the contractor must identify construction activities and resources and complete a schedule for the project.

Earthwork is generally the most problematic part of the job, as it involves many variables for design and construction activities. CERM Table 86.4 provides a good example of the construction sequencing for a project that involves earthwork activities, specifically involving erosion and sedimentation activities.

Table 86.4 Construction Activities and ESC Measures

construction activity	ESC measures
designate site access	Stabilize exposed areas with gravel and/or temporary vegetation. Immediately apply stabilization to areas exposed throughout site development.
protect runoff outlets and conveyance systems	Install principal sediment traps, fences, and basins prior to grading; stabilize stream banks and install storm drains, channels, etc.
land clearing	Mark trees and buffer areas for preservation.
site grading	Install additional ESC measures as needed during grading.
site stabilization	Install temporary and permanent seeding, mulching, sodding, riprap, etc.
building construction and utilities installation	Install additional ESC measures as needed during construction.
landscaping and final site stabilization*	Remove all temporary control measures; install topsoil, trees and shrubs, permanent seeding, mulching, sodding, riprap, etc.; stabilize all open areas, including borrow and spoil areas.

*This is the last construction phase.

Fast Tracking

CERM: Sec. 86.3

The fast-track method requires close coordination between the architect, contractor, subcontractors, owner, and others. It can save time and money; however, if communication on a project is poor, then activities may take longer and cost more.

Ordering long-lead materials and equipment is the practice of procuring equipment and materials for a project before they are needed. For example, ordering the steel for a steel building and having it fabricated while civil (earthwork) is being performed means that the steel arrives on site and is ready to be erected as the concrete finishes curing.

Fast tracking is really just taking advantage of critical path and noncritical activities.

Braced cuts
   Pressure distribution

sand

soft to
med clay

stiff
clay

Shear reinforcement
  = stirrups
      ↳ closed ties
        ↳ hoops

2 Means and Methods

Content in blue refers to the *NCEES Handbook*.

A. Construction Loads.. 2-1
B. Construction Methods....................................... 2-2
C. Temporary Structures and Facilities 2-5

The knowledge area of Means and Methods makes up between three and five questions out of the 80 questions on the PE Civil exam. The organization of this chapter follows the order of subtopics given by NCEES for this knowledge area. Each subtopic is covered in the following sections.

A. CONSTRUCTION LOADS

Key concepts: These key concepts are important for answering exam questions in subtopic 2.A, Construction Loads.

- allowable stress design method
- dynamic loads
- environmental loads
- lateral loads
- load and resistance factor design method
- structural integrity

NCEES Handbook: The *Handbook* does not contain any material relevant to this subtopic. Be sure to study the listed sections in CERM.

PE Civil Reference Manual **(CERM):** Study these sections in CERM that either relate directly to this subtopic or provide background information.

- Section 39.2: Braced Cuts
- Section 39.14: Cofferdams
- Section 45.1: Allowable Stress Design
- Section 45.2: Ultimate Strength Design
- Section 50.21: Shear Reinforcement
- Section 51.13: Concrete Deck Systems in Bridges
- Section 54.2: Bearing Walls: Empirical Method
- Section 62.1: Introduction
- Section 68.13: ASD Wall Design: Shear
- Section 69.4: Minimum Column Eccentricity

The following concepts are relevant for subtopic 2.A, Construction Loads.

Structural Integrity

CERM: Sec. 45.1, Sec. 45.2

Structural integrity is the ability of a structure or structural component to resist loads applied to it. Various criteria may be used in defining structural integrity, such as strength and serviceability.

Construction loads that may impact structural integrity include loads from temporary structures or scaffolding, crane loads, equipment loads, and shoring loads. Also, structural elements that are relied on to provide stability may not be present during construction, and this must be included in evaluations of structural integrity during that time. For more about loads from temporary structures or scaffolding, see subtopic 2.C, Temporary Structures and Facilities, in this chapter.

Lateral Loads

CERM: Sec. 39.2, Sec. 50.21, Sec. 54.2, Sec. 62.1, Sec. 68.13, Sec. 69.4

Lateral loading can be caused by construction activities (e.g., construction vehicles colliding with the structure), the eccentricity of vertical loads, seismic loading, or wind loading. Lateral loads should be considered during construction. The most common lateral load during construction is wind.

Allowable Strength Design Method

CERM: Sec. 45.1

The allowable strength design (ASD) method is also known as the allowable stress design method. The ASD method applies a factor of safety to the material strength of a structural component. This reduced strength of the material is known as the allowable strength. The ASD method is used extensively in steel design and almost exclusively in geotechnical design. The ASD method uses service-level loads—that is, the loads the structure would be expected to encounter if used as designed. For more information on the ASD method, see the entry "Allowable Strength Design Method" in Chap. 25 of this book).

Load and Resistance Factor Design Method

CERM: Sec. 45.2

The load and resistance factor design (LRFD) method is also known as the ultimate strength design method, load factor design method, plastic design method, or strength design method. The LRFD method does not use allowable stresses at all. Instead, the member is designed so that its actual nominal strength exceeds the required ultimate strength. For more information on the LRFD method, see the entry "Load and Resistance Factor Design Method" in Chap. 25 of this book.

B. CONSTRUCTION METHODS

Key concepts: These key concepts are important for answering exam questions in subtopic 2.B, Construction Methods.

- common methods used in calculating areas
- dewatering and pumping
- Dupuit's formula
- raft foundations
- soil stabilization
- trenching and excavation

NCEES Handbook: To prepare for this subtopic, familiarize yourself with these sections in the *Handbook*.

- Cross-Section Methods
- Cross-Sectional End Areas
- Borrow Pit Grid Method
- Earthwork Area Formulas
- Dewatering and Pumping
- Soil Stabilization Methods
- Dupuit's Formula

PE Civil Reference Manual (CERM): Study these sections in CERM that either relate directly to this subtopic or provide background information.

- Section 21.11: Well Drawdown in Aquifers
- Section 36.13: General Considerations for Rafts
- Section 39.1: Excavation
- Section 78.32: Areas Bounded by Irregular Boundaries
- Section 80.14: Average End Area Method
- Section 80.15: Prismoidal Formula Method
- Section 80.16: Borrow Pit Geometry
- Section 83.6: Trenching and Excavation

The following equations and figures are relevant for subtopic 2.B, Construction Methods.

Average End Area Method

Handbook: Cross-Section Methods; Cross-Sectional End Areas

CERM: Sec. 80.14

$$V = L\left(\frac{A_1 + A_2}{2}\right)$$

The average end area method calculates the volume between two consecutive cross sections as the average of their areas multiplied by the distance between the two areas. This method assumes the fill is positive and the cut is negative. See *Handbook* figure Cross-Sectional End Areas for examples of fill and cut diagrams.

Prismoidal Formula

Handbook: Cross-Section Methods; Cross-Sectional End Areas

CERM: Sec. 80.15

$$V = L\left(\frac{A_1 + 4A_m + A_2}{6}\right)$$

The prismoidal formula is the preferred cross-sectional end area formula when two end areas differ greatly or the ground surface is irregular. The volume calculated using this method is generally smaller than that calculated using the average end area method, and so favors the owner and developer in earthwork cost estimating. See *Handbook* figure Cross-Sectional End Areas for examples of fill and cut diagrams.

Borrow Pit Grid Method

Handbook: Borrow Pit Grid Method

CERM: Sec. 80.16

It is often necessary to borrow earth from an adjacent area to construct embankments. Normally, the borrow pit area is laid out in a rectangular grid with 10 ft (3 m), 50 ft (15 m), or even 100 ft (30 m) squares. Elevations are determined at the corners of each square by leveling before and after excavation so that the cut at each corner can be computed. See the illustration in *Handbook* section Borrow Pit Grid Method or CERM Fig. 80.7.

Coordinate Method

Handbook: Earthwork Area Formulas

$$A = \tfrac{1}{2}\big(X_A(Y_B - Y_N) + X_B(Y_C - Y_A) \\ + X_C(Y_D - Y_B) + \cdots + X_N(Y_A - Y_{N-1})\big)$$

In this equation, X_A, X_B, ..., X_N are widths, and Y_A, Y_B, ..., Y_N are heights. This equation is used to calculate the area of earthwork to be removed from an excavation.

Trapezoidal Method

Handbook: Earthwork Area Formulas

$$A = w\big(\tfrac{1}{2}(h_1 + h_n) + h_2 + h_3 + h_4 + \cdots + h_{n-1}\big)$$

CERM: Sec. 78.32

$$A = d\left(\frac{h_1 + h_n}{2} + \sum_{i=2}^{n-1} h_i\right) \quad 78.40$$

The length of the common interval is w in the *Handbook* equation and d in the CERM equation.

Areas of sections with irregular boundaries cannot be determined precisely, and approximation methods must be used. If the irregular side can be divided into a series of cells and each cell is fairly straight, the *trapezoidal rule* can be used.

Simpson's Rule

Handbook: Earthwork Area Formulas

$$A = \left(\frac{1}{3}\right)\begin{pmatrix}\text{first value} + \text{last value} + \\ (4 \cdot \text{sum of odd-numbered values}) \\ + (2 \cdot \text{sum of even-numbered values})\end{pmatrix}$$
$$\cdot \text{ length of interval}$$

CERM: Sec. 78.32

$$A = \frac{d}{3}\left(h_1 + h_n + 2\sum_{\text{odd}} h_i + 4\sum_{\text{even}} h_i\right) \quad 78.41$$
$$= \frac{d}{3}(h_1 + 4h_2 + 2h_3 + 4h_4 + \ldots + h_n)$$

Areas of sections with irregular boundaries cannot be determined precisely, and approximation methods must be used. If the irregular side can be divided into a series of cells and each cell is curved or parabolic, *Simpson's rule* (sometimes referred to as *Simpson's 1/3 rule*) can be used.

CERM Eq. 78.41 and the *Handbook* equation in Earthwork Area Formulas are equivalent. The difference is that the CERM equation is written using classic summation nomenclature, while the *Handbook* equation is written using words to explain the terms of the equation.

Dewatering and Pumping

Handbook: Dewatering and Pumping; Dupuit's Formula

Dewatering (sometimes known as *unwatering*) is the removal of water from the job site, typically by lowering the water table. Submersible pumps and vacuum well-point methods are used to remove water. The following equations relate the rate of pumping to the level of the water table at any given distance from the well.

For artesian wells,

$$Q = \frac{2\pi K f (H - h)}{\ln \dfrac{R_O}{r_w}}$$

For water-table aquifers,

$$Q = \frac{\pi K (H^2 - h^2)}{\ln \dfrac{R_O}{r_w}}$$

In these equations,

f	original aquifer depth	ft
h	aquifer depth at radial distance r_w	ft
H	aquifer depth at radial distance R_O	ft
K	coefficient of permeability	ft/sec
Q	pumping quantity	ft^3/sec
r_w	radius of well	ft
R_O	radial distance from well	ft

Dupuit's formula can be used to predict the steady-state drawdown and volumetric flow from a deep-penetrating well. (Dupuit's formula is also equivalent to the previous equation.)

$$Q = \frac{\pi K (h_2^2 - h_1^2)}{\ln \dfrac{r_2}{r_1}}$$

In Dupuit's formula,

h_1	aquifer depth at radial distance r_1 (i.e., at perimeter of well)	ft
h_2	aquifer depth at radial distance r_2	ft
r_1	radius of well	ft
r_2	radial distance from well	ft

However, as long as h_1 and h_2 correspond to r_1 and r_2, respectively, Dupuit's formula will give the same result whether the well radius is taken as r_1 or r_2.

CERM: Sec. 21.11

For artesian wells,

$$Q = \frac{2\pi KY(y_1 - y_2)}{\ln \dfrac{r_1}{r_2}} \qquad 21.27$$

For deep-penetrating wells,

$$Q = \frac{\pi K(y_1^2 - y_2^2)}{\ln \dfrac{r_1}{r_2}} \qquad 21.25$$

CERM Eq. 21.25 is Dupuit's formula again. The nomenclature in the two CERM equations is similar to those from the *Handbook*, but r_1 and r_2 switch meanings (i.e., r_2 is the well radius). Also, y and Y are used instead of h and f, respectively.

y_1	aquifer depth at radial distance r_1	ft
y_2	aquifer depth at radial distance r_2 (i.e., at perimeter of well)	ft
Y	original aquifer depth	ft

Visual representations of the dimensions used in the *Handbook* equations can be found in *Handbook* figures **Confined Aquifer** and **Water-Table Aquifer**. Visual representations of the dimensions used in the CERM equations can be found in CERM Fig. 21.2. The *Handbook* and CERM assign the subscripts 1 and 2 differently to the larger and smaller radial distances; however, it doesn't matter which subscript is assigned to which as long as the aquifer depths also correspond correctly.

Excavation by Soil Type

Handbook: Soil Stabilization Methods

CERM: Sec. 83.6

Handbook table **Soil Stabilization Methods** shows the differences between stabilization methods for portland cement and lime.

CERM Fig. 83.3 illustrates the different excavations that can be performed in the three different types of soils, and also provides the required slopes and maximum dimensions for the excavations.

Trenching and Excavation

CERM: Sec. 83.6

Excavations deeper than 5 ft in all types of earth must be protected from cave-in and collapse unless they are excavations entirely performed in stable rock. Timber shoring, aluminum shoring, and trench shields may be used in excavations up to 20 ft. See CERM Fig. 83.2 for slope and shield configurations.

Figure 83.2 *Slope and Shield Configurations*

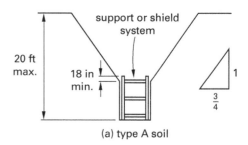

(a) type A soil

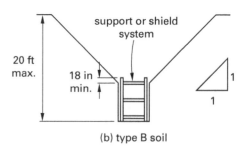

(b) type B soil

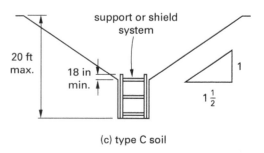

(c) type C soil

Sloping and benching the trench walls may be substituted for shoring. Sloped walls in excavations deeper than 20 ft (6 m) must be designed by a professional engineer. CERM Table 83.3 provides maximum slopes for excavations less than 20 ft deep.

Raft Foundations

CERM: Sec. 36.13

A raft foundation is a combined footing slab that usually covers the entire area beneath a building and supports all walls and columns. A raft foundation should be used whenever the individual footings would constitute half or more of the area beneath a building.

C. TEMPORARY STRUCTURES AND FACILITIES

Key concepts: These key concepts are important for answering exam questions in subtopic 2.C, Temporary Structures and Facilities.

- formwork
- formwork pressure
- scaffolding

NCEES Handbook: The *Handbook* does not contain any material relevant to this subtopic. Be sure to study the listed sections in CERM.

PE Civil Reference Manual (**CERM**): Study these sections in CERM that either relate directly to this subtopic or provide background information.

- Section 49.15: Formwork
- Section 49.16: Lateral Pressure on Formwork
- Section 83.15: Scaffolds
- Section 83.16: Temporary Structures

The following equations, figures, and concepts are relevant for subtopic 2.C, Temporary Structures and Facilities.

Formwork

CERM: Sec. 49.15

Formwork is the system of boards, ties, and bracing required to construct the mold in which wet concrete is placed.

Forms are constructed out of a variety of materials. Plywood is the most common formwork material, usually 3/4 in thick and coated on one side with oil, water-resistant glue, or plastic. This prevents adhesion to the concrete so that the forms are easier to remove. The plywood is supported by solid wood framing, which is braced or shored. CERM Fig. 49.1 shows two typical wood-framed forms.

Figure 49.1 Concrete Framework

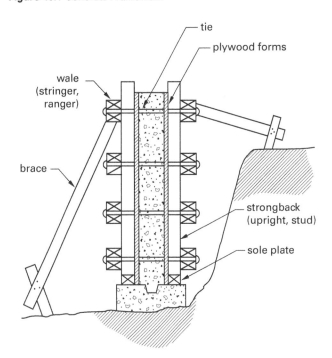

(a) wall formwork

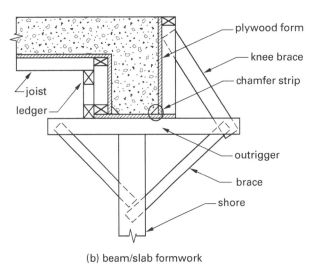

(b) beam/slab formwork

For exposed architectural surfaces, consideration must be given to the method and design of the formwork, as the joints and form ties are visible when the framework is removed. *Form ties* are metal wires or rods used to hold opposite sides of the form together and prevent their collapse. When the forms are removed, the wire remains in the concrete, and the excess is twisted or cut off.

Formwork Pressure

CERM: Sec. 49.16

$$p_{\max,\text{kPa}} = 30C_w \leq C_w C_c \left(7.2 + \frac{785 R_{\text{m/h}}}{T_{°\text{C}} + 17.8°}\right)$$
$$\leq \rho g h \quad \text{[SI]} \quad 49.3(a)$$
$$\begin{bmatrix} \text{columns: } h \leq 1.2 \text{ m} \\ \text{walls: } h \leq 4.2 \text{ m} \\ R \leq 2.1 \text{ m/h} \end{bmatrix}$$

$$p_{\max,\text{psf}} = 600C_w \leq C_w C_c \left(150 + \frac{9000 R_{\text{ft/hr}}}{T_{°\text{F}}}\right)$$
$$\leq \gamma h \quad \text{[U.S.]} \quad 49.3(b)$$
$$\begin{bmatrix} \text{columns: } h \leq 4 \text{ ft} \\ \text{walls: } h \leq 14 \text{ ft} \\ R \leq 7 \text{ ft/hr} \end{bmatrix}$$

$$p_{\max,\text{kPa}} = 30C_w \leq C_w C_c$$
$$\cdot \left(7.2 + \frac{1156 + 244 R_{\text{m/h}}}{T_{°\text{C}} + 17.8°}\right) \leq \rho g h \quad \text{[SI]} \quad 49.4(a)$$
$$\begin{bmatrix} \text{walls: } h > 4.2 \text{ m}, R \leq 2.1 \text{ m/h} \\ \text{walls: } 2.1 \text{ m/h} < R \leq 4.5 \text{ m/h} \end{bmatrix}$$

$$p_{\max,\text{psf}} = 600C_w \leq C_w C_c$$
$$\cdot \left(150 + \frac{43{,}400 + 2800 R_{\text{ft/hr}}}{T_{°\text{F}}}\right) \quad \text{[U.S.]} \quad 49.4(b)$$
$$\leq \gamma h$$
$$\begin{bmatrix} \text{walls: } h > 14 \text{ ft}, R \leq 7 \text{ ft/hr} \\ \text{walls: } 7 \text{ ft/hr} < R \leq 15 \text{ ft/hr} \end{bmatrix}$$

Formwork must be strong enough to withstand hydraulic loading from the concrete during pouring and setting. The hydraulic load is greatest immediately after pouring. CERM Eq. 49.3 and Eq 49.4 can be used to find the lateral pressure. The pressure from wet concrete is given by CERM Eq. 49.2.

$$p = \rho g h \quad \text{[SI]} \quad 49.2(a)$$

$$p = \rho g h \quad \text{[U.S.]} \quad 49.2(b)$$

Scaffolding

CERM: Sec. 83.15

A scaffold is any temporary elevated platform and its supporting structure used for supporting employees, materials, or both. The construction and use of scaffolds are regulated by the Occupational Safety and Health Administration (OSHA) in the Code of Federal Regulations, Title 29, Part 1926, Subpart L.

Per OSHA Sec. 1926.451(g)(1), fall protection must be provided for each employee on a scaffold more than 10 ft above a lower level. This differs from the 6 ft threshold for fall protection for other walking and working surfaces because scaffolds are temporary structures that provide a work platform for employees who are constructing or demolishing other structures.

3 Soil Mechanics

Content in blue refers to the *NCEES Handbook*.

A. Lateral Earth Pressure 3-1
B. Soil Consolidation 3-5
C. Effective and Total Stresses 3-8
D. Bearing Capacity 3-10
E. Foundation Settlement 3-12
F. Slope Stability 3-14

The knowledge area of Soil Mechanics makes up between five and eight questions out of the 80 questions on the PE Civil exam. The organization of this chapter follows the order of subtopics given by NCEES for this knowledge area. Each subtopic is covered in the following sections.

A. LATERAL EARTH PRESSURE

Key concepts: These key concepts are important for answering exam questions in subtopic 3.A, Lateral Earth Pressure.

- backfill soil types
- earth pressure coefficients
- Rankine earth pressure theory
- types of earth pressure
- types of retaining wall structures

NCEES Handbook: To prepare for this subtopic, familiarize yourself with these sections in the *Handbook*.

- Stress States on a Soil Element Subjected Only to Body Stresses
- At-Rest Coefficients
- Development of Rankine Active and Passive Failure Zones for a Smooth Retaining Wall
- Failure Surfaces, Pressure Distribution and Forces
- Lateral Pressure Due to Surcharge Loadings

PE Civil Reference Manual (CERM): Study these sections in CERM that either relate directly to this subtopic or provide background information.

- Section 37.3: Earth Pressure
- Section 37.4: Vertical Soil Pressure
- Section 37.5: Active Earth Pressure
- Section 37.6: Passive Earth Pressure
- Section 37.7: At-Rest Soil Pressure
- Section 37.10: Surcharge Loading
- Section 37.11: Effective Stress
- Section 37.12: Cantilever Retaining Walls: Analysis

The following equations, figures, and tables are relevant for subtopic 3.A, Lateral Earth Pressure.

Stress States on a Soil Element

Handbook: Stress States on a Soil Element Subjected Only to Body Stresses

Handbook figure Stress States on a Soil Element Subjected Only to Body Stresses is useful for determining the forces and pressures acting on soil at depth and on retaining walls. These forces are important for retaining wall design, especially when analyzing the failure modes of retaining wall structures.

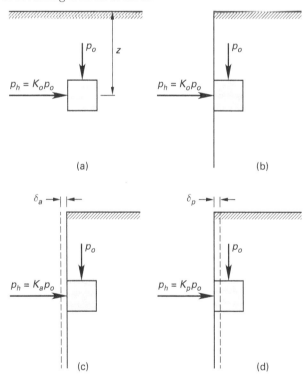

Source: Federal Highway Administration. National Highway Institute. *Soils and Foundations, Reference Manual.* Vol. I. FHWA-NHI-06-088. Washington, DC: U.S. Department of Transportation, December 2006, Fig. 2-19, p. 2-43. www.fhwa.dot.gov/engineering/geotech/pubs/nhi06088.pdf.

K_a	active earth pressure coefficient	–
K_o	at-rest earth coefficient	–
K_p	passive earth coefficient	–
p_h	horizontal soil pressure	lbf/ft^2
p_o	vertical soil pressure	lbf/ft^2
z	depth	ft
δ_a	wall movement away from original wall location	ft
δ_p	wall movement toward retained soil	in

Part (a) shows vertical and horizontal stresses acting on a soil element in situ and the calculation for the pressure on a soil element at rest. Part (b) shows the same soil element stressed against an infinitely rigid wall and the calculation for the pressure under those conditions. Part (c) shows an active earth pressure condition in which the wall moves away from retained soil when subjected to both vertical and horizontal stresses and the calculation for active earth pressure. (p_h is shown in the negative direction in this case.) Part (d) shows a passive earth pressure condition in which the wall moves into retained soil subjected to the same stresses.

Exam tip: These formulas assume granular soil, which has a cohesion of 0.

At-Rest Coefficients

Handbook: At-Rest Coefficients

CERM: Sec. 37.7

For normally consolidated soils,

$$K_o = 1 - \sin \phi'$$

For overconsolidated soils,

$$K_o = (1 - \sin \phi')(\text{OCR})^\Omega$$

$$\Omega = \sin \phi'$$

The *Handbook* includes an additional equation for overconsolidated soils that is not included in CERM.

At-rest soil is defined as completely confined soil that cannot move. The term is appropriate to use for soil next to bridge abutments, basement walls, walls restrained at their tops, walls bearing on rock, and walls with soft-clay backfill, as well as for sand deposits of infinite depth and extent.

The horizontal pressure at rest can be calculated using the at-rest earth pressure coefficient, K_o. CERM Table 37.2 shows typical ranges of earth pressure coefficients, which vary from 0.4 to 0.5 for untamped sand, 0.5 to 0.7 for normally consolidated clays, and 1.0 and up for overconsolidated clays.

Table 37.2 Typical Range of Earth Pressure Coefficients

condition	granular soil	cohesive soil
active	0.20–0.33	0.25–0.5
passive	3–5	2–4
at rest	0.4–0.6	0.4–0.8

Rankine Active and Passive Failure Zones

Handbook: Development of Rankine Active and Passive Failure Zones for a Smooth Retaining Wall

CERM: Sec. 37.3

Handbook figure Development of Rankine Active and Passive Failure Zones for a Smooth Retaining Wall indicates the location and geometry of passive and active failure zones at a smooth retaining wall. For a more detailed illustration, see CERM Fig. 37.2.

The angle α can be determined from CERM Eq. 37.1.

$$\alpha = 45° + \frac{\phi}{2} \quad \text{[Rankine]} \qquad 37.1$$

In this equation,

α	angle of failure plane	deg
ϕ	angle of internal friction	deg

Exam tip: Rankine theory neglects friction between the retaining wall and the soil.

Failure Surfaces, Pressure Distribution, and Forces

Handbook: Failure Surfaces, Pressure Distribution and Forces

The active pressure at depth z is

$$p_a = K_a \gamma z$$

The active force within depth z is

$$P_a = \frac{K_a \gamma z^2}{2}$$

The passive pressure at depth z is

$$p_p = K_p \gamma z$$

The passive force within depth z is

$$P_p = \frac{K_p \gamma z^2}{2}$$

CERM: Sec. 37.4, Sec. 37.5, Sec. 37.6

$$p_a = p_v k_a - 2c\sqrt{k_a} \qquad 37.4$$

$$p_p = p_v k_p + 2c\sqrt{k_p} \qquad 37.13$$

Vertical soil pressure, p_v, is caused by the soil's own weight and calculated in the same manner as a fluid column.

There are notable differences between the *Handbook* equations and the CERM equations. CERM refers to depth as H, while the *Handbook* refers to it as z. Pressure equations in CERM include a value for cohesion, c, while the *Handbook* assumes the soil is granular, which means $c = 0$.

Handbook figure Failure Surfaces, Pressure Distribution and Forces shows pressure distributions and failure surfaces behind a vertical wall. Because wall friction is neglected in Rankine theory, the triangular resultant pressure acts at $H/3$ from the base and is normal to the wall.

In CERM Eq. 37.3, p_v is calculated as

$$p_v = \rho g H \qquad \text{[SI]} \qquad 37.3(a)$$

$$p_v = \gamma H \qquad \text{[U.S.]} \qquad 37.3(b)$$

Exam tip: For granular soils cohesion, $c = 0$.

Lateral Pressure Due to Point Load Surcharges

Handbook: Lateral Pressure Due to Surcharge Loadings

For $\overline{m} > 0.4$,

$$p_h \left(\frac{H^2}{Q_p} \right) = \frac{1.77 \overline{m}^2 \overline{n}^2}{(\overline{m}^2 + \overline{n}^2)^3}$$

For $\overline{m} \leq 0.4$,

$$p_h \left(\frac{H^2}{Q_p} \right) = \frac{0.28 \overline{n}^2}{(0.16 + \overline{n}^2)^3}$$

CERM: Sec. 37.10

$$p_q = \frac{1.77 V_q m^2 n^2}{H^2 (m^2 + n^2)^3} \qquad [m > 0.4] \qquad 37.33$$

$$p_q = \frac{0.28 V_q n^2}{H^2 (0.16 + n^2)^3} \qquad [m \leq 0.4] \qquad 37.34$$

$$R_q \approx \frac{0.78 V_q}{H} \qquad [m = 0.4] \qquad 37.35$$

$$R_q \approx \frac{0.60 V_q}{H} \qquad [m = 0.5] \qquad 37.36$$

$$R_q \approx \frac{0.46 V_q}{H} \qquad [m = 0.6] \qquad 37.37$$

In the *Handbook* equation,

H	wall height	ft
$\overline{m}$	ratio of horizontal distance of point load from inside face of retaining wall in relation to depth of soil retained	–
$\overline{n}$	ratio of depth at which pressure due to surcharge load acts on wall in relation to depth of soil retained	–
p_h	horizontal pressure	lbf/ft^2
Q_p	point load	lbf

In the CERM equations, nomenclature is identical to the *Handbook* nomenclature except as shown.

p_q	horizontal pressure	lbf/ft^2
R_q	resultant surcharge load	lbf
V_q	point load	lbf
y	vertical distance to resultant horizontal force from bottom of wall	ft

CERM Eq. 37.35, Eq. 37.36, and Eq. 37.37 give the resultant, or total force applied, from the point load surcharge in units of pounds-force.

For horizontal pressures caused by point load Q_p at different depths and distances from the inner face of the wall, see *Handbook* figure Lateral Pressure Due to Surcharge Loadings.

Lateral Pressure Due to Line Load Surcharges

Handbook: Lateral Pressure Due to Surcharge Loadings

For $\overline{m} > 0.4$,

$$p_h\left(\frac{H}{Q_l}\right) = \frac{1.28\overline{m}^2\overline{n}}{(\overline{m}^2 + \overline{n}^2)^2}$$

$$P_h = \frac{0.64 Q_l}{\overline{m}^2 + 1}$$

For $\overline{m} \leq 0.4$,

$$p_h\left(\frac{H}{Q_l}\right) = \frac{0.20\overline{n}}{(0.16 + \overline{n}^2)^2}$$

$$P_h = 0.55 Q_l$$

CERM: Sec. 37.10

$$p_q = \frac{4L_q m^2 n}{\pi H (m^2 + n^2)^2} \quad [m > 0.4] \qquad 37.40$$

$$p_q = \frac{0.203 L_q n}{H(0.16 + n^2)^2} \quad [m \leq 0.4] \qquad 37.42$$

In the *Handbook* equations,

H	wall height	ft
$\overline{m}$	ratio of horizontal distance of point load from inside face of retaining wall in relation to depth of soil retained	–
$\overline{n}$	ratio of depth at which pressure due to surcharge load acts on wall in relation to depth of soil retained	–
p_h	horizontal pressure	lbf/ft²
Q_l	line load	lbf

In the CERM equations, all nomenclature is identical to the *Handbook* nomenclature except as shown.

p_q	horizontal pressure	lbf/ft²
L_q	line load	lbf
y	vertical distance to resultant horizontal force from bottom of wall	ft

Handbook figure Lateral Pressure Due to Surcharge Loadings gives equations for horizontal pressures caused by the line load (Q_l in the *Handbook*; L_q in CERM) at different depths and distances from the inner face of the wall, and for the resultant, or total pressure, due to line load surcharges. The equations vary depending on the value of $\bar{m}$, which is the ratio of the horizontal distance and the wall height, and the value of $\bar{n}$, which is the ratio of the vertical distance and the wall height.

Lateral Pressure Due to Strip Load Surcharges

Handbook: Lateral Pressure Due to Surcharge Loadings

$$p_h = \left(\frac{2q}{\pi}\right)(\beta - \sin\beta\cos 2\alpha)$$

In the *Handbook* equation,

p_h	horizontal pressure	lbf
q	strip load	lbf/ft
α	angle between center of strip and vertical wall face	rad
β	angle between front and back boundaries of strip load	rad

Handbook figure Lateral Pressure Due to Surcharge Loadings shows the applied pressure due to a strip load, p_h, at a given location along the wall.

Lateral Pressure Due to Uniform Load Surcharges

Handbook: Lateral Pressure Due to Surcharge Loadings

$$-p_h(\text{due to } q) = qK$$

CERM: Sec. 37.10

$$p_q = k_a q \qquad 37.31$$

In the *Handbook* equation,

K	earth pressure coefficient	–
p_h	horizontal pressure	lbf
q	uniform load	lbf/ft

In CERM Eq. 37.31, the nomenclature is identical except that the earth pressure coefficient is k.

Handbook figure Lateral Pressure Due to Surcharge Loadings shows the horizontal pressures caused by the uniform surcharge load q_p and gives the equation for finding them. The appropriate coefficient, whether passive, active, or at rest, is indicated by K in the *Handbook* equation and k in the CERM equation.

The resultant, or total force applied, from the uniform surcharge across the distance, H, is given by CERM Eq. 37.32.

$$R_q = k_a q H \qquad 37.32$$

Types of Retaining Wall Structures

CERM: Sec. 37.3

CERM Fig. 37.1 illustrates types of retaining walls.

A *gravity wall* is a high-bulk structure that relies on self-weight. A *buttress wall* depends on compression ribs between the stem and the toe to resist flexure and overturning. *Counterfort walls* depend on tension ribs between the stem and the heel to resist flexure and overturning. *Cantilever walls* resist overturning through a combination of the soil weight over the heel and the resisting pressure under the base.

B. SOIL CONSOLIDATION

Key concepts: These key concepts are important for answering exam questions in subtopic 3.B, Soil Consolidation.

- average time factor
- clay consolidation and settlement
- coefficients of consolidation
- consolidated compression index
- settlement time
- types of settlement
- vertical effective pressure

NCEES Handbook: To prepare for this subtopic, familiarize yourself with these sections in the *Handbook*.

- Construction of Field Virgin Consolidation Relationships
- Normally Consolidated Soils
- Typical Consolidation Curve for Normally Consolidated Soil
- Overconsolidated Soils
- Typical Consolidation Curve for Overconsolidated Soil
- Diagram Illustrating Consolidation of a Layer of Clay Between Two Pervious Layers
- Time Rate of Settlement
- Average Degree of Consolidation, U, Versus Time Factor, T_v, for Uniform Initial Increase in Pore Water Pressure
- Settlement Ratio for Overconsolidation

PE Civil Reference Manual (CERM): Study these sections in CERM that either relate directly to this subtopic or provide background information.

- Section 40.4: Settling
- Section 40.5: Clay Condition
- Section 40.6: Consolidation Parameters
- Section 40.7: Primary Consolidation
- Section 40.8: Primary Consolidation Rate
- Section 40.9: Secondary Consolidation

The following equations, figures, and tables are relevant for subtopic 3.B, Soil Consolidation.

Consolidation

Handbook: Construction of Field Virgin Consolidation Relationships

CERM: Sec. 40.4

Handbook figure Construction of Field Virgin Consolidation Relationships graphically shows the relationship between soil void ratio and vertical applied pressure.

Settling is generally due to consolidation (i.e., a decrease in void fraction) of the supporting soil. There are three types of consolidation. *Immediate settling* occurs immediately after a structure is built and typically occurs in sandy soils. *Primary consolidation* of clayey soils occurs due to the extrusion of water from the void spaces. *Secondary consolidation* of clayey soils occurs due to plastic readjustment of the soil grains, taking place at a much slower rate.

Settlement for Normally Consolidated Soils

Handbook: Normally Consolidated Soils

$$S_C = \sum_{1}^{n} \left(\frac{C_c}{1+e_o}\right) H_o \log_{10} \frac{p_f}{p_o}$$

$$S_C = \sum_{1}^{n} C_{c\varepsilon} H_o \log_{10} \frac{p_f}{p_o}$$

CERM: Sec. 40.7

$$S_{\text{primary}} = \frac{H \Delta e}{1+e_o} = \frac{H C_c \log_{10} \dfrac{p_o' + \Delta p_v'}{p_o'}}{1+e_o} \qquad 40.16$$

$$= H(\text{CR}) \log_{10} \frac{p_o' + \Delta p_v'}{p_o'}$$

[normally consolidated]

In the *Handbook* equations,

C_c	coefficient of consolidation (void ratio)	–
$C_{c\varepsilon}$	coefficient of consolidation (strain)	–
e_o	initial void ratio	–
H_o	thickness of compressible layer	ft
p_f	final overburden pressure	lbf/ft²
p_o	initial overburden pressure	lbf/ft²
S_C	settlement from consolidation	ft

In the CERM equation,

C_c	compression index	–
CR	compression ratio	–
e_o	void ratio	–
Δe	change in void ratio	–
H	thickness of clay layer	in
p'_o	original effective pressure	lbf/ft²
$\Delta p'_v$	increase in vertical pressure	lbf/ft²
S_{primary}	primary settlement	in

In CERM Eq. 40.16, $\Delta p'_v$ is the increase in effective vertical pressure, and p'_o is the original effective pressure (effective stress) at the midpoint of the clay layer and directly below the foundation.

p_f in the *Handbook* equation is the final overburden pressure, represented in the CERM equation as $p'_o + \Delta p'_v$.

The coefficient of consolidation, $C_{c\varepsilon}$, also called the compression ratio, is defined in the *Handbook* as

$$C_{c\varepsilon} = \frac{C_c}{1+e_o}$$

This equation corresponds to CERM Eq. 40.12.

CERM Eq. 40.8 through Eq. 40.11 can be used to determine the compression index.

Consolidation Curves for Normally Consolidated Soils

Handbook: Typical Consolidation Curve for Normally Consolidated Soil

CERM: Sec. 40.6, Sec. 40.7

Primary consolidation occurs when a clay layer is loaded and water is squeezed from the voids. This happens over a long period of time, and water loss decreases with time. *Handbook* figure Typical Consolidation Curve for Normally Consolidated Soil shows the relationship between soil void ratio and vertical effective stress in part (a) and between vertical strain and vertical effective stress in part (b).

In this figure,

C_c	coefficient of consolidation (void ratio)	–
$C_{c\varepsilon}$	coefficient of consolidation (strain)	–
e_C	initial void ratio	–
e_f	final void ratio	–
e_o	initial void ratio	–
Δp	increase in effective stress	lbf/ft²
p_c	preconsolidation pressure	lbf/ft²
p_f	final overburden pressure	lbf/ft²
p_o	initial overburden pressure	lbf/ft²
ε_{vc}	initial vertical strain	–
ε_{vf}	final vertical strain	–
ε_{vo}	initial vertical strain	–

This graph gives a visual representation of all the values required for the corresponding settlement equation. The final overburden pressure is equal to the sum of the initial overburden pressure and the increase in the effective stress.

Exam tip: For normally consolidated soils, initial overburden pressure and preconsolidation pressure are equal.

Overconsolidated Soils

Handbook: Normally Consolidated Soils; Overconsolidated Soils

$$S = \sum_1^n \left(\frac{H_o}{1+e_o}\right)\left(C_r \log_{10} \frac{p_c}{p_o} + C_c \log_{10} \frac{p_f}{p_c}\right)$$

$$S = \sum_1^n H_o\left(C_{r\varepsilon} \log_{10} \frac{p_c}{p_o} + C_{c\varepsilon} \log_{10} \frac{p_f}{p_c}\right)$$

CERM: Sec. 40.5, Sec. 40.6, Sec. 40.7

$$S_{\text{primary}} = \frac{H\Delta e}{1+e_o} = \frac{HC_r \log_{10} \frac{p'_o + \Delta p'_v}{p'_o}}{1+e_o} \quad 40.15$$

$$= H(\text{RR})\log_{10} \frac{p'_o + \Delta p'_v}{p'_o}$$

[overconsolidated]

Clay that has previously experienced a stress that is no longer there is known as overconsolidated or preconsolidated clay.

In these equations, S is the settlement in feet. In CERM Eq. 40.15, $\Delta p'_v$ is the increase in effective vertical pressure, and p'_o is the original effective pressure (effective stress) at the midpoint of the clay layer and directly below the foundation. The nomenclature in these equations is otherwise identical, except for the final overburden pressure, represented in the *Handbook* equation as p_f and in CERM Eq. 40.15 as $p'_o + \Delta p'_v$ (i.e., the sum of the initial overburden pressure and the increase in the effective stress).

The coefficient of consolidation, $C_{c\varepsilon}$, also called the compression ratio, is defined in *Handbook* section Normally Consolidated Soils.

$$C_{c\varepsilon} = \frac{C_c}{1+e_o}$$

This equation corresponds to CERM Eq. 40.12.

CERM Eq. 40.8 through Eq. 40.11 can be used to determine the compression index.

Typical Consolidation Curve for Overconsolidated Soil

Handbook: Typical Consolidation Curve for Overconsolidated Soil

CERM: Sec. 40.5 through Sec. 40.7

In *Handbook* figure Typical Consolidation Curve for Overconsolidated Soil, the curve in part (a) shows the relationship between void ratio and vertical effective stress for overconsolidated soil, and the curve in part (b) shows the relationship between vertical strain and vertical effective stress in overconsolidated soil.

In this figure,

C_c	coefficient of consolidation (void ratio)	–
$C_{c\varepsilon}$	coefficient of consolidation (strain)	–
C_r	coefficient of reconsolidation (void ratio)	–
$C_{r\varepsilon}$	coefficient of reconsolidation (strain)	–
e_C	initial void ratio	–
e_f	final void ratio	–
e_o	initial void ratio	–
Δp	increase in effective stress	lbf/ft^2
p_c	preconsolidation pressure	lbf/ft^2
p_f	final overburden pressure	lbf/ft^2
p_o	initial overburden pressure	lbf/ft^2
ε_{vc}	initial vertical strain	–
ε_{vf}	final vertical strain	–
ε_{vo}	initial vertical strain	–

The beginning of the curve shows the overconsolidated segment and is also known as the *recompression curve*. The flatter portion of the curve, before the breakpoint, shows the stress that has previously been applied to the clay. The steeper portion of the curve shows the newly applied stress.

Consolidation of a Layer of Clay Between Two Pervious Layers

Handbook: Diagram Illustrating Consolidation of a Layer of Clay Between Two Pervious Layers

Handbook figure Diagram Illustrating Consolidation of a Layer of Clay Between Two Pervious Layers shows the consolidation behavior of a clay layer between an upper and lower layer of pervious soil.

The straight line from *a* to *b* in the *Handbook* figure shows the initial hydrostatic pressure in the clay layer. As the water pressure changes over time, the hydrostatic pressure is not linear, but curved, as shown on lines C_1 and line C_2 in the figure.

It is important to remember that this figure assumes that drainage is equal in both directions (i.e., both upper and lower materials are equally permeable).

> *Exam tip*: For normally consolidated soils, initial overburden pressure and preconsolidation pressure are equal.

Time Rate of Settlement

Handbook: Time Rate of Settlement

CERM: Sec. 40.8

$$t = \frac{T_v H_d^2}{c_v}$$

In this equation, H_d is the full thickness of the consolidated layer if there is one drainage surface, and half the layer thickness if there is drainage from both the top and bottom of the consolidated layer.

T_v is a dimensionless time factor that depends on the degree of consolidation, U_z. See the entry "Average Degree of Consolidation Versus Time Factor" in this chapter for information on approximate time factors.

The coefficient of consolidation, c_v, with typical units of ft²/day (m²/d), is assumed to remain constant over small variations in the void ratio, e.

$$C_v = \frac{K(1 + e_o)}{a_v \gamma_{\text{water}}} \qquad 40.21$$

Average Degree of Consolidation Versus Time Factor

Handbook: Average Degree of Consolidation, U, Versus Time Factor, T_v, for Uniform Initial Increase in Pore Water Pressure

CERM: Sec. 40.8

Handbook table Average Degree of Consolidation, U, Versus Time Factor, T_v, for Uniform Initial Increase in Pore Water Pressure shows the relationship between degree of consolidation and time factor. The degree of consolidation is the fraction of the total consolidation that is expected.

CERM Eq. 40.23 can be used to calculate the exact time factor. U_z in the CERM equations is the same as U in the *Handbook* table.

$$T_v = \frac{1}{4}\pi U_z^2 \quad [U_z < 0.60] \qquad \text{[SI]} \quad 40.23(a)$$

$$T_v = 1.781 - 0.933 \log\bigl(100(1 - U_z)\bigr) \qquad \text{[U.S.]} \quad 40.23(b)$$
$$[U_z \geq 0.60]$$

Settlement Ratio for Overconsolidation

Handbook: Settlement Ratio for Overconsolidation

The figure in *Handbook* section Settlement Ratio for Overconsolidation shows the relationship between the settlement ratio and overconsolidation ratio for different values of the uniform load and the thickness of the clay stratum.

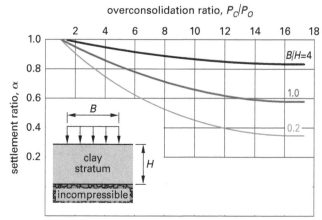

Source: United Facilities Criteria (UFC), *Soil Mechanics*, UFC 3-220-10N. Washington DC: U.S. Department of Defense, June 2005, Fig. 8, p. 225. wbdg.org/FFC/DOD/UFC/ufc_3_220_10n_2005.pdf

B	uniform load	lbf/ft
H	depth of clay stratum	ft
P_C	preconsolidation pressure	lbf/ft²
P_O	initial overburden pressure	lbf/ft²

The *Handbook* figure illustrates how the settlement ratio decreases as the ratio of uniform load, B, to thickness, H, decreases. In other words, the lower the load compared to the thickness of the clay, the lower the settlement ratio will be. In addition, the higher the overconsolidation ratio for a given load and thickness, the lower the settlement ratio will be.

C. EFFECTIVE AND TOTAL STRESSES

Key concepts: These key concepts are important for answering exam questions in subtopic 3.C, Effective and Total Stresses.

- effective versus total stress
- shear strength
- soil phases

NCEES Handbook: To prepare for this subtopic, familiarize yourself with these sections in the *Handbook*.

- Shear Strength of (a) Cohesionless Soils and (b) Cohesive Soils
- Shear Strength Effective Stress

- Relationship Between the Ratio of Undrained Shear Strength to Effective Overburden Pressure and Plasticity Index for Normally Consolidated and Overconsolidated Clays
- Drained Shear Strength of Clays
- Unified Soil Classification System (USCS)

PE Civil Reference Manual (**CERM**): Study these sections in CERM that either relate directly to this subtopic or provide background information.

- Section 35.7: Effective Stress
- Section 35.17: Direct Shear Test
- Section 35.18: Triaxial Stress Test

The following equations and figures are relevant for subtopic 3.C, Effective and Total Stresses.

Shear Strength: Total Stress

Handbook: Shear Strength of (a) Cohesionless Soils and (b) Cohesive Soils

CERM: Sec. 35.17

The direct shear test is used to determine the relationship of shear strength to consolidation stress. In this test, a disc of soil is inserted into a direct shear box. The box has a top half and a bottom half that can slide laterally with respect to each other. A normal stress is applied vertically, and then one half of the box is moved laterally relative to the other at a constant rate. Measurements of vertical and horizontal displacement and horizontal shear load are taken. The test is usually repeated at three different vertical normal stresses.

Handbook figure Shear Strength of (a) Cohesionless Soils and (b) Cohesive Soils shows the relationship between soil shear strength and consolidation (normal) stress. The lines on the graphs represent the failure envelopes given by Coulomb's equation. See CERM Fig. 35.13 for another example.

Effective Shear Strength

Handbook: Shear Strength Effective Stress

$$\tau' = c' + (\sigma_n - u)\tan\phi' = c' + \sigma'_n \tan\phi'$$

CERM: Sec. 35.18

$$s = c' + \sigma'\tan\phi' \quad \quad 35.43$$

The *Handbook* equation and CERM Eq. 35.43 are identical except that the *Handbook* uses τ' for shear stress while CERM uses s.

The first form of the *Handbook* equation includes the calculation for effective stress, σ'_n, which is the portion of the total stress that is supported through grain contact and is equal to the difference between the normal stress and the pore pressure.

$$\sigma' = \sigma - u$$

If soil is saturated, an increase in load may be supported by the soil at soil-grain contact points or through increases in pore pressure. The amount each mode contributes depends on the rate of loading and the permeability of the soil.

Undrained Shear Strength of Clays

Handbook: Relationship Between the Ratio of Undrained Shear Strength to Effective Overburden Pressure and Plasticity Index for Normally Consolidated and Overconsolidated Clays; Unified Soil Classification System (USCS)

Handbook figure Relationship Between the Ratio of Undrained Shear Strength to Effective Overburden Pressure and Plasticity Index for Normally Consolidated and Overconsolidated Clays shows the variation of the undrained shear strength/effective overburden pressure ratio versus the plasticity index of clayey soils. The data was collected from four different researchers, Skempton, Kenney, "young" Bjerrum, and "aged" Bjerrum, as named in the figure.

p_c is the preconsolidation pressure, and p_o is the in situ overburden pressure. The plasticity index, PI, is determined using the equation in Unified Soil Classification System (USCS).

$$PI = LL - PL$$

The data from Skempton provides a good average estimate of values and allows for a conceptual, ballpark check of results obtained in the laboratory on individual soil samples. As shown by all four curves, the undrained shear strength/effective overburden pressure ratio increases as the plasticity index increases. It can also be observed that the ratio increases with an increase in the overconsolidation ratio (OCR).

Drained Shear Strength of Clays

Handbook: Drained Shear Strength of Clays

Handbook figure Drained Shear Strength of Clays shows how the friction angle decreases as the plasticity index increases in various clays.

The plasticity index, PI, is determined using the equation from Unified Soil Classification System (USCS).

$$PI = LL - PL$$

A higher PI indicates a greater amount of clay in the soil, which corresponds to a lower friction angle. As shown by the individual scatter plot points, the friction angle is highly variable when simply compared to PI, and this relationship should be used for conceptual estimates only.

D. BEARING CAPACITY

Key concepts: These key concepts are important for answering exam questions in subtopic 3.D, Bearing Capacity.

- bearing capacities of sand versus clay
- footing shapes
- frost depth
- Terzaghi bearing capacity factors
- Terzaghi-Meyerhof equation

NCEES Handbook: To prepare for this subtopic, familiarize yourself with these sections in the *Handbook*.

- Bearing Capacity Theory
- Bearing Capacity Equation for Concentrically Loaded Strip Footings
- Bearing Capacity Equation for Concentrically Loaded Square or Rectangular Footings
- Frost Depth

PE Civil Reference Manual **(CERM):** Study these sections in CERM that either relate directly to this subtopic or provide background information.

- Section 36.3: General Considerations for Footings
- Section 36.5: General Bearing Capacity Equation
- Section 36.6: Selecting a Bearing Capacity Theory
- Section 36.7: Bearing Capacity of Clay
- Section 36.8: Bearing Capacity of Sand
- Section 36.9: Shallow Water Table Correction
- Section 36.11: Effects of Water Table on Footing Design
- Section 36.14: Rafts on Clay
- Section 36.15: Rafts on Sand

The following equations and figures are relevant for subtopic 3.D, Bearing Capacity.

Bearing Capacity Theory

Handbook: Bearing Capacity Theory

CERM: Sec. 36.6

The figure in *Handbook* section Bearing Capacity Theory presents modes of bearing capacity failures and the relationship between soil settlement and load for each case. CERM Sec. 36.6 covers bearing capacity theories, their considerations, where each is appropriate, and how they may affect the calculated bearing capacity value.

Terzaghi's shallow foundation bearing capacity theory predicts shear failure along known planes and is derived from level strip footings with footing depths of $D_f < B$.

The Meyerhof model includes correction factors for footing shape, eccentricity, load inclination, and foundation depth. It varies from Terzaghi's theory in that it takes shear strength of soil above the footing into account so the benefit of surcharge is included.

The Hansen model considers footing depth, shape, and load inclination.

The Vesic model closely follows the Meyerhof and Hansen models but addresses the concern that local shear failure leads to lower bound estimates of ultimate bearing capacity.

Bearing Capacity Equation: Concentrically Loaded Strip Footings

Handbook: Bearing Capacity Equation for Concentrically Loaded Strip Footings

$$q_{ult} = cN_c + qN_q + 0.5\gamma B_f N_\gamma$$

$$q = q_{appl} + \gamma_a D_f$$

$$N_q = e^{\pi \tan \phi} \tan^2\left(45° + \frac{\phi}{2}\right)$$

$$N_c = 2 + \pi = 5.14 \quad [\text{for } \phi = 0°]$$

$$N_c = (N_q - 1)\cot\phi = 5.14 \quad [\text{for } \phi > 0°]$$

$$N_\gamma = 2(N_q - 1)\tan\phi$$

CERM: Sec. 36.5

$$q_{ult} = \tfrac{1}{2}\rho g B N_\gamma + cN_c + (p_q + \rho g D_f)N_q \quad [\text{SI}] \quad 36.1(a)$$

$$q_{ult} = \tfrac{1}{2}\gamma B N_\gamma + cN_c + (p_q + \gamma D_f)N_q \quad [\text{U.S.}] \quad 36.1(b)$$

In these equations,

B_f	width of footing	ft
c	cohesion of the soil	lbf/ft²
q	total surcharge at the base of the footing	lbf/ft
q_{ult}	ultimate bearing capacity	lbf/ft²
N_c	bearing capacity factor for cohesion	–
N_q	bearing capacity factor for surcharge	–
N_γ	bearing capacity factor for unit weight	–
γ	unit weight of soil	lbf/ft³

The equation for the ultimate bearing capacity for a shallow continuous footing, known as the Terzaghi-Meyerhof equation, is valid for both sandy and clayey soils. The *Handbook* refers to this equation as the bearing capacity equation, and its version of the equation and CERM Eq. 36.1(b) are equivalent with a few minor differences in nomenclature. In the *Handbook* equation, the total surcharge at the base of the footing, q, is found from the equation shown. In CERM Eq. 36.1(b), the relationship between applied surcharge (represented by p_q instead of q, as in the *Handbook*), unit weight of soil above base of footing, and depth of footing is included in the equation.

Handbook section Bearing Capacity Equation for Concentrically Loaded Strip Footings provides additional equations for determining the various bearing capacity factors. The bearing capacity factors N_c, N_q, and N_γ can also be found using *Handbook* table Bearing Capacity Factors. CERM Fig. 36.2 and Table 36.2 are equivalent resources to the *Handbook* table. Bearing capacity factors are covered in more detail in Chap. 25 of this book.

Bearing Capacity Equation: Concentrically Loaded Square or Rectangular Footings

Handbook: Bearing Capacity Equation for Concentrically Loaded Square or Rectangular Footings

$$q_{ult} = cN_c s_c + qN_q s_q + 0.5\gamma B_f N_\gamma s_\gamma$$

CERM: Sec. 36.5

$$q_{ult} = \tfrac{1}{2}\rho g B N_\gamma + cN_c + (p_q + \rho g D_f)N_q \quad \text{[SI]} \qquad 36.1(a)$$

$$q_{ult} = \tfrac{1}{2}\gamma B N_\gamma + cN_c + (p_q + \gamma D_f)N_q \quad \text{[U.S.]} \qquad 36.1(b)$$

In these equations,

B_f	width of footing	ft
c	cohesion of the soil	lbf/ft²
N_c	bearing capacity factor for cohesion	–
N_q	bearing capacity factor for surcharge	–
N_γ	bearing capacity factor for unit weight	–
q	total surcharge at the base of the footing	lbf/ft
q_{ult}	ultimate bearing capacity	lbf/ft²
s_c	shape factor for cohesion	–
s_q	shape factor for surcharge	–
s_γ	shape factor for unit weight	–
γ	unit weight of soil	lbf/ft³

The *Handbook* equation for the ultimate bearing capacity for concentrically loaded square or rectangular footings and CERM Eq. 36.1(b) are similar with a few exceptions. In the *Handbook* equation, the total surcharge at the base of the footing, q, is found from the equation shown in section Bearing Capacity Equation for Concentrically Loaded Strip Footings. In CERM Eq. 36.1(b), the relationship between applied surcharge (represented by p_q instead of q, as in the *Handbook*), unit weight of soil above base of footing, and depth of footing is included in the equation. The other exception is the shape factor, s_q, which is only present in the *Handbook* equation.

This ultimate bearing capacity equation is appropriate for a foundation in a continuous wall footing. The bearing capacity factors N_c and N_γ are multiplied by the appropriate shape factors when used in CERM Eq. 36.1(b).

See *Handbook* table Shape Correction Factors for shape factor terms. The bearing capacity factors N_c, N_q, and N_γ can be found using *Handbook* table Bearing Capacity Factors. CERM Fig. 36.2 and Table 36.2 are equivalent resources to the *Handbook* table.

Bearing capacity factors and shape correction factors are covered in more detail in Chap. 25 of this book.

Frost Depth

Handbook: Frost Depth

Handbook figure Frost Depth shows the approximate frost depth across the United States. The frost depth is the maximum depth below the ground surface above which water in the soil will freeze and create frost heave.

Frost depth is important when it comes to foundation design problems. The frost depth can be used to determine the frost line, which is the minimum depth at which foundations should be designed to avoid frost heave. Frost heave is a phenomenon that occurs when moisture within soil freezes and expands.

E. FOUNDATION SETTLEMENT

Key concepts: These key concepts are important for answering exam questions in subtopic 3.E, Foundation Settlement.

- Boussinesq contour chart
- Boussinesq's equation
- influence chart
- methods of determining settlement
- pressure from applied loads
- zone of influence

NCEES Handbook: To prepare for this subtopic, familiarize yourself with these sections in the *Handbook*.

- Vertical Stress Contours (Isobars) Based on Boussinesq's Theory for Continuous and Square Footings
- Distribution of Vertical Stress by the 2:1 Method
- Settlement (Elastic Method)
- Shape and Rigidity Factors, C_d, for Calculating Settlements of Points on Loaded Areas at the Surface of a Semi-Infinite, Elastic Half Space
- Elastic Constants of Various Soils
- Settlement (Schmertmann's Method)
- Simplified Vertical Strain Influence Factor Distributions

PE Civil Reference Manual **(CERM):** Study these sections in CERM that either relate directly to this subtopic or provide background information.

- Section 40.1: Pressure from Applied Loads: Boussinesq's Equation
- Section 40.2: Pressure from Applied Loads: Zone of Influence
- Section 40.3: Pressure from Applied Loads: Influence Chart

The following equations and figures are relevant for subtopic 3.E, Foundation Settlement.

Vertical Stress Contours for Continuous and Square Footings

Handbook: Vertical Stress Contours (Isobars) Based on Boussinesq's Theory for Continuous and Square Footings

CERM: Sec. 40.1

The stress distribution for various depths based on Boussinesq's theory (also known as stress contours) is shown in *Handbook* figure **Vertical Stress Contours (Isobars) Based on Boussinesq's Theory for Continuous and Square Footings.**

The increase in vertical pressure caused by the application of a point load at the surface can be found from CERM Eq. 40.1, Boussinesq's equation.

$$\Delta p_v = \frac{3h^3 P}{2\pi s^5}$$

$$= \left(\frac{3P}{2\pi h^2}\right)\left(\frac{1}{1+\left(\frac{r}{h}\right)^2}\right)^{5/2} \quad [h > 2B] \quad 40.1$$

CERM Eq. 40.1 assumes the footing width, B, is small compared to the depth, h, and that the soil is semi-infinite, elastic, isotropic, and homogenous. CERM Eq. 40.1 can be solved using a Boussinesq contour chart.

Zone of Influence

Handbook: Distribution of Vertical Stress by the 2:1 Method

CERM: Sec. 40.2

The zone of influence is the area of the horizontal plane enclosed by the influence boundaries, as shown in *Handbook* figure **Distribution of Vertical Stress by the 2:1 Method.** CERM Fig. 40.2 is a similar figure shown here as an example. It illustrates how the 2:1 method is used to determine the zone of influence of the footing.

Figure 40.2 *Zone of Influence*

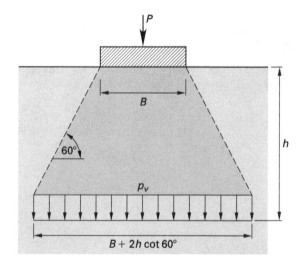

In the CERM figure and following equations,

A	area	ft^2
B	footing width	ft
h	distance below water table	ft
L	length	ft
p_v	vertical pressure	lbf/ft^2
P	load	lbf

The 2:1 method is used to calculate the area of the zone of influence of the horizontal plane enclosed by the influence boundaries.

CERM Eq. 40.2 is the complete form of the equation shown in CERM Fig. 40.2.

$$A = \big(B + 2(h \cot 60°)\big)\big(L + 2(h \cot 60°)\big) \quad 40.2$$

However, for simplicity, the 60° angle is assumed to be 63.4° for 2V:1H to make calculations easier. The equation then simplifies to CERM Eq. 40.3, which shows how to calculate the rectangular area of influence.

$$A = (B + h)(L + h) \quad 40.3$$

CERM Eq. 40.2 and Eq. 40.3 assume a rectangular footing of width B and length L using the 60° or 2:1 method to calculate the area of influence at depth d.

CERM Eq. 40.4 calculates the increase in pressure at the specified depth.

$$\Delta p_v = \frac{P}{A} \quad 40.4$$

Elastic Method of Settlement

Handbook: Settlement (Elastic Method); Shape and Rigidity Factors, C_d, for Calculating Settlements of Points on Loaded Areas at the Surface of a Semi-Infinite, Elastic Half Space; Elastic Constants of Various Soils

$$\delta_v = \frac{C_d \Delta p B_f (1 - v^2)}{E_m}$$

The vertical settlement at the surface of a soil, also known as the elastic settlement, is found from the equation shown. *Handbook* table Shape and Rigidity Factors, C_d, for Calculating Settlements of Points on Loaded Areas at the Surface of a Semi-Infinite, Elastic Half Space shows the shape and rigidity factors, and *Handbook* table Elastic Constants of Various Soils shows typical values for the Young's modulus and Poisson's ratio for different soil types.

Schmertmann's Method

Handbook: Settlement (Schmertmann's Method)

$$S_i = C_1 C_2 \Delta p \sum_{i=1}^{n} \Delta H_i$$

$$\Delta H_i = H_c \left(\frac{I_z}{XE_s} \right)$$

Schmertmann's method is used to estimate the immediate elastic settlement of the soil under the footing. ΔH_i is the thickness of the ith sublayer. See the entry "Vertical Strain Influence Factor Distributions" in this chapter for an explanation of how the influence factor, I_z, is determined.

Schmertmann's method was developed primarily for spread footings. See CERM Fig. 36.3 for the dimensions of a spread footing.

Figure 36.3 Spread Footing Dimensions

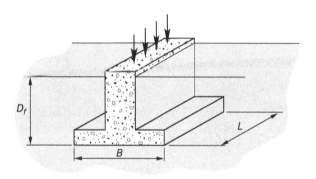

Vertical Strain Influence Factor Distributions

Handbook: Simplified Vertical Strain Influence Factor Distributions

Handbook figure Simplified Vertical Strain Influence Factor Distributions shows the relationship between the rigid footing vertical strain influence factor and the relative depth below the footing and gives the equation for the depth to peak strain influence factor, I_{zp}.

$$I_{zp} = 0.5 + 0.1 \left(\frac{\Delta p}{p_{op}} \right)^{0.5}$$

In the *Handbook* equation,

I_{zp}	peak strain influence factor	–
Δp	change in pressure	lbf/ft^2
p_{op}	effective overburden pressure at depth of peak strain	lbf/ft^2

Influence Chart

CERM: Sec. 40.3

CERM Fig. 40.3 contains an example of an influence chart. The implied scale in the influence chart is determined by making the indicated distance A-B equal to the depth at which the pressure is wanted. Using this scale, a plan view of the mat foundation is drawn and the number of squares under the footing is counted, with partial squares counted as fractions. The number is then multiplied by the influence value I and the applied pressure to calculate the pressure at depth.

$$\Delta p_v = I(\text{no. of squares})p_{\text{app}} \quad 40.5$$

F. SLOPE STABILITY

Key concepts: These key concepts are important for answering exam questions in subtopic 3.F, Slope Stability.

- slope stability guidelines for design
- stability of clay versus stability of sand
- Taylor slope stability
- types of slope failures

NCEES Handbook: To prepare for this subtopic, familiarize yourself with these sections in the *Handbook*.

- Stability Charts for $\phi = 0$ Soils
- Taylor's Chart for Soils with Friction Angle
- Taylor's Chart for $\phi' = 0$ Conditions for Slope Angles (β) Less than 54°
- Stability Analysis of Transitional Failure
- Stability of Rock Slope
- Infinite Slope
- Infinite Slope Failure in Dry Sand
- Slice for Ordinary Method of Slices with External Water Loads
- Typical Slice and Forces for Ordinary Method of Slices
- Slope Stability Guidelines for Design

PE Civil Reference Manual **(CERM):** Study these sections in CERM that either relate directly to this subtopic or provide background information.

- Section 40.10: Slope Stability in Saturated Clay

The following equations, figures, and concepts are relevant for subtopic 3.F, Slope Stability.

Slope Stability Guidelines for Design

Handbook: Stability Charts for $\phi = 0$ Soils; Slope Stability Guidelines for Design

CERM: Sec. 40.10

The angle of internal friction is the maximum slope for cuts in a cohesionless drained sand. A Taylor's slope stability chart can be used to determine the maximum slope angle for cohesive soils. Taylor's charts relate a dimensionless depth factor, d, with a dimensionless stability number, N_o. See *Handbook* figure Stability Charts for $\phi = 0$ Soils and CERM Fig. 40.6 for examples.

Figure 40.6 Taylor Slope Stability (undrained, cohesive soils; $\phi = 0°$)

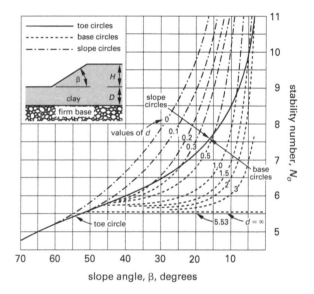

Source: *Soil Mechanics*, NAVFAC Design Manual DM-7.1, 1986, Fig. 2, p. 7.1-319.

For more information on Taylor's slope stability charts, review the entry "Taylor's Chart for Slope Angles or Friction Angles" in this chapter.

Design guidelines are available in *Handbook* table Slope Stability Guidelines for Design.

Taylor's Chart for Slope Angles or Friction Angles

Handbook: Taylor's Chart for Soils with Friction Angle; Taylor's Chart for $\phi' = 0$ Conditions for Slope Angles (β) Less than 54°

CERM: Sec. 40.10

Handbook figure Taylor's Chart for $\phi' = 0$ Conditions for Slope Angles (β) Less than 54° indicates the different failure modes for soils with a slope angle less than 54°. *Handbook* figure Taylor's Chart for Soils with Friction Angle indicates the different failure modes for soils with a slope angle greater than 54°.

The chart assumes there is no open water outside of the slope, there are no surcharges or tension cracks, shear strength is only due to cohesion, and failure takes place as a rotation around an arc.

The depth factor used in a Taylor's chart is found using the vertical distance of the slope to the firm base, D, and the slope height, H, as shown in CERM Eq. 40.27.

$$d = \frac{D}{H} \qquad 40.27$$

Exam tip: The Taylor's stability chart shows that toe circle failures will always occur in slopes steeper than or equal to 54°.

Translational Failure

Handbook: Stability Analysis of Transitional Failure

The top third of *Handbook* figure Stability Analysis of Transitional Failure shows the potential sliding surfaces within different soil strata at different positions along the slopes. The middle and lower thirds of the *Handbook* figure show the resultant horizontal force for a wedge sliding along P_α and a resultant horizontal force for a wedge sliding along P_β, respectively.

Rock Slope Failure

Handbook: Stability of Rock Slope

Handbook figure Stability of Rock Slope shows the free-body diagram of a mass of soil on an inclined plane of rock, as well as the forces acting on said mass of soil.

The factor of safety, F_s, can be determined using the equation shown at the bottom of the figure.

$$F_s = \frac{cL + \begin{pmatrix} W\cos\alpha - kW\sin\alpha \\ -P_{\text{WR}} - P_{\text{WJ}}\sin\alpha \end{pmatrix}\tan\phi}{W\sin\alpha + kW\cos\alpha + P_{\text{WJ}}\cos\alpha}$$

In this equation,

c	cohesive strength of failure surface	lbf/ft^2
F_s	factor of safety	–
k	seismic coefficient to account for dynamic horizontal force	–
L	length	ft
P_{WJ}	water force on the joint reaction, $\gamma_w h_w^2/2$	lbf/ft
P_{WR}	water force on the assumed failure plane, $\gamma_w h_w L/2$	lbf/ft
W	weight of wedge	lbf
α	angle	deg
ϕ	angle	deg

Infinite Slope Failure in Dry Sand

Handbook: Infinite Slope; Infinite Slope Failure in Dry Sand

Handbook figure Infinite Slope Failure in Dry Sand shows the forces acting on a mass of dry sand and the potential slope failure surface.

The factor of safety can be computed for infinite slope either with (top) or without the pore water pressure (bottom).

$$F = \left(\frac{\tan\phi'}{\tan\beta}\right)(1 - r_u(1 + \tan^2\beta))$$

$$F = \frac{\tan\phi'}{\tan\beta}$$

Exam tip: The pore water pressure is sometimes referred to as the neutral stress because it is equal in all directions.

Forces for Ordinary Method of Slices

Handbook: Slice for Ordinary Method of Slices with External Water Loads; Typical Slice and Forces for Ordinary Method of Slices

Handbook figure Slice for Ordinary Method of Slices with External Water Loads shows the free-body diagram for soil on a circular failure surface. The method of slices involves breaking the sloped soil into segments of equal width. The free-body diagram of each slice can then be treated as a statics problem and solved for the unknowns. The overall stability calculation is an iterative, guess-and-check procedure using a summation of forces from each slice within the failure zone. *Handbook* figure Typical Slice and Forces for Ordinary Method of Slices shows the forces acting on each individual slice.

Structural Mechanics

Content in blue refers to the *NCEES Handbook*.

A. Dead and Live Loads 4-1
B. Trusses 4-2
C. Bending 4-4
D. Shear 4-6
E. Axial 4-7
F. Combined Stresses 4-8
G. Deflection 4-9
H. Beams 4-10
I. Columns 4-13
J. Slabs 4-15
K. Footings 4-17
L. Retaining Walls 4-17

The knowledge area of Structural Mechanics makes up between five and eight questions out of the 80 questions on the PE Civil exam. The organization of this chapter follows the order of subtopics given by NCEES for this knowledge area. Each subtopic is covered in the following sections.

A. DEAD AND LIVE LOADS

Key concepts: These key concepts are important for answering exam questions in subtopic 4.A, Dead and Live Loads.

- allowable strength design (ASD) versus load and resistance factor design (LRFD)
- ASCE/SEI 7 specifications for dead and live loads
- concentrated live loads
- live load design
- load combinations
- partial loading on a structure
- partition loads

NCEES Handbook: To prepare for this subtopic, familiarize yourself with these sections in the *Handbook*.

- Influence Lines for Beams and Trusses

PE Civil Reference Manual (CERM): Study these sections in CERM that either relate directly to this subtopic or provide background information.

- Section 41.25: Influence Lines for Reactions
- Section 46.10: Influence Diagrams
- Section 50.4: Allowable Stress Design Versus Strength Design
- Section 50.5: Service Loads, Factored Loads, and Load Combinations

The following concepts are relevant for subtopic 4.A, Dead and Live Loads.

Influence Lines

Handbook: Influence Lines for Beams and Trusses

CERM: Sec. 41.25

An influence line is a graph of the magnitude of a reaction as a function of load placement. The x-axis corresponds to the position of the load along the member span, and the y-axis corresponds to the magnitude of the load.

Influence lines show the variation of load, shear force, and moment as a function of the load location, and they can be used to determine the load position causing the maximum shear and moment in a beam or truss.

Dead Loads

CERM: Sec. 50.4

Dead loads are loads within a building system that are considered to be permanently in place and not transient. Dead loads consist of the combined weight of all materials of construction incorporated into the building, including but not limited to walls, floors, roofs, ceilings, stairways, built-in partitions, finishes, cladding, and other similarly incorporated architectural and structural items; as well as fixed service equipment, including the weight of cranes.

Minimum Design Loads for Buildings and Other Structures (ASCE/SEI 7) is the most commonly used standard for dead and live load estimation. It also covers snow, wind, seismic, and other loads. These topics are outside the scope of this knowledge area.

Live Loads

CERM: Sec. 50.4

Live loads are loads within a building system that are considered to be transient or impermanent. These include people and furniture. Mechanical equipment is typically considered to be a live load due to moving parts and the vibrations caused during normal operation.

Live load reduction is particularly useful for the design of columns and foundations of high-rise buildings. Live load reduction applies only to uniformly distributed live loads; it is not permitted for live loads exceeding 100 lbf/ft^2 (4.79 kN/m^2) or for passenger vehicle garages.

Concentrated live loads can have a greater impact on a structure than uniformly distributed loads. When considering concentrated live loads or uniformly distributed loads, design for whichever load will cause the greatest action.

Allowable Stress Design Versus Allowable Strength Design

CERM: Sec. 50.4

Structural members can be sized using one of two alternative design procedures. In allowable strength design (ASD), the stresses induced in a member are estimates made with the assumption that the behavior is linearly elastic, and the member is sized so that the computed stresses do not exceed certain predetermined values. In load and resistance factor design (LRFD), a member is chosen with adequate strength to support appropriately factored loads.

The term *service loads* refers to the loads expected to be imposed on the structure during its service life. Service loads are taken from applicable building codes. The term *factored loads* refers to service loads that have been increased by various load factors. Factored loads can also be thought of as the required strength.

For more on ASD and LRFD, see the entries "Allowable Strength Design Method" and "Load and Resistance Factor Design Method" in Chap. 25 of this book.

B. TRUSSES

Key concepts: These key concepts are important for answering exam questions in subtopic 4.B, Trusses.

- equilibrium
- free-body diagrams
- method of joints
- method of sections
- reactions in two dimensions for determinate structures
- special types of trusses
- zero-force members

NCEES Handbook: To prepare for this subtopic, familiarize yourself with these sections in the *Handbook*.

- Resolution of a Force
- Systems of Forces
- Plane Truss: Method of Joints
- Plane Truss: Method of Sections

PE Civil Reference Manual (**CERM**): Study these sections in CERM that either relate directly to this subtopic or provide background information.

- Section 41.17: Conditions of Equilibrium
- Section 41.22: Free-Body Diagrams
- Section 41.31: Trusses
- Section 41.32: Determinate Trusses
- Section 41.33: Zero-Force Members
- Section 41.34: Method of Joints
- Section 41.36: Method of Sections

The following equations, figures, and concepts are relevant for subtopic 4.B, Trusses.

Free-Body Diagrams

Handbook: Resolution of a Force; Systems of Forces

CERM: Sec. 41.22

A free-body diagram is a representation of a body in equilibrium. It shows all applied forces, moments, and reactions but does not consider the internal structure or construction of the body. Free-body diagrams are fundamental to structural engineering, and understanding them is important for the exam.

CERM Fig. 41.10 shows the difference between a drawing of a structure (on the left) and a free-body diagram (on the right). Note the section cut on the truss.

Figure 41.10 Bodies and Free Bodies

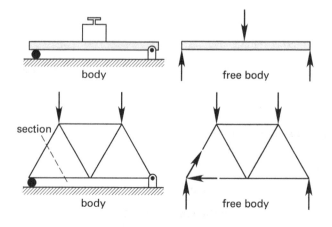

In a free-body diagram, the body is typically isolated from its physical supports in order to evaluate the reaction forces. (Including supports in a free-body diagram is a common mistake.) In other cases, the body may be sectioned (that is, cut) in order to determine the forces at the section. The equations in *Handbook* sections Resolution of a Force and Systems of Forces can be used to calculate components of the forces.

Drawing a free-body diagram including the coordinate axis and the direction of rotation can be useful for solving exam problems. Establishing the positive direction of rotation (e.g., clockwise) can make it easier to find moments in the structure. The system of axes and positive moment direction is entirely arbitrary, but when establishing a convenient set of coordinate axes, it is generally preferred that one axis be along the direction of the structure.

Equilibrium

Handbook: Plane Truss: Method of Joints

$$\sum F_H = 0$$

$$\sum F_V = 0$$

CERM: Sec. 41.17

$$\mathbf{F}_R = \sum \mathbf{F} = 0 \qquad 41.35$$

$$F_R = \sqrt{F_{R,x}^2 + F_{R,y}^2 + F_{R,z}^2} = 0 \qquad 41.36$$

$$\mathbf{M}_R = \sum \mathbf{M} = 0 \qquad 41.37$$

$$M_R = \sqrt{M_{R,x}^2 + M_{R,y}^2 + M_{R,z}^2} = 0 \qquad 41.38$$

The *Handbook* equations and CERM Eq. 41.35 through Eq. 41.38 show that, if a force system is in equilibrium, the forces in the vertical direction and those in the horizontal direction must each sum to zero.

An object is *static* when it is stationary. To be stationary, all forces and moments on the object must be in equilibrium.

A typical exam problem covers forces in two directions, such as in a truss. The methods used to solve two-dimensional problems are the same methods used to solve three-dimensional problems.

Method of Joints

Handbook: Plane Truss: Method of Joints

$$\sum F_H = 0$$

$$\sum F_V = 0$$

CERM: Sec. 41.34

The method of joints is one of the two main ways to find the internal forces in each member of a truss. For more about the method of sections, see the entry "Method of Sections" in this chapter.

The method of joints directly applies the equations of equilibrium in the x- and y-directions. Start by finding the reactions supporting the truss. Next, evaluate the joint at one of the reactions, which identifies all member forces framing into the joint. Then, using the knowledge of one or more of the member forces from the previous step, analyze an adjacent joint. Repeat the process until all unknown quantities are found.

This method is most useful in finding the forces in all the members of a truss. If only one member is being solved for, especially a member in the middle of the span, the method of joints is inefficient, as it requires calculating the forces all the way from the support.

Method of Sections

Handbook: Plane Truss: Method of Sections

CERM: Sec. 41.36

The method of sections is one of the two main ways to find the internal forces in each member of a truss. For more about the method of joints, see the entry "Method of Joints" in this chapter.

The method of sections is a direct approach to finding the forces in any truss member. First, the support reactions are found. Then, a cut is made through the truss passing through the unknown member. Finally, all three conditions of equilibrium are applied as needed to the remaining truss portion. This method is convenient when only a few truss member forces are unknown.

Truss Basics

CERM: Sec. 41.31

A truss is a set of pin-connected axial members. Member weights are neglected, and loads are applied only at connection points, referred to as *joints*. Each closed loop of members is referred to as a *structural cell*. For a truss to be stable (i.e., to be a rigid truss), all structural cells must be triangles. CERM Fig. 41.13 illustrates the parts of a bridge truss.

Figure 41.13 Parts of a Bridge Truss

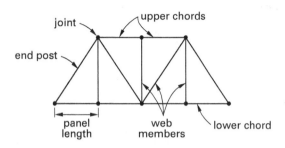

Trusses are considered to act only in the plane of the truss, and they are analyzed as two-dimensional structures. Forces in truss members hold the various parts together and are known as *internal forces*. Internal forces are found by applying equations of equilibrium to appropriate free-body diagrams. Free-body diagrams are typically made of the joints of a truss.

See CERM Fig. 41.14 for examples of special types of trusses.

Statically Determinate Trusses

CERM: Sec. 41.32

$$\text{no. of members} = 2(\text{no. of joints}) - 3 \quad 41.48$$

A truss is *statically determinate* if the number of members in the truss is equal to twice the number of joints in the truss minus three, as CERM Eq. 41.48 shows. If the value of the left-hand side of the equation is greater than the value of the right-hand side (i.e., there are redundant members), the truss is statically indeterminate. If the value of the left-hand side of the equation is less than the value of the right-hand side, the truss is unstable and will collapse under certain types of loading.

Zero-Force Members

CERM: Sec. 41.33

A third member framing into a joint that already connects two collinear members carries no internal force unless there is a load applied at that joint. In this case, this third member is referred to as a *zero-force member*. Similarly, both members forming an apex of a truss are zero-force members unless there is a load applied at the apex. See CERM Fig. 41.15 for an illustration of this concept.

Figure 41.15 Zero-Force Members

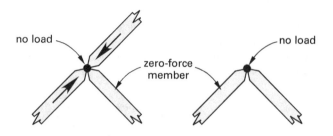

Where there are zero-force members in a truss, the forces in the truss can sometimes be determined by inspection. This rule can be proven by applying the equations of equilibrium at the joints.

Exam tip: Take some time to learn about recognizing zero-force members. They can be used to simplify solving a truss.

C. BENDING

Key concepts: These key concepts are important for answering exam questions in subtopic 4.C, Bending.

- bending stress
- bending stress distribution in a beam
- moment at a point
- section modulus
- shear and moment diagrams

NCEES Handbook: To prepare for this subtopic, familiarize yourself with these sections in the *Handbook*.

- Area Moment of Inertia (Table)
- Shearing Force and Bending Moment Sign Conventions
- Stresses in Beams
- Moment, Shear, and Deflection Diagrams

PE Civil Reference Manual (CERM): Study these sections in CERM that either relate directly to this subtopic or provide background information.

- Section 44.11: Shear and Moment
- Section 44.12: Shear and Bending Moment Diagrams
- Section 44.14: Bending Stress in Beams

The following equations and figures are relevant for subtopic 4.C, Bending.

Bending Moment Diagrams

Handbook: Shearing Force and Bending Moment Sign Conventions; Moment, Shear, and Deflection Diagrams

$$w(x) = -\frac{dV(x)}{dx}$$

$$V = \frac{dM(x)}{dx}$$

$$V_2 - V_1 = \int_{x1}^{x2} -w(x)\,dx$$

$$M_2 - M_1 = \int_{x1}^{x2} V(x)\,dx$$

CERM: Sec. 44.11, Sec. 44.12

$$V = \frac{dM}{dx} \qquad 44.26$$

$$M = \int V dx \qquad 44.27$$

Handbook section Shearing Force and Bending Moment Sign Conventions contains tips for understanding the relationship between the load, shear, and moment in beams. *Handbook* section Moment, Shear, and Deflection Diagrams contains a list of variables used in drawing moment diagrams, as well as an exhaustive list of tables that include various moment diagrams. CERM Eq. 44.26 and Eq. 44.27 are equivalent to the *Handbook* equations for shear and moment.

The moment diagram is often drawn on the compression face of concrete beams, so the engineer can remember which face the reinforcement needs to be applied to. CERM Fig. 44.9 illustrates this concept.

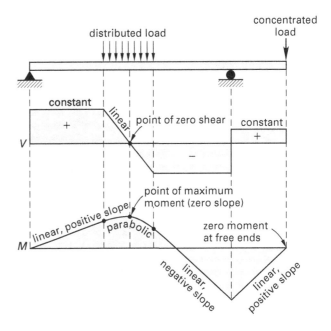

Figure 44.9 Drawing Shear and Moment Diagrams

For more on bending moments, see the entry "Bending Moments" in this chapter.

Exam tip: On exam problems, students may be asked to choose the appropriate shear or bending moment diagram. Become familiar with how different loading and boundary conditions can affect the overall shear and bending moment diagrams.

Bending Stresses

Handbook: Area Moment of Inertia (Table); Stresses in Beams

$$\sigma = -\frac{My}{I}$$

$$\sigma = \pm\frac{Mc}{I}$$

CERM: Sec. 44.14

These two *Handbook* equations calculate, first, the normal stress in a beam due to bending and, second, the maximum normal stress in a beam due to bending. Although bending stress is a normal stress, the term *bending stress* (or *flexural stress*) is used to indicate the cause of the stress.

Normal stress occurs in a bending beam where the beam is acted upon by a transverse force, as shown in CERM Fig. 44.12. Bending stress varies with location (depth) within the beam. The dashed line in the figure indicates the neutral plane (at which there is neither tension nor compression).

Figure 44.12 Normal Stress Due to Bending

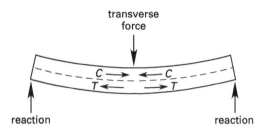

Equations for simple and complex shapes can be found in *Handbook* table Area Moment of Inertia (Table).

For information about normal stresses in regard to axial members, see the entry "Axial Members" in this chapter.

Exam tip: Be aware of which side of the beam is in compression or tension. The negative sign indicating compression is commonly omitted, so the absence of a negative sign may not indicate the absence of a compression force.

Bending Moments

CERM: Sec. 44.11, Sec. 44.14

$$M = \sum_{\substack{\text{point to} \\ \text{one end}}} F_i d_i + \sum_{\substack{\text{point to} \\ \text{one end}}} C_i \qquad 44.25$$

Moment at a point is the total bending moment acting on an object. In the case of a beam, the moment, M, will be the algebraic sum of all moments and couples located between the investigation point and one of the beam ends, as shown in CERM Eq. 44.25. As with shear, the number of calculations required to calculate the moment can be minimized by careful choice of which beam end to use as a reference point.

Since the beam ends are higher than the midpoint, it is commonly said that "a positive moment will make the beam smile." In other words, moment is taken as positive when the upper surface of a beam is in compression and the lower surface is in tension.

D. SHEAR

Key concepts: These key concepts are important for answering exam questions in subtopic 4.D, Shear.

- average shear stress
- average shear stress for flanged beams
- shear at a point on a beam
- shear force diagrams
- shear stress distribution
- shear stresses in beams

NCEES Handbook: To prepare for this subtopic, familiarize yourself with these sections in the *Handbook*.

- Shearing Force and Bending Moment Sign Conventions
- Stresses in Beams
- Moment, Shear, and Deflection Diagrams

PE Civil Reference Manual (CERM): Study these sections in CERM that either relate directly to this subtopic or provide background information.

- Section 44.11: Shear and Moment
- Section 44.12: Shear and Bending Moment Diagrams
- Section 44.13: Shear Stress in Beams

The following equations and figures are relevant for subtopic 4.D, Shear.

Shear at a Point on a Beam

Handbook: Shearing Force and Bending Moment Sign Conventions

CERM: Sec. 44.11

$$V = \sum_{\substack{\text{point to} \\ \text{one end}}} F_i \qquad 44.24$$

Shear at a point is the sum of all vertical forces acting on an object. Shear is not the same as shear stress, since the area of the object is not considered.

Handbook section Shearing Force and Bending Moment Sign Conventions explains the sign conventions for shear in a beam. The sign is typically positive when the net force to the left of a point is upward and negative when it is downward. CERM Eq. 44.24 is a generalized equation for shear.

Shear Force Diagrams

Handbook: Shearing Force and Bending Moment Sign Conventions; Moment, Shear, and Deflection Diagrams

$$w(x) = -\frac{dV(x)}{dx}$$

$$V = \frac{dM(x)}{dx}$$

$$V_2 - V_1 = \int_{x1}^{x2} -w(x)\,dx$$

$$M_2 - M_1 = \int_{x1}^{x2} V(x)\,dx$$

CERM: Sec. 44.11

$$V = \sum_{\substack{\text{point to}\\ \text{one end}}} F_i \qquad 44.24$$

Shear at a point on a beam, V, is defined as the sum of all vertical forces between the point and one of the ends, as shown in the *Handbook* equations and CERM Eq. 44.24.

Shear is taken as positive when the net force to the left of a point is upward and negative when it is downward. It is convenient to describe the shear function graphically. See CERM Fig. 44.9 for an example.

Figure 44.9 Drawing Shear and Moment Diagrams

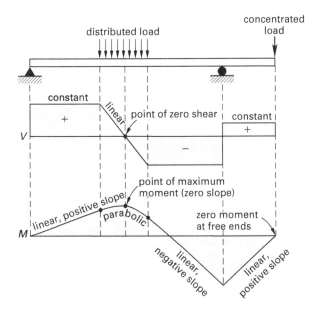

Reactions are really applied loads to a beam, so they can be treated as such. Distributed loads can be drawn similarly to point loads, but they change gradually over the length to which the load is applied. See the tables in *Handbook* section Moment, Shear, and Deflection Diagrams.

Shear Stress in Beams

Handbook: Stresses in Beams

$$\tau_{xy} = \frac{VQ}{Ib}$$

$$Q = A'\bar{y}'$$

CERM: Sec. 44.13

$$\tau = \frac{V}{A} \qquad 44.28$$

$$\tau = \frac{V}{t_w d} \qquad 44.29$$

The *Handbook* equation for transverse shear stress is essentially the same as CERM Eq. 44.29, except that the *Handbook* equation uses the moment of the area and the moment of inertia instead of the depth of the beam, and the *Handbook* equation uses the variable b to represent the depth instead of d.

The average shear stress experienced at a point along the length of a beam depends on the shear, V, at that point and the total area, A, of the beam. CERM Eq. 44.28 illustrates this, and Eq. 44.29 and the *Handbook* equation show the calculation using the individual dimensions of the beam rather than the area.

Shear stress is generally not the limiting factor in design. However, it often controls the design (or is limited by codes and standards) in thin tubes and in wood, masonry, and concrete beams.

In most cases, the entire area of the beam is used in calculating the average shear stress. However, in flanged beam calculations, it is assumed that only the web carries the average shear stress; the flanges carry the flexural stresses in the beam.

Exam tip: It is always a good idea to perform a shear check and a deflection check when required.

E. AXIAL

Key concepts: These key concepts are important for answering exam questions in subtopic 4.E, Axial.

- axial members
- internal forces and stresses of trusses

NCEES Handbook: The *Handbook* does not contain any material relevant to this subtopic. Be sure to study the listed sections in CERM.

PE Civil Reference Manual (CERM): Study these sections in CERM that either relate directly to this subtopic or provide background information.

- Section 41.18: Two- and Three-Force Members
- Section 41.29: Axial Members
- Section 41.31: Trusses
- Section 41.32: Determinate Trusses

The following concepts are relevant for subtopic 4.E, Axial.

Axial Members

CERM: Sec. 41.29

An axial member is commonly identified by its endpoints, and the force in a member is designated by the two endpoints. For example, the axial force in a member connecting points C and D may be written as F_{CD}.

Axial loads in a member only create normal stresses in the member. Since the ends are assumed to be pinned (and therefore rotation free), an axial member cannot support moments.

A column is not necessarily an axial member. This is because a column does not necessarily have pinned ends; it can support moments. Also, columns are not necessarily loaded only at their ends.

Exam tip: The self-weight of axial members is typically ignored for problems in this knowledge area.

Internal Forces and Stresses of a Truss

CERM: Sec. 41.18, Sec. 41.31, Sec. 41.32

The force on a joint at the edge of a member will be in the opposite direction of the force in the axial member. This is because the force at the joint is a reaction to the force in the member. With typical bridge trusses supported at the ends and loaded downward at the joints, the upper chords are almost always in compression, and the end panels and lower chords are almost always in tension. It is common practice to label forces in axial members with (T) for tension and (C) for compression.

Exam tip: When solving problems involving a simple truss, the answers can be checked by verifying whether the chord forces would make sense if they were replaced by a beam. It's not impossible for a truss to have a tension or compression member that does not follow this rule of thumb, but it would definitely be the exception to the rule.

F. COMBINED STRESSES

Key concepts: These key concepts are important for answering exam questions in subtopic 4.F, Combined Stresses.

- axial stress
- beam-columns
- bending stress
- combined stresses
- concentric loading
- eccentric loading
- normal stresses

NCEES Handbook: To prepare for this subtopic, familiarize yourself with these sections in the *Handbook*.

- Uniaxial Loading and Deformation
- Stresses in Beams

PE Civil Reference Manual (CERM): Study these sections in CERM that either relate directly to this subtopic or provide background information.

- Section 44.14: Bending Stress in Beams
- Section 44.16: Eccentric Loading of Axial Members

The following equations, figures, and concepts are relevant for subtopic 4.F, Combined Stresses.

Normal Stress

Handbook: Uniaxial Loading and Deformation; Stresses in Beams

$$\sigma = \frac{P}{A}$$

$$\sigma = \pm \frac{Mc}{I}$$

CERM: Sec. 44.16

$$\sigma_{\text{max,min}} = \frac{F}{A} \pm \frac{Mc}{I_c} \quad\quad 44.42$$

$$\sigma_{\text{max,min}} = \frac{F}{A} \pm \frac{Fec}{I_c} \quad\quad 44.43$$

These equations calculate the maximum and minimum combined stresses caused by axial and bending stresses. c is the distance from the outermost tension or compression fiber to the neutral axis, and P is the axial force. The nomenclature for CERM Eq. 44.42 and

Eq. 44.43 is the same as the nomenclature for the *Handbook* equations, with the addition of *e* for eccentricity and the use of *F* for the axial force instead of *P*.

Exam tip: Shear stresses are a different type of stress from normal stresses, and the two cannot be directly added. The interaction of shear and normal stresses is outside of the scope of this knowledge area.

Bending Stresses

CERM: Sec. 44.16

$$\sigma_{\max,\min} = \frac{F}{A} \pm \frac{Fec}{I_c} \qquad 44.43$$

In CERM Eq. 44.43, *c* is the distance from the outermost tension or compression fiber to the neutral axis, and *F* is the axial force.

Bending stresses can be caused by loads applied perpendicular to the axis of the member or by axial loads not applied directly over the centroid of the beam. The distance from the centroid to the point where the load is applied is known as the *eccentricity*.

Concentric Loading

CERM: Sec. 44.16

In concentric loading, the load is applied through the centroid of a tension or compression member's cross section.

Eccentric Loading

CERM: Sec. 44.16

Eccentric loading is different from concentric loading. If an axial member is loaded eccentrically, it will bend and experience bending stress in the same manner as a beam. Because the member experiences both axial stress and bending stress, it is known as a beam-column. CERM Fig. 44.14 shows eccentric loading of an axial member; *c* is the distance from the outermost tension or compression fiber to the neutral axis, and *e* is the eccentricity.

Figure 44.14 Eccentric Loading of an Axial Member

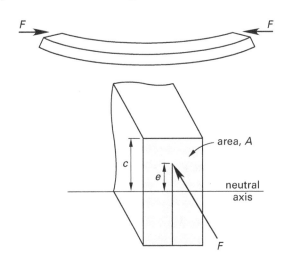

G. DEFLECTION

Key concepts: These key concepts are important for answering exam questions in subtopic 4.G, Deflection.

- long-term deflections
- serviceability
- stiffness

NCEES Handbook: To prepare for this subtopic, familiarize yourself with these sections in the *Handbook*.

- Shears, Moments, and Deflections

PE Civil Reference Manual (**CERM**): Study these sections in CERM that either relate directly to this subtopic or provide background information.

- Section 44.5: Stiffness and Rigidity
- Section 44.22: Beam Deflection: Table Lookup Method
- Section 50.14: Cracked Moment of Inertia
- Section 50.15: Serviceability: Deflections
- Section 50.16: Long-Term Deflections

The following equations, tables, and concepts are relevant for subtopic 4.G, Deflection.

Beam Deflections

Handbook: Shears, Moments, and Deflections

CERM: Sec. 44.22

Handbook table Shears, Moments, and Deflections includes commonly used deflection formulas for simple beams, fixed beams, cantilevered beams, overhanging beams, and continuous beams. These formulas should never need to be derived and should be used whenever possible. They are particularly useful in calculating deflections due to multiple loads using the principle of superposition. These formulas are also available in CERM App. 44.A.

Stiffness and Rigidity

CERM: Sec. 44.5

$$k = \frac{F}{\delta} \quad \text{[general form]} \qquad \text{[SI]} \qquad 44.7(a)$$

$$k = \frac{AE}{L_o} \quad \text{[normal stress form]} \qquad \text{[U.S.]} \qquad 44.7(b)$$

Stiffness is the amount of force required to cause a unit of deformation. Stiffness is often referred to as the *spring constant*. Typical units are pounds per inch and newtons per meter. The stiffness of a spring or other structure can be calculated from the deformation equation, CERM Eq. 44.7. The deformation equation is valid for tensile and compressive normal stress. For torsion and bending, the equation will depend on how the deflection is calculated.

When more than one spring or resisting member shares a load, the relative stiffnesses are known as *rigidities*. Rigidities have no units, and individual rigidity values have no significance. Rigidity is proportional to the reciprocal of deflection.

CERM Table 44.1 gives equations for deflection and stiffness for various systems.

Exam tip: The magnitude of the deflection of a structural member is dependent on the rigidity of the member. The more rigid a member is, the less it will deflect.

Beam Deflections: Moment of Inertia

CERM: Sec. 50.14

Reinforced concrete beam deflections have two components: an immediate deflection that occurs as a result of the strains induced by the applied loads and a long-term deflection that develops due to creep and shrinkage of the concrete. Immediate (instantaneous) deflections are computed using any available general method based on elastic linear response. However, the effect of cracking on the flexural stiffness of the beam must be considered. See the entry "Deflection" in this chapter for more in-depth information.

Exam tip: In concrete beams, the total deflection is the summation of the immediate deflections and the long-term deflections.

Long-Term Deflections

CERM: Sec. 50.16

$$\Delta_a = \lambda \Delta_i \qquad 50.46$$

$$\lambda = \frac{i}{1 + 50\rho'} \qquad 50.47$$

Long-term deflection is caused by creep and shrinkage of the concrete. A long-term deflection develops rapidly at first and then slows down and is largely complete after approximately five years.

Long-term deflection is calculated using CERM Eq. 50.46. As specified by *Building Code Requirements for Structural Concrete* (ACI 318) Sec. 24.2.4.1, the long-term deformation, Δ_a, is the product of the immediate deflection produced by the portion of the load that is sustained, Δ_i, and an amplification factor, λ. When calculating the sustained part of the immediate deflection, judgment is needed to decide what fraction of the prescribed service live load can be assumed to be acting continuously.

The amplification factor is calculated using CERM Eq. 50.47. ρ' is the compression steel ratio, which is calculated at midspan for simply supported and continuous beams and at supports for cantilevers. i is an empirical factor that accounts for the rate of the additional deflection. i is 1.0 at three months, 1.2 at six months, 1.4 at one year, and 2 at five years or more.

Exam tip: Long-term deflection is dependent on the amount of compression reinforcement in the concrete beam.

H. BEAMS

Key concepts: These key concepts are important for answering exam questions in subtopic 4.H, Beams.

- allowable strength design for steel beams
- bending stresses in beams
- concrete beams
- determinate beams
- load and resistance factor design for steel beams

STRUCTURAL MECHANICS

- reinforced concrete beams
- shear stresses in beams
- steel W-shaped beams
- types of steel cross sections

NCEES Handbook: To prepare for this subtopic, familiarize yourself with these sections in the *Handbook*.

- Stresses in Beams
- Simply Supported Beam Slopes and Deflections
- Cantilevered Beam Slopes and Deflections
- Shears, Moments, and Deflections

PE Civil Reference Manual (**CERM**): Study these sections in CERM that either relate directly to this subtopic or provide background information.

- Section 41.21: Types of Determinate Beams
- Section 44.13: Shear Stress in Beams
- Section 44.14: Bending Stress in Beams
- Section 44.26: Modes of Beam Failure
- Section 46.3: Indeterminate Beams
- Section 50.2: Steel Reinforcing
- Section 50.3: Types of Beam Cross Sections
- Section 50.6: Design Strength and Design Criteria
- Section 59.1: Types of Beams
- Section 59.3: Flexural Strength in Steel Beams

The following equations, figures, tables, and concepts are relevant for subtopic 4.H, Beams.

Shear Stresses in Beams

Handbook: Stresses in Beams

CERM: Sec. 44.13

$$\tau_{xy} = \frac{VQ}{Ib}$$

$$Q = A'\bar{y}'$$

The *Handbook* equation for transverse shear stress and CERM Eq. 44.30 are identical except for some minor differences in arrangement and subscripts.

Shear stress is not the limiting factor in most designs. However, it can control the design in thin tubes and in wood, masonry, and concrete beams. The average shear stress experienced at a point along the length of a beam depends on the shear at that point, V, and the area of the beam, A, as shown in CERM Eq. 44.28.

$$\tau = \frac{V}{A} \qquad 44.28$$

In most cases, the entire area of the beam is used in calculating the average shear stress. However, in flanged beam calculations, it is assumed that only the web carries the average shear stress. Flanges are not included in shear stress calculations, as shown in CERM Eq. 44.29. See CERM Fig. 44.10 for an illustration of how the thickness of the web and the depth of the flanged beam are measured.

$$\tau = \frac{V}{t_w d} \qquad 44.29$$

Figure 44.10 Web of a Flanged Beam

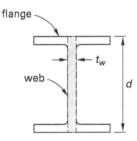

Bending Stresses in Beams

Handbook: Stresses in Beams

$$\sigma = -\frac{My}{I}$$

$$\sigma = \pm\frac{Mc}{I}$$

$$\sigma = \frac{M}{s}$$

$$s = \frac{I}{c}$$

CERM: Sec. 44.14

$$\sigma_b = \frac{-My}{I_c} \qquad 44.36$$

$$\sigma_{b,\max} = \frac{Mc}{I_c} \qquad 44.37$$

$$\sigma_{b,\max} = \frac{M}{S} \qquad 44.38$$

$$S = \frac{I_c}{c} \qquad 44.39$$

The *Handbook* equations for normal stress in a beam due to bending and for the elastic section modulus of a beam due to bending are identical to CERM Eq. 44.36 through Eq. 44.39 except for slight differences in nomenclature.

Normal stress occurs in a bending beam, as shown in CERM Fig. 44.12, where the beam is acted on by a transverse force. Although it is a normal stress, the term *bending stress* (or *flexural stress*) is used to indicate the cause of the stress.

Figure 44.12 Normal Stress Due to Bending

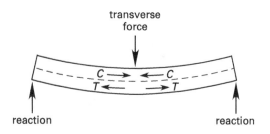

Bending stress varies with location (depth) within the beam. It is zero at the neutral axis and increases linearly with distance from the neutral axis, as modeled by the equations for bending stress. See CERM Fig. 44.13 for an illustration of this principle.

Figure 44.13 Bending Stress Distribution in a Beam

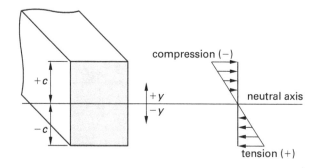

CERM Eq. 44.37 and its equivalent *Handbook* equation show that the maximum bending stress will occur where the moment along the length of the beam is maximum.

Types of Determinate Beams

Handbook: Simply Supported Beam Slopes and Deflections; Cantilevered Beam Slopes and Deflections; Shears, Moments, and Deflections

CERM: Sec. 41.21

There are three types of determinate beams, as illustrated in CERM Fig. 41.9.

Figure 41.9 Types of Determinate Beams

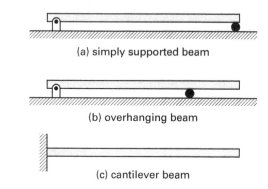

Handbook table Shears, Moments, and Deflections shows diagrams and equations for simply supported beams, overhanging beams, and cantilevered beams. The manner in which a body is supported determines the type, location, and direction of the reactions. For beams, the two most common types of supports are roller supports and the pinned supports.

A roller support, shown in CERM Fig. 41.9 as a cylinder supporting the beam, supports vertical forces only. Rather than supporting a horizontal force, a roller support simply rolls into a new equilibrium position. Only one equilibrium equation (i.e., the sum of vertical forces) is needed at a roller support. Generally, the terms *simple support* and *simply supported* refer to a roller support.

A pinned support supports both vertical and horizontal forces. Two equilibrium equations are needed.

Types of Indeterminate Beams

Handbook: Shears, Moments, and Deflections

CERM: Sec. 46.3

Three common configurations of beams can easily be recognized as statically indeterminate. These are continuous beams, propped cantilever beams, and fixed-end beams, as shown in CERM Fig. 46.1.

Figure 46.1 Types of Indeterminate Beams

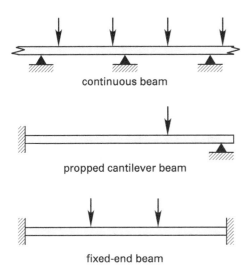

Handbook table Shears, Moments, and Deflections shows diagrams and equations for continuous beams, propped cantilevered beams, and fixed-end beams.

Design Strength and Criteria

CERM: Sec. 50.6

$$\text{design strength} = \phi(\text{nominal strength}) \quad 50.8$$

$$\phi M_n \geq M_u \quad [\text{SI}] \quad 50.9(a)$$

$$\phi V_n \geq V_u \quad [\text{U.S.}] \quad 50.9(b)$$

The design strength is the result of multiplying the nominal strength by a strength reduction factor (also known as a capacity reduction factor), as shown in CERM Eq. 50.8. The design criteria for all sections in a beam are given by CERM Eq. 50.9.

Building Code Requirements for Structural Concrete (ACI 318) uses the subscript n to indicate a "nominal" quantity. A nominal value can be interpreted as being in accordance with theory for the specified dimensions and material properties. The nominal moment strength and shear strength are designated M_n and V_n, respectively.

Beam Shape

CERM: Sec. 59.1

Each beam shape has a specific purpose or situation for which it is best suited. For example, W shapes carry the flexure in the flanges and shear in the webs, L shapes (angles) are useful as ledgers to support vertical loads framing into a wall, and square HSS sections are stable columns because they are equally strong in both directions.

Each shape has some disadvantages that must be accounted for. For example, channels tend to roll when loaded on the flange, due to the shear center not being in the same spot as the centroid of the member. As a result, channels often require horizontal bracing or some other lateral support to keep them from rolling when loaded.

Exam tip: Being aware of the benefits of each shape can help you narrow down answer options on the exam.

I. COLUMNS

Key concepts: These key concepts are important for answering exam questions in subtopic 4.I, Columns.

- available compressive stress
- eccentricity
- effective length factors
- Euler's column buckling theory
- minor axis buckling and bracing
- slenderness ratio
- steel columns

NCEES Handbook: To prepare for this subtopic, familiarize yourself with these sections in the *Handbook*.

- Columns

PE Civil Reference Manual (**CERM**): Study these sections in CERM that either relate directly to this subtopic or provide background information.

- Section 52.1: Introduction
- Section 52.4: Design for Small Eccentricity
- Section 61.2: Euler's Column Buckling Theory
- Section 61.3: Effective Length
- Section 61.4: Geometric Terminology
- Section 61.5: Slenderness Ratio

The following equations, figures, tables, and concepts are relevant for subtopic 4.I, Columns.

Slenderness Ratio

Handbook: Columns

CERM: Sec. 61.5

$$\frac{KL}{r} = \text{effective slenderness ratio for the column}$$

The *Handbook* equation and CERM Eq. 61.5 are identical except for differences in nomenclature.

Steel columns are categorized as either long columns or intermediate columns depending on their slenderness ratios.

For unsymmetrical members, there are two values, corresponding to the x- and y-directions, of the radius of gyration, effective length factor, and length. As a result, there are also two slenderness ratios.

> *Exam tip*: The slenderness ratio of compressive members should preferably not exceed 200, per AISC *Specification for Structural Steel Buildings* Sec. E2. This limit is partially due to constructability concerns.

Effective Length

Handbook: Columns

CERM: Sec. 61.3

The restraints placed on a column's ends greatly affect its stability. For this reason, an *effective length factor*, K (also known as an *end-restraint factor*), is used to modify the unbraced length. The product of the effective length factor and the unbraced length (i.e., KL) is known as the *effective length* of the column. The effective length approximates the length over which a column actually buckles.

The effective length can be longer or shorter than the actual unbraced length. For braced columns (sidesway inhibited), $K \leq 1$, whereas for unbraced columns (sidesway uninhibited), $K > 1$. *Handbook* section Columns gives theoretical effective length factors for various columns. CERM Table 61.1 also lists recommended values of K for use with steel columns. The values of K in Table 61.1 do not require knowing the column size or shape.

> *Exam tip*: The effective length factor is based on the end conditions of a given column. These end conditions dictate the buckled shape of the column.

Euler's Column Buckling Theory

Handbook: Columns

$$P_{cr} = \frac{\pi^2 EI}{(KL)^2}$$

E	modulus of elasticity	kips/in^2
I	moment of inertia	in^4
K	effective length factor	–
L	length	in
P_{cr}	critical axial load	lbf

CERM: Sec. 61.3

$$F_e = \frac{\pi^2 E}{\left(\dfrac{KL}{r}\right)^2} \quad [\textit{AISC Specification } \text{Eq. E3-4}] \quad 61.3$$

The difference between the *Handbook* equation and CERM Eq. 61.3 is that the *Handbook* equation is used to calculate the critical axial load in a column as a force, while CERM Eq. 61.3 is used to calculate the critical axial load in a column as a stress. This is why the *Handbook* equation includes the moment of inertia, I, and CERM Eq. 61.3 does not.

An ideal column is initially perfectly straight, isotropic, and free of residual stresses. When loaded to the buckling (Euler) load, a column will fail by sudden buckling (bending). The Euler column buckling load for an ideal, pin-ended, concentrically loaded column is given by CERM Eq. 61.1. The presence of the modulus of elasticity, E, implies that Eq. 61.1 is valid as long as the loading remains in the elastic region.

$$P_e = \frac{\pi^2 EI}{L^2} \quad 61.1$$

Equation 61.2 is convenient for design and is valid for all types of cross sections and all grades of steel.

$$F_e = \frac{P_e}{A} = \frac{\pi^2 E}{\left(\dfrac{L}{r}\right)^2} \quad 61.2$$

> *Exam tip*: The Euler load theory applies to all materials. However, it is applied in a slightly different way for each material, as per the specifications governing the design of the column.

Geometric Characteristics

CERM: Sec. 61.4

CERM Fig. 61.3 shows a W shape used as a column. In this column, the moment of inertia in the y-direction is smaller than the moment of inertia in the x-direction. In this case, the column is said to buckle about the minor axis.

Figure 61.3 Minor Axis Buckling and Bracing

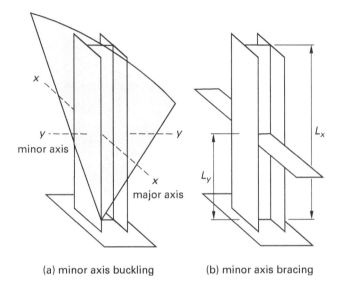

(a) minor axis buckling (b) minor axis bracing

Since buckling about the minor axis is the expected buckling mode, bracing is usually provided for the minor axis at intermediate locations to reduce the unbraced length. Associated with each column are two unbraced lengths, L_x and L_y, as shown in CERM Fig. 61.3(b). L_x is the full column height, and L_y is half the column height, assuming that the brace is placed at midheight.

Another important geometric characteristic is the radius of gyration. Since there are two moments of inertia, there are also two radii of gyration. Because I_y is smaller than I_x, r_y will be smaller than r_x. r_y is known as the least radius of gyration.

Exam tip: A column will buckle in the direction in which the moment of inertia is the smallest.

Design for Small Eccentricity

CERM: Sec. 52.4

The relative importance of bending and axial compression in the design of a column is measured by the normalized eccentricity. The normalized eccentricity is the ratio of the maximum moment acting on the column to the product of the axial load and the dimension of the column perpendicular to the axis of bending.

Exam tip: An eccentric load on a column induces an additional moment into the column and can significantly increase the design load in the column.

J. SLABS

Key concepts: These key concepts are important for answering exam questions in subtopic 4.J, Slabs.

- one-way slabs
- slab design for flexure and shear
- two-way slabs

NCEES Handbook: The *Handbook* does not contain any material relevant to this subtopic. Be sure to study the listed sections in CERM.

PE Civil Reference Manual (**CERM**): Study these sections in CERM that either relate directly to this subtopic or provide background information.

- Section 51.2: One-Way Slabs
- Section 51.6: Slab Design for Flexure
- Section 51.7: Slab Design for Shear
- Section 51.8: Two-Way Slabs

The following figures and concepts are relevant for subtopic 4.J, Slabs.

One-Way Slab

CERM: Sec. 51.2

One-way slabs are slabs designed under the assumption that bending takes place in only one direction. Floor slabs are typically supported on all four sides. If the slab length is more than twice the slab width, a uniform load will produce a deformed surface that has little curvature in the direction parallel to the long dimension. Given that moments are proportional to curvature, bending moments will be significant only in the short direction.

In a one-way slab, the internal forces are calculated by taking a strip of unit width and treating it like a beam. The torsional restraint introduced by the supporting beams is typically neglected. For example, the moments per unit width in the short direction for the floor system shown in CERM Fig. 51.1(a) would be obtained by analyzing the four-span continuous beam depicted in CERM Fig. 51.1(b).

Figure 51.1 One-Way Slab

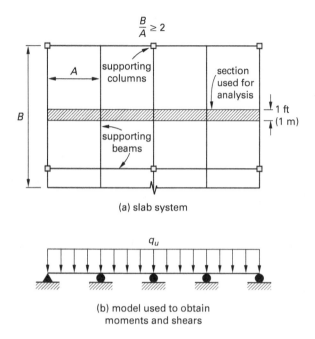

(a) slab system

(b) model used to obtain moments and shears

Figure 51.2 Two-Way Slabs

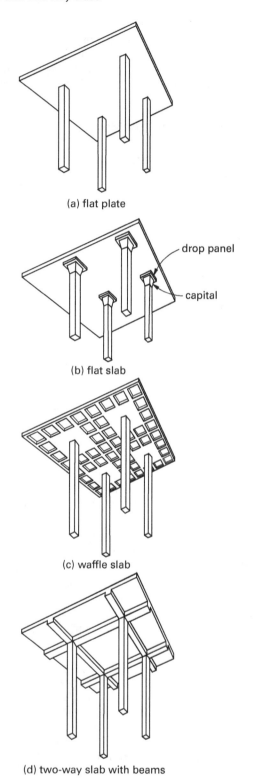

(a) flat plate

(b) flat slab

(c) waffle slab

(d) two-way slab with beams

Two-Way Slabs

CERM: Sec. 51.8

A *two-way slab* is a slab for which the ratio of long sides to short sides is no greater than two. A two-way slab supported on a column grid without the use of beams is known as a *flat plate*. (See CERM Fig. 51.2(a).)

A modified version of a flat plate, where the shear capacity around the columns is increased by thickening the slab in those regions, is known as a *flat slab*. (See CERM Fig. 51.2(b).) The thickened part of the flat plate is known as a drop panel.

Another two-way slab, often used without beams between column lines, is a *waffle slab*. (See CERM Fig. 51.2(c).) In a waffle slab, the forms used to create the voids are omitted around the columns to increase resistance to punching shear.

A two-way slab system that is supported on beams is illustrated in Fig. 51.2(d).

The moments used to design the reinforcement in a two-way slab, whether with or without beams, are obtained in the same manner as for beams and one-way slabs: the slab system and supporting columns are reduced to a series of one-dimensional frames running in both directions. As CERM Fig. 51.3 illustrates, the beams in the frames are wide elements whose edges are defined by cuts midway between the columns.

K. FOOTINGS

Key concepts: These key concepts are important for answering exam questions in subtopic 4.K, Footings.

- column footings
- wall footings

NCEES Handbook: The *Handbook* does not contain any material relevant to this subtopic. Be sure to study the listed sections in CERM.

***PE Civil Reference Manual* (CERM):** Study these sections in CERM that either relate directly to this subtopic or provide background information.

- Section 55.1: Introduction
- Section 55.2: Wall Footings
- Section 55.3: Column Footings

The following equations and figures are relevant for subtopic 4.K, Footings.

Design Considerations

CERM: Sec. 55.1

$$q_s = \frac{P_s}{A_f} + \gamma_c h + \gamma_s(H-h) \pm \frac{M_s\left(\frac{B}{2}\right)}{I_f} \leq q_a \quad 55.1$$

$$I_f = \tfrac{1}{12}LB^3 \quad 55.2$$

A footing is designed in two steps: (a) selecting the footing area and (b) selecting the footing thickness and reinforcement. The footing area is chosen so that the soil contact pressure is within limits. The footing thickness and the reinforcement are chosen to keep the shear and bending stresses in the footing within permissible limits. Although the footing area is obtained from the unfactored service loads, the footing thickness and reinforcement are calculated from factored loads.

CERM Fig. 55.1 illustrates the general case of a footing, with dimensions L and B carrying a vertical (downward) axial service load, P_s, and a service moment, M_s. For such cases, the maximum and minimum service soil pressures are given by CERM Eq. 55.1. I_f is the moment of inertia of the footing, which can be found from CERM Eq. 55.2.

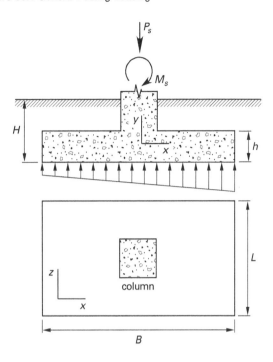

Figure 55.1 General Footing Loading

Exam tip: A footing spreads the load from a column or wall to a pressure that is less than the allowable pressure of the soil it bears on.

L. RETAINING WALLS

Key concepts: These key concepts are important for answering exam questions in subtopic 4.L, Retaining Walls.

- retaining wall design
- types of retaining wall structures

NCEES Handbook: To prepare for this subtopic, familiarize yourself with these sections in the *Handbook*.

- At-Rest Coefficients
- Development of Rankine Active and Passive Failure Zones for a Smooth Retaining Wall
- Rankine Active and Passive Coefficients (Friction Only)
- Failure Surfaces, Pressure Distribution and Forces
- General Distribution of Combined Active Earth Pressure and Water Pressure
- Rankine Active and Passive Coefficients (Friction and Cohesion)

PE Civil Reference Manual (**CERM**): Study these sections in CERM that either relate directly to this subtopic or provide background information.

Section 37.1: Types of Retaining Wall Structures

Section 37.3: Earth Pressure

Section 54.5: Design of Retaining Walls

The following concepts are relevant for subtopic 4.L, Retaining Walls.

Design of Retaining Walls

Handbook: At-Rest Coefficients; Development of Rankine Active and Passive Failure Zones for a Smooth Retaining Wall; Rankine Active and Passive Coefficients (Friction Only); Failure Surfaces, Pressure Distribution and Forces; General Distribution of Combined Active Earth Pressure and Water Pressure; Rankine Active and Passive Coefficients (Friction and Cohesion)

CERM: Sec. 54.5

A retaining wall is essentially a vertical cantilever beam with the additional complexities of nonuniform loading and soil-to-concrete contact. The wall is designed to resist moments from the lateral earth pressure. For more information on the design of retaining walls, see Chap. 24 of this book.

Types of Retaining Wall Structures

CERM: Sec. 37.1

CERM Fig. 37.1 illustrates types of retaining walls. A *gravity wall* is a high-bulk structure that relies on self-weight. A *buttress wall* depends on compression ribs between the stem and the toe to resist flexure and overturning. *Counterfort walls* depend on tension ribs between the stem and the heel to resist flexure and overturning. *Cantilever walls* resist overturning through a combination of the soil weight over the heel and the resisting pressure under the base.

Figure 37.1 Types of Retaining Walls

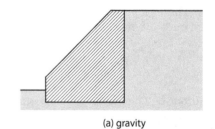

(a) gravity

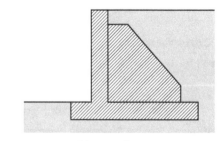

(b) buttress

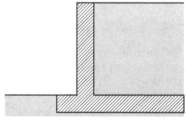

(c) counterfort

(d) cantilever

5 Hydraulics and Hydrology

Content in blue refers to the *NCEES Handbook*.

A. Open-Channel Flow 5-1
B. Stormwater Collection and Drainage 5-5
C. Storm Characteristics.. 5-5
D. Runoff Analysis 5-10
E. Detention/Retention Ponds 5-11
F. Pressure Conduit 5-12
G. Energy and/or Continuity Equation 5-15

The knowledge area of Hydraulics and Hydrology makes up between six and nine questions out of the 80 questions on the PE Civil exam. The organization of this chapter follows the order of subtopics given by NCEES for this knowledge area. Each subtopic is covered in the following sections.

A. OPEN-CHANNEL FLOW

Key concepts: These key concepts are important for answering exam questions in subtopic 5.A, Open-Channel Flow.

- Chezy equation
- flow measurement with weirs
- friction losses
- friction slope
- Hazen-Williams equation
- hydraulic radius
- Manning equation
- open channel flow properties
- section factor
- uniform flow
- weir shapes

NCEES Handbook: To prepare for this subtopic, familiarize yourself with these sections in the *Handbook*.

- Principles of One-Dimensional Fluid Flow: Energy Equation
- Hydraulic Radius
- Rectangular Weirs: Sharp-Crested Weirs
- Triangular (V-Notch) Weirs
- Hazen-Williams Equation
- Manning's Equation
- Conveyance
- Chezy Equation
- Geometric Elements of Channel Sections
- Channel Section Critical Depths
- Best Hydraulic Efficient Sections Without Freeboard
- Hydraulic Classification of Slopes
- Gradually Varied Flow Profile Diagrams
- Composite Slopes Channel Profiles

PE Civil Reference Manual (**CERM**): Study these sections in CERM that either relate directly to this subtopic or provide background information.

- Section 16.8: Hydraulic Radius
- Section 19.4: Velocity Distribution
- Section 19.5: Parameters Used in Open Channel Flow
- Section 19.6: Governing Equations for Uniform Flow
- Section 19.8: Hazen-Williams Velocity
- Section 19.10: Energy and Friction Relationships
- Section 19.14: Flow Measurement with Weirs
- Section 19.15: Triangular Weirs
- Section 19.16: Trapezoidal Weirs
- Section 19.20: Uniform and Nonuniform Steady Flow

The following equations and figures are relevant for subtopic 5.A, Open-Channel Flow.

Hydraulic Radius

Handbook: Hydraulic Radius

$$R_H = \frac{A}{P}$$

CERM: Sec. 16.8, Sec. 19.5

The hydraulic radius is the ratio of the cross-sectional area of flow to the wetted perimeter and is used primarily in the Hazen-Williams, Chezy, and Manning equations. The wetted perimeter is the length of the line representing the interface between the fluid and the pipe or channel. It does not include the free surface length (i.e., the interface between fluid and atmosphere). For a circular channel flowing either full or half-full, the hydraulic radius is one-fourth of the hydraulic diameter, $D_h/4$. The hydraulic radii of other channel shapes are easily calculated from the basic definition. CERM Table 19.2 summarizes parameters for the basic shapes.

Exam tip: For channels where the width is significantly greater than the depth, the hydraulic radius is approximately equal to the depth.

Section Factor

Handbook: Channel Section Critical Depths; Geometric Elements of Channel Sections; Best Hydraulic Efficient Sections Without Freeboard

CERM: Sec. 19.5

$$\text{section factor} = AR^{2/3} \quad \text{[general uniform flow]} \quad 19.4$$

$$\text{section factor} = A\sqrt{D_h} \quad \text{[critical flow only]} \quad 19.5$$

The general uniform flow section factor is central to the Manning equation and is found in *Handbook* table Channel Section Critical Depths. CERM Eq. 19.4 can be used to calculate the section factor for general uniform flow, and CERM Eq. 19.5 can be used to calculate the section factor for critical flow. The uniform flow section factor represents a frequently occurring variable group. The section factor is often evaluated against depth of flow when working with discharge from irregular cross sections.

Section factors are listed in *Handbook* tables Geometric Elements of Channel Sections and Best Hydraulic Efficient Sections Without Freeboard.

Slope

Handbook: Hydraulic Classification of Slopes; Gradually Varied Flow Profile Diagrams; Composite Slopes Channel Profiles

CERM: Sec. 19.5, Sec. 19.20

Three definitions of slope exist for open channel flow: the slope of the channel bottom, the slope of the water surface, and the slope of the energy gradient line, which is more commonly known as the friction slope. CERM Fig. 19.12 illustrates these three definitions.

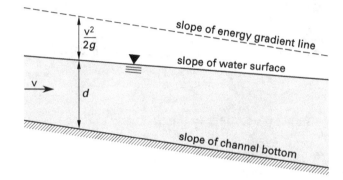

Figure 19.12 Slopes Used in Open Channel Flow

For small slopes, as shown in CERM Eq. 19.6, typical of almost all natural waterways, the channel length and horizontal run are essentially identical.

$$S = \frac{dE}{dL} \quad 19.6$$

In general, the friction slope can be calculated from the Bernoulli equation as the energy loss per unit length of channel. If the flow is uniform, the slope of the energy line will parallel the water surface and channel bottom, and the energy gradient will equal the geometric slope, as shown in CERM Eq. 19.7.

$$S_0 = \frac{\Delta z}{L} \approx S \quad \text{[uniform flow]} \quad 19.7$$

Manning Equation

Handbook: Manning's Equation

$$Q = \left(\frac{1.486}{n}\right) A R_H^{2/3} S^{1/2}$$

CERM: Sec. 19.6

$$\mathrm{v} = \left(\frac{1}{n}\right) R^{2/3} \sqrt{S} \qquad \text{[SI]} \quad 19.12(a)$$

$$\mathrm{v} = \left(\frac{1.49}{n}\right) R^{2/3} \sqrt{S} \qquad \text{[U.S.]} \quad 19.12(b)$$

$$Q = vA = \left(\frac{1.49}{n}\right)AR^{2/3}\sqrt{S} \quad \text{[U.S.]} \quad 19.13(b)$$
$$= K\sqrt{S}$$

The Manning equation is used to calculate flow rate in open channels. The *Handbook* equation and CERM Eq. 19.12(b), used for U.S. units, are nearly identical, with only a minor rounding difference. CERM Eq. 19.12(a) is used for SI units. Note that the *Handbook* uses R_H to represent the hydraulic radius, while CERM uses R. CERM Eq. 19.13(b) combines all the coefficients and constants into the conveyance, K. (See the entry "Conveyance" in this chapter for more information.) Roughness coefficient values can be found in *Handbook* table Approximate Values of Manning's Roughness Coefficient.

Chezy Equation

Handbook: Chezy Equation

CERM: Sec. 19.6

$$v = C\sqrt{R_H S}$$

The most common equation used to calculate the flow velocity in open channels is the Chezy equation, given in *Handbook* section Chezy Equation and by CERM Eq. 19.9. When the channel is large and the flow is fully turbulent, the constant, C, can be evaluated using the 1888 Manning formula, given in *Handbook* section Chezy Equation and by CERM Eq. 19.11. The *Handbook* equation and CERM Eq. 19.11(a) are used for SI units and are nearly identical, with the *Handbook* using R_H and CERM using R to represent the hydraulic radius. CERM Eq. 19.11(b) is used for U.S. units.

$$C = \left(\frac{1}{n}\right)R_H^{1/6}$$

$$C = \left(\frac{1.49}{n}\right)R^{1/6} \quad \text{[U.S.]} \quad 19.11(b)$$

Combining the Chezy equation and 1888 Manning formula yields the Chezy-Manning equation, more commonly referred to as the Manning equation. See the entry "Manning Equation" in this chapter for more information.

Conveyance

Handbook: Conveyance

CERM: Sec. 19.6

$$K = \left(\frac{1.486}{n}\right)AR_H^{2/3}$$

$$K = \frac{Q}{\sqrt{S}}$$

The coefficients and constants in the Manning equation may be combined into the conveyance, K. The *conveyance* of a channel section is a measure of the carrying capacity of the channel section per unit longitudinal slope.

Hazen-Williams Velocity

Handbook: Hazen-Williams Equation

$$v = k_1 CR_H^{0.63} S^{0.54}$$

CERM: Sec. 19.8

$$v = 0.85 CR^{0.63} S_0^{0.54} \quad \text{[SI]} \quad 19.14(a)$$

$$v = 1.318 CR^{0.63} S_0^{0.54} \quad \text{[U.S.]} \quad 19.14(b)$$

The *Handbook* and CERM equations are nearly identical. For the *Handbook* equation, k_1 is 0.849 for SI units and 1.318 for U.S. units. The equation can be converted into a flow equation with the addition of the cross-sectional area of the pipe or channel.

$$Q = k_1 CAR_H^{0.63} S^{0.54}$$

CERM displays the Hazen-Williams velocity equation slightly differently than the *Handbook* by using two separate equations, CERM Eq. 19.14(a) and (b), to distinguish between SI and U.S. units. The variable C in the equations is used to characterize the roughness of the channel. See *Handbook* table Values of Hazen-Williams Coefficient C for values used for various pipe materials.

Friction Losses

Handbook: Principles of One-Dimensional Fluid Flow: Energy Equation

CERM: Sec. 19.10

Friction loss describes the energy lost in a fluid system as a result of friction between the fluid and the wetted surface of the conveyance system. The Bernoulli equation is an expression for the conservation of energy along a fluid streamline, incorporating friction losses into the equation.

$$\frac{p_1}{\rho g} + z_1 + \frac{v_1^2}{2g} = \frac{p_2}{\rho g} + z_2 + \frac{v_2^2}{2g} + h_f$$

$$\frac{p_1}{\gamma} + z_1 + \frac{v_1^2}{2g} = \frac{p_2}{\gamma} + z_2 + \frac{v_2^2}{2g} + h_f$$

The Bernoulli equation can also be written for two points along the bottom of an open channel.

Sharp-Crested Weirs

Handbook: Rectangular Weirs: Sharp-Crested Weirs

$$Q = CLH^{3/2}$$

CERM: Sec. 19.14

$$Q = \tfrac{2}{3} C_1 b \sqrt{2g}\, H^{3/2} \qquad 19.49$$

Sharp-crested weirs are constructed with a sharp weir edge that is used for measurements. The equations shown are basic equations used to calculate flows from rectangular sharp-crested weirs. The *Handbook* equation is slightly different from the CERM equation because the *Handbook* incorporates different design conditions and assumptions into the coefficient, C, as shown in the Rehbock coefficient of the discharge equation.

$$C = 3.27 + 0.4\left(\frac{H}{h}\right) \quad [\text{U.S.}]$$

For any given width of weir opening (referred to as the *weir length*), the discharge will be a function of the head over the weir. The head (or sometimes surface elevation) can be determined by a standard *staff gauge* mounted adjacent to the weir. The full channel flow usually goes over the weir. It is possible, however, to divert a small portion of the total flow through a measurement channel. The full channel flow rate can be extrapolated from a knowledge of the split fractions.

Velocity Distribution in Open Channels

Handbook: Principles of One-Dimensional Fluid Flow: Continuity Equation

CERM: Sec. 19.4

Due to the adhesion between the wetted surface of the channel and the water, the flow velocity will not be uniform within the cross section but will vary similar to the cross section shown in CERM Fig. 19.1.

Figure 19.1 Velocity Distribution in an Open Channel

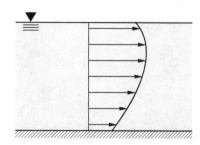

Open channel flow velocity refers to the mean velocity. As indicated by the continuity equation (given in *Handbook* section Principles of One-Dimensional Fluid Flow: Continuity Equation and by CERM Eq. 19.1), the mean velocity, when multiplied by the cross-sectional area, gives the volumetric flow rate at that cross section.

$$Q = Av$$

Triangular (V-Notch) Weirs

Handbook: Triangular (V-Notch) Weirs

CERM: Sec. 19.15, Sec. 19.16

general V-notch

$$Q = \tfrac{8}{15} C_d \sqrt{2g} \left(\tan\frac{\theta}{2}\right) H^{5/2}$$

90° V-notch (cone equation)

$$Q = CH^{5/2}$$

The general equation for triangular or V-notch weirs in the *Handbook* is fundamentally the same as CERM Eq. 19.55. Triangular weirs should be used when small flow rates are to be measured. The flow coefficient over a triangular weir depends on the notch angle, θ, and generally varies from 0.58 to 0.61. For a 90° weir, $C_2 \approx 0.593$. See CERM Fig. 19.8 for an example of a triangular weir.

Figure 19.8 Triangular Weir

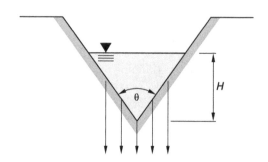

A trapezoidal weir is essentially a rectangular weir with a triangular weir on either side. An example of a trapezoidal weir is shown in CERM Fig. 19.9.

Figure 19.9 Trapezoidal Weir

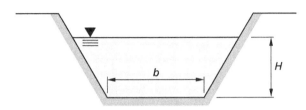

B. STORMWATER COLLECTION AND DRAINAGE

Key concepts: These key concepts are important for answering exam questions in subtopic 5.B, Stormwater Collection and Drainage.

- culverts
- curb and gutter flow
- street inlets

NCEES Handbook: To prepare for this subtopic, familiarize yourself with these sections in the *Handbook*.

- Culvert Flow Types (USGS)
- Culvert Head Loss, Total
- Curb and Gutter as Triangular Channels

PE Civil Reference Manual (**CERM**): Study these sections in CERM that either relate directly to this subtopic or provide background information.

- Section 17.26: Pressure Culverts
- Section 19.37: Culverts
- Section 19.38: Determining Type of Culvert Flow
- Section 19.39: Culvert Design
- Section 28.10: Street Inlets

The following equations and concepts are relevant for subtopic 5.B, Stormwater Collection and Drainage.

Culverts

Handbook: Culvert Flow Types (USGS); Culvert Head Loss, Total

CERM: Sec. 17.26, Sec. 19.37, Sec. 19.38, Sec. 19.39

The *Handbook* classifies culvert flows into six different types. The flow equations in the *Handbook* for each type of culvert may appear at first to be fundamentally different from those used in CERM, but the culvert types are not numbered in the same order. For instance, submerged flow is Type 1 in the *Handbook* and Type 4 in CERM.

Exam tip: Become familiar with the culvert types as they are presented in *Handbook* section Culvert Flow Types (USGS) to prepare for the test.

Street Gutter Flow

Handbook: Curb and Gutter as Triangular Channels

$$Q = \left(\frac{K_u}{n}\right) S_x^{1.67} S_L^{0.5} T^{2.67}$$

$$Q = \left(\frac{K_u}{n}\right)\left(\frac{S_L^{0.5}}{S_x}\right) d^{8/3}$$

$$T = \left(\frac{Qn}{K_u S_x^{1.67} S_L^{0.5}}\right)^{0.375}$$

The curb and gutter within street systems are often used to convey flows to stormwater inlets or other channelized stormwater conveyance systems. Flows within the curb and gutter can be calculated based on a geometry-specific form of the Manning equation, as shown in the *Handbook* equations. *Handbook* section Curb and Gutter as Triangular Channels includes a figure that illustrates these concepts.

Street Inlets

CERM: Sec. 28.10

Street inlets are required at all low points where ponding could occur, and they should be placed no more than 600 ft (180 m) apart. A limit of 300 ft (90 m) is preferred. A common practice is to install three inlets in a sag vertical curve, one at the lowest point and one on each side, with an elevation of 0.2 ft (60 mm) above the center inlet. Depth of gutter flow is found using the Manning equation. Inlet capacities of street inlets have traditionally been calculated from partly empirical formulas.

C. STORM CHARACTERISTICS

Key concepts: These key concepts are important for answering exam questions in subtopic 5.C, Storm Characteristics.

- hydrologic cycle
- precipitation
- probabilities of flooding events

- rainfall intensity
- rain gauges and hyetographs
- sheet flow
- time of concentration
- unit hydrograph
- water balance equation

NCEES Handbook: To prepare for this subtopic, familiarize yourself with these sections in the *Handbook*.

- Storm/Flood Frequency Probabilities: General Probability
- Probability of Single Occurrence in a Given Storm Year
- Risk (Probability of Exceedance)
- Frequency Factor for Normal Distribution
- Point Precipitation
- Time of Concentration: SCS Lag Formula
- Time of Concentration: Sheet Flow Formula
- Time of Concentration: Inlet Flow Formula
- Unit Hydrograph
- Graphical Representation of a Unit Hydrograph
- Snyder Synthetic Unit Hydrograph
- SCS (NRCS) Unit Hydrograph
- Clark Unit Hydrograph
- Rainfall Gauging Stations: Precipitation Gauge Analysis
- Rainfall Gauging Stations: Thiessen Polygon Method
- Rainfall Gauging Stations: Isohyetal Method
- National Weather Service IDF Curve Creation
- Distance Weighting
- Rainfall Gauging Stations: Arithmetic/Station Average Method
- Rainfall Gauging Stations: Normal-Ratio Method
- Surface Water System Hydrologic Budget

PE Civil Reference Manual **(CERM)**: Study these sections in CERM that either relate directly to this subtopic or provide background information.

- Section 20.1: Hydrologic Cycle
- Section 20.3: Precipitation Data
- Section 20.4: Estimating Unknown Precipitation
- Section 20.5: Time of Concentration
- Section 20.6: Rainfall Intensity
- Section 20.7: Floods
- Section 20.10: Unit Hydrograph
- Section 20.11: NRCS Synthetic Unit Hydrograph
- Section 20.12: NRCS Synthetic Unit Triangular Hydrograph

The following equations and concepts are relevant for subtopic 5.C, Storm Characteristics.

Hydrologic Cycle

CERM: Sec. 20.1

The *hydrologic cycle* is the full life cycle of water. The cycle begins with precipitation, which encompasses all of the hydrometric forms, including rain, snow, sleet, and hail from a storm. Precipitation can

- fall on vegetation and structures and evaporate back into the atmosphere
- be absorbed into the ground and either make its way to the water table or be absorbed by plants, after which it evaporates and spires back into the atmosphere
- travel as surface water to a depression, watershed, or creek from which it either evaporates back into the atmosphere, infiltrates into the groundwater system, or flows via streams and rivers to an ocean or lake

The cycle is completed when lake and ocean water evaporates into the atmosphere.

Water Balance Equation

Handbook: Surface Water System Hydrologic Budget

$$P + Q_{in} - Q_{out} + Q_g - E_s - T_s - I = \Delta S_S$$

CERM: Sec. 20.1

$$P = Q + E + \Delta S$$

The water balance equation is the balance between water moving into and out of a watershed, lake, or other surface water system. The *Handbook* presents the equation in a slightly different manner than CERM, but the equations are essentially the same: the change in water storage is the difference between the inflows and outflows.

The total amount of water that is intercepted (and subsequently evaporates) and absorbed into groundwater before runoff begins is known as the *initial abstraction*. Even after runoff begins, the soil continues to absorb some infiltrated water. Initial abstraction and infiltration do not contribute to surface runoff. CERM Eq. 20.1 can be restated as CERM Eq. 20.2.

$$\frac{\text{total}}{\text{precipitation}} = \text{initial abstraction} + \text{infiltration} \quad 20.2$$
$$+ \text{surface runoff}$$

Estimating Precipitation

Handbook: Rainfall Gauging Stations: Precipitation Gauge Analysis; Rainfall Gauging Stations: Thiessen Polygon Method; Rainfall Gauging Stations: Isohyetal Method; Rainfall Gauging Stations: Arithmetic/Station Average Method; Rainfall Gauging Stations: Normal-Ratio Method; National Weather Service IDF Curve Creation; Distance Weighting

CERM: Sec. 20.3, Sec. 20.4

The *Handbook* includes many of the methods discussed in CERM Sec. 20.3 for estimating rainfall based on gauge data. If a precipitation measurement at a location is unknown, it may still be possible to estimate the value by one of these procedures or by using the method outlined in CERM Sec. 20.4. CERM Eq. 20.2 gives a general equation for calculating precipitation.

$$\frac{\text{total}}{\text{precipitation}} = \text{initial abstraction} + \text{infiltration} \quad 20.2$$
$$+ \text{surface runoff}$$

Time of Concentration

Handbook: Time of Concentration: SCS Lag Formula; Time of Concentration: Inlet Flow Formula

$$t_L = 0.000526 \, L^{0.8} \left(\frac{1000}{\text{CN}} - 9\right)^{0.7} S^{-0.5}$$

$$t_c = \frac{5}{3} t_L$$

CERM: Sec. 20.5

$$t_{c,\min} = 1.67 \, t_{\text{watershed lag time, min}}$$

$$= \frac{(1.67)\left(60 \, \frac{\min}{\text{hr}}\right) L_{o,\text{ft}}^{0.8} (S_{\text{in}} + 1)^{0.7}}{1900 \sqrt{S_{\text{percent}}}} \quad 20.11$$

$$= \frac{(1.67)\left(60 \, \frac{\min}{\text{hr}}\right) L_{o,\text{ft}}^{0.8} \left(\frac{1000}{\text{CN}} - 9\right)^{0.7}}{1900 \sqrt{S_{\text{percent}}}}$$

Time of concentration, t_c, is defined as the time of travel from the hydraulically most remote (timewise) point in the watershed to the watershed outlet or other design point. The most prominent method for calculating the time of concentration in the *Handbook* is the Natural Resource Conservation Service (NRCS) lag formula. In the *Handbook*, this is called the SCS lag formula (the NRCS was formerly known as the Soil Conservation Service).

CERM also includes the NRCS lag method in Eq. 20.11. However, where the *Handbook* presents the equation in two parts and combines constants to form decimals and fractions, showing time of concentration in terms of lag time, CERM Eq. 20.11 embeds the lag time equation into the time of concentration equation while keeping the constants in the equation separated out.

For storm drain inlets, *Handbook* section Time of Concentration: Inlet Flow Formula suggests a different approach to calculating the time of concentration.

$$t = t_i + t_s$$

For points (e.g., manholes) along storm drains being fed from a watershed, time of concentration is taken as the largest combination of overland flow time (sheet flow), swale or ditch flow (shallow concentrated flow), and storm drain, culvert, or channel time, as shown by CERM Eq. 20.5.

$$t_c = t_{\text{sheet}} + t_{\text{shallow}} + t_{\text{channel}} \quad 20.5$$

CERM also covers the Federal Aviation Administration (FAA) formula and the kinematic wave formula methods for calculating the time of concentration. These methods are not included in the *Handbook*.

Exam tip: It is unusual for time of concentration to be less than 0.1 hr (6 min) when using the NRCS method or less than 10 min when using the rational method.

Sheet Flow

Handbook: Time of Concentration: Sheet Flow Formula

CERM: Sec. 20.5

Sheet flow, also known as *overland flow*, is the flow of water overland prior to concentration. Sheet flow travel time contributes to the time of concentration and can be calculated using the equation given in *Handbook* section Time of Concentration: Sheet Flow Formula.

$$T_{ti} = \left(\frac{K_u}{I^{0.4}}\right)\left(\frac{nL}{\sqrt{S}}\right)^{0.6}$$

CERM Eq. 20.10 is equivalent, but it is formatted differently. It gives 0.94 as the value of K_u, which is slightly different from the value of 0.933 given in the *Handbook*.

$$t_{c,\min} = \frac{0.94 L_{o,\text{ft}}^{0.6} n^{0.6}}{I_{\text{in/hr}}^{0.4} S_{\text{decimal,ft/ft}}^{0.3}} \quad 20.10$$

For irregularly shaped drainage areas, it may be necessary to evaluate several distances related to sheet flow. For example, most natural drainages are long and tend to become narrow at the downstream end of the watershed where flows enter into channelized waterways. CERM Fig. 20.3 shows a rough approximation of such a drainage area.

Figure 20.3 Irregular Drainage Area

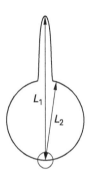

Unit Hydrograph

Handbook Unit Hydrograph; Graphical Representation of a Unit Hydrograph; Snyder Synthetic Unit Hydrograph; SCS (NRCS) Unit Hydrograph; Clark Unit Hydrograph

The *Handbook* explores several different methods of developing a unit hydrograph. A general example is shown in *Handbook* figure Graphical Representation of a Unit Hydrograph. Other examples include *Handbook* figures Snyder Synthetic Unit Hydrograph, SCS (NRCS) Unit Hydrograph, and Clark Unit Hydrograph. The *Handbook* includes the NRCS unit hydrograph method, but you should become familiar with all the methods in preparing for the test. The general equation for determining flow using peak discharge values is given in *Handbook* section Unit Hydrograph.

$$Q = \tfrac{1}{2} q_p T = \tfrac{1}{2} q_p (t_p + t_r)$$

Peak Flow

Handbook: SCS (NRCS) Unit Hydrograph

$$q_p = \frac{K_p A_m Q_D}{t_p}$$

CERM: Sec. 20.11

$$Q_p = \frac{484 A_{d,\text{mi}^2}}{t_p} \quad 20.25$$

The peak flow calculation as presented in the *Handbook* is the same as that presented in CERM Eq. 20.25. In the *Handbook*, the equation cites a peaking constant, K_p, which is 484 when the area of the watershed is calculated in square miles. It also includes the variable Q_D, which is the volume of direct runoff. For a unit hydrograph, where the precipitation is equal to 1 in, Q_D is also 1 in. The CERM equation accordingly drops this variable entirely.

Time to Peak

Handbook: SCS (NRCS) Unit Hydrograph

CERM: Sec. 20.11, Sec. 20.12

$$t_p = \tfrac{2}{3} t_c$$

The time to peak in the *Handbook* is an approximation based on the time to concentration. It is the same estimation method called out in CERM Eq. 20.28. The lag time as shown in *Handbook* section SCS (NRCS) Unit Hydrograph is the same as that expressed in CERM Eq. 20.23 and is based on the potential maximum retention.

$$t_L = \frac{L^{0.8}(S+1)^{0.7}}{1900 \, Y^{0.5}}$$

This is another expression of the SCS lag formula for time to concentration as expressed in *Handbook* section Time of Concentration: SCS Lag Formula, which utilizes the curve number of the watershed.

SCS (NRCS) Unit Hydrograph

Handbook: SCS (NRCS) Unit Hydrograph

CERM: Sec. 20.10, Sec. 20.11

The NRCS developed a synthetic unit hydrograph based on the *curve number*, CN. See *Handbook* figure SCS (NRCS) Unit Hydrograph for an example. In order to draw the NRCS synthetic unit hydrograph, it is necessary to calculate the time to peak flow, t_p, and the peak discharge, Q_p.

$$t_p = 0.5t_R + t_1 \qquad 20.22$$

t_R in CERM Eq. 20.22 is the storm duration (i.e., the duration of the rainfall), and t_1 is the lag time (i.e., the time from the centroid of the rainfall distribution to the peak discharge). Lag time can be determined from correlations with geographical region and drainage area or calculated from CERM Eq. 20.23. S in CERM Eq. 20.23 is the soil water storage capacity in inches, computed as a function of the curve number.

$$t_{1,\text{hr}} = \frac{L_{o,\text{ft}}^{0.8}(S+1)^{0.7}}{1900\sqrt{S_{\text{percent}}}} \qquad 20.23$$

The peak runoff is calculated as

$$Q_p = \frac{0.756 A_{d,\text{ac}}}{t_p} \qquad 20.24$$

$$Q_p = \frac{484 A_{d,\text{mi}^2}}{t_p} \qquad 20.25$$

Q_p and t_p contribute only one point to the construction of the unit hydrograph. To construct the remainder, CERM Table 20.3 must be used. Using time as the independent variable, selections of time different from t_p are arbitrarily made, and the ratio t/t_p is calculated. The curve is then used to obtain the ratio of Q_t/Q_p.

Rainfall Intensity

Handbook: Point Precipitation; National Weather Service IDF Curve Creation

$$i = \frac{c}{T_d^e + f}$$

$$i = \frac{cT^m}{T_d + f}$$

$$i = \frac{cT^m}{T_d^e + f}$$

CERM: Sec. 20.6

$$I = \frac{K' F^a}{(t+b)^c} \qquad 20.13$$

Rainfall intensity is the amount of precipitation per hour. The instantaneous intensity changes throughout the storm. However, it may be averaged over short time intervals or over the entire storm duration. Rainfall intensity is a function of local and regional climate conditions and the storm or hydrologic conditions for which the engineer is designing. CERM Fig. 20.6 shows typical intensity-duration-frequency curves.

Figure 20.6 Typical Intensity-Duration-Frequency Curves

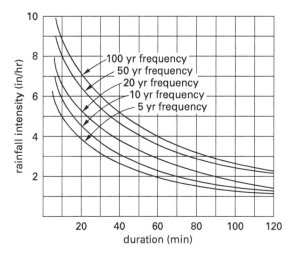

Handbook section National Weather Service IDF Curve Creation also includes an equation for intensity as it is calculated for intensity-duration-frequency curves. This equation is not necessarily accurate for all regions of the United States or the world.

$$i = \frac{60P}{T_d}$$

$$P = \frac{T_d + 20}{100}$$

Probability of Flooding

Handbook: Storm/Flood Frequency Probabilities: General Probability; Probability of Single Occurrence in a Given Storm Year; Risk (Probability of Exceedance); Frequency Factor for Normal Distribution

CERM: 20.7 Floods

$$P(x \geq x_T) = p = \frac{1}{T}$$

$$P(x \geq x_T \text{ at least once in } n \text{ years}) = 1 - \left(1 - \frac{1}{T}\right)^n$$

$$x_T = \bar{x} + K_T s$$

The *Handbook* provides several statistical equations for the probability or risk of a storm event in a given year or over a specified period of time. The equation in *Handbook* section Probability of Single Occurrence in a Given Storm Year is the same as that shown in CERM Eq. 20.19, with T in the *Handbook* replacing F in CERM. Similarly, the equation in *Handbook* section Risk (Probability of Exceedance) is the same as CERM Eq. 20.20. *Handbook* table Frequency Factor for Normal Distribution illustrates the calculation of flood frequency assuming a normal distribution of frequency factors.

The average number of years between storms of a given intensity is known as the *frequency of occurrence*.

D. RUNOFF ANALYSIS

Key concepts: These key concepts are important for answering exam questions in subtopic 5.D, Runoff Analysis.

- base flow
- detention and retention ponds
- hydrograph histograms
- hydrographs
- NRCS curve number method
- peak runoff using the rational method
- stream flow measurements
- stream hydrographs

NCEES Handbook: To prepare for this subtopic, familiarize yourself with these sections in the *Handbook*.

- Runoff Analysis: Rational Formula Method
- Runoff Coefficients for Rational Formula
- NRCS (SCS) Rainfall Runoff Method: Peak Discharge Method
- Minimum Infiltration Rates for the Various Soil Groups
- Coefficients for SCS Peak Discharge Method
- Hydrograph Estimate of Stream Flow

PE Civil Reference Manual **(CERM):** Study these sections in CERM that either relate directly to this subtopic or provide background information.

- Section 20.8: Total Surface Runoff from Stream Hydrograph
- Section 20.9: Hydrograph Separation
- Section 20.15: Peak Runoff from the Rational Method

- Section 20.16: NRCS Curve Number
- Section 20.17: NRCS Graphical Peak Discharge Method

The following equations and concepts are relevant for subtopic 5.D, Runoff Analysis.

Stream Hydrograph

Handbook: Hydrograph Estimate of Stream Flow

CERM: Sec. 20.8

$$Q_t = Q_0 K^t$$

After a rainfall event, runoff and groundwater increase stream flow. A plot of the stream discharge versus time is known as a *hydrograph*. Hydrograph periods may be very short (e.g., hours) or very long (e.g., days, weeks, or months). A typical stream hydrograph is shown in CERM Fig. 20.7.

Figure 20.7 *Stream Hydrograph*

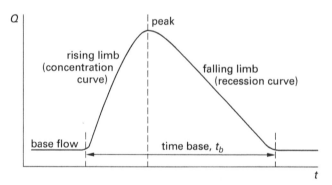

The time base is the length of time that the stream flow exceeds the original base flow. The flow rate increases on the rising limb (concentration curve) and decreases on the falling limb (recession curve).

Rational Method

Handbook: Runoff Analysis: Rational Formula Method; Runoff Coefficients for Rational Formula

CERM: Sec. 20.15

$$Q = CIA$$

The *Handbook* equation is nearly identical to CERM Eq. 20.36; the primary difference lies in the method used to determine the runoff coefficient, C. In contrast to the *Handbook*, CERM calculates the runoff coefficient, C, as a function of impervious area and the slope of the basin or watershed. The intensity, I, used in the equation depends on the time of concentration and the degree of protection desired (i.e., the recurrence interval).

Handbook table Runoff Coefficients for Rational Formula provides a range of runoff coefficients based on the land cover and, in some cases, slope. The *Handbook* also includes an equation that can be used to calculate a weighted runoff coefficient for watersheds or drainage basins that include multiple types of land cover.

$$C_w = \frac{A_1 C_1 + A_2 C_2 + \cdots + A_n C_n}{A_1 + A_2 + \cdots + A_n}$$

NRCS Peak Discharge Method

Handbook: NRCS (SCS) Rainfall Runoff Method: Peak Discharge Method; Coefficients for SCS Peak Discharge Method

$$q_p = q_u A_m Q$$

$$q_u = 10^{C_0 + C_1 \log t_c + C_2 (\log t_c)^2}$$

CERM: Sec. 20.17

$$Q_p = q_u A_{mi^2} Q_{in} F_p \qquad 20.45$$

Equations related to the NRCS rainfall runoff method are outlined in the *Handbook* and correspond to CERM Eq. 20.45. Where CERM includes a factor for the presence of ponds and swamps in the watershed, F_p, the *Handbook* equation assumes no ponds or swamps. *Handbook* table Coefficients for SCS Peak Discharge Method includes the TR-55 coefficients C_1, C_2, and C_3 (from NRCS Technical Release 55).

NRCS Curve Number Method

Handbook: NRCS (SCS) Rainfall Runoff Method: Runoff; Minimum Infiltration Rates for the Various Soil Groups

CERM: Sec. 20.16

$$Q = \frac{(P - 0.2S)^2}{P + 0.8S}$$

$$S = \frac{1000}{\text{CN}} - 10$$

$$\text{CN} = \frac{1000}{S + 10}$$

Handbook section NRCS (SCS) Rainfall Runoff Method: Runoff provides equations related to calculating runoff. The first and second *Handbook* equations correspond, respectively, to CERM Eq. 20.44 and Eq. 20.43. The NRCS method classifies the land use and soil type by a single parameter called the *curve number*. This method can be used for any size of homogeneous watershed with a known percentage of imperviousness. Infiltration rates for soil group types are found in *Handbook* table Minimum Infiltration Rates for the Various Soil Groups.

Hydrograph Separation

CERM: Sec. 20.9

The stream discharge consists of both surface runoff and subsurface groundwater flows. A procedure known as *hydrograph separation* or *hydrograph analysis* is used to separate runoff (surface flow, net flow, or overland flow) and base flow. There are several methods of separating base flow from runoff. Three methods that are easily carried out manually are presented in CERM Sec. 20.9.

E. DETENTION/RETENTION PONDS

Key concepts: These key concepts are important for answering exam questions in subtopic 5.E, Detention/Retention Ponds.

- detention and retention ponds
- modified puls method
- reservoir routing
- reservoir sizing

NCEES Handbook: To prepare for this subtopic, familiarize yourself with these sections in the *Handbook*.

- Detention and Retention: Rational Method
- Detention and Retention: Routing Equation
- Detention and Retention: Modified Puls Routing Method

PE Civil Reference Manual (**CERM**): Study these sections in CERM that either relate directly to this subtopic or provide background information.

- Section 20.18: Reservoir Sizing: Modified Rational Method
- Section 20.19: Reservoir Sizing: Nonsequential Drought Method
- Section 20.20: Reservoir Sizing: Reservoir Routing
- Section 20.21: Reservoir Routing: Stochastic Simulation

The following equations are relevant for subtopic 5.E, Detention/Retention Ponds.

Detention and Retention: Rational Method

Handbook: Detention and Retention: Rational Method

$$V_{\text{in}} = i \sum A C t$$

$$V_{\text{out}} = Q_0 t$$

The *Handbook* includes the basic equations for using the rational method to calculate the storage required for a detention or retention pond, infiltration system, or swale. The volume of a reservoir needed to hold stream flow from a single storm is simply the total area of the hydrograph. Similarly, when comparing two storms, the incremental volume needed is simply the difference in the areas of their two hydrographs.

Detention and Retention: Routing

Handbook: Detention and Retention: Routing Equation

$$\tfrac{1}{2}(I_1 + I_2)\Delta t + (S_1 - \tfrac{1}{2}O_1 \Delta t) = (S_2 + \tfrac{1}{2}O_2 \Delta t)$$

Large reservoirs must be sized considering the inflows and outflows within the system over longer time frames or more refined time steps. The *Handbook* includes the basic equations for routing flows through a reservoir.

> *Exam tip:* This equation calculates the volume of water coming into a reservoir and leaving it over a given time period, requiring that any instantaneous flow rates be converted into the average volume of water entering or exiting the system within that time.

Detention and Retention: Modified Puls Routing Method

Handbook: Detention and Retention: Modified Puls Routing Method

CERM: Sec. 20.19, Sec. 20.21

$$(I_1 + I_2) + \left(\frac{2S_1}{\Delta t} - O_1\right) = \frac{2S_2}{\Delta t} + O_2$$

In contrast to the traditional routing equation, the modified puls method calculates the inflow and outflow in terms of volume over time.

CERM includes discussion of data-intensive processes such as stochastic simulation (CERM Sec. 20.21), which is based on Monte Carlo simulation processes, and the nonsequential drought method (CERM Sec. 20.19).

F. PRESSURE CONDUIT

Key concepts: These key concepts are important for answering exam questions in subtopic 5.F, Pressure Conduit.

- Darcy equation
- Darcy friction factor
- friction losses
- Hagen-Poiseuille equation
- laminar and turbulent flow
- minor losses
- Moody friction factor chart
- pipes in series and parallel
- pitot-static gauges
- relative and specific roughness
- Reynolds number

NCEES Handbook: To prepare for this subtopic, familiarize yourself with these sections in the *Handbook*.

- Reynolds Number—Circular Pipes
- Head Loss Due to Flow (Darcy-Weisbach Equation)
- Pitot Tube
- Darcy-Weisbach Equation (Head Loss)
- Minor Losses in Pipe Fittings, Contractions, and Expansions
- Entrance and Exit Head Losses
- Multiple Pipes in Parallel
- Pipe Network Analysis: Continuity

PE Civil Reference Manual **(CERM):** Study these sections in CERM that either relate directly to this subtopic or provide background information.

- Section 16.7: Pitot Tube
- Section 16.10: Reynolds Number
- Section 16.11: Laminar Flow
- Section 16.12: Turbulent Flow
- Section 17.6: Relative Roughness
- Section 17.7: Friction Factor
- Section 17.8: Energy Loss Due to Friction: Laminar Flow
- Section 17.9: Energy Loss Due to Friction: Turbulent Flow
- Section 17.28: Series Pipe Systems
- Section 17.29: Parallel Pipe Systems
- Section 17.33: Pitot-Static Gauge

The following equations, figures, and concepts are relevant for subtopic 5.F, Pressure Conduit.

Reynolds Number

Handbook: Reynolds Number—Circular Pipes

CERM: Sec. 16.10, Sec. 17.7

$$\text{Re} = \frac{\text{v}D\rho}{\mu} = \frac{\text{v}D}{\nu}$$

The *Handbook* includes a few different iterations of the equation for the Reynolds number, which are the same as CERM Eq. 16.16(a) and Eq. 16.17. Water is considered a Newtonian fluid, so the Reynolds number for power law fluids does not apply. The Reynolds number can communicate the relative level of turbulence in a pipe, which often dictates which equations apply when determining other fluid dynamic properties. Flow through a closed pressurized pipe is generally characterized as

- laminar flow: Re < 2100
- transitional flow: 2100 < Re < 4000
- fully turbulent flow: 4000 < Re

The Reynolds number, Re, is a dimensionless number interpreted as the ratio of inertial forces to viscous forces in the fluid, as shown in CERM Eq. 16.15. The inertial forces are proportional to the flow diameter, velocity, and fluid density. (Increasing these variables will increase the momentum of the fluid in flow.)

$$\text{Re} = \frac{\text{inertial forces}}{\text{viscous forces}} \quad\quad 16.15$$

The friction factor can also be calculated using CERM Eq. 17.16. While the friction factor diagram shown in CERM Fig. 17.3 is more accurate, this equation will establish an approximate friction factor when other data is not available.

$$f = \frac{64}{\text{Re}} \quad [\text{circular pipe}] \quad\quad 17.16$$

Friction Factor

Handbook: Reynolds Number—Circular Pipes; Head Loss Due to Flow (Darcy-Weisbach Equation); Darcy-Weisbach Equation (Head Loss)

CERM: Sec. 17.6, Sec. 17.7

The *Darcy friction factor*, f, is one of the parameters used to calculate friction loss. The friction factor is not constant but decreases as the Reynolds number (fluid velocity) increases, up to a certain point known as fully turbulent flow (or rough-pipe flow). Once the flow is fully turbulent, the friction factor remains constant and depends only on the relative roughness, not the Reynolds number. A graphical representation of the friction factor is shown in CERM Fig. 17.3.

Figure 17.3 Friction Factor as a Function of Reynolds Number

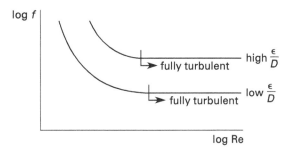

The *Handbook* does not include a specific section on the friction factor, but the friction factor is central to a number of other related sections. *Handbook* figure Moody, Darcy, or Stanton Friction Factor Diagram can be used with a known Reynolds number, material, and diameter to determine the friction factor in a pipe. This factor, in turn, is used to calculate head loss in pipes. CERM Table 17.2 includes the specific roughness of several pipe materials.

Parallel Pipe Systems

Handbook: Multiple Pipes in Parallel; Pipe Network Analysis: Continuity

Pipes in parallel can be two or more pipes, and can split from the same node, merge into the same node, and be in series and parallel at the same time. CERM Fig. 17.17 shows a simple illustration of a parallel pipe system.

Figure 17.17 Parallel Pipe System

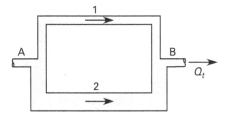

Adding a second pipe in parallel with a first is a standard method of increasing the capacity of a line. A *pipe loop* is a set of two pipes placed in parallel, both originating and terminating at the same junction. The two pipes are referred to as branches or legs. *Handbook* section Pipe Network Analysis: Continuity includes an equation to determine the total flow rate in parallel pipes.

$$\sum Q_{\text{in}} - \sum Q_{\text{out}} = Q_c I$$

Three principles govern the distribution of flow between the two branches.

- The flow divides in such a manner as to make the head loss in each branch the same.
- The head loss between junctions A and B is the same as the head loss in branches 1 and 2.
- The total flow rate is the sum of the flow rates in the two branches.

If the pipe diameters are known, CERM Eq. 17.95 and Eq. 17.97 can be solved simultaneously for the branch velocities. See the entry "Pipes in Parallel" in Chap. 46 of this book for more information.

$$h_{f,1} = h_{f,2} \quad 17.95$$

$$\dot{V}_t = \dot{V}_1 + \dot{V}_2 \quad 17.97$$

Minor Losses

Handbook: Minor Losses in Pipe Fittings, Contractions, and Expansions

$$\frac{p_1}{\gamma} + z_1 + \frac{v_1^2}{2g} = \frac{p_2}{\gamma} + z_2 + \frac{v_2^2}{2g} + h_f + h_{f,\text{fitting}}$$

$$\frac{p_1}{\rho g} + z_1 + \frac{v_1^2}{2g} = \frac{p_2}{\rho g} + z_2 + \frac{v_2^2}{2g} + h_f + h_{f,\text{fitting}}$$

$$h_{f,\text{fitting}} = C\left(\frac{v^2}{2g}\right)$$

$$\frac{v^2}{2g} = 1 \text{ velocity head}$$

Head losses occur as a fluid flows through pipe fittings (elbows, valves, couplings, etc.) and as it encounters sudden pipe contractions and expansions. *Handbook* sections **Entrance and Exit Head Losses** and **Minor Head Loss Coefficients** respectively include a figure and a table containing characteristic values for the coefficient C for different fittings, bends, and exits. See Chap. 46 in this book for more in-depth information.

Pitot-Static Gauge

Handbook: Pitot Tube

$$v = \sqrt{\left(\frac{2}{\rho}\right)(p_0 - p_s)} = \sqrt{\frac{2g(p_0 - p_s)}{\gamma}}$$

CERM: Sec. 16.7, Sec. 17.33

$$v_1 = \sqrt{\frac{2(p_2 - p_1)}{\rho}} \quad [\text{SI}] \quad 16.11(a)$$

$$v_1 = \sqrt{\frac{2g_c(p_2 - p_1)}{\rho}} \quad [\text{U.S.}] \quad 16.11(b)$$

The *Handbook* includes an equation for fluid velocity that is similar to CERM Eq. 16.11. Additionally, CERM Sec. 17.33 contains CERM Eq. 17.142, an equation for the velocity within the pipe.

$$\frac{v^2}{2} = \frac{p_t - p_s}{\rho} = hg \quad [\text{SI}] \quad 17.142(a)$$

$$\frac{v^2}{2g_c} = \frac{p_t - p_s}{\rho} = h \cdot \frac{g}{g_c} \quad [\text{U.S.}] \quad 17.142(b)$$

The main difference between the *Handbook* equation and CERM Eq. 17.142 is the form in which each is given. The *Handbook* equation can be used to calculate the velocity within the pipe, and CERM Eq. 17.142 can be used to calculate the velocity head as a function of depth.

A pitot tube measures the total energy within a channel or pipe. If the pressure energy within the system is known at the point of measurement, velocity head can be calculated easily. The upstream velocity can be calculated if the static and stagnation pressures are known. Piezometer tubes and wall taps are used to measure static pressure energy. The difference between the total and static energies is the kinetic energy of the flow. CERM Fig. 17.22 illustrates a comparative method of directly measuring the velocity head for an incompressible fluid.

Figure 17.22 Comparative Velocity Head Measurement

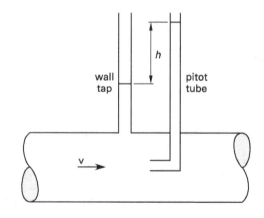

Exam tip: The energy measured by a pitot tube is also referred to as the energy grade line in pipe flow. In unpressurized systems, velocity head can be calculated as the additional depth of water above the surface water elevation within the pitot tube.

Laminar Flow

CERM: Sec. 16.11

If all the particles in a fluid move in paths parallel to the overall flow direction (i.e., in layers, or *laminae*), the flow is said to be *laminar*. (The terms *viscous flow* and *streamline flow* are also used.) This occurs in pipeline flow when the Reynolds number is less than approximately 2100. Laminar flow is typical when the flow channel is small, the velocity is low, and the fluid is viscous. Viscous forces are dominant in laminar flow.

See the entry "Energy Loss Due to Friction: Laminar Flow" in Chap. 46 in this book for equations and methods for calculating energy loss for fluids experiencing laminar flow.

Turbulent Flow

CERM: Sec. 16.12

A fluid is said to be in *turbulent flow* if the Reynolds number is greater than approximately 4000. Turbulent flow is characterized by three-dimensional movement of the fluid particles superimposed on the overall direction of motion. A stream of dye injected into a turbulent flow will quickly disperse and uniformly mix with the surrounding flow. Inertial forces dominate in turbulent flow. At very high Reynolds numbers, the flow is said to be fully turbulent.

See the entry "Energy Loss Due to Friction: Turbulent Flow" in Chap. 46 in this book for equations and methods for calculating energy loss for fluids experiencing turbulent flow.

Series Pipe Systems

CERM: Sec. 17.28

The *Handbook* does not include information specifically dedicated to pipes in series. A system of pipes in series consists of two or more lengths of pipes of different diameters connected end to end. In the case of the series pipe from a reservoir discharging to the atmosphere shown in CERM Fig. 17.16, the available head will be split between the velocity head and the friction loss.

Figure 17.16 Series Pipe System

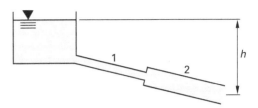

If the flow rate or velocity in any part of the system is known, the friction loss can easily be found as the sum of the friction losses in the individual sections, as given by CERM Eq. 17.89.

$$h_{f,t} = h_{f,a} + h_{f,b} \qquad 17.89$$

$$A_a \mathrm{v}_a = A_b \mathrm{v}_b \qquad 17.90$$

Note that CERM Eq. 17.90 suggests that the flows in each section of pipe are the same. This equation assumes that there are no inflows or outflows related to stormwater inlets, drinking water laterals, and so on, and it assumes that losses due to leaking in pipes are minimal. See the entry "Pipes in Series" in Chap. 46 of this book for more information.

G. ENERGY AND/OR CONTINUITY EQUATION

Key concepts: These key concepts are important for answering exam questions in subtopic 5.G, Energy and/or Continuity Equation.

- Bernoulli equation
- Bernoulli equation assumptions
- continuity of flow
- energy and hydraulic grade lines
- extended Bernoulli equation
- impulse-momentum principle

NCEES Handbook: To prepare for this subtopic, familiarize yourself with these sections in the *Handbook*.

- Principles of One-Dimensional Fluid Flow: Continuity Equation
- Principles of One-Dimensional Fluid Flow: Energy Equation
- Impulse-Momentum Principle
- Momentum Equation

PE Civil Reference Manual (**CERM**): Study these sections in CERM that either relate directly to this subtopic or provide background information.

- Section 16.5: Bernoulli Equation
- Section 16.15: Energy Grade Line
- Section 17.2: Conservation of Mass
- Section 17.19: Extended Bernoulli Equation
- Section 17.20: Energy and Hydraulic Grade Lines with Friction
- Section 17.38: Impulse-Momentum Principle

The following equations and figures are relevant for subtopic 5.G, Energy and/or Continuity Equation.

Impulse-Momentum Principle

Handbook: Impulse-Momentum Principle

$$\sum F = \sum Q_2 \rho_2 v_2 - \sum Q_1 \rho_1 v_1$$

CERM: Sec. 17.38

$$F = \dot{m}\Delta v \qquad \text{[SI]} \qquad 17.171(a)$$

$$F = \frac{\dot{m}\Delta v}{g_c} \qquad \text{[U.S.]} \qquad 17.171(b)$$

The *Handbook* and CERM equations are equivalent once you consider that the mass flow rate equals flow rate times density.

The momentum of a moving object is a vector quantity defined as the product of the object's mass and velocity. Force is the product of mass and acceleration. In this case, the product of flow, density, and velocity produces the same results.

> *Exam tip*: The momentum equation is used to calculate total force from flows. If the flow vectors are meeting, the total force will indicate which direction the flow will go.

Bernoulli Equation

Handbook: Principles of One-Dimensional Fluid Flow: Energy Equation

$$\frac{p_1}{\gamma} + z_1 + \frac{v_1^2}{2g} = \frac{p_2}{\gamma} + z_2 + \frac{v_2^2}{2g} + h_f$$

$$\frac{p_1}{\rho g} + z_1 + \frac{v_1^2}{2g} = \frac{p_2}{\rho g} + z_2 + \frac{v_2^2}{2g} + h_f$$

The Bernoulli equation follows from the principle of conservation of energy. Energy (or head) can be broken down into three components: pressure energy (pressure head), potential energy (gravitational head), and kinetic energy (velocity head).

Head loss, h_f, has two components; major losses (fluid friction losses) and minor losses (turbulence or changes in flow direction at valves, bends, or tees). Consider flow running through a pipe. As the water slides through the pipe, a portion of the water "sticks" to the walls. This "sticky" water drags the rest of the water running through the pipe. The higher the water's velocity through the pipe, the bigger the drag is. This drag is the friction loss.

> *Exam tip*: The equation is an energy balance, meaning both sides must be equal. Unless otherwise noted, variables without a subscript number should be equal on both sides, leaving pressure, velocity, and water height as the only variables that change.

Continuity Equation

Handbook: Principles of One-Dimensional Fluid Flow: Continuity Equation

CERM: Sec. 17.2

$$A_1 v_1 = A_2 v_2$$
$$Q = A v$$

Because mass cannot be created or destroyed, it is a constant throughout pipelines and open channels. Flow rates have many different units. Units of volumetric flow rate include gal/min, gal/day, and m^3/s. Flow rate measures the velocity of a moving volume.

A classic example of using the continuity equation is to evaluate two segments within the same pipe or channel to determine the velocity, friction loss, or other flow properties within the channel or pipeline. A more powerful application is the one shown in the illustration, for pipe distribution networks. The flow along the main line will equal the sum of the flows along the distribution lines (pipes in parallel), as shown.

> *Exam tip*: Total flow in must equal total flow out, as with conservation of mass and momentum. However, the fluid velocity at the inlet can be different from the fluid velocity in the discharge pipes.

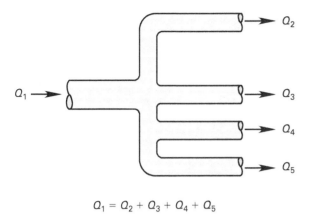

$$Q_1 = Q_2 + Q_3 + Q_4 + Q_5$$

Momentum Equation

Handbook: Momentum Equation

$$M = \frac{Q^2}{gA} + Ay$$

Momentum has a directional vector. The most common way to solve it is to divide the vector into the Cartesian coordinates x and y. One of the main applications of the momentum equation is in the design of pipe restraints and thrust blocks. Two acting forces, the pressure and momentum forces, develop the reaction force.

Exam tip: The required bearing area of a thrust block is dependent upon the reaction force and the soil bearing pressure so that the soil is not "punched" in. The allowable soil bearing pressure (i.e., capacity) must be determined by geotechnical testing and will usually be provided in a problem.

Geometrics

Content in blue refers to the *NCEES Handbook*.

A. Basic Circular Curve Elements............................ 6-1
B. Basic Vertical Curve Elements 6-3
C. Traffic Volume... 6-5

The knowledge area of Geometrics makes up between three and five questions out of the 80 questions on the PE Civil exam. The organization of this chapter follows the order of subtopics given by NCEES for this knowledge area. Each subtopic is covered in the following sections.

A. BASIC CIRCULAR CURVE ELEMENTS

Key concepts: These key concepts are important for answering exam questions in subtopic 6.A, Basic Circular Curve Elements.

- degree of curve
- geometric design of horizontal circular curves
- horizontal circular curves
- stationing on a horizontal curve

NCEES Handbook: To prepare for this subtopic, familiarize yourself with these sections in the *Handbook*.

- Parts of a Circular Curve
- Deflection Angles on a Simple Circular Curve

PE Civil Reference Manual (**CERM**): Study these sections in CERM that either relate directly to this subtopic or provide background information.

- Section 78.16: Stationing
- Section 79.1: Horizontal Curves
- Section 79.2: Degree of Curve
- Section 79.3: Stationing on a Horizontal Curve

The following equations, figures, and concepts are relevant for subtopic 6.A, Basic Circular Curve Elements.

Horizontal Circular Curves

Handbook: Parts of a Circular Curve

CERM: Sec. 79.1

A *horizontal circular curve* is the circular arc between two straight lines known as tangents, and is designed to change a highway's direction of travel from one tangent to another. The tangent that the driver encounters before the curve is the *back tangent* (or *approach tangent*), and the tangent after the curve is the *forward tangent* (*departure tangent*). *Handbook* figure Parts of a Circular Curve illustrates horizontal circular curve elements. CERM Fig. 79.1 is an equivalent figure. The CERM figure uses the variable I for the intersection angle, whereas the *Handbook* figure uses Δ.

Figure 79.1 Horizontal Curve Elements

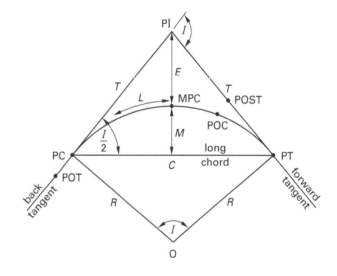

Exam tip: The nomenclature for horizontal circular curves is often used in equations on the exam, so it is important to understand.

Degree of Curve

Handbook: Deflection Angles on a Simple Circular Curve

$$D = \frac{100\left(\frac{180}{\pi}\right)}{R} = \frac{18{,}000}{\pi R}$$

CERM: Sec. 79.2

$$D = \frac{(360°)(100 \text{ ft})}{2\pi R} = \frac{5729.578 \text{ ft}}{R} \quad \text{[arc basis]} \quad 79.8$$

The *Handbook* equation and CERM Eq. 79.8 are equivalent equations, with the values arranged and combined differently.

In the United States, curvature of city streets, highways, and railways can be specified by either the radius or the degree of curve. In most highway work, the length of the curve is understood to be the actual curved arc length, and the degree of the curve is the angle subtended by an arc of 100 ft. When the degree of curve is related to an arc of 100 ft, it is said to be calculated on an arc basis.

Curve Radius

Handbook: Deflection Angles on a Simple Circular Curve

$$R = \frac{5729.6}{D}$$

$$R = \frac{50}{\sin \frac{D}{2}}$$

CERM: Sec. 79.1

$$R = \frac{5729.578 \frac{\text{ft}}{\text{deg}}}{D} \quad \text{[U.S.; arc definition]} \quad 79.1$$

$$R = \frac{50 \text{ ft}}{\sin \frac{D}{2}} \quad \text{[U.S.; chord definition]} \quad 79.2$$

The radius of a curve is a function of the deflection angle, as shown in the equations in *Handbook* section Deflection Angles on a Simple Circular Curve. CERM Eq. 79.1 and Eq. 79.2 are equivalent equations, with the major difference being that the CERM equations include units for the constants.

Length of Curve

Handbook: Deflection Angles on a Simple Circular Curve

$$L = \frac{R\Delta\pi}{180}$$

CERM: Sec. 79.1

$$L = \frac{2\pi RI}{360°} = RI_{\text{radians}} = \frac{(100 \text{ ft})I}{D} \quad \text{[U.S.]} \quad 79.3$$

The length of a curve is a function of the curve radius. The *Handbook* equation and CERM Eq. 79.3 are equivalent equations, with the values arranged and combined differently. The variables I and Δ both represent the same value.

External Distance

Handbook: Deflection Angles on a Simple Circular Curve

$$E = R \sec \frac{\Delta}{2} - R = T \tan \frac{\Delta}{4} = R \left(\frac{1}{\cos \frac{\Delta}{2}} - 1 \right)$$

CERM: Sec. 79.1

$$E = R\left(\sec \frac{I}{2} - 1\right) = R \tan \frac{I}{2} \tan \frac{I}{4} \quad 79.5$$

The *external distance* of a curve is a function of the radius and the intersection angle. The *Handbook* equation and CERM Eq. 79.5 are similar equations arranged differently. The variables I and Δ both represent the same value.

Middle Ordinate

Handbook: Deflection Angles on a Simple Circular Curve

$$M = R - R\cos \frac{\Delta}{2}$$
$$= R\left(1 - \cos \frac{\Delta}{2}\right)$$

CERM: Sec. 79.1

$$M = R\left(1 - \cos \frac{I}{2}\right) = \frac{C}{2} \tan \frac{I}{4} \quad 79.6$$

The *middle ordinate* is the distance from the midpoint of the curve to the long chord. It is a function of the radius and the intersection angle. The *Handbook* equation and CERM Eq. 79.6 are similar equations arranged differently, with the CERM equation also finding the middle ordinate as a function of the long chord. The variables I and Δ both represent the intersection angle.

Chord Length

Handbook: Deflection Angles on a Simple Circular Curve

$$C = 2R \sin \frac{\Delta}{2}$$

CERM: Sec. 79.1

$$C = 2R \sin \frac{I}{2} = 2T \cos \frac{I}{2} \qquad 79.7$$

The chord length is a function of the radius and the intersection angle. The *Handbook* equation and CERM Eq. 79.7 are similar equations arranged differently, with the CERM equation also finding the chord length as a function of the tangent distance. The variables I and Δ both represent the intersection angle.

Stationing on a Circular Curve

CERM: Sec. 78.16, Sec. 79.3

$$\text{sta PT} = \text{sta PC} + L \qquad 79.11$$

$$\text{sta PC} = \text{sta PI} - T \qquad 79.12$$

In route surveying, lengths are divided into 100 ft or 100 m sections called *stations*. The word *station* can mean both a location and a distance. A stake is normally set at each station, at the point of curve (PC), where the back tangent ends and the curve begins, and the point of tangent (PT), where the curve ends and the forward tangent begins. The PT station is equal to the PC station plus the curve length, as shown in CERM Eq. 79.11. Likewise, the PC station is equal to the point of intersection (PI) station minus the tangent length, as shown in Eq. 79.12.

Normal curve layout follows the convention of showing increased stationing (ahead stationing) from left to right.

B. BASIC VERTICAL CURVE ELEMENTS

Key concepts: These key concepts are important for answering exam questions in subtopic 6.B, Basic Vertical Curve Elements.

- curve length
- elevation along a curve
- horizontal distances along a curve
- PVC, PVI, and PVT locations
- low and high points on vertical curves
- types of vertical curves
- vertical clearance
- vertical curve properties

NCEES Handbook: To prepare for this subtopic, familiarize yourself with these sections in the *Handbook*.

- Vertical Curve Formulas
- Symmetrical Vertical Curve Formula

PE Civil Reference Manual (**CERM**): Study these sections in CERM that either relate directly to this subtopic or provide background information.

- Section 79.10: Superelevation
- Section 79.17: Vertical Curves
- Section 79.18: Vertical Curves Through Points
- Section 79.19: Vertical Curve to Pass Through Turning Point

Transportation Depth Reference Manual (**CETR**): Study these sections in CETR that either relate directly to this subtopic or provide background information.

- Section 4.9: Vertical Curves

The following equations, figures, and concepts are relevant for subtopic 6.B, Basic Vertical Curve Elements.

Rate of Grade Change per Station

Handbook: Vertical Curve Formulas; Symmetrical Vertical Curve Formula

$$r = \frac{g_2 - g_1}{L}$$

CERM: Sec. 79.17

$$R = \frac{G_2 - G_1}{L} \quad \text{[may be negative]} \qquad 79.46$$

The *Handbook* equation and CERM Eq. 79.46 are the same equation, with CERM Eq. 79.46 using uppercase variables instead of lowercase.

The *rate of grade change per station* is a measure of how quickly or sharply a curve changes direction. It is a function of the grade of the back tangent, the grade of the forward tangent, and the length of the curve.

Curve Elevation at Any Point on a Vertical Curve

Handbook: Vertical Curve Formulas; Symmetrical Vertical Curve Formula

$$\text{curve elevation} = Y_{\text{PVC}} + g_1 x + ax^2$$
$$= Y_{\text{PVC}} + g_1 x + x^2 \left(\frac{g_2 - g_1}{2L} \right)$$

CERM: Sec. 79.17

$$\text{elev}_x = \frac{Rx^2}{2} + G_1 x + \text{elev}_{\text{BVC}} \quad \quad 79.47$$

CERM Eq. 79.47 is a rearrangement of the *Handbook* equation that uses the variable elev for elevation instead of Y_{PVC} and an uppercase G for grade instead of a lowercase g. The calculation of half the rate of change, $R/2$, is equivalent to the parabola constant, a, in the *Handbook*.

The elevation at any point along the curve can be found if the elevation of the point of vertical curvature (PVC), incoming grade, and rate of change of the curve are known.

Vertical Curves

Handbook: Symmetrical Vertical Curve Formula

CERM: Sec. 79.17

A *vertical curve* is used to change from one highway grade to another. Most vertical curves are equal-tangent parabolas. Vertical curves are designed to have specified clearances from vertical objects. For example, a sag curve that goes under an overpass may require 16 ft to allow for certain trucks to pass under the overpass.

The vertex of a vertical curve is the point where the tangent slopes intersect. In a symmetrical vertical curve, the sections of the road at the beginning and end of the vertical curve have equal slopes. The length of a vertical curve is the straight-line distance between the beginning and end of the curve, while the length of a horizontal curve is the distance along the curve (or along the 100 ft chord lengths for the chord basis). For symmetrical curves, the intersection point of the two tangents (i.e., the vertex) divides the curve in half.

Handbook section Symmetrical Vertical Curve Formula contains a list of nomenclature terms used for vertical curves, along with a diagram; Y is elevation, which is not defined in the list. A similar illustration is shown in CERM Fig. 79.10.

Sag and Crest Curves

Handbook: Symmetrical Vertical Curve Formula

CERM: Sec. 79.10

CETR: Sec. 4.9

Sag curves are concave, upward curves, as shown in CETR Fig. 4.17. They are most often used to transition from negative to positive vertical grades.

Figure 4.17 *Three Conditions of Sag Vertical Curves*

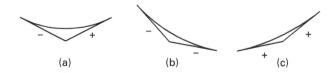

Crest curves are concave, downward curves, as shown in CETR Fig. 4.18. They are most often used to transition from positive to negative vertical grades.

Figure 4.18 *Three Conditions of Crest Vertical Curves*

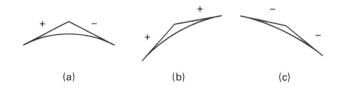

Vertical Curves Through Points

CERM: Sec. 79.18

$$s = \sqrt{\frac{\text{elev}_\text{E} - \text{elev}_\text{G}}{\text{elev}_\text{E} - \text{elev}_\text{F}}} \quad \quad 79.50$$

$$L = \frac{2d(s+1)}{s-1} \quad \quad 79.51$$

If a curve is to have some minimum clearance from an obstruction, the curve length will generally not be known in advance. If the station and elevation of the point of vertical curve (or the vertex) and the gradient values are known, then CERM Eq. 79.51 can be used to find the length of a vertical curve with an obstruction. The constant s used in this equation can be found using CERM Eq. 79.50.

CERM Fig. 79.11 shows a vertical curve with an obstruction.

Figure 79.11 Vertical Curve with an Obstruction

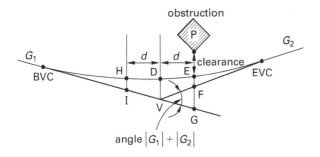

C. TRAFFIC VOLUME

Key concepts: These key concepts are important for answering exam questions in subtopic 6.C, Traffic Volume.

- annual vehicle miles traveled (VMT)
- average annual daily traffic (AADT)
- design hour volume
- free-flow speed
- levels of service
- peak hour factors
- rate of flow
- traffic density
- traffic flow rates
- traffic volume abbreviations

NCEES Handbook: To prepare for this subtopic, familiarize yourself with these sections in the *Handbook*.

- Traffic Flow, Density, Headway, and Speed Relationships
- Headway
- Space Mean Speed
- Greenshields Maximum Flow Rate Relationship
- Speed-Density Model
- Flow-Density Model
- Speed-Flow Model
- Time Mean Speed
- Average Speed (Mean Speed)
- Average Annual Daily Traffic Estimation

PE Civil Reference Manual (**CERM**): Study these sections in CERM that either relate directly to this subtopic or provide background information.

- Section 73.2: Abbreviations and Units
- Section 73.4: Design Vehicles
- Section 73.5: Levels of Service
- Section 73.6: Speed Parameters
- Section 73.8: Volume Parameters
- Section 73.10: Speed, Flow, and Density Relationships

The following equations, figures, tables, and concepts are relevant for subtopic 6.C, Traffic Volume.

Traffic Flow Rate

Handbook: Traffic Flow, Density, Headway, and Speed Relationships

$$q = \frac{n}{t}$$

$$q = \frac{n(3600)}{t}$$

Traffic flow rate is defined as the ratio of the number of vehicles passing some designated point in the roadway to the time interval in which those vehicles pass that point. Traffic flow rate should not be confused with *traffic density*, which is the ratio of the number of vehicles in the roadway to the roadway length.

Headway

Handbook: Headway

$$t = \sum_{i=1}^{n} h_i$$

$$q = \frac{n}{\sum_{i=1}^{n} h_i}$$

$$q = \frac{1}{\bar{h}}$$

CERM: Sec. 73.10

Headway is the time between successive vehicles, calculated as a function of the headway of the ith vehicle in a stream of i vehicles and the number of measured headways taken at a designated point. The average hourly flow can be found from the same values.

A concept related to headway is *gap*, or the time interval between vehicles in a traffic stream. Gap is important for pedestrians trying to cross a traffic flow.

Space Mean Speed

Handbook: Space Mean Speed

$$u_s = \frac{q}{k}$$

$$u_s = \frac{nL}{\sum_{i=1}^{n} t_i}$$

Space mean speed is calculated by dividing the total distance traveled by all vehicles by the total of the travel times of all vehicles. It can also be found by dividing the traffic flow by the traffic density.

Greenshields Maximum Flow Rate Relationship

Handbook: Greenshields Maximum Flow Rate Relationship

$$q_{\max} = \frac{k_j u_f}{4}$$

The maximum flow rate can be found using the free-flow speed and the jam density.

Speed-Density Model

Handbook: Speed-Density Model

$$u_s = u_f \left(1 - \frac{k}{k_j}\right)$$

CERM: Sec. 73.10

$$S = S_f \left(1 - \frac{D}{D_j}\right) \quad 73.9$$

The *Handbook* equation and CERM Eq. 73.9 are the same equation with different nomenclature: S is the same as u_s, S_f is the same as u_f, and k is the same as D. The speed-density relationship is shown graphically in CERM Fig. 73.1.

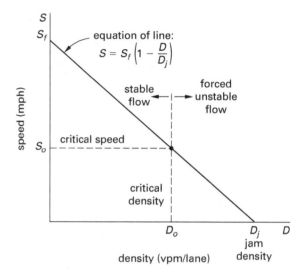

Figure 73.1 Space Mean Speed Versus Density

As traffic density increases, traffic can become interrupted, and the speed on the roadway will no longer be the free-flow speed. The density of the roadway may eventually increase to the *jam density*, the density at which the vehicles are all at a standstill. Knowing the current density of the roadway, the jam density, and the free-flow speed can allow for calculation of the speed of the roadway when traffic is not in the free-flow condition.

Flow-Density Model

Handbook: Flow-Density Model

$$q = u_f \left(k - \frac{k^2}{k_j}\right)$$

$$q_{\text{cap}} = u_f \left(\frac{k_j}{4}\right)$$

CERM: Sec. 73.10

The *flow-density relationship* is the relationship between the flow rate, the free-flow speed, and the traffic density. It is used to calculate the maximum traffic volume at the peak of a curve. This relationship can be mathematically described using the *Handbook* equations shown, and is illustrated in CERM Fig. 73.3.

Speed-Flow Model

Handbook: Speed-Flow Model

$$k = k_j\left(1 - \frac{u_s}{u_f}\right)$$

$$q = k_j\left(u_s - \frac{u_s^2}{u_f}\right)$$

CERM: Sec. 73.10

The speed-flow relationship can be used to find the critical speed of the system, as shown in the *Handbook* equations. CERM Fig. 73.2 illustrates this relationship.

Time Mean Speed

Handbook: Time Mean Speed

$$\overline{u_t} = \frac{\sum_{i}^{t} u_i}{n}$$

Time mean speed (*average spot speed*) is the arithmetic mean of the instantaneous speed of all cars at a particular point.

Average Speed (Mean Speed)

Handbook: Average Speed (Mean Speed)

$$\overline{x} = \frac{\sum n_i S_i}{N}$$

Average travel speed (*mean speed*) is the speed over a specified section of highway, including operational delays such as stops for traffic signals.

Traffic Estimation

Handbook: Average Annual Daily Traffic Estimation

CERM: Sec. 73.8

$$\text{AADT} = (V_{24ij})(\text{DF}_i)(\text{MF}_j)$$

$$\text{DF} = \frac{V_{\text{avg}}}{V_{\text{day}}}$$

$$\text{MF}_i = \frac{\text{AADT}}{\text{ADT}_i}$$

The *average annual daily traffic*, AADT, is a function of the 24-hour traffic volume for a given day in a given month; the daily adjustment factor, DF, for that day; and the monthly adjustment factor, MF, for that month. DF is the ratio of the average daily count of vehicles for all days of the week to the average daily count for a given day of the week. The monthly adjustment factor is a function of AADT and the average daily traffic, ADT, for that month. For purposes of calculating MF, AADT can be estimated as the average of 12 monthly ADTs.

Traffic Volume Abbreviations

CERM: Sec. 73.2

The units and abbreviations shown in CERM Table 73.1 are derived from the *Highway Capacity Manual* (HCM) and are used in standard practice.

Table 73.1 Abbreviations

abbreviation	meaning
ln	lane
km/h	kilometers per hour
mph	miles per hour
p	people, person, or pedestrian
pc	passenger car
pcph	passenger cars per hour
pcphg	passenger cars per hour of green signal
pcphpl	passenger cars per hour per lane
pcphgpl	passenger cars per hour of green signal per lane
pc/km/ln	passenger cars per kilometer per lane
pcpmpl	passenger cars per mile per lane
ped	pedestrian
pers	person
veh	vehicle
vph	vehicles per hour
veh/km	vehicles per kilometer
vph/lane	vehicles per hour per lane
vpk	vehicles per kilometer
vpm	vehicles per mile

Some HCM abbreviations differ when SI units are used. For example, while the HCM still uses "pcph" for calculations in English units, it has adopted "pc/h" for calculations in SI units, and "ln" is used for "lane" in SI calculations.

Design Vehicles

CERM: Sec. 73.4

Design vehicles represent a variety of common vehicles. They ensure that geometric features accommodate all commonly sized vehicles in that category. They are given specific designations in *A Policy on Geometric Design of Highways and Streets* (AASHTO *Green Book*), and CERM Table 73.3 shows a list of standard design vehicles.

Vehicle equivalents are calculated using factors obtained from the *Highway Capacity Manual* (HCM) and are influenced by many variables, including speed, size, and density.

> *Exam tip*: For exams other than the PE Civil Transportation exam, design vehicle information will be provided. If it is not provided, then the term *design vehicle* can be assumed to be synonymous with *vehicle*.

Level of Service

CERM: Sec. 73.5

Level of service (LOS) is a user's quality of service through or over a specific facility, such as a highway, intersection, or crosswalk. The designations run from A, unimpeded flow, to F, gridlock. A lower LOS (where A is highest and F is lowest) typically indicates less safety afforded by the roadway.

LOS is a function of vehicle speed and flow rate on the roadway. The relationship between the flow rate and the speed is also known as *vehicular density*. Vehicular density is calculated by dividing the flow rate (in vehicles per hour per lane) by the speed (in miles per hour). The result is in units of vehicles per mile per lane.

Evaluations for level of service focus on 15 minutes of peak flow. This peak is obtained from the hourly peak traffic observed in the field.

It is desirable to have LOS A on every roadway and street, but this is unrealistic in urban areas, and only possible when the volume of traffic is small. The benefits of designing for LOS A may not be enough to justify the cost of construction.

7 Materials

Content in blue refers to the *NCEES Handbook*.

A. Soil Classification and Boring Log
   Interpretation.. 7-1
B. Soil Properties... 7-4
C. Concrete.. 7-6
D. Structural Steel.. 7-7
E. Material Test Methods and Specification
   Conformance.. 7-7
F. Compaction... 7-8

The knowledge area of Materials makes up between five and eight questions out of the 80 questions on the PE Civil exam. The organization of this chapter follows the order of subtopics given by NCEES for this knowledge area. Each subtopic is covered in the following sections.

A. SOIL CLASSIFICATION AND BORING LOG INTERPRETATION

Key concepts: These key concepts are important for answering exam questions in subtopic 7.A, Soil Classification and Boring Log Interpretation.

- Atterberg limits
- boring logs
- classes of soil
- classification of soil particle sizes
- coefficient of curvature
- cone penetrometer test
- grain and particle size distribution
- group index
- Hazen uniformity coefficient
- liquid limit
- organic material
- plasticity index
- plastic limit
- sieve analysis
- standard penetration test

NCEES Handbook: To prepare for this subtopic, familiarize yourself with these sections in the *Handbook*.

- Guidelines for Minimum Number of Exploration Points and Depth of Exploration
- Soil Classification Chart
- AASHTO Soil Classification System
- AASHTO Classification System
- Atterberg Limits
- Conceptual Changes in Soil Phases as a Function of Water Content
- Concepts of Soil Phase, Soil Strength, and Soil Deformation Based on Liquidity Index
- U.S. Standard Sieve Sizes and Corresponding Opening Dimension
- Sample Grain Size Distribution Curves
- Gradation Tests
- Gradation Based on C_u and C_c Parameters

PE Civil Reference Manual (**CERM**): Study these sections in CERM that either relate directly to this subtopic or provide background information.

- Section 35.1: Soil Particle Size Distribution
- Section 35.3: AASHTO Soil Classification
- Section 35.4: Unified Soil Classification
- Section 35.9: Standard Penetration Test
- Section 35.14: Atterberg Limit Tests

The following equations, figures, tables, and concepts are relevant for subtopic 7.A, Soil Classification and Boring Log Interpretation.

Boring Log

Handbook: Guidelines for Minimum Number of Exploration Points and Depth of Exploration

Handbook table Guidelines for Minimum Number of Exploration Points and Depth of Exploration provides guidelines for taking soil samples based on application.

A *boring log* is a record of a soil profile measured from the ground surface to a predefined logging depth. Completing a boring log typically involves taking soil samples and performing various field tests. A number of techniques can be used to extract soil samples at different depths, such as wash boring and rotary boring.

The depth of the water table should be noted. Sometimes initial and final water tables are given in the same boring log (i.e., the water table can be seasonal). Depth is often given as elevation instead of depth below surface. A boring log should also indicate the location where samples were taken.

A boring log can be used to find the properties of a given layer of soil and to predict how different soil properties will interact. It is very useful when determining the foundation type that is needed in an area.

Specifications for soil borings (e.g., number, depth, location, and type) are given in a project geotechnical report, which is produced by a licensed geotechnical engineer.

USCS Soil Classification

Handbook: Soil Classification Chart

CERM: Sec. 35.4

The criteria for the Unified Soil Classification System (USCS) are provided in *Handbook* table Soil Classification Chart. This criteria is also shown in CERM Table 35.5.

The USCS is based on soil particle size distribution, liquid limit, and the plasticity index. Each USCS group is designated by group symbols and a corresponding group name (e.g., GW is the group symbol for well-graded gravel). Each group symbol contains two letters, the first representing particle sizes, and the second being a descriptive modifier, such as W for well-graded or fairly clean soil and C for soil with significant amounts of clay.

AASHTO Soil Classification System

Handbook: AASHTO Soil Classification System; AASHTO Classification System

$$GI = (F-35)(0.2 + 0.005(LL-40)) + 0.01(F-15)(PI-10)$$

CERM: Sec. 35.3

$$I_g = (F_{200} - 35)\bigl(0.2 + 0.005(LL-40)\bigr) \\ + 0.01(F_{200} - 15)(PI-10) \quad 35.3$$

The *Handbook* equation and CERM Eq. 35.3 are the same except for the variable used to represent the group index (GI in the *Handbook*, I_g in CERM) and the use of the subscript 200 on the variable F for the percentage of soil that passes through a no. 200 sieve.

The American Association of State Highway and Transportation Officials (AASHTO) soil classification system is based on sieve analysis, liquid limit, and the plasticity index. It is used most often to classify soils supporting roadways. The AASHTO classification system categorizes soils into groups and subgroups, and some further with a group index, which compares soils within the same group.

Handbook table AASHTO Soil Classification System and CERM Table 35.4 display the AASHTO soil classification chart.

Exam tip: For the A-2-6 and A-2-7 subgroups, only the second term of the equation, $0.01(F-15)(PI-10)$, is used.

Atterberg Limits

Handbook: Atterberg Limits; Conceptual Changes in Soil Phases as a Function of Water Content; Concepts of Soil Phase, Soil Strength, and Soil Deformation Based on Liquidity Index

CERM: Sec. 35.14

$$LI = \frac{w - PL}{PI}$$

Handbook figure Conceptual Changes in Soil Phases as a Function of Water Content illustrates soil phases based on water content. The water content values corresponding to the transitions between solid, semisolid, plastic solid, and liquid are known as the *Atterberg limits*. The most commonly used Atterberg limits are liquid limit, LL, and plastic limit, PL.

The *consistency* of a clay is the water content relative to the Atterberg limits. The consistency can be represented by the liquidity index, LI. A liquidity index between zero and one indicates a water content between the plastic limit and liquid limit, and a liquidity index greater than one indicates that the water content is above the liquid limit. *Handbook* table Concepts of Soil Phase, Soil Strength, and Soil Deformation Based on Liquidity Index provides soil phases and strengths for various liquidity index values.

A third Atterberg limit, the shrinkage limit, is less commonly used. The shrinkage limit is defined as the water content at which further drying out of the soil does not decrease the volume of soil. Below the shrinkage limit, air enters the voids, and water content decreases are not accompanied by decreases in volume. The shrinkage index, SI, is defined as the range in moisture content over which the soil is in a semisolid condition, as shown in CERM Eq. 35.24.

$$SI = PL - SL \quad 35.24$$

Plasticity Index

Handbook: Atterberg Limits

CERM: Sec. 35.14

$$PI = LL - PL$$

The plasticity index, PI, is the range of moisture content at which a soil is in a plastic condition. Soil in a plastic condition can be deformed and will hold together without crumbling. As shown in the *Handbook* equation and CERM Eq. 35.23, the plasticity index is the difference between the liquid limit and plastic limit.

The plasticity index also correlates with strength and deformation properties of soil. A high PI is typically found in clay soils. A medium PI is typically found in silty soils that contain some clay. Soils with a PI at or close to zero are considered nonplastic (NP). These low PIs are typically found in sands and silts with little or no clay.

Sieve Analysis

Handbook: U.S. Standard Sieve Sizes and Corresponding Opening Dimension

CERM: Sec. 35.1

In sieve analysis, sieves with holes of various diameters are stacked in order, with the largest openings on top and the smallest on the bottom. Each sieve is weighed when empty and weighed again after the soil has been sieved. The weight of the soil retained by each sieve is found by subtracting the sieve's empty weight from its full weight. The weight of the soil retained on each sieve is then plotted against the size of the sieve's openings, as shown in CERM Fig. 35.1. The percentage by weight of soil passing each sieve is plotted on the vertical axis. The grain diameter (log scale) is plotted on the horizontal axis.

Figure 35.1 Typical Particle Size Distribution

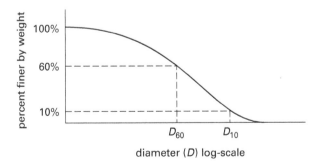

Handbook table U.S. Standard Sieve Sizes and Corresponding Opening Dimension lists standard sieve numbers and properties. There are sieves with smaller openings than a no. 200 sieve, but they are less commonly used. The smallest sieve passes only the fines (silt and clay). A hydrometer test is used to distinguish between silt and clay.

The grain diameter is typically described in terms of the percentage of soil that passes through a sieve of a given diameter, as shown in CERM Fig. 35.1. D_{10}, D_{30}, and D_{60} are the sieve diameters that 10%, 30%, and 60% of the soil will pass through, respectively. D_{10} is also known as the *effective grain size* of the soil sample. These values are used to define various soil distribution parameters, namely the soil uniformity and curvature coefficients.

Particle Size Distribution

Handbook: Sample Grain Size Distribution Curves

CERM: Sec. 35.1

To determine a soil's type, its grain size distribution is estimated. The grain size distribution for a coarse soil is found from a sieve analysis test. It is less common to determine the grain size distribution for the finer particles. *Handbook* figure Sample Grain Size Distribution Curves shows particle size distributions for gap-graded, well-graded, and uniformly graded soils.

For more about sieve analysis, see the entry "Sieve Analysis" in this chapter.

Hazen Uniformity Coefficient

Handbook: Gradation Tests; Gradation Based on C_u and C_c Parameters

CERM: Sec. 35.1

$$C_u = \frac{D_{60}}{D_{10}}$$

The Hazen uniformity coefficient, found using the equation shown, indicates the general shape of the particle size distribution. The higher the coefficient, the more well graded the soil is; a Hazen uniformity coefficient of 10 or more indicates a well-graded soil.

D_{60} is the diameter that 60% of the soil is finer than; D_{10} is the diameter that 10% of the soil is finer than.

Handbook table Gradation Based on C_u and C_c Parameters provides typical Hazen uniformity coefficients and coefficients of curvature for different soil types. CERM Table 35.3 provides typical Hazen uniformity coefficients for different soil types.

Coefficient of Curvature

Handbook: Gradation Tests; Gradation Based on C_u and C_c Parameters

CERM: Sec. 35.1

$$C_c = \frac{D_{30}^2}{D_{60}D_{10}}$$

The *coefficient of curvature*, also known as the *coefficient of gradation*, is an index for the uniformity of the soil, much like the Hazen uniformity coefficient. The *Handbook* equation and CERM Eq. 35.2 are identical except for the variable used for the coefficient of curvature.

D_{10} is the diameter that 10% of the soil is finer than, D_{30} is the diameter that 30% of the soil is finer than, and D_{60} is the diameter that 60% of the soil is finer than.

Handbook table Gradation Based on C_u and C_c Parameters provides typical Hazen uniformity coefficients and coefficients of curvature for different soil types.

B. SOIL PROPERTIES

Key concepts: These key concepts are important for answering exam questions in subtopic 7.B, Soil Properties.

- degree of saturation
- mass-volume relationships
- moisture content
- permeability

NCEES Handbook: To prepare for this subtopic, familiarize yourself with these sections in the *Handbook*.

- A Unit of Soil Mass and Its Idealization
- Volume and Weight Relationships
- Weight-Volume Relationships
- Laboratory Permeability Tests
- Hazen's Equation for Permeability
- Darcy's Law

PE Civil Reference Manual **(CERM):** Study these sections in CERM that either relate directly to this subtopic or provide background information.

- Section 35.5: Mass-Volume Relationships
- Section 35.15: Permeability Tests

The following equations and figures are relevant for subtopic 7.B, Soil Properties.

Phase Diagram

Handbook: A Unit of Soil Mass and Its Idealization

CERM: Sec. 35.5

A phase diagram is a schematic representation of the solids-water-air percentages in terms of mass and volume. Solids-water-air percentages are used to calculate the aggregate properties of the soil. A phase diagram can be used to understand and quickly calculate other soil testing parameters, like porosity or void ratio. See *Handbook* figure A Unit of Soil Mass and Its Idealization or CERM Fig. 35.4 for an example of a phase diagram.

Exam tip: It is best to draw the phase diagram to solve problems dealing with soil phases.

Porosity and Void Ratio

Handbook: Volume and Weight Relationships

CERM: Sec. 35.5

$$e = \frac{V_v}{V_s}$$

$$n = \frac{V_v}{V}$$

The porosity, e, is the ratio of the volume of voids to the total volume. The void ratio, n, is the ratio of the volume of voids to the volume of solids.

The main difference between void ratio and porosity is that the porosity is used to define the amount of free space that is in the sample, and the void ratio is used to relate the amount of free space to filled space. Either of these values can be used to draw a soil phase diagram.

The *Handbook* often treats weight and mass as equivalent; other sources may use mass in place of weight in these equations. Unit weight calculations vary based on water content. *Handbook* table Weight-Volume Relationships provides formulas for calculation of unit weight.

Moisture Content and Degree of Saturation

Handbook: Volume and Weight Relationships

$$w = \frac{W_w}{W_s}$$

$$S = \frac{V_w}{V_v}$$

CERM: Sec. 35.5

$$w = \frac{m_w}{m_s} \cdot 100\% \qquad 35.6$$

$$S = \frac{V_w}{V_v} \cdot 100\%$$
$$= \frac{V_w}{V_g + V_w} \cdot 100\% \qquad 35.7$$

The *Handbook* equations and CERM Eq. 35.6 and Eq. 35.7 are identical, except that CERM Eq. 35.6 specifies that the water content is a percentage, and Eq. 35.7 expands the equation to include a calculation on the basis of both volume of water and volume of air. As with many other *Handbook* equations, the *Handbook* treats weight and mass as equivalent. *Handbook* table Weight-Volume Relationships provides formulas for calculation of unit weight.

The water content (or moisture content), w, is the ratio of the mass of water to the mass of solids. The degree of saturation, S, is the ratio of the volume of water to the total volume of voids, usually expressed as a percentage. The water content and the degree of saturation both indicate how much of the void space is filled with water.

If all the voids are filled with water, then the volume of air is zero, the degree of saturation is 100%, and the sample is said to be fully saturated. Fully saturated soil, which typically occurs below the water table, weighs more than soil that is only partially saturated. This affects the differences between the total and effective stresses of a soil sample. See *Handbook* table Volume and Weight Relationships or CERM Table 35.7 for a summary of mass-volume relationships in soil phases.

Unit Weight

Handbook: Volume and Weight Relationships

$$\gamma_{\text{sat}} = \frac{W_s + W_w}{V_s + V_w}$$

$$\gamma_d = \frac{W_s}{V_s + V_w}$$

CERM: Sec. 35.5

$$\rho = \frac{m_t}{V_t} = \frac{m_w + m_s}{V_g + V_w + V_s} \qquad 35.8$$

$$\rho_d = \frac{m_s}{V_t} = \frac{m_s}{V_g + V_w + V_s} \qquad 35.10$$

The *Handbook* equations are identical to CERM Eq. 35.8 and Eq. 35.10 except for the use of unit weight and weight in the *Handbook* equations and density and mass in the CERM equations. The *Handbook* often treats weight and mass as equivalent. *Handbook* table Weight-Volume Relationships provides formulas for calculation of unit weight.

The *total unit weight* is the ratio of the total weight to the total volume. The total unit weight is the *moist unit weight* if the soil is not fully saturated, or the *saturated unit weight* if the soil is fully saturated (i.e., if it falls below the water table).

The *dry unit weight* is the ratio of the solid weight (i.e., the weight of the soil grains) to the total volume. Dry unit weight is important because it is used for the estimation of relative density and relative compaction.

Permeability of Soils

Handbook: Laboratory Permeability Tests

$$k = \frac{QL}{tAh}$$

$$k = \frac{aL}{A(t_1 - t_2)} \ln \frac{h_1}{h_2}$$

CERM: Sec. 35.15

$$K = \frac{VL}{hAt} \quad \text{[constant head]} \qquad 35.29$$

$$K = \frac{A'L}{At} \ln \frac{h_i}{h_f} \quad \text{[falling head]} \qquad 35.30$$

The equations shown are used to calculate the coefficient of permeability. The *Handbook* equations are equivalent to CERM Eq. 35.29 and Eq. 35.30. The differences are K versus k for the coefficient of permeability, Q versus V for the volume of water, and t versus $t_1 - t_2$ for the difference in time.

Permeability implies constant voids in the soil sample, not just the presence of voids in the soil; all permeable soils will have relatively large void ratios, but not all soils with large void ratios will be permeable. Landfill liners are made from clay with low permeability, while gravelly soils are used where high permeability is desired, such as draining water away from structures.

Hazen's Equation for Permeability

Handbook: Hazen's Equation for Permeability

$$k = CD_{10}^2$$

CERM: Sec. 35.15

$$K_{\text{cm/s}} \approx CD_{10,\text{mm}}^2 \qquad 35.28$$

The *Handbook* version of Hazen's equation and CERM Eq. 35.28 are equivalent except for k versus K for the coefficient of permeability and the subscripts indicating units. The CERM equation is an approximation in metric units. Otherwise, it is equivalent to the *Handbook* equation.

The permeability of soils is the ability to transfer water through soil voids. The permeable material supports a flow of water. It can be estimated using Hazen's formula for uniform soils. Hazen's formula provides only a crude estimate; actual values can vary by up to two orders of magnitude.

Darcy's Law

Handbook: Darcy's Law

$$Q = kiA$$

CERM: Sec. 35.15

$$Q = \text{v} A_{\text{gross}} \quad \quad 35.26$$

$$\text{v} = Ki \quad \quad 35.27$$

The *Handbook* version of Darcy's law is a combination of CERM Eq. 35.26 and Eq. 35.27. The variables K and k both stand for permeability.

The flow of water through a permeable aquifer or soil is given by Darcy's law. Darcy's law uses testing to calculate the time for water to travel through a soil. Darcy's law can be rearranged to solve for the permeability, which is common when conducting laboratory and field testing.

C. CONCRETE

Key concepts: These key concepts are important for answering exam questions in subtopic 7.C, Concrete.

- compressive strength
- concrete
- material testing

***NCEES Handbook*:** To prepare for this subtopic, familiarize yourself with these sections in the *Handbook*.

- Concrete Testing
- Design Provisions: Definitions
- Concrete Exposure Categories and Classes

***PE Civil Reference Manual* (CERM):** Study these sections in CERM that either relate directly to this subtopic or provide background information.

- Section 48.1: Concrete
- Section 48.2: Cementitious Materials
- Section 48.8: Compressive Strength

The following concepts are relevant for subtopic 7.C, Concrete.

Concrete Compressive Strength

Handbook: Concrete Testing; Design Provisions: Definitions

CERM: Sec. 48.8

Handbook section Concrete Testing contains an overview of material properties and testing for concrete. *Handbook* section Design Provisions: Definitions contains common nomenclature and definitions for concrete design.

Concrete compressive strength is the maximum stress a concrete specimen can sustain in compressive axial loading. Concrete compressive strength is determined by testing concrete cylinders.

Construction documents provide minimum concrete compressive strengths that must be met by the project concrete suppliers. Typically, the design strength of the concrete is specified at 28 days, because while the compressive strength increases with time, the increase levels off at approximately 28 days. However, cylinders are often tested at three and/or seven days. A typical seven-day concrete break strength is 75% of the 28-day break strength.

Concrete Overview

Handbook: Concrete Exposure Categories and Classes

CERM: Sec. 48.1, Sec. 48.2

Handbook section Concrete Exposure Categories and Classes contains categories, conditions, and classes for concrete exposure.

Concrete is a mixture of cementitious materials, aggregates, water, and air. It is relatively strong in compression, but weak in tension. It is an economical choice for many structural applications, because its primary volume is made up of aggregates that can be sourced virtually anywhere in the world.

During the concrete curing process, microcracks that form in the crystalline structure are forced together in compression and pulled apart in tension, which explains why concrete behaves well in compression but needs steel reinforcement to resist tension. The behavior and design of concrete are driven by these facts.

Various admixtures can be used in concrete to improve numerous properties of concrete. Admixtures can be chemical or cementitious.

D. STRUCTURAL STEEL

Key concepts: These key concepts are important for answering exam questions in subtopic 7.D, Structural Steel.

- material properties
- stress-strain behavior
- structural steel
- tensile strength

NCEES Handbook: To prepare for this subtopic, familiarize yourself with these sections in the *Handbook*.

- Stress-Strain Curve for Mild Steel

PE Civil Reference Manual (**CERM**): Study these sections in CERM that either relate directly to this subtopic or provide background information.

- Section 58.2: Types of Structural Steel and Connecting Elements
- Section 58.3: Steel Properties
- Section 58.4: Structural Shapes

The following figures, tables, and concepts are relevant for subtopic 7.D, Structural Steel.

Stress-Strain Curves for Steel

Handbook: Stress-Strain Curve for Mild Steel

CERM: Sec. 58.3

Handbook figure Stress-Strain Curve for Mild Steel shows a stress-strain curve for mild steel. CERM Fig. 58.1 shows some stress-strain curves for various steels.

A stress-strain curve is, as the name suggests, a graph of the stress and strain a sample of steel can resist when subjected to a tension test. The slope of the initial straight-line portion of the stress-strain curve is the *modulus of elasticity*, which is taken as 29,000 ksi for all structural steel design calculations.

The yield strength is the unit tensile stress at which the stress-strain curve exhibits a well-defined increase in strain without a large increase in stress.

Properties of Structural Steel

CERM: Sec. 58.3

Structural steel is produced to specified minimum mechanical properties and American Society for Testing and Materials (ASTM) designations. The mechanical properties of structural steel are generally determined from tension tests on small specimens in accordance with standard ASTM procedures. CERM Table 58.1 lists some typical properties of structural steel.

The modulus of elasticity and the density of steel are independent of the type of steel used, while the tensile strength and yield stress are dependent on the type of steel used. Tensile strength and yield stress also depend on the size and thickness of the steel piece.

Three types of hot-rolled structural steel are used in buildings: carbon steel, high-strength low-alloy steel, and quenched tempered alloy steel. Most structural steel is made from carbon steel. W shapes are typically made from high-strength, low-alloy steels.

Structural Shape Designations

CERM: Sec. 58.4

Each steel structural shape is ideally used in a specific application (beam, column, etc.) based on the inertia in the x- and y-directions. Usually, sections with similar or close inertia in both directions are used as columns. Sections with a clear difference in inertia in both directions are used as beams.

CERM Fig. 58.2 illustrates several structural shapes, and CERM Table 58.2 provides nomenclature used for shape designations. CERM Fig. 58.3 shows how some sections can be combined.

E. MATERIAL TEST METHODS AND SPECIFICATION CONFORMANCE

Key concepts: These key concepts are important for answering exam questions in subtopic 7.E, Material Test Methods and Specification Conformance.

- compressive strength
- concrete mixture proportioning
- concrete testing
- material properties and testing

NCEES Handbook: To prepare for this subtopic, familiarize yourself with these sections in the *Handbook*.

- Concrete Testing
- Mixture Proportioning

PE Civil Reference Manual (**CERM**): Study these sections in CERM that either relate directly to this subtopic or provide background information.

- Section 48.8: Compressive Strength
- Section 49.6: Absolute Volume Method

The following equations are relevant for subtopic 7.E, Material Test Methods and Specification Conformance.

ASTM Material Testing

Handbook: Concrete Testing
CERM: Sec. 48.8

$$f'_c = \frac{P}{A} \qquad 48.1$$

American Society for Testing and Materials (ASTM) C39 is a compressive test specimen of a mold cylinder. The ASTM C39 cylinder has a 6 in (150 mm) diameter and a 12 in (300 mm) height. The concrete in the cylinder is cured for a specific amount of time at a specific temperature, and then the specimen is axially loaded to failure at a specific rate. The specific amount of time that the concrete is cured could be three days, a week, 28 days, or more.

CERM Eq. 48.1 gives the compressive strength of an ASTM C39 concrete specimen. P is the axial load on the sample, and A is the area of the sample.

Absolute Volume Method

Handbook: Mixture Proportioning

$$\text{yield} = \frac{\text{total mass of batched materials}}{\text{density of freshly mixed concrete}}$$

$$\frac{\text{absolute}}{\text{volume}} = \frac{\text{mass of loose material}}{\text{relative density (or specific gravity)} \atop \text{of material} \cdot \text{density of water}}$$

CERM: Sec. 49.6

$$V_{\text{absolute}} = \frac{m}{(\text{SG})\rho_{\text{water}}} \qquad [\text{SI}] \qquad 49.1(a)$$

$$V_{\text{absolute}} = \frac{W}{(\text{SG})\gamma_{\text{water}}} \qquad [\text{U.S.}] \qquad 49.1(b)$$

CERM Eq. 49.1 is the same as the *Handbook* equation, except it uses variables instead of words to represent the values calculated.

The absolute volume method can be used to find the absolute volume that a given loose material occupies in a unit volume of concrete. The absolute volume is found from the mass of the loose material, the relative density (specific gravity) of the material, and the density of water. To use the absolute volume method, it is necessary to know the solid densities of the constituents.

Yield is the volume of wet concrete produced in a batch (in cubic yards). The yield is found from the total mass of the batched materials and the density of the freshly mixed concrete.

Exam tip: Relative densities in mix design calculations are based on saturated surface-dry (SSD) conditions.

F. COMPACTION

Key concepts: These key concepts are important for answering exam questions in subtopic 7.F, Compaction.

- field compaction methods
- field density
- nuclear gauge method
- sand cone method

NCEES Handbook: To prepare for this subtopic, familiarize yourself with these sections in the *Handbook*.

- Nondestructive Test Methods
- Nondestructive Test Methods for Concrete
- Field Density Testing
- Relative Density, Relative Compaction, and Void Ratio Concepts
- Acceptance Criteria for Compaction
- Compactors Recommended for Various Types of Soil and Rock
- Summary of Compaction Criteria

***PE Civil Reference Manual* (CERM):** Study these sections in CERM that either relate directly to this subtopic or provide background information.

- Section 35.11: Proctor Test
- Section 35.13: In-Place Density Tests

The following equations, figures, tables, and concepts are relevant for subtopic 7.F, Compaction.

Nuclear Gauge Method

Handbook: Nondestructive Test Methods; Nondestructive Test Methods for Concrete

CERM: Sec. 35.13

Handbook sections Nondestructive Test Methods and Nondestructive Test Methods for Concrete cover the variety of purposes and properties of nondestructive test methods.

Nuclear gauges use the rate of radiation penetration through the soil to determine both the wet density and the water content of the soil. These devices have a probe containing radioactive material and are inserted into a hole punched into the compacted soil. The relative compaction and/or field density of a compacted soil are output from the nuclear gauge. These values are used to compare the field conditions to the maximum dry density and optimum water content found in the lab using one of the Proctor tests.

Relative Compaction

Handbook: Field Density Testing; Relative Density, Relative Compaction, and Void Ratio Concepts; Acceptance Criteria for Compaction

$$\text{RC} = \frac{\gamma_{d,\text{field}}}{\gamma_{d,\text{max}}} \cdot 100\%$$

CERM: Sec. 35.11

$$\text{RC} = \frac{\rho_d}{\rho_d^*} \cdot 100\% \qquad 35.19$$

The *Handbook* equation and CERM Eq. 35.19 are identical except that the *Handbook* equation uses unit weight and the CERM equation uses density. The *Handbook* treats weight and mass as equivalent. *Handbook* table Weight-Volume Relationships provides formulas for calculation of unit weight.

Relative compaction, RC, compares the density or unit weight of compacted soil in the field to the maximum density or unit weight obtainable in a lab. It is the percentage of the maximum value determined in the laboratory. Project specifications usually indicate the required range of water content and the minimum density in the field. The maximum relative compaction is less than 100% because the maximum dry unit weight occurs when all the voids are filled with water (i.e., the maximum water content).

Relative compaction is different from relative density (also known as specific gravity), which is the ratio of a soil's density to that of water. *Handbook* figure Relative Density, Relative Compaction, and Void Ratio Concepts illustrates relative compaction percentages versus relative densities. *Handbook* figure Acceptance Criteria for Compaction illustrates desired ranges of compaction used for initial testing and pretesting of backfill.

Field Compaction Methods

Handbook: Compactors Recommended for Various Types of Soil and Rock; Summary of Compaction Criteria

CERM: Sec. 35.11

Compaction equipment densifies soil using different mechanisms, such as static loading, impact, vibration, and kneading. Each type of equipment is most suitable for a specific soil type. *Handbook* figure Compactors Recommended for Various Types of Soil and Rock includes several types of compactors for different zones of application. See *Handbook* table Summary of Compaction Criteria for a summary of compaction criteria for different soil types. For more in-depth information on field compaction, see subtopic 11.A, Excavation and Embankment, Borrow Source Studies, Laboratory and Field Compaction, in Chap. 19 of this book.

Sand Cone Method

The sand cone method can be used to find the in situ density and unit weight of soils. The basics of the sand cone method include filling a hole with a known amount of sand in order to find the volume of the hole, and then calculating the density of the sample excavated from the hole. The main components of the testing device are a balloon initially filled with standard sand of a known density, a valve, and a cone.

As with all field testing, the sand cone method should be performed at various points in the field. The sand cone test is only able to measure and calculate the density and unit weight of the soil that was in the excavated hole. The in situ densities are typically the important densities for compaction efforts, but at times the excavated density can also be important (e.g., when working with cuts and fills).

For more information on the sand cone method, see ASTM Standard D1556.

Site Development

Content in blue refers to the *NCEES Handbook*.

A. Excavation and Embankment 8-1
B. Construction Site Layout and Control................. 8-4
C. Temporary and Permanent Soil Erosion and
    Sediment Control... 8-5
D. Impact of Construction on Adjacent Facilities 8-7
E. Safety 8-7

The knowledge area of Site Development makes up between four and six questions out of the 80 questions on the PE Civil exam. The organization of this chapter follows the order of subtopics given by NCEES for this knowledge area. Each subtopic is covered in the following sections.

A. EXCAVATION AND EMBANKMENT

Key concepts: These key concepts are important for answering exam questions in subtopic 8.A, Excavation and Embankment.

- average end area method
- borrow pits
- braced cuts
- earthwork
- excavation
- grade points and grade lines
- mass diagrams
- net volume calculations
- prismoidal formula method
- profile diagrams
- sheet piling
- soil shrinkage and expansion

NCEES Handbook: To prepare for this subtopic, familiarize yourself with these sections in the *Handbook*.

- Excavation and Embankment
- Cross-Section Methods
- Cross-Sectional End Areas
- Borrow Pit Grid Method
- Earthwork Area Formulas

PE Civil Reference Manual (**CERM**): Study these sections in CERM that either relate directly to this subtopic or provide background information.

- Section 39.1: Excavation
- Section 39.2: Braced Cuts
- Section 39.3: Braced Cuts in Sand
- Section 39.4: Braced Cuts in Stiff Clay
- Section 39.5: Braced Cuts in Soft Clay
- Section 39.6: Braced Cuts in Medium Clay
- Section 39.7: Stability of Braced Excavations in Clay
- Section 39.8: Stability of Braced Excavations in Sand
- Section 39.9: Sheet Piling
- Section 39.10: Analysis/Design of Braced Excavations
- Section 78.30: Traverse Area: Method of Coordinates
- Section 78.32: Areas Bounded by Irregular Boundaries
- Section 80.3: Swell and Shrinkage
- Section 80.5: Cut and Fill
- Section 80.7: Cross Sections
- Section 80.9: Typical Sections
- Section 80.10: Distance Between Cross Sections
- Section 80.13: Earthwork Volumes
- Section 80.14: Average End Area Method
- Section 80.15: Prismoidal Formula Method
- Section 80.16: Borrow Pit Geometry
- Section 80.17: Mass Diagrams

The following equations, figures, and concepts are relevant for subtopic 8.A, Excavation and Embankment.

Loose and Compacted Volume

Handbook: Excavation and Embankment

$$V_L = \left(1 + \frac{S_w}{100\%}\right) V_B$$

$$V_C = \left(1 - \frac{S_h}{100\%}\right) V_B$$

CERM: Sec. 80.3

$$V_l = \left(\frac{100\% + \% \text{ swell}}{100\%}\right) V_b = \frac{V_b}{\text{LF}} \qquad 80.1$$

$$V_c = \left(\frac{100\% - \% \text{ shrinkage}}{100\%}\right) V_b \qquad 80.2$$

The *Handbook* equation for a soil's loose volume is equivalent to CERM Eq. 80.1, and the *Handbook* equation for the compacted volume is equivalent to CERM Eq. 80.2. The differences are in the nomenclature and the arrangement of the equation. CERM Eq. 80.1 additionally includes a different form of the equation that finds the loose volume using the load factor.

When earth is excavated, it increases in volume because of an increase in voids. The change in volume of earth from its natural state to its loose state is known as *swell*.

The decrease in volume of earth from its natural state to its compacted state is known as *shrinkage*. Shrinkage is expressed as a percentage decrease from a soil's natural state. Swell and shrinkage vary with soil type. For more in-depth information, see the entries "Material Volume Change Characteristics," "Shrinkage Factor," and "Swell and Swell Factor" in Chap. 10 of this book.

Swell and Shrinkage

Handbook: Excavation and Embankment

$$\text{shrinkage factor} = \frac{\text{compacted unit weight}}{\text{bank unit weight}}$$

$$S_h = \frac{\text{compacted unit weight} - \text{bank unit weight}}{\text{compacted unit weight}} \cdot 100\%$$

$$\text{swell factor} = \frac{\text{loose unit weight}}{\text{bank unit weight}}$$

$$S_w = \left(\frac{\frac{W}{V_B}}{\frac{W}{V_L}} - 1\right) \cdot 100\%$$

Swell, S_w, is expressed as a percentage of the natural volume. A soil's load factor in a particular excavation environment is the inverse of the *swell factor* (also known as the *bulking factor*), which is the sum of 1.0 and the swell expressed as a decimal.

The decrease in volume of earth from its natural state to its compacted state is known as *shrinkage*. Shrinkage is expressed as a percentage decrease from a soil's natural state. See CERM Sec. 80.3 for more details on swell and shrinkage.

In the *Handbook* equations, the compacted unit weight, bank unit weight, and loose unit weight are all dry unit weights.

Average End Area Method

Handbook: Cross-Section Methods; Cross-Sectional End Areas

CERM: Sec. 80.14

$$V = L\left(\frac{A_1 + A_2}{2}\right)$$

The average end area method calculates the volume between two consecutive cross sections as the average of their areas multiplied by the distance between the two areas. This method assumes the fill is positive and the cut is negative. *Handbook* figure Cross-Sectional End Areas shows examples of fill and cut diagrams. See the entry "Average End Area Method" in Chap. 10 of this book for more information.

Prismoidal Formula Method

Handbook: Cross-Section Methods; Cross-Sectional End Areas

CERM: Sec. 80.15

$$V = L\left(\frac{A_1 + 4A_m + A_2}{6}\right)$$

The prismoidal formula is the preferred cross-sectional area formula when two end areas differ greatly or the ground surface is irregular. See *Handbook* figure Cross-Sectional End Areas for examples of fill and cut diagrams, and see the entry "Prismoidal Formula" in Chap. 10 of this book for more information.

Borrow Pit Grid Method

Handbook: Borrow Pit Grid Method

CERM: Sec. 80.16

It is often necessary to borrow earth from an adjacent area to construct embankments. Normally, the borrow pit area is laid out in a rectangular grid, and elevations are determined at the corners of each square so that the cut at each corner can be computed. The illustration in *Handbook* section Borrow Pit Grid Method and CERM Fig. 80.4 show this method. See the entry "Borrow Pit Grid Method" in Chap. 10 of this book for more information.

Simpson's Rule

Handbook: Earthwork Area Formulas

$$A = \frac{1}{3}\begin{pmatrix} \text{first value} + \text{last value} \\ + 4(\text{sum of odd-numbered values}) \\ + 2(\text{sum of even-numbered values}) \\ \cdot \text{length of interval} \end{pmatrix}$$

CERM: Sec. 78.32

$$A = \frac{d}{3}\left(h_1 + h_n + 2\sum_{\text{odd}} h_i + 4\sum_{\text{even}} h_i\right) \quad 78.41$$
$$= \frac{d}{3}(h_1 + 4h_2 + 2h_3 + 4h_4 + \ldots + h_n)$$

When the irregular side of each cell is curved, Simpson's rule (sometimes referred to as Simpson's 1/3 rule) can be used to determine the area. The difference between the *Handbook* equation and CERM Eq. 78.41 is that CERM Eq. 78.41 uses classic summation nomenclature, while the *Handbook* equation uses words to explain the terms of the equation.

Coordinate Method

Handbook: Earthwork Area Formulas

$$A = \tfrac{1}{2}\big(X_A(Y_B - Y_N) + X_B(Y_C - Y_A) + X_C(Y_D - Y_B) + \cdots + X_N(Y_A - Y_{N-1})\big)$$

CERM: Sec. 78.30

$$A = \frac{1}{2}\left|\begin{array}{l}\sum \text{of full line products} \\ -\sum \text{of broken line products}\end{array}\right| \quad 78.37$$

The area calculation is simplified in CERM Eq. 78.37, but the *Handbook* and CERM equations are fundamentally the same. The area of a simple traverse can be found by dividing the traverse into a number of geometric shapes and summing their areas. If the coordinates of the traverse leg end points are known, the coordinate method can be used. The coordinates can be x-y coordinates referenced to some arbitrary set of axes, or they can be sets of departure and latitude. See CERM Sec. 78.30 for more details on this method.

Trapezoidal Method

Handbook: Earthwork Area Formulas

$$A = w\big(\tfrac{1}{2}(h_1 + h_n) + h_2 + h_3 + h_4 + \cdots + h_{n-1}\big)$$

In this equation,

A	area	ft^2
h	height	ft
w	length of common interval	ft

CERM: Sec. 78.32

$$A = d\left(\frac{h_1 + h_n}{2} + \sum_{i=2}^{n-1} h_i\right) \quad 78.40$$

The nomenclature for CERM Eq. 78.40 is identical to the *Handbook* nomenclature, except for the use of d instead of w to represent the length of the common interval. The formulas are similar, except that CERM Eq. 78.40 uses a summation to simplify the second half of the equation.

Exam tip: If the irregular side of each cell is fairly straight, the trapezoidal rule can be used.

Cut and Fill

CERM: Sec. 80.5

Earthwork that is to be excavated is known as *cut*. Excavation that is placed in an embankment is known as *fill*.

The volume of cut and fill is calculated from the cross-sectional areas at various intervals and the distance between the areas. Net volume is the difference between the cut and the fill. A net cut results in material needing to be removed from the project, while a net fill results in material needing to be brought in from a borrow site.

Braced Cuts

CERM: Sec. 39.2 to Sec. 39.10

A *braced cut* is an excavation in which the active earth pressure from one bulkhead is used to support the facing bulkhead. The method used to analyze braced cuts varies depending on the type of soil.

B. CONSTRUCTION SITE LAYOUT AND CONTROL

Key concepts: These key concepts are important for answering exam questions in subtopic 8.B, Construction Site Layout and Control.

- survey leveling
- survey stake markings
- survey staking

NCEES Handbook: To prepare for this subtopic, familiarize yourself with these sections in the *Handbook*.

- Site Layout and Control

PE Civil Reference Manual **(CERM):** Study these sections in CERM that either relate directly to this subtopic or provide background information.

- Section 81.1: Staking
- Section 81.2: Stake Markings
- Section 81.3: Establishing Slope Stake Markings

The following equations, figures, tables, and concepts are relevant for subtopic 8.B, Construction Site Layout and Control.

Stake Location

Handbook: Site Layout and Control

$$\text{elevation of BM} + \text{BS} = \text{HI}$$

$$\text{elevation of TP} = \text{HI} - \text{FS}$$

CERM: Sec. 81.3

Locating a construction stake properly requires knowing the elevation of the ground and the height of instrument, HI, which is the elevation of the instrument above the reference datum. The *Handbook* equations can be used to determine HI from the elevation of the benchmark, BM, and the backsight measurement, BS; and the elevation of the turning point, TP, from HI and the foresight measurement, FS.

CERM Eq. 81.1 finds the height of the instrument given the ground elevation and either the instrument height of the ground or the ground rod, which is slightly different from the *Handbook* equation.

$$\text{HI} = \text{elev}_{\text{ground}} + \text{instrument height above the ground} \quad 81.1$$
$$= \text{elev}_{\text{ground}} + \text{ground rod}$$

The use of HI to determine stake location is illustrated in CERM Fig. 81.3.

Figure 81.3 Determining Stake Location

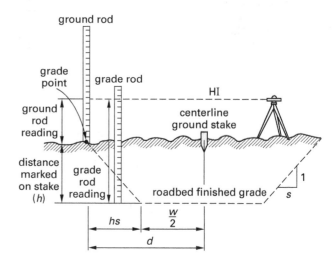

The grade rod is calculated from the planned grade elevation and the height of the instrument, as shown in CERM Eq. 81.2.

$$\text{grade rod} = \text{HI} - \text{elev}_{\text{grade}} \quad 81.2$$

The distance, h, marked on a construction stake is the distance between the natural and finished grade elevations (i.e., the cut or fill) and is easily calculated from the ground and grade rods, as shown in CERM Eq. 81.3.

$$h = \text{grade rod} - \text{ground rod} \quad 81.3$$

The horizontal distance, d, from the grade point to the centerline stake of the finished surface is found from the width of the finished surface, w, the side slope ratio, s, and the cut or fill at the grade point, h, as shown in CERM Eq. 81.4.

$$d = \frac{w}{2} + hs \quad 81.4$$

Staking

CERM: Sec. 81.1

Surveying markers are referred to as construction stakes, alignment stakes, offset stakes, grade stakes, or slope stakes, depending on their purpose. The distances on stakes are measured in feet or meters. Stakes can relate a lot of information to personnel on a job site, including the station at which the engineer is located, so it is important that engineers understand the specific stake markings for a given project.

Precision of staking depends on the type of feature being documented. As shown in CERM Table 81.1, bridges and buildings require much higher accuracy in survey staking than pipes or telephone poles. The values in this table apply to new construction and when locating the different components of a project.

Stake Markings

CERM: Sec. 81.2

Stake markings vary greatly from agency to agency. The abbreviations shown in CERM Table 81.2 are typical but not universal.

Typically, the front (the side facing the construction) and back of the stake are marked permanently using pencil, carpenter's crayon, or permanent ink markers. Stakes are read from top to bottom. The front of the stake is marked with header information (e.g., offset distance) and cluster information (e.g., horizontal and vertical measurements, slope ratio). The header is separated from the first cluster by a double horizontal line. Multiple clusters are separated by single horizontal lines. All cluster information is measured in the same direction from the same point.

CERM Fig. 81.1 illustrates a construction stake for a storm drain.

Figure 81.1 Construction Stake for Storm Drain

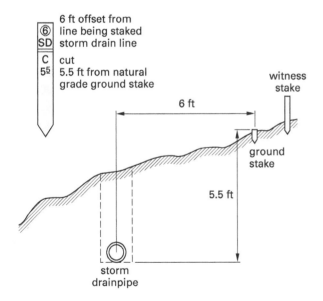

Slope Stakes

CERM: Sec. 81.3

Slope stakes indicate *grade points*, points where the cuts and fills begin, and the planned side slopes intersect the natural ground surface. Slope stakes are marked "SS" to indicate their purpose and are typically placed with a 10 ft (3 m) offset. In addition to indicating the grade point, the fronts of slope stakes are marked to indicate the nature of the earthwork (C for cut and F for fill), the offset distance, the type of line being staked, the distance from the centerline or control line, the slope to finished grade, and the elevation difference between the grade point and the finished grade. Distances from the centerline can be marked L or R to indicate whether the stake points are to the left or the right of the centerline when looking up-station. The station is marked on the stake back.

CERM Fig. 81.2 illustrates the use of slope stakes along three adjacent sections of a proposed highway.

C. TEMPORARY AND PERMANENT SOIL EROSION AND SEDIMENT CONTROL

Key concepts: These key concepts are important for answering exam questions in subtopic 8.C, Temporary and Permanent Soil Erosion and Sediment Control.

- erosion and sediment control permits
- mechanics of erosion
- mechanics of sedimentation
- slope and erosion control features
- types of control measures

NCEES Handbook: The *Handbook* does not contain any material relevant to this subtopic. Be sure to study the listed sections in CERM.

PE Civil Reference Manual (**CERM**): Study these sections in CERM that either relate directly to this subtopic or provide background information.

- Section 26.13: Sedimentation Removal Efficiency: Unmixed Basins
- Section 80.20: Slope and Erosion Control Features

The following figures and concepts are relevant for subtopic 8.C, Temporary and Permanent Soil Erosion and Sediment Control.

Sedimentation

CERM: Sec. 26.13

Erosion is the action of particles moving from a surface. *Sedimentation* is the result of those particles being deposited somewhere by a flowing stream.

A low velocity of flow can be ensured through various mitigation techniques. The idea behind each technique is to slow the flow rate down so that particles can sediment on the construction site.

Temporary Berms

CERM: Sec. 80.20

A *temporary berm* is constructed of compacted soil, with or without a shallow ditch, either at the top of fill slopes or transverse to the centerline on fills. Berms are temporarily used at the top of newly constructed slopes to prevent excessive erosion until permanent controls are installed or slopes are stabilized.

Temporary berm design may be subject to specifications set forth by governing agencies. They typically have a minimum width of 24 in at the top and a minimum height of 12 in.

CERM Fig. 80.12 illustrates a temporary berm.

Slope Drain

CERM: Sec. 80.20

Slope drains carry water down slopes. Common slope drain materials include stone, concrete, and asphalt gutters; fiber mats; plastic sheets; pipe; and sod.

Slope drains intercept runoff before it can affect other parts of a project, and they reduce erosion by carrying accumulating water down slopes prior to the installation of permanent facilities or the growth of adequate ground cover on the slopes.

Slope drains must be anchored to slopes to prevent disruption by the force of the water flowing in the drains. Energy dissipators (rock dumps or basins) are needed at the outlet ends of slope drains to prevent erosion downstream and to collect sediment. Slope drains are shown in CERM Fig. 80.12.

Silt Fence

CERM: Sec. 80.20

A *silt fence* is a sediment trap that passes water and holds soil fines. A silt fence uses wire mesh or other material attached to posts.

For best results, erosion control and silt fences should be implemented together. Silt fences should be used where siltation is expected, such as at the bottoms of fill slopes. Silt fences should be constructed to retain the suspended particles; the filter fabric should be on the upstream side of the fence, and the fabric should rest on the natural ground. An example of silt fencing is shown in CERM Fig. 80.12.

Sediment Structure

CERM: Sec. 80.20

A *sediment structure* is an energy-dissipating rock dump, basin, pond, or trap that catches and stores sediment from upstream erodible areas. Its purpose is to protect properties and stream channels, including culverts and pipe structures, below a construction site from excessive siltation. A sediment structure should be at least twice as long as it is wide.

Check Dam

CERM: Sec. 80.20

A *check dam* is a barrier composed of logs and poles, large stones, or other material placed across a natural or constructed drainway to retard stream flow and catch small sediment loads. A check dam with a minimum depth of 2 ft should be keyed into the sides and bottom of the channel.

Temporary Seeding and Mulching

CERM: Sec. 80.20

Temporary seeding and mulching consists of seeding, mulching, fertilizing, and matting used to reduce erosion. Typically, all cut and fill slopes, including waste sites and borrow pits, should be seeded.

Erosion control fabric is used on steep slopes to prevent erosion of soil and mulch. Mulch on slopes exceeding a 3:1 V:H ratio should be held in place by erosion control fabric. This type of fabric is shown in CERM Fig. 80.12.

Additional Erosion and Sediment Control Measures

CERM: Sec. 80.20

A *brush barrier* consists of brush, tree trimmings, shrubs, plants, and other refuse from a clearing and grubbing operation. A brush barrier should be placed on natural ground at the bottom of a fill slope to restrain sedimentation particles.

Baled hay or straw erosion checks are temporary measures to control erosion and prevent siltation. Bales may be either hay or straw containing 5 ft^3 or more of material. They should be used where the existing ground slopes toward or away from the embankment along the toes of slopes, in ditches, or in other areas where siltation erosion or runoff is a problem.

Ditch lining fabric can be used to prevent erosion of ditches by fastening the fabric to the ground with stakes, U-staples, or wire.

Erosion and Sediment Control Permits

CERM: Sec. 80.20

Erosion and sediment control permits are often referred to as *environmental permits*. One of the most common permits is the National Pollutant Discharge Elimination System (NPDES) permit, often called a *stormwater permit*. It is a permit from the U.S. Environmental Protection Agency (EPA) that regulates the flow of water over a construction site. The permit may dictate the mitigation activities required for construction activities, such as using a temporary berm or a baled hay erosion check.

Other permits are often required by the agency in charge of the project. Sometimes adjacent agencies may have input on the permitting process. For example, a road being built near a waterway may require permitting from the Army Corps of Engineers.

D. IMPACT OF CONSTRUCTION ON ADJACENT FACILITIES

Key concepts: These key concepts are important for answering exam questions in subtopic 8.D, Impact of Construction on Adjacent Facilities.

- impacts of excavation on existing structures
- mitigation measures

NCEES Handbook: To prepare for this subtopic, familiarize yourself with these sections in the *Handbook*.

- Work Zone and Public Safety
- Determination of Soil Type
- Slope and Shield Configurations
- Allowable Slopes
- Slope Configurations: Excavations in Layered Soils
- Excavations Made in Type A Soil
- Excavations Made in Type B Soil

PE Civil Reference Manual (CERM): Study these sections in CERM that either relate directly to this subtopic or provide background information.

- Section 80.20: Slope and Erosion Control Features
- Section 83.5: Soil Classification

The following equations, figures, tables, and concepts are relevant for subtopic 8.D, Impact of Construction on Adjacent Facilities.

Traffic and Noise

Handbook: Work Zone and Public Safety

$$D = 100\% \cdot \sum \frac{C_i}{T_i}$$

Noise from a construction site can impact nearby facilities and people. The *Handbook* equation can be used to calculate noise exposure from a construction site.

Construction traffic can affect local traffic patterns and may cause extra wear and tear on roadways. Additionally, it may cause safety concerns for the local population.

Trenching and Excavation

Handbook: Determination of Soil Type; Slope and Shield Configurations; Allowable Slopes; Slope Configurations: Excavations in Layered Soils; Excavations Made in Type A Soil; Excavations Made in Type B Soil

CERM: Sec. 83.5

The most dangerous part of construction is excavation, particularly deep excavation. Excavations need to be protected from caving in by scaffolds or slopes.

Soil classification is the controlling factor in the proper design of an excavation protection system. *Handbook* section Determination of Soil Type provides OSHA categories for soil and rock deposits.

Sloping is typically cheaper than benching or shoring, but is not always possible due to the tight space around many construction sites. *Handbook* figure Slope and Shield Configurations illustrates slope and shield configurations for each type of soil, and *Handbook* table Allowable Slopes gives values for allowable slopes based on soil type. *Handbook* figure Slope Configurations: Excavations in Layered Soils provides slope configuration diagrams for excavations in layered soils, and *Handbook* figures Excavations Made in Type A Soil and Excavations Made in Type B Soil provide diagrams specific to excavations made in type A and B soils, respectively.

Settlement, Dust, and Runoff

CERM: Sec. 80.20

Settlement can cause issues for construction sites as well as existing buildings. Settlement analysis should be performed prior to construction activities, and site surveys may be utilized for settlement during a project.

Dust and stormwater runoff can affect neighboring projects and the water quality of nearby water bodies. Erosion and sedimentation control (ESC) measures should be taken on construction sites. Erosion mitigation techniques such as temporary silt fences or erosion control fabric can be employed.

E. SAFETY

Key concepts: These key concepts are important for answering exam questions in subtopic 8.E, Safety.

- electrical safety
- fall and impact protection
- incidence rate
- OSHA guidelines

- recordable injury
- safe practices for crane operation
- scaffolds
- trenching and excavation

NCEES Handbook: To prepare for this subtopic, familiarize yourself with these sections in the *Handbook*.

- Telescoping Boom Industrial Cranes
- Telescoping Boom Rough-Terrain Cranes
- Telescoping Boom Carrier- and Crawler-Mounted Cranes
- Telescoping Boom Truck-Mounted Cranes
- Telescoping Boom Crane Components
- Lattice Boom Cranes
- Lattice Boom Crane Components
- Safety Incidence Rate
- Experience Modification Rate
- Work Zone and Public Safety
- Work Zone and Public Safety: Permissible Noise Exposure (OSHA)
- Determination of Soil Type
- Slope and Shield Configurations
- Allowable Slopes
- Slope Configurations: Excavations in Layered Soils
- Excavations Made in Type A Soil
- Excavations Made in Type B Soil

PE Civil Reference Manual (**CERM**): Study these sections in CERM that either relate directly to this subtopic or provide background information.

- Chapter 83: Construction and Job Site Safety

The following equations, figures, tables, and concepts are relevant for subtopic 8.E, Safety.

Cranes

Handbook: Telescoping Boom Industrial Cranes; Telescoping Boom Rough-Terrain Cranes; Telescoping Boom Carrier- and Crawler-Mounted Cranes; Telescoping Boom Truck-Mounted Cranes; Telescoping Boom Crane Components; Lattice Boom Cranes; Lattice Boom Crane Components

CERM: Sec. 83.17

Handbook sections Telescoping Boom Industrial Cranes, Telescoping Boom Rough-Terrain Cranes, Telescoping Boom Carrier- and Crawler-Mounted Cranes, Telescoping Boom Truck-Mounted Cranes, and Lattice Boom Cranes illustrate several different types of cranes. *Handbook* section Telescoping Boom Crane Components provides illustrations of telescoping boom crane components, and *Handbook* section Lattice Boom Crane Components illustrates lattice boom crane components.

A hydraulic truck crane consists of a boom, a jib, a rotex gear, outriggers, counterweights, wire rope, and a hook. A tower crane consists of a base, a mast, and a slewing unit, which itself consists of the gear and motor that allow the crane to rotate. A tower crane is depicted in CERM Fig. 83.5.

Safety Incidence Rate

Handbook: Safety Incidence Rate

$$\text{IR} = \frac{N \cdot 200{,}000}{T}$$

CERM: Sec. 83.2, Sec. 83.3

The *Handbook* equation finds the safety incidence rate, IR, based on the number of injuries, illnesses, and fatalities and the total hours worked by all employees during a given period of time. CERM Sec. 83.3 provides a similar equation that refers to the safety incidence rate as the recordable injury incidence rate. Both equations represent the number of injuries per 200,000 hours, or the equivalent of 100 employees working 40 hours per week, 50 weeks per year.

Only actual, recordable, nonfatal injuries and illnesses are included in the injury incidence rate calculation. The types of incidents that OSHA considers recordable include illnesses and injuries that result in a loss of consciousness, restriction of work or motion, or permanent transfer to another job within the company, or that require some type of medical treatment beyond first aid.

Experience Modification Rate

Handbook: Experience Modification Rate

$$\text{EMR} = \frac{B + H + EW + (1 - W)F}{D + H + FW + (1 - W)F}$$

The *experience modification rate* is an annual adjustment for workers' compensation insurance premiums based on actual loss experienced for the previous three full years.

Permissible Noise Exposure

Handbook: Work Zone and Public Safety; Work Zone and Public Safety: Permissible Noise Exposure (OSHA)

$$D = 100\% \cdot \sum \frac{C_i}{T_i}$$

CERM: Sec. 83.14

Permissible noise exposure is defined by OSHA as the given time spent at a specified sound pressure level. The *Handbook* equation and CERM Eq. 83.5 are the same.

Handbook table Work Zone and Public Safety: Permissible Noise Exposure (OSHA) provides permissible times (in hours) for common noise levels in A-weighted decibels. CERM Table 83.5 is an equivalent table.

Soil Classification and Soil Excavations

Handbook: Determination of Soil Type; Slope and Shield Configurations; Allowable Slopes; Slope Configurations: Excavations in Layered Soils; Excavations Made in Type A Soil; Excavations Made in Type B Soil

CERM: Sec. 83.5, Sec. 83.6, Sec. 83.7

OSHA categorizes soil and rock deposits into four types: A, B, C, and D. *Handbook* section Determination of Soil Type provides definitions of the four types, and CERM Fig. 83.1 is a diagram of soil types organized by geotechnical qualities.

Handbook figure Slope and Shield Configurations illustrates slope and shield configurations for each type of soil, and *Handbook* table Allowable Slopes provides values for allowable slopes based on soil type. *Handbook* figure Slope Configurations: Excavations in Layered Soils provides slope configuration diagrams for excavations in layered soils, and *Handbook* figures Excavations Made in Type A Soil and Excavations Made in Type B Soil provide diagrams specific to excavations made in type A and B soils, respectively.

OSHA Regulations for Crane Use

CERM: Sec. 83.18

CERM Sec. 83.18 explains crane use and safety and lists the OSHA regulations for crane safety.

Some states have their own safety standards. A qualified person may conduct annual inspections of equipment, and a competent person may conduct work shift and monthly equipment inspections.

Crane operation must be performed only by qualified and trained personnel. The ground underneath the crane must be firm, stable, and within 1% of level, especially when using truck cranes.

Crane Loading

CERM: Sec. 83.19

Crane loading is typically governed by stability issues (i.e., overturning of the crane during service). Crane stability can be affected by many issues, such as the weight distribution of a load, the angle of the crane boom, and soil conditions under the crane. Engineers need to take all of these factors into consideration when dealing with cranes.

Every crane has its own load chart that specifies the crane's features, its dimensions, and how its lifting capacity varies with configuration. These charts should be read conservatively; for example, if a boom angle is used that is not listed on the load chart, the next lower angle noted on the chart should be used to determine the capacity of the crane. Likewise, when using a particular radius that is not listed on the load chart, the next larger radius measurement should be used for determining the crane capacity.

A *lift capacity chart*, or *lift table*, consists of a table of maximum loads. A *lift range chart*, also called a *range diagram* or *range table*, illustrates how much boom length is needed to pick up and lift a load at given distances and vertical lift heights. A *lift angle chart* illustrates the maximum lift with luffing and fixed jibs. A *crane in motion chart* defines the lift capacity for a pick and carry operation.

Electrical Safety

CERM: Sec. 83.10, Sec. 83.11

Electrical safety hazards include shock, arc flash, explosion, and fire. CERM Table 83.4 includes the effects of current on humans.

Clearance requirements can be calculated for voltages up to 1000 kV using CERM Eq. 83.2. Values should be rounded up to the nearest 5 ft.

$$\text{line clearance} = 3 \text{ m} + (10.2 \text{ mm}) \cdot (V_{\text{kV}} - 50 \text{ kV}) \quad [\text{SI}] \quad 83.2(a)$$

$$\text{line clearance} = 10 \text{ ft} + (0.4 \text{ in}) \cdot (V_{\text{kV}} - 50 \text{ kV}) \quad [\text{U.S.}] \quad 83.2(b)$$

NIOSH Lifting Equation

CERM: Sec. 83.13

$$LI = \frac{L}{RWL} \quad \text{83.3}$$

$$RWL = (LC)(HM)(VM)(DM)(AM)(FM)(CM) \quad \text{83.4}$$

The NIOSH lifting equation predicts the relative risk of a task that involves lifting objects, and consists of the calculation of two values. The lifting index, LI, is the ratio of the actual load to the recommended weight limit, RWL, which can be calculated from the load constant, LC, and a set of multiplying factors. The load constant is the recommended weight for lifting under optimal conditions, and is taken to be 51 lbm (23 kg). The multipliers vary depending on the individual task and the person performing it.

A lifting index greater than 3.0 exposes most workers to a high risk of developing low-back pain and injury. The goal is to design lifting tasks so that LI is less than or equal to 1.0.

Scaffolds and Temporary Structures

CERM: Sec. 83.12, Sec. 83.15, Sec. 83.16

A *scaffold* is any temporary elevated platform used for supporting employees, materials, or both. Construction and use of scaffolds are regulated in detail by OSHA 1926.451. This includes fall protection.

Temporary structures are those that exist for only a period of time during construction and are removed prior to project completion.

Engineering Economics

Content in blue refers to the *NCEES Handbook*.

Engineering Economics..9-1

An engineer requires a basic understanding of economics to determine the best value of various engineering solutions. While the NCEES exam specifications do not specifically include engineering economics analysis for every civil discipline, some problems may require knowledge of engineering economics.

ENGINEERING ECONOMICS

Key concepts: These key concepts are important for answering exam questions on engineering economics and cost estimating.

- cash flow calculations
- cash flow diagrams
- comparison of economic alternatives
- conventions of basic engineering economics calculations
- nomenclature used in engineering economics
- various methods of depreciation

NCEES Handbook: To prepare for this subtopic, familiarize yourself with these sections and tables in the *Handbook*.

- Engineering Economics
- Nonannual Compounding
- Breakeven Analysis
- Inflation
- Depreciation
- Book Value
- Capitalized Costs
- Rate-of-Return
- Benefit-Cost Analysis
- Interest Rate Tables

PE Civil Reference Manual **(CERM):** Study these sections in CERM that either relate directly to this subtopic or provide background information.

- Chapter 87: Engineering Economic Analysis

The following equations and tables are relevant for knowledge area Engineering Economics.

Economic Factor Conversions

Handbook: Engineering Economics

CERM: Table 87.1

The complete list of factor names and corresponding formulas are available in *Handbook* table Engineering Economics, *Handbook* section Engineering Economics: Nomenclature and Definitions, and CERM Chap. 87.

A method of remembering the notation is to interpret the factors algebraically. The (F/P) factor could be thought of as the fraction F/P. Algebraically, the (F/P) factor would be as shown in CERM Eq. 87.6.

$$F = P\left(\frac{F}{P}\right) \qquad 87.6$$

Single Payment Compound Amount

Handbook: Engineering Economics

CERM: Table 87.1

$$(F/P, i\%, n) = (1+i)^n$$

For the appropriate interest rate, i, and number of years, n, multiply the present value, P, of a sum of money by this factor to obtain its future value, F.

CERM Eq. 87.2 is similar but includes the variables F and P.

Values of this factor for common values of i and n can be found using *Handbook* table Interest Rate Tables or CERM App. 87.B.

Single Payment Present Worth

Handbook: Engineering Economics

CERM: Table 87.1

$$(P/F, i\%, n) = (1+i)^{-n}$$

For the appropriate interest rate, i, and number of years, n, multiply the future value, F, of a sum of money by this factor to obtain its present value, P.

Values of this factor for common values of i and n can be found using *Handbook* table Interest Rate Tables or CERM App. 87.B.

Uniform Series Sinking Fund

Handbook: Engineering Economics

CERM: Table 87.1

$$(A/F, i\%, n) = \frac{i}{(1+i)^n - 1}$$

Multiply the future value, F, of a sum of money by this factor to obtain its annual value, A.

The A/F factor is called the sinking fund factor.

Values of this factor for common values of i and n can be found using *Handbook* table Interest Rate Tables or CERM App. 87.B.

Capital Recovery

Handbook: Engineering Economics

CERM: Table 87.1

$$(A/P, i\%, n) = \frac{i(1+i)^n}{(1+i)^n - 1}$$

Multiply the present value, P, of a sum of money by this factor to obtain its annual value, A.

Values of this factor for common values of i and n can be found using *Handbook* table Interest Rate Tables or CERM App. 87.B.

Uniform Series Compound Amount

Handbook: Engineering Economics

CERM: Table 87.1

$$(F/A, i\%, n) = \frac{(1+i)^n - 1}{i}$$

Multiply the annual value, A, of a sum of money by this factor to obtain its future value, F.

Values of this factor for common values of i and n can be found using *Handbook* table Interest Rate Tables or CERM App. 87.B.

Uniform Series Present Worth

Handbook: Engineering Economics

CERM: Table 87.1

$$(P/A, i\%, n) = \frac{(1+i)^n - 1}{i(1+i)^n}$$

Multiply the annual value, A, of a sum of money by this factor to obtain its present value, P.

Values of this factor for common values of i and n can be found using *Handbook* table Interest Rate Tables or CERM App. 87.B.

Uniform Gradient Present Worth

Handbook: Engineering Economics

CERM: Table 87.1

$$(P/G, i\%, n) = \frac{(1+i)^n - 1}{i^2(1+i)^n} - \frac{n}{i(1+i)^n}$$

Multiply a gradient cash flow, G, by this factor to obtain its present value, P.

Values of this factor for common values of i and n can be found using *Handbook* table Interest Rate Tables or CERM App. 87.B.

Uniform Gradient Future Worth

Handbook: Engineering Economics

CERM: Table 87.1

$$(F/G, i\%, n) = \frac{(1+i)^n - 1}{i^2} - \frac{n}{i}$$

Multiply a gradient cash flow, G, by this factor to obtain its future value, F.

Values of this factor for common values of i and n can be found using *Handbook* table Interest Rate Tables or CERM App. 87.B.

Uniform Gradient Uniform Series

Handbook: Engineering Economics

CERM: Table 87.1

$$(A/G, i\%, n) = \frac{1}{i} - \frac{n}{(1+i)^n - 1}$$

Multiply a gradient cash flow, G, by this factor to obtain its annual value, A.

Values of this factor for common values of i and n can be found using *Handbook* table Interest Rate Tables or CERM App. 87.B.

Non-Annual Compounding

Handbook: Nonannual Compounding

$$i_e = \left(1 + \frac{r}{m}\right)^m - 1$$

r is the interest rate. m is the number of time intervals in a given year for which the compounding occurs.

Inflation

Handbook: Inflation

$$d = i + f + (i \cdot f)$$

CERM: Sec. 87.58

$$i' = i + e + ie \qquad 87.72$$

In the *Handbook* equation, the term d is the interest rate period adjusted (deflated) for inflation rate f per interest rate period of i. It is used to calculate the present worth, P, of the money.

In CERM Eq. 87.72, the effective annual interest rate, i, is replaced with a value corrected for inflation, i'. The term e is the decimal inflation rate.

Assuming that inflation is constant, cash flows can be adjusted to $t = 0$ by dividing cash flows by the following term.

$$(1 + e)^n$$

e is the decimal inflation rate, and n is the year (or years) of the cash flow(s).

Straight-Line Depreciation

Handbook: Depreciation

CERM: 87.36

$$D_j = \frac{C - S_n}{n}$$

This is depreciation in year j, or D_j, which depends upon the cost, C, and salvage value in year n, S_n.

The *Handbook* equation is equivalent to CERM Eq. 87.25.

Book Value

Handbook: Book Value

CERM: Sec. 87.58

$$BV = \text{initial cost} - \sum D_j$$

BV is the book value. The book value is determined by subtracting the depreciation at year j from the initial cost, which is represented by the term C in CERM Eq. 87.40.

Benefit-Cost Analysis

Handbook: Benefit-Cost Analysis

$$B - C \geq 0, \text{ or } B/C \geq 1$$

CERM: Sec. 87.58

$$\frac{B_2 - B_1}{C_2 - C_1} \geq 1 \text{ [alternative 2 superior]} \qquad 87.22$$

The *Handbook* equation compares benefits, B, to costs, C, of a single alternative.

The CERM equation compares two alternatives. If the value is greater than 1, then alternative 2 is superior; if not, alternative 1 is superior.

Modified Accelerated Cost Recovery System (MACRS)

CERM: Sec. 87.58

$$D_j = C \cdot \text{factor} \qquad 87.33$$

Under MACRS, the cost recovery amount in the jth year of an asset's cost recovery period is calculated by multiplying the initial cost by a factor. The factor used depends on the asset's cost recovery period. (See CERM Table 87.4.)

Topic II: Construction

Chapter

10. Earthwork Construction and Layout
11. Estimating Quantities and Costs
12. Construction Operations and Methods
13. Scheduling
14. Material Quality Control and Production
15. Temporary Structures
16. Health and Safety

10 Earthwork Construction and Layout

Content in blue refers to the *NCEES Handbook*.

A. Excavation and Embankment 10-1
B. Borrow Pit Volumes 10-6
C. Site Layout and Control 10-8
D. Earthwork Mass Diagrams and Haul
   Distance.. 10-9
E. Site and Subsurface Investigations 10-11

The knowledge area of Earthwork Construction and Layout makes up between five and eight questions out of the 80 questions on the PE Civil Construction exam. The organization of this chapter follows the order of subtopics given by NCEES for this knowledge area. Each subtopic is covered in the following sections.

A. EXCAVATION AND EMBANKMENT

Key concepts: These key concepts are important for answering exam questions in subtopic 9.A, Excavation and Embankment.

- bank versus compacted versus loose cubic yards
- cross-sectional volume calculations
- cut and fill
- fundamental soil relationships
- saturation
- soil indexing formulas
- soil phases and structure
- swell and load factors
- swelling and shrinkage

NCEES Handbook: To prepare for this subtopic, familiarize yourself with these sections in the *Handbook*.

- Excavation and Embankment
- Method of Compaction Categorized by Soil Type
- Cross-Section Methods
- Cross-Sectional End Areas
- Borrow Pit Grid Method
- Earthwork Area Formulas
- Mass Diagrams and Profile Diagrams
- Method of Compaction Categorized by Soil Type

PE Civil Reference Manual (**CERM**): Study these sections in CERM that either relate directly to this subtopic or provide background information.

- Section 7.2: Areas with Irregular Boundaries
- Section 35.5: Mass-Volume Relationships
- Section 35.11: Proctor Test
- Section 35.12: Modified Proctor Test
- Section 78.30: Traverse Area: Method of Coordinates
- Section 78.32: Areas Bounded by Irregular Boundaries
- Section 80.3: Swell and Shrinkage
- Section 80.5: Cut and Fill
- Section 80.8: Original and Final Cross Sections
- Section 80.14: Average End Area Method
- Section 80.15: Prismoidal Formula Method
- Section 80.16: Borrow Pit Geometry

Construction Depth Reference Manual (**CECN**): Study these sections in CECN that either relate directly to this subtopic or provide background information.

- Section 1.1: Earthmoving and Soil Fundamentals
- Section 1.2: Cross-Sectional Volume Calculations

The following equations, figures, and tables are relevant for subtopic 9.A, Excavation and Embankment.

Material Volume Change Characteristics

Handbook: Excavation and Embankment; Mass Diagrams and Profile Diagrams

$$V_L = \left(1 + \frac{S_w}{100\%}\right) V_B$$

$$V_C = \left(1 - \frac{S_h}{100\%}\right) V_B$$

$$V_B = \left(\frac{\gamma_F}{\gamma_B}\right)V_F + \frac{W_L}{\gamma_B}$$

CERM: Sec. 80.3

$$V_l = \left(\frac{100\% + \%\,\text{swell}}{100\%}\right)V_b = \frac{V_b}{\text{LF}} \quad \text{80.1}$$

$$V_c = \left(\frac{100\% - \%\,\text{shrinkage}}{100\%}\right)V_b \quad \text{80.2}$$

The *Handbook* relates the volume of loose soil, V_L, and the volume of compacted soil, V_C, to the volume of the undisturbed, or bank, soil, V_B. The *Handbook* equation uses the swell factor, S_w, and shrinkage factor, S_h, when relating the loose soil or compacted soil volume with the undisturbed soil volume. Unit weights, γ, are also given in terms of loose, compacted, and bank soil. The equations provided in CERM also use the terms *loose*, *compacted*, and *bank soil* when relating volume and unit weights. However, they also include a load factor, LF, when determining the volume of loose soil.

CECN Eq. 1.11, Eq. 1.12, and Eq. 1.14 are used to determine the change in unit weight of a material and contain variables for bank cubic yards (BCY), loose cubic yards (LCY), and compacted cubic yards (CCY). The total volume increase or decrease is then calculated using CECN Eq. 1.16.

$$\gamma_{d,b} = \frac{W_{s,b}}{V_{\text{BCY}}} \quad \text{1.11}$$

$$\gamma_{d,l} = \frac{W_s}{V_{\text{LCY}}} \quad \text{1.12}$$

$$\gamma_{d,c} = \frac{W_{s,c}}{V_{\text{CCY}}} \quad \text{1.14}$$

$$V_{\text{BCY}}\gamma_{d,b} = V_{\text{LCY}}\gamma_{d,l} = V_{\text{CCY}}\gamma_{d,c} \quad \text{1.16}$$

Equations for the unit weight of loose soil and compacted soil are also given in the *Handbook*.

$$\gamma_L = \frac{\gamma_B}{1 + \dfrac{S_w}{100\%}}$$

$$\gamma_C = \frac{\gamma_B}{1 - \dfrac{S_h}{100\%}}$$

The *Handbook* equations for unit weight correspond to CECN Eq. 1.6, with the major difference being that the *Handbook* equation distinguishes between loose and compacted soil, and accounts for shrinkage and swell factors.

Factors from CERM Table 80.2 can be used to estimate the loose volume and compacted volume, which are useful in determining the number of vehicles needed to transport the soil and the volume of soil that must be transported from the borrow site to the construction fill site.

Handbook section Mass Diagrams and Profile Diagrams provides sample mass and profile diagrams and explains some common terms and concepts used in earthwork.

Relative Soil Compaction

Handbook: Excavation and Embankment; Method of Compaction Categorized by Soil Type

$$\text{RC} = \frac{\gamma_{d,\,\text{field}}}{\gamma_{d,\,\text{max}}} \cdot 100\%$$

CERM: Sec. 35.11

$$\text{RC} = \frac{\rho_d}{\rho_d^*} \cdot 100\% \quad \text{35.19}$$

The specification given to the grading contractor sets forth the minimum acceptable density, as well as a range of acceptable water content values. The minimum density is specified as the relative compaction, which is the percentage of the maximum value determined in the laboratory.

The *Handbook* equation and CERM Eq. 35.19 are the same except that the *Handbook* equation is based on unit weight and the CERM equation is based on density. γ and ρ both represent the density.

Handbook table Method of Compaction Categorized by Soil Type includes classifications and definitions of soil compaction techniques.

Shrinkage Factor

Handbook: Excavation and Embankment

$$\text{shrinkage factor} = \frac{\text{compacted unit weight}}{\text{bank unit weight}}$$

CECN: Sec. 1.1

$$\text{shrinkage factor} = \frac{\gamma_{d,b}}{\gamma_{d,c}} \quad \text{1.20}$$

EARTHWORK CONSTRUCTION AND LAYOUT

The volume of a loose pile of excavated earth will be greater than the original, in-place natural volume. If the earth is compacted after it is placed, the volume may be less than its original volume.

Swell and Swell Factor

Handbook: Excavation and Embankment

$$S_w = \left(\frac{\frac{W}{V_B}}{\frac{W}{V_L}} - 1 \right) \cdot 100\%$$

$$\text{swell factor} = \frac{\text{loose unit weight}}{\text{bank unit weight}}$$

CECN: Sec. 1.1

$$\text{swell} = \left(\frac{\gamma_{d,b}}{\gamma_{d,l}} - 1 \right) \cdot 100\% \qquad 1.17$$

$$\text{LF} = \frac{\gamma_{d,l}}{\gamma_{d,b}} \qquad 1.18$$

The amount that a given material increases in volume when excavated is expressed as a *swell percentage*. When the dry unit weight of a material in both the bank and loose states is known, use the first *Handbook* equation, which is equivalent to CECN Eq. 1.17, to find the swell percentage of the material.

The *swell factor* is the ratio of the unit weight of loose material to the unit weight of bank material. The *Handbook* equation for the swell factor is equivalent to CECN Eq. 1.18, which uses the term *load factor*, LF.

CECN Eq. 1.16 can be used to calculate the total volume increase (swell) or total volume decrease (shrinkage) in a material.

$$V_{\text{BCY}}\gamma_{d,b} = V_{\text{LCY}}\gamma_{d,l} = V_{\text{CCY}}\gamma_{d,c} \qquad 1.16$$

Material Volume Change During Earthmoving

Handbook: Excavation and Embankment

CECN: Sec. 1.1

CECN Fig. 1.2 and the equation included therein show the relationship of bank cubic yards (BCY) to loose cubic yards (LCY) and compacted cubic yards (CCY).

Figure 1.2 Typical Material Volume Change During Earthmoving

1.0 BCY in natural condition (bank cubic yards) = 1.25 LCY after digging (loose cubic yards) = 0.90 CCY after compaction (compacted cubic yards)

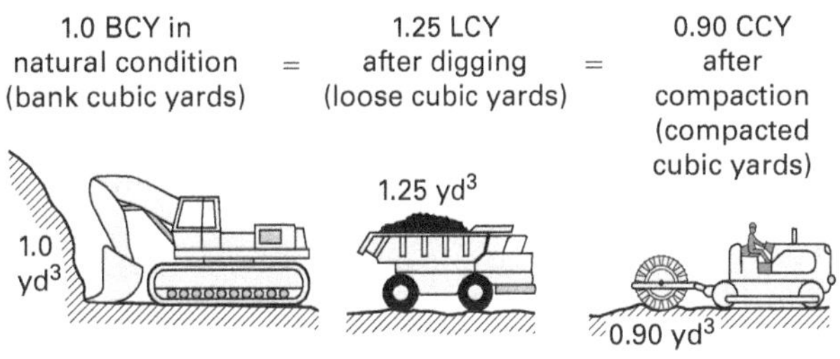

CERM Table 80.1 shows the relationships between soil factors and bank, loose, or cubic yard volumes.

Exam tip: Soil is described as *bank soil* when in an undisturbed state, *loose soil* after it has been disturbed, and *compacted soil* is after it has been placed and compacted.

Average End Area Method

Handbook: Cross-Section Methods; Cross-Sectional End Areas

CERM: Sec. 80.8

CECN: Sec. 1.2

$$V = L \left(\frac{A_1 + A_2}{2} \right)$$

The average end area method calculates the volume between two consecutive cross sections as the average of their areas multiplied by the distance between the two areas. This method assumes the fill is positive and the cut is negative. *Handbook* figure Cross-Sectional End Areas include examples of fill and cut diagrams.

This method disregards the slopes and orientations of the ends and sides, but it is sufficiently accurate for most earthwork calculations. When the end area is complex, it may be necessary to use a planimeter or to plot the area on fine grid paper and simply count the squares. The average end area method usually overestimates the actual soil volume.

Exam tip: The precision obtained from the average end area method is generally sufficient unless one of the end areas is very small or zero. In that case, the volume should be computed as a pyramid, or an alternative method should be used.

Coordinate Method

Handbook: Earthwork Area Formulas

$$A = \tfrac{1}{2}\big(X_A(Y_B - Y_N) + X_B(Y_C - Y_A) \\ + X_C(Y_D - Y_B) + \cdots + X_N(Y_A - Y_{N-1})\big)$$

CERM: Sec. 78.30

$$A = \tfrac{1}{2}\left|\begin{array}{c}\sum \text{of full line products} \\ -\sum \text{of broken line products}\end{array}\right| \quad 78.37$$

The area calculation is simplified in the CERM equation, but the *Handbook* and CERM equations are fundamentally the same. The area of a simple traverse can be found by dividing the traverse into a number of geometric shapes and summing their areas. If the coordinates of the traverse leg endpoints are known, the coordinate method can be used. The coordinates can be x-y coordinates referenced to some arbitrary set of axes, or they can be sets of departure and latitude. See CERM Sec. 78.30 for more details on this method.

Prismoidal Formula

Handbook: Cross-Section Methods

CERM: Sec. 80.15

$$V = L\left(\frac{A_1 + 4A_m + A_2}{6}\right)$$

This is the preferred formula when two end areas differ greatly or when the ground surface is irregular. It generally produces a smaller volume than the average end area method and favors the owner-developer in earthwork cost estimating. See *Handbook* figure Cross-Sectional End Areas for examples of fill and cut diagrams.

> *Exam tip*: The prismoidal method should be used when there is a significant difference between the two ends, as it is more accurate than the average end area method.

Borrow Pit Grid Method

Handbook: Borrow Pit Grid Method

CERM: Sec. 80.16

It is often necessary to borrow earth from an adjacent area to construct embankments. Normally, the *borrow pit* area is laid out in a rectangular grid divided into squares. Elevations are determined at the corners of each square by leveling before and after excavation so that the cut at each corner can be calculated. CERM Fig. 80.7 and the figure shown in *Handbook* section Borrow Pit Grid Method show the sample layout.

Figure 80.7 Depth of Excavation (Cut) in a Borrow Pit Area

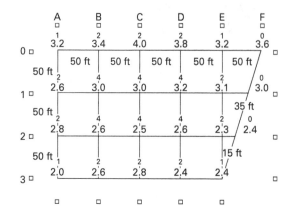

Handbook section Borrow Pit Grid Method includes an equation to calculate the total volume.

$$\begin{array}{l}\text{volume of material} \\ \text{in one grid square}\end{array} = \tfrac{1}{4}(a+b+c+d) \\ \cdot (\text{area of grid square})$$

For partial grid squares at edges of an excavation, the amount of material may be estimated using standard volume formulas for three-dimensional shapes, as shown in these two equations from *Handbook* section Borrow Pit Grid Method.

$$V = \tfrac{1}{2}bhl$$

$$V = \tfrac{1}{12}\pi r^2 h$$

> *Exam tip*: The PE Civil exam is not likely to include a problem involving a full squares wedge.

Trapezoidal Method

Handbook: Earthwork Area Formulas

$$A = w\big(\tfrac{1}{2}(h_1 + h_n) + h_2 + h_3 + h_4 + \cdots + h_{n-1}\big)$$

CERM: Sec. 78.32

$$A = d\left(\frac{h_1 + h_n}{2} + \sum_{i=2}^{n-1} h_i\right) \quad 78.40$$

If the irregular side of each cell is fairly straight, the trapezoidal rule can be used. To denote the length of the common interval, the *Handbook* uses w and CERM uses d.

Simpson's Rule

Handbook: Earthwork Area Formulas

$$A = \tfrac{1}{3}\begin{pmatrix} \text{first value} + \text{last value} \\ +4(\text{sum of odd-numbered values}) \\ +2(\text{sum of even-numbered values}) \end{pmatrix}$$
$$\cdot (\text{length of interval})$$

CERM: Sec. 7.2, Sec. 78.32

$$A = \frac{d}{3}\left(h_1 + h_n + 2\sum_{\text{odd}} h_i + 4\sum_{\text{even}} h_i\right) \quad 78.41$$
$$= \frac{d}{3}(h_1 + 4h_2 + 2h_3 + 4h_4 + \ldots + h_n)$$

When the irregular side of each cell is curved or parabolic, *Simpson's rule* (sometimes referred to as *Simpson's 1/3 rule*) can be used to determine the area. CERM uses classic summation nomenclature, while the *Handbook* equation uses words to explain the terms of the equation.

Soil Phases

CECN: Sec. 1.1

The volume between the solid particles in a sample represents the voids. This void space can be filled with air, water, or gas. CECN Fig. 1.1 illustrates soil phases.

Figure 1.1 Soil Phases

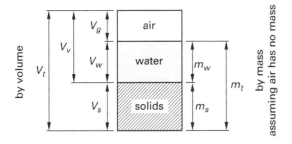

V_g	volume of gas in soil	ft³/m³
V_s	volume of solids in soil	ft³/m³
V_t	total volume of soil	ft³/m³
V_v	volume of voids in soil	ft³/m³
V_w	volume of water in soil	ft³/m³

The equation to calculate the total volume of the soil sample is

$$V_t = V_v + V_s = V_g + V_w + V_s$$

V_v is the volume of voids in the soil sample, which is calculated as

$$V_v = V_g + V_w$$

CECN Sec. 1.1 includes more information on the differences of the volumes and masses shown in the figure and the relationships between them.

Soil Indexing Formulas

CERM: Sec. 35.5

CERM Table 35.7 shows the equations for various soil indexing formulas. In the equations from the table, the *porosity*, n, is the ratio of the volume of voids to the total volume. The *void ratio*, e, is the ratio of the volume of voids to the volume of solids. The *moisture content* (or *water content*), w, is the ratio of the mass of water to the mass of solids.

The *degree of saturation*, S, is the percentage of the volume of water to the total volume of voids. This indicates how much of the void space is filled with water. If all the voids are filled with water, then the volume of air is zero, and the sample's degree of saturation is 100%.

The *density*, ρ, is the ratio of the total mass to the total volume. The total density may be referred to as the *moist density* (or *wet density*) above the water table and as the *saturated density* below the water table. The *dry density*, ρ_d, is the ratio of the solid mass to the total volume. The *buoyant density* (or *submerged density*), ρ_b, is the difference between the total density and the density of water.

The density of the solid constituents, ρ_s, is the ratio of the mass of the solids to the volume of the solids. This would also be the density of the soil if there were no voids. The *percent pore space* (PPS) is the ratio of the volume of the voids to the total volume, as a percentage.

B. BORROW PIT VOLUMES

Key concepts: These key concepts are important for answering exam questions in subtopic 9.B, Borrow Pit Volumes.

- average end area method
- borrow pit grid method
- grid method
- pit excavations
- point and block diagram of excavation sites
- prismoidal method
- roadway cross-sectional volumes
- Simpson's rule
- spoil banks
- trapezoidal rule
- trenches

NCEES Handbook: To prepare for this subtopic, familiarize yourself with these sections in the *Handbook*.

- Borrow Pit Grid Method

PE Civil Reference Manual **(CERM):** Study these sections in CERM that either relate directly to this subtopic or provide background information.

- Section 78.32: Areas Bounded by Irregular Boundaries
- Section 80.16: Borrow Pit Geometry
- Section 83.6: Trenching and Excavation

Construction Depth Reference Manual **(CECN):** Study these sections in CECN that either relate directly to this subtopic or provide background information.

- Section 1.2: Cross-Sectional Volume Calculations
- Section 1.3: Borrow Pit Volumes and Soil Banks

The following equations, figures, and concepts are relevant for subtopic 9.B, Borrow Pit Volumes.

Borrow Pit Grid Method

Handbook: Borrow Pit Grid Method

CERM: Sec. 80.16

It is often necessary to borrow earth from an adjacent area to construct embankments. Normally, the *borrow pit* area is laid out in a rectangular grid divided into squares. Elevations are determined at the corners of each square by leveling before and after excavation so that the cut at each corner can be calculated. CERM Fig. 80.7 and the figure shown in *Handbook* section Borrow Pit Grid Method show the sample layout.

Figure 80.7 Depth of Excavation (Cut) in a Borrow Pit Area

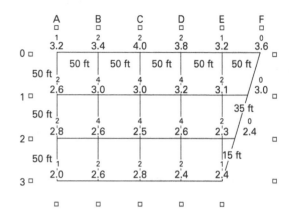

Handbook section **Borrow Pit Grid Method** includes an equation to calculate the total volume.

$$\begin{matrix}\text{volume of material}\\ \text{in one grid square}\end{matrix} = \tfrac{1}{4}(a+b+c+d) \cdot (\text{area of grid square})$$

For partial grid squares at the edges of an excavation, the amount of material may be estimated using standard volume formulas for three-dimensional shapes, as shown in these two equations from *Handbook* section **Borrow Pit Grid Method**.

$$V = \tfrac{1}{2}bhl$$

$$V = \tfrac{1}{12}\pi r^2 h$$

Average Depth of Pit Excavation

CERM: Sec. 80.16

CECN: Sec. 1.3

$$D_{\text{ave}} = \frac{1\sum\binom{\text{depth of points}}{\text{part of 1 block}} + 2\sum\binom{\text{depth of points}}{\text{part of 2 blocks}} + 3\sum\binom{\text{depth of points}}{\text{part of 3 blocks}} + 4\sum\binom{\text{depth of points}}{\text{part of 4 blocks}}}{4(\text{number of blocks})}$$

1.27

D_{ave} is the average depth of the overall excavation.

Exam tip: Multiply the average depth by the total area of the site to determine the volume of material to be excavated.

Trench Excavations

CECN: Sec. 1.3

$$V = LDw \qquad 1.28$$

$$V_l = \frac{LDw}{2} \qquad 1.29$$

Typically, trenches on a construction site are either rectangles or triangles. CECN Eq. 1.28 calculates the volume of a square trench, and CECN Eq. 1.29 calculates the volume of a triangular trench. CECN Fig. 1.7 and Fig. 1.8 respectively illustrate a rectangular and a triangular trench.

Figure 1.7 Rectangular Trenches

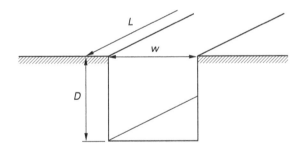

Figure 1.8 Triangular Trenches

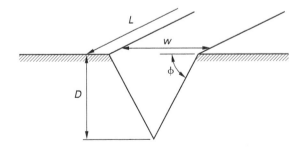

Triangular Spoil Banks

CECN: Sec. 1.3

$$V_{\text{LCY}} = \frac{Lhw}{2} \qquad 1.30$$

$$w = \sqrt{\frac{4V_l}{L \tan \phi}} \qquad 1.31$$

$$h = \frac{w \tan \phi}{2} \qquad 1.32$$

The volume of a spoil bank is measured in loose cubic yards (LCY). Calculation of the volume requires knowing the *soil angle of repose*, the angle that the sides naturally form with the horizon. Triangular spoil banks are common adjacent to trenches. CECN Fig. 1.9 shows a triangular spoil bank.

Figure 1.9 Triangular Spoil Banks

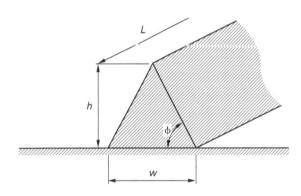

Conical Spoil Banks

CECN: Sec. 1.3

$$V_{\text{LCY}} = \tfrac{1}{3} A_{\text{base}} h \qquad 1.33$$

$$D = \sqrt[3]{\frac{24 V_l}{\pi \tan \phi}} \qquad 1.34$$

$$h = \frac{D \tan \phi}{2} \qquad 1.35$$

The volume of a spoil bank is measured in loose cubic yards (LCY). Calculation of the volume requires knowing the *soil angle of repose*, the angle that the sides naturally form with the horizon. Conical spoil banks are typically produced in aggregate production operations as crushed aggregate falls from the end of a conveyor belt.

C. SITE LAYOUT AND CONTROL

Key concepts: These key concepts are important for answering exam questions in subtopic 9.C, Site Layout and Control.

- grade rod
- ground rod
- ground stake
- highway sections
- slope stake markings
- survey staking

NCEES Handbook: To prepare for this subtopic, familiarize yourself with these sections in the *Handbook*.

- Site Layout and Control

Construction Depth Reference Manual **(CECN):** Study these sections in CECN that either relate directly to this subtopic or provide background information.

- Section 1.5: Staking
- Section 1.6: Stake Markings
- Section 1.7: Establishing Slope Stake Markings

The following equations, figures, and concepts are relevant for subtopic 9.C, Site Layout and Control.

Construction Stakes for Storm Drains

CECN: Sec. 1.6

Stakes are read from top to bottom. The front of the stake is marked with header information (e.g., RPSS, offset distance) and cluster information (e.g., horizontal and vertical measurements, slope ratio). All cluster information is measured in the same direction from the same point.

Hub stakes are located, identified, and protected by witness stakes or guard stakes. A *witness stake* calls attention to a hub stake but does not itself locate a specific point. A *guard stake* may be driven at an angle with its top over the flush-driven hub stake.

The back of a witness stake is used to record the station and other information, including literal descriptions (e.g., "at ramp"). Actual elevations, when included, are marked on the thin edge of the stake. CECN Fig. 1.15 illustrates how a construction stake would be marked to identify a trench for a storm drain.

Figure 1.15 Construction Stake for Storm Drain

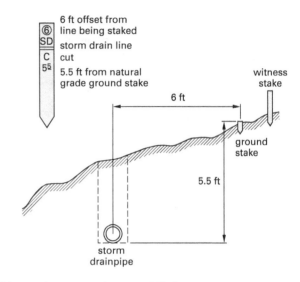

Slope Stakes Along a Highway

CECN: Sec. 1.6

Slope stakes indicate grade points—points where the cuts and fills begin and the planned side slopes intersect the natural ground surface. Typically, they are placed with a 10 ft offset. CECN Fig. 1.16 illustrates the use of slope stakes along three adjacent sections of a proposed highway.

In addition, the fronts of slope stakes are marked to indicate the nature of the earthwork (i.e., C for cut and F for fill), the offset distance, the type of line being staked, the distance from the centerline or control line, the slope to finished grade, and the elevation difference between the grade point and the finished grade.

Common Stake Marking Abbreviations

CECN: Sec. 1.6

Stake markings vary greatly from agency to agency. See CECN Table 1.5 for a list of common stake marking abbreviations.

Exam tip: Reviewing a construction dictionary may help in becoming familiar with technical terms.

Determining Stake Location

CECN: Sec. 1.7

The actual steps taken to calculate the stake marking depend on whether the earthwork is a cut or a fill and whether the instrument is above or below the finished grade. Sketching a diagram will help clarify the algebraic steps and prevent sign errors. An example is shown in CECN Fig. 1.17.

Figure 1.17 Determining Stake Location

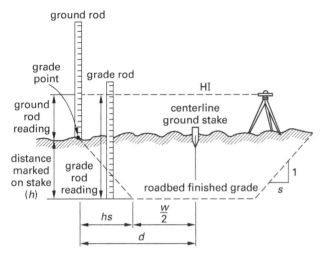

The distance marked on the construction stake, h, or the cut or fill stake is the distance between natural and finished grade elevations, and is easily calculated from the ground and grade rods. The cut or fill stake marking is the difference between grade and ground rod elevations.

Establishing Slope Stake Markings

CECN: Sec. 1.7

$$\text{HI} = \text{elev}_{\text{ground}} + \begin{array}{l}\text{instrument height}\\ \text{above the ground}\end{array} \quad 1.36$$

$$= \text{elev}_{\text{ground}} + \text{ground rod}$$

$$\text{grade rod} = \text{HI} - \text{elev}_{\text{grade}} \quad 1.37$$

$$h = \text{grade rod} - \text{ground rod} \quad 1.38$$

The markings on a construction stake are determined from a survey of the natural ground surface. This survey requires two individuals: the leveler and the rod person. The leveler, or instrument person, works with the instrument, and the rod person holds the leveling rod.

The elevation of the instrument (or the elevation of the ground at the instrument location) and the height of the instrument above the ground must be known if actual elevations are to be marked on a stake. The height of the instrument is the elevation of the instrument above the reference datum.

Survey Leveling

Handbook: Site Layout and Control

$$\text{elevation of BM} + \text{BS} = \text{HI}$$
$$\text{elevation of TP} = \text{HI} - \text{FS}$$

Benchmark is the common name given to permanent monuments of known vertical positions. The elevations of temporary benchmarks are generally found in field notes and local official filings. With direct leveling, a level is set up at a point approximately midway between the two points whose difference in elevation is desired. The vertical backsight (plus sight) and foresight (minus sight) are read directly from the rod. HI is the height of the instrument above the ground. CERM Sec. 78.12 includes more information on benchmarks, and CERM Sec. 78.19 includes more information on the measurement of elevation.

D. EARTHWORK MASS DIAGRAMS AND HAUL DISTANCE

Key concepts: These key concepts are important for answering exam questions in subtopic 9.D, Earthwork Mass Diagrams and Haul Distance.

- average haul distance
- balance lines
- balance point
- characteristics of mass diagrams
- constructing mass diagrams
- equipment capacity
- freehaul
- overhaul
- using mass diagrams

NCEES Handbook: To prepare for this subtopic, familiarize yourself with these sections in the *Handbook*.

- Mass Diagrams and Profile Diagrams
- Freehaul and Overhaul
- Production Rate for Loading and Hauling Earthwork

PE Civil Reference Manual (CERM): Study these sections in CERM that either relate directly to this subtopic or provide background information.

- Section 80.17: Mass Diagrams
- Section 80.18: Capacities of Earth-Handling Equipment

Construction Depth Reference Manual (CECN): Study these sections in CECN that either relate directly to this subtopic or provide background information.

- Section 1.4: Mass Diagrams

The following figures and concepts are relevant for subtopic 9.D, Earthwork Mass Diagrams and Haul Distance.

Profile Mass Diagrams

Handbook: Mass Diagrams and Profile Diagrams

CECN: Sec. 1.4

The figure in *Handbook* section Mass Diagrams and Profile Diagrams shows how to interpret a mass diagram, showing original versus finish grade, and how to correlate that to balancing of site (cut vs. fill). CECN Fig. 1.11 shows a similar figure.

Figure 1.11 Baseline and Centerline Profile Mass Diagram

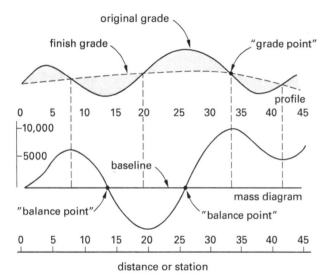

Excess material will be present at the end of the cut/fill operation. The area above the finish grade line indicates a cut, and the area below the finish grade line indicates a fill. The local minima and maxima identify the grade points.

Balance Line Between Two Points

CECN: Sec. 1.4

CECN Fig. 1.12 shows how to determine if you are cutting or filling based on finished grade profile. A balance line has been drawn that intersects the mass diagram at two points. These points represent the inclusive stations for which the cut and fill volumes are equal.

Figure 1.12 Balance Line Between Two Points

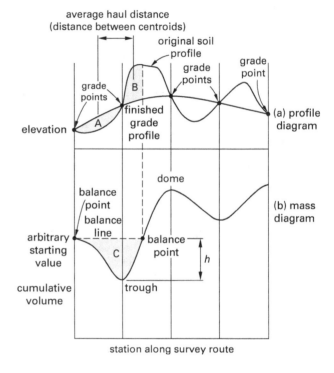

Freehaul and Overhaul

Handbook: Freehaul and Overhaul

CECN: Sec. 1.4

The *freehaul distance* is the maximum distance a contractor is expected to transport earth without asking for additional payment. The typical distance is between 500 ft and 1000 ft. Any soil transported farther than the freehaul distance is considered *overhaul*. Freehaul is expected, but overhaul results in additional costs.

The haul distance should be known because this is the total distance that the contractor is expected to haul material over the site. The figure in *Handbook* section Freehaul and Overhaul illustrates the freehaul distance versus overhaul volume. CECN Fig. 1.14 shows another example of this relationship.

Figure 1.14 Freehaul and Overhaul

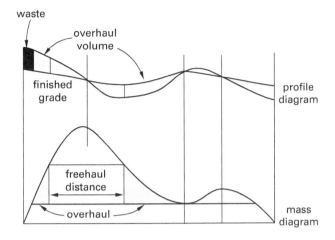

Capacities of Earth-Handling Equipment

Handbook: Production Rate for Loading and Hauling Earthwork

CERM: Sec. 80.18

Handbook section Production Rate for Loading and Hauling Earthwork provides methods to determine cycle times for moving material. CERM Table 80.4 provides typical capacities for various equipment and is helpful in differentiating between equipment capacity types. The exam may require you to determine how long it will take to move a certain volume of material.

E. SITE AND SUBSURFACE INVESTIGATIONS

Key concepts: These key concepts are important for answering exam questions in subtopic 9.E, Site and Subsurface Investigations.

- Atterberg limits
- cone penetration test
- field vane test
- in-place density
- permeability
- pressuremeter test
- seismic refraction
- soil classification
- standard penetration test
- subsurface investigation techniques

NCEES Handbook: To prepare for this subtopic, familiarize yourself with these sections in the *Handbook*.

- Particle Size Distribution Curves
- Soil Classification Chart
- AASHTO Soil Classification System
- Atterberg Limits
- Volume and Weight Relationships

PE Civil Reference Manual (**CERM**): Study these sections in CERM that either relate directly to this subtopic or provide background information.

- Section 35.1: Soil Particle Size Distribution
- Section 35.2: Soil Classification
- Section 35.3: AASHTO Soil Classification
- Section 35.4: Unified Soil Classification
- Section 35.8: Standardized Soil Testing Procedures
- Section 35.9: Standard Penetration Test
- Section 35.11: Proctor Test
- Section 35.12: Modified Proctor Test
- Section 35.13: In-Place Density Tests
- Section 35.14: Atterberg Limit Tests
- Section 35.15: Permeability Tests
- Section 35.17: Direct Shear Test

Construction Depth Reference Manual (**CECN**): Study these sections in CECN that either relate directly to this subtopic or provide background information.

- Section 1.1: Earthmoving and Soil Fundamentals

The following equations, figures, and tables are relevant for subtopic 9.E, Site and Subsurface Investigations.

Particle Size Distribution

Handbook: Particle Size Distribution Curves

CERM: Sec. 35.4

Handbook figure Particle Size Distribution Curves shows particle size distribution curves. The particle size can indicate the status of grading, particle size, and percent passing a given sieve. CERM Fig. 35.3 gives another example of particle size distribution curves. When sieve data is incomplete, the needed values can be interpolated by plotting known data.

AASHTO Soil Classification System

Handbook: AASHTO Soil Classification System

CERM: Sec. 35.3

$$GI = (F - 35)(0.2 + 0.005(LL - 40)) + 0.01(F - 15)(PI - 10)$$

The sieve analysis, liquid limit, and plasticity index contribute to how soils are classified in the American Association of State Highway Transportation Officials (AASHTO) classification. Soils are graded between A-1 (the most suitable soil for roadway use) to A-8 (not suitable for roadway use). See *Handbook* table **AASHTO Soil Classification System**, which is similar to CERM Table 35.4.

The *group index* is another method of soil classification, but it is used to compare soils within a group, not between groups. Group index is measured on a scale between zero and 20 (or higher), with zero representing good subgrade material within the group, and 20 or higher representing poor subgrade material.

Exam tip: Be sure to review the second footnote to the *Handbook* table. The exam may offer two A-4 answers to choose from, but with different group index values.

Typical Soil Characteristics

Handbook: Volume and Weight Relationships

CECN: Sec. 1.1

$$e = \frac{V_v}{V_s}$$

$$n = \frac{V_v}{V}$$

$$W_s = \frac{W_t}{1 + w}$$

The preceding *Handbook* equations calculate the void ratio of a soil, the porosity of a soil, and the weight of solids in a soil sample, respectively. The equation for the void ratio compares the volume of voids (air) to the volume of solids. This is different from the percent pore space, which is the ratio of the volume of voids to the total volume expressed as a percentage. The equation for porosity, n, defines it as the ratio of the volume of voids to the total soil volume.

CERM Table 35.6 shows typical soil characteristics for a variety of soils.

Table 35.6 Typical Soil Characteristics

description	n	e	w_{sat}	ρ_d (lbm/ft^3 (kg/m^3))	ρ_{sat} (lbm/ft^3 (kg/m^3))
sand, loose and uniform	0.46	0.85	0.32	90 (1440)	118 (1890)
sand, dense and uniform	0.34	0.51	0.19	109 (1750)	130 (2080)
sand, loose and mixed	0.40	0.67	0.25	99 (1590)	124 (1990)
sand, dense and mixed	0.30	0.43	0.16	116 (1860)	135 (2160)
glacial clay, soft	0.55	1.20	0.45	76 (1200)	110 (1760)
glacial clay, stiff	0.37	0.60	0.22	106 (1700)	125 (2000)

(Multiply lbm/ft^3 by 16.02 to obtain kg/m^3.)

Unified Soil Classification System

Handbook: Soil Classification Chart

CERM: Sec. 35.4

Handbook table **Soil Classification Chart** shows the soil classification based on the Unified Soil Classification System (USCS). CERM Table 35.5 shows a similar table. The grain size distribution, liquid limit, and plasticity index contribute to how soils are classified in the USCS. Coarse-grained soils are divided into two categories: gravel soils and sand soils. See CERM Fig. 83.1 for a diagram of soil type by geotechnical qualities.

Plasticity Index

Handbook: Atterberg Limits

CERM: Sec. 35.14

$$PI = LL - PL$$

The equation shown is used to calculate the *plasticity index*, which is the difference between liquid and plastic limits. The plasticity index indicates the range of moisture content over which the soil is in a plastic condition. Large plasticity indexes (greater than 20) show that a considerable amount of water can be added to the soil before it becomes liquid.

The *liquidity index* is the water content relative to the Atterberg limits.

$$LI = \frac{w - PL}{PI}$$

When the liquidity index is between zero and one, the water content is between the plastic limit and the liquid limit. When the liquidity index is greater than one, the water content is above the liquid limit.

11 Estimating Quantities and Costs

Content in blue refers to the *NCEES Handbook*.

A. Quantity Takeoff Methods.............................. 11-1
B. Cost Estimating 11-3
C. Cost Analysis for Resource Selection................ 11-4
D. Work Measurement and Productivity 11-5

The knowledge area of Estimating Quantities and Costs makes up between five and eight questions out of the 80 questions on the PE Civil Construction exam. The organization of this chapter follows the order of subtopics given by NCEES for this knowledge area. Each subtopic is covered in the following sections.

A. QUANTITY TAKEOFF METHODS

Key concepts: These key concepts are important for answering exam questions in subtopic 10.A, Quantity Takeoff Methods.

- average end area method
- borrow pit method
- prismoidal formula method
- quantity takeoffs for asphalt
- quantity takeoffs for concrete and concrete formwork
- quantity takeoffs for reinforcing and structural steel
- Simpson's rule
- trapezoidal method

NCEES Handbook: To prepare for this subtopic, familiarize yourself with these sections in the *Handbook*.

- Cross-Section Methods
- Borrow Pit Grid Method
- Earthwork Area Formulas
- Cost Estimate Classification Matrix for Building and General Construction Industries

PE Civil Reference Manual (**CERM**): Study these sections in CERM that either relate directly to this subtopic or provide background information.

- Section 48.23: Reinforcing Steel
- Section 78.18: Elevation Measurement
- Section 78.32: Areas Bounded by Irregular Boundaries
- Section 80.15: Prismoidal Formula Method
- Section 86.2: Budgeting

Construction Depth Reference Manual (**CECN**): Study these sections in CECN that either relate directly to this subtopic or provide background information.

- Section 1.2: Cross-Sectional Volume Calculations
- Section 1.3: Borrow Pit Volumes and Spoil Banks
- Section 2.3: Engineering Economics

The following equations, tables, and concepts are relevant for subtopic 10.A, Quantity Takeoff Methods.

Quantity and Cost Estimates

Handbook: Cost Estimate Classification Matrix for Building and General Construction Industries

CERM: Sec. 86.2

CECN: Sec. 2.3

Estimators compile and analyze data on factors that can influence costs, including materials, labor, location, project duration, and special machinery. The quantity takeoff process can be manual (using a printed copy of the plan, a red pen, and a clicker) or electronic (using a digitizer). The user can take measurements from paper bid documents or with an integrated takeoff viewer program that interprets electronic bid documents. A sample takeoff report is shown in CERM Table 86.1.

The *Handbook* does not give much information regarding cost estimating, but *Handbook* table Cost Estimate Classification Matrix for Building and General Construction Industries gives insights into different estimate classes and how to build an estimate based on design status.

> *Exam tip*: Become comfortable with organizing quantities of different items. You may be given a situation (e.g., four walls of a certain size to be built), one or more materials, and a unit cost for each material. Keeping track of quantities and costs is critical. To solve, determine how much of each material is needed, and then calculate the total cost of materials. Remember that you will sometimes need multiple sides (e.g., formwork for a footing or grade beam).

Average End Area Method

Handbook: Cross-Section Methods

CECN: Sec. 1.2

$$V = L\left(\frac{A_1 + A_2}{2}\right)$$

The cross-sectional volume equations assume that the volume between two consecutive cross sections is the average of their areas multiplied by the distance between them. In cases where areas of two surfaces are given, perform the calculation for A_1 and A_2, then A_2 and A_3, then A_3 and A_4, and so forth.

> *Exam tip*: The PE Civil exam might include a problem asking for the total volume of an excavation. For example, the problem may give the cross-sectional areas at station 1+00, station 1+50, and station 2+00. Find the volume between areas 1 and 2 by taking the average of the two cross-sectional areas and multiplying by the distance between them (in this case, 50 ft). Repeat to find the volume between areas 2 and 3. Finally, add the calculated volumes to determine the total volume.

Prismoidal Formula

Handbook: Cross-Section Methods

CERM: Sec. 80.15

$$V = L\left(\frac{A_1 + 4A_m + A_2}{6}\right)$$

The prismoidal formula is the preferred method of estimating earthwork volume when the two end areas differ greatly or when the ground surface is irregular. It generally produces a smaller volume than the average end area method and favors the owner-developer in earthwork cost estimating.

> *Exam tip*: Using either the average end area method or the prismoidal method on the exam will usually produce the same answer. Be sure to multiply the average area by the length to obtain volume.

Borrow Pit Grid Method

Handbook: Borrow Pit Grid Method

CECN: Sec 1.3

This method can be used to estimate the volume of material taken from an irregularly shaped borrow pit. Lay an imaginary grid of equal squares over the pit, and then multiply each square's average depth by its area to obtain the volume of material within that square. (Equal rectangles can also be used.) Once all volumes are determined, add them to get the total volume of material.

Coordinate Method

Handbook: Earthwork Area Formulas

$$A = \frac{1}{2}\begin{pmatrix} X_A(Y_B - Y_N) + X_B(Y_C - Y_A) \\ + X_C(Y_D - Y_B) + \cdots + X_N(Y_A - Y_{N-1}) \end{pmatrix}$$

This equation can be used to calculate the area of earthwork to be removed from an excavation. Because of its complexity, however, this equation is not likely to be needed on the exam.

Trapezoidal Method

Handbook: Earthwork Area Formulas

$$A = w(\tfrac{1}{2}(h_1 + h_n) + h_2 + h_3 + h_4 + \cdots + h_{n-1})$$

CERM: Sec. 78.32

$$A = d\left(\frac{h_1 + h_n}{2} + \sum_{i=2}^{n-1} h_i\right) \quad \text{78.40}$$

Cross-sectional areas with irregular boundaries, such as those of creek banks, cannot be determined precisely, and approximation methods must be used. In this equation, the length of the common interval is multiplied by an averaged depth. The equation accounts for the trapezoidal shape of the area.

The trapezoidal method can give a more accurate value than would be obtained by simply averaging the separate given areas. If the slope of the mass profile curve is relatively small, the trapezoidal method should be used. If the slope is high, Simpson's rule will provide a more accurate value. If the irregular side can be divided into a series of cells of width d, either the trapezoidal rule or Simpson's rule can be used.

The *Handbook* equation and CERM Eq. 78.40 are equivalent. The length of the common interval is w in the *Handbook*, while CERM uses d. The *Handbook* spells out the second term of the equation while CERM uses the symbol for summation.

Simpson's Rule

Handbook: Earthwork Area Formulas

$$A = \left(\frac{1}{3}\right)\begin{pmatrix} \text{first value} + \text{last value} \\ +(4 \cdot \text{ sum of odd-numbered values}) \\ +(2 \cdot \text{ sum of even-numbered values}) \end{pmatrix} \cdot \text{length of interval}$$

CERM: Sec. 78.18, Sec. 78.32

$$A = \frac{d}{3}\left(h_1 + h_n + 2\sum_{\text{odd}} h_i + 4\sum_{\text{even}} h_i\right) \qquad 78.41$$
$$= \frac{d}{3}(h_1 + 4h_2 + 2h_3 + 4h_4 + \ldots + h_n)$$

Cross-sectional areas with irregular boundaries, such as those of creek banks, cannot be determined precisely, and approximation methods must be used. If the irregular side of each cell is curved or parabolic, then Simpson's rule (sometimes referred to as Simpson's 1/3 rule) can be used. Elevations are taken at equal intervals. The *Handbook* equation for Simpson's rule and CERM Eq. 78.41 are equivalent, but CERM uses mathematical notation while the *Handbook* uses words.

Exam tip: Because of its relative complexity, Simpson's rule is less likely than the trapezoidal method to be needed on the exam. If the irregular side can be divided into a series of cells of width d, either the trapezoidal method or Simpson's rule can be used.

B. COST ESTIMATING

Key concepts: These key concepts are important for answering exam questions in subtopic 10.B, Cost Estimating.

- direct and indirect costs
- project delivery methods
- types of cost items
- types of estimates

NCEES Handbook: To prepare for this subtopic, familiarize yourself with these sections in the *Handbook.*

- Cost Estimate Classification Matrix for Building and General Construction Industries
- Cost Indexes

PE Civil Reference Manual (CERM): Study these sections in CERM that either relate directly to this subtopic or provide background information.

- Section 86.2: Budgeting
- Section 87.46: Rate and Period Changes

The following equations and tables are relevant for subtopic 10.B, Cost Estimating.

Cost Estimating

Handbook: Cost Estimate Classification Matrix for Building and General Construction Industries

CERM: Sec. 86.2

Handbook table Cost Estimate Classification Matrix for Building and General Construction Industries provides a generic and acceptable classification system that can be used as a guideline to compare and assess a firm's standards with generally accepted cost practices. The matrix breaks down cost estimation classes based on the maturity of the project, end usage, the type of estimating applicable to each class, and the expected accuracy of each class. See subtopic 1.B, Cost Estimating, in Chap. 1 of this book for more information on using the cost estimate classification matrix.

Cost Index

Handbook: Cost Indexes

CERM: Sec. 86.2

$$\text{current \$} = (\text{cost in year } M)\left(\frac{\text{current index}}{\text{index in year } M}\right)$$

A cost index is an indicator of the average cost movement over time of goods and services in the construction industry. The *Handbook* equation determines the current cost for an item of equipment in year M given the historical purchase cost and the current index value. Tables of index values are available from government and industry sources. The *Handbook* does not include a table of index values, so any index values needed will be given in the problem.

Rate and Period Changes

CERM: Sec. 87.46

The equation for the effective annual rate, i, is given in CERM Eq. 87.51.

$$i = (1 + \phi)^k - 1$$
$$= \left(1 + \frac{r}{k}\right)^k - 1 \qquad 87.51$$

Effective interest rate computations can be used to determine the total cost of a loan. This is important when determining overall total cost.

C. COST ANALYSIS FOR RESOURCE SELECTION

Key concepts: These key concepts are important for answering exam questions in subtopic 10.C, Cost Analysis for Resource Selection.

- annual cost ratio
- benefit-cost ratio
- earned value method
- engineering economics
- value engineering

NCEES Handbook: To prepare for this subtopic, familiarize yourself with these sections in the *Handbook*.

- Earned-Value Analysis

PE Civil Reference Manual (**CERM**): Study these sections in CERM that either relate directly to this subtopic or provide background information.

- Section 87.25: Choice of Alternatives: Comparing an Alternative with a Standard

Construction Depth Reference Manual (**CECN**): Study these sections in CECN that either relate directly to this subtopic or provide background information.

- Section 2.3: Engineering Economics
- Section 2.4: Value Engineering and Costing

The following equations, tables, and concepts are relevant for subtopic 10.C, Cost Analysis for Resource Selection.

Benefit-Cost Ratio

Handbook: Earned-Value Analysis

CERM: Sec. 87.25

A *benefit-cost ratio* compares the benefits of an alternative (for all beneficiaries) with its cost. The benefit-cost ratio method is often used in municipal project evaluations where benefits and costs accrue to different segments of the community.

$$\text{B/C} = \frac{\Delta \text{ user benefits}}{\Delta \text{ investment cost} + \Delta \text{ maintenance} - \Delta \text{ residual value}} \quad 87.21$$

The end benefits are compared with the total cost after subtracting any residual value.

Exam tip: Pay attention to signs for specific variables.

Earned Value Method

Handbook: Earned-Value Analysis

CECN: Sec. 2.4

$$\text{ACWP} = Q_{\text{actual}} C_{\text{actual}} \quad 2.38$$

$$\text{BCWP} = Q_{\text{actual}} C_{\text{budgeted}} \quad 2.39$$

$$\text{CPI} = \frac{\text{BCWP}}{\text{ACWP}} \quad 2.40$$

The *earned value method* is a project management technique that correlates actual project value with earned value. The actual cost of work performed (ACWP) can be represented by CECN Eq. 2.38. The budgeted cost of work performed (BCWP) can be represented by CECN Eq. 2.39, where Q represents the quantity and C represents the cost.

The *cost performance index* (CPI) is the ratio of the BCWP to the ACWP, as shown in CECN Eq. 2.40. The CPI indicates the quality of the original estimate. If the CPI is greater than 1, then the estimate was good.

Cost Variance

Handbook: Earned-Value Analysis

CECN: Sec. 2.4

$$\text{CV} = \text{BCWP} - \text{ACWP}$$

The *cost variance* (CV) for a work item is the amount by which the item has come in over or under budget. The CV is equal to the actual cost of work performed (ACWP) subtracted from the budgeted cost of work performed (BCWP) for the work item. The *Handbook* equation is the same as CECN Eq. 2.41.

The CV gives the status of the budget in terms of remaining funds. A positive CV at the end of work means that budgeted funds are still available, and the budget was good. A negative CV value is cause for further analysis.

Value Engineering

CECN: Sec. 2.4

Value engineering involves a methodical process of planning to arrive at the best design. The best design minimizes cost while achieving high quality.

$$\text{value} = \frac{\sum \text{benefits}}{\sum \text{costs}}$$

This metric can be used during the generation stage of the value engineering process to evaluate the economic feasibility of a design alternative. *Value* is the same thing as *benefit-cost ratio*, and as in a benefit-cost analysis, the benefits and costs can be reviewed to evaluate each possible design alternative. The greater the benefit-cost ratio, the greater the value.

Engineering Economics

CECN: Sec. 2.3

CECN Table 2.6 and Table 2.7 define the terms for specific costs. It is important to be familiar with these terms and their meanings. All of these can be represented as costs when applying values toward determining a benefit-cost ratio.

Table 2.6 Equipment Owning Costs

depreciation	decline in value of assets
investment interest cost	annual cost (converted to an hourly cost) of the capital invested in a machine
insurance cost	cost of fire, theft, accident, and liability insurance for the equipment
taxes	cost of property tax and licenses for the equipment
storage cost	cost of rent and maintenance for equipment storage yards and facilities, the wages of guards and employees involved in handling equipment in and out of storage, and associated direct overhead

Table 2.7 Equipment Operating Costs

fuel cost	cost of fuel for every hour the equipment is operated
service cost	cost of oil, hydraulic fluids, grease, and filters in addition to the labor required to perform the maintenance service
repair cost	cost of all equipment repair and non-routine maintenance
replacement cost	cost of equipment and part replacement

One common economic concern for construction engineers is equipment cost. Decisions about the equipment cost, such as whether to rent or purchase a piece of equipment, or when a piece of equipment should be replaced, are central to proper economic management of a construction firm.

Exam tip: Use these definitions of costs when calculating benefit-cost ratios.

D. WORK MEASUREMENT AND PRODUCTIVITY

Key concepts: These key concepts are important for answering exam questions in subtopic 10.D, Work Measurement and Productivity.

- crew hours
- equipment productivity and selection
- labor costs
- wage rates

NCEES Handbook: The *Handbook* does not contain any material relevant to this subtopic. Be sure to study the listed sections in CERM and CECN.

PE Civil Reference Manual **(CERM):** Study these sections in CERM that either relate directly to this subtopic or provide background information.

- Section 87.51: Accounting Costs and Expense Terms

Construction Depth Reference Manual **(CECN):** Study these sections in CECN that either relate directly to this subtopic or provide background information.

- Section 2.2: Elements of a Cost Estimate
- Section 3.1: Equipment Productivity and Selection

The following equations, tables, and concepts are relevant for subtopic 10.D, Work Measurement and Productivity.

Labor and Crew Hour Rates

CERM: Sec. 87.51

CECN: Sec. 2.2

Each type of laborer will have a different hourly rate, which depends on the work location and/or union agreements.

$$\text{labor hour rate} = \frac{\text{crew hour rate}}{\text{no. of workers}}$$

$$\text{crew hour rate} = \sum (\text{no. of workers})(\text{hour rate})$$

Estimated labor cost is the cost of labor required to complete the work for a particular task (e.g., constructing a retaining wall or painting a room), as shown in CECN Eq. 2.20, Eq. 2.21, and Eq. 2.22.

$$\text{labor productivity rate} = \frac{Q_w}{t_w}$$

$$\text{labor rate} = \frac{C_w}{t_w}$$

$$C_{l,t} = Q_{\text{workers}} t_w (\text{labor rate})$$

Total cost is determined using a labor productivity rate (time for a certain amount of production), a labor rate (cost per unit time), and the number of workers.

Exam tip: Pay attention to whether the problem statement gives you a productivity rate per single person or per crew.

Wage Rates

The *wage rate*, also called the *labor rate*, is used to determine labor costs. The minimum federal wage rate is determined by the U.S. Department of Labor (DOL), but it can also be set by unions. All the trade wages on a federally funded construction project must comply with DOL minimum wage rates. The federal government often requires agencies to do periodic checks to ensure that contractors are complying with federal wage rates.

Exam tip: For exam purposes, you will need to know that minimum wages exist and where to obtain them, but you will not need to know the actual minimum wages for various kinds of labor. These will generally be provided in the exam problems as needed.

Equipment Productivity and Selection

CECN: Sec. 3.1

CECN Table 3.1 shows the major pieces of equipment used in excavating and loading operations. Equipment productivity rates can be used in a similar manner as labor rates. CECN Sec. 3.1 gives productivity rates for several kinds of equipment.

Exam tip: Exam problems may ask you to supply a specific equipment productivity rate or to use a given rate to determine costs.

Crew Hours

Crew labor rate is the total labor rate for the total number of workers. If the individual labor rate is $12.00/hr and there are five workers on a crew, then the crew labor rate is the product of $12.00/hr and 5, or $60.00/hr.

Exam tip: Pay attention to the number of workers on a given crew.

12 Construction Operations and Methods

Content in blue refers to the *NCEES Handbook*.

A. Lifting and Rigging 12-1
B. Crane Stability 12-3
C. Dewatering and Pumping 12-6
D. Equipment Operations 12-8
E. Deep Foundation Installation 12-10

The knowledge area of Construction Operations and Methods makes up between six and nine questions out of the 80 questions on the PE Civil Construction exam. The organization of this chapter follows the order of subtopics given by NCEES for this knowledge area. Each subtopic is covered in the following sections.

A. LIFTING AND RIGGING

Key concepts: These key concepts are important for answering exam questions in subtopic 11.A, Lifting and Rigging.

- center of gravity of a load
- crane operations personnel
- load in a sling
- moments
- multiple lift rigging technique
- OSHA requirements for hoisting and rigging sling length
- tensile and compressive forces on cranes and rigging
- trigonometry

NCEES Handbook: To prepare for this subtopic, familiarize yourself with these sections in the *Handbook*.

- Trigonometry
- Moments (Couples)
- Systems of Forces

PE Civil Reference Manual (**CERM**): Study these sections in CERM that either relate directly to this subtopic or provide background information.

- Section 6.3: Triangles
- Section 6.4: Right Triangles
- Section 6.5: Circular Transcendental Functions
- Section 41.5: Moments
- Section 41.6: Moment of a Force About a Point

Construction Depth Reference Manual (**CECN**): Study these sections in CECN that either relate directly to this subtopic or provide background information.

- Section 3.3: Lifting and Rigging

Codes and standards: To prepare for this subtopic, familiarize yourself with these sections from codes and standards.

Occupational Safety and Health Act (OSHA)

- 29 CFR 1926 Subpart H: Materials Handling, Storage, Use, and Disposal
- 29 CFR 1926 Subpart N Section 552: Material Hoists, Personnel Hoists, and Elevators
- 29 CFR 1926 Subpart R: Steel Erection
- 29 CFR 1926 Subpart CC: Cranes and Derricks in Construction

The following equations, figures, and concepts are relevant for subtopic 11.A, Lifting and Rigging.

Lifting Tensile and Compressive Forces

Handbook: Trigonometry; Moments (Couples); Systems of Forces

CECN: Sec. 3.3

The tensile and compressive forces should be evaluated for each lift. CECN Fig. 3.8 demonstrates the lifting of a single beam. The tensile force in the two lifting cables and the compressive force in the beam will vary depending on the length of the cables used.

Figure 3.8 Lifting a Single Beam

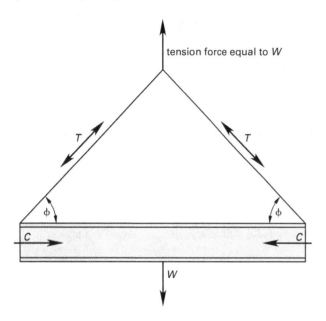

Review *Handbook* section Trigonometry to familiarize yourself with the trigonometric concepts that apply to the calculation of tensile and compressive forces. CECN Eq. 3.24 gives the tension in lifting cables.

$$T = \frac{W}{2\sin\phi} \quad 3.24$$

CECN Eq. 3.25 gives the compression in a beam.

$$C = \frac{W}{2\tan\phi} \quad 3.25$$

Handbook sections Moments (Couples) and Systems of Forces provide information needed for solving for unknown parameters.

Multiple Lift Rigging

Handbook: Moments (Couples); Systems of Forces

CECN: Sec. 3.3

OSHA: 29 CFR 1926 Subpart R

The multiple lift rigging technique is used when multiple loads must be lifted at one time. A spreader beam is used to attach the multiple loads to the rigging. The lifting point is always directly above the center of gravity. The center of gravity, load in the cables, and bending moment capacity of the spreader beam should all be evaluated. OSHA standards in 29 CFR 1926 Subpart R have specific requirements for multiple lift rigging.

Multiple lift rigging requires the calculation of the centers of gravity of the load alone and of the load in skewed cables of variable sling lengths, as well as the evaluation of the bending moment capacity of the spreader beam. CECN Fig. 3.9 illustrates the dimensions and values used to perform the calculations.

Figure 3.9 Multiple Lift Rigging Technique

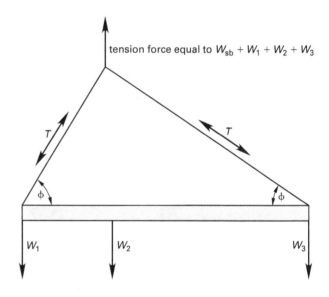

Handbook sections Moments (Couples) and Systems of Forces provide information necessary to solve for unknowns.

> *Exam tip*: To begin a problem on multiple lift rigging, assess what information you are given (weights, distances, forces, etc.) and what you will ultimately be solving. Take a moment about one end of the spreader beam to start solving for unknowns. Take note of where each weight is situated on the spreader beam. You may need to solve for the distance to the center of gravity.

OSHA Regulations for Hoisting and Rigging

OSHA: 29 CFR 1926 Sec. 251, Sec. 552, Subpart CC

A maximum of five members can be present per lift, and loads must be rigged by a qualified rigger. All rigging must be inspected prior to each use and must have permanent markings that display the safe working load; any custom-designed hooks, clamps, and so on must be marked with the safe working load and be proof tested at 125% of the safe working load.

The crane regulations are detailed in OSHA 1926 Subpart CC. OSHA 1926 Subpart H further identifies regulations for rigging equipment such as slings and hooks. Additional information on rigging can be found in OSHA Sec. 1926.251, and additional information on hoists can be found in OSHA Sec. 1926.552.

B. CRANE STABILITY

Key concepts: These key concepts are important for answering exam questions in subtopic 11.B, Crane Stability.

- crane components
- crane selection process
- crane specifications
- crane types
- moments
- range diagrams
- load charts
- static basis for cranes

NCEES Handbook: To prepare for this subtopic, familiarize yourself with these sections in the *Handbook*.

- Telescoping Boom Industrial Cranes
- Telescoping Boom Rough-Terrain Cranes
- Telescoping Boom Carrier- and Crawler-Mounted Cranes
- Telescoping Boom Truck-Mounted Cranes
- Telescoping Boom Crane Components
- Lattice Boom Cranes
- Lattice Boom Crane Components

PE Civil Reference Manual (**CERM**): Study these sections in CERM that either relate directly to this subtopic or provide background information.

- Section 41.5: Moments
- Section 41.6: Moment of a Force About a Point
- Section 83.17: Truck and Tower Cranes
- Section 83.19: Crane Load Charts

Construction Depth Reference Manual (**CECN**): Study these sections in CECN that either relate directly to this subtopic or provide background information.

- Section 3.2: Cranes

Codes and standards: To prepare for this subtopic, familiarize yourself with these sections from codes and standards.

Occupational Safety and Health Act (OSHA)

- 29 CFR 1926 Section 1401: Definitions

The following equations, figures, tables, and concepts are relevant for subtopic 11.B, Crane Stability.

Mobile Cranes and Their Components

CECN: Sec. 3.2

OSHA: 29 CFR 1926 Sec. 1401

CECN Fig. 3.4 shows the static basis for a crane.

Figure 3.4 *Static Basis for a Mobile Crane*

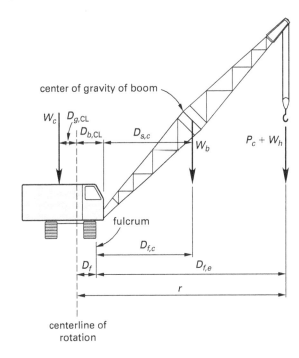

A mobile crane typically has a quicker setup time than a tower crane does, can be moved easily around a site, and offers a great deal of versatility.

A fly jib can be added to the top of a jib boom to extend the crane's operating range. The use of outriggers on a crane also increases its range.

The distance from the end of the superstructure to the center of the boom can be found using CECN Eq. 3.18.

$$D_{s,c} = \frac{r - D_{b,\text{CL}}}{2} \quad \quad 3.18$$

The distance from the fulcrum to the center of the boom can be found using CECN Eq. 3.19.

$$D_{f,c} = D_{s,c} + D_{b,\text{CL}} - D_j \quad 3.19$$

The distance from the fulcrum to the end of the boom can be found using CECN Eq. 3.20.

$$D_{f,e} = r - D_j \quad 3.20$$

The following *Handbook* figures illustrate various types of mobile cranes and their components:

- Telescoping Boom Industrial Cranes
- Telescoping Boom Rough-Terrain Cranes
- Telescoping Boom Carrier- and Crawler-Mounted Cranes
- Telescoping Boom Truck-Mounted Cranes
- Telescoping Boom Crane Components
- Lattice Boom Cranes
- Lattice Boom Crane Components

Exam tip: OSHA 1926 Sec. 1401 includes definitions for crane terminology.

Tipping Load for a Mobile Crane

CECN: Sec. 3.2

The tipping load for a mobile crane can be found using CECN Eq. 3.21.

$$P_c = \frac{W_c(D_{g,\text{CL}} + D_f) - W_b D_{f,c}}{D_{f,e}} - W_h \quad 3.21$$

CECN Fig. 3.4 shows the static basis of a mobile crane.

Exam tip: A crane's tipping load can be determined by calculating the moment about the crane's centerline of rotation.

Tipping Load for a Horizontal Boom Tower Crane

CECN: Sec. 3.2

The tipping load for a horizontal boom tower crane can be found using CECN Eq. 3.22.

$$P_c = \frac{W_{c,1} D_{c,1} + W_{c,2} D_{c,2} - W_b D_{b,c}}{D_{l,c}} - W_h \quad 3.22$$

CECN Fig. 3.5 illustrates a horizontal boom tower crane.

Figure 3.5 Horizontal Boom Tower Crane

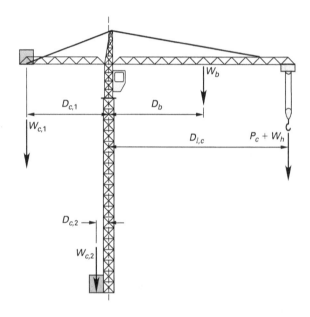

Exam tip: A crane's tipping load can be determined by calculating the moment about the crane's centerline of rotation.

Tipping Load for a Lifting Boom Tower Crane

CECN: Sec. 3.2

The tipping load for a lifting boom tower crane can be found using CECN Eq. 3.23.

$$P_c = \frac{W_c D_c}{D_{l,c}} - \frac{W_b D_l}{2 D_{l,c}} - W_h \quad 3.23$$

Figure 3.6 Lifting Boom Tower Crane

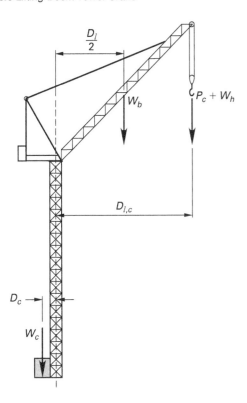

CECN Fig. 3.6 illustrates a lifting boom tower crane.

Exam tip: A crane's tipping load can be determined by calculating the moment about the crane's centerline of rotation.

Crane Selection Process

CECN: Sec. 3.2

A crane should be selected to maximize the production of all equipment available. The selected crane should have sufficient load capacity for the project loads, and it should fit on site. The crane should be chosen according to the types of lifting, should not hurt the environment, and should not interfere with the surrounding structures.

The crane should be appropriately sized, as oversizing can lead to higher expenses. Crane size is a function of

- the size of the loads to be moved
- the distance that the loads must be moved
- the amount of space available to assemble, disassemble, and maneuver the loads and crane
- efficiency

Crane Specifications

CECN: Sec. 3.2

A crane comes with a data sheet and a manufacturer's load chart. The crane data sheet includes specifications such as dimensions, weights, and working range. The manufacturer's load chart includes specifications such as load capacity and tipping load.

A crane that lifts safely will do so without becoming unstable and moving uncontrollably. The weight that can be lifted safely is usually less than 75% of the tipping load. The load capacity and tipping load are functions of crane configuration and positioning, and can be determined from the manufacturer's load chart.

Load Charts

CECN: Sec. 3.2

To account for unintentional rotation and possible overloading of the crane, assume a 360° rotation. If outriggers are to be used in the manner described in the crane's load chart manual, choose the outriggers with the highest load capacity. Jib booms should not be used for carrying loads, as they may damage the crane and make it unstable. CECN Table 3.7 shows a sample load chart.

The load radius is equal to the distance from the centerline of the crane to the hook. *Over front* loading refers to loads that will remain within 0° to 5° of the front of the crane. The capacities listed for 360° rotation are usually lower than those listed for over front loading.

Exam tip: There is controversy as to whether interpolating in load charts is allowed. For safety concerns, do not interpolate; use the most conservative value. The values in the load chart can be assumed to include the factor of safety.

Range Diagrams

CECN: Sec. 3.2

Range diagrams are unique to each crane manufacturer. CECN Fig. 3.2 shows a sample range diagram.

Figure 3.2 Sample Range Diagram (26 ft to 61 ft boom)

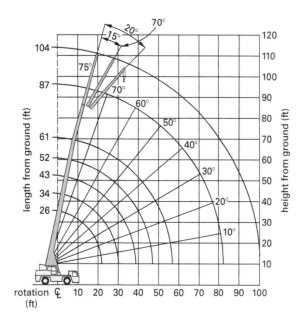

To use a range diagram chart to find the required boom length, first sum the following to find the total height from the ground.

- distance the load is to be lifted above the ground
- height of the load
- sling distance
- hook block (hoisting tackle) distance

Then enter the chart on the right y-axis at the corresponding height from the ground, proceed to the correct distance from the centerline of the crane to the load (bottom axis), and finally read the left axis as the required boom length.

C. DEWATERING AND PUMPING

Key concepts: These key concepts are important for answering exam questions in subtopic 11.C, Dewatering and Pumping.

- cavitation
- drawdown curves
- pump design curves and affinity laws
- surface runoff
- temporary erosion control
- total dynamic head
- types of aquifers and pumps
- types of wells

NCEES Handbook: To prepare for this subtopic, familiarize yourself with these sections in the *Handbook*.

- Confined Aquifer
- Water-Table Aquifer
- Total Dynamic Pumping Head
- Centrifugal Pump Characteristics
- Pump Brake Horsepower
- Runoff Analysis: Rational Formula Method
- Dupuit's Formula
- Thiem Equation

PE Civil Reference Manual (**CERM**): Study these sections in CERM that either relate directly to this subtopic or provide background information.

- Section 17.18: Introduction to Pumps and Turbines
- Section 17.19: Extended Bernoulli Equation
- Section 18.10: Pumping Power
- Section 18.11: Pumping Efficiency
- Section 18.15: Specific Speed
- Section 18.23: Operating Point
- Section 20.15: Peak Runoff from the Rational Method
- Section 21.1: Aquifers
- Section 21.11: Well Drawdown in Aquifers

Construction Depth Reference Manual (**CECN**): Study these sections in CECN that either relate directly to this subtopic or provide background information.

- Section 3.5: Construction Dewatering and Pumping

The following equations, figures, tables, and concepts are relevant for subtopic 11.C, Dewatering and Pumping.

Aquifers

Handbook: Confined Aquifer; Water-Table Aquifer

CERM: Sec. 21.1

CECN: Sec. 3.5

Underground or subsurface water is contained in saturated geological formations known as *aquifers*. A *confined aquifer* is bounded on top and bottom. It has an impenetrable layer above that prevents water from rising to the ground surface. An *unconfined aquifer* (or *water-table aquifer*) is bounded only at the bottom. It is usually covered by sand or is open to the atmosphere. Illustrations of aquifers are provided in *Handbook* figures Confined Aquifer and Water-Table Aquifer. CECN Fig. 3.19 illustrates an unconfined aquifer.

Figure 3.19 Unconfined Aquifer

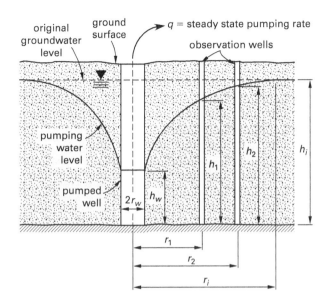

Well Discharge

Handbook: Dupuit's Formula; Thiem Equation

CERM: Sec. 21.11

Dupuit's formula is

$$Q = \frac{\pi K(h_2^2 - h_1^2)}{\ln\left(\dfrac{r_2}{r_1}\right)}$$

Dupuit's formula can be used to determine the drawdown level of a well if the following are known.

- radius of influence of the well
- hydraulic conductivity of the soil
- soil flow rate of the water

CERM Eq. 21.25 is equivalent, with only minor differences in nomenclature. The *Handbook* and CERM assign the subscripts differently to the larger and smaller radial distances and the larger and smaller aquifer depths; however, it doesn't matter which subscript is assigned to which as long as the aquifer depths also correspond correctly. If an area of excavation needs to be dried, a construction engineer must calculate how far the groundwater table can be lowered.

The Thiem equation is used to determine the flow rate from a pump in an unconfined aquifer. CERM Eq. 21.27 is equivalent, again with only minor differences in nomenclature.

$$Q = \frac{2\pi T(h_2 - h_1)}{\ln\left(\dfrac{r_2}{r_1}\right)}$$

Rational Formula Method

Handbook: Runoff Analysis: Rational Formula Method

CERM: Sec. 20.15

$$Q = CIA$$

The rational formula for peak discharge is applicable for small areas (less than several hundred acres), but is rarely used for areas greater than 2 mi^2.

The rational method makes various assumptions.

- Rainfall occurs at a constant rate.
- The recurrence interval of the peak flow is the same as for the design storm.
- The runoff coefficient is constant.
- The rainfall is spatially uniform over the drainage area.

Total Dynamic Head

Handbook: Total Dynamic Pumping Head

CERM: Sec. 18.10

$$TDH = H_L + H_F + H_v$$

$$H_v = \frac{v^2}{2g}$$

The energy (head) added by a pump can be determined from the difference in total energy on either side of the pump. In most applications, the change in velocity and potential heads is either zero or small in comparison to the increase in pressure head.

Total dynamic head is represented by TDH. The velocity head is usually small and therefore often neglected when calculating TDH. Minor losses, including those caused by components such as valves, bends, tees, and other appurtenances in the system, may need to be considered.

Pump Design

Handbook: Centrifugal Pump Characteristics; Pump Brake Horsepower

CERM: Sec. 17.18, Sec. 18.10, Sec. 18.11, Sec. 18.23

A pump usually has an allowable operating range characterized by a range of discharge over which the pump will operate with acceptable vibration. *Handbook* figure **Centrifugal Pump Characteristics** includes pump curves that illustrate the relationship between head, power, and efficiency.

The difference between the system capacity curve and the pump head curve is evident in CERM Fig. 18.13. The pump curve is obtained from the pump manufacturer and is plotted along with the system head curve. The intersection of the two curves represents the pump's *operating point*. The operating point defines the system head and system flow rate.

Figure 18.13 Extreme Operating Points

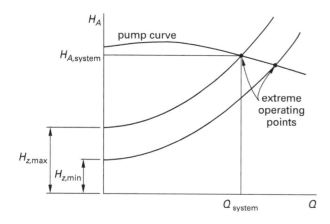

Pump brake horsepower can be calculated using the following *Handbook* equations.

$$\text{BHP} = \frac{\text{HP}}{\eta}$$

$$\text{BHP} = \frac{Q\gamma H}{\left(550 \, \dfrac{\text{ft-lbf}}{\text{hp-sec}}\right)\eta}$$

The relationship between brake horsepower and input power in kilowatts is given by CERM Eq. 18.14.

$$P_{m,\text{kW}} = \frac{0.7457 \text{BHP}}{\eta_m} \qquad 18.14$$

If the mass flow rate is known, the specific energy added by the pump can be calculated from the input power to the pump using CERM Eq. 17.60.

$$E_A = \frac{\left(1000 \, \dfrac{\text{W}}{\text{kW}}\right) P_{\text{kW,input}} \eta_{\text{pump}}}{\dot{m}} \qquad \text{[SI]} \qquad 17.60(a)$$

$$E_A = \frac{\left(550 \, \dfrac{\text{ft-lbf}}{\text{sec-hp}}\right) P_{\text{hp,input}} \eta_{\text{pump}}}{\dot{m}} \qquad \text{[U.S.]} \qquad 17.60(b)$$

See CERM Table 18.5 for a summary of hydraulic horsepower equations.

Types of Wells

CERM: Sec. 21.1

A *monitor well* (or *observation well*) is a well drilled to monitor the quality and quantity of water. A *relief well* is a well drilled to dewater soil. A *hard rock well* is a well drilled in a hard rock formation.

An *artesian well* is a well whose water flows to the surface naturally. An artesian well is measured by the distance the water reaches above the ground. A non-artesian well is measured by the distance from the ground surface to the water in the well.

To avoid pumping of solids and pump damage, the use of one or more screens is recommended for both artesian and non-artesian wells. Multiple screens may be needed if the well penetrates multiple aquifer zones of fine and coarse material.

D. EQUIPMENT OPERATIONS

Key concepts: These key concepts are important for answering exam questions in subtopic 11.D, Equipment Operations.

- equipment production estimates
- excavation load measurements
- paver speed estimates
- roller speed estimates
- types of construction equipment

NCEES Handbook: To prepare for this subtopic, familiarize yourself with these sections in the *Handbook*.

- Production Rate for Soil Compaction

PE Civil Reference Manual (**CERM**): Study these sections in CERM that either relate directly to this subtopic or provide background information.

- Section 80.3: Swell and Shrinkage

Construction Depth Reference Manual (CECN): Study these sections in CECN that either relate directly to this subtopic or provide background information.

- Section 3.1: Equipment Productivity and Selection

The following equations, tables, and concepts are relevant for subtopic 11.D, Equipment Operations.

Compaction Equipment

Handbook: Production Rate for Soil Compaction

CECN: Sec. 3.1

The production capacity of compaction equipment is equal to the production capacity of excavating and hauling equipment. Compaction equipment should be selected based on the soil properties, soil lift thickness, type of work to be performed, construction specifications, and project size. CECN Table 3.4 gives a summary of the descriptions and uses of various kinds of compaction equipment.

For rollers and vibratory plates, production can be estimated as

$$\text{compacted cubic yards per hour} = \frac{1}{n}(16.3\,WSL)(\text{efficiency})$$

Types of Construction Equipment

CECN: Sec. 3.1

Excavation and loading equipment is used to excavate material for large earthwork cuts, install underground utilities, and construct foundations. Excavation equipment production capacities vary from 1 yd³ to 80 yd³. See the entry "Excavation and Loading Equipment" in this chapter for more information.

Hauling and placing equipment is used to transport soil from one location on a site to another. Production capacity of hauling and placing equipment is measured in transport distance as well as bucket load capacity. Transportation distances vary per the equipment but can range up to 10,000 ft or more for trucks. See the entry "Hauling and Placing Equipment" in this chapter for more information.

Compaction equipment uses a variety of forces to mechanically increase the density of soil. The selection of compaction equipment depends largely on soil compaction properties and soil layer thickness. Compaction equipment production capacity is measured in cubic yards per minute. See the entry "Compaction Equipment" in this chapter for more information.

Finishing and grading equipment is used to bring earthwork to the shape and elevation specified by the project. Production capacity of finishing and grading equipment is measured in feet per minute. See the entry "Finishing and Grading Equipment" in this chapter for more information.

Excavation and Loading Equipment

CECN: Sec. 3.1

Excavation and loading equipment is used for rough grading and for the loading of hauling equipment.

Soil that has been excavated is measured in *loose cubic yards* (LCY), in-place soil before excavation is measured in *bank cubic yards* (BCY), and soil compacted in place after placement is measured in *compact cubic yards* (CCY).

Backhoes are typically used for small jobs, as they fit in confined areas and are easy to transport. A backhoe has a typical load capacity between 1 yd³ and 3 yd³. Given the efficiency, capacity, fill factor, and cycle time, the production of a backhoe can be estimated as

$$p_{\text{backhoe}} = E\left(\frac{c_h F}{t}\right) \qquad 3.3$$

Excavators have higher capacities than backhoes but require more maintenance. The typical load capacity of an excavator is between 1 yd³ and 5 yd³.

Draglines require longer site setup times and are generally used for dredging. They typically have high load capacities, between 40 yd³ and 80 yd³.

Dozers are used for pushing material but can have ripper attachments fitted. They are not designed for loading and do not have any load capacity. The production of a dozer can be estimated as

$$p_{\text{dozer}} = \frac{c_{\text{blade}}}{t} \qquad 3.6$$

Loaders are for loading haul equipment or moving material around job sites. The typical load capacity of a loader is between 1 yd³ and 9 yd³. The production of a loader can be estimated as

$$p_{\text{loader}} = \frac{c_{\text{bucket}} F E}{t} \qquad 3.7$$

Load capacity is generally measured as *heaped capacity*, the capacity when material is mounded above the rim of the bucket at a 2:1 slope. If material is filled only to the rim of the bucket, it is measured as *struck capacity*.

Hauling and Placing Equipment

CECN: Sec. 3.1

CECN Table 3.2 lists the most common types of hauling and placing equipment. CECN Table 3.3 describes the types of scrapers in more detail, along with reasons for and against using a pusher. CECN Eq. 3.8 can be used to find the production capability of scrapers without pushers, and CECN Eq. 3.9 can be used to find the production capability of scrapers with pushers.

$$p_{\text{scraper}} = \left(\frac{c_h E}{t}\right)(\text{number of scrapers}) \quad 3.8$$

$$p_{\text{scraper}} = \frac{c_h E}{t} \quad 3.9$$

CECN Eq. 3.10 gives the production capability of a truck.

$$p_{\text{truck}} = \left(\frac{c_h E}{t}\right)(\text{number of trucks}) \quad 3.10$$

Finishing and Grading Equipment

CECN: Sec. 3.1

Finishing and grading equipment is used to attain finished grades that are specified by construction plans. Minor changes can be made to the site elevations during the grading stage. Grader production can be estimated as

$$p_{\text{grader}} = \text{v} w_{\text{gr}} E \quad 3.13$$

Finishing equipment is used for final preparation of a subgrade, establishing roadway cross slopes, and preparing drainage ditches. Finishing equipment can move only a limited amount of material. Paver production can be estimated as

$$\text{production} = \gamma_{\text{asphalt}} t_l w \quad 3.14$$

Paver speed can be estimated as

$$\text{v}_p = \frac{p_{\text{plant}}}{\text{production}} \quad 3.15$$

E. DEEP FOUNDATION INSTALLATION

Key concepts: These key concepts are important for answering exam questions in subtopic 11.E, Deep Foundation Installation.

- constant rate of penetration tests
- drop hammers
- load capacity
- load test interpretation
- maintained load tests
- pile drivers
- pile installation
- pile load tests
- pile types
- static load tests

NCEES Handbook: To prepare for this subtopic, familiarize yourself with these sections in the *Handbook*.

- Hammer-Pile-Soil System
- Modified Engineering-News Formula
- Gates Formula
- Typical Wave Equation Models
- Summary Output for Wave Equation Analysis
- Maximum Allowable Stresses in Pile for Top-Driven Piles

PE Civil Reference Manual **(CERM):** Study these sections in CERM that either relate directly to this subtopic or provide background information.

- Section 38.1: Introduction
- Section 38.2: Pile Capacity from Driving Data
- Section 38.3: Theoretical Point-Bearing Capacity
- Section 38.5: H-Piles

Construction Depth Reference Manual **(CECN):** Study these sections in CECN that either relate directly to this subtopic or provide background information.

- Section 3.8: Deep Foundations

The following equations, figures, tables, and concepts are relevant for subtopic 11.E, Deep Foundation Installation.

Allowable Pile Load

Handbook: Modified Engineering-News Formula

$$P_{\text{allow}} = \frac{R}{6} = \frac{2E_n}{S+k}$$

This formula is used to determine the allowable pile load. The total soil resistance must be known, or else the driving energy (calculated by multiplying the weight of the hammer with the hammer dropping height), the pile penetration per blow, and a constant, k, based on the hammer type, must all be known. Each variable has a specified unit; if values are given in other units, they must be converted to the units specified.

Ultimate Pile Capacity

Handbook: Gates Formula

$$R_u = 1.75\sqrt{E_r}(\log_{10}10N_b) - 100$$

The Gates formula calculates the ultimate pile capacity given the rated hammer energy at field-observed ram stroke and the number of hammer blows per inch at final penetration. Each variable has a specified unit; if values are given in other units, they must be converted to the units specified.

Pile Load Tests

Handbook: Maximum Allowable Stresses in Pile for Top-Driven Piles

CECN: Sec. 3.8

There are several methods available for testing piles. The most common method is to measure the duration that a force is applied to a pile and the strain induced in the pile.

In a *maintained load test* (MLT), the load is applied to the pile in discrete increments, and the resulting pile movement and settlement are monitored. In a *constant rate of penetration* (CRP) test, the load needed to cause a pile to penetrate into the ground at a constant rate is monitored until either the maximum specified test load is achieved or failure of the pile occurs.

Handbook table Maximum Allowable Stresses in Pile for Top-Driven Piles provides design stress and driving stress equations that align with the different pile types (steel H-piles, unfilled steel pipe piles, concrete-filled steel pipe piles, precast prestressed concrete piles, conventionally reinforced concrete piles, and timber piles).

Pile-Driving Hammers

Handbook: Hammer-Pile-Soil System; Typical Wave Equation Models; Summary Output for Wave Equation Analysis

CECN: Sec. 3.8

A *drop hammer* is a hammer, approximately the same weight as the pile, that is raised to a suitable height in a back guide and released to strike the pile head. A single-acting hammer has a massive weight in the shape of a cylinder inside a hollow piston rod. A double-acting hammer can be driven by steam or compressed air.

Pile driving can also be done using electrically powered or hydraulically powered vibrating hammers.

Handbook figure Hammer-Pile-Soil System illustrates the compressive and tensile force pulses from a hammer-pile driving system, given a ram and drive head driving a pile into a dense layer of soil beneath a soft layer of soil.

Handbook figures Typical Wave Equation Models and Wave Equation Bearing Graph show the results of an air-steam hammer. The diagram in figure Typical Wave Equation Models shows the assembly of the hammer, the soil model, and the representation of the soil model. The graph in figure Wave Equation Bearing Graph shows a summary of the stroke, compressive stress, tensile stress, and driving capacity versus the blow count.

Auger Pile Installation

CECN: Sec. 3.8

As shown in the illustration, piles can be installed by driving, jetting, or augering. The installation type chosen is a function of soil conditions, bearing capacity, available equipment, and local practice.

A continuous flight auger uses an augered hole. As the auger is removed, concrete is poured, and reinforcement is inserted into the wet concrete.

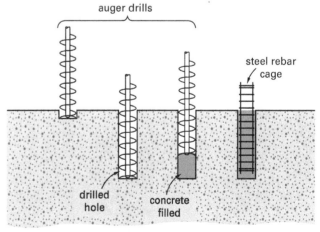

continuous flight auger installation

Continuous flight augering has many advantages. It eliminates vibration and eliminates the need to drill in bentonite mud, reduces disposal of waste materials and noise, and avoids having an open hole. There are some disadvantages as well: it is not applicable to all soil types, the pile quality is dependent on the workmanship, and it makes it difficult to achieve batter.

Ultimate Bearing Capacity

CERM: Sec. 38.1

$$Q_{\text{ult}} = Q_p + Q_f \qquad 38.1$$

The *ultimate static bearing capacity* of a single pile is the sum of its point-bearing and skin friction capacities. The allowable capacity, Q_a, of a pile depends on the factor of safety, which is typically two to three for both compression and tension piles. The lower value of Q_a is used when the capacity can be verified by pile load tests.

$$Q_a = \frac{Q_{\text{ult}}}{F} \qquad 38.2$$

Theoretical Point-Bearing Capacity

CERM: Sec. 38.3

$$Q_p = A_p\left(\tfrac{1}{2}\gamma B N_\gamma + c N_c + \gamma D_f N_q\right) \quad \text{[U.S.]} \qquad 38.8(b)$$

The theoretical point-bearing capacity (also known as *tip resistance* or *point capacity*), Q_p, of a single pile can be calculated in much the same manner as for a footing. Since the pile size, B, is small, the ½ $\gamma B N_\gamma$ term is generally omitted.

13 Scheduling

Content in blue refers to the *NCEES Handbook*.

A. Construction Sequencing.................................. 13-1
B. Activity Time Analysis..................................... 13-3
C. Critical Path Method Network Analysis............ 13-4
D. Resource Scheduling and Leveling 13-5
E. Time-Cost Trade-Off.. 13-7

The knowledge area of Scheduling makes up between five and eight questions out of the 80 questions on the PE Civil Construction exam. The organization of this chapter follows the order of subtopics given by NCEES for this knowledge area. Each subtopic is covered in the following sections.

A. CONSTRUCTION SEQUENCING

Key concepts: These key concepts are important for answering exam questions in subtopic 12.A, Construction Sequencing.

- design sequencing
- fast tracking
- Gantt charts
- precedence diagrams

***NCEES Handbook*:** To prepare for this subtopic, familiarize yourself with these sections in the *Handbook*.

- CPM Precedence Relationships

***PE Civil Reference Manual* (CERM):** Study these sections in CERM that either relate directly to this subtopic or provide background information.

- Section 86.3: Scheduling
- Section 86.4: Resource Leveling

***Construction Depth Reference Manual* (CECN):** Study these sections in CECN that either relate directly to this subtopic or provide background information.

- Section 4.1: Introduction to Scheduling
- Section 4.2: CPM Network Analysis Testing

The following equations, figures, and concepts are relevant for subtopic 12.A, Construction Sequencing.

Precedence Diagrams

Handbook: CPM Precedence Relationships

CECN: Sec. 4.2

Construction sequencing can be performed using precedence diagrams. This method uses boxes to represent activities. A precedence diagram can be used to represent many more relationships between activities than can be represented using a network diagram. CECN Fig. 4.9 shows the NCEES format for representing the values used in precedent diagrams.

Figure 4.9 Precedence Diagram Values

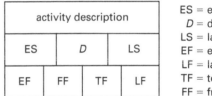

ES = earliest start
D = duration
LS = latest start
EF = earliest finish
LF = latest finish
TF = total float
FF = free float

There are three possible relationships between a preceding activity and a succeeding activity.

finish-to-start: In a finish-to-start relationship, the succeeding activity can begin only after the preceding activity has finished.

start-to-start: In a start-to-start relationship, the succeeding activity can start only after the preceding activity has started.

finish-to-finish: In a finish-to-finish relationship, the succeeding activity can finish only after the preceding activity has finished.

With any of these relationships, a lag time may also be required (as illustrated in CECN Fig. 4.7 and Fig. 4.8).

Figure 4.7 Start-to-Start Diagram

The start of B depends on the start of A, plus lag time.

Figure 4.8 Finish-to-Finish Diagram

The finish of B depends on the finish of A, plus lag time.

The early finish (EF) is the earliest possible finish date of an activity. It is calculated as part of a forward pass and is the earliest start plus the activity's duration.

$$EF = ES + D$$

The late start (LS) is the latest possible start of an activity. It is calculated as part of a backward pass. The LS will indicate the latest time an activity can be started so that the total project schedule is not affected. It is the latest finish minus the activity's duration.

$$LS = LF - D$$

Gantt Charts

CECN: Sec. 4.1

A Gantt chart allows a project manager to identify the activities in a project and the durations of these activities. A project manager will often use computer software to build a Gantt chart for a project. See CERM Fig. 86.3 for an example of a Gantt chart.

One advantage of a Gantt chart is that the chart is simple to make and understand. A disadvantage is that the chart cannot show all the sequences and dependencies of activities.

The process for working with a Gantt chart is as follows.

1. Use a work breakdown structure (WBS) to list activities.
2. Identify task relationships.
   - Sequential (linear) tasks are dependent activities.
   - Parallel tasks can be done concurrently.
3. Input activities into the chart.
4. Estimate the duration of activities.
5. Monitor progress.
6. Update the chart as needed.

Exam tip: Problems on the exam that involve Gantt charts will require students to understand the general concepts of a Gantt chart.

Design Sequencing Phases

CECN: Sec. 4.1

Design sequencing refers to a scheduling method in which construction on a project can begin as soon as the designs for a portion of the project are finished. The phases of design are

1. schematic design
2. design development
3. construction documents
4. bidding
5. construction

Construction sequencing is the responsibility of a project's construction engineer and is the process of creating a work schedule in which activities that disturb the land on the job site (e.g., earthwork) are timed properly with the installation of structures used for erosion and sedimentation control.

Earthwork is generally the most problematic part of the job, as it involves many variables for design and construction activities. CERM Table 86.4 provides a good example of the construction sequencing for a project that involves earthwork activities, specifically involving erosion and sedimentation control activities.

Fast Tracking

CERM: Sec. 86.3

Fast tracking is a project delivery strategy in which construction is begun before the design is complete. The purpose is to shorten the time to completion, which reduces the overall costs of the construction project. It requires close coordination between the architect, contractor, subcontractors, owner, and others. It can save time and money. For example, ordering the steel for a steel building and having it fabricated while the earthwork is being performed means that the steel arrives on site and is ready to be erected as the concrete finishes curing.

Fast tracking is really just taking advantage of critical path and noncritical activities. Benefits of fast tracking include

- shorter schedules, which reduce time to completion, costs of construction, and overhead costs
- project costs that are less likely to be affected by inflation

The risks of fast tracking include

- more complex management than the traditional design-bid-build process
- less certainty concerning final project cost

- early design decisions potentially conflicting with later needs of the project
- potential added time and cost due to reduced communication during the project

B. ACTIVITY TIME ANALYSIS

Key concepts: These key concepts are important for answering exam questions in subtopic 12.B, Activity Time Analysis.

- activity-on-arrow diagrams
- activity-on-node diagrams
- calculating scheduling information
- precedence diagrams

NCEES Handbook: To prepare for this subtopic, familiarize yourself with these sections in the *Handbook*.

- CPM Precedence Relationships

PE Civil Reference Manual (CERM): Study these sections in CERM that either relate directly to this subtopic or provide background information.

- Section 86.5: Activity-on-Node Networks
- Section 86.6: Solving a CPM Problem
- Section 86.7: Activity-on-Arc Networks

Construction Depth Reference Manual (CECN): Study these sections in CECN that either relate directly to this subtopic or provide background information.

- Section 4.2: CPM Network Analysis Testing

The following figures and concepts are relevant for subtopic 12.B, Activity Time Analysis.

Activity-on-Node Diagram

Handbook: CPM Precedence Relationships

CERM: Sec. 86.5, Sec. 86.6

CECN: Sec. 4.2

The most common type of network diagram is the activity-on-node diagram, which shows the activity duration on the node. *Handbook* section CPM Precedence Relationships contains activity-on-node diagrams for start-to-start, finish-to-finish, and finish-to-start precedence relationships, as well as the components that an activity has related to its position on the critical path. CECN Fig. 4.5 contains a more complex activity-on-node diagram.

Figure 4.5 Activity-on-Node Network Diagram

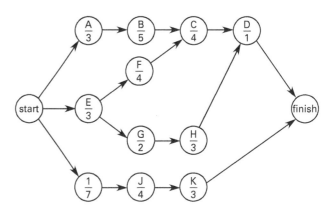

As shown in CECN Fig. 4.5, an activity is represented by a node. The node (activity) has a duration and an activity description. The arrow indicates only dependency and does not have any duration. The start node indicates the start of the project, and it does not have any duration. The finish node indicates the finish of the project, and it also does not have any duration. These nodes are also known as *dummy nodes*.

Activity-on-Arrow Diagram

Handbook: CPM Precedence Relationships

CERM: Sec. 86.7

CECN: Sec. 4.2

Activity-on-arrow diagrams are most commonly used to show a schedule and time calculations graphically. Each activity is represented by an arrow with an associated description given above the tail of the arrow and an expected duration given as a number below the arrow. *Handbook* section CPM Precedence Relationships contains an activity-on-arrow diagram. The event chronology is shown by the arrow direction. Nodes indicate milestones. Arrows have duration, but the nodes do not. CECN Fig. 4.2 shows a more complex example of an activity-on-arrow diagram.

Figure 4.2 Activity-on-Arrow Terminology

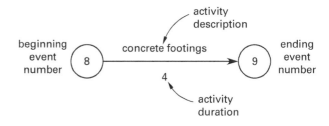

C. CRITICAL PATH METHOD NETWORK ANALYSIS

Key concepts: These key concepts are important for answering exam questions in subtopic 12.C, Critical Path Method Network Analysis.

- activity-on-node
- critical path conditions
- free float
- precedence relationships
- total float

NCEES Handbook: To prepare for this subtopic, familiarize yourself with these sections in the *Handbook*.

- CPM Precedence Relationships
- Lead and Lag Relationships
- Resource Scheduling and Leveling

PE Civil Reference Manual (**CERM**): Study these sections in CERM that either relate directly to this subtopic or provide background information.

- Section 86.4: Resource Leveling
- Section 86.5: Activity-on-Node Networks
- Section 86.6: Solving a CPM Problem

Construction Depth Reference Manual (**CECN**): Study these sections in CECN that either relate directly to this subtopic or provide background information.

- Section 4.2: CPM Network Analysis Testing
- Section 4.4: Resource Scheduling

The following equations and concepts are relevant for subtopic 12.C, Critical Path Method Network Analysis.

Network Diagrams

Handbook: CPM Precedence Relationships

CERM: Sec. 86.5

CECN: Sec. 4.2

Network diagrams are used on projects with complex schedules in which multiple tasks can occur simultaneously. The main advantage of network diagrams is that they show the relationship between activities. Network diagrams are used to schedule activities, monitor construction progress, show interrelationships between activities, and show the interdependencies between activities. There are two types: activity-on-node diagrams and activity-on-arrow diagrams. *Handbook* section **CPM Precedence Relationships** illustrates how these network diagrams work within a critical path method network analysis. See the entry "Activity Time Analysis" in this chapter for more information.

Total Float

Handbook: CPM Precedence Relationships

CERM: Sec. 86.5

CECN: Sec. 4.2

$$TF = LF - EF$$
$$TF = LS - ES$$

The total float is the difference between the latest date an activity can finish and the earliest date it can finish. The total float value for an activity is the amount of extra time available to complete that activity without delaying project completion.

Free Float

Handbook: CPM Precedence Relationships

CECN: Sec. 4.2

$$FF = ES - EF$$

The free float value is the amount of time an activity can be delayed without taking float away from its successor activities. The free float of the final activity is always zero days, because it has no successor activities. If there is a lag time between the given activity and successor activities, that lag time is subtracted from the free float value.

Critical Path

Handbook: CPM Precedence Relationships; Lead and Lag Relationships

CERM: Sec. 86.6

CECN: Sec. 4.2

The critical path is the longest continuous path with zero float. It shows the minimum project duration. The critical path always includes the first and last activities in the project. An activity being delayed (with free float) will not necessarily delay the project, but an activity that runs out of total float will delay the project. *Handbook* section **CPM Precedence Relationships** contains diagrams and equations useful to showing the critical path in a schedule and how to mathematically determine activities on the critical path. For an activity to be on the critical path, CECN Eq. 4.8 through Eq. 4.10 must all be true.

$$ES_c = LS_c \quad 4.8$$

$$EF_c = LF_c \quad 4.9$$

$$TF_c = FF_c = 0 \quad 4.10$$

Exam tip: Critical path activities are those activities that determine the minimum overall duration of the schedule.

Resource Leveling

Handbook: Resource Scheduling and Leveling

CERM: Sec. 86.4

CECN: Sec. 4.4

Resource leveling is used to address overallocation. The limited resources are usually people and equipment, but project funding may also be limited if it becomes available in stages. If resource leveling is used with activities on a project's critical path, the project's completion date will inevitably be delayed.

D. RESOURCE SCHEDULING AND LEVELING

Key concepts: These key concepts are important for answering exam questions in subtopic 12.D, Resource Scheduling and Leveling.

- construction activity sequencing
- crew distribution chart
- linear scheduling
- resource leveling tools

NCEES Handbook: To prepare for this subtopic, familiarize yourself with these sections in the *Handbook*.

- Resource Scheduling and Leveling
- Lead and Lag Relationships

PE Civil Reference Manual (**CERM**): Study these sections in CERM that either relate directly to this subtopic or provide background information.

- Section 86.3: Scheduling
- Section 86.4: Resource Leveling

Construction Depth Reference Manual (**CECN**): Study these sections in CECN that either relate directly to this subtopic or provide background information.

- Section 4.4: Resource Scheduling

The following figures and concepts are relevant for subtopic 12.D, Resource Scheduling and Leveling.

Resource Leveling

Handbook: Resource Scheduling and Leveling

CERM: Sec. 86.4

CECN: Sec 4.4

Resource leveling is a way of determining when activities should be performed based on materials, labor, and equipment. It smooths demand for resources and avoids exceeding given resource limits. It is used to address overallocation (i.e., situations that demand more resources than are available). Two common approaches are used to level resources.

- Tasks can be delayed until resources become available.
- Tasks can be split so that the parts are completed when planned, and the remainders can be completed when resources become available.

Resource Scheduling

Handbook: Resource Scheduling and Leveling

CERM: Sec. 86.4

CECN: Sec 4.4

Resource scheduling is the process of tabulating project resource demands over time. Resource scheduling examines resource requirements during specific periods of the project and attempts to minimize the variations in resource demand, either to improve efficiency or to reflect reality.

Many variables can affect construction time. Most can be controlled, but others, such as weather, are beyond anyone's control. Material delivery times and labor availability have a large impact on construction schedules. Start and end dates that can be adjusted are those for the different activities and not necessarily for the entire project's dates. Adjustment of the project's start and completion dates may be necessary, however, based on availability of resources, weather, and so on.

Lead and Lag Relationships

Handbook: Lead and Lag Relationships

In lead and lag relationships, activities A and B have a finish-to-start relationship with lead or lag time. The figure in *Handbook* section Lead and Lag Relationships shows how scheduling works for these relationships.

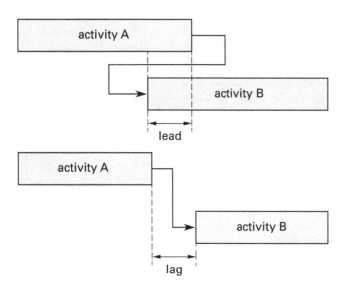

If the lag time is zero, activity B can be started immediately upon completion of activity A. If the lag time between activity A and activity B is positive, activity B can be started once activity A is completed and the lag time has passed. If the lag time between activity A and activity B is negative, activity B can be started prior to the completion of activity A. The *Handbook* refers to negative lag time between the start of two activities as lead time, as shown in the *Handbook* figure.

Crew Distribution Charts

CECN: Sec. 4.4

A *crew distribution chart* is a bar graph that shows the number of workers required for each activity in the project. A crew distribution chart shows the number of crew members required on any given day based on the number of members needed for each activity on that day. CECN Fig. 4.12 shows a time-scaled network and its corresponding crew distribution chart.

Figure 4.12 Time-Scaled Network and Crew Distribution Chart

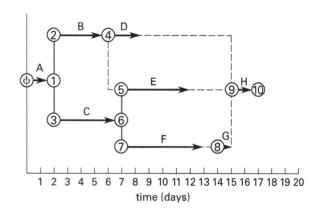

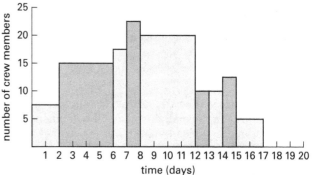

A time-scaled network represents the duration of each activity in relation to the project's start; that is, the cumulative time since the project's start using the duration of each activity. Crew distribution charts are commonly used in the construction industry.

Linear Scheduling

A *linear schedule* graphs the project schedule (duration) versus the length of the project, whether it is stationing (such as a highway) or height (such as a high-rise building). The linear schedule is also known as a *time versus distance diagram* (T-D chart).

Construction Activity Sequencing

CERM: Sec. 86.3

A project manager should confirm that the general construction schedule and the construction sequencing schedule are compatible. Key construction activities and associated erosion and sedimentation control (ESC) measures are shown in CERM Table 86.4. This table describes the process for construction activities. See the entry "Construction Sequencing" in this chapter for more detailed information.

E. TIME-COST TRADE-OFF

Key concepts: These key concepts are important for answering exam questions in subtopic 12.E, Time-Cost Trade-Off.

- cost per unit of time reduction
- schedule crashing
- schedule versus cost relationship

NCEES Handbook: To prepare for this subtopic, familiarize yourself with these sections in the *Handbook*.

- Time-Cost Trade-Off

PE Civil Reference Manual **(CERM):** Study these sections in CERM that either relate directly to this subtopic or provide background information.

- Section 86.3: Scheduling

Construction Depth Reference Manual **(CECN):** Study these sections in CECN that either relate directly to this subtopic or provide background information.

- Section 4.2: CPM Network Analysis Testing
- Section 4.3: Earned Value Method
- Section 4.5: Time-Cost Trade-off

The following equations and concepts are relevant for subtopic 12.E, Time-Cost Trade-Off.

Time-Cost Trade-Off

Handbook: Time-Cost Trade-Off

CERM: Sec. 86.3

CECN: Sec. 4.5

The essence of a time-cost trade-off is that cost increases as project time is decreased and vice versa. The project cost is the sum of the direct and indirect costs. Direct costs are connected to a specific construction activity. Indirect costs cannot be directly attributed to a construction activity. Instead, they are costs that help maintain the functioning of a company.

Trade-offs include liquidated damages and completion incentives. Liquidated damages are penalties a contractor pays for not completing the project or a certain milestone on time. Completion incentives are usually financial.

Crashing the Schedule

CECN: Sec. 4.5

Compressing or *crashing* a project schedule is a method of accelerating project activities to reduce project time. The *crash time* of an activity is the least amount of time in which a project can be completed. Shortening the time to complete an activity results in additional costs (e.g., overtime pay), but it may help reduce overall project costs. A schedule is crashed only after alternatives are analyzed to determine how to get the maximum schedule compression for the least additional cost.

Schedule acceleration can be achieved by reducing activity durations or adjusting the sequencing of activities. Usually, schedule acceleration comes with a higher cost.

Cost per Unit of Time Reduction

CECN: Sec. 4.5

$$s = \frac{C_{cr} - C_n}{t_n - t_{cr}} \qquad 4.18$$

The slope of each cost-over-time trade-off for an activity can be determined using CECN Eq. 4.18.

Schedule Variance

CECN: Sec. 4.3

$$SV = BCWP - BCWS \qquad 4.12$$

Schedule variance (SV) is the difference between the value of work accomplished for a given period and the value of the work planned. It measures how much a project is ahead of or behind schedule. Schedule variance is measured in value (e.g., dollars), not time.

14 Material Quality Control and Production

Content in blue refers to the *NCEES Handbook*.

A. Material Properties and Testing....................... 14-1
B. Weld and Bolt Installation.............................. 14-2
C. Quality Control Process 14-5
D. Concrete Proportioning and Placement............. 14-6
E. Concrete Maturity and Early Strength
    Evaluation 14-8

The knowledge area of Material Quality Control and Production makes up between five and eight questions out of the 80 questions on the PE Civil Construction exam. The organization of this chapter follows the order of subtopics given by NCEES for this knowledge area. Each subtopic is covered in the following sections.

A. MATERIAL PROPERTIES AND TESTING

Key concepts: These key concepts are important for answering exam questions in subtopic 13.A, Material Properties and Testing.

- asphalt-concrete properties
- concrete types and testing
- Marshall mix design method
- soil testing

***NCEES Handbook*:** To prepare for this subtopic, familiarize yourself with these sections in the *Handbook*.

- Concrete Testing

***PE Civil Reference Manual* (CERM):** Study these sections in CERM that either relate directly to this subtopic or provide background information.

- Section 35.2: Soil Classification
- Section 48.1: Concrete
- Section 48.2: Cementitious Materials
- Section 48.3: Aggregate
- Section 48.4: Water
- Section 48.5: Admixtures
- Section 48.6: Slump
- Section 48.7: Density
- Section 48.8: Compressive Strength

- Section 76.13: Hot Mix Asphalt Concrete Mix Design Methods
- Section 76.14: Marshall Mix Design
- Section 76.15: Marshall Mix Test Procedure
- Section 76.16: Hveem Mix Design
- Section 76.17: Superpave

***Construction Depth Reference Manual* (CECN):** Study these sections in CECN that either relate directly to this subtopic or provide background information.

- Section 5.1: Quality Control
- Section 5.2: Material Testing

The following equations, figures, and concepts are relevant for subtopic 13.A, Material Properties and Testing.

Concrete Testing

Handbook: Concrete Testing

CERM: Sec. 48.1, Sec. 48.8

CECN: Sec. 5.1

Handbook section Concrete Testing provides information on concrete testing, with particular focus on how strength test samples are taken and results are interpreted. As shown in CERM Eq. 48.1, the compressive strength of concrete of a sample can be found from the maximum axial load and the cross-sectional area of the sample.

$$f'_c = \frac{P}{A} \qquad 48.1$$

Soil Testing

CERM: Sec. 35.2

CECN: Sec. 5.2

Soil tests are performed on soils to determine their plasticity, dry density, water content, and compaction, and to ensure that the required soil specifications are met.

The sand-cone test measures the dry density and water content, and is only applicable for soil without appreciable amounts of rock or coarse materials of a diameter greater than 1.5 in. The water content of a sample can be determined using CECN Eq. 5.1.

$$w = \frac{W_t - W_d}{W_d} \quad \quad 5.1$$

The weight of the sand is converted to volume using CECN Eq. 5.2. The dry density of a sample can be determined using CECN Eq. 5.3.

$$V_h = \frac{W_{\text{test}} g_c}{\rho_{\text{test}} g} \quad \quad 5.2$$

$$\rho = \frac{W_s g_c}{V_h g} \quad \quad 5.3$$

The Proctor test measures the compaction; it is the only laboratory compaction test accepted by highway departments and other local agencies. The test is performed in accordance with *Standard Test Methods for Laboratory Compaction Characteristics of Soil Using Modified Effort* (ASTM Standard D1557) or *Standard Method of Test for Moisture-Density Relations of Soils Using a 4.54-kg (10-lb) Rammer and a 457-mm (18-in.) Drop* (AASHTO T180). The nuclear gauge field measures the water content and density by using a portable nuclear gauge. This method is illustrated in CECN Fig. 5.1.

Figure 5.1 Nuclear Gauge Field Test

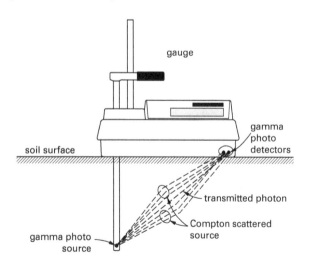

Asphalt-Concrete

CERM: Sec. 76.13, Sec. 76.14, Sec. 76.15, Sec. 76.16, Sec. 76.17

CECN: Sec. 5.2

Asphalt-concrete (AC) is a material mix used as a flexible pavement; it is also referred to as *hot mix asphalt* (HMA). It consists of well-graded aggregates, filler (usually fine aggregate), bituminous binder (4–8%), air voids, and high-temperature mix and placement. The aggregates in AC are chosen for friction properties and durability and to prevent rutting.

AC can be mixed using one of three common methods: the Marshall mix design method, the Hveem mix design method, and the Superpave method.

The Marshall mix design method consists of three basic steps: aggregate selection, asphalt binder selection, and determination of the optimum asphalt binder content. The Hveem mix design method is similar to the Marshall method, with the addition of measuring resistance to shear and considering asphalt absorption by aggregates. The Superpave method is a volumetric mix design process developed to characterize asphalt binders more accurately than the Marshall method, based on the idea that a binder's properties should be related to the conditions where it will be used, including expected climatic conditions and aging considerations.

All asphalt mixes used for infrastructure construction projects must be inspected prior to placement. The roller pass/control strip method and the determination of in-place density are two ways to test asphalt in the field. Visual observations and testing for density are used to determine in-place density, using the guidelines in *Standard Test Method for Density of Bituminous Concrete in Place by Nuclear Methods* (ASTM Standard D2950).

B. WELD AND BOLT INSTALLATION

Key concepts: These key concepts are important for answering exam questions in subtopic 13.B, Weld and Bolt Installation.

- bolt and weld inspection procedures
- bolt testing
- recurring problems with welds
- types of welded joints
- types of welds
- weld symbols
- welding methods

NCEES Handbook: To prepare for this subtopic, familiarize yourself with these sections in the *Handbook*.

- Basic Welding Symbols

***PE Civil Reference Manual* (CERM)**: Study these sections in CERM that either relate directly to this subtopic or provide background information.

- Section 45.11: Bolts
- Section 45.15: Fillet Welds
- Section 66.1: Introduction
- Section 66.2: Types of Welds and Joints
- Section 66.3: Fillet Welds

***Construction Depth Reference Manual* (CECN)**: Study these sections in CECN that either relate directly to this subtopic or provide background information.

- Section 5.3: Welding and Bolt Testing

Codes and standards: To prepare for this subtopic, familiarize yourself with these sections from codes and standards.

Steel Construction Manual (AISC)

- Part 8: Design Considerations for Welds
- Part 16.1: Specification for Structural Steel Buildings
- Part 16.2: Specification for Structural Joints Using High-Strength Bolts
- Section J2.2: Fillet Welds
- Section J2.4: Strength

The following equations, figures, and tables are relevant for subtopic 13.B, Weld and Bolt Installation.

Weld Symbols

Handbook: Basic Welding Symbols

CERM: Sec. 66.2

AISC: Part 8

Handbook table Basic Welding Symbols illustrates the different types of weld symbols and locations specified by the American Welding Society and provides explanations of how to read weld symbols and labels. This table is directly copied from AISC *Steel Construction Manual* Table 8-2. CERM Fig. 66.3 is an equivalent illustration.

Weld Designations

CERM: Sec. 66.1

Two processes are most commonly used for structural welding, each with its own designations.

The designations for shielded metal arc welding (SMAW) take the form E60XX, E70XX, E80XX, E90XX, E100XX, or E110XX. The E denotes an electrode, the number denotes the electrode's ultimate tensile strength in units of ksi, and the XX indicates how the electrode is used. The first X indicates the welding position the electrode can be used in, and the second X indicates the coating, penetration, and current type used.

The designations for submerged arc welding (SAW) are a combination of a flux and an electrode classification like those used in SMAW. For example, the designation F7X-E7XX is for a granular flux where the tensile strengths of the weld and the electrode are both 70 ksi.

Welds and Joints

CERM: Sec. 66.2

Different weld types are used for different applications. The weld type is usually determined by the joint that needs to be welded. CERM Fig. 66.1 illustrates the types of welds.

Figure 66.1 Types of Welds

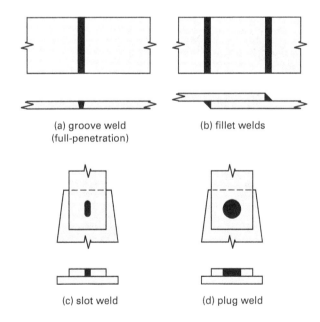

(a) groove weld (full-penetration)

(b) fillet welds

(c) slot weld

(d) plug weld

There are also different types of welded joints (see CERM Fig. 66.2). The joint type used depends on different factors, such as loading. A lap joint is most common because of the ease of fitting the joint.

Figure 66.2 Types of Welded Joints

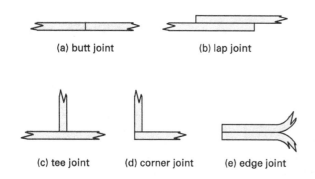

Nondestructive Weld Inspections

CECN Sec. 5.3

Nondestructive examinations are performed after welding and visual inspections are complete. These tests include liquid penetrant inspection (PT), magnetic particle inspection (MT), radiographic inspection (RT), and ultrasonic inspection (UT). CECN Table 5.4 lists the different nondestructive inspection methods, along with the coordinating equipment required, problems detected, advantages, limitations, and additional remarks.

Bolt Testing

CERM: Sec. 45.11

CECN: Sec. 5.3

AISC: Sec. 16.2

Before being installed, bolts need to be verified. There are three methods for testing bolts used in construction: the calibrated wrench method, the turn-of-nut method, and the load-indicating bolt method. AISC Part 16.2 explains the procedures and requirements for each bolt test.

Fillet Welds

CERM: Sec. 45.15, Sec. 66.3

AISC: Part 16.1 Sec. J2.2, Part 16.1 Sec. J2.4

CERM Fig. 66.4 illustrates a fillet weld.

Figure 66.4 Fillet Weld

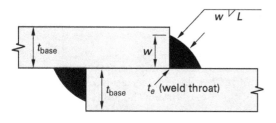

The effective weld throat thickness, t_e, for a shielded metal arc welding (SMAW) process can be found using CERM Eq. 66.1.

$$t_e = 0.707w \quad \quad 66.1$$

The available shear strength on the weld throat can be found from AISC Table J2.5 and CERM Eq. 66.2.

$$F_v = 0.60 F_{u,\text{rod}} \quad \quad 66.2$$

The shear stress in a concentrically loaded fillet weld group can be found from CERM Eq. 66.3. Shear and tensile strengths of the weld may not be greater than the nominal strength per unit area of the base metal, or the nominal strength of the weld metal per unit area times the area of the weld, per AISC Part 16.1 Sec. J2.4.

$$f_v = \frac{P}{A_w} = \frac{P}{l_w t_e} \quad \quad 66.3$$

The shear resistance per unit length of the weld is found from CERM Eq. 66.4.

$$R_w = 0.30 t_e F_{u,\text{rod}} \quad \quad 66.4$$

Special restrictions apply to fillet welds. These are specified in AISC Table J2.4 and summarized in CERM Table 66.1.

Table 66.1 Minimum Fillet Weld Size

thickness of thinner part joined (in)	minimum weld size,* w (in)
to $\frac{1}{4}$ inclusive	$\frac{1}{8}$
over $\frac{1}{4}$ to $\frac{1}{2}$ inclusive	$\frac{3}{16}$
over $\frac{1}{2}$ to $\frac{3}{4}$ inclusive	$\frac{1}{4}$
over $\frac{3}{4}$	$\frac{5}{16}$

(Multiply in by 25.4 to obtain mm.)
*Leg dimension of fillet weld. Single pass welds must be used.

Source: From *AISC Specification* Table J2.4.

C. QUALITY CONTROL PROCESS

Key concepts: These key concepts are important for answering exam questions in subtopic 13.C, Quality Control Process.

- analysis and trending
- documentation
- inspection categories
- quality control processes
- root cause analysis
- testing types
- tracking

NCEES Handbook: To prepare for this subtopic, familiarize yourself with these sections in the *Handbook*.

- Nondestructive Test Methods
- Nondestructive Test Methods for Concrete

Construction Depth Reference Manual (CECN): Study these sections in CECN that either relate directly to this subtopic or provide background information.

- Section 5.1: Quality Control

The following tables and concepts are relevant for subtopic 13.C, Quality Control Process.

Testing for Quality Assurance

Handbook: Nondestructive Test Methods; Nondestructive Test Methods for Concrete

CECN: Sec. 5.1

Testing for quality assurance includes in-place testing of structural and architectural features, destructive testing of samples or portions of the installation, performance or service testing of equipment and systems, and nondestructive testing of welds and concrete.

In-place testing of structural and architectural features verifies that the proper construction methods are followed and that the installations have the capacities required to perform as needed when construction is complete.

Performance and service testing verifies the integrity or performance of equipment and systems.

Destructive testing is used to measure the point of failure of materials or installations, or to determine the chemical and physical properties of materials.

Nondestructive testing of welds and concrete detects cracks, voids, surface imperfections, and other structural issues that may not be obvious from visual inspection of the installation. *Handbook* section Nondestructive Test Methods explains reasons for and goals of using nondestructive tests, and *Handbook* table Nondestructive Test Methods for Concrete provides a summary of primary and secondary nondestructive testing methods for concrete.

Quality and Quality Control

CECN: Sec. 5.1

Quality is subjective; the quality required on a project depends on the contractor's policies and core values as well as the requirements from the client. The client's requirements can be obtained from the project specifications and industry standards, and also from discussions with the client to understand their expectations.

Quality assurance includes all planned and systematic actions necessary to provide confidence that a product or facility will perform satisfactorily in its role. It is an all-encompassing term that includes quality control, independent assurance, and acceptance.

Quality control (also known as *process control*) is the set of actions and considerations necessary to assess production and construction processes and control the level of quality in an end product. It typically includes sampling and testing by the contractor.

Exam tip: Be familiar with the differences between quality, quality control, and quality assurance.

Quality Control Stages

CECN: Sec. 5.1

The stages of the quality control (QC) process are startup, planning, execution, and closeout. During startup and planning, the project quality program is developed, and the requirements for the project are determined from specifications, plan sheets, preactivity meetings, and so on. During project execution, all quality-related documentation is gathered, and the owner agrees that the project adheres to the construction documentation and any agreed-on modifications. Material receiving and storage are performed to ensure that all ordered materials are free from damage. During closeout, punch lists are finalized, reporting documentation is gathered, and the project is turned over to the client.

Inspection

CECN Sec. 5.1

Inspection is a series of examinations meant to determine whether a material, project, or facility conforms to requirements, including material types and quantities, construction methods and processes, and dimensional checks.

Other Quality Measures

CECN: Sec. 5.1

Tracking verifies the use of appropriate materials in specific portions of the installation and requires identification and tagging of materials by item or lot.

Documenting provides a final record of quality, and generally includes all test results and an indication of acceptance by an authorized individual.

Analysis and *trending* include documenting and developing numerical summaries of problems, as well as providing statistical analysis of test results, inspection results, or problems with specific operations.

D. CONCRETE PROPORTIONING AND PLACEMENT

Key concepts: These key concepts are important for answering exam questions in subtopic 13.D, Concrete Proportioning and Placement.

- absolute volume method
- concrete mix design
- mix proportions by weight
- volumetric method
- weight method

NCEES Handbook: To prepare for this subtopic, familiarize yourself with these sections in the *Handbook*.

- Mixture Proportioning

PE Civil Reference Manual **(CERM):** Study these sections in CERM that either relate directly to this subtopic or provide background information.

- Section 49.4: Water-Cement Ratio and Cement Content
- Section 49.5: Proportioning Mixes
- Section 49.6: Absolute Volume Method
- Section 49.7: Adjustments for Water and Air

Construction Depth Reference Manual **(CECN):** Study these sections in CECN that either relate directly to this subtopic or provide background information.

- Section 5.4: Concrete Mix Design

The following equations and concepts are relevant for subtopic 13.D, Concrete Proportioning and Placement.

Concrete Mix Design

Handbook: Mixture Proportioning

$$\begin{aligned}\text{yield} &= \text{volume of fresh concrete produced in a batch} \\ &= \frac{\text{total mass of batched materials}}{\text{density of freshly mixed concrete}} \\ &= \text{sum of absolute volumes of concrete ingredients}\end{aligned}$$

CECN: Sec. 5.4

$$y_c = \frac{W_{m,t}}{\gamma_c} \qquad 5.5$$

CECN Eq. 5.5 is similar to the equation provided in the *Handbook*, except that the CECN equation uses variables instead of plain text descriptions of the values.

A concrete mix design determines the most economical and practical combination of available materials to produce a concrete that will satisfy the performance requirements for a construction project under the anticipated conditions of use. The strength of a concrete mix is largely defined by its water-to-cement ratio, the weight of water in the mix divided by the weight of cement in the mix. The yield is calculated from the total weight of materials used in the concrete mix and the unit weight of the mix, as shown in the *Handbook* equation and CECN Eq. 5.5.

The amount of cement used for a construction project is measured in sacks. One sack of cement is 1 ft³ of cement weighing 94 lbf. The dry weight of the coarse or fine aggregates, or both, in a mix can be found using CECN Eq. 5.6.

$$W_d = \frac{V_{a,t}(\text{percentage of composition})\text{SG}_a \gamma_w}{100\%} \qquad 5.6$$

The weight of the water in the aggregates of a mix can be found from CECN Eq. 5.8, given the batch weight of the aggregates in the mix.

$$W_w = \left(\frac{w_{\text{fine}}}{1+w_{\text{fine}}}\right) W_{\text{batch,fine}} \\ + \left(\frac{w_{\text{coarse}}}{1+w_{\text{coarse}}}\right) W_{\text{batch,coarse}} \qquad 5.8$$

Exam tip: For purposes of concrete mix calculations, the unit weight of water is always assumed to be 62.4 lbf/ft^3 or 8.34 lbf/gal.

Exam tip: Make sure that you use the correct units for the density or specific gravity of water, given what data is available for the problem.

Absolute Volume Method

Handbook: Mixture Proportioning

$$\text{absolute volume} = \frac{\text{mass of loose material}}{\begin{pmatrix} \text{relative density or} \\ \text{specific gravity of material} \\ \cdot \text{ density of water} \end{pmatrix}}$$

CERM: Sec. 49.6

$$V_{\text{absolute}} = \frac{m}{(\text{SG})\rho_{\text{water}}} \quad [\text{SI}] \quad 49.1(a)$$

$$V_{\text{absolute}} = \frac{W}{(\text{SG})\gamma_{\text{water}}} \quad [\text{U.S.}] \quad 49.1(b)$$

CECN: Sec. 5.4

$$V_{c,\text{ab}} = \frac{W_c}{\gamma_w \text{SG}_c} \quad 5.12$$

$$V_{w,\text{ab}} = \frac{W_w}{\gamma_w \text{SG}_w} \quad 5.13$$

CERM Eq. 49.1 is similar to the *Handbook* equation, but it is shown using nomenclature instead of plain text and includes separate equations for SI and customary U.S. units. CECN Eq. 5.12 and Eq. 5.13 are likewise very similar to the *Handbook*, but are used to calculate the absolute volume of cement and absolute volume of water in a mix, respectively.

The standard practice for selecting proportions for a concrete mix is the absolute volume method. As shown in the *Handbook* equation, the absolute volume of a concrete mix is the mass of loose material in the mix divided by the product of the relative density or specific gravity of the material and the density of water.

Mix Proportions by Weight

Handbook: Mixture Proportioning

$$\text{total moisture (\%)} = \frac{\text{wet weight} - \text{oven dry weight}}{\text{oven dry weight}} \cdot 100\%$$

$$\text{absorbed moisture (\%)} = \frac{\text{SSD weight} - \text{oven dry weight}}{\text{oven dry weight}} \cdot 100\%$$

$$\text{free moisture (\%)} = \text{total moisture (\%)} - \text{absorbed moisture (\%)}$$

$$\text{water-cementitious material ratio} = \frac{w}{cm}$$
$$= \frac{\text{mass of water}}{\text{mass of cementitious materials}}$$

CERM: Sec. 49.5

CECN: Sec. 5.4

A design specification will often be given in terms of mix proportion. The mix proportions are indicated as a ratio, with the numbers indicating the weight of cement, the dry weight of fine aggregate, and the dry weight of coarse aggregate. For example, a mix proportion of 1:2.5:3.25 indicates that for every 1 lbf of cement, there are 2.5 dry lbf of fine aggregate and 3.25 dry lbf of coarse aggregate.

The saturated, surface-dry (SSD) condition occurs when the aggregate holds as much water as it can without trapping any free water between the aggregate particles.

Proportioning by weight requires knowledge of the aggregate's absorption, the amount of water the aggregate will absorb during the mixing process, and the free moisture/water content. The total weight of aggregates in a mix, including the dry weight of aggregates and the free moisture, can be found using CECN Eq. 5.17.

$$W_{t,a} = (1 + w) W_{d,a} \quad 5.17$$

The weight of free moisture in a mix can be determined using CECN Eq. 5.18.

$$W_w = W_{t,a} - W_{d,a} \qquad 5.18$$

To calculate the effect of absorption and determine the amount of water the aggregate will absorb, CECN Eq. 5.19 can be used.

$$W_{w,\text{ab}} = W_{d,a} w \qquad 5.19$$

The total weight of the aggregate can also be calculated using the dry weight of the aggregate and the weight of the free moisture, as shown in CECN Eq. 5.20.

$$W_{t,a} = W_{d,a} + W_{w,\text{ab}} \qquad 5.20$$

E. CONCRETE MATURITY AND EARLY STRENGTH EVALUATION

Key concepts: These key concepts are important for answering exam questions in subtopic 13.E, Concrete Maturity and Early Strength Evaluation.

- percentage of strength calculations
- strength-gain curves

NCEES Handbook: To prepare for this subtopic, familiarize yourself with these sections in the *Handbook*.

- Concrete Testing
- Concrete Maturity

PE Civil Reference Manual **(CERM):** Study these sections in CERM that either relate directly to this subtopic or provide background information.

- Section 48.8: Compressive Strength

Construction Depth Reference Manual **(CECN):** Study these sections in CECN that either relate directly to this subtopic or provide background information.

- Section 5.5: Concrete Maturity and Early Strength Evaluation

The following equations and figures are relevant for subtopic 13.E, Concrete Maturity and Early Strength Evaluation.

Concrete Strength Gain Factors

Handbook: Concrete Testing

CERM: Sec. 48.8

$$f'_c = \frac{P}{A} \qquad 48.1$$

CECN: Sec. 5.5

$$f'_c = \frac{P_{c,\max}}{A_c} \qquad 5.21$$

The water-cement ratio, the aggregate strength, and the bond all influence the rate at which concrete gains its strength. The strength the concrete has reached at certain time intervals is determined by testing two cylinders of each concrete mix to find the maximum axial load on each cylinder, and then using CERM Eq. 48.1 to calculate the strength. CECN Eq. 5.21 is an equivalent equation.

Concrete Maturity: Time-Temperature Factor Method

Handbook: Concrete Maturity

$$M = \sum_0^t (T - T_0) \Delta t$$

The *time-temperature factor method*, also known as the Nurse-Saul method, directly relates the strength of concrete to the temperature of the concrete. This method uses a specific datum temperature to calculate the time-temperature factor (also known as the maturity factor), M, as a function of that datum temperature, the elapsed time since the concrete began maturing, and the average temperature in the concrete sample during that time. This time-temperature factor is then related to the measured concrete flexural or compressive strength of the concrete to establish an estimated concrete strength at that time-temperature factor that can be used in future design stages.

Concrete Strength-Gain Curve

CECN: Sec. 5.5

In general, concrete gains strength as it cures. Concrete reaches about 50% strength at 3 days, 75% at 7 days, and about 90% in the first 28 days. It is important to keep concrete moist to allow for proper curing. See CECN Fig. 5.5 for an illustration of a concrete strength-gain curve.

Figure 5.5 Strength-Gain Curve

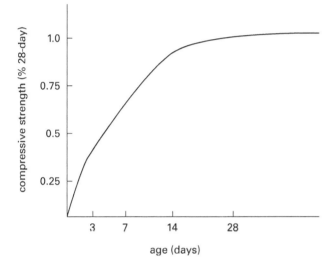

15 Temporary Structures

Content in blue refers to the *NCEES Handbook*.

A. Construction Loads, Codes, and Standards 15-1
B. Formwork 15-3
C. Falsework and Scaffolding............................... 15-7
D. Shoring and Reshoring 15-8
E. Bracing and Anchorage for Stability 15-9
F. Temporary Support of Excavation 15-11

The knowledge area of Temporary Structures makes up between six and nine questions out of the 80 questions on the PE Civil Construction exam. The organization of this chapter follows the order of subtopics given by NCEES for this knowledge area. Each subtopic is covered in the following sections.

A. CONSTRUCTION LOADS, CODES, AND STANDARDS

Key concepts: These key concepts are important for answering exam questions in subtopic 14.A, Construction Loads, Codes, and Standards.

- allowable strength design
- construction loads
- load and resistance factor design
- load combinations and load factors for strength design

NCEES Handbook: The *Handbook* does not contain any material relevant to this subtopic. Be sure to study the listed sections in CERM and CECN and from codes and standards.

***PE Civil Reference Manual* (CERM):** Study these sections in CERM that either relate directly to this subtopic or provide background information.

- Section 58.7: Methods of Design

***Construction Depth Reference Manual* (CECN):** Study these sections in CECN that either relate directly to this subtopic or provide background information.

- Section 6.1: Codes and Standards
- Section 6.2: Construction Loads
- Section 6.3: Environmental Loads

Codes and standards: To prepare for this subtopic, familiarize yourself with these sections from codes and standards.

Design Loads on Structures During Construction (ASCE 37)

- Section 2.2.3: Combinations Using Strength Design
- Section 2.3.1: Additive Combinations
- Section 6.2: Wind
- Section 6.3: Thermal Loads
- Section 6.4: Snow Loads
- Section 6.5: Earthquake
- Section 6.6: Rain
- Section 6.7: Ice

The following equations, tables, and concepts are relevant for subtopic 14.A, Construction Loads, Codes, and Standards.

Allowable Stress Design (ASD)

CERM: Sec. 58.7

CECN: Sec. 6.1

ASCE 37: Sec. 2.2.3

Allowable stress design (ASD) is one of two design approaches for temporary structures, the other being *load and resistance factor design* (LRFD). (See the entry "Load and Resistance Factor Design (LRFD)" in this chapter). ASD is the older method and assumes that members remain elastic.

ASD is also known as working stress design. ASD is usually preferred for use with temporary structures and wooden structures. One factor of safety is used, and the predicted stress is greater than the allowable stress. The factor of safety depends on building codes. The nominal load is the total load per load combinations prescribed by codes.

The maximum design load effect is obtained by comparing and analyzing the greatest of the prescribed load combinations. The equations are shown in ASCE 37 Sec. 2.2.3 and CECN Table 6.1. The factors applied to each load depend on the uncertainty of the load and are prescribed by codes. In LRFD design, material loads are divided into fixed material loads, P_{FML}, and variable material loads, P_{VML}. However, in ASD, material loads are lumped into one category, P_{ML}.

Table 6.1 ASD Basic Safety Factor Formulas

$P_D + P_d + P_{FML} + P_{VML} + P_L$

$P_D + P_d + P_{FML} + P_{VML} + P_p + P_h + P_L$

$P_D + P_d + P_{FML} + P_{VML} + 0.6W + P_p + P_L$

$P_D + P_d + P_{FML} + P_{VML} + 0.7E + P_p + P_L$

$P_D + P_d + (0.7E \text{ or } 0.6W)$

Adapted from *ASCE 37: Design Loads on Structures During Construction*, American Society of Civil Engineers, copyright © 2014.

Load and Resistance Factor Design (LRFD)

CERM: Sec. 58.7

CECN: Sec. 6.1

ASCE 37: Sec. 2.2.3

Load and resistance factor design (LRFD) is one of two design approaches for temporary structures, the other being *allowable stress design* (ASD). (See the entry "Allowable Stress Design (ASD)" in this chapter.)

LRFD is the more recently developed design approach and is used for a variety of materials, although it is typically used with concrete. LRFD uses two factors of safety: a resistance factor and a load factor.

The equations to calculate factored loads are shown in ASCE 37 Sec. 2.2.3 and CECN Table 6.2. The strength that the member is required to support is the greatest of the basic load combinations shown. Codes may require other combinations to be considered. The factors applied to each load depend on the uncertainty of the load; these factors are prescribed by codes. In LRFD, material loads are divided into fixed material loads, P_{FML}, and variable material loads, P_{VML}.

Table 6.2 LRFD Basic Load Factor Formulas

$1.4P_D + 1.4P_d + 1.2P_{FML} + 1.4P_{VML}$

$1.2P_D + 1.2P_d + 1.2P_{FML} + 1.4P_{VML} + 1.6P_L$

$1.2P_D + 1.2P_d + 1.2P_{FML} + 1.4P_{VML}$
$\quad + 1.6P_p + 1.6P_H + 0.5P_L$

$1.2P_D + 1.2P_d + 1.2P_{FML} + 1.4P_{VML}$
$\quad + 1.0W + 0.5P_p + 0.5P_L$

$1.2P_D + 1.2P_d + 1.2P_{FML} + 1.4P_{VML}$
$\quad + 1.0E + 0.5P_p + 0.5P_L$

$0.9P_D + 0.9P_d + (1.0W \text{ or } 1.0E)$

Adapted from *ASCE 37: Design Loads on Structures During Construction*, American Society of Civil Engineers, copyright © 2014.

Construction Loads

CECN: Sec. 6.1, Sec. 6.2

Construction loads consist of dead loads, material loads, personnel loads, equipment loads, horizontal construction loads, lateral earth loads, and general dynamic loads. Dead loads include the weight of the temporary structure and do not change over time. The weight of the permanent structure does not count when considering construction loads. Live loads include people, movable furniture, and automobiles and can change over time.

Concrete is a special case. While being poured, it is a temporary load; once it is cured and can support its own weight, it becomes a permanent dead load.

Personnel and equipment loads are loads induced by workers and equipment on a structure. CECN Eq. 6.1 can be used to estimate personnel and equipment loads with respect to weight and area.

$$p_t = \frac{W_t}{A_t} \quad \quad 6.1$$

Horizontal loads are induced on a structure by moving loads. For wheeled vehicles, the horizontal load is typically calculated as 20% of a fully loaded single vehicle or 10% of the sum of all vehicles. Lateral earth loads apply whenever construction operations require a temporary excavation or trench. Active lateral earth pressure can be calculated using CECN Eq. 6.2, where k_a is the active earth pressure coefficient given by CECN Eq. 6.4. Passive earth pressure can be found using CECN Eq. 6.3, where k_p is the passive earth pressure coefficient, which can be found using CECN Eq. 6.5.

$$p_{EM,a} = \gamma_s h k_a \quad \quad 6.2$$

$$p_{EM,p} = \gamma_s h k_p \quad \quad 6.3$$

$$k_a = \tan^2\left(45° - \frac{\phi}{2}\right) \quad \quad 6.4$$

$$k_p = \tan^2\left(45° + \frac{\phi}{2}\right) \quad \quad 6.5$$

Dynamic loads are loads that move from place to place on or in a structure. They are increased during design to account for dynamic forces, and this amplified load is called *equivalent static load*. Dynamic loads include personnel and equipment.

Environmental Loads

CECN: Sec. 6.3

ASCE 37: Sec. 6.2, Sec. 6.3, Sec. 6.4, Sec. 6.5, Sec. 6.6, Sec. 6.7

ASCE 37 Chap. 6 describes environmental loads, which include wind loads, thermal loads, snow loads, earthquake loads, rain loads, and ice loads. Wind loads must be modeled if there is any significant chance of exposure to wind. ASCE 37 Sec. 6.2 states that design wind loading must meet the requirements of ASCE/SEI 7. CECN Sec. 6.3 details the method in ASCE/SEI 7 for determining the *main wind force resisting system* (MWFRS).

Thermal loads are considered in the design of long structures with more than a 4:1 length-to-width ratio subjected to significant temperature swings during their limited life. To find the thermal load, you can multiply the pertinent length of the structure by the largest of the following temperature values.

- the highest average daily temperature minus the lowest average daily temperature
- the expected average temperature for the period of structure use minus the lowest average daily temperature
- the highest average temperature minus the expected average temperature for the period of structure use

As with wind speed, ground snow loads are based on a 2% probability with a 50-year mean recurrence, and reduction of these ground loads is allowed to account for shorter durations. However, there is only one common reduction: a factor of 0.8 (20% reduction) for a construction period less than five years. Where snow loads are significant, complete calculations for applicable snow loads include provisions for exposure, surface slope, drift, loading imbalance, ice accumulation at edges, and impact from snow falling from above. These load calculations can be found in ASCE/SEI 7 and in Sec. 1608 of the *International Building Code* (IBC).

Earthquake loads can be applied based on judgment. Guidance can be found in ASCE 37 Sec. 6.5 but is beyond the scope of the PE Civil Construction exam.

Exam tip: Rain and ice loads are rarely considered in construction.

B. FORMWORK

Key concepts: These key concepts are important for answering exam questions in subtopic 14.B, Formwork.

- formwork components
- formwork pressures
- lateral pressure
- pour rates
- sizing of formwork
- tributary area calculations

NCEES Handbook: The *Handbook* does not contain any material relevant to this subtopic. Be sure to study the listed sections in CERM and CECN and from codes and standards.

PE Civil Reference Manual (CERM): Study these sections in CERM that either relate directly to this subtopic or provide background information.

- Section 49.15: Formwork
- Section 49.16: Lateral Pressure on Formwork

Construction Depth Reference Manual (CECN): Study these sections in CECN that either relate directly to this subtopic or provide background information.

- Section 6.4: Formwork for Concrete
- Appendix 6.B: Representative Base Design Stresses, Normal Load Duration, Visually Graded Dimension Lumber at 19% Moisture, and Plywood Used Wet

Code and standards: To prepare for this subtopic, familiarize yourself with these sections from codes and standards.

Formwork for Concrete (ACI SP-4)

- Section 4.2: Engineered Wood Products
- Section 5.4.3: Lateral Pressure of Concrete Equations
- Section 5.4.5: Column Forms
- Section 5.4.6: Wall Forms
- Section 7.4: Design Criteria for Wood Beams
- Section 7.7: Slab Form Design
- Section 7.8: Beam Form Design
- Section 7.9.1: Types of Column Forms
- Section 7.9.2: General Design Procedure
- Section 11.5: Wall Forms
- Section 11.8.1: Beam and Slab Construction

Guide to Formwork for Concrete (ACI 347R)

- Section 4.2: Loads
- Section 7.4: Design

The following equations and figures are relevant for subtopic 14.B, Formwork.

Walls

CERM: Sec. 49.16

CECN: Sec. 6.4

ACI SP-4: Sec. 5.4.3, Sec. 5.4.6, Sec. 11.5

ACI 347R: Sec. 4.2.2

ACI SP-4 Sec. 5.4.3 provides equations for the lateral pressure of concrete. Sec. 5.4.6 and Sec. 11.5 provide information and equations for wall forms.

To design elements of vertical formwork for columns 4 ft or less in height and walls 14 ft or less in height, use CERM Eq. 49.3 to calculate the maximum lateral pressure.

$$p_{\max,\text{kPa}} = 30 C_w \leq C_w C_c \left(7.2 + \frac{785 R_{\text{m/h}}}{T_{°\text{C}} + 17.8°}\right)$$
$$\leq \rho g h \quad \text{[SI]} \quad 49.3(a)$$
$$\begin{bmatrix} \text{columns: } h \leq 1.2 \text{ m} \\ \text{walls: } h \leq 4.2 \text{ m} \\ R \leq 2.1 \text{ m/h} \end{bmatrix}$$

$$p_{\max,\text{psf}} = 600 C_w \leq C_w C_c \left(150 + \frac{9000 R_{\text{ft/hr}}}{T_{°\text{F}}}\right)$$
$$\leq \gamma h \quad \text{[U.S.]} \quad 49.3(b)$$
$$\begin{bmatrix} \text{columns: } h \leq 4 \text{ ft} \\ \text{walls: } h \leq 14 \text{ ft} \\ R \leq 7 \text{ ft/hr} \end{bmatrix}$$

The temperature and the pouring rate are the controlling factors. CERM Eq. 49.3 is based on ACI 347R Sec. 4.2.2.1a and Sec. 4.2.2.1b and on standards in ACI SP-4 Sec. 5.4.3 for the maximum lateral pressure for regular concrete with no greater than 4 in slump, ordinary work, and internal vibration.

When concrete has a slump of 7 in or less and is placed with internal vibration, CERM Eq. 49.4 can be used for walls when filled to a depth exceeding 14 ft with a placement rate of 7 ft/hr or less and for all walls with placement rates of 7 ft/hr to 15 ft/hr. Unit weight coefficients, C_w, can be found in CERM Table 49.4, and chemistry coefficients can be found in CERM Table 49.5. Similar information can also be found in ACI SP-4 Table 5.4 and Table 5.5.

$$p_{\max,\text{kPa}} = 30 C_w \leq C_w C_c$$
$$\cdot \left(7.2 + \frac{1156 + 244 R_{\text{m/h}}}{T_{°\text{C}} + 17.8°}\right) \leq \rho g h \quad \text{[SI]} \quad 49.4(a)$$
$$\begin{bmatrix} \text{walls: } h > 4.2 \text{ m}, R \leq 2.1 \text{ m/h} \\ \text{walls: } 2.1 \text{ m/h} < R \leq 4.5 \text{ m/h} \end{bmatrix}$$

$$p_{\max,\text{psf}} = 600 C_w \leq C_w C_c$$
$$\cdot \left(150 + \frac{43{,}400 + 2800 R_{\text{ft/hr}}}{T_{°\text{F}}}\right) \quad \text{[U.S.]} \quad 49.4(b)$$
$$\leq \gamma h$$
$$\begin{bmatrix} \text{walls: } h > 14 \text{ ft}, R \leq 7 \text{ ft/hr} \\ \text{walls: } 7 \text{ ft/hr} < R \leq 15 \text{ ft/hr} \end{bmatrix}$$

The major components of formwork for concrete walls are illustrated in CECN Fig. 6.2. For all wall formworks, the minimum internal lateral pressure that the formwork should be built to resist is defined in CECN Eq. 6.8. Values for C_w can be found in ACI 347R Table 4.2.2.1a(c) and CECN Table 6.4.

$$p_{c,\min} = C_w \left(600 \frac{\text{lbf}}{\text{ft}^2}\right) \qquad 6.8$$

CECN Eq. 6.9 determines the lateral pressure on the formwork for walls less than 14 ft tall with a concrete placement rate less than 7 ft/hr. Values for chemistry coefficients, C_c, can be found in ACI 347R Table 4.2.2.1a(b) and CECN Table 6.3.

$$p_{c,w} = C_w C_c \left(150 \frac{\text{lbf}}{\text{ft}^3} + 9000 \left(\frac{r}{T}\right)\right) \qquad 6.9$$

For walls less than 14 ft tall with a concrete placement rate between 7 ft/hr and 15 ft/hr, or for walls taller than 14 ft with a placement rate less than 7 ft/hr, the internal lateral pressure on the formwork is found from CECN Eq. 6.10.

$$p_{c,w} = C_w C_c \left(150 \frac{\text{lbf}}{\text{ft}^3} + \frac{43{,}400}{T} + 2800 \left(\frac{r}{T}\right)\right) \qquad 6.10$$

Wall formwork must be braced laterally to allow for accidental eccentric loads or wind loads prescribed by local building codes. ACI recommends all formwork designs assume a minimum of a 100 lbf/ft wind load applied at the top of the wall or a 15 lbf/ft² wind pressure applied to the top half of the wall, whichever is greater. This is represented in CECN Eq. 6.11(a) and (b) and CECN Fig. 6.3.

$$M = \left(100 \ \frac{\text{lbf}}{\text{ft}}\right)h \qquad \qquad 6.11(a)$$

$$M = \left(15 \ \frac{\text{lbf}}{\text{ft}^2}\right)\left(\frac{1}{2}h\right)\left(\frac{3}{4}h\right) \qquad 6.11(b)$$

For all wall formworks, the minimum internal lateral pressure the formwork should be built to resist can be computed by ACI 347R Eq. 4.2.2.1a(a), with a minimum of (600 lbf/ft²)C_w.

$$C_{\text{CP}} = wh \quad [\text{ACI Eq. 4.2.2.1a(a)}]$$

For walls less than 14 ft tall with a concrete placement rate less than 7 ft/hr, the internal lateral pressure on the formwork is found from ACI 347R Eq. 4.2.2.1a(b) (or CECN Eq. 6.9). The equation already includes the conversion factor, so it is dimensionally inconsistent.

$$C_{\text{CP max}} = C_c C_w \left(150 + \frac{9000R}{T}\right) \quad [\text{ACI Eq. 4.2.2.1a(b)}]$$

For walls less than 14 ft tall with a concrete placement rate between 7 ft/hr and 15 ft/hr, the internal lateral pressure on the formwork is found from ACI 347R Eq. 4.2.2.1a(c) (or CECN Eq. 6.10).

$$C_{\text{CP max}} = \left(150 + \frac{43{,}400}{T} + \frac{2800R}{T}\right) \quad [\text{ACI Eq. 4.2.2.1a(c)}]$$

Design of Concrete Formwork

CERM: Sec. 49.15

CECN Sec. 6.4

ACI 347R: Sec. 4.2, Sec. 7.4

Formwork refers to the system of boards, ties, and bracing required to construct the mold in which wet concrete is placed. It must be strong enough to withstand the weight and pressure created by the wet concrete and must also be easy to erect and remove.

Formwork is required for the construction of walls, columns, beams, and slabs and must be easy to erect and remove. Most importantly, it must be designed so it will support all vertical and lateral loads until the dead load and construction loads can be carried by the structure.

ACI 347R is the most widely used guide for the design of formwork, specifically Sec. 4.2 and Sec. 7.4. Formwork design is based upon the maximum allowable bending stress and the maximum shear.

CECN Eq. 6.15 is used to calculate the safe support distance based on the maximum allowable bending stress. CECN Eq. 6.16 is used to calculate the safe support distance based on the maximum allowable shear stress.

$$l_{\text{in}} = 10.95 \sqrt{\frac{F'_{b,\text{lbf/in}^2} S_{\text{in}^3}}{w_{\text{lbf/ft}}}} \qquad 6.15$$

$$l_{\text{in}} = \frac{40 F'_{h,\text{lbf/in}^2} b_{\text{in}} d_{\text{in}}}{3 w_{\text{lbf/ft}}} + 2d_{\text{in}} \qquad 6.16$$

The maximum shear stress for a given support spacing is found using CECN Eq. 6.17, in which Ib/Q is the rolling shear constant.

$$l_{\text{in}} = \left(\frac{20 F'_{s,\text{lbf/in}^2}}{w_{\text{lbf/ft}}}\right)\left(\frac{Ib}{Q}\right)_{\text{in}^2} + 1.5 \qquad 6.17$$

Deflection occurs when the depth of a member is small, such as with plywood. The deflection tolerance for a member can be found using CECN Eq. 6.18. Formwork design for a deflection tolerance of $1/360$ in is governed by Eq. 6.19, and formwork design for a deflection tolerance of $1/400$ in is governed by Eq. 6.20.

$$\Delta_{\text{max}} = \left(\frac{1}{145}\right)\left(\frac{w_{\text{lbf/ft}}}{12}\right)\left(\frac{l_{\text{in}}^4}{E'_{\text{lbf/in}^2} I_{\text{in}^4}}\right) \qquad 6.18$$

$$l_{\text{in}} = 1.69 \sqrt[3]{\frac{E'_{\text{lbf/in}^2} I_{\text{in}^4}}{w_{\text{lbf/ft}}}} \qquad 6.19$$

$$l_{\text{in}} = 1.63 \sqrt[3]{\frac{E'_{\text{lbf/in}^2} I_{\text{in}^4}}{w_{\text{lbf/ft}}}} \qquad 6.20$$

Beams

CECN Sec. 6.4

ACI SP-4: Sec. 7.8

Formwork for beams requires supporting the concrete on both sides of the beam and the bottom. Although lateral pressures are not great, the load that must be supported can be significant, depending on the weight and depth of the beam being supported. Major components of formwork for beams include sheathing, chamfer, strips, and spreaders, as shown in CECN Fig. 6.4.

Figure 6.4 Beam Formwork

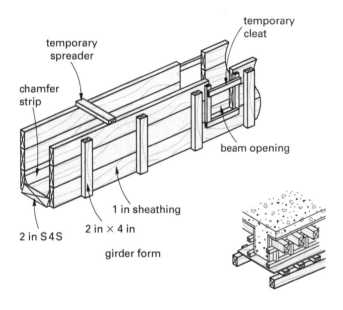

CECN Eq. 6.12 gives the maximum pressure capacity. Beam formwork is never designed to resist internal lateral pressure less than $(600 \text{ lbf/ft}^2) C_w$, and the maximum pressure capacity can never exceed $150h$.

$$p_c = 150 \, \frac{\text{lbf}}{\text{ft}^3} + 9000 \left(\frac{r}{T}\right) \qquad 6.12$$

To determine the vertical loads on beam formwork, the weight of the concrete and the construction live load can be added together.

Columns

CECN: Sec. 6.4

ACI SP-4: Sec. 5.4.5, Sec. 7.9.1, Sec. 7.9.2

ACI SP-4 Sec. 5.4.5 gives a brief description of column forms. Section 7.9.1 describes types of column forms, and Sec. 7.9.2 describes the general design procedure for forms. The design of formwork for columns depends on the height of the column and the rate of placement of the concrete to be used during construction. Major components of formwork for concrete columns include sheathing, clamps or ties, and guys. These are illustrated in CECN Fig. 6.5.

Figure 6.5 Formwork for Columns

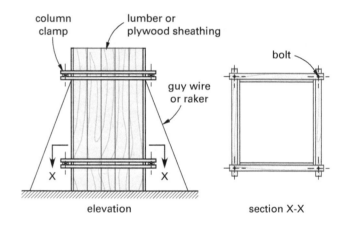

Column formwork is built to resist a minimum internal lateral pressure of $(600 \text{ lbf/ft}^2) C_w$, and columns must be designed for a maximum internal concrete pressure of 3000 lbf/ft^2, per ACI 347R.

Slabs

CECN: Sec. 6.4

ACI SP-4: Sec. 7.7, Sec. 11.8.1

ACI SP-4 Sec. 7.7 describes slab form design, and Sec. 11.8.1 provides information on beam and slab construction. Further information on different types of slabs can be found in the remainder of Sec. 11.8.

Major components of formwork for concrete slabs include sheathing, joists, stringers, and shores, which are illustrated in CECN Fig. 6.6. CECN Eq. 6.13 is used to calculate concrete pressure when the height does not exceed 18 ft.

$$p_c = 150 \, \frac{\text{lbf}}{\text{ft}^3} + 9000 \left(\frac{r}{T}\right) \qquad 6.13$$

p_c should not exceed 3000 psi nor be less than 600 psi, and in no case be greater than $150h$. The vertical pressure on slab formwork can be found using CECN Eq. 6.14, given the weight of the slab formwork and the live pressure on the slab.

$$p_{c,V} = p_c + p_L \quad\quad 6.14$$

Elevated slab formwork is supported by posts. The tributary area by a given group of posts is dependent on the spacing of the posts and the spacing of the stringers, as shown in CECN Eq. 6.24.

$$A_{tr} = l_{stringer} l_p \quad\quad 6.24$$

Once the tributary area of the post is known, the total load on the post can be found using CECN Eq. 6.25.

$$P_p = p_c A_{tr} \quad\quad 6.25$$

The allowable compressive stress for a circular post of height h_p and diameter d is found from CECN Eq. 6.26, which is ACI 347's version of the Euler formula for buckling stress.

$$F'_c = \frac{0.3 E'_{lbf/in^2}}{\left(\dfrac{h_{p,in}}{d_{in}}\right)^2} \quad\quad 6.26$$

The required cross-sectional area of a given post is found in CECN Eq. 6.27.

$$A_{tr,req} = \frac{P_p}{F'_c} \quad\quad 6.27$$

Bearings

CECN: Sec. 6.4, App. 6.B

The bearing of members is checked to ensure that members will support all loads without being crushed. The tributary area is calculated using CECN Eq. 6.28.

$$A_{tr,1st\,member} = l_{1st\,member} l_{supporting\,member} \quad\quad 6.28$$

The total load at the area of contact can be found using CECN Eq. 6.29.

$$P_{1st\,member,supporting\,member} = p_c A_{tr,1st\,member} \quad\quad 6.29$$

The stress at the area of contact p can be found using CECN Eq. 6.30. A is the area of contact between the two members and is calculated using CECN Eq. 6.31.

$$p_{1st\,member,\,supporting\,member} = \frac{P_{1st\,member,\,supporting\,member}}{A_{1st\,member,\,supporting\,member}} \quad\quad 6.30$$

$$A_{1st\,member,\,supporting\,member} = w_{1st\,member} \cdot w_{supporting\,member} \quad\quad 6.31$$

For members used horizontally, the bearing stress capacity is equal to the maximum compressive stress perpendicular to the grain. For members used vertically, the bearing stress capacity is equal to the maximum compressive stress parallel to the grain.

C. FALSEWORK AND SCAFFOLDING

Key concepts: These key concepts are important for answering exam questions in subtopic 14.C, Falsework and Scaffolding.

- allowable load ratings
- OSHA load factors

NCEES Handbook: The *Handbook* does not contain any material relevant to this subtopic. Be sure to study the listed sections in CERM and CECN and from codes and standards.

***PE Civil Reference Manual* (CERM):** Study these sections in CERM that either relate directly to this subtopic or provide background information.

- Section 83.15: Scaffolds

***Construction Depth Reference Manual* (CECN):** Study these sections in CECN that either relate directly to this subtopic or provide background information.

- Section 6.5: Falsework and Scaffolding

Codes and standards: To prepare for this subtopic, familiarize yourself with these sections from codes and standards.

Occupational Safety and Health Act (OSHA)

- 29 CFR 1926 Section 451(a)(1): General Requirements

The following tables and concepts are relevant for subtopic 14.C, Falsework and Scaffolding.

Scaffolding

CERM: Sec. 83.15

CECN: Sec. 6.5

OSHA: 29 CFR 1926 Sec. 451(a)(1)

Scaffolds are any temporary elevated platform and its supporting structure used for supporting employees, materials, or both. Construction and use of scaffolds are regulated by OSHA Sec. 1926.451.

Fall protection must be provided for each employee on a scaffold more than 10 ft above a lower level. This differs from the 6 ft threshold for fall protection for other walking/working surfaces in construction because scaffolds are temporary structures that provide a work platform for employees who are constructing or demolishing other structures.

According to OSHA Sec. 1926.451(a)(1), scaffolds must be capable of supporting their own weight and at least four times the maximum intended load without failure. For this reason, a load factor of 4.0 is applied to the personnel and equipment load, fixed material load, and variable material load. A load factor of 1.0 is applied to the construction dead load. Capacity reduction factors of 1.0 are used with these load factors.

Scaffolding load ratings are found in CECN Table 6.5. This table provides load ratings for light-duty, medium-duty, and heavy-duty loadings.

Table 6.5 Scaffolding Load Ratings

light-duty loading	25 lbf/ft^2 maximum working load for the support of labor and small hand tools (no equipment or material storage on the platform)
medium-duty loading	50 lbf/ft^2 maximum working load for labor, small hand tools, and limited staging of materials, often used for bricklaying and plaster work
heavy-duty loading	75 lbf/ft^2 maximum working load for labor, small hand tools, and the staging of materials, often used for stone masonry work

Falsework

CECN: Sec. 6.5

Falsework is a temporary structure used to support spanning structures, arched structures, or concrete formwork. Falsework holds a component in place until its construction is advanced enough to support itself.

D. SHORING AND RESHORING

Key concepts: These key concepts are important for answering exam questions in subtopic 14.D, Shoring and Reshoring.

- Euler buckling formula
- load in shoring and reshoring

NCEES Handbook: To prepare for this subtopic, familiarize yourself with these sections in the *Handbook*.

- Columns

***Construction Depth Reference Manual* (CECN):** Study these sections in CECN that either relate directly to this subtopic or provide background information.

- Section 6.6: Shoring and Reshoring

Codes and Standards: To prepare for this subtopic, familiarize yourself with these sections from codes and standards.

Formwork for Concrete (ACI SP-4)

- Section 11.9.1 Shore Layout and Installation
- Section 12.6: Reshoring

The following equations and concepts are relevant for subtopic 14.D, Shoring and Reshoring.

Shoring

CECN: Sec. 6.6

ACI SP-4: Sec. 11.9.1

ACI SP-4 Sec. 11.9.1 provides information on shore layout and installation, including illustrations.

Shoring may be wood, aluminum, or steel and supports the forms, workers, and fresh concrete at the top level. It distributes loads from the form to the slab below (the top of the reshoring system). During concrete construction, temporary shoring is used to support freshly placed concrete slabs and their formwork. Shoring behaves as an integrated structural system.

Allowable Compressive Load

Handbook: Columns

$$P_{cr} = \frac{\pi^2 EI}{(KL)^2}$$

CECN: Sec. 6.6

The Euler buckling formula, given in *Handbook* section Columns and CECN Eq. 6.32, calculates the allowable compressive load for shoring. Self-weight must be considered. Field testing is usually used to determine compressive strength before shoring can be put into use.

Reshoring

CECN: Sec. 6.6

ACI SP-4: Sec. 12.6

ACI SP-4 Sec. 12.6 focuses on reshoring operations.

During concrete construction, temporary shoring is used to support freshly placed concrete slabs and their formwork. The loads are transferred to the shore from the formwork and freshly placed concrete. Shores are designed similarly to formwork for concrete slabs. Upon removal of concrete formwork, concrete has yet to obtain its full strength, so reshoring is performed at key locations (e.g., under beams or slabs) to facilitate construction above the freshly placed concrete.

For buildings or facilities with multiple floors, shoring is supported by the floors below it. Proper construction technique is to construct formwork and shoring for the first floor, pour the concrete, and allow it to achieve full strength before reshoring the concrete for the first floor and beginning construction of formwork and shoring for the second floor.

Reshoring posts can help prevent cracks in the previously poured slabs. Reshoring behaves as an integrated structural system.

E. BRACING AND ANCHORAGE FOR STABILITY

Key concepts: These key concepts are important for answering exam questions in subtopic 14.E, Bracing and Anchorage for Stability.

- stress on braces and anchors
- tilt-up construction
- wind load

NCEES Handbook: The *Handbook* does not contain any material relevant to this subtopic. Be sure to study the listed sections in CERM and CECN and from codes and standards.

PE Civil Reference Manual (**CERM**): Study these sections in CERM that either relate directly to this subtopic or provide background information.

- Section 39.7: Stability of Braced Excavations in Clay

Construction Depth Reference Manual (**CECN**): Study these sections in CECN that either relate directly to this subtopic or provide background information.

- Section 6.7: Bracing
- Section 6.8: Anchorage

Codes and standards: To prepare for this subtopic, familiarize yourself with these sections from codes and standards.

Formwork for Concrete (ACI SP-4)

- Section 5.5.2: Wind Loads
- Section 5.5.6: Wind Loads on Elevated Slab Formwork
- Section 8.6.2: Design of Braces
- Section 8.8.1: Deadmen
- Section 8.8.2: Concrete Anchors
- Section 8.8.3: Ground Anchors
- Section 8.8.4: Anchors

Minimum Design Loads for Buildings and Other Structures (ASCE/SEI 7)

- Section 26.5.1: Basic Wind Speed

The following equations and figures are relevant for subtopic 14.E, Bracing and Anchorage for Stability.

Bracing

CECN: Sec. 6.7

ACI SP-4: Sec. 8.6.2

Bracing is the process of supporting a wall to prevent the collapse of the wall or the structure it supports. A brace consists of one or more members sloping between the face of the wall it is applied to and the ground.

Bracing can be evaluated by determining the compressive force on the brace. To do this, take the sum of the moments about a convenient location and factor in all applicable loads. The allowable compressive stress in a member can be found with CECN Eq. 6.37.

$$F'_{c,b} = \frac{0.3E'}{\left(\dfrac{l}{d}\right)^2} \quad 6.37$$

l/d is the unbraced length of the compression member divided by the net dimension of the buckling dimension being considered. (As members in formwork are generally considered to be pinned at both ends, K is almost always equal to 1.0 and can be ignored.) Note that $l_e/d < 50$ is an appropriate limit for formwork compression members.

Tilt-up construction frequently uses bracing through precast members that are braced and anchored in place. Bracing equations and principles apply. Brace location should be placed above the wall's mass center of gravity. CECN Fig. 6.9 illustrates members used to construct a brace, as well as their placement when used to brace a wall.

Figure 6.9 *Bracing*

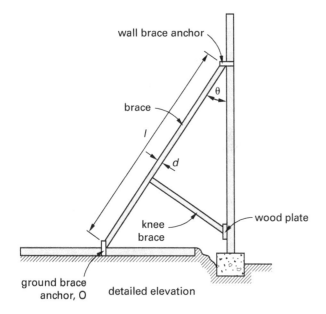

Wind Force

CERM: Sec. 39.7

CECN: Sec. 6.7

ACI SP-4: Sec. 5.5.2, Sec. 5.5.6

ASCE/SEI 7: Sec. 26.5.1

For short-duration formwork, construction period factors can be used to reduce the probable maximum wind velocity. These factors can be found in ACI SP-4 Table 5.10. The basic wind speed velocity can be found using the maps in ASCE/SEI 7 Fig. 26.5-1. Typically, formwork is considered "other structures" and is classified as a category II building. The design velocity pressure can be determined from ACI SP-4 Eq. 5.16.

$$q_z = 0.00256 K_z K_{zt} K_d V_C^2 \quad \text{[ACI SP-4 Eq. 5.16]}$$

$$V_C = (V)(\text{construction period factor})$$

CECN Fig. 6.10 depicts the values that must be known to verify that a brace is adequate to support a structure against wind load.

Figure 6.10 *Wind Load*

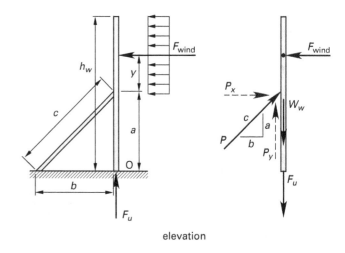

Wind is assumed to act on only the upper half of a wall, so a factor of 1/2 is assumed in CECN Eq. 6.33.

$$F_{\text{wind}} = \frac{p_{\text{wind}} A_{w,\text{tr}}}{2} \quad 6.33$$

The uplift force can be determined using CECN Eq. 6.34.

$$F_u = \frac{F_{\text{wind}}(a+y) - W_w b}{b} \quad 6.34$$

Compressive stress in the brace member can be determined by taking the moment about point O, as shown in CECN Eq. 6.36. If the compressive stress is less than the allowable compressive stress in the brace member, then the brace will prevent collapse due to wind load. This can be determined by CECN Eq. 6.37.

$$\sum M_O = 0 = P_x a - P_y b + F_u b + W_w b \quad 6.36$$
$$- F_{\text{wind}}(a+y)$$

$$F'_{c,b} = \frac{0.3 E'}{\left(\dfrac{l}{d}\right)^2} \quad 6.37$$

Anchorage

CECN: Sec. 6.8

ACI SP-4: Sec. 8.8.1, Sec. 8.8.2, Sec. 8.8.3, Sec. 8.8.4

ACI SP-4 Sec. 8.8 explains that braces for wall and column forms are typically anchored at their base to a ground stake, to a surrounding slab, or to a temporary surface deadman or buried deadman. The anchoring element must be able to resist the horizontal and vertical components of the brace force simultaneously.

Anchor strength is measured in terms of both shear strength and tension strength. Shear strength measures the vertical load that an anchor can resist before the bolt tears or breaks off flush with a wall.

Tension strength measures the force required to pull the bolt out of a hole. It can be found by taking moments around the outermost edge of a plate, as shown in CECN Fig. 6.11.

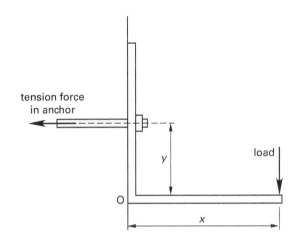

Figure 6.11 Anchor Strength

F. TEMPORARY SUPPORT OF EXCAVATION

Key concepts: These key concepts are important for answering exam questions in subtopic 14.F, Temporary Support of Excavation.

- calculations for cofferdams
- calculations for soldier piles and lagging
- calculations for tieback systems
- types of support of excavation (SOEs)

NCEES Handbook: To prepare for this subtopic, familiarize yourself with these sections in the *Handbook*.

- General Distribution of Combined Active Earth Pressure and Water Pressure

PE Civil Reference Manual (**CERM**): Study these sections in CERM that either relate directly to this subtopic or provide background information.

- Section 39.14: Cofferdams

Construction Depth Reference Manual (**CECN**): Study these sections in CECN that either relate directly to this subtopic or provide background information.

- Section 6.9: Temporary Excavation Support Systems

The following equations and figures are relevant for subtopic 14.F, Temporary Support of Excavation.

Cofferdams

Handbook: General Distribution of Combined Active Earth Pressure and Water Pressure

$$p_w = \gamma_w z_w$$

CERM: Sec. 39.14

CECN: Sec. 6.9

Cofferdams are temporary walls or enclosures for protecting an excavation. Double-wall cofferdams consist of two lines of sheet piles tied to each other with the space between filled with sand. An illustration of this type of cofferdam is provided in CERM Fig. 39.7.

Figure 39.7 Double-Wall Cofferdam

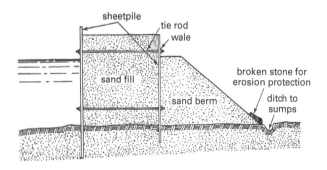

A cellular cofferdam is composed of relatively wide units and is typically used in the construction of dams, locks, wharves, and bridge piers. This type of cofferdam is illustrated in CERM Fig. 39.8.

Single-wall cofferdams form an enclosure with only one line of sheeting and may be built with soldier piles if there will be no water pressure on the sheeting.

Cofferdams are designed to support hydrostatic pressures inside and outside. The inside hydrostatic force acts against the outside hydrostatic force, so it can be subtracted from the outside force. CECN Fig. 6.13 shows an illustration of a cofferdam and the forces exerted on it.

Figure 6.13 Cofferdam

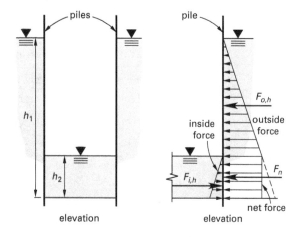

The outside hydrostatic force exerted on a cofferdam, the inside hydrostatic force exerted on a cofferdam, and the net hydrostatic force exerted on a cofferdam are presented in CECN Eq. 6.38, CECN Eq. 6.39, and CECN Eq. 6.40, respectively.

$$F_{o,h} = \frac{\gamma_w h_1^2}{2} \quad 6.38$$

$$F_{i,h} = \frac{\gamma_w (h_2^3)}{2} \quad 6.39$$

$$F_n = \frac{\gamma_w (h_1^2 - h_2^2)}{2} \quad 6.40$$

The equation included in *Handbook* figure General Distribution of Combined Active Earth Pressure and Water Pressure gives the pressure over a cofferdam's area. To find the net force, multiply the original force by the height of the wall and divide by 2.

Soldier Piles and Lagging

CECN: Sec. 6.9

Soldier piles are long steel members inserted into pre-augured holes. Lagging refers to timber or steel plates placed horizontally between the piles. Soldier piles are usually spaced 6 ft to 10 ft apart. CECN Fig. 6.14 shows steel and wood members placed in accordance with the soldier pile and lagging technique.

Figure 6.14 Soldier Pile and Lagging Technique (overhead view)

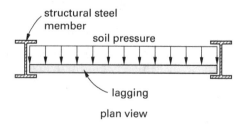

A uniform load on a soldier pile is derived from the soil pressure carried by the soldier pile and the tributary width between adjacent soldier piles. This is illustrated in CECN Fig. 6.15.

Figure 6.15 Uniform Load on a Soldier Pile

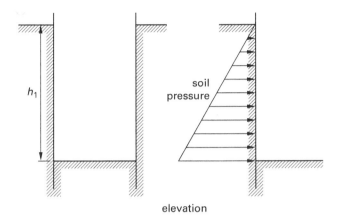

elevation

Figure 6.16 Raking Shore

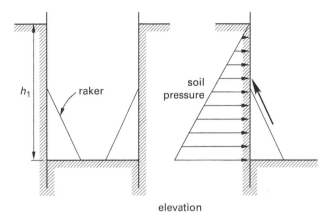

elevation

The maximum moment for lagging of length L and a uniform load w is found from CECN Eq. 6.41. The maximum shear for lagging is found from CECN Eq. 6.42.

Wales and struts are similar to rakers in that they help support walls, but wales and struts use opposing walls for stabilization. Wales are modeled as beams with point loads. CECN Fig. 6.17 shows a typical strut and wale configuration.

Figure 6.17 Struts and Wales

$$M_{\text{maximum}} = \frac{wL^2}{8} \qquad 6.41$$

$$V_{\text{maximum}} = \frac{wL}{2} \qquad 6.42$$

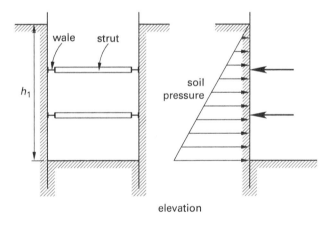

elevation

To select an appropriate cross section for the piles, determine the maximum moment, $M_{\max}$, on the piles and use CECN Eq. 6.43.

$$Z_x = \frac{M_{\max}}{\phi_b F_y} \qquad 6.43$$

Tieback systems are an alternative to struts and wales or rakers. These systems allow for better movement of equipment and personnel within the excavation. The angle of tiebacks ranges between 10° to 20°. CECN Fig. 6.19 illustrates a typical tieback system.

Excavation Supports

CECN: Sec. 6.9

Rakers are supports placed against the ground and a wall to help support the walls from collapse or cave-in. The best support is obtained when the raker meets the wall at an angle between 60° and 70°. Rakers are typically used in deep and wide excavations where the raker can adequately fit. This is illustrated in CECN Fig. 6.16.

Figure 6.19 Tieback System

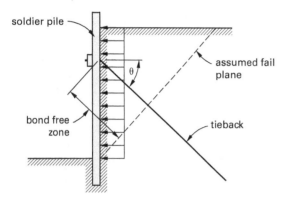

CECN Eq. 6.44 gives the total load on a tieback system, P_t. θ is the insertion angle of the tiebacks, given in degrees. P_h is the horizontal load on the tiebacks, found using CECN Eq. 6.45. d is the depth of the excavation, l is the horizontal spacing between tiebacks, and p_h is the horizontal soil pressure on the support system.

$$P_t = \frac{P_h}{\cos \theta} \qquad 6.44$$

$$P_h = dlp_h \qquad 6.45$$

$$P_{\text{tieback}} = \frac{P_h}{\theta} \qquad 6.46$$

16 Health and Safety

Content in blue refers to the *NCEES Handbook*.

A. OSHA Regulations and Hazard Identification/
   Abatement 16-1
B. Safety Management and Statistics 16-3
C. Work Zone and Public Safety 16-5

The knowledge area of Health and Safety makes up between three and five questions out of the 80 questions on the PE Civil Construction exam. The organization of this chapter follows the order of subtopics given by NCEES for this knowledge area. Each subtopic is covered in the following sections.

A. OSHA REGULATIONS AND HAZARD IDENTIFICATION/ABATEMENT

Key concepts: These key concepts are important for answering exam questions in subtopic 15.A, OSHA Regulations and Hazard Identification/Abatement.

- compliance inspections
- OSHA regulations
- trenching and excavation

NCEES Handbook: To prepare for this subtopic, familiarize yourself with these sections in the *Handbook*.

- Work Zone and Public Safety
- Work Zone and Public Safety: Permissible Noise Exposure (OSHA)

PE Civil Reference Manual **(CERM):** Study these sections in CERM that either relate directly to this subtopic or provide background information.

- Section 83.1: Introduction
- Section 83.6: Trenching and Excavation

Construction Depth Reference Manual **(CECN):** Study these sections in CECN that either relate directly to this subtopic or provide background information.

- Section 6.2: Construction Loads
- Section 7.1: Safety Management

Codes and standards: To prepare for this subtopic, familiarize yourself with these sections from codes and standards.

Occupational Safety and Health Act (OSHA)

- Section 1926.652(b): Design of Sloping and Benching Systems

The following equations, figures, tables, and concepts are relevant for subtopic 15.A, OSHA Regulations and Hazard Identification/Abatement.

OSHA Standards Overview

CERM: Sec. 83.1

The Occupational Safety and Health Administration (OSHA) is a part of the U.S. Department of Labor and ensures safe working conditions. OSHA regulations are divided into two parts that are contained in the *Code of Federal Regulations* (CFR). CERM Table 83.1 lists OSHA 1910 subjects for general industry, including emergency action plans, walking and working surfaces, and ergonomics. CERM Table 83.2 lists OSHA 1926 subjects for the construction industry, including personal protective and life-safety equipment, fall protection, and concrete and masonry construction.

State divisions (such as Cal/OSHA) are charged with enforcing federal and state safety regulations. Surface and underground mines are regulated by the federal Mine Safety and Health Act (MSHA).

OSHA incorporates the standards set by other agencies and organizations, including

- American National Standards Institute (ANSI)
- American Society for Testing and Materials (ASTM) International
- Federal Highway Administration (FHWA)
- National Fire Protection Association (NFPA)

Exam tip: Familiarize yourself with the table of contents of OSHA 29 CFR 1926 (Safety and Health Regulations for Construction). The code is provided as a PDF during the exam, and knowing where to look for information about a given topic can save time on the exam.

OSHA 1910

CERM: Sec. 83.1

OSHA 1910 gives standards for general industry, and CERM Table 83.1 lists subjects for general industry covered by OSHA 1910. For information on standards for the construction industry, see the entry "OSHA 1926" in this chapter.

Table 83.1 OSHA 1910 Subjects for General Industry

Safety and Health Programs
Recordkeeping
Hazard Communication
Exit Routes
Emergency Action Plans
Fire Prevention
Fire Detection and Protection
Electrical
Flammable and Combustible Liquids
Lockout/Tagout
Machine Guarding
Walking and Working Surfaces
Welding, Cutting, and Brazing
Material Handling
Ergonomics
Permit-Required Confined Spaces
Personal Protective Equipment (PPE)
Industrial Hygiene
Blood-Borne Pathogens
Hand and Portable Power Tools and Equipment

OSHA 1926

CERM: Sec. 83.1

OSHA 1926 gives standards for the construction industry, and CERM Table 83.2 lists subjects for the construction industry covered by OSHA 1926. For information on standards for general industry, see the entry "OSHA 1910" in this chapter.

Table 83.2 OSHA 1926 Subjects for the Construction Industry

Health Hazards in Construction
Hazard Communication
Fall Protection
Signs, Signals, and Barricades
Cranes
Rigging
Excavations and Trenching
Tools
Material Handling
Scaffolds
Walking and Working Surfaces
Stairways and Ladders
Hand and Power Tools
Welding and Cutting
Electrical
Fire Prevention
Concrete and Masonry
Confined Space Entry
Personal Protective Equipment (PPE)
Motor Vehicles

Compliance Inspections

CECN: Sec. 7.1

Compliance inspections are completed by a compliance officer. These inspections can be random, the result of an incident where a fatality or hospitalization of more than five workers occurred, due to a threat of imminent danger, a result of a worker complaint of violation, or a follow-up to a previous inspection.

Events can prompt a compliance inspection, but OSHA can also perform unscheduled inspections.

Compliance officers have the right of entry to worksites provided they schedule an inspection. For an unscheduled inspection, a compliance officer needs a search warrant.

Depending on the findings during an inspection, a compliance officer may impose a fine, request that the employer immediately correct the violation, or even cancel the work operation for a period of time (abatement). A follow-up inspection may be necessary to verify that a violation has been corrected.

Permissible Noise Exposure

Handbook: Work Zone and Public Safety; Work Zone and Public Safety: Permissible Noise Exposure (OSHA)

$$D = 100\% \cdot \sum \frac{C_i}{T_i}$$

The *Handbook* equation is used to calculate permissible noise exposure in accordance with OSHA regulations. *Handbook* table Work Zone and Public Safety: Permissible Noise Exposure (OSHA) provides permissible time in hours for various noise levels in decibels.

Trenching and Excavation

CERM: Sec. 83.6

OSHA: Sec. 1926.652(b)

Except for excavations entirely in stable rock, excavations deeper than 5 ft (1.5 m) in all types of earth must be protected from cave-in and collapse. Specifications are provided in OSHA Sec. 1926.652. Excavations less than 5 ft (1.5 m) deep are usually exempt, but they may need to be protected when inspection indicates that hazardous ground movement is possible.

Excavations by Soil Type

CERM: Sec. 83.6

CECN: Sec. 6.2

OSHA: Sec. 1926.652(b)

In many soil structures, soil failure modes are such that the probability of collapse in an excavation that is maintaining current stability is fairly low. However, the probability of collapse increases with time and with changing environmental conditions, such as rain or vibration. Even in relatively shallow excavations, a collapse can be fatal. Because of this, safety authorities such as OSHA have concentrated enforcement efforts on construction in excavations.

CERM Table 83.3 shows maximum sloping based on depth of excavation and soil type. Slope and shield configurations and excavations by soil type are given in CERM Fig. 83.2 and Fig. 83.3. Additional information on the design of sloping and benching systems can be found in OSHA Sec. 1926.652(b).

Table 83.3 Maximum Allowable Slopes

soil or rock type	maximum allowable slopes (H:V)[a] for excavations less than 20 ft deep[b]
stable rock	vertical (90°)
type A[c]	3/4:1 (53°)
type B	1:1 (45°)
type C[d]	1 1/2:1 (34°)

[a]Numbers shown in parentheses next to maximum allowable slopes are angles expressed in degrees from the horizontal. Angles have been rounded off.
[b]Sloping or benching for excavations greater than 20 ft (6 m) deep must be designed by a registered professional engineer.
[c]A short-term maximum allowable slope of 1/2 H:1V (63°) is allowed in excavations in type A soil that are 12 ft (3.67 m) or less in depth. Short-term maximum allowable slopes for excavations greater than 12 ft (3.67 m) in depth must be 3/4 H:1V (53°).
[d]These slopes must be reduced 50% if the soil shows signs of distress.
Source: OSHA 1926 Subpart P App. B

Excavations in Layered Soils

CERM: Sec. 83.6

CECN: Sec. 6.2

In many soil structures, soil failure modes are such that the probability of collapse in an excavation that is maintaining current stability is fairly low. However, the probability of collapse increases with time and with changing environmental conditions, such as rain or vibration. Even in relatively shallow excavations, a collapse can be fatal. Because of this, safety authorities such as OSHA have concentrated enforcement efforts on construction in excavations.

CERM Fig. 83.4 shows sloping requirements based on depth of excavation and layering of soil types.

B. SAFETY MANAGEMENT AND STATISTICS

Key concepts: These key concepts are important for answering exam questions in subtopic 15.B, Safety Management and Statistics.

- experience modification rates
- first aid
- incidence rate
- recordable injuries and illnesses
- safety recordkeeping

NCEES Handbook: To prepare for this subtopic, familiarize yourself with these sections in the *Handbook*.

- Safety Incidence Rate
- Experience Modification Rate

PE Civil Reference Manual (**CERM**): Study these sections in CERM that either relate directly to this subtopic or provide background information.

- Section 83.2: Occupational Injuries and Illnesses
- Section 83.3: OSHA Recordable Injury Incidence Rate

Construction Depth Reference Manual (**CECN**): Study these sections in CECN that either relate directly to this subtopic or provide background information.

- Section 7.2: Safety Statistics
- Section 7.3: OSHA Regulations

The following equations, tables, and concepts are relevant for subtopic 15.B, Safety Management and Statistics.

Incidence Rate

Handbook: Safety Incidence Rate

$$\text{IR} = \frac{N \cdot 200{,}000}{T}$$

CERM: Sec. 83.3

CECN: Sec. 7.2

$$R = \frac{(\text{no. of injuries})(200{,}000 \text{ hr})}{\text{no. of hours worked}} \qquad 83.1$$

The *Handbook* equation for the safety incidence rate and CERM Eq. 83.1 can be used to calculate the incidence rate for a company.

An incidence rate is useful for monitoring the effects of safety measures and comparing a company's safety performance against state and national averages. OSHA uses a recordable injury incidence rate that represents the number of injuries per 200,000 hours, the equivalent of 100 employees working 40 hours per week, 50 weeks per year. Only actual recordable nonfatal injuries and illnesses are included in the injury incidence rate calculation.

CECN Eq. 7.1 is also used to calculate the incidence rate and is very similar to the safety incidence rate equation in the *Handbook*.

$$= \frac{\begin{array}{c}\text{incidence rate (total number of} \\ \text{injuries and illnesses)} \\ \cdot (200{,}000 \text{ hr})\end{array}}{\text{hours worked by all employees}} \qquad 7.1$$

Exam tip: Make sure you are familiar with what is considered an injury or illness.

Experience Modification Rate

Handbook: Experience Modification Rate

$$\text{EMR} = \frac{B + H + EW + (1 - W)F}{D + H + FW + (1 - W)F}$$

CERM: Sec. 83.2

The experience modification rate (EMR) is a number that insurance companies use to measure past costs of injuries and risks of future injuries. It is based on a three-year average ending the year before the current policy expires.

An average company will have an EMR of 1.0. An EMR of more than 1.0 will cost the company higher insurance premiums. The class rate is determined by comparing a company with similar companies. If the rate is high, insurance companies will assume the future rate will also be high and will therefore charge a higher premium.

Exam tip: Familiarize yourself with the industry average of 1.0 and make sure you know what it means to have a higher EMR.

Recordable Injury or Illness

CERM: Sec. 83.3

According to OSHA 1904, an employer must record any work-related fatality, and any injury or illness that results in loss of consciousness, days away from work, restricted work, or transfer to another job. An employer must also record any injury or illness requiring medical treatment beyond first aid, and diagnosed cases of cancer, irreversible diseases, fractured bones, and punctured eardrums.

The log of an incident shows the time of the incident, name of the injured worker, regular job of the worker, department that the worker normally works in, type of injury or illness, and a one- to two-line description of the injury.

Safety Recordkeeping

CERM: Sec. 83.2

OSHA 1904 describes the requirements for recording and reporting injuries and illnesses. A company not listed in the "industries partially exempt" table and that has more than 10 employees at any time during the calendar year must report all work-related recordables.

OSHA Form 300 is a log of work-related injuries and illnesses and must be filled out when an accident occurs.

OSHA Form 300A is a summary of work-related injuries and illnesses and must be filled out yearly regardless of whether any incidents have occurred.

OSHA Form 301 is an injury and illness incident report to be filled out when an incident occurs.

OSHA Construction Standards

CECN: Sec. 7.3

CECN Table 7.2 lists OSHA sections, subparts, and topics and can help you quickly locate a particular subject within OSHA 29 CFR 1926.

OSHA regulations refer to and define authorized persons and competent persons. An *authorized person* is one approved by an employer to perform a specific type of duty or duties, or to work at a specific location on the job site. A *competent person* is one who, from training or experience, is capable of identifying workplace hazards or working conditions that are unsanitary or hazardous to employees, and who has authorization to take prompt measures to eliminate them.

C. WORK ZONE AND PUBLIC SAFETY

Key concepts: These key concepts are important for answering exam questions in subtopic 15.C, Work Zone and Public Safety.

- buffer widths
- taper types and lengths
- temporary traffic control zones and applications
- work zone definition and areas

NCEES Handbook: The *Handbook* does not contain any material relevant to this subtopic. Be sure to study the listed sections in CERM and CECN and from codes and standards.

PE Civil Reference Manual (**CERM**): Study these sections in CERM that either relate directly to this subtopic or provide background information.

- Section 73.37: Temporary Traffic Control Zones

Construction Depth Reference Manual (**CECN**): Study these sections in CECN that either relate directly to this subtopic or provide background information.

- Section 6.10: Traffic and Work Zone Safety

Codes and standards: To prepare for this subtopic, familiarize yourself with these sections from codes and standards.

Manual on Uniform Traffic Control Devices for Streets and Highways (MUTCD)

- Section 6C.04: Advance Warning Area
- Section 6C.08: Tapers

The following equations, figures, tables, and concepts are relevant for subtopic 15.C, Work Zone and Public Safety.

Work Zone

CERM: Sec. 73.37

CECN: Sec. 6.10

MUTCD: Sec. 6C.08

A *work zone* is an area of highway with construction, maintenance, or utility work activities. It is typically marked by signs, channelizing devices, or the equivalent. The work zone extends from the first warning sign to the "END ROAD WORK" sign.

A work zone is different from an incident zone. An *incident zone* is an area of highway where temporary traffic control is imposed by local officials due to a temporary traffic incident.

Channelizing devices include barriers, pavement markings, and other means to direct traffic on a specified route, usually away from the work zone.

Information on tapers, including a diagram similar to CECN Fig. 6.20, can be found in MUTCD Sec. 6C.08. For the areas within the work zone, see the entry "Areas of a Work Zone" in this chapter.

Areas of a Work Zone

CERM: Sec. 73.37

CECN: Sec. 6.10

CECN Fig. 6.20 shows the safety measures for traffic and work zones.

In an *advanced warning area*, drivers are informed of what to expect. Advance warning may vary from a single sign or flashing lights on a vehicle to a series of signs in advance of a temporary traffic control zone transition area. On freeways and expressways, where driver speed is generally in a higher range (45 mph or more), signs may be placed from 500 ft to 0.5 mi or more before the temporary traffic control zone.

A *transition area* redirects users out of the normal path, typically by means of a taper. It is an important component in the safe design of traffic control planning.

An *activity area* is where the work is taking place. It includes traffic space, a buffer space, and the work space.

A *termination area* starts where the actual work stops and ends after the "END ROAD WORK" sign. It returns users to the normal path.

Figure 6.20 Traffic and Work Zone Safety Measures

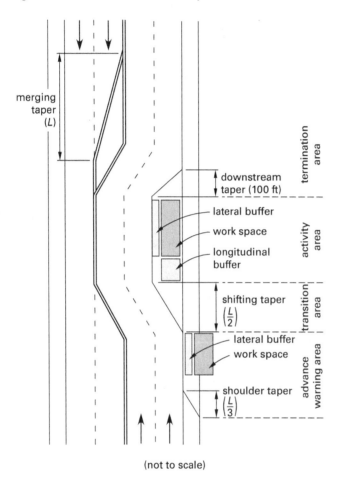

(not to scale)

Lane Tapers

CERM: Sec. 73.37

CECN: Sec. 6.10

MUTCD: Sec. 6C.08

$$L_{\text{ft}} = \frac{W_{\text{ft}} S_{\text{mph}}^2}{60} \quad [S \leq 40 \text{ mph}] \qquad 6.47$$

$$L_{\text{ft,min}} = W_{\text{ft}} S_{\text{mph}} \quad [S \geq 45 \text{ mph}] \qquad 6.48$$

For taper length calculation, L depends on the highest speed, as shown. CECN Eq. 6.47 is used when the speed is 40 mph or less; CECN Eq. 6.48 is used when the speed is 45 mph or more.

In CECN Eq. 6.47 and Eq. 6.48,

L	length of taper	ft
S	the highest of the posted speed, the off-peak 85th percentile speed prior to work starting, or the anticipated operating speed	mph
W	width of offset	ft

Tapers can be used in transition and termination areas. They can be created with different types of channelizing devices or pavement markings.

MUTCD Sec. 6C.08 incudes additional information on tapers, including diagrams and similar equations.

Buffer Widths

CECN: Sec. 6.10

A buffer space is a space that separates traffic flow from the work activity or from a potentially hazardous area. A buffer space also provides recovery space for a vehicle that strays over the edge of the traffic space. Neither work activity nor storage of equipment, vehicles, or material should occur in this space. A buffer space may be positioned longitudinally or laterally with respect to the direction of traffic flow.

Lateral buffer widths depend on the approach speed. Typical values are shown in CECN Table 6.6, but engineering judgment should be used to determine the actual width to be used.

Table 6.6 Lateral Buffer Widths

approach speed of traffic (posted speed limit)	minimum lateral buffer width
over 45 mph	within 6 ft of a traffic lane, but not on a traffic lane
45 mph and under	within 3 ft of a traffic lane, but not on a traffic lane

Suggested Advance Warning Sign Spacing

CERM: Sec. 73.37

MUTCD: Sec. 6C.04

An advance warning area is where road users learn of an upcoming traffic control zone. Warning may be given via signage, flashing light trailers, or rotating lights and strobes on parked vehicles. (In situations where the activity area does not interfere with normal traffic, there may be no advance warning area.)

CERM Table 73.19 provides the suggested distance between advance warning signs (shown as A, B, and C in CERM Fig. 73.18), based on road type.

MUTCD Sec. 6C.04 includes additional information on advance warning areas.

Figure 73.18 Sections of a Temporary Traffic Control Zone

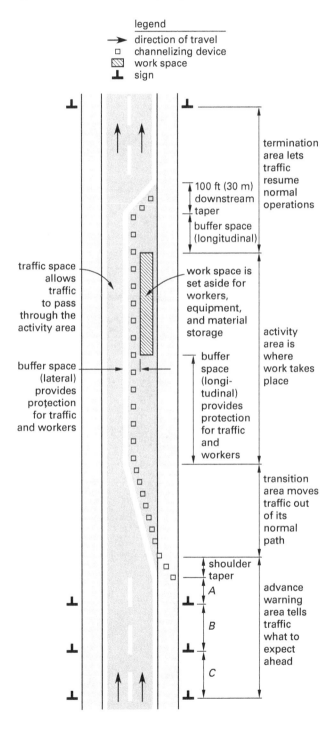

Reprinted with permission from Fig. 6C-1 of the *Manual on Uniform Traffic Control Devices*, 2009 ed., U.S. Department of Transportation, Federal Highway Administration, 2009.

Topic III: Geotechnical

Chapter

17. Site Characterization
18. Soil Mechanics, Laboratory Testing, and Analysis
19. Field Materials Testing, Methods, and Safety
20. Earthquake Engineering and Dynamic Loads
21. Earth Structures
22. Groundwater and Seepage
23. Problematic Soil and Rock Conditions
24. Earth Retaining Structures (ASD or LRFD)
25. Shallow Foundations (ASD or LRFD)
26. Deep Foundations (ASD or LRFD)

17 Site Characterization

Content in blue refers to the *NCEES Handbook*.

A. Interpretation of Available Existing Site Data and Proposed Site Development Data 17-1
B. Subsurface Exploration Planning 17-3
C. Geophysics 17-4
D. Drilling Techniques 17-5
E. Sampling Techniques 17-6
F. In Situ Testing 17-8
G. Description and Classification of Soils 17-9
H. Rock Classification and Characterization 17-9
I. Groundwater Exploration, Sampling, and Characterization 17-10

The knowledge area of Site Characterization makes up between four and six questions out of the 80 questions on the PE Civil Geotechnical exam. The organization of this chapter follows the order of subtopics given by NCEES for this knowledge area. Each subtopic is covered in the following sections.

A. INTERPRETATION OF AVAILABLE EXISTING SITE DATA AND PROPOSED SITE DEVELOPMENT DATA

Key concepts: These key concepts are important for answering exam questions in subtopic 9.A, Interpretation of Available Existing Site Data and Proposed Site Development Data.

- desktop study
- evaluating site impacts on the project
- material (soil and rock) classification
- sample collection and laboratory testing

NCEES Handbook: To prepare for this subtopic, familiarize yourself with these sections in the *Handbook*.

- Guidelines for Minimum Number of Exploration Points and Depth of Exploration
- Commonly Performed Laboratory Tests on Soils
- Corrosion of Buried Steel
- Frost Susceptibility Classification by Percentage of Mass
- In Situ Tests for Subsurface Exploration in Pavement Design and Construction
- Effect of Resistivity on Corrosion

PE Civil Reference Manual (**CERM**): Study these sections in CERM that either relate directly to this subtopic or provide background information.

- Section 35.1: Soil Particle Size Distribution
- Section 35.2: Soil Classification
- Section 35.3: AASHTO Soil Classification
- Section 35.4: Unified Soil Classification
- Section 35.5: Mass-Volume Relationships
- Section 35.6: Swell
- Section 35.9: Standard Penetration Test
- Section 35.10: Cone Penetrometer Test
- Section 35.11: Proctor Test
- Section 35.12: Modified Proctor Test
- Section 35.13: In-Place Density Tests
- Section 35.14: Atterberg Limit Tests
- Section 35.19: Vane-Shear Test
- Section 35.21: Sensitivity
- Section 35.23: Plate Bearing Value Test
- Section 35.26: Characterizing Rock Mass Quality
- Section 40.17: Liquefaction

Codes and standards: To prepare for this subtopic, familiarize yourself with these sections from codes and standards.

Geotechnical Site Characterization (FHWA NHI-16-072)

- Section 3.4: Influence of Number of Measurements
- Chapter 5: Identification and Characterization of Problematic Soil and Rock

The following figures, tables, and concepts are relevant for subtopic 9.A, Interpretation of Available Existing Site Data and Proposed Site Development Data.

Site Characterization

Handbook: Guidelines for Minimum Number of Exploration Points and Depth of Exploration

FHWA NHI-16-072: Sec. 3.4

The goal of site characterization is to create a virtual three-dimensional site and surroundings on which a project can be overlaid and against which it can be compared. Site characterization can reveal geotechnical features, materials, and processes that drive project stability and performance and help evaluate site impacts on the project.

The type and extent of subsurface investigation depend on the type of structure planned at a given location on the project site. *Handbook* table Guidelines for Minimum Number of Exploration Points and Depth of Exploration provides the recommended frequency and depth of exploration based on structure type. The structures listed are

- retaining walls
- embankment foundations
- cut slopes
- shallow foundations
- deep foundations

Exam tip: FHWA NHI-16-072 Sec. 3.4 provides in-depth recommendations for subsurface exploration and planning.

Site Characterization Tasks

Handbook: Guidelines for Minimum Number of Exploration Points and Depth of Exploration; Commonly Performed Laboratory Tests on Soils; In Situ Tests for Subsurface Exploration in Pavement Design and Construction

CERM: Sec. 35.1 through Sec. 35.6, Sec. 35.9 through Sec. 35.14, Sec. 35.19, Sec. 35.23, Sec. 35.26

Tasks that create the site characterization include

- desktop study
- in situ testing and instrumentation
- material (soil and rock) classification
- subsurface exploration planning and execution
- sample collection and laboratory testing

Each task is important and complements the other tasks. The need for any given task and the extent to which the task is explored depend on an understanding of project risk. The project is impacted by experience with similar sites, projects, design requirements, geotechnical uncertainty, and cost considerations.

CERM Sec. 35.1 through Sec. 35.6, Sec. 35.11, Sec. 35.12, and Sec. 35.14 provide details of many types of geotechnical testing, including laboratory testing. CERM Sec. 35.9, Sec. 35.10, Sec. 35.13, Sec. 35.19, and Sec. 35.23 provide details related to the in situ testing. Descriptions of various types of sampling techniques are shown in CERM Sec. 35.9 and Sec. 35.26.

Handbook table Guidelines for Minimum Number of Exploration Points and Depth of Exploration provides the recommended frequency and depth of exploration, based on structure type.

Handbook tables In Situ Tests for Subsurface Exploration in Pavement Design and Construction and Commonly Performed Laboratory Tests on Soils provide different test designations and parameters for various soils tests.

Project Plans

Handbook: Guidelines for Minimum Number of Exploration Points and Depth of Exploration

In addition to the typical design phase players (civil engineer, structural engineer, owner, developer), contractors may also be involved if a construction approach has already been pursued or if specialty services are anticipated. Timing, sequencing, materials, and methods may all impact the geotechnical scope of services.

In some cases, the geotechnical engineer may need to discuss changes to the exploration scope offered. This includes not only the number of explorations but also exploration locations, types of explorations, exploration depths, and sampling intervals.

Handbook table Guidelines for Minimum Number of Exploration Points and Depth of Exploration provides the recommended frequency and depth of exploration based on structure type.

Identification and Characterization of Problematic Soil and Rock

Handbook: Effect of Resistivity on Corrosion; Corrosion of Buried Steel; Frost Susceptibility Classification by Percentage of Mass

CERM: Sec. 35.6, Sec. 35.21, Sec. 40.17

FHWA NHI-16-072: Chap. 5

FHWA NHI-16-072 Chap. 5 is dedicated to problematic rock and soil conditions. The soil conditions discussed are collapsible soils, expansive and shrinking soils, organic soils, dispersive soils, liquefiable soils, colluvium and talus, degradable rock, corrosive soils, cemented sands, sensitive quick clays, high sulfate soils, pyritic/acid rock, unsaturated soils, and permafrost. *Handbook* figure Frost Susceptibility Classification by Percentage of Mass provides a graph for determining the amount of expected heave in soil. The *Handbook* also provides equations for corrosion in section Corrosion of Buried Steel and provides the aggressiveness of soils in table Effect of Resistivity on Corrosion. CERM discusses the sensitivity of clays in Sec. 35.21 and swell in Sec. 35.6. Liquefaction typically occurs in sand during seismic events and is discussed in CERM Sec. 40.17.

B. SUBSURFACE EXPLORATION PLANNING

Key concepts: These key concepts are important for answering exam questions in subtopic 9.B, Subsurface Exploration Planning.

- drainage
- past site use
- topography
- vegetation
- visible soils and rocks

NCEES Handbook: To prepare for this subtopic, familiarize yourself with these sections in the *Handbook*.

- Hammer Efficiency
- Depth Adjustment

PE Civil Reference Manual (**CERM**): Study these sections in CERM that either relate directly to this subtopic or provide background information.

- Section 35.9: Standard Penetration Test
- Section 35.10: Cone Penetrometer Test
- Section 35.26: Characterizing Rock Mass Quality

Codes and standards: To prepare for this subtopic, familiarize yourself with these sections from codes and standards.

Geotechnical Site Characterization (FHWA NHI-16-072)

- Section 3.3: Site Reconnaissance
- Section 3.4: Influence of Number of Measurements
- Section 3.5: Influence of Type of Measurements
- Section 3.9: Considerations for Different Levels of Site Characterization
- Section 12.2: Karst Hazards
- Section 12.3: Underground Mine Hazards
- Section 12.6: Landslide and Rockfall Hazards

The following equations and concepts are relevant for subtopic 9.B, Subsurface Exploration Planning.

Site Reconnaissance Overview

FHWA NHI-16-072: Sec. 3.3

Sound subsurface exploration plans reflect not only project details and performance requirements, but also geologic setting.

Site reconnaissance completes what can be called the project discovery phase. At this point, there is a need to collect, examine, test, and analyze the physical materials within and below the site. Geotechnical engineers should have a good idea of where the project is headed relative to design and construction challenges, and their expectations should be based on exploration, sampling, testing, site instrument development, and construction recommendations.

FHWA NHI-16-072 covers preliminary site investigation in Chap. 1 through Chap. 3, and site reconnaissance is specifically addressed in Sec. 3.3. The initial site visit for reconnaissance helps to confirm what was shown in the desktop study. Site features such as soft surface soils, improperly represented topography, and fences or gates that can impede construction may not be apparent in the desktop study, but can be easily identified during site reconnaissance.

Site Reconnaissance Primary Site Features

FHWA NHI-16-072: Sec. 3.4, Sec. 3.5, Sec. 3.9

When evaluating a site, several features of interest include topography, drainage, visible soil and rocks, vegetation, current site uses, and evidence of past site uses.

Topography includes the overall relief, the slope height and gradient, and the natural and manmade site topography.

Drainage includes general runoff, concentrated surface flow, seepage, sources, and impedances. Visible soil and rocks include natural and artificial rocks and soils, coarse-grained and fine-grained soils, outcroppings, and rock structures.

Site reconnaissance not only helps address logistical issues like access but can reveal much about how a site and its surroundings have been used in the past, how historic structures have performed in turn, and to what risks future projects might be exposed.

Site reconnaissance is critical to ensure that the topography used in the desktop study has not been modified by construction or natural processes. FHWA NHI-16-072 discusses the influence of the number of measurements in Sec. 3.4, the influence of the type of measurement in Sec. 3.5, and the influence of the scope of the investigation in Sec. 3.9.

Site Reconnaissance Secondary Site Features

FHWA NHI-16-072: Sec. 12.2, Sec. 12.3, Sec. 12.6

When evaluating a site, several features of interest include geological hazards, structural or grade distress, and environmental hazards.

- Geological hazards include slope failure, erosion, rockfall, and karst.
- Structural or grade distress includes grade settlement, retaining walls, and pavement stress.
- Environmental hazards include buried structures, monitoring wells, and debris piles.

Descriptions of karst hazards, underground mines, landslides, and rockfalls, along with other geological and environmental hazards, are provided in FHWA NHI-16-072 Sec. 12.2, Sec. 12.3, and Sec. 12.6.

Material Sampling

Handbook: Hammer Efficiency; Depth Adjustment

CERM: Sec. 35.9, Sec. 35.10, Sec. 35.26

Sample collection is an important part of confirming the composition and origin of materials encountered through exploration. It includes aspects such as how these materials vary with breadth and depth throughout a site and how they may affect design and construction.

A *standard penetration test* (SPT) measures the soil's resistance to penetration by a standard split-spoon sampler. As the sample is being collected, the sampler is driven into the ground by blows of a measured force, and the number of blows—called the *standard penetration resistance* or *N-value*—needed to advance the sampler by a specified distance is recorded. *N*-values obtained in the field must be adjusted using equations in *Handbook* sections Hammer Efficiency and Depth Adjustment.

Rock samples obtained during the coring process can be measured to provide a rock quality designation (RQD). CERM Eq. 35.48 shows how to calculate RQD percentage.

$$\text{RQD} = \frac{\sum_{L_i > 4 \text{ in}} L_i}{L_{\text{core}}} \cdot 100\% \qquad 35.48$$

C. GEOPHYSICS

Key concepts: These key concepts are important for answering exam questions in subtopic 9.C, Geophysics.

- electrical resistivity tomography (ERT)
- ground-penetrating radar (GPR)
- indirect in situ testing
- seismic reflection and refraction

***NCEES Handbook*:** The *Handbook* does not contain any material relevant to this subtopic. Be sure to study the listed sections in CERM and from codes and standards.

Codes and standards: To prepare for this subtopic, familiarize yourself with these sections from codes and standards.

Geotechnical Site Characterization (FHWA NHI-16-072)

- Section 8.4.1: Intrusive Field Methods for Stress-Wave Measurements
- Section 12.2.2: Identification and Characterization of Karst Hazards

The following concepts are relevant for subtopic 9.C, Geophysics.

Indirect In Situ Testing

FHWA NHI-16-072: Sec. 12.2.2

Indirect in situ testing and measurement methods are geophysical. The methods include electrical resistivity tomography (ERT), ground-penetrating radar (GPR), and seismic reflection and refraction.

Of the indirect methods, ERT and GPR are used extensively to fill gaps and define trends between discrete exploration locations, and to discover geologic anomalies (e.g., karst) and urban anomalies (e.g., utilities). Indirect methods also provide a safe means of exploring subsurface conditions in cases where subsurface environmental hazards exist.

FHWA NHI-16-072 provides the level of preference of various geophysical methods for voids and sinkholes in Table 12-1 and depth to bedrock in Table 12-2. GPR is preferred for both cases.

Electrical Resistivity Tomography (ERT)

FHWA NHI-16-072: Sec. 12.2.2

Electrical resistivity tomography (ERT) is a technique used for indirect in situ testing. ERT testing analyzes soils, sand, and rocks and profiles bedrock and groundwater.

ERT is applied to a wide range of geotechnical, structural, and environmental problems. ERT is most often used for larger-scale and deeper geotechnical profiling of earth materials, groundwater, and environmental conditions; for smaller-scale and shallower profiling of structural elements and buried structures, ground-penetrating radar (GPR) is more often used. One disadvantage of ERT surveys is that they require staking and then disassembling electrode arrays that can be several hundred feet long; the equipment for GPR is more portable. See the entry "Ground-Penetrating Radar (GPR)" in this chapter for more information.

FHWA NHI-16-072 Fig. 12-11 shows how ERT can be used to identify air-filled voids, clay-filled voids, and bedrock depth. The wider the spacing of the electrodes, the deeper the penetration of the mapping. Smaller voids, however, are hard to identify when a wide electrode spacing is used, especially when they are deep below the ground surface.

Ground-Penetrating Radar (GPR)

FHWA NHI-16-072: Sec. 12.2.2

Ground-penetrating radar (GPR) is a technique used for indirect in situ testing. GPR testing analyzes soils, rocks, and pavement and locates shallow objects such as utilities and bedrock.

GPR is applied to a wide range of geotechnical, structural, and environmental problems. GPR is most often used for smaller-scale and shallower profiling of structural elements and buried structures; for larger-scale and deeper geotechnical profiling of earth materials, groundwater, and environmental conditions, electrical resistivity tomography (ERT) is more often used. One advantage of GPR is the portability of the equipment; ERT surveys require staking and then disassembling electrode arrays that can be several hundred feet long.

FHWA NHI-16-072 Sec. 12.2.2 discusses the use of GPR to identify the potential and location of karst conditions. GPR uses the reflection of electromagnetic waves from material interfaces to determine voids in the subsurface. FHWA NHI-16-072 Fig. 12-7 shows a cross section illustrating how GPR is used to determine voids below the ground surface.

Crosshole Seismic Tomography

FHWA NHI-16-072: Sec. 8.4.1, Sec. 12.2.2

Crosshole methods are an intrusive investigative technique in which instruments are placed into pre-drilled holes in the ground. FHWA NHI-16-072 Fig. 8-4(a) illustrates crosshole testing. The instruments, consisting of a receiver and a source, are placed at the same depth below the ground surface. The source is typically a hammer that produces a shear wave that can be read by the receiver in another borehole. The spacing of the boreholes is generally 10 ft to 15 ft to provide enough distance to remove near-field effects from the hammer, but also close enough that other vibrations do not create static for the receiver.

Multiple sets of receivers and sources may be used to create a more thorough subsurface profile. FHWA NHI-16-072 Fig. 12-13 and Fig. 12-14 provide a borehole cross section and subsurface cross section showing the wave paths from the source to the receiver.

D. DRILLING TECHNIQUES

Key concepts: These key concepts are important for answering exam questions in subtopic 9.D, Drilling Techniques.

- boring
- creating boreholes from which monitoring devices (e.g., wells, piezometers, slope inclinometer casings) can be installed
- creating uncased or cased boreholes from which material samples can be retrieved
- generating large sample volumes for examination and testing
- limited reliability for short-term groundwater observations
- sampling methods favorable for determination of material origin (e.g., filled vs. natural, clay vs. organic)
- variable ability to penetrate highly consolidated soils
- variable ability to penetrate soft rock

NCEES Handbook: To prepare for this subtopic, familiarize yourself with these sections in the *Handbook*.

- Guidelines for Minimum Number of Exploration Points and Depth of Exploration
- Soil Classification Chart
- Rock Quality Designation (RQD)
- Basic Pavement Structure

PE Civil Reference Manual (CERM): Study these sections in CERM that either relate directly to this subtopic or provide background information.

- Section 35.8: Standardized Soil Testing Procedures
- Section 35.9: Standard Penetration Test
- Section 35.10: Cone Penetrometer Test
- Section 35.26: Characterizing Rock Mass Quality
- Section 80.16: Borrow Pit Geometry

The following equations and concepts are relevant for subtopic 9.D, Drilling Techniques.

Boring

Handbook: Guidelines for Minimum Number of Exploration Points and Depth of Exploration; Soil Classification Chart; Basic Pavement Structure

CERM: Sec. 35.9

Boring creates uncased or cased boreholes from which material samples can be retrieved and monitoring devices such as wells, piezometers, and slope inclinometer casings can be installed.

The sampling methods are favorable for the determination of material origin (e.g., filled vs. natural, clay vs. organic). Each boring technique is generally intended to produce

- material samples for examination (at a minimum)
- possible laboratory index testing
- laboratory strength, compressibility, and/or hydraulic conductivity testing (depending on the sampling technique employed)

When employed to qualify projects with high stability and performance expectations, these techniques need to be augmented with a sampling program capable of retrieving the appropriate types of samples for the project's testing requirements.

Test Pits

CERM: Sec. 80.16

Test pits are generally practical to about 20 ft in depth; the length varies but typically matches the depth, and the width varies based on the need for entry. Test pits provide a relatively inexpensive means of qualifying and quantifying project impacts due to unfavorable near-surface conditions, as well as a means of confirming whether those impacts go deeper than anticipated.

Test pits are basic but helpful in providing a visual description of the below-grade material. They also provide a large number of disturbed samples that can be further tested in the laboratory, and they create an open area for potential further undisturbed testing in the bottom or sides of the pit.

Exam tip: Test pits do not provide an objective value, but only a visual, subjective description of the soil.

Rock Cores

Handbook: Rock Quality Designation (RQD)

CERM: Sec. 35.26

Rock cores are needed to confirm the presence, lithology, and quality of in-place bedrock. Through knowledge of the local bedrock sequence and structure, cores can be examined and conclusions drawn regarding origin and suitability for structure support. *Handbook* section Rock Quality Designation (RQD) includes the following equation, which is the percentage of the core run that consists of pieces longer than 4 in. CERM Eq. 35.48 is equivalent.

$$\text{RQD} = \frac{\sum \text{length of sound core pieces} > 4 \text{ in}}{\text{total core run length}}$$

Although the RQD is a measurement of joints in the rock core, a single core will not provide the angle of the joint.

Exam tip: The correct method to measure the length of the continuous rock core is to measure along the centerline of the sample (not the edges).

E. SAMPLING TECHNIQUES

Key concepts: These key concepts are important for answering exam questions in subtopic 9.E, Sampling Techniques.

- index tests
  - Atterberg limits
  - gradation/hydrometer
  - moisture content
  - unit density
- strength, compressibility, hydraulic, and other tests
  - consolidation tests
  - hydraulic conductivity tests
  - pavement subgrade (California bearing ratio (CBR), *R*-value) tests with companion Proctor density tests
  - shear strength tests (direct shear, triaxial shear)

NCEES Handbook: To prepare for this subtopic, familiarize yourself with these sections in the *Handbook*.

- Commonly Performed Laboratory Tests on Soils
- Methods for Index Testing of Soils
- Methods for Performance Testing of Soils
- Consolidation Testing

PE Civil Reference Manual (CERM): Study these sections in CERM that either relate directly to this subtopic or provide background information.

- Section 35.8: Standardized Soil Testing Procedures
- Section 35.9: Standard Penetration Test
- Section 35.11: Proctor Test
- Section 35.12: Modified Proctor Test
- Section 35.13: In-Place Density Tests
- Section 35.15: Permeability Tests
- Section 35.16: Consolidation Tests
- Section 35.17: Direct Shear Test
- Section 35.18: Triaxial Stress Test
- Section 35.20: Unconfined Compressive Strength Test
- Section 35.21: Sensitivity
- Section 35.22: California Bearing Ratio Test
- Section 83.5: Soil Classification

The following tables and concepts are relevant for subtopic 9.E, Sampling Techniques.

Sampling Considerations

Handbook: Commonly Performed Laboratory Tests on Soils; Methods for Index Testing of Soils; Methods for Performance Testing of Soils; Consolidation Testing

CERM: Chap. 35

Sampling is dependent on testing requirements and disturbed/undisturbed soil. Geotechnical specialty tests include

- unconfined compression tests
- direct shear and triaxial shear strength tests
- consolidation tests
- hydraulic conductivity tests
- pavement technology tests such as California bearing ratio (CBR), *R*-value, and resilient modulus

Specialty testing can be performed on a variety of soil types in a remolded state. It is generally performed only on cohesive soils in their intact state.

Exam tip: Pay attention to the description in the problem statement regarding the state of the soil sample (undisturbed vs. disturbed).

Disturbed Samples

Handbook: Commonly Performed Laboratory Tests on Soils; Methods for Index Testing of Soils; Methods for Performance Testing of Soils; Consolidation Testing

CERM: Sec. 35.9

Disturbed samples are those whose structure has been compromised in the sampling process, and whose probable response to material (shear or consolidation) loads or hydraulic (infiltration) loads can no longer be reasonably modeled or simulated in a laboratory environment. For example, auger cuttings, generated by displacement as auger borings are advanced, are disturbed.

Disturbed samples are acceptable for numerous tests, including those for

- moisture content
- organic content
- Atterberg limits
- particle size distribution
- pH
- resistivity

If the soil is placed in a recompacted condition in the field, an undisturbed sample is not helpful.

Exam tip: Many laboratory tests can be performed on both disturbed and undisturbed samples. Pay attention to the sample condition listed in the problem statement.

Undisturbed Samples

Handbook: Commonly Performed Laboratory Tests on Soils; Methods for Index Testing of Soils; Methods for Performance Testing of Soils; Consolidation Testing

CERM: Sec. 35.16, Sec. 35.21, Sec. 83.5

Undisturbed samples are those whose structure is *not* considered to be unfavorably compromised by sampling, extraction, and handling. Shelby tube samples represent one type of undisturbed sample. Shelby tube samples are obtained in tubes by direct push through a hollow-stem auger.

Exam tip: Many laboratory tests can be performed on both disturbed and undisturbed samples. Make sure to pay attention to the sample condition listed in the problem statement.

Sample Handling and Storage

CERM: Sec. 35.8

Disturbed samples should, at a minimum, be protected from moisture changes and freezing temperatures. Undisturbed samples should be handled as infrequently as possible and secured when not being handled to limit the impact that careless handling, transportation, and storage might have on laboratory test results.

The level of care required for a soil needed for a mass fill operation is much less than the level of care required for an undisturbed sample needed for testing of a mass excavation or deep foundation.

F. IN SITU TESTING

Key concepts: These key concepts are important for answering exam questions in subtopic 9.F, In Situ Testing.

- direct/invasive methods
  - cone penetrometer test (CPT)
  - flat-plate dilatometer test
  - Iowa borehole shear test
  - pump/slug tests
  - standard penetration test (SPT)
- indirect/noninvasive methods
  - ground-penetrating radar (GPR)
  - resistivity
  - seismic reflection/refraction

NCEES Handbook: To prepare for this subtopic, familiarize yourself with these sections in the *Handbook*.

- Hammer Efficiency
- Instruments for Measuring Piezometric Pressure
- In Situ Tests for Subsurface Exploration in Pavement Design and Construction

PE Civil Reference Manual **(CERM):** Study these sections in CERM that either relate directly to this subtopic or provide background information.

- Section 35.9: Standard Penetration Test
- Section 35.10: Cone Penetration Test

The following equations and concepts are relevant for subtopic 9.F, In Situ Testing.

Standard Penetration Test (SPT)

Handbook: Hammer Efficiency; In Situ Tests for Subsurface Exploration in Pavement Design and Construction

CERM: Sec. 35.9

$$N_{60} = \left(\frac{E_{\text{eff}}}{60}\right) N_{\text{meas}}$$

The standard penetration test (SPT) is an empirical means of estimating material properties. An auger used with an SPT provides an undisturbed sample. A split spoon, used to determine the *N*-value, provides a minimally disturbed (nearly undisturbed) column of soil. Other tests can be performed at the same time as the SPT in the same borehole.

Exam tip: In the *Handbook* equation given, make sure the raw *N*-value is converted to an N_{60} value before using it in another equation.

Cone Penetrometer Test (CPT)

Handbook: In Situ Tests for Subsurface Exploration in Pavement Design and Construction

CERM: Sec. 35.10

The cone penetrometer test (CPT) is one of the most versatile subsurface exploration tools. The ability to collect material samples, even if only for examination and index testing purposes, completes this strong technology. Cone penetrometer testing technology is particularly suited to

- fine-grained soils
- finely layered soils
- weak and/or compressible soils that are likely to be sensitive to loading

CPT does not provide a visual or a sample of the soil that is being tested; the results from the cone are the sole measurement. CPT provides continuous output from its tip and can identify layers of soil that a standard penetration test (SPT) may not be able to find due to gaps in the sampling. (See also the entry "Standard Penetration Test (SPT)" in this chapter.)

Vibrating-Wire Piezometers

Handbook: Instruments for Measuring Piezometric Pressure

Vibrating-wire piezometers can be installed easily. One or more devices are simply fixed to a sacrificial casing, the casing is lowered into a completed borehole, and the borehole is then grouted to the surface. Vibrating-wire piezometers produce data more quickly than conventional wells and standpipes, as the latter need to be bailed and developed, and groundwater levels within need to be allowed to restabilize, which can take days when installed in finer-grained materials.

G. DESCRIPTION AND CLASSIFICATION OF SOILS

Key concepts: These key concepts are important for answering exam questions in subtopic 9.G, Description and Classification of Soils.

- Atterberg limits
- plasticity

NCEES Handbook: To prepare for this subtopic, familiarize yourself with these sections in the *Handbook*.

- Unified Soil Classification System (USCS)
- AASHTO Soil Classification System
- Gradation Tests

PE Civil Reference Manual **(CERM):** Study these sections in CERM that either relate directly to this subtopic or provide background information.

- Section 35.1: Soil Particle Size Distribution
- Section 35.2: Soil Classification
- Section 35.3: AASHTO Soil Classification
- Section 35.4: Unified Soil Classification

The following equations and figures are relevant for subtopic 9.G, Description and Classification of Soils.

Unified Soil Classification

Handbook: Unified Soil Classification System (USCS); Gradation Tests

CERM: Sec. 35.4

The Unified Soil Classification System (USCS) is the most complex of the three soil classification systems used in geotechnical engineering. If the soil is coarse grained (more than 50% retained on a #200 sieve), it is categorized further on the basis of particle size distribution. If the soil is fine grained, the focus is on liquid limit and plasticity index.

Exam tip: The axes of the plasticity chart are plasticity index (PI) and liquid limit (LL). There are different USCS flowcharts provided in CERM and the *Handbook*; they will give the same result.

AASHTO Soil Classification

Handbook: AASHTO Soil Classification System; Gradation Tests

CERM: Sec. 35.3

The AASHTO soil classification system, as shown in the *Handbook*, is based more on behavior and performance than other systems. AASHTO soil classification is based on particle size distribution and plasticity index. There is generally only one chart for the AASHTO classification system.

Exam tip: The number designation provides the quality of the soil for use as roadway subgrades. A-1 is excellent, indicating a very granular soil made of stone and rock fragments, while A-8 is unsatisfactory, indicating a highly organic soil.

USDA Textural Classification

CERM: Sec. 35.2

The United States Department of Agriculture (USDA) textural classification system is based on the particle sizes (sand, silt, and clay) that the USDA believes most impact soil behavior. The USDA system sets the boundary between gravel- and sand-sized particles at 2 mm to emphasize the importance of finer-grained particle sizes on soil classification.

The chart for the USDA classification is triangular and has three axes. The intersection of the percentages of sand, silt, and clay determines the soil classification.

Exam tip: The USDA classification is the only classification discussed in this book that uses the term *loam*.

H. ROCK CLASSIFICATION AND CHARACTERIZATION

Key concepts: These key concepts are important for answering exam questions in subtopic 9.H, Rock Classification and Characterization.

- lithology
  - igneous, sedimentary, metamorphic
- rock core descriptors
  - discontinuities (joints, faults, shear zones)
  - rock mass (degree of weathering, hardness, texture)

- rock structure (bedding thickness, degree of fracturing (jointing), dip of bed or fracture)
- quality of rock mass (MR)
- rock mass quality (Q)
- rock quality index (QI)
- rock structure rating (RSR)
- simplified rock mass rating (R)

NCEES Handbook: To prepare for this subtopic, familiarize yourself with these sections in the *Handbook*.

- Rock Quality Designation (RQD)
- Rock Groups and Types

PE Civil Reference Manual (**CERM**): Study these sections in CERM that either relate directly to this subtopic or provide background information.

- Section 35.25: Classification of Rocks
- Section 35.26: Characterizing Rock Mass Quality

The following equations, tables, and concepts are relevant for subtopic 9.H, Rock Classification and Characterization.

Rock Quality Designation (RQD)

Handbook: Rock Quality Designation (RQD)

$$\text{RQD} = \frac{\sum \text{length of sound core pieces} > 4 \text{ in}}{\text{total core run length}}$$

CERM: Sec. 35.26

$$\text{RQD} = \frac{\sum_{L_i > 4 \text{ in}} L_i}{L_{\text{core}}} \cdot 100\% \qquad 35.48$$

Rock quality designation (RQD) is a rough measure of rock quality based on core recovery. The RQD is defined as the total length of all recovered core pieces greater than 4 in long divided by the intended length (not the recovered length) of a single core run.

Rock quality is characterized as very good (RQD > 90%), good (90%–75%), fair (75%–50%), poor (50%–25%), or very poor (RQD < 25%).

Exam tip: The measurement for determining RQD should be located at the center of the core sample, not the edge.

Rock Classification Principles

Handbook: Rock Groups and Types

CERM: Sec. 35.25

Unlike soil classification, which is based primarily on texture, rock classification is based primarily on structure. The type of rock referenced by the exam may be hard rock, crystalline rock, or particulate rock that is sufficiently consolidated and cemented (lithified) to require coring (instead of auguring) to penetrate.

Rock grades that use Roman numerals I through VI describe the amount of weathering that has occurred to the minerals that compose the rock. *Handbook* table **Rock Groups and Types** includes some additional information on rock classification principles.

Lithology

Handbook: Rock Groups and Types

CERM: Sec. 35.25

The likelihood of rock being of poor quality, or of being vulnerable to processes that reduce its quality, is consistent to some extent with the type of rock and source of placement or deposition.

Igneous rock is formed from the cooling of magma into massive crystalline deposits (e.g., granite) or lava (e.g., basalt) into structured or layered deposits. *Sedimentary rock* is formed from the deposition, consolidation, and cementation of soil particles (e.g., sandstone, siltstone, shale) or chemical precipitate (e.g., limestone, dolomite). *Metamorphic rock* consists of igneous and sedimentary rocks that have been altered by extreme pressure and/or temperature (e.g., shale to slate, granite to gneiss, limestone to marble).

Exam tip: The *Handbook* provides a very useful table titled **Rock Groups and Types** that lists many types of rocks and their formation history.

I. GROUNDWATER EXPLORATION, SAMPLING, AND CHARACTERIZATION

Key concepts: These key concepts are important for answering exam questions in subtopic 9.I, Groundwater Exploration, Sampling, and Characterization.

- Darcy's law
- flow nets
- groundwater flow
- infiltration
- permeability

NCEES Handbook: To prepare for this subtopic, familiarize yourself with these sections in the *Handbook*.

- Dewatering and Pumping
- Typical Flow Net Showing Basic Requirements and Computations
- Seepage Through Embankment and Foundation
- Darcy's Law
- Approximate Coefficient of Permeability for Various Sands
- Concepts of Flow Paths Through a Soil Column
- Flow Net Concepts
- Gravity Dam on Pervious Foundation of Finite Depth
- Minimum Infiltration Rates for the Various Soil Groups
- Infiltration: Richards Equation
- Infiltration: Horton Model
- Green-Ampt Equation
- Unconfined Aquifers
- Dupuit's Formula
- Thiem Equation
- Cooper Jacobs Equations

PE Civil Reference Manual (CERM): Study these sections in CERM that either relate directly to this subtopic or provide background information.

- Section 21.1: Aquifers
- Section 21.2: Aquifer Characteristics
- Section 21.3: Permeability
- Section 21.4: Darcy's Law
- Section 21.14: Flow Nets
- Section 21.15: Seepage from Flow Nets
- Section 21.16: Hydrostatic Pressure Along Flow Path
- Section 21.17: Infiltration

Codes and standards: To prepare for this subtopic, familiarize yourself with these sections from codes and standards.

Geotechnical Site Characterization (FHWA NHI-16-072)

- Section 10.1: Uses for Hydraulic Properties and Groundwater Conditions for Design and Construction

The following equations, figures, and tables are relevant for subtopic 9.I, Groundwater Exploration, Sampling, and Characterization.

Groundwater Through an Aquifer

Handbook: Darcy's Law; Concepts of Flow Paths Through a Soil Column; Approximate Coefficient of Permeability for Various Sands; Unconfined Aquifers; Dupuit's Formula; Thiem Equation; Cooper Jacobs Equations

$$Q = kiA$$

CERM: Sec. 21.1 through Sec. 21.4

$$Q = -KiA_{\text{gross}} = -\text{v}_e A_{\text{gross}} \qquad 21.10$$

Movement of groundwater through an aquifer is given by Darcy's law. The hydraulic gradient may be specified in either ft/ft (m/m) or ft/mi (m/km), depending on the units of area used.

Handbook figure Concepts of Flow Paths Through a Soil Column provides a visual of groundwater flow through soil. *Handbook* table Approximate Coefficient of Permeability for Various Sands provides the permeability of various sands. The table provides the coefficient of permeability, k, in both cm/sec and ft/min. Note that the permeability of a fine-grained soil is much less than the permeability of a coarse-grained soil.

Handbook sections Unconfined Aquifers, Dupuit's Formula, Thiem Equation, and Cooper Jacobs Equations provide equations of flow rates for unconfined and confined aquifers. *Handbook* section Dewatering and Pumping also provides flow rate calculations for groundwater related to dewatering and pumping in various aquifers.

> *Exam tip*: Darcy's law is applicable only when the Reynolds number is less than one.

Flow Nets

Handbook: Typical Flow Net Showing Basic Requirements and Computations; Flow Net Concepts; Gravity Dam on Pervious Foundation of Finite Depth

CERM: Sec. 21.14

Groundwater seepage is from locations of high hydraulic head to locations of lower hydraulic head. Relatively complex two-dimensional problems may be evaluated using a graphical technique that shows the decrease in hydraulic head along the flow path. The resulting graphic representation of pressure and flow path is called a *flow net*.

Handbook section Flow Net Concepts includes a simple diagram of a flow net. *Handbook* figure Gravity Dam on Pervious Foundation of Finite Depth provides a more complex cross section of a flow net and also provides a formula to determine the flow rate for the system. *Handbook* figure Typical Flow Net Showing Basic Requirements and Computations takes the concept even further and provides example calculations for the system. CERM Sec. 21.14 provides a helpful list of rules on how to construct a flow net.

Exam tip: Follow the rules of flow net construction to quickly create a flow net cross section. The level of size or detail makes only a small difference in the final calculation. Three to five streamlines will be enough to calculate a correct answer.

Seepage from Flow Nets

Handbook: Gravity Dam on Pervious Foundation of Finite Depth; Seepage Through Embankment and Foundation

$$Q = kh\left(\frac{N_f}{N_d}\right)$$

CERM: Sec. 21.15

$$Q = KH\left(\frac{N_f}{N_p}\right) \quad \text{[per unit width]} \quad 21.33$$

Once a flow net is drawn, it can be used to calculate the seepage. First count the number of flow channels, N_f, between the streamlines. Then count the number of equipotential drops, N_d, between equipotential lines. The total hydraulic head, h, is determined as a function of the water surface levels.

Handbook figure Seepage Through Embankment and Foundation shows four different flow nets for four different embankment configurations. All four flow net diagrams provide the corresponding calculations for seepage based on the varying number of flow channels and equipotential drops.

Infiltration

Handbook: Infiltration: Horton Model; Green-Ampt Equation; Infiltration: Richards Equation; Minimum Infiltration Rates for the Various Soil Groups

CERM: Sec. 21.17

$$f = (f_0 - f_c)e^{-kt} + f_c$$

$$F = f_c t_p + \left(\frac{f_0 - f_c}{k}\right)(1 - e^{-kt_p})$$

An aquifer and the groundwater table can be refilled in a number of natural and artificial ways. When the rainfall supply exceeds the infiltration capacity, infiltration decreases exponentially over time. The cumulative infiltration will not correspond to the increase in water table elevation due to the effects of porosity.

The Horton equation, given in the *Handbook* and in CERM Eq. 21.39, provides a lower bound on the infiltration capacity, f. The cumulative infiltration over time, F, is shown in the second equation above, as in CERM Eq. 21.42.

Another model for calculation of infiltration is found in *Handbook* section Green-Ampt Equation. It's a more practical version of the equation in *Handbook* section Infiltration: Richards Equation. The Green-Ampt method requires more inputs than the Horton method, but it provides what many consider to be a more accurate measurement of the infiltration rate. Minimum infiltration rates for various soil types are provided in *Handbook* table Minimum Infiltration Rates for the Various Soil Groups.

18 Soil Mechanics, Laboratory Testing, and Analysis

Content in blue refers to the *NCEES Handbook*.

A. Index Properties and Testing 18-1
B. Strength Testing of Soil and Rock 18-4
C. Stress-Strain Testing of Soil and Rock 18-6
D. Permeability Testing Properties of Soil and Rock 18-9
E. Effective and Total Stresses 18-11

The knowledge area of Soil Mechanics, Laboratory Testing, and Analysis makes up between four and six questions out of the 80 questions on the PE Civil Geotechnical exam. The organization of this chapter follows the order of subtopics given by NCEES for this knowledge area. Each subtopic is covered in the following sections.

A. INDEX PROPERTIES AND TESTING

Key concepts: These key concepts are important for answering exam questions in subtopic 10.A, Index Properties and Testing.

- effective and total stress conditions
- index
  - basic physical properties
  - general classification
- permeability
  - constant head
  - falling head
- strength
  - direct shear
  - triaxial shear
  - unconfined compression
- stress-strain
  - consolidation

NCEES Handbook: To prepare for this subtopic, familiarize yourself with these sections in the *Handbook*.

- Typical Consolidation Curve for Normally Consolidated Soil
- Shear Strength of (a) Cohesionless Soils and (b) Cohesive Soils
- Shear Strength Effective Stress
- Relationship Between the Ratio of Undrained Shear Strength to Effective Overburden Pressure and Plasticity Index for Normally Consolidated and Overconsolidated Clays
- Drained Shear Strength of Clays
- A Unit of Soil Mass and Its Idealization
- Weight-Volume Relationships
- Volume and Weight Relationships
- Hazen's Equation for Permeability
- Darcy's Law

PE Civil Reference Manual **(CERM):** Study these sections in CERM that either relate directly to this subtopic or provide background information.

- Section 21.3: Permeability
- Section 35.5: Mass-Volume Relationships
- Section 35.7: Effective Stress
- Section 35.15: Permeability Tests
- Section 35.16: Consolidation Tests
- Section 35.17: Direct Shear Test
- Section 35.18: Triaxial Stress Test
- Section 35.19: Vane-Shear Test
- Section 35.20: Unconfined Compressive Strength Test
- Section 35.21: Sensitivity
- Section 35.22: California Bearing Ratio Test
- Section 37.11: Effective Stress
- Section 40.5: Clay Condition
- Section 40.6: Consolidation Parameters

- Section 40.7: Primary Consolidation
- Section 40.8: Primary Consolidation Rate
- Section 40.9: Secondary Consolidation

The following equations and figures are relevant for subtopic 10.A, Index Properties and Testing.

Index Properties

Handbook: A Unit of Soil Mass and Its Idealization; Weight-Volume Relationships; Volume and Weight Relationships

CERM: Sec. 35.5

Index tests—including water content, unit density, particle size distribution, Atterberg limits, organic content, and others—confirm visual and manual classifications, provide behavior-based parameters, and allow empirical correlations to be made with performance-based parameters.

Unit density and water content tests are the most common; these parameters facilitate the development of in situ stress profiles. Available shear strength, bearing capacity, settlement, and lateral earth pressure are determined from the stress profiles. Void ratio is a key property in settlement computations. Specific gravity is also needed to derive the sum of available mass- and volume-related parameters for a given material.

Soil is composed of three types of particles: air, water, and solids. See *Handbook* figure A Unit of Soil Mass and Its Idealization for an example of a soil phase diagram. This same graph is shown in CERM Fig. 35.4.

Figure 35.4 *Soil Phases*

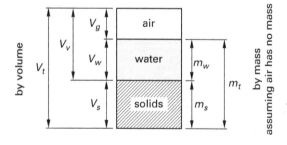

Handbook table Volume and Weight Relationships provides many weight-volume relationships and is useful for determining many different soil properties if only a few of these values are known. Weight is commonly used in soil calculations instead of mass. The corresponding density is called *unit weight*, γ. *Handbook* table Weight-Volume Relationships provides several unit weight relationships.

Materials Strength Testing

Handbook: Shear Strength of (a) Cohesionless Soils and (b) Cohesive Soils; Relationship Between the Ratio of Undrained Shear Strength to Effective Overburden Pressure and Plasticity Index for Normally Consolidated and Overconsolidated Clays; Drained Shear Strength of Clays

CERM: Sec. 35.17 through Sec. 35.22

Strength tests give the performance-based parameters used to compute bearing capacity, estimate lateral earth pressures, and determine slope stability factors of safety; these parameters are all about what the ground can hold. For situations in which the impact of loadings on in situ materials is limited, design can often proceed based on presumptive design parameters. The tests performed to determine soil strength are

- direct shear test
- triaxial stress test
- vane shear test
- unconfined compression test

The following *Handbook* figures demonstrate shear strength relationships.

- Shear Strength of (a) Cohesionless Soils and (b) Cohesive Soils
- Relationship Between the Ratio of Undrained Shear Strength to Effective Overburden Pressure and Plasticity Index for Normally Consolidated and Overconsolidated Clays
- Drained Shear Strength of Clays

Exam tip: Depending on the type of soil and the test performed, the strength values may be reported as friction angle, cohesion, or drained or undrained shear strengths.

Stress-Strain Testing

Handbook: Typical Consolidation Curve for Normally Consolidated Soil

CERM: Sec. 35.16 through Sec. 35.18, Sec. 40.5 through Sec. 40.9

Stress-strain tests predict how much earth materials will give in response to stress changes (just because the ground holds does not mean it will not yield). Consolidation tests are the primary tests from this category (in situ testing offers more and better options).

SOIL MECHANICS, LABORATORY TESTING, AND ANALYSIS

Handbook figure Typical Consolidation Curve for Normally Consolidated Soil shows a typical consolidation curve. See also CERM Fig. 40.4 for a typical consolidation curve for clay.

Figure 40.4 Consolidation Curve for Clay

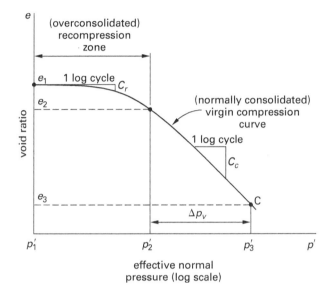

Exam tip: These values will depend on whether the soil is normally consolidated or overconsolidated. Pay attention to the soil description in the problem statement.

Permeability Testing

Handbook: Hazen's Equation for Permeability; Darcy's Law

$$Q = kiA$$

CERM: Sec. 35.15

$$Q = \mathrm{v} A_{\text{gross}} \qquad 35.26$$

$$\mathrm{v} = Ki \qquad 35.27$$

The flow of water through a permeable aquifer or soil is given by Darcy's Law in the *Handbook* or CERM Eq. 35.26. Permeability tests, like strength and stress-strain tests, measure a performance property directly, which is preferred to estimating empirically. The permeability of a soil is a measure of continuous voids. In addition to a soil having large voids, the voids must also be connected for water to flow through them. A permeable material supports a flow of water. The *Handbook* provides Hazen's Equation for Permeability, the equation for calculating the value of permeability, k.

$$k = C(D_{10})^2$$

CERM Sec. 35.15 provides typical values for the coefficient of permeability for different soil types.

Exam tip: Permeability is also referred to as hydraulic conductivity, so look for that term in problem statements.

Effective Stresses

Handbook: Shear Strength Effective Stress

$$\tau' = c' + (\sigma_n - u)\tan\phi' = c' + \sigma'_n \tan\phi'$$

CERM: Sec. 35.18

$$s = c' + \sigma'\tan\phi' \qquad 35.43$$

Stress is everywhere in engineering analyses; stress-dependent qualities include shear strength as determined by

- friction angle
- bearing capacity
- settlement
- earth pressure

The effective stress, σ', is the portion of the total stress that is supported through grain contact. It is the difference between the total stress, σ, and the pore water pressure, u. The pore water pressure is sometimes called the neutral stress because it is equal in all directions; that is, it has no shear stress component.

Exam tip: The effective stress condition is typically a long-term condition when the water pressure has had time to dissipate from the system. A total stress condition includes the pore water pressure and is typically used for an analysis condition during construction or immediately after construction.

B. STRENGTH TESTING OF SOIL AND ROCK

Key concepts: These key concepts are important for answering exam questions in subtopic 10.B, Strength Testing of Soil and Rock.

- California bearing ratio (CBR)
- direct shear test
- rock quality designation (RQD)
- triaxial stress test
- unconfined compressive strength test

NCEES Handbook: To prepare for this subtopic, familiarize yourself with these sections in the *Handbook*.

- Rock Quality Designation (RQD)
- Commonly Performed Laboratory Tests on Soils
- Methods for Performance Testing of Soils
- Determination of Soil Type
- In Situ Tests for Subsurface Exploration in Pavement Design and Construction
- Typical CBR Values

PE Civil Reference Manual (**CERM**): Study these sections in CERM that either relate directly to this subtopic or provide background information.

- Section 35.17: Direct Shear Test
- Section 35.18: Triaxial Stress Test
- Section 35.20: Unconfined Compressive Strength Test
- Section 35.22: California Bearing Ratio Test
- Section 35.26: Characterizing Rock Mass Quality

The following equations, figures, and tables are relevant for subtopic 10.B, Strength Testing of Soil and Rock.

Unconfined Compressive Strength Test

Handbook: Commonly Performed Laboratory Tests on Soils; Determination of Soil Type

CERM: Sec. 35.20

Handbook table Commonly Performed Laboratory Tests on Soils provides the test designation for the unconfined compressive strength of cohesive soils.

Unconfined compression tests are a relatively inexpensive option for determining the undrained shear strength of cohesive material samples of moderate to high strength. The results give an idea of the absolute strength of the material being tested but do not simulate field or design stresses.

The unconfined compressive strength is calculated using CERM Eq. 35.44. The undrained shear strength is calculated using CERM Eq. 35.45 as one-half of the unconfined compressive strength. The equations are typically used for Type A soils such as clays or silts (not sands, gravels, or rock). See *Handbook* section Determination of Soil Type for a description of Type A soils.

$$S_{uc} = \frac{P}{A} \qquad 35.44$$

$$s_u = \frac{S_{uc}}{2} \qquad 35.45$$

Exam tip: CERM Eq. 35.45 is a helpful equation to determine the undrained shear strength, a common input to many other equations.

Direct Shear Test

Handbook: Methods for Performance Testing of Soils

CERM: Sec. 35.17

The *direct shear test* is used to determine the relationship of shear strength to consolidation stress. Details of the test can be found in CERM Sec. 35.17. See *Handbook* table Methods for Performance Testing of Soils for an overview of the procedure and applicable soil types. The test is usually repeated at three different vertical normal stresses. (See CERM Fig. 35.13.)

Representative values of typical strength characteristics ϕ and c are given in CERM Table 35.12. ϕ is the *angle of internal friction*, and c is the *cohesion intercept*, a characteristic of cohesive soils.

A line drawn through all the test values is called the *failure envelope* (or *failure line* or *rupture line*). The equation for the failure envelope is given by Coulomb's equation, CERM Eq. 35.37, which relates the strength of the soil, S, to the normal stress on the failure plane.

$$S = \tau = c + \sigma \tan \phi \qquad 35.37$$

Exam tip: This test is typically performed on sands and gravels (not silts or clays).

Figure 35.13 Graphing Direct Shear Test Results

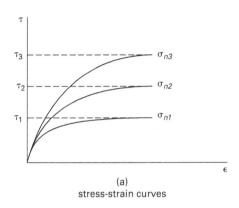

(a)
stress-strain curves

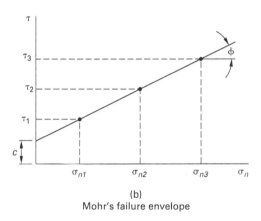

(b)
Mohr's failure envelope

Triaxial Stress Test

Handbook: Methods for Performance Testing of Soils; Commonly Performed Laboratory Tests on Soils

CERM: Sec. 35.18

See *Handbook* table Methods for Performance Testing of Soils and CERM Sec. 35.18 for an overview of the triaxial stress test and applicable soil types. Results of a triaxial test at a given chamber pressure are plotted as a stress-strain curve. Two such examples are illustrated in CERM Fig. 35.14.

Figure 35.14 Triaxial Test Stress-Strain Curves

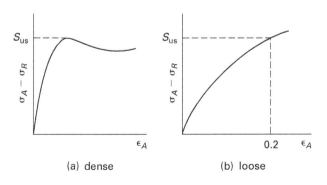

The normal and shear stresses on any plane can be found from the combined stress equations. The angle α is measured counterclockwise from the horizontal (i.e., counterclockwise from the plane that σ_1 acts on). Compression is considered positive. CERM Eq. 35.38 and Eq. 35.39 generate points on Mohr's circle, which can be constructed once σ_A and σ_R are known. *Handbook* table Commonly Performed Laboratory Tests on Soils provides the AASHTO and ASTM designations for the various types of triaxial stress test.

$$\sigma_\alpha = \tfrac{1}{2}(\sigma_A + \sigma_R) + \tfrac{1}{2}(\sigma_A - \sigma_R)\cos 2\alpha \quad 35.38$$

$$\tau_\alpha = \tfrac{1}{2}(\sigma_A - \sigma_R)\sin 2\alpha \quad 35.39$$

Exam tip: For saturated clays in quick (undrained) shear, it is commonly assumed that $\phi = 0$. For this condition only, the undrained shear strength can be found using CERM Eq. 35.42.

$$S_{\text{us}} = c = \frac{\sigma_D}{2} \quad 35.42$$

California Bearing Ratio

Handbook: In Situ Tests for Subsurface Exploration in Pavement Design and Construction; Typical CBR Values

CERM: Sec. 35.22

The *California bearing ratio* (CBR) test is used to determine the suitability of a soil for use as a subbase in pavement sections. It is most reliable when used to test cohesive soils. See *Handbook* table In Situ Tests for Subsurface Exploration in Pavement Design and Construction for properties that can be determined from field CBR tests. The ratio of actual load to a standard load derived from a sample of crushed stone is multiplied by 100, and the result is reported without a percentage symbol.

$$\text{CBR} = \frac{\text{actual load}}{\text{standard load}} \cdot 100 \quad 35.47$$

Figure 35.17 Plotting CBR Test Data

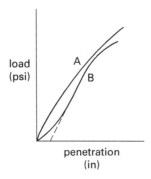

The resulting data will be in the form of inches of penetration versus load. This data can be plotted as shown in CERM Fig. 35.17. If the plot is concave upward (curve B), the steepest slope is extended downward to the x-axis. This point is taken as the zero penetration point, and all penetration values are adjusted accordingly. *Handbook* table Typical CBR Values gives typical CBR values.

Exam tip: Sometimes in pavement design, the soil resistance value (R-value) is required. The R-value is a function of the CBR. Be sure to use the appropriate equation to make the conversion.

Rock Quality Designation (RQD)

Handbook: Rock Quality Designation (RQD)

$$\text{RQD} = \frac{\sum \text{length of sound core pieces} > 4 \text{ in}}{\text{total core run length}}$$

CERM: Sec. 35.26

$$\text{RQD} = \frac{\sum\limits_{L_i > 4 \text{ in}} L_i}{L_\text{core}} \cdot 100\% \qquad 35.48$$

The *Handbook* equation for Rock Quality Designation (RQD) and CERM Eq. 35.48 are equivalent, with only minor differences in nomenclature. The suitability of certain rock exposures is characterized by its RQD. RQD is a rough measure of the degree of fracturing (jointing) in a rock mass, measured as a percentage of drill core lengths that are unfragmented. The RQD is the percentage of the core run that consists of pieces longer than 4 in. CERM Table 35.18 provides recommended allowable bearing pressure for footings.

C. STRESS-STRAIN TESTING OF SOIL AND ROCK

Key concepts: These key concepts are important for answering exam questions in subtopic 10.C, Stress-Strain Testing of Soil and Rock.

- coefficient of consolidation
- consolidation test
- modulus of subgrade reaction
- resilient modulus

NCEES Handbook: To prepare for this subtopic, familiarize yourself with these sections in the *Handbook*.

- Normally Consolidated Soils
- Overconsolidated Soils
- Logarithm-of-Time Method for Determination of c_v
- Square-Root-of-Time Method for Determination of c_v
- Resilient Modulus Under Cyclic Loading
- Subgrade Resilient Modulus
- Correlations between Subgrade Resilient Modulus and Other Soil Properties
- Default M_R Values for Unbound Granular and Subgrade Materials at Unsoaked Optimum Moisture Content and Density Conditions
- Coefficient of Subgrade Reaction k

PE Civil Reference Manual (CERM): Study these sections in CERM that either relate directly to this subtopic or provide background information.

- Section 35.16: Consolidation Tests
- Section 35.23: Plate Bearing Value Test
- Section 40.7: Primary Consolidation
- Section 40.8: Primary Consolidation Rate
- Section 76.25: Layer Strengths
- Section 77.4: Layer Material Strengths

The following equations, figures, and tables are relevant for subtopic 10.C, Stress-Strain Testing of Soil and Rock.

Resilient Modulus

Handbook: Resilient Modulus Under Cyclic Loading; Subgrade Resilient Modulus; Correlations between Subgrade Resilient Modulus and Other Soil Properties; Default M_R Values for Unbound Granular and Subgrade Materials at Unsoaked Optimum Moisture Content and Density Conditions

CERM: Sec. 76.25

The resilient modulus, M_R, is the most important parameter used for the design of flexible primary-road pavements. It is sensitive to varying moisture and temperature conditions within the subgrade. A composite value is typically used, per CERM Fig. 76.9, which considers fluctuations in the parameter based on seasonal moisture and temperature conditions that impact subgrade strength.

The resilient modulus is the same as the modulus of elasticity, E, of the soil. It is not the same as the *modulus of subgrade reaction*, k, used in rigid pavement design, although the two are related. For positive values of the resilient modulus, $M_R \approx 19.4k$.

The following *Handbook* figures and tables provide correlations and relationships to determine the resilient modulus.

- Resilient Modulus Under Cyclic Loading
- Subgrade Resilient Modulus
- Correlations between Subgrade Resilient Modulus and Other Soil Properties
- Default M_R Values for Unbound Granular and Subgrade Materials at Unsoaked Optimum Moisture Content and Density Conditions

Exam tip: Make sure to pay attention to the inputs and requirements for use to ensure you are using the appropriate correlation equation to determine the resilient modulus.

Modulus of Subgrade Reaction

Handbook: Coefficient of Subgrade Reaction k

CERM: Sec. 35.23, Sec. 77.4

Slab-on-grade design is based mainly on the *modulus of subgrade reaction*, a parameter most often correlated to other physical or strength properties, including classification, relative density, and unconfined compressive strength.

A *plate bearing value test* is performed on compacted soil in the field and provides an indication of the shear strength of pavement components. This test is graphed in *Handbook* figure Coefficient of Subgrade Reaction k. (See also CERM Fig. 35.18.) The bearing value is the interpolated load that would produce a deflection of 0.5 in (12 mm). The *subgrade modulus* (*modulus of subgrade reaction*), k, is the slope of the line (in psi per inch) in the loading range encountered by the soil. Typical values are given in CERM Table 35.17.

Table 35.17 Typical Values of the Subgrade Modulus

group symbol	range of subgrade modulus, k (psi/in (kPa/mm))
GW	300–500 (80–140)
GP	250–400 (68–110)
GM	100–400 (27–110)
GC	100–300 (27–80)
SW	200–300 (54–80)
SP	200–300 (54–80)
SM	100–300 (27–80)
SM-SC	100–300 (27–80)
SC	100–300 (27–80)
ML	100–200 (27–54)
ML-CL	—
CL	50–200 (14–54)
OL	50–100 (14–27)
MH	50–100 (14–27)
CH	50–150 (14–41)
OH	25–100 (6.8–27)

(Multiply psi/in by 0.2714 to obtain kPa/mm.)

CERM Sec. 77.4 provides many correlations between various soil relationships, including CBR, subgrade modulus, and the R-value of the soil. These values can be used to assist in pavement design and determining subbase thickness.

Consolidation and Normally Consolidated Soils

Handbook: Normally Consolidated Soils

CERM: Sec. 40.7

Compression of cohesive materials and resulting structure settlement can vary significantly based on in situ stress, the anticipated change in stress, and the relationship between that bandwidth of stress and historic stress. The laboratory consolidation test evaluates load-induced compression by establishing a benchmark—the preconsolidation pressure—to which in situ stresses and post-construction stresses are compared, allowing the significance of the load increases to be quantified.

Recompression requires that the sum of the in situ stress and any additional stresses, $\Delta p'_v$, do not exceed the historic maximum (therefore, p'_o and $\Delta p'_v$ need to be known).

The significance of compression and settlement in cases where stresses push into the virgin compression zone depends on where the preconsolidation pressure is situated with respect to the in situ and post-construction stresses.

$$S_C = \sum_1^n \left(\frac{C_c}{1+e_o}\right) H_o \log_{10} \frac{p_f}{p_o}$$

$$S_C = \sum_1^n C_{c\varepsilon} H_o \log_{10} \frac{p_f}{p_o}$$

$$C_{c\varepsilon} = \frac{C_c}{1+e_o}$$

These equations from Normally Consolidated Soils in the *Handbook* apply exclusively to situations where in situ and post-construction stresses remain below or lie entirely beyond the preconsolidation pressure, respectively, and compression and settlement are governed only by the recompression index or compression index. The first equations are equivalent to CERM Eq. 40.16.

$$S_{\text{primary}} = \frac{H \Delta e}{1+e_o} = \frac{H C_c \log_{10} \frac{p'_o + \Delta p'_v}{p'_o}}{1+e_o} \quad\quad 40.16$$

$$= H(\text{CR}) \log_{10} \frac{p'_o + \Delta p'_v}{p'_o}$$

[normally consolidated]

For overconsolidated soils, see the entry "Overconsolidated Soils" in this chapter.

Overconsolidated Soils

Handbook: Overconsolidated Soils

CERM: Sec. 40.7

$$S = \sum_1^n \left(\frac{H_o}{1+e_o}\right)\left(C_r \log_{10} \frac{p_c}{p_o} + C_c \log_{10} \frac{p_f}{p_c}\right)$$

$$S = \sum_1^n H_o \left(C_{r\varepsilon} \log_{10} \frac{p_c}{p_o} + C_{c\varepsilon} \log_{10} \frac{p_f}{p_c}\right)$$

These equations show how even overconsolidated soils can experience appreciable amounts of compression if the gap between the in situ and preconsolidation pressure is limited and/or the magnitude of the anticipated stress increase is large; the sum of the in situ and post-construction stresses must always be compared to the preconsolidation pressure. These equations are equivalent to CERM Eq. 40.15.

$$S_{\text{primary}} = \frac{H \Delta e}{1+e_o} = \frac{H C_r \log_{10} \frac{p'_o + \Delta p'_v}{p'_o}}{1+e_o} \quad\quad 40.15$$

$$= H(\text{RR}) \log_{10} \frac{p'_o + \Delta p'_v}{p'_o}$$

[overconsolidated]

For normally consolidated soils, see the entry "Consolidation and Normally Consolidated Soils" in this chapter.

Overconsolidation Ratio

CERM: Sec. 35.16

$$\text{OCR} = \frac{p'_{\text{max}}}{p'_o} \quad\quad 35.33$$

The *overconsolidation ratio*, OCR, is defined by CERM Eq. 35.33. p'_o is the present or in situ overburden pressure, and p'_{max} is the maximum past pressure or *preconsolidation pressure*.

An overconsolidation ratio of 1 means that the soil is normally consolidated; an overconsolidation ratio greater than 1 means that the soil is overconsolidated. In rare circumstances, such as during construction or rapid underwater deposition, a soil may be underconsolidated; that is, it has not yet come to equilibrium with its present load.

Coefficient of Consolidation

Handbook: Logarithm-of-Time Method for Determination of c_v; Square-Root-of-Time Method for Determination of c_v

Logarithm-of-Time Method for Determination of c_v

$$c_v = \frac{0.197 H_d^2}{t_{50}}$$

Square-Root-of-Time Method for Determination of c_v

$$c_v = \frac{0.848 H_d^2}{t_{90}}$$

CERM: Sec. 40.8

$$t = \frac{T_v H_d^2}{C_v} \qquad 40.20$$

The time for a single layer to reach a specific consolidation is given by the *Handbook* equations and by CERM Eq. 40.20. The differences in the constant applied to the c_v equations are due to the difference in the time at which c_v is being calculated. One is based on t_{50}, the time at which 50% consolidation has occurred; the other is based on t_{90}, the time at which 90% consolidation has occurred. The square-root-of-time method often provides a slightly higher value of c_v than the logarithm-of-time method. The C_v value is used in CERM Eq. 40.20 to calculate the estimated time of settlement.

In problems dealing with the time rate of consolidation, mistakes are often made due to the units. For example, the coefficient of consolidation may be expressed in square feet per day, but the answer may require time to be expressed in years. The time factor, T_v, is similar numerically to the degree of consolidation being calculated. The drainage path length, H_d, resembles layer thickness, H, but could amount to half that value where there is two-way drainage.

Exam tip: There are many variables related to consolidation that begin with a C (C_c, C_{ec}, C_s, etc.). Pay attention to the subscript to make sure you are using the correct term and/or equation.

D. PERMEABILITY TESTING PROPERTIES OF SOIL AND ROCK

Key concepts: These key concepts are important for answering exam questions in subtopic 10.D, Permeability Testing Properties of Soil and Rock.

- bearing, lateral loading, and steady-state slope stability
  - direct shear
  - triaxial shear
  - unconfined compression
  - unit weight
- constant- and falling-head tests
- construction
  - Atterberg limits
  - corrosion testing
  - gradation
  - moisture content
- deformation and transient slope stability
  - consolidation
  - permeability
- infiltration and steady-state seepage
  - gradation
  - permeability
- pavement design
  - Atterberg limits
  - California bearing ratio
  - gradations
  - moisture content

NCEES Handbook: To prepare for this subtopic, familiarize yourself with these sections in the *Handbook*.

- Taylor's Chart for Soils with Friction Angle
- Taylor's Chart for $\phi' = 0$ Conditions for Slope Angles (β) Less than 54°
- Slope Stability Guidelines for Design
- Hazen's Equation for Permeability
- Determination of Soil Type
- Darcy's Law
- Approximate Coefficient of Permeability for Various Sands
- Geotechnical Influences on Major Distresses in Flexible Pavements
- Geotechnical Influences on Major Distresses in Rigid Pavements
- Typical Zone of Influence for an Asphalt Pavement Section
- Typical CBR Values
- Subgrade Resilient Modulus
- Correlations between Subgrade Resilient Modulus and Other Soil Properties
- Default M_R Values for Unbound Granular and Subgrade Materials at Unsoaked Optimum Moisture Content and Density Conditions

PE Civil Reference Manual (CERM): Study these sections in CERM that either relate directly to this subtopic or provide background information.

- Section 35.15: Permeability Tests
- Section 35.27: Geotechnical Safety and Risk Mitigation
- Section 40.10: Slope Stability in Saturated Clay
- Section 77.7: Pavement Design Methodology

The following equations and figures are relevant for subtopic 10.D, Permeability Testing Properties of Soil and Rock.

Constant- and Falling-Head Tests

Handbook: Hazen's Equation for Permeability; Darcy's Law; Approximate Coefficient of Permeability for Various Sands

$$k = C(D_{10})^2$$

CERM: Sec. 35.15

$$K = \frac{VL}{hAt} \quad \text{[constant head]} \qquad 35.29$$

$$K = \frac{A'L}{At}\ln\frac{h_i}{h_f} \quad \text{[falling head]} \qquad 35.30$$

The equation in *Handbook* section Hazen's Equation for Permeability, which calculates the permeability constant, k, provides a crude estimate of permeability, and actual values can vary by two orders of magnitude. Accurate values are calculated using CERM Eq. 35.29 and Eq. 35.30 from controlled permeability tests using constant- or falling-head permeameters. (See CERM Fig. 35.8.)

In a *constant-head test*, shown in CERM Eq. 35.29, the volume, V, of water percolating through the soil over time is measured. The test is applicable to coarse-grained soils with $K > 10^{-3}$ cm/s. In a *falling-head test*, shown in CERM Eq. 35.30, the change in head over time is measured as the water percolates through the soil. The test is applicable to fine-grained soils with $K < 10^{-3}$ cm/s.

Darcy's law is provided in *Handbook* section Darcy's Law; this equation is also useful in calculating the permeability, as are diagrams for constant- and falling-head test setups.

> *Exam tip*: Typical values of permeability are provided in *Handbook* table Approximate Coefficient of Permeability for Various Sands. Use this table to make sure your calculations are in the right range.

Steady-State Slope Stability

Handbook: Slope Stability Guidelines for Design; Taylor's Chart for Soils with Friction Angle; Taylor's Chart for $\phi' = 0$ Conditions for Slope Angles (β) Less than 54°

CERM: Sec. 40.10

Handbook table Slope Stability Guidelines for Design provides extensive guidance regarding which slope stability analysis should be used, depending on soil type.

Long-term steady-state solutions are often limiting as they are static snapshots of stress and stability conditions after construction is complete. Short-term stresses on earth materials during construction are often more critical.

The maximum slope for cuts in cohesionless drained sand is the angle of internal friction, ϕ. For saturated clay with ϕ equal to 0°, the Taylor slope stability chart (shown in the *Handbook* in Taylor's Chart for Soils with Friction Angle and Taylor's Chart for $\phi' = 0$ Conditions for Slope Angles (β) Less than 54°, and in CERM Fig. 40.6) can be used to determine the factor of safety against slope failure. The Taylor chart makes a set of assumptions, which are detailed in CERM Sec. 40.10.

Figure 40.6 Taylor Slope Stability (undrained, cohesive soils; $\varphi = 0°$)

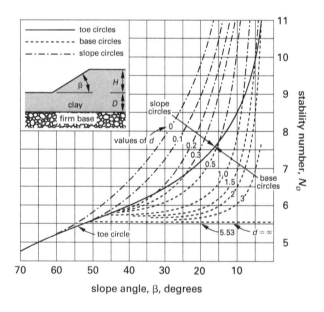

Source: *Soil Mechanics*, NAVFAC Design Manual DM-7.1, 1986, Fig. 2, p. 7.1-319.

CERM Eq. 40.28 gives the relationship between the stability number and the cohesive factor of safety.

$$F_{\text{cohesive}} = \frac{N_o c}{\gamma_{\text{eff}} H} \qquad 40.28$$

The effective specific weight is used when the clay is submerged. Otherwise, the bulk density is used. Generally, the saturated bulk density is used to account for worst-case loading. The minimum acceptable factor of safety is approximately 1.3 to 1.5.

Infinite slopes are calculated similarly to a traditional sliding block analysis (if no water is present). The method of slices presented in the *Handbook* is useful for understanding the concept.

> *Exam tip*: *Handbook* table Slope Stability Guidelines for Design is a great starting point to determine the next steps in slope stability analysis.

Pavement Design

Handbook: Geotechnical Influences on Major Distresses in Flexible Pavements; Geotechnical Influences on Major Distresses in Rigid Pavements; Typical Zone of Influence for an Asphalt Pavement Section; Typical CBR Values; Subgrade Resilient Modulus; Correlations between Subgrade Resilient Modulus and Other Soil Properties; Default M_R Values for Unbound Granular and Subgrade Materials at Unsoaked Optimum Moisture Content and Density Conditions

CERM: Sec. 77.7

Slab thickness is determined from CERM Fig. 77.2 for each k-value and then rounded to the nearest 0.5 in (13 mm). The following variables are required: (a) effective k-value; (b) estimated future traffic in equivalent single axle loads (ESALs) using the rigid pavement equivalency factors; (c) reliability level, R; (d) overall standard deviation, S_o; (e) design serviceability loss, ΔPSI; (f) concrete modulus of elasticity, E_c; (g) concrete modulus of rupture, S'_c; (h) load transfer coefficient, J; and (i) drainage.

The *Handbook* provides causes of problems in pavement in the Geotechnical Influences on Major Distresses in Flexible Pavements and Geotechnical Influences on Major Distresses in Rigid Pavements charts. The *Handbook* chart Typical Zone of Influence for an Asphalt Pavement Section shows how material below the pavement's surface affects the pavement's design. The Typical CBR Values chart in the *Handbook* can be used to check calculated CBRs prior to subsequent calculations, and the Subgrade Resilient Modulus chart and Correlations between Subgrade Resilient Modulus and Other Soil Properties table allow the user to translate between different input values. Typical values and ranges of M_R are provided in the table Default M_R Values for Unbound Granular and Subgrade Materials at Unsoaked Optimum Moisture Content and Density Conditions.

Construction

Handbook: Determination of Soil Type

CERM: Sec. 35.27

The materials placed on the project site have to be the right materials placed in the right way. This aspect of testing is also beneficial to engineers thinking ahead to how a project might be built, and what that work means to their analysis of stability and performance. The determination of soil and rock is critical to construction safety and the angle of temporary excavations. CERM Sec. 35.27 details some of the subjects affected by soil characteristics. *Handbook* section Determination of Soil Type provides the OSHA categories of soil and rock and the construction that is allowed in each type.

E. EFFECTIVE AND TOTAL STRESSES

Key concepts: These key concepts are important for answering exam questions in subtopic 10.E, Effective and Total Stresses.

- geologic profiles
- profiling
  - pore water pressure
  - total stresses (earth and water)
  - undrained shear strength

NCEES Handbook: To prepare for this subtopic, familiarize yourself with these sections in the *Handbook*.

- Diagram Illustrating Consolidation of a Layer of Clay Between Two Pervious Layers
- Shear Strength of (a) Cohesionless Soils and (b) Cohesive Soils
- Hammer Efficiency

PE Civil Reference Manual (CERM): Study these sections in CERM that either relate directly to this subtopic or provide background information.

- Section 35.7: Effective Stress
- Section 35.9: Standard Penetration Test
- Section 35.10: Cone Penetrometer Test

The following equations, figures, and concepts are relevant for subtopic 10.E, Effective and Total Stresses.

Profiling

Handbook: Diagram Illustrating Consolidation of a Layer of Clay Between Two Pervious Layers

CERM: Sec. 35.7

Stress is everywhere in engineering analyses; stress-dependent qualities include

- shear strength as determined by friction angle
- bearing capacity
- settlement
- earth pressure

Total stress analyses—employing undrained shear strength parameters—have their place but limit the engineer to only a glimpse of immediate or ultimate engineering stability or performance.

Effective stress analyses—employing drained shear strength parameters—allow the engineer to explore and even manipulate structure stability and performance at the same endpoints that total stress analyses do, but also at an infinite number of intermediate steps to fully characterize geologic profile and structure response to stress changes.

CERM Sec. 35.7 provides several examples of stress analyses and pore pressure diagrams.

Even if pore water pressures cannot be determined or estimated with accuracy, their use as estimated values still provides insight into the impact of water on stability and performance.

The straight line *ab* in *Handbook* figure Diagram Illustrating Consolidation of a Layer of Clay Between Two Pervious Layers shows the initial hydrostatic pressure in the clay layer. As the water pressure changes over time, the hydrostatic pressure is not linear, but curved, as shown on lines C_1 and C_2 in the figure. It is also important to remember that this figure assumes that drainage is equal in both directions (both upper and lower materials are equally permeable).

> *Exam tip*: On the exam, pay attention to the ϕ and c designations in the problem statement and equations used to solve the problem. If there is a prime mark next to the variable (ϕ' or c'), that designates the effective friction angle or the effective cohesion. No such mark indicates the total friction angle or the total cohesion value.

Geologic Profiles

Handbook: Hammer Efficiency

CERM: Sec. 35.7, Sec. 35.9, Sec. 35.10

$$N_{60} = \left(\frac{E_{\text{eff}}}{60}\right) N_{\text{meas}}$$

The independent determination and development of total stress, pore water pressure, and effective stress profiles are the most rigorous means of qualifying a geologic profile. Effective stresses in materials extending below the groundwater surface can be determined directly by using buoyant material weight rather than total or saturated unit weight. The use of buoyant unit weight to compute effective stresses within materials below the groundwater surface saves multiple analytical steps in the determination process.

See CERM Sec. 35.7, Sec. 35.9, and Sec. 35.10 for more information on effective stress and penetration tests.

Handbook section Hammer Efficiency provides adjustment terms for raw blow counts from a standard penetration test (SPT) hammer in order to determine the N_{60} value. The N_{60} value is used in many formulas to determine other soil properties.

> *Exam tip*: SPTs and cone penetrometer tests (CPTs) are commonly used to prepare subsurface profiles.

Effective Stress

Handbook: Shear Strength of (a) Cohesionless Soils and (b) Cohesive Soils

CERM: Sec. 35.7

$$\sigma' = \sigma - u \qquad \text{35.17}$$

The *effective stress*, σ', is the portion of the total stress that is supported through grain contact. It is calculated using CERM Eq. 35.17 as the difference between the *total stress*, σ, and the *pore water pressure*, u. (The pore water pressure is sometimes called the *neutral stress* because it is equal in all directions; that is, it has no shear stress component.) The effective stress is the average stress on a plane through the soil, not the actual contact stress between two soil particles (which can be much higher). *Handbook* figure Shear Strength of (a) Cohesionless Soils and (b) Cohesive Soils provides a description and graph of the total stress shear strength for both cohesive and purely frictional soils. The effective stress condition is similar, but the graphed line is altered to remove the pore water pressure.

19 Field Materials Testing, Methods, and Safety

Content in blue refers to the *NCEES Handbook*.

A. Excavation and Embankment, Borrow Source Studies, Laboratory and Field Compaction....... 19-1
B. Trench and Construction Safety 19-3
C. Geotechnical Instrumentation 19-4

The knowledge area of Field Materials Testing, Methods, and Safety makes up between three and five questions out of the 80 questions on the PE Civil Geotechnical exam. The organization of this chapter follows the order of subtopics given by NCEES for this knowledge area. Each subtopic is covered in the following sections.

A. EXCAVATION AND EMBANKMENT, BORROW SOURCE STUDIES, LABORATORY AND FIELD COMPACTION

Key concepts: These key concepts are important for answering exam questions in subtopic 11.A, Excavation and Embankment, Borrow Source Studies, Laboratory and Field Compaction.

- dynamic or static cone penetrometer
- hand auger
- pocket penetrometer or vane shear
- sand cone or nuclear density gauge

NCEES Handbook: To prepare for this subtopic, familiarize yourself with these sections in the *Handbook*.

- Slope Stability Guidelines for Design
- Guidelines for Minimum Number of Exploration Points and Depth of Exploration
- Evaluation of the Consistency of Fine-Grained Soils
- Methods for Index Testing of Soils
- Methods for Performance Testing of Soils
- Determination of Soil Type
- Slope and Shield Configurations
- Allowable Slopes
- Slope Configurations: Excavations in Layered Soils
- Excavations Made in Type A Soil
- Excavations Made in Type B Soil
- Instruments for Measuring Piezometric Pressure
- Categories of Instruments for Measuring Deformation
- In Situ Tests for Subsurface Exploration in Pavement Design and Construction

PE Civil Reference Manual (CERM): Study these sections in CERM that either relate directly to this subtopic or provide background information.

- Section 35.2: Soil Classification
- Section 35.8: Standardized Soil Testing Procedures
- Section 35:11: Proctor Test
- Section 35.12: Modified Proctor Test
- Section 39.1: Excavation
- Section 39.7: Stability of Braced Excavations in Clay
- Section 39.10: Analysis/Design of Braced Excavations
- Section 80.4: Classification of Materials
- Section 83.5: Soil Classification
- Section 83.6: Trenching and Excavation

Codes and standards: To prepare for this subtopic, familiarize yourself with these sections from codes and standards.

Occupational Safety and Health Act (OSHA)

- 29 CFR 1926 Subpart P App. A: Soil Classification

The following tables and concepts are relevant for subtopic 11.A, Excavation and Embankment, Borrow Source Studies, Laboratory and Field Compaction.

Embankment Construction

Handbook: Slope Stability Guidelines for Design; Guidelines for Minimum Number of Exploration Points and Depth of Exploration; Instruments for Measuring Piezometric Pressure; Categories of Instruments for Measuring Deformation

Handbook table Slope Stability Guidelines for Design gives guidance for the construction of embankments on a cohesive foundation. It is important that the embankment be constructed (and analyzed) in stages, with waiting periods to allow water to seep out of the foundation, consolidation to take place, and the strength of the clay to improve slowly as the load is applied. The strength properties of the embankment and/or its foundation may change over time due to the water and loading conditions associated with the construction of the embankment. Consolidation testing is needed to estimate the length of each waiting period.

Handbook table Guidelines for Minimum Number of Exploration Points and Depth of Exploration provides recommendations for the amount, location, and depth of the field explorations. *Handbook* tables Instruments for Measuring Piezometric Pressure and Categories of Instruments for Measuring Deformation list different types of instruments that can be used to measure water pressure and deformation to assist with the necessary field consolidation.

Material Testing Tools

Handbook: Evaluation of the Consistency of Fine-Grained Soils; Methods for Index Testing of Soils; Methods for Performance Testing of Soils

CERM: Sec. 83.5

OSHA: 29 CFR 1926 Subpart P App. A

Certain tools make observations easier. Dynamic or static *cone penetrometers* allow the geotechnical representative to both classify and determine the relative strength of soils (via resistance to advancement). *Pocket penetrometer tests* and *shear vane tests* can be performed on excavation banks to measure undrained shear strength and evaluate compliance with OSHA requirements for excavation stability.

CERM Sec. 83.5 provides insight into the various types of tests that OSHA requires to determine soil type. This section describes the thumb penetration, pocket penetrometer, and torvane shear tests. The pocket penetrometer and shear vane tests provide numerical readings. The thumb penetration test is more subjective but still permissible by the OSHA classification requirements.

Handbook table Evaluation of the Consistency of Fine-Grained Soils provides a list of manual manipulations that can be performed on a soil specimen to determine the unconfined compressive strength. The "tools" used to perform these tests are the technician's fingers and thumbs. The Methods for Index Testing of Soils and Methods for Performance Testing of Soils tables in the *Handbook* provide a list of soil tests and their associated procedures. The tools used in these tests are sometimes included in the name of the test performed (in the case of the sieve tests).

Grading Activities

Handbook: Determination of Soil Type; Slope and Shield Configurations; Allowable Slopes; Slope Configurations: Excavations in Layered Soils; Excavations Made in Type A Soil; Excavations Made in Type B Soil

CERM: Sec. 35.11, Sec. 35.12, Sec. 80.4, Sec. 83.5, Sec. 83.6

The design of foundations, slabs, pavements, walls, and other earth-supported structures is generally based on a limited amount of subsurface information and assumes no inconsistencies in the composition or competence of the earth materials present. In reality, however, subsurface conditions vary.

Excavated material is usually classified as common excavation or rock excavation. *Common excavation* is the excavation of soil. In highway construction, *common road excavation* is soil found in the roadway. *Common borrow* is soil found outside the roadway and brought into it. Borrow is necessary where there is not enough material in the roadway excavation to provide for the embankment.

Handbook sections Slope Configurations: Excavations in Layered Soils, Determination of Soil Type, and Excavations Made in Type B Soil show various configurations for cuts. The various layered soil conditions are type A, B, or C, as defined by *Handbook* table Allowable Slopes. Allowable excavations in a pure type A soil are shown in section Excavations Made in Type A Soil and allowable excavations in a pure type B soil are shown in section Excavations Made in Type B Soil. These guidelines are provided by OSHA (see *Handbook* figure Slope and Shield Configurations); excavations in type A soil are allowed to be steeper than excavations in type B soil. Soil types are listed in *Handbook* section Determination of Soil Type.

Material Testing and Observation

Handbook: In Situ Tests for Subsurface Exploration in Pavement Design and Construction

CERM: Sec. 35.8, Sec. 35.12

Grading contractors are required only to excavate or fill to specified depths or elevations, to procure materials that meet project specifications, and to meet compaction requirements.

Excavations need to be clear of debris and groundwater and be safely sloped or shored to facilitate the evaluation of materials at and below the excavation bottoms.

Groundwater makes evaluating excavation bottoms more difficult and can corrupt earth and construction materials being placed in the excavations.

Handbook table In Situ Tests for Subsurface Exploration in Pavement Design and Construction describes a variety of tests, including what type of soil each test is applicable for and what can be determined in regard to pavement design.

CERM Table 35.8 provides a thorough list of ASTM standard soil tests. Some are laboratory tests, and others are tests performed at the time of construction in the field. The results from laboratory tests must be confirmed and evaluated in the field at the time of construction to confirm agreement with the lab results.

CERM Table 35.10 provides a range of the optimum moisture content for a complete range of soil types. This moisture content and compaction density must be confirmed in the field at the time of construction.

B. TRENCH AND CONSTRUCTION SAFETY

Key concepts: These key concepts are important for answering exam questions in subtopic 11.B, Trench and Construction Safety.

- disturbance
- fissuring
- groundwater
- layering
- soil composition and strength
- vibrations

NCEES Handbook: To prepare for this subtopic, familiarize yourself with these sections in the *Handbook*.

- Determination of Soil Type
- Slope Configurations: Excavations in Layered Soils
- Excavations Made in Type A Soil
- Excavations Made in Type B Soil

PE Civil Reference Manual **(CERM):** Study these sections in CERM that either relate directly to this subtopic or provide background information.

- Section 83.5: Soil Classification
- Section 83.6: Trenching and Excavation

Codes and standards: To prepare for this subtopic, familiarize yourself with these sections from codes and standards.

Occupational Safety and Health Act (OSHA)

- 29 CFR 1926 Section 652(b): Design of sloping and benching systems
- 29 CFR 1926 Subpart P App. B: Sloping and Benching

The following figures and concepts are relevant for subtopic 11.B, Trench and Construction Safety.

Geotechnical OSHA Regulations

Handbook: Determination of Soil Type; Slope Configurations: Excavations in Layered Soils

CERM: Sec. 83.5

OSHA: 29 CFR 1926 Sec. 652(b), 29 CFR 1926 Subpart P App. B

Handbook figure Slope Configurations: Excavations in Layered Soils provides a comprehensive overview of OSHA regulations for the establishment and maintenance of excavations, with helpful illustrations.

For conventional sloped excavations, the regulations given in OSHA 29 CFR 1926 Subpart P App. B are commonly used. A large percentage of excavations terminate at depths shallower than 20 ft, and compliance with these regulations eliminates issues with procuring and gaining approval for an alternative design. The regulations can be applied as a decision tree (as shown in CERM Fig. 83.1) or by using the process of elimination.

Layered soils occur when an excavation is made into soil that is not homogeneous. *Handbook* section Determination of Soil Type provides a definition of the various soil types. *Handbook* figure Slope Configurations: Excavations in Layered Soils shows how the angle of excavation must be modified based on the different soil types encountered. It is important to recognize the order of the soil layering.

Exam tip: If the stronger soil is above the weaker soil, the weaker soil excavation angle must be followed for the entire excavation. If the stronger soil is below the weaker soil, the excavation angle may be increased (made steeper) within the layer of stronger soil.

Slope Configuration: Type A

Handbook: Slope Configurations: Excavations in Layered Soils; Excavations Made in Type A Soil

CERM: Sec. 83.6

The type A classification is the most favorable, allowing the steepest slopes for soil. The type A classification is limited to cohesive soils of appreciable strength but has a number of exclusions based on soil structure (fissures, disturbance, and steeply oriented layer boundaries) and adjacent activities (vibrations).

CERM Fig. 83.3 shows the different types of slope excavations in soil type A. *Handbook* figure Excavations Made in Type A Soil provides cross sections showing the allowable excavation angles and depths for type A soil. Short-term excavations are at the steepest angle. Note the location of the overall angle measurement (the dashed line) to ensure that an excavation is not too steep. Pay close attention to the description of each cross section. These descriptions determine the angle and height of the excavation. Type A soil excavations are allowed to be steeper and deeper than type B excavations.

Handbook section Slope Configurations: Excavations in Layered Soils provides guidance for excavations in layered soils. When type A soil is above type B or type C soil, the guidance for the weaker material controls the slope angle of the type A soil as well. Although the type A soil alone could be excavated at a steeper angle, it must be kept at a flatter angle if it lays on top of a weaker soil.

Slope Configuration: Type B

Handbook: Slope Configurations: Excavations in Layered Soils; Excavations Made in Type B Soil

CERM: Sec. 83.6

Type B is the intermediate soil classification and includes a variety of soils, including granular soils containing angular stone, silt, or clay fines.

The type B classification also includes type A soils that have been downgraded due to fissuring, disturbance, or exposure to vibrations. Unstable dry rock is included in the type B classification as well.

CERM Fig. 83.3 shows the different types of slope excavations in type B soil. *Handbook* figure Slope Configurations: Excavations in Layered Soils shows only the second and third cross sections from CERM Fig. 83.3. These two cross sections correspond to the two types of excavations that are permitted in cohesive type B soil. The first cross section shown in the *Handbook* for type B soil corresponds to the first cross section in the *Handbook* for type A soil but shows a flatter angle for type B soil and applies to all type B soils.

Handbook section Excavations Made in Type B Soil shows excavations made in soil that is purely type B. The angle of these excavations can be no steeper than 45°.

Slope Configuration: Type C

Handbook: Determination of Soil Type; Slope Configurations: Excavations in Layered Soils

CERM: Sec. 83.6

The type C classification is the least favorable of the three classifications. Type C soils include the weakest of the cohesive soils, the balance of the granular soils, soil profiles with steeply dipping layers, and all soils impacted by groundwater. See CERM Fig. 83.3 for an example of a simple slope excavation with type C soil.

Handbook figure Slope Configurations: Excavations in Layered Soils shows the different types of slope excavations in layered soils.

Characteristics of type C soil are described in *Handbook* section Determination of Soil Type. *Handbook* figure Slope Configurations: Excavations in Layered Soils shows how type C soil interacts with and is treated differently than type A and type B soil. Type C soil cannot be excavated any steeper than 34° from horizontal, and this angle must be reduced by 50% if signs of distress appear.

C. GEOTECHNICAL INSTRUMENTATION

Key concepts: These key concepts are important for answering exam questions in subtopic 11.C, Geotechnical Instrumentation.

- deformation monitoring
- pore water pressure
- settlement monitoring
- vibration monitoring

NCEES Handbook: To prepare for this subtopic, familiarize yourself with these sections in the *Handbook*.

- Categories of Instruments for Measuring Deformation
- Measurements and Instruments for Long-Term Performance Monitoring

***PE Civil Reference Manual* (CERM):** Study these sections in CERM that either relate directly to this subtopic or provide background information.

- Section 85.1: Accuracy
- Section 85.2: Precision
- Section 85.3: Stability
- Section 85.4: Calibration
- Section 85.18: Strain Gauges
- Section 85.23: Stress Measurements in Known Directions

Codes and standards: To prepare for this subtopic, familiarize yourself with these sections from codes and standards.

Design and Construction of Driven Pile Foundations – Volume I (FHWA NHI-16-009)

- Section 3.5.4: Condition Surveys and Vibration Monitoring of Adjacent Facilities

The following equations, figures, tables, and concepts are relevant for subtopic 11.C, Geotechnical Instrumentation.

Settlement Monitoring

Handbook: Categories of Instruments for Measuring Deformation; Measurements and Instruments for Long-Term Performance Monitoring

CERM: Sec. 85.1, Sec. 85.2, Sec. 85.4

In general, settlement is not a major construction or post-construction performance concern. However, for large earthen roadway embankments, rail embankments, earthen levees and dams, and thick-grade fills, settlement can be appreciable and take a significant length of time.

Various devices are used to monitor settlement. The simplest is a *settlement plate*, consisting of a wooden or metal platform to which a vertical standpipe is fixed. One or more settlement plates are installed on surfaces that will receive fill so that the standpipes will extend above the fill; the standpipes can be extended upward as needed as the fill increases. The elevation of the top of each standpipe is then carefully measured and monitored over time.

Settlement cells accomplish the same thing but track the settlement pneumatically and avoid the possibility of the standpipes being damaged during filling. A *shape array* measures changes in grade through the deflection of horizontally oriented inclinometers that can profile settlement over the width of an earth fill instead of just at one point.

Handbook table Categories of Instruments for Measuring Deformation provides a list of instruments that measure vertical deformation (settlement). Many of these instruments also measure horizontal, axial, and rotational deformation. *Handbook* table Measurements and Instruments for Long-Term Performance Monitoring shows that surveying techniques and Global Positioning System (GPS) instruments are recommended for surface vertical deformations.

> *Exam tip*: Review CERM Sec. 85.1 through Sec. 85.4 to understand the keywords associated with measurement reporting (accuracy, precision, stability, calibration, etc.).

Deformation Monitoring

Handbook: Categories of Instruments for Measuring Deformation; Measurements and Instruments for Long-Term Performance Monitoring

In some cases, how far a structure moves laterally is as important as how much it moves vertically. For example, an earth retention system may support a nearby building that cannot be allowed to deflect more than $\frac{1}{2}$ in.

Shape arrays and *slope inclinometer casings* allow lateral deformation to be profiled along a line of interest and over time, so that actual movements can be compared with estimated or threshold movements. *Extensometers* track horizontal movement between two points at a single location over time.

Horizontal deformation can be measured by six of the seven categories listed in *Handbook* table Categories of Instruments for Measuring Deformation. Recommended instruments for the measurement of surface lateral deformations are provided in *Handbook* table Measurements and Instruments for Long-Term Performance Monitoring. Surveying methods, extensometers, and inclinometers are the most common and effective means to measure lateral deformation and displacement.

> *Exam tip*: In exam problems concerning deformation monitoring, be sure to take note of not just the magnitude but also the direction of movement given in the problem statement. Calculations and recommendations for lateral movement are completely different from those for vertical movement.

Strain Gauges

CERM: Sec. 85.18, Sec. 85.23

A bonded *strain gauge* is a metallic resistance device that is cemented to the surface of an unstressed member. CERM Fig. 85.5 shows how strain gauges work.

Strain gauges are the most frequently used tool for determining the stress in a member. Stress can be calculated from strain, or the measurement circuitry can be calibrated to give the stress directly.

For stress in only one direction, only one strain gauge is needed, and the stress can be calculated from Hooke's law. When a surface experiences simultaneous stresses in two directions (biaxial stress), strain in one direction affects the strain in the other direction. Two strain gauges are required, even if determining the stress in only one direction is needed. The strains measured by the gauges are known as the *net strains*.

Figure 85.5 Strain Gauge

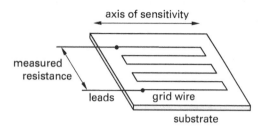

(a) folded-wire strain gauge

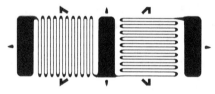

(b) commercial two-element rosette

CERM Sec. 85.18 describes how strain gauges are used in measuring the strain inside a member. Hooke's law relates stress to strain and is shown in CERM Eq. 85.24.

$$\sigma = E\epsilon \quad \quad 85.24$$

If strain in two directions is present, Hooke's law is applied to both directions, and the stresses are solved for simultaneously, as shown in CERM Eq. 85.25 and Eq. 85.26.

$$\epsilon_x = \frac{\sigma_x - \nu\sigma_y}{E} \quad \quad 85.25$$

$$\epsilon_y = \frac{\sigma_y - \nu\sigma_x}{E} \quad \quad 85.26$$

Vibration Monitoring

FHWA NHI-16-009: Sec. 3.5.4

Construction activities can

- cause cracking of plastered and gypsum walls in structures near construction sites
- cause slopes to deform and fail
- impact the curing of concrete and masonry if performed near such materials in their unconsolidated state

Threshold peak particle velocities for these and other situations are well established (by the U.S. Bureau of Mines) and are often used to gauge the severity of ground shaking near construction sites. Portable seismographs are installed at locations of interest to support such evaluations. This type of monitoring is often accompanied by condition survey work that allows the condition of potentially vulnerable structures to be assessed in advance of construction and damage claims.

Neither the *Handbook* nor CERM provides guidance related to vibration monitoring. However, in FHWA NHI-06-009 Sec. 3.5.4, it is recommended that the internal and external condition of nearby structures be documented with pictures and videos prior to the start of construction. In-depth documentation must be performed for existing cracks, and crack monitors may need to be installed, depending on the sensitivity and proximity of the structure to the proposed construction.

20 Earthquake Engineering and Dynamic Loads

Content in blue refers to the *NCEES Handbook*.

A. Liquefaction Analysis and Mitigation Techniques 20-1
B. Seismic Site Characterization 20-2
C. Pseudostatic Analysis and Earthquake Loads 20-4

The knowledge area of Earthquake Engineering and Dynamic Loads makes up between two and four questions out of the 80 questions on the PE Civil Geotechnical exam. The organization of this chapter follows the order of subtopics given by NCEES for this knowledge area. Each subtopic is covered in the following sections.

A. LIQUEFACTION ANALYSIS AND MITIGATION TECHNIQUES

Key concepts: These key concepts are important for answering exam questions in subtopic 12.A, Liquefaction Analysis and Mitigation Techniques.

- evaluating liquefaction potential
- factor of safety for liquefaction

NCEES Handbook: The *Handbook* does not contain any material relevant to this subtopic. Be sure to study the listed sections in CERM and from codes and standards.

PE Civil Reference Manual **(CERM):** Study these sections in CERM that either relate directly to this subtopic or provide background information.

- Section 40.17: Liquefaction

Codes and standards: To prepare for this subtopic, familiarize yourself with these sections from codes and standards.

LRFD Seismic Analysis and Design of Transportation Geotechnical Features and Structural Foundations Reference Manual (FHWA NHI-11-032)

- Section 6.3.1: Hazard Description and Initial Screening

The following equations, figures, and concepts are relevant for subtopic 12.A, Liquefaction Analysis and Mitigation Techniques.

Liquefaction

CERM: Sec. 40.17

Liquefaction is a sudden drop in soil's shear strength. It occurs when cyclic shear reduces effective overburden stresses, resulting in loss of shear strength. Liquefaction can cause bearing capacity failure, structure settlement, and lateral spreading.

Liquefaction is a potentially catastrophic consequence of ground shaking. When compressing a susceptible soil structure, cyclic shear induces an intergranular pore water pressure that, in soils with sufficient fines, cannot dissipate as quickly as it accumulates. When excess pore water pressure is built up, the effective stress within the susceptible soil layer can be reduced to zero, at which point it possesses no shear strength. This can cause settlement of nonliquefied soil layers and structures above it and, if near to such structures, bearing capacity failure.

Factor of Safety: Liquefaction

CERM: Sec. 40.17

FHWA NHI-11-032: Sec. 6.3.1

Liquefaction potential can be quantified in terms of a factor of safety that roughly relates the shear stress required to cause liquefaction to the shear stress induced by the design earthquake. The factor of safety against liquefaction is determined by dividing the cyclic resistance ratio (CRR) by the cyclic stress ratio (CSR).

$$F_s = \frac{\text{CRR}}{\text{CSR}}$$

In general, free-draining gravels are less susceptible to liquefaction when compared to sands, and dense, more granular soils are less likely to liquefy than loose soils. Clean granular soils are more likely to liquefy than soils with fines, especially when clay particles are included. FHWA NHI-11-032 Sec. 6.3.1 provides a checklist for when an evaluation of liquefaction should be assessed. The determination criteria are based on the seismic hazard level and mean earthquake magnitude.

Cyclic Stress Ratio (CSR)

CERM: Sec. 40.17

$$\frac{\tau_{h,\text{ave}}}{\sigma'_o} \approx 0.65\left(\frac{a_{\max}}{g}\right)\left(\frac{\sigma_o}{\sigma'_o}\right)r_d \qquad 40.37$$

In CERM Eq. 40.37,

$a_{\max}$	peak horizontal acceleration at the ground surface	m/s²
r_d	stress reduction factor (between 1.0 and 0.9)	–
σ_o	total overburden stress	kPa
σ'_o	effective overburden stress within suspect layer	kPa
$\tau_{h,\text{ave}}$	average cyclic shear stress due to earthquake	kPa

The cyclic stress ratio (CSR) is given by CERM Eq. 40.37. This ratio compares the average cyclic shear stress developed on the horizontal surfaces of the sand due to the earthquake loading with the initial vertical effective stress before the earthquake forces were applied. The CSR considers initial overburden stresses in addition to applicable seismic parameters, including cyclic shear stress and peak horizontal acceleration. A stress reduction factor between 1.0 and 0.9 is also applied (although it is conservatively assumed to be 1.0).

In cases where vulnerable layers are close to structures, the effective stress need not be reduced to zero to cause damage to the structures, but only to a value small enough to reduce the available shear strength and associated bearing capacity to less than the applied bearing pressure.

Cyclic Resistance Ratio (CRR)

CERM: Sec. 40.17

The cyclic resistance ratio (CRR) is determined graphically based on standard penetration resistance values corrected for overburden, $(N_1)_{60}$. Ignoring the boundary between the liquefaction and nonliquefaction zones, the CRR is the threshold value at which liquefaction otherwise occurs. The liquefaction potential for a magnitude 7.5 event is shown.

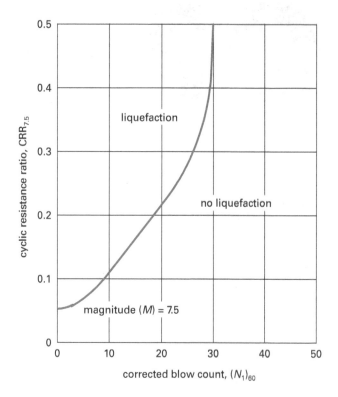

From NAVFAC DM 7.3, *Soil Dynamics, Deep Stabilization, and Special Geotechnical Construction*, Chap. 2, Fig. 24, © 1983, Naval Facilities Engineering Command.

When the ratio of CRR to the cyclic stress ratio (CSR) drops below 1.1, the potential for liquefaction is expected and should be considered in the site development plans and design. A ratio greater than 1.1 means that liquefaction is not likely to occur.

B. SEISMIC SITE CHARACTERIZATION

Key concepts: These key concepts are important for answering exam questions in subtopic 12.B, Seismic Site Characterization.

- epicenter
- epicenter distance
- fault
- focal depth
- hypocenter
- site class
    - A: hard rock
    - B: rock
    - C: very dense soil and soft rock
    - D: stiff soil
    - E: soft clay soil

NCEES Handbook: The *Handbook* does not contain any material relevant to this subtopic. Be sure to study the listed sections from codes and standards.

Codes and standards: To prepare for this subtopic, familiarize yourself with these sections from codes and standards.

LRFD Seismic Analysis and Design of Transportation Geotechnical Features and Structural Foundations Reference Manual (FHWA NHI-11-032)

- Section 2.2: Earthquake Sources
- Section 2.3: Seismic Waves
- Section 2.5: Parameters Describing Earthquake Size and Location
- Section 2.6: Parameters Describing Faulting
- Section 2.7: Parameters Describing Ground Shaking

Minimum Design Loads for Buildings and Other Structures (ASCE/SEI 7)

- Section 20.3: Site Class Definitions
- Section 20.3.3: Site Classes C, D, and E
- Section 20.3.4: Shear Wave Velocity for Site Class B
- Section 20.3.5: Shear Wave Velocity for Site Class A
- Section 20.4.1: $\bar{v}_s$, Average Shear Wave Velocity
- Section 20.4.2: $\bar{N}$, Average Field Standard Penetration Resistance and $\bar{N}_{ch}$, Average Standard Penetration Resistance for Cohesionless Soil Layers
- Section 20.4.3: $\bar{s}_u$, Average Undrained Shear Strength

The following equations, figures, and concepts are relevant for subtopic 12.B, Seismic Site Characterization.

Earthquake Principles

FHWA NHI-11-032: Sec. 2.2, Sec. 2.3, Sec. 2.5, Sec. 2.6, Sec. 2.7

Structures do not commonly bear on rock but rather on soil deposits formed in place or deposited atop the rock. As a result, structures in seismically active areas can experience strong ground motions, as shear wave amplitude is generally higher, and frequency generally lower, in soil than in rock.

Most analyses of seismic ground motions assume that motions are controlled by upward propagation of shear waves from the event hypocenter at depth. Wave energy attenuates away from the event epicenter (the surface projection of the event hypocenter) as it approaches structures.

FHWA NHI-11-032 Fig. 2-11 shows various distance measures used in earthquake engineering.

FHWA NHI-11-032 Chap. 2 is devoted to earthquake fundamentals and engineering seismology. Section 2.2 discusses earthquake sources, including plate tectonics, fault movements, and fault activity. Section 2.3 discusses seismic waves. Section 2.5 introduces parameters describing earthquake size and location, including magnitude, location, and recurrence. Section 2.6 covers parameters describing faulting, and Sec. 2.7 discusses parameters describing ground shaking, including intensity, peak ground motion, energy content, duration, and response spectrum.

Seismic Site Classification

ASCE/SEI 7: Sec. 20.3

In addition to fault movement and other seismic event characteristics, ground motion is impacted by physical soil and rock properties related to the relative density, stiffness, or hardness of the wave propagation media. ASCE/SEI 7 Table 20.3-1 provides site class ratings based primarily on shear wave velocity, standard penetration resistance, and undrained shear strength, all of which generally increase in proportion to density, stiffness, or hardness.

Site classes A and B are typically only found in the western United States. These sites are indicated by rock and hard rock near the ground surface. Site classes C, D, and E are more common in the central and eastern United States and are described by various soil conditions near the ground surface. Site class F is undesirable and requires special site analysis.

For more information on seismic site classification, refer to the entry "Seismic Site Class Categories" in this chapter.

Seismic Site Class Categories

ASCE/SEI 7: Sec. 20.3.3, Sec. 20.3.4, Sec. 20.3.5, Sec. 20.4.1, Sec. 20.4.2, Sec. 20.4.3

Site class categories are ranked A through E based on averaging the soil shear wave velocity, $\bar{v}_s$, standard penetration resistance, $\bar{N}$, and undrained shear strength, $\bar{s}_u$, over the upper 100 ft of subsurface profile.

$$\bar{v}_s = \frac{\sum_{i=1}^{n} d_i}{\sum_{i=1}^{n} \frac{d_i}{v_{si}}} \quad \text{[ASCE/SEI 7 Eq. 20.4-1]}$$

$$\bar{N} = \frac{\sum_{i=1}^{n} d_i}{\sum_{i=1}^{n} \frac{d_i}{N_i}} \quad \text{[ASCE/SEI 7 Eq. 20.4-2]}$$

$$\bar{s}_u = \frac{d_c}{\sum_{i=1}^{k} \frac{d_i}{s_{ui}}} \quad \text{[ASCE/SEI 7 Eq. 20.4-4]}$$

The site class categories are based on average conditions to a depth of 100 ft below the local ground surface. When multiple material types comprise that profile, average shear wave velocities, standard penetration resistance values, and undrained shear strength values can be computed using ASCE/SEI 7 Eq. 20.4-1, Eq. 20.4-2, and Eq. 20.4-4, respectively.

Only one of the equations is needed to determine seismic site class C, D, or E, per ASCE/SEI 7 Sec. 20.3.3. The determination of site class B is measured on site or estimated as described in ASCE/SEI 7 Sec. 20.3.4. Site class A is supported by shear wave velocities on site or from similar rock formations with similar weathering conditions according to ASCE/SEI 7 Sec. 20.3.5.

For more information on seismic site class categories, refer to the entry "Seismic Site Classification" in this chapter.

C. PSEUDOSTATIC ANALYSIS AND EARTHQUAKE LOADS

Key concepts: These key concepts are important for answering exam questions in subtopic 12.C, Pseudostatic Analysis and Earthquake Loads.

- Mononobe-Okabe Method
- seismic coefficient
- seismic slope stability

NCEES Handbook: To prepare for this subtopic, familiarize yourself with these sections in the *Handbook*.

- Pseudostatic Analysis and Earthquake Loads
- Forces Behind a Gravity Wall
- Stability of Rock Slope

Codes and standards: To prepare for this subtopic, familiarize yourself with these sections from codes and standards.

LRFD Seismic Analysis and Design of Transportation Geotechnical Features and Structural Foundations Reference Manual (FHWA NHI-11-032)

- Section 11.4.2: Displacement-Based Seismic Coefficient

The following equations, figures, and concepts are relevant for subtopic 12.C, Pseudostatic Analysis and Earthquake Loads.

Mononobe-Okabe Method

Handbook: Pseudostatic Analysis and Earthquake Loads; Forces Behind a Gravity Wall

$$P_{AE} = \tfrac{1}{2} K_{AE} \gamma H^2 (1 - k_v)$$

$$P_{PE} = \tfrac{1}{2} K_{PE} \gamma H^2 (1 - k_v)$$

$$K_{AE} = \frac{\cos^2(\phi - \theta - \beta)}{\cos\theta (\cos^2\beta) \cos(\beta + \delta + \theta) D}$$

$$D = \left(1 + \sqrt{\frac{\sin(\phi + \delta)\sin(\phi - \theta - i)}{\cos(\delta + \beta + \theta)\cos(i - \beta)}}\right)^2$$

$$K_{PE} = \frac{\cos^2(\phi - \theta + \beta)}{\cos\theta (\cos^2\beta) \cos(\delta - \beta + \theta) D'}$$

$$D' = \left(1 + \sqrt{\frac{\sin(\phi + \delta)\sin(\phi + i - \theta)}{\cos(\delta - \beta + \theta)\cos(i - \beta)}}\right)^2$$

$$\theta = \tan^{-1} \frac{k_h}{1 - k_v}$$

The equations used in the Mononobe-Okabe method are based on the Coulomb earth pressure theory. The equations for P_{AE} and P_{PE} are the same as that for static earth pressure but use the value of K_{AE} to determine the horizontal earth pressure. The other modification is the $1 - k_v$ term. The value of k_v is commonly taken as zero in practice. The basis of the Mononobe-Okabe theory is built into the calculation of K_{AE}.

Handbook figure Forces Behind a Gravity Wall provides a cross section of a retaining wall free-body diagram with seismic forces applied in a pseudostatic condition. The seismic forces have been added through the inclusion of the k_h and k_v terms, which represent the horizontal and vertical seismic coefficients and are multiplied by the weight of the soil to calculate the increase in soil pressure during a seismic event.

Slope Stability

Handbook: Stability of Rock Slope

Slopes subject to ground shaking experience an effective increase in driving force based on a seismic coefficient, k. The seismic coefficient is multiplied by the static normal and tangential forces computed for mass or by the slice weight alone.

Static slope stability is evaluated by computing normal and tangential forces on theoretical or known sliding surfaces and dividing the shear strength available from the normal forces by the tangential forces to determine the factor of safety. Dynamic slope stability involves adding (or subtracting) a dynamic force component to (or from) each static force, which is based on the gravitational acceleration developed during the seismic event. The acceleration will always increase the driving force, but it may reduce, increase, or leave the resisting force unchanged, depending on failure surface or slice orientation.

See the equation given in *Handbook* section Stability of Rock Slope. This section provides a free-body diagram showing the forces acting on the rock slope and gives a visual description for each term in the equation.

$$F_s = \frac{cL + (W \cos \alpha - kW \sin \alpha - P_{WR} - P_{WJ} \sin \alpha) \tan \phi}{W \sin \alpha + kW \cos \alpha + P_{WJ} \cos \alpha}$$

During a static analysis, the kW term does not exist. This term is the product of the ground acceleration and the weight of the wedge, which equates to the inertial force of the wedge itself.

Earthquake-Induced Ground Displacement

FHWA NHI-11-032: Sec. 11.4.2

Not all seismic events cause slopes to fail, but many cause slopes to deform. Unfavorable deformation can cause cracking or other slope integrity issues that ultimately lead to failure. In the case of retention structures, unfavorable deformation can lead to piping or breaching and failure.

FHWA NHI-11-032 Sec. 11.4.2 provides an allowance for a retaining wall to displace during a seismic event in exchange for a reduction in the seismic coefficient used for design. Much of this discussion is based on Sec. 6.2.3 of the same document, where the background and concepts of displacement are discussed.

21 Earth Structures

Content in blue refers to the *NCEES Handbook*.

A. Slab on Grade 21-1
B. Ground Improvement 21-3
C. Geosynthetic Applications 21-5
D. Slope Stability and Slope Stabilization 21-7
E. Earth Dams, Levees, and Embankments 21-9
F. Landfills and Caps 21-11
G. Pavement Structures......................................21-13
H. Settlement 21-14

The knowledge area of Earth Structures makes up between four and six questions out of the 80 questions on the PE Civil Geotechnical exam. The organization of this chapter follows the order of subtopics given by NCEES for this knowledge area. Each subtopic is covered in the following sections.

A. SLAB ON GRADE

Key concepts: These key concepts are important for answering exam questions in subtopic 13.A, Slab on Grade.

- compaction methods
- drainage
- excavation
- material compaction
- material types and soils
- moisture control
- post-construction stabilization
- recompaction
- resilient modulus
- scarification
- stabilization methods
- subgrade preparation

NCEES Handbook: To prepare for this subtopic, familiarize yourself with these sections in the *Handbook*.

- Compactors Recommended for Various Types of Soil and Rock
- Summary of Compaction Criteria
- Grain Size Ranges Considered for Different Stabilization Methods
- Grouping of Soils for Dynamic Compaction
- Geotechnical Influences on Major Distresses in Rigid Pavements
- Subgrade Improvement Methods
- Guide for Selection of Admixture Stabilization Method(s)

PE Civil Reference Manual **(CERM):** Study these sections in CERM that either relate directly to this subtopic or provide background information.

- Section 35.11: Proctor Test
- Section 35.12: Modified Proctor Test
- Section 35.13: In-Place Density Tests
- Section 77.4: Layer Material Strengths

Codes and standards: To prepare for this subtopic, familiarize yourself with these sections from codes and standards.

Foundations & Earth Structures Design Manual 7.02 (NAVFAC DM-7.02)

- Section 4.4: Mat and Continuous Beam Foundations
- Section 4.5: Foundations on Engineered Fill
- Section 4.6: Foundations on Expansive Soils

Guide to Design of Slabs-on-Ground (ACI 360R)

- Section 3.2: Slab Types
- Section 3.3: General Comparison of Slab Types
- Section 4.3: Subgrade Classification
- Section 4.4: Modulus of Subgrade Reaction
- Section 4.5: Design of Slab-Support System

The following figures, tables, and concepts are relevant for subtopic 13.A, Slab on Grade.

Subgrade Compaction

Handbook: Compactors Recommended for Various Types of Soil and Rock; Summary of Compaction Criteria; Grain Size Ranges Considered for Different Stabilization Methods; Grouping of Soils for Dynamic Compaction

CERM: Sec. 35.11, Sec. 35.12, Sec. 35.13

Surface compaction typically involves

1. scarifying the subgrade to a specified depth
2. wetting or drying the disturbed materials so that their moisture content falls within a specified optimum range
3. recompacting the materials to a specified percentage of their maximum laboratory Proctor density

Compaction subcuts merely extend the depth to which subgrade soils are improved.

The incorporation of a pavement subbase below the pavement section typically involves the removal of a specified thickness of in situ subgrade material and replacement with a material of greater strength. The surface on which the subbase material is placed is surface compacted and may also be subjected to a compaction subcut to further increase the effective depth of improvement and strength of the improved subgrade.

Many variables go into proper compaction. The type of compaction, amount of compaction, compaction equipment, and type of soil affect the density of the material. CERM Fig. 35.5 shows different types of compaction equipment, and *Handbook* table Summary of Compaction Criteria provides general guidance regarding proper compaction for various types of soil and equipment.

Handbook figure Grain Size Ranges Considered for Different Stabilization Methods provides options for compaction as a function of soil type and particle size. Dynamic compaction is detailed further in *Handbook* figure Grouping of Soils for Dynamic Compaction. *Handbook* figure Compactors Recommended for Various Types of Soil and Rock provides a list of compactors to use based on the type of soil being compacted.

Exam tip: Compaction is a measurement of density. Make sure to match the correct unit weight and Proctor density to the material in question.

Stabilization Methods

Handbook: Grain Size Ranges Considered for Different Stabilization Methods; Subgrade Improvement Methods; Guide for Selection of Admixture Stabilization Method(s)

Chemical stabilization can improve subgrade conditions in several ways. Hydrated lime, fly ash, and cement all decrease plasticity and increase strength. Cement and calcium chloride temper strength loss though freeze-thaw cycles in cold climates. Bitumen binds the particles and increases water resistance.

Chemical agents are generally applied dry and mixed into disked or pulverized subgrade materials, with subgrade moisture providing the hydrating effect. The effectiveness of any chemical agent is dependent on matching to subgrade mineralogy and plasticity, proportioning, and mixing. Compressive strength testing and freeze-thaw durability testing in cold-weather environments are needed to demonstrate long-term benefits.

The use of chemical stabilization measures is common in expansive soils. Geotextiles are often used in subgrade improvement for weak soils. *Handbook* table Subgrade Improvement Methods shows the applicability of geotextiles, geogrid, and GG-GT composites in roadway subgrade design. *Handbook* figure Guide for Selection of Admixture Stabilization Method(s) provides a list of suitable stabilization options based on grain size and plasticity index (PI) values. *Handbook* figure Grain Size Ranges Considered for Different Stabilization Methods lists different stabilization methods based on the grain size of the soil.

Subgrade Preparation

Handbook: Geotechnical Influences on Major Distresses in Rigid Pavements; Subgrade Improvement Methods

CERM: Sec. 77.4

ACI 360R: Sec. 3.2, Sec. 3.3, Sec. 4.3, Sec. 4.4, Sec. 4.5

NAVFAC DM-7.02: Sec. 4.4, Sec. 4.5, Sec. 4.6

A variety of subgrade preparation alternatives are available depending on performance tolerances, budget constraints, material availability, and risk. *Handbook* tables Geotechnical Influences on Major Distresses in Rigid Pavements and Subgrade Improvement Methods provide charts with recommendations to prevent problems.

ACI 360R Sec. 3.2 and Sec. 3.3 discuss and compare different slab types. ACI 360R provides detail regarding the subgrade classification in Sec. 4.3, the modulus of subgrade reaction in Sec. 4.4, and the design of slab-support systems in Sec. 4.5.

NAVFAC DM-7.02 provides guidance for foundation design. Mat and continuous beam foundations are covered in Sec. 4.4, foundations on engineered fill are covered in Sec. 4.5, and foundations on expansive soils are discussed in Sec. 4.6.

If the design of the slab itself becomes undesirable due to cost, time, or size, subgrade improvements should be considered in order to reduce the design (and/or size) of the foundation itself. CERM Table 77.6 and Table 77.7 provide descriptions of various types of subgrade materials and their corresponding design values.

Exam tip: Slabs on grade are also referred to as *rafts*. Do not confuse a slab on grade with a floor slab that is not supported by the earth.

Post-Construction Stabilization

Mudjacking and grinding are used for temporary mitigation of settlement or heave. *Mudjacking* is commonly used to raise a slab that has settled (but without doing anything to densify the compressible material that is causing the settlement). *Grinding* is used to reduce the grade differential between a slab that has heaved and an adjacent slab or structure that has not (but without mitigating the mineralogical or moisture-related sources that are causing the slab to heave).

Underpinning is a more robust, longer-term form of settlement mitigation. Underpinning is feasible only when there are materials at depth that are capable of either isolating the slab from continued compression of the subgrade materials beneath it or restraining the slab from continued heaving. See subtopic 16.F, Underpinning, in Chap. 24 of this book for details.

Post-construction stabilization (or repair) is always more expensive than stabilization at the time of construction.

B. GROUND IMPROVEMENT

Key concepts: These key concepts are important for answering exam questions in subtopic 13.B, Ground Improvement.

- bearing capacity/subgrade strength
- dry and wet soil mixing
- dynamic deep compaction
- excavation stability
- exposure to environmental hazards
- liquefaction
- magnitude and time rate of settlement
- preconsolidation and surcharging
- rammed aggregate piers
- seepage control
- vibrocompaction and replacement
- void propagation and collapse

NCEES Handbook: To prepare for this subtopic, familiarize yourself with these sections in the *Handbook*.

- Grain Size Ranges Considered for Different Stabilization Methods
- Physical Properties of Chemical Grouts
- Range of Soil Types Treated by Vibrocompaction
- Grouping of Soils for Dynamic Compaction
- Grout Curtain or Cutoff Trench Around an Excavation
- Guide for Selection of Admixture Stabilization Method(s)

PE Civil Reference Manual (CERM): Study these sections in CERM that either relate directly to this subtopic or provide background information.

- Section 35.1: Soil Particle Size Distribution
- Section 35.2: Soil Classification
- Section 35.11: Proctor Test
- Section 35.12: Modified Proctor Test

Codes and standards: To prepare for this subtopic, familiarize yourself with these sections from codes and standards.

Geotechnical Aspects of Pavements (FHWA NHI-05-037)

- Section 7.6.5: Admixture Stabilization

Soil Dynamics and Special Design Aspects (NAVFAC DM-7.03)

- Section 1.5.3: Dynamic Compaction

The following figures and concepts are relevant for subtopic 13.B, Ground Improvement.

Soil Size

Handbook: Grain Size Ranges Considered for Different Stabilization Methods

CERM: Sec. 35.1, Sec. 35.2

Ground improvement options provide tools for mitigating structural issues such as foundation bearing and settlement, as well as global site issues such as settlement due to large fills, liquefaction potential, and karst. A conventional excavation and backfill approach can be taken when there are no excavation constraints (e.g., adjacent structures) that would otherwise require a substantial investment in retention. With a suitable ground improvement option, excavations can potentially be eliminated.

Ground improvement provides alternatives to what would otherwise be simple excavation and backfill work when the materials involved aren't impacted by environmental contaminants or hazardous materials. Combined with a minimum amount of site restoration based on regulatory requirements, ground improvement can limit the volume of material that has to be exported to a landfill.

See *Handbook* figure Grain Size Ranges Considered for Different Stabilization Methods, which provides recommended ground improvement methods based on soil type.

The type of soil and particle size affect the type of compaction that can be effective. For larger soil particles such as sand and gravel, vibration is effective. For smaller soil particles such as silt and clay, the use of water is effective. Depending on the particle size, various grouting techniques can be used to improve the foundation.

Vibrocompaction

Handbook: Grain Size Ranges Considered for Different Stabilization Methods; Range of Soil Types Treated by Vibrocompaction

Vibrocompaction densifies loose granular soils through horizontal vibratory action on a grid pattern. It is effective above and below groundwater and to depths up to about 100 ft.

Vibrocompaction takes dynamic deep compaction further, being effective to greater depths and also below groundwater. Like soil mixing, vibrocompaction and replacement can be applied locally to support discrete structural elements or in closely spaced arrays to support broader loads associated with buildings and grade raises. As with dynamic deep compaction, the cavities created by vibrocompaction must be filled, and final grades must be reestablished with compacted fill. Filling the voids created by the vibrator with concrete increases strength and stiffness well beyond what densification does, and filling with free-draining aggregate helps mitigate the potential for liquefaction. *Handbook* figure Range of Soil Types Treated by Vibrocompaction shows that vibrocompaction is effective for coarse-grained materials (sand and gravel), but vibroreplacement is needed for finer-grained soils (silt and clay).

Handbook figure Grain Size Ranges Considered for Different Stabilization Methods plots the average particle size for various types of ground improvement.

Dynamic Deep Compaction

Handbook: Grouping of Soils for Dynamic Compaction

CERM: Sec. 35.11, Sec. 35.12

NAVFAC DM-7.03: Sec. 1.5.3

Dynamic deep compaction is an effective soil compaction method for large parcels being redeveloped or reclaimed for building construction. Suitable sites include those where removing and recompacting or replacing the fill deposits is impractical, such as abandoned urban fill sites and mining reclamation projects. This method is not suited for submerged materials, but successful densification can be achieved if a material is only partially submerged. The densification process leaves large craters. On completing the treatment, anywhere from 3 ft to 6 ft of disturbed material is removed and recompacted, and additional fill is imported and compacted to offset the grade loss due to compaction.

The depth of the compaction influence in feet is half of the weight in tons multiplied by the height in feet that the weight is dropped from. The weight used is commonly a minimum of 10 tons and can be more than 40 tons. The weight is usually dropped from a minimum of 50 ft and a maximum of 130 ft.

Handbook figure Grouping of Soils for Dynamic Compaction plots the percentage of finer and coarser soils against sieve numbers.

Exam tip: Although not provided for use by NCEES during the exam, Sec. 1.5.3 of NAVFAC DM-7.03 is a great resource for dynamic compaction and should be reviewed before taking the exam.

Seepage Control

Handbook: Physical Properties of Chemical Grouts; Grout Curtain or Cutoff Trench Around an Excavation

Containment of seepage can be achieved by jet grouting or chemical grouting, and collection of seepage can be performed with vibroreplacement columns or rammed aggregate piers.

Both jet grouting and chemical grouting can be used to build strength and stiffness, but jet grouting is more often used for seepage control, and chemical grouting is more often used for excavation support. Jet grouting involves in situ high-velocity fluid erosion of a soil column and mixing with grout. Chemical grouting involves injecting permeable soils with a polymer to fill and seal voids.

Both vibroreplacement columns and rammed aggregate piers, when filled with free-draining aggregate, provide reservoirs capable of collecting steady-state seepage (associated with a slope failure) or relieving load-induced pore water pressure.

Handbook figure Grout Curtain or Cutoff Trench Around an Excavation shows a cross section of an excavation into sand underlain with an impermeable layer. The grout curtain is used to prevent seepage through the sand into the excavation. Pumps and piezometers are used to monitor and remove any water that is detected on the excavated side of the grout curtain wall. Properties of the grout, including class, viscosity, and unconfined compressive strength, can be found in *Handbook* table Physical Properties of Chemical Grouts.

Exam tip: These concepts can be used to minimize seepage of a contaminant as well as groundwater seepage.

Dry and Wet Soil Mixing

Handbook: Guide for Selection of Admixture Stabilization Method(s)

FHWA NHI-05-037: Sec. 7.6.5

Where densification cannot be performed via impact methods, the materials in question can be disturbed with large augers and mixed with cementitious materials or grout, depending on the consistency of the material to be mixed. Mixing does not rely on densification to provide enhanced strength and stiffness, but instead relies on the hardening effect of the additive.

Mixing can also be used to reinforce known or potential slope failure planes. In such cases, shear strength along the failure surface in question is determined based on the fraction of a unit area occupied by mixed columns.

It is common for lime, cement, or fly ash to be mixed with the soil to improve the soil quality. FHWA NHI-05-037 Sec. 7.6.5 provides a detailed description of how each can be added to the soil for subbase improvement.

Handbook figure Guide for Selection of Admixture Stabilization Method(s) gives the plasticity index (PI) for different forms of stabilizations.

C. GEOSYNTHETIC APPLICATIONS

Key concepts: These key concepts are important for answering exam questions in subtopic 13.C, Geosynthetic Applications.

- clogging and migration of particles leading to voids
- geosynthetic grids or textiles
- punching failure
- shear resistance

NCEES Handbook: To prepare for this subtopic, familiarize yourself with these sections in the *Handbook*.

- Identification of Usual Primary Function Versus Type of Geosynthetics
- Soil Retention Criteria for Geotextile Filter Design Using Steady-State Flow Conditions
- Recommended Strength-Reduction Factor Values for Use in Allowable Tensile Strength Equation
- Recommended Reduction Factor Values for Use in Determining Allowable Tensile Strength of Geogrids
- Geotextile-Reinforced Walls
- Earth Pressure Concepts and Theory for Geotextile Wall Design
- Elements of Geogrid (or Geotextile) Reinforced Wall Design
- Reinforced Slopes with Geotextiles
- Steep, Reinforced, Soil-Slope Design Charts for Zero-Pore Water Pressure
- Limit Equilibrium Forces Involved in a Finite Length Slope Analysis for a Uniformly Thick Cover Soil
- Peak Friction Values and Efficiencies of Various Geosynthetic Interfaces
- Solid-Waste Containment System with High Geosynthetic Utilization
- Subgrade Improvement Methods

PE Civil Reference Manual (CERM): Study these sections in CERM that either relate directly to this subtopic or provide background information.

- Section 40.14: Geotextiles

The following figures, tables, and concepts are relevant for subtopic 13.C, Geosynthetic Applications.

Separation

Handbook: Identification of Usual Primary Function Versus Type of Geosynthetics; Limit Equilibrium Forces Involved in a Finite Length Slope Analysis for a Uniformly Thick Cover Soil; Peak Friction Values and Efficiencies of Various Geosynthetic Interfaces; Solid-Waste Containment System with High Geosynthetic Utilization; Subgrade Improvement Methods

CERM: Sec. 40.14

Separation contains the embankment fill over soft ground and prevents punching failure of the soil underneath. The separation contains an aggregate base and subbase material placed on soft subgrades.

Separation is critical to establish and maintain the structural integrity of structural fill materials placed upon potentially weak or compressible foundation or subgrade soils.

The placement of a geotextile over fine-grained soils in advance of granular fill placement provides the tensile strength needed to facilitate compaction of the granular soil and helps distribute fill and traffic loads more uniformly. As the fine-grained foundation soils compress under the weight of the rising embankment, the embankment fill becomes more secure and the geotextile assumes a concave upward shape. The settlement reaches a peak value beneath the embankment centerline and tapers to a lesser value at the embankment toes.

Handbook table Subgrade Improvement Methods provides guidance on what type of geosynthetic is suitable for use with specific types of soil. *Handbook* table Identification of Usual Primary Function Versus Type of Geosynthetics is also a helpful resource for the type of geosynthetic useful in specific applications.

When a geosynthetic is placed at an angle, as shown in *Handbook* figure Limit Equilibrium Forces Involved in a Finite Length Slope Analysis for a Uniformly Thick Cover Soil, a force balance with an appropriate factor of safety must occur to prevent a sliding failure. This diagram provides all the forces acting on the free-body diagram. *Handbook* table Peak Friction Values and Efficiencies of Various Geosynthetic Interfaces provides values to use in the force balance equations. These friction angles are a function of both the soil type and the geosynthetic type. *Handbook* figure Solid-Waste Containment System with High Geosynthetic Utilization shows uses of various geosynthetics.

Exam tip: Geosynthetics may have multiple purposes, such as separation and filtration.

Filtration

Handbook: Soil Retention Criteria for Geotextile Filter Design Using Steady-State Flow Conditions

CERM: Sec. 40.14

Filtration prevents the migration of particles between materials of varying grain size distribution, and the resulting formation of voids. It prevents clogging and maintains flow.

Filtration is needed where water will flow between materials of varying grain size distributions or opening sizes. If the disparity in particle size distribution or opening size is too great, finer-grained soil may migrate into the coarser-grained fill, which may create a void within the finer-grained soil or clog the coarser-grained fill. The migration of fine-grained soil into the coarse-grained soil (or previously gap-graded soil) will not only create a well-graded soil, but also result in a depression at the ground surface as the fine-grained material moves into the openings of the previously open material.

Tensile strength is not critical for filtration, but it can benefit situations where there is a significant strength differential between the materials being placed on top of the geosynthetic and the materials present beneath the geosynthetic.

Handbook figure Soil Retention Criteria for Geotextile Filter Design Using Steady-State Flow Conditions is helpful in determining the type of soil and filter to use.

Exam tip: Filtration geotextiles may be placed at any orientation, depending on the configuration between the dissimilar materials.

Drainage

Handbook: Soil Retention Criteria for Geotextile Filter Design Using Steady-State Flow Conditions

CERM: Sec. 40.14

Synthetic panels used for drainage are advantageous where space or economics favor them. The panels have raised "buttons" to create flow corridors. The geotextile covers will filter the earth materials on the "wet" side of the panels.

In situations where flowing water reaches a stopping point, the water must be collected and transported away from the impacting structure so as not to alter the assumed design loading conditions. This is often accomplished with a free-draining soil layer, but it can also be done with a geosynthetic to save time and space. Drainage panels can be wrapped around a drain pipe located along the base of the wall or terminated at weep holes extended into the front of the wall.

Geosynthetics used for drainage are often a combination of two materials (commonly HDPE and a filter fabric) and are usually thicker than a typical geosynthetic. The thicker drain board material provides a large void space for the water to flow, and the outer filter fabric prevents soil from entering the drainage void.

Handbook figure Soil Retention Criteria for Geotextile Filter Design Using Steady-State Flow Conditions shows how to navigate filter criteria when using property soil tests.

Exam tip: Drainage composites are commonly used in place of free-draining soil because drainage composites have an equal or better flow rate but take up less space than free-draining soil.

Stability of Walls and Slopes

Handbook: Recommended Strength-Reduction Factor Values for Use in Allowable Tensile Strength Equation; Recommended Reduction Factor Values for Use in Determining Allowable Tensile Strength of Geogrids; Geotextile-Reinforced Walls; Earth Pressure Concepts and Theory for Geotextile Wall Design; Elements of

Geogrid (or Geotextile) Reinforced Wall Design; Reinforced Slopes with Geotextiles; Steep, Reinforced, Soil-Slope Design Charts for Zero-Pore Water Pressure

Geosynthetics are used with retaining walls to create a block of soil called the *reinforced zone* that acts as a single unit, similar to how a gravity retaining wall acts. The strength of the geosynthetic reinforcement interacts with the soil to provide tensile capacity to the soil. The strength of the reinforcement is combined with the strength of the soil to provide a stable structure. *Handbook* figure Elements of Geogrid (or Geotextile) Reinforced Wall Design shows the potential failure modes of a retaining wall that is reinforced with geosynthetics. *Handbook* section Geotextile-Reinforced Walls provides equations to calculate the necessary length, spacing, and strength of the reinforcement.

Handbook figure Earth Pressure Concepts and Theory for Geotextile Wall Design shows how the reinforcement spacing is used to resist the forces of the soil, surcharge, and live load pressures applied to the wall system.

Geosynthetics are added to reinforced slopes to increase the factor of safety. The strength of the reinforcement is added to the resisting forces that are provided by the strength of the soil. *Handbook* section Reinforced Slopes with Geotextiles provides equations showing how the tensile strength of the reinforcement is added to resisting forces and increases the factor of safety. *Handbook* figure Steep, Reinforced, Soil-Slope Design Charts for Zero-Pore Water Pressure provides minimum reinforcement lengths for slopes of angles from 30° to 90°.

Exam tip: Use *Handbook* tables Recommended Reduction Factor Values for Use in Determining Allowable Tensile Strength of Geogrids and Recommended Strength-Reduction Factor Values for Use in Allowable Tensile Strength Equation to determine the design value for the geosynthetic.

D. SLOPE STABILITY AND SLOPE STABILIZATION

Key concepts: These key concepts are important for answering exam questions in subtopic 13.D, Slope Stability and Slope Stabilization.

- factor of safety
- fill slopes
- pore water pressure increase
- saturation
- seepage forces
- weathering

NCEES Handbook: To prepare for this subtopic, familiarize yourself with these sections in the *Handbook*.

- Stability Charts for $\phi = 0$ Soils
- Taylor's Chart for Soils with Friction Angle
- Taylor's Chart for $\phi' = 0$ Conditions for Slope Angles (β) Less than 54°
- Stability Analysis of Transitional Failure
- Stability of Rock Slope
- Infinite Slope
- Slice for Ordinary Method of Slices with External Water Loads
- Typical Slice and Forces for Ordinary Method of Slices
- Slope Stability Guidelines for Design
- Determination of Soil Type
- Slope and Shield Configurations
- Reinforced Slopes with Geotextiles
- Details of Circular Arc Slope Stability Analysis for (c, ϕ) Shear Strength Soils

PE Civil Reference Manual (**CERM**): Study these sections in CERM that either relate directly to this subtopic or provide background information.

- Section 40.10: Slope Stability in Saturated Clay
- Section 40.15: Soil Nailing
- Section 83.5: Soil Classification

The following equations, figures, and tables are relevant for subtopic 13.D, Slope Stability and Slope Stabilization.

Slope Stability Factor of Safety

Handbook: Stability Charts for $\phi = 0$ Soils; Taylor's Chart for Soils with Friction Angle; Taylor's Chart for $\phi' = 0$ Conditions for Slope Angles (β) Less than 54°; Stability Analysis of Transitional Failure; Stability of Rock Slope; Infinite Slope; Slice for Ordinary Method of Slices with External Water Loads; Typical Slice and Forces for Ordinary Method of Slices; Slope Stability Guidelines for Design; Details of Circular Arc Slope Stability Analysis for (c, ϕ) Shear Strength Soils

CERM: Sec. 40.10

The factor of safety indicates, on a percentage basis, how much greater the resisting force is than the driving force. When the factor of safety approaches 1.0, the resisting and driving forces are balanced, and failure is assumed to be imminent, needing only a small reduction in strength or a slight increase in stress.

The minimum factor of safety targeted by designers is 1.3 for typical cases, 1.125 for temporary cases (or seismic conditions), and 1.5 for critical slopes. The *Handbook* provides guidance for translational failures in figure Stability Analysis of Transitional Failure, infinite slope failures in section Infinite Slope, and rotational/circular failures using the method of slices in section Slice for Ordinary Method of Slices with External Water Loads. Rock slope failures are shown in *Handbook* figure Stability of Rock Slope. There are also design charts to assist in stability verification in *Handbook* figures Stability Charts for $\phi = 0$ Soils, Taylor's Chart for Soils with Friction Angle, and Taylor's Chart for $\phi' = 0$ Conditions for Slope Angles (β) Less than 54°. The concepts used in Taylor's charts are represented in *Handbook* figures Details of Circular Arc Slope Stability Analysis for (c, ϕ) Shear Strength Soils and Typical Slice and Forces for Ordinary Method of Slices. *Handbook* table Slope Stability Guidelines for Design provides a list of the types of analysis that should be performed based on soil type, along with the strength parameters required for analysis and comments regarding the method of analysis that should be used.

Exam tip: If manually calculating a factor of safety for slope stability, use the stability charts to check your answer.

Slope Stability Failure

Handbook: Stability Analysis of Transitional Failure; Stability of Rock Slope; Infinite Slope; Slice for Ordinary Method of Slices with External Water Loads

CERM: Sec. 40.10

An increase in driving force or a decrease in resisting force can cause failure. Loading at the top, unloading at the toe, saturation, weathering, an increase in pore water pressure, and seepage forces associated with groundwater flow can all contribute to failure.

The slope stability factor of safety can fall due to an increase in the driving force, a decrease in the resisting force, or a combination of both. Events that increase the driving force include

- load application at the top of the slope (filling, building)
- load removal at the slope's toe (excavation, erosion)
- seepage through the slope

Events that decrease the resisting force include

- shear strength reduction (weathering, softening)
- confining stress reduction (saturation, submergence)

In many cases, water proves to be the critical factor.

If a failure plane is already defined (or assumed), as shown in *Handbook* figure Stability of Rock Slope, the factor of safety can be calculated using the equation shown below the cross section. A complete free-body diagram is also provided in the cross section. This calculation can be thought of as a traditional sliding block analysis, but the geotechnical aspects of the problem are highlighted in these diagrams.

When considering the stability of a slope used in construction, an effective stress condition should be checked along with a total stress condition. An effective stress analysis (using the effective friction angle of the soil) will generally provide a lower factor of safety compared to the total stress analysis (using the undrained shear strength of the soil). However, the total stress condition may be applicable for a temporary condition or the initial construction phase.

The *Handbook* provides guidance for translational failures in figure Stability Analysis of Transitional Failure, infinite slope failures in section Infinite Slope, and rotational/circular failures using the method of slices in section Slice for Ordinary Method of Slices with External Water Loads.

Soil and Rock Slopes

Handbook: Stability of Rock Slope; Infinite Slope; Stability Analysis of Transitional Failure; Determination of Soil Type; Slope and Shield Configurations

CERM: Sec. 83.5

Soil and rock slope stability is typically governed by local geomorphology. The material properties and geologic origin and structure are major contributing factors.

Even slopes made of robust materials experience erosion from uncontrolled or concentrated drainage. This can result in the loss of vegetation at the least, and downcutting or slumping if left unattended.

Geological layering contributes to the stability (or instability) of a soil and rock slope. If the angle of layering is known, a cut should be made as close to perpendicular as possible. A cut angle equal to (or close to equal to) the layering angle will result in a sliding failure. The example shown in *Handbook* figure Stability of Rock Slope provides an equation for the factor of safety of a rock slope failure, but this equation is simply a specialized version of the general translational failure equation provided in *Handbook* section Infinite Slope. Both are simplified versions of *Handbook* figure Stability Analysis of Transitional Failure. Any of these types of failures can be thought of as a typical sliding block analysis.

Handbook sections Determination of Soil Type and Slope and Shield Configurations provide general guidance regarding stable slope angles based on the soil and rock type being excavated.

Exam tip: Drawing a free-body diagram of the sliding block is helpful for visualizing all the forces acting on the sliding mass.

Retention Structures

Handbook: Reinforced Slopes with Geotextiles

CERM: Sec. 40.15

$$\text{FS} = \frac{\sum_{i=1}^{n}(N_i \tan\phi + c\Delta l_i)R + \sum_{i=1}^{m} T_i y_i}{\sum_{i=1}^{n}(W_i \sin\theta_i)R}$$

$$N_i = W_i \cos\theta_i$$

$$\text{FS} = \frac{\sum_{i=1}^{n}(\overline{N}_i \tan\overline{\phi} + \overline{c}\Delta l_i)R + \sum_{i=1}^{m} T_i y_i}{\sum_{i=1}^{n}(\overline{W}_i \sin\theta_i)R}$$

$$\overline{N}_i = N_i - u_i \Delta x_i$$

$$u_i = h_i \gamma_w$$

From a stability standpoint, slopes also include timber walls, block walls, and piled walls whose elements contribute to the balance or imbalance of driving and resisting forces. The steeper and more abrupt geometry established by such structures increases the risk of failure, a risk that often requires foundation improvement to mitigate.

Retention structures supported on weak foundation soils produce concentrated driving forces that can precipitate a failure through the foundation.

Structures with insufficient embedment on sloping grades can precipitate failures along underlying parallel sloping strata or fail in response to the weakening and mobilization of shallow, strength-sensitive materials.

The principles that govern general slope stability still apply whether a retention structure is located at the bottom, middle, or top of the slope. If properly designed, the retention structure will add to the stability of the slope. The equations given in *Handbook* section Reinforced Slopes with Geotextiles apply not only to slopes but to any system with geosynthetics. *Soil nailing*, which is discussed in CERM Sec. 40.15, is a slope-stabilization method similar to the geosynthetic stabilization methods shown in the *Handbook*, but is more common in a cut slope or landslide repair. Geosynthetics are common in fill slopes.

Reinforced Slopes with Geotextiles

Handbook: Reinforced Slopes with Geotextiles

$$\text{FS} = \frac{\sum_{i=1}^{n}(N_i \tan\phi + c\Delta l_i)R + \sum_{i=1}^{m} T_i y_i}{\sum_{i=1}^{n}(W_i \sin\theta_i)R}$$

Geotextiles are commonly used in fill slope applications. The geotextile is placed in horizontal layers as the fill soil is compacted around it. The geotextile creates a stronger mass than if the soil were placed alone and allows for a steeper slope, a taller slope, heavier loading at the top of the slope, or a combination of these.

The resisting forces in the slope are a summation of the inherent soil strength combined with the additional strength provided by the geotextile. The factor-of-safety equation shown in *Handbook* section Reinforced Slopes with Geotextiles is a summation of the strength of the soil (friction and cohesion values) and the geotextile strength, each multiplied by their moment arms, divided by the driving force multiplied by its moment arm.

E. EARTH DAMS, LEVEES, AND EMBANKMENTS

Key concepts: These key concepts are important for answering exam questions in subtopic 13.E, Earth Dams, Levees, and Embankments.

- earth dam failure modes
- embankment failure modes
- levee components
- stability relative to strong and weak foundations
- Taylor slope stability chart

NCEES Handbook: To prepare for this subtopic, familiarize yourself with these sections in the *Handbook*.

- Entrance, Discharge, and Transfer Conditions of Seepage Line
- Seepage Through Embankment and Foundation

PE Civil Reference Manual (CERM): Study these sections in CERM that either relate directly to this subtopic or provide background information.

- Section 15.3: Hydrostatic Pressure
- Section 15.13: Hydrostatic Forces on a Dam
- Section 19.30: Controls on Flow
- Section 19.31: Flow Choking
- Section 19.32: Varied Flow

- Section 19.33: Hydraulic Jump
- Section 19.36: Erodible Channels
- Section 21.14: Flow Nets

The following figures and concepts are relevant for subtopic 13.E, Earth Dams, Levees, and Embankments.

Retention Structures

Handbook: Entrance, Discharge, and Transfer Conditions of Seepage Line

CERM: Sec. 15.3, Sec. 15.13

Retention structures consist of grade-relief structures, seepage elements, and erosion protection.

Earth dams, impoundments, and levees can be more complicated than either conventional earth slopes or earth slopes with structural elements. Water is a factor in most slope failures, but in the case of dams, impoundments, and levees, water is a constant agent of potential instability. The release of water from these structures is by itself likely to be many times more destructive than the movement of the earth or the structures that are retaining it. Retention structures are complex in design, and thus require great attention to detail when it comes to construction. Unfortunately, such structures may be built without regulation, observation, or testing.

In the *Handbook* and in CERM, dams are most often shown as earthen with a trapezoidal shape, such as in *Handbook* figure Entrance, Discharge, and Transfer Conditions of Seepage Line. This style of dam can be composed of soil, rock, or concrete. Other components of the dam are included in the design to prevent failure, as just described. Dams can be any shape or size as long as they properly retain the required volume of water. Even a cast-in-place concrete wall or grouted boulder wall can act as a dam.

Exam tip: Make sure you understand the type of dam used in the problem to ensure you are using the appropriate equations.

Failure Modes

Handbook: Entrance, Discharge, and Transfer Conditions of Seepage Line; Seepage Through Embankment and Foundation

CERM: Sec. 15.13, Sec. 21.14

The failure of an earth dam, impoundment, or levee may be closely related to construction means and methods, but in any case, it will be strongly tied to the flow and control of water through, below, and/or over the structure. To qualify how the flow of water will affect a structure, two additional phases of analysis are needed. One of these focuses on flow through or below the structure, which destabilizes the structure from within, and the other focuses on flow over the structure, which compromises it from above. These exercises bring the geotechnical engineer into the realm of hydrogeology.

The design of a dam must account for all potential failure modes. Some failure modes are external and encompass the entire dam; others are internal, resulting in failure through the inside of the dam. Examples of external failures are overturning and sliding, both described in CERM Sec. 15.13. An example of internal failure is piping (also known as seepage), shown in CERM Fig. 21.3 and *Handbook* figures Seepage Through Embankment and Foundation and Entrance, Discharge, and Transfer Conditions of Seepage Line.

Failure by Piping

Handbook: Entrance, Discharge, and Transfer Conditions of Seepage Line

CERM: Sec. 21.14

A dam, impoundment, or levee can fail due to piping. Piping occurs when water exerts enough force to mobilize the material particles within or below the structure, causing voids to form. The particle movement may begin at the surface of the structure, or, when the disparity in the gradation of differing materials is too great to restrict it, the particle movement may begin within the structure. Where such conditions exist, filter analyses should be performed to confirm that the coarser materials can effectively block the movement of the finer materials while still allowing water to move through the retention system.

A *flow net* provides the easiest way to perform a seepage analysis of a dam, but the lines of the flow net and the seepage path will vary depending on how many different materials are included within the dam. *Handbook* figure Entrance, Discharge, and Transfer Conditions of Seepage Line shows how the seepage path changes as the water encounters different materials at different angles. Following the figure, the *Handbook* also provides equations for both homogeneous dams and anisotropic sections. See CERM Fig. 21.3 for examples of flow nets for dams on impervious foundations, as well as for when seepage through the foundation occurs.

Exam tip: It is much easier to evaluate a dam with isotropic properties than one composed of multiple materials.

Failure by Toe Heave

CERM: Sec. 21.14

The uplift of grade at or beyond the landside toe of a structure can occur if confinement is removed from permeable strata under pressure. Uplift can lead to loss of confinement, the release of water under pressure, and slope failure, structure failure, or failure by piping.

Toe heave is unlike the excavation heave that can occur in the case of a retained excavation. Heave involves mobilizing masses of soil, which in turn can lead to failure by piping. The loss of stabilizing earth at the toe of an earth dam, impoundment slope, or levee or floodwall, however, can by itself lead to slope failure long before piping would act to cause such a failure.

Preventing toe heave may consist of purposely penetrating a confining layer or "blanket" with relief wells to allow seepage forces to dissipate through the controlled release of water in the impacted area, or building a berm along and out from the structure toe to help contain/balance the upward-acting seepage forces.

The uplift force is less for impervious materials in the foundation; as the permeability of the foundation increases, so does the uplift force. To complicate the uplift calculation, impermeable clay in the foundation can still create an uplift force due to pressure beneath the clay layer. Drains with appropriate filtration can be installed to minimize the uplift pressure, as shown in CERM Fig. 21.3.

Failure by Breaching

CERM: Sec. 19.30 through Sec. 19.33, Sec. 19.36, Sec. 21.14

Uncontrolled overtopping can weaken and erode materials at crest elevation. Lower crests enable higher velocity and more erosive flow. Failure leads to structure collapse and interior or downstream flooding.

Breaching is the most frequent cause of dam, impoundment, or levee failure. Breaching involves the uncontrolled progression of erosion in the crest of a dam, impoundment, or levee, or the undermining of the spillway in a dam or impoundment. Most dams and impoundments with "permanent" reservoirs are equipped with spillways or intake structures that allow excessively high or sudden rises in water levels to be managed before the water reaches the top of the structure and flows over the crest. Even spillways and intake structures, however, can be agents of instability if water is allowed to erode particles in contact with those structures and create voids along, around, or beneath those structures.

CERM Sec. 19.30 through Sec. 19.33 and Sec. 19.36 describe various conditions related to the water flow. If the type of flow is not properly accounted for at the outfall of the dam, then erosion can occur, and, ultimately, a breach. Proper erosion control and flow management are critical to dam safety. Hydraulic jumps and drops, along with flow choking and open channel flow, are all part of dam design.

F. LANDFILLS AND CAPS

Key concepts: These key concepts are important for answering exam questions in subtopic 13.F, Landfills and Caps.

- base liners
- compacted waste
- final covers
- landfill design
- leachate

NCEES Handbook: To prepare for this subtopic, familiarize yourself with these sections in the *Handbook*.

- Solid-Waste Containment System with High Geosynthetic Utilization
- Various Leachate Removal Designs for Primary Leachate Collection Systems
- Design Model and Related Forces Used to Calculate Geomembrance Thickness
- Darcy's Law

PE Civil Reference Manual (CERM): Study these sections in CERM that either relate directly to this subtopic or provide background information.

- Section 31.2: Landfills
- Section 31.3: Landfill Capacity
- Section 31.5: Clay Liners
- Section 31.6: Flexible Membrane Liners
- Section 31.7: Landfill Caps
- Section 31.8: Landfill Siting
- Section 31.10: Ultimate Landfill Disposition
- Section 31.12: Landfill Leachate
- Section 31.13: Leachate Migration from Landfills
- Section 31.14: Groundwater Dewatering
- Section 31.15: Leachate Recovery Systems
- Section 31.16: Leachate Treatment

The following figures and concepts are relevant for subtopic 13.F, Landfills and Caps.

Landfill Design

Handbook: Solid-Waste Containment System with High Geosynthetic Utilization; Various Leachate Removal Designs for Primary Leachate Collection Systems

CERM: Sec. 31.2, Sec. 31.3, Sec. 31.8

Landfills require additional stability and performance considerations because they contain materials that

- are highly variable in density and strength
- have sloping interfaces of variable strength
- are impacted by leachate (water with various other elements associated with decomposition)

This is especially true for large landfills, as they can exert great stresses on foundation materials of limited strength and high compressibility. In some situations, large landfills are also built with steep slopes to economize on space.

CERM Sec. 31.8 discusses the planning required for proper selection of a landfill site.

Calculation of a landfill's capacity is difficult due to the variations mentioned. CERM Eq. 31.1 through Eq. 31.3 provide calculations that account for the population that contributes to the landfill, the amount of municipal solid waste that is generated per day, and the ratio of soil to municipal solid waste that is used in the landfill construction and waste placement. These equations can be used to determine the daily increase in landfill volume and ultimately the capacity of the landfill.

Handbook figures Solid-Waste Containment System with High Geosynthetic Utilization and Various Leachate Removal Designs for Primary Leachate Collection Systems contain graphic representations of different landfills.

Landfill Stability and Performance

Handbook: Solid-Waste Containment System with High Geosynthetic Utilization

CERM: Sec. 31.3, Sec. 31.10

Although a landfill is not built on, the performance of a landfill's drainage system can nevertheless be affected by unfavorable amounts of compression within the earth materials on which the landfill is built. CERM Sec. 31.10 lists some acceptable uses of a landfill once it is filled and closed, and describes the large amount of settlement that can occur at the top of the cap.

Despite their unique design and construction elements, landfills are generally evaluated just like any other embankment or slope. The settlement and stability analyses performed for landfills, however, are prone to much greater uncertainties. For one, material consistency varies much more than with other earth structures; further, it can be hard to model analytically how leachate level fluctuations will affect the varied materials and structure. Additional failure modes are generated by the presence of liner and cover membranes and by layer boundaries that are oriented parallel to the below- and above-grade landfill geometries.

Handbook figure Solid-Waste Containment System with High Geosynthetic Utilization shows how geogrid reinforcement can be used on the inside of a landfill berm to strengthen and potentially steepen the berm to increase the capacity of the landfill. Geosynthetics can also be used along the outside of a landfill to steepen its outer walls and increase its capacity.

Landfill Liners and Caps

Handbook: Solid-Waste Containment System with High Geosynthetic Utilization; Design Model and Related Forces Used to Calculate Geomembrance Thickness

CERM: Sec. 31.5, Sec. 31.6, Sec. 31.7

Landfill liners commonly fail from drying out and cracking (a process called *desiccation*) or from chemical interactions involving the waste inside the landfill. *Handbook* figure Design Model and Related Forces Used to Calculate Geomembrance Thickness shows the various properties needed to calculate the liner thickness for a landfill.

A flexible membrane liner (FML) is sometimes combined with a clay layer to create a geocomposite liner. FML materials include polyvinyl chloride (PVC); chlorinated polyethylene (CPE); low-density polyethylene (LDPE); linear low-density polyethylene (LLDPE); very low-density polyethylene (VLDPE); high-density polyethylene (HDPE); chlorosulfonated polyethylene, also known as Hypalon (CSM or CSPE); ethylene propylene diene monomer (EP or EPDM); polychloroprene, also known as neoprene (CR); isobutylene isoprene, also known as butyl (IIR); and oil-resistant polyvinyl chloride (ORPVC). HDPE is the most commonly used FML material due to its high tensile strength, high resistance to tearing and puncturing, and ease of seaming.

Once the landfill has been filled, a clay or FML cap is placed on top. *Handbook* figure Solid-Waste Containment System with High Geosynthetic Utilization shows how geosynthetics and FMLs can be used both as caps and as liners at landfill sites.

Leachate

Handbook: Various Leachate Removal Designs for Primary Leachate Collection Systems; Darcy's Law

CERM: Sec. 31.12 through Sec. 31.16

Leachate is liquid waste containing dissolved and finely suspended solid matter and microbial waste produced in a landfill. Leachate is a mixture of precipitation with whatever liquids and contaminants have been disposed of in the landfill.

Over time, the leachate percolates down to the bottom of the landfill. In a natural landfill, the leachate will contaminate the surrounding soil and groundwater. In a lined landfill, the leachate must be removed before it builds up and does damage to the liner system.

Handbook figure Various Leachate Removal Designs for Primary Leachate Collection Systems provides three different configurations for leachate removal systems. If the leachate is not properly removed, or if the liner is defective, the rate of leachate migration through the soil can be calculated using Darcy's law (see *Handbook* section Darcy's Law or CERM Eq. 31.13 and Eq. 31.14).

The flow of leachate is a function of a liner's permeability. A flexible membrane liner (FML) provides much lower and more consistent permeability than a clay liner, whose permeability can change over time.

G. PAVEMENT STRUCTURES

Key concepts: These key concepts are important for answering exam questions in subtopic 13.G, Pavement Structures.

- aggregate properties
- binder and aggregate
- flexible pavement
- rigid pavement

NCEES Handbook: To prepare for this subtopic, familiarize yourself with these sections in the *Handbook*.

- Basic Pavement Structure
- Common Pavement Systems
- Geotechnical Influences on Major Distresses in Flexible Pavements
- Geotechnical Influences on Major Distresses in Rigid Pavements
- Specific Gravity of Aggregates
- Specific Gravity of Fine Aggregates
- Asphalt Mixture Volumetrics
- Asphalt Concrete Volumetric Terms and Definitions Using Phase Diagram
- Structural Layer Coefficients

PE Civil Reference Manual **(CERM):** Study these sections in CERM that either relate directly to this subtopic or provide background information.

- Section 76.1: Asphalt Concrete Pavement
- Section 76.3: Asphalt Grades
- Section 76.4: Aggregate
- Section 76.12: Characteristics of Asphalt Concrete
- Section 76.26: Pavement Structural Number
- Section 77.1: Rigid Pavement
- Appendix 76.A: Axle Load Equivalency Factors for Flexible Pavements (single axles and p_t of 2.5)
- Appendix 76.B: Axle Load Equivalency Factors for Flexible Pavements (tandem axles and p_t of 2.5)
- Appendix 76.C: Axle Load Equivalency Factors for Flexible Pavements (triple axles and p_t of 2.5)
- Appendix 77.A: Axle Load Equivalency Factors for Rigid Pavements (single axles and p_t of 2.5)
- Appendix 77.B: Axle Load Equivalency Factors for Rigid Pavements (double axles and p_t of 2.5)
- Appendix 77.C: Axle Load Equivalency Factors for Rigid Pavements (triple axles and p_t of 2.5)

Codes and standards: To prepare for this subtopic, familiarize yourself with these sections from codes and standards.

Geotechnical Aspects of Pavements (FHWA NHI-05-037):

- Section 5.5.1: 1993 AASHTO Guide

The following equations, figures, and tables are relevant for subtopic 13.G, Pavement Structures.

Pavement Design Categories

Handbook: Basic Pavement Structure; Common Pavement Systems; Asphalt Concrete Volumetric Terms and Definitions Using Phase Diagram

CERM: Sec. 76.1, Sec. 76.4, Sec. 77.1

Flexible pavement is a mixture of asphalt binder and mixed aggregate. It is also called *asphalt*, *bituminous asphalt*, or *hot mix asphalt*. Failure occurs from excessive horizontal tensile stress at the base asphalt.

Rigid pavement, also called *concrete* or *portland cement concrete*, carries loads across concrete slabs and transfers loads across the joints. Failure occurs at joints or from cracking at corners and mid-panel.

Aggregate pavement is a dense-graded compacted aggregate on a prepared subgrade. The load transfer and design principles are similar to flexible pavement.

Handbook figure Asphalt Concrete Volumetric Terms and Definitions Using Phase Diagram shows a phase diagram for asphalt concrete.

Handbook figure Common Pavement Systems shows six different cross sections of pavement. The differences between systems are in the components and relative thicknesses of the layers. However, all systems begin at the natural subgrade, have a compacted subgrade layer at the system's base, and are finished with a layer of asphalt at the surface. *Handbook* section Basic Pavement Structure provides a more detailed description of each of the layered components for an asphalt system.

Asphalt Service Life

Handbook: Geotechnical Influences on Major Distresses in Rigid Pavements; Geotechnical Influences on Major Distresses in Rigid Pavements

CERM: Sec. 77.1

Poor construction practices and/or quality control measures can easily sabotage a good design and set up a pavement for premature failure. *Handbook* tables Geotechnical Influences on Major Distresses in Flexible Pavements and Geotechnical Influences on Major Distresses in Rigid Pavements provide insight into common problems and failures associated with pavement systems. These tables can help the engineer prevent a specific problem or determine the cause of a problem once it has occurred.

Pavement Thickness

Handbook: Structural Layer Coefficients

CERM: Sec. 76.26, App. 76.A, App. 76.B, App. 76.C, App. 77.A, App. 77.B, App. 77.C

FHWA NHI-05-037: Sec. 5.5.1

CERM Eq. 76.33 is the AASHTO *layer-thickness equation*, which combines pavement layer properties and thickness into one variable, the design *structural number*.

$$SN = D_1 a_1 + D_2 a_2 m_2 + D_3 a_3 m_3 \quad \quad 76.33$$

Layer thicknesses are determined in top-down fashion, with actual layer thicknesses potentially varying from what is required to satisfy the structural number based on minimum thickness requirements, as well as placement, compaction, and cost considerations.

For the CERM equation,

a	layer coefficients	–
D_1	thickness of surface	in
D_2	thickness of base	in
D_3	thickness of subbase	in
m	drainage coefficients (base and subbase only)	–

The a-value can be obtained from *Handbook* table Structural Layer Coefficients. This value depends on the layer (wearing surface, base, subbase) and the type of material used (concrete, stone, cement, etc.). Structural numbers for flexible pavements are available in CERM App. 76.A, App. 76.B, and App. 76.C and for rigid pavement in App. 77.A, App. 77.B, and App. 77.C. Drainage coefficients are provided in FHWA NHI-05-037 Table 5-49.

Exam tip: The AASHTO layer-thickness equation should not be used to determine the structural number when the base or subbase modulus of resilience is greater than 40,000 psi.

Aggregates

Handbook: Specific Gravity of Aggregates; Specific Gravity of Fine Aggregates; Asphalt Mixture Volumetrics

CERM: Sec. 76.4, Sec. 76.12

The mineral aggregate component of the asphalt mixture comprises 90–95% of the weight and 75–85% of the mixture volume. The size and grading of the aggregate are important because the minimum lift thickness depends on the maximum aggregate size. The minimum lift thickness should generally be at least three times the nominal maximum aggregate size. Therefore, the maximum-size aggregate can be as much as 80% of the lift thickness.

Coarse aggregate is retained above a no. 8 sieve, and fine aggregate passes through a no. 8 sieve. Dense-graded aggregate mixtures contain a wide range of particle sizes so that there is minimal void space in the mixture. Open-graded aggregate is most common in hot mix asphalt design and does not include fine or sand content. Gap-graded mixtures are a combination of two distinct sizes that are on opposite ends of the size spectrum.

Handbook sections Specific Gravity of Aggregates and Specific Gravity of Fine Aggregates provide calculations for the specific gravities of different materials that are used in the asphalt mix. *Handbook* section Asphalt Mixture Volumetrics uses these values to further the asphalt design process.

H. SETTLEMENT

Key concepts: These key concepts are important for answering exam questions in subtopic 13.H, Settlement.

- lightweight fill materials
- prefabricated vertical drains (PVD)
- vertical strip drains (VSD)
- wick drains

NCEES Handbook: To prepare for this subtopic, familiarize yourself with these sections in the *Handbook*.

- Typical Consolidation Curve for Normally Consolidated Soil
- Typical Consolidation Curve for Overconsolidated Soil
- Physical Property Requirements of Geofoam

PE Civil Reference Manual (**CERM**): Study these sections in CERM that either relate directly to this subtopic or provide background information.

- Section 36.1: Shallow Foundations
- Section 40.4: Settling
- Section 40.7: Primary Consolidation
- Section 40.9: Secondary Consolidation

The following equations, figures, and tables are relevant for subtopic 13.H, Settlement.

Primary Consolidation

Handbook: Typical Consolidation Curve for Normally Consolidated Soil; Typical Consolidation Curve for Overconsolidated Soil

CERM: Sec. 40.7

When pressure is applied to soil, it squeezes water from the soil below the load. As the water is lost, the soil consolidates and the ground surface settles. The long-term consolidation due to water loss is the primary consolidation, S_{primary}. CERM Eq. 40.15 and Eq. 40.16 calculate the primary consolidation for an overconsolidated soil and a normally consolidated soil, respectively.

$$S_{\text{primary}} = \frac{H \Delta e}{1 + e_o} = \frac{HC_r \log_{10} \frac{p_o' + \Delta p_v'}{p_o'}}{1 + e_o} \quad 40.15$$

$$= H(\text{RR}) \log_{10} \frac{p_o' + \Delta p_v'}{p_o'}$$

[overconsolidated]

$$S_{\text{primary}} = \frac{H \Delta e}{1 + e_o} = \frac{HC_c \log_{10} \frac{p_o' + \Delta p_v'}{p_o'}}{1 + e_o} \quad 40.16$$

$$= H(\text{CR}) \log_{10} \frac{p_o' + \Delta p_v'}{p_o'}$$

[normally consolidated]

Handbook figures Typical Consolidation Curve for Normally Consolidated Soil and Typical Consolidation Curve for Overconsolidated Soil plot the void ratio to the vertical effective stresses for consolidated and overconsolidated soils, respectively.

Settlement

Handbook: Physical Property Requirements of Geofoam

CERM: Sec. 36.1, Sec. 40.4

Settlement is the tendency of soils to deform under applied loads. Since structures can tolerate only a limited amount of settlement, foundation design will often be controlled by settlement criteria because soil usually deforms significantly before it fails in shear.

To determine the total amount of settlement, all phases of anticipated consolidation must be summed. Settlement that occurs during the elastic phase is reversible. The settlement that has occurred in this range can easily be reversed by removing the applied load. However, once the settlement moves into the other phases of consolidation, the settlement becomes permanent, since the soil properties—most importantly, the void ratio—have changed.

The load applied to the soil must include the weight of any fill that is placed during construction. *Geofoam* is a common lightweight fill material that keeps the overburden pressure to a minimum and thus minimizes the amount of settlement.

Handbook table Physical Property Requirements of Geofoam gives various requirements for geofoam based on ASTM D6817.

> *Exam tip*: The equations to determine settlement are different for normally consolidated and overconsolidated soil. Make sure to use the correct equations for the type of soil in the problem statement.

Secondary Consolidation

CERM: Sec. 40.9

Secondary consolidation is a gradual continuation of consolidation that continues long after the cessation of primary consolidation. Secondary consolidation does not occur in granular soils; however, it is a major contribution to the total settlement with inorganic clays and silts, as well as for highly organic soil (although highly organic soils are not typically used in structural construction). Secondary consolidation is a much slower process than primary consolidation. The coefficient of secondary consolidation, C_α, is given by CERM

Eq. 40.25. The original void ratio is often used as an approximation of the average void ratio over the range of interest.

$$C_\alpha = \frac{\alpha}{1 + e_o} \quad \quad 40.25$$

The secondary consolidation during the period t_4 to t_5 is given by CERM Eq. 40.26.

$$S_{\text{secondary}} = C_\alpha H \log_{10} \frac{t_5}{t_4} \quad \quad 40.26$$

22 Groundwater and Seepage

Content in blue refers to the *NCEES Handbook*.

A. Seepage Analysis/Groundwater Flow 22-1
B. Dewatering Design, Methods, and Impact on Nearby Structures... 22-3
C. Drainage Design/Infiltration........................... 22-5
D. Grouting and Other Methods of Reducing Seepage... 22-7

The knowledge area of Groundwater and Seepage makes up between three and five questions out of the 80 questions on the PE Civil Geotechnical exam. The organization of this chapter follows the order of subtopics given by NCEES for this knowledge area. Each subtopic is covered in the following sections.

A. SEEPAGE ANALYSIS/GROUNDWATER FLOW

Key concepts: These key concepts are important for answering exam questions in subtopic 14.A, Seepage Analysis/Groundwater Flow.

- Darcy's law
- flow nets
- hydraulic gradient
- uplift

NCEES Handbook: To prepare for this subtopic, familiarize yourself with these sections in the *Handbook*.

- Flow Rates
- Recommended Flow-Reduction Factor Values for Use in Allowable Flow Rates Equation
- Transformation Method for Analysis of Anisotropic Embankments
- Flow Net Concepts
- Typical Flow Net Showing Basic Requirements and Computations
- Seepage Through Embankment and Foundation
- Darcy's Law
- Gravity Dam on Pervious Foundation of Finite Depth

PE Civil Reference Manual (**CERM**): Study these sections in CERM that either relate directly to this subtopic or provide background information.

- Section 21.2: Aquifer Characteristics
- Section 21.4: Darcy's Law
- Section 21.11: Well Drawdown in Aquifers
- Section 21.14: Flow Nets
- Section 21.15: Seepage from Flow Nets
- Section 21.16: Hydrostatic Pressure Along Flow Path

The following equations and figures are relevant for subtopic 14.A, Seepage Analysis/Groundwater Flow.

Gradient and Uplift

Handbook: Flow Rates; Recommended Flow-Reduction Factor Values for Use in Allowable Flow Rates Equation

CERM: Sec. 21.16

When a point being investigated is along the bottom of a structure, the hydrostatic pressure is referred to as *uplift pressure*. All uplift is due to the neutral pore pressure. The actual uplift force on a surface area, A, is given by CERM Eq. 21.37. N is the *neutral stress coefficient*, the fraction of the surface that is exposed to the hydrostatic pressure.

$$U = N p_u A \qquad 21.37$$

The gradient factor of safety qualifies a geologic column subject to seepage. The gradient factor of safety applies to flotation or uplift potential. When the upward forces on a geologic column equal the downward forces, flotation occurs. The gradient is considered critical.

$$i_c = \frac{h}{L} = \frac{g_{\text{buoy}}}{g_{\text{H}_2\text{O}}}$$

The factor of safety for flotation is the ratio of critical gradient, i_c, to computed gradient, i_e, which is generally the exit gradient.

$$\text{FS}_i = \frac{i_c}{i_e}$$

CERM Eq. 21.38 is similar.

$$(FS)_{heave} = \frac{\text{downward pressure}}{\text{uplift pressure}} \quad 21.38$$

Exit gradients are generally highest (and gradient factors of safety are generally lowest) near grade transitions along the downstream sides of structures. High exit gradients commonly occur where flow through granular, high-permeability strata is restricted by overlying cohesive, low-permeability blanket strata.

Subgrade drainage can be used to modify the flow rate or gradient to help facilitate an acceptable factor of safety. *Handbook* section Flow Rates shows equations for flow rate calculations.

$$q_{allow} = q_{ult}\left(\frac{1}{(RF_{SCB})(RF_{CR})(RF_{IN})(RF_{CC})(RF_{BC})}\right)$$

$$q_{allow} = q_{ult}\left(\frac{1}{\prod RF}\right)$$

Handbook table Recommended Flow-Reduction Factor Values for Use in Allowable Flow Rates Equation provides ranges of reduction factor values for various filters and drains used to modify the flow rate.

Flow Nets

Handbook: Transformation Method for Analysis of Anisotropic Embankments; Typical Flow Net Showing Basic Requirements and Computations; Seepage Through Embankment and Foundation; Flow Net Concepts; Gravity Dam on Pervious Foundation of Finite Depth

CERM: Sec. 21.14, Sec. 21.15

For complicated, real-life models, flow nets are commonly used. *Handbook* figures Flow Net Concepts, Gravity Dam on Pervious Foundation of Finite Depth, Seepage Through Embankment and Foundation, Typical Flow Net Showing Basic Requirements and Computations, and Transformation Method for Analysis of Anisotropic Embankments are representations of how flow nets are used to calculate seepage through an embankment. Although the cross sections change in shape and configuration, the concepts are the same for each diagram.

CERM Sec. 21.14 lists the rules for drawing a flow net, and CERM Fig. 21.3 shows examples of flow nets. The rules given in CERM are followed in the *Handbook*, and many flow nets are provided as examples.

Once the flow net is drawn, CERM Eq. 21.33 can be used to determine the seepage through the system.

Handbook figures Flow Net Concepts, Gravity Dam on Pervious Foundation of Finite Depth, and Seepage Through Embankment and Foundation provide explanations of flow nets through common embankment configurations.

Handbook figure Typical Flow Net Showing Basic Requirements and Computations provides a good example of how to use the measurements taken from the flow net.

Handbook figure Transformation Method for Analysis of Anisotropic Embankments provides an example of how to convert a complicated embankment composed of multiple materials into an embankment of a single material equivalent in which the principles in the previously mentioned diagrams apply.

Flow Rate

Handbook: Darcy's Law

$$Q = kiA$$

CERM: Sec. 21.4, Sec. 21.11

$$Q = -KiA_{gross}$$
$$= -v_e A_{gross} \quad 21.10$$

The flow rate, Q, through a porous medium is defined by Darcy's law, given in the *Handbook* and CERM Eq. 21.10. Flow is important because drainage collection elements need to have adequate capacity for probable flows. Gradient, on the other hand, is important from stability and performance standpoints; gradient is related to the balance of earth and hydraulic forces where flow is occurring and is used to qualify the risk of seepage-based failure hazards.

Note that flow is expressed as a positive value in the *Handbook* equation, even though by convention it is expressed as a negative value (as in CERM Eq. 21.10) because head loss (and thus gradient) is negative in the direction of flow.

Darcy's law is repurposed in Dupuit's formula (CERM Eq. 21.25) to determine the gravity well discharge from unconfined aquifers and in the Thiem equation (CERM Eq. 21.27) to determine the artesian well discharge from an unconfined aquifer.

Handbook section Darcy's Law provides a cross section of a basic setup used to describe the rate of water flow through a soil with permeability k. With known values of distance and area, the flow rate, Q, can be determined.

Exam tip: Darcy's law is applicable only when the Reynolds number is less than 1.0.

Hydraulic Gradient

Handbook: Darcy's Law

$$i = \frac{\Delta h}{\Delta S}$$

CERM: Sec. 21.2

$$i = \frac{\Delta H}{L} \quad\quad 21.6$$

CERM Eq. 21.6 can be used to calculate the hydraulic gradient, i, which is the change in hydraulic head over a particular distance. The hydraulic head is determined as the piezometric head at observation wells. Gradient is related to forces within a flow regime. The drag associated with flow through a porous medium creates *seepage forces*, acting in the direction of flow, that may

- add to or reduce the gravitational forces acting on a particular subsurface profile
- impose additional driving loads on potential failure masses
- increase or reduce local shear strength

These forces expose structures impacted by flow to unique failure hazards.

The *Handbook* equation shows the hydraulic gradient as the quotient of Δh and ΔS. Although the *Handbook* equation uses different variables, it is equivalent to CERM Eq. 21.6. A visual illustration of the hydraulic gradient is shown in *Handbook* section Darcy's Law.

B. DEWATERING DESIGN, METHODS, AND IMPACT ON NEARBY STRUCTURES

Key concepts: These key concepts are important for answering exam questions in subtopic 14.B, Dewatering Design, Methods, and Impact on Nearby Structures.

- piezometers
- sumps and pumps
- wells

NCEES Handbook: To prepare for this subtopic, familiarize yourself with these sections in the *Handbook*.

- Instruments for Measuring Piezometric Pressure
- Drainage of an Open Deep Cut by Means of a Multistage Wellpoint System
- Vacuum Wellpoint System
- Jet-Eductor Wellpoint System for Dewatering a Shaft
- Deep-Well System for Dewatering an Excavation in Sand
- Single Dewatering Well
- Grout Curtain or Cutoff Trench Around an Excavation

PE Civil Reference Manual **(CERM):** Study these sections in CERM that either relate directly to this subtopic or provide background information.

- Section 17.33: Pitot-Static Gauge
- Section 21.9: Wells
- Section 21.11: Well Drawdown in Aquifers
- Section 40.7: Primary Consolidation
- Section 80.21: Site Dewatering

The following equations, figures, tables, and concepts are relevant for subtopic 14.B, Dewatering Design, Methods, and Impact on Nearby Structures.

Piezometers

Handbook: Instruments for Measuring Piezometric Pressure; Deep-Well System for Dewatering an Excavation in Sand

CERM: Sec. 17.33, Sec. 40.7

Piezometers are used to determine groundwater elevation. This information is helpful prior to construction to determine if the groundwater elevation must be lowered (and if so, how much) and to select the specific type of dewatering system used.

Handbook table Instruments for Measuring Piezometric Pressure provides a list of different piezometers and the advantages of each. Piezometers are shown in many *Handbook* figures as a tool to measure the groundwater elevation. A construction piezometer is shown in *Handbook* figure Deep-Well System for Dewatering an Excavation in Sand. In this example, the piezometer is used to measure the groundwater elevation at the centerline of an excavation. The turbine pump installed elsewhere on the site is tasked with lowering the groundwater table at the piezometer location.

Exam tip: The piezometer is installed at the location where groundwater elevation is critical. The pump is installed at the location where it will have the greatest effect on the groundwater table.

Wells

Handbook: Drainage of an Open Deep Cut by Means of a Multistage Wellpoint System; Vacuum Wellpoint System; Jet-Eductor Wellpoint System for Dewatering a Shaft; Single Dewatering Well

$$Q_w = \frac{\pi k (H^2 - h_w^2)}{\ln \frac{R}{r_w}}$$

$$Q_w = \frac{\pi k (H^2 - h^2)}{\ln \frac{R}{r}}$$

CERM: Sec. 21.9, Sec. 21.11

$$Q = \frac{\pi K (y_1^2 - y_2^2)}{\ln \frac{r_1}{r_2}} \quad \text{21.25}$$

$$Q = \frac{2\pi T (s_2 - s_1)}{\ln \frac{r_1}{r_2}} \quad \text{21.26}$$

Wells are typically required where water needs to be drawn down in advance of construction or where larger volumes of water need to be managed over the long term.

CERM Eq. 21.25 is for unconfined aquifers, while CERM Eq. 21.26 is for confined aquifers and unconfined aquifers of great thickness. The equations can be rearranged to solve for drawdown depth or distance from the well if the other parameters are known. The equations in the *Handbook* and CERM are equivalent but use different variables. *Handbook* variables H and h are equivalent to y_1 and y_2 in CERM and represent the depth of the aquifer. K in CERM is the same as k in the *Handbook* and represents the hydraulic conductivity. R in the *Handbook* is the same as r_1 in CERM and is the depth of the aquifer at the corresponding radial distance H or y_1. (r in the *Handbook* and r_2 in CERM are similar).

A vacuum wellpoint system is needed for wells in fine-grained soil. See *Handbook* figure Vacuum Wellpoint System for an example. This type of system is most effective at depths of 15 ft or shallower.

Handbook figure Jet-Eductor Wellpoint System for Dewatering a Shaft highlights a type of wellpoint best used for deep excavations (up to 100 ft). A system using multiple pipes is constructed to create a higher-pressure system to dewater soils with a very low permeability.

Handbook figure Jet-Eductor Wellpoint System for Dewatering a Shaft illustrates another method for reducing the water table (in this case, by 15 ft or more).

With this system, multiple wellpoints may be used at varying elevations to lower the water table if an excavation can be made.

Exam tip: In these equations, the variable y or h is the height of the initial or post-dewatering water column. This can be confusing, as many questions ask for the depth to the water surface. Note that variables y_1 and (or H and h) are equal to the height of the initial or lowered water column, not the depth to the water surface.

Sumps and Pumps

Handbook: Deep-Well System for Dewatering an Excavation in Sand; Grout Curtain or Cutoff Trench Around an Excavation

CERM: Sec. 80.21

Sumps and pumps deal effectively with low flows through low-permeability materials. Groundwater can be drawn down below construction subgrades and grade-supported structures. Aggregate surfacing can serve as a working platform and drainage medium. Drawdown helps mitigate the risk of subgrade heave and seepage-related piping.

Sumps and pumps are used where the volume of water needing to be collected is limited or where operation or stability requirements do not warrant the use of wells.

Sumps are typically filled with gravel, partly to provide a reservoir from which water can be pumped, and partly to limit the clogging of pumps with finer-grained soils. Spread atop a surface through which water might seep, the gravel provides a stable working platform. Sloping excavations gently over large areas helps encourage seepage to flow to the sumps.

While Chap. 3 of the *Handbook* does not address pumps directly, it references them in figures Deep-Well System for Dewatering an Excavation in Sand and Grout Curtain or Cutoff Trench Around an Excavation. The specifics of actual pumps are discussed in detail in *Handbook* Chap. 6. A geotechnical engineer should understand the concepts of flow rate and the need for pumping.

Subsidence

CERM: Sec. 40.7

Subsidence is the lowering of the ground surface and can occur when groundwater is removed from the soil. Dewatering removes the buoyant effect that groundwater has on earth materials, resulting in an increase in unit weight from buoyant to saturated over the range of depths or elevations subject to dewatering. Effective stresses in potentially compressible material layers will thus increase by an amount equal to the unit weight of water times the change in water surface elevation. Analysis of this situation proceeds in a manner consistent with any other loading situation, except that the applied load is just the increase in in situ density from drawdown.

Depending on the amount and elevation of initial groundwater compared to the amount and elevation of groundwater after pumping, the ground elevation may be lowered. This is most common in low-lying areas where the groundwater is close to the ground surface.

C. DRAINAGE DESIGN/INFILTRATION

Key concepts: These key concepts are important for answering exam questions in subtopic 14.C, Drainage Design/Infiltration.

- surface drain control
  - edge drains
  - swales
  - trench drains
- underground filtration systems
  - stored water contributing to expansion or frost heave
  - travel of stored water to adjacent slope
  - unanticipated infiltration rates if system not evaluated

NCEES Handbook: To prepare for this subtopic, familiarize yourself with these sections in the *Handbook*.

- Identification of Usual Primary Function Versus Type of Geosynthetics
- Soil Retention Criteria for Geotextile Filter Design Using Steady-State Flow Conditions
- Solid-Waste Containment System with High Geosynthetic Utilization
- Criteria for Filters
- Perforated Pipe
- Runoff Analysis: Rational Formula Method
- Minimum Infiltration Rates for the Various Soil Groups
- Unit Hydrograph
- Infiltration: Richards Equation
- Infiltration: Horton Model
- Green-Ampt Equation

PE Civil Reference Manual (CERM): Study these sections in CERM that either relate directly to this subtopic or provide background information.

- Section 20.10: Unit Hydrograph
- Section 20.15: Peak Runoff from the Rational Method
- Section 20.16: NRCS Curve Number
- Section 21.17: Infiltration
- Section 32.35: Rainwater Runoff from Highways
- Section 40.14: Geotextiles
- Section 76.31: Subgrade Drainage

The following equations, figures, and tables are relevant for subtopic 14.C, Drainage Design/Infiltration.

Subsurface Drainage

Handbook: Identification of Usual Primary Function Versus Type of Geosynthetics

CERM: Sec. 40.14, Sec. 76.31

Below-grade walls are generally fitted with a drainage medium and drainage collection pipes. Walls retaining layered soils and sedimentary rock are likely to be exposed to water flowing from layer boundaries or from bedding planes or joints, either periodically or seasonally.

Slabs do not typically require subfloor drainage unless there is risk of the groundwater surface rising or other conditions warranting such a system.

Subsurface drainage is important for exterior grade-supported structures. Even with a sound surface drainage plan, water can pass through material interface gaps and joints, accumulating within aggregate and granular subbase layers and potentially precipitating or magnifying expansion or frost heave.

CERM Sec. 76.31 describes the criteria for when a subgrade drain is needed in pavement design and provides details regarding the location of the drain. CERM Sec. 40.14 provides a description of geotextile uses, including subsurface drainage. CERM Table 40.6 provides typical Department of Transportation (DOT) minimum geotextile specifications for use with subsurface drainage.

Handbook section Identification of Usual Primary Function Versus Type of Geosynthetics lists types of geosynthetics used for drainage. These geosynthetics are commonly used in retaining wall drainage, as described previously.

Filters

Handbook: Soil Retention Criteria for Geotextile Filter Design Using Steady-State Flow Conditions; Solid-Waste Containment System with High Geosynthetic Utilization; Criteria for Filters; Perforated Pipe

CERM: Sec. 32.35, Sec. 40.14

Filters have many uses, depending on what material is being filtered out of what medium. CERM Sec. 32.35 gives a good discussion of filters that clean pollutants out of highway runoff. The uses of filter berms, filter chambers, and sand filter beds are provided. *Handbook* table Criteria for Filters can be used to size filters for base soils containing particles finer than 4.75 mm.

Geotextiles may be used as low-profile filters. CERM Sec. 40.14 discusses different types of geotextiles and what materials each type may filter out. Geotextiles provide a filtering function and prevent the mixing of dissimilar materials (e.g., an aggregate base and soil). *Handbook* figure Soil Retention Criteria for Geotextile Filter Design Using Steady-State Flow Conditions provides the guidance needed for selection of a geotextile based on the particle size of the soil. *Handbook* figure Solid-Waste Containment System with High Geosynthetic Utilization shows many different geosynthetic membranes, including filters that allow water and leachate to pass through. *Handbook* section Perforated Pipe gives the equation to prevent infiltration of filter material into perforated pipe, screens, and so on.

$$\frac{\text{minimum 50\% size of filter material}}{\text{hole diameter or slot width}} \geq 1.0$$

Surface Drainage

Handbook: Runoff Analysis: Rational Formula Method; Unit Hydrograph

$$Q = CIA$$

CERM: Sec. 20.10, Sec. 20.15

$$Q_p = CIA_d \qquad 20.36$$

Finished grades should be sloped away from buildings and grade-supported structures. Site features exposed to concentrated or high-velocity flow should be protected from erosion. Surface drainage should not be directed toward or over slopes or walls unless the structures have been designed to accommodate the drainage.

The equation in *Handbook* section Runoff Analysis: Rational Formula Method, also given by CERM Eq. 20.36, provides a runoff calculation of peak discharge based on the runoff coefficient, rainfall intensity, and watershed area. The runoff coefficient is based on the type of drainage area and varies between 0.05 in flat, sandy soil and 0.95 for roofs, concrete and asphaltic streets, and downtown business areas.

Unit hydrographs such as those shown in *Handbook* section Unit Hydrograph and CERM Fig. 20.13 are graphical representations of peak discharge versus time. The graphs show how the discharge increases to the peak discharge value as the duration of the rain event increases. After the peak is met, the graphs show the time needed for recession back to the original state.

Figure 20.13 Unit Hydrograph

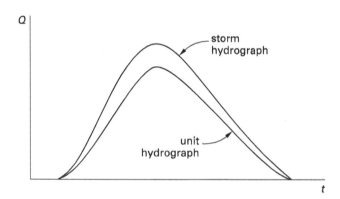

Exam tip: Most surface drainage discussion and equations are in *Handbook* Chap. 6, not Chap. 3.

Infiltration

Handbook: Minimum Infiltration Rates for the Various Soil Groups; Infiltration: Richards Equation; Infiltration: Horton Model; Green-Ampt Equation

$$f = (f_0 - f_c)e^{-kt} + f_c$$

$$F = f_c t_p + \left(\frac{f_0 - f_c}{k}\right)(1 - e^{-kt_p})$$

CERM: Sec. 20.16, Sec. 21.17

$$f_t = f_c + (f_0 - f_c)e^{-kt} \qquad 21.39$$

$$F_t = f_c t + \left(\frac{f_0 - f_c}{k}\right)(1 - e^{-kt}) \qquad 21.42$$

The groundwater table can be refilled in several ways. When the rainfall supply exceeds the infiltration capacity, infiltration decreases exponentially over time. The cumulative infiltration will not correspond to the increase in water table elevation due to the effects of porosity.

Infiltration rates for intermediate, silty soils are approximately 50% of those for sand. Clay infiltration rates are approximately 10% of those for sand. The depth of water stored in a layer is equal to the total depth of infiltration minus the depth of water that has leaked out the bottom of the layer. Except in very permeable soils, water leaks out the bottom of the layer slowly.

The Horton equation given in *Handbook* section Infiltration: Horton Model and CERM Eq. 21.39 provides a lower bound on the infiltration capacity.

The cumulative infiltration over time is provided by CERM Eq. 21.42 (also provided in *Handbook* section Infiltration: Horton Model).

Another model for calculation of infiltration is in *Handbook* section Green-Ampt Equation. This is a more practical version of the equation in *Handbook* section Infiltration: Richards Equation. The Green-Ampt method requires more input than the Horton method, but it provides a more accurate measurement of the infiltration rate. Minimum infiltration rates for various soil types are provided in *Handbook* figure Minimum Infiltration Rates for the Various Soil Groups.

D. GROUTING AND OTHER METHODS OF REDUCING SEEPAGE

Key concepts: These key concepts are important for answering exam questions in subtopic 14.D, Grouting and Other Methods of Reducing Seepage.

- collection elements
  - chimney drains
  - interceptor pipes
  - sand blanket drains
  - trench drains
- flow prevention to vulnerable structures
- particle filtration
- seepage berms

NCEES Handbook: To prepare for this subtopic, familiarize yourself with these sections in the *Handbook*.

- Physical Properties of Chemical Grouts
- Rough Guide for Grouting Pressure
- Identification of Usual Primary Function Versus Type of Geosynthetics

- Soil Retention Criteria for Geotextile Filter Design Using Steady-State Flow Conditions
- Grout Curtain or Cutoff Trench Around an Excavation

PE Civil Reference Manual **(CERM):** Study these sections in CERM that either relate directly to this subtopic or provide background information.

- Section 21.10: Design of Gravel Screens and Porous Filters
- Section 30.23: Sludge Disposal
- Section 67.14: Grout
- Section 67.15: Grout Slump
- Section 67.16: Grout Compressive Strength
- Section 80.20: Slope and Erosion Control Features

The following equations, figures, tables, and concepts are relevant for subtopic 14.D, Grouting and Other Methods of Reducing Seepage.

Grouting

Handbook: Physical Properties of Chemical Grouts; Rough Guide for Grouting Pressure

CERM: Sec. 67.14, Sec. 67.15, Sec. 67.16

Grouting can mitigate unfavorable geotechnical conditions contributing to seepage-based stability and performance risks. Slurry, soil mix, and jet grouting can create effective cutoffs. Injection grouting can target open bedding planes and joints and fractures in rock. Chemical grouting can cement permeable media.

Depending on the technology employed, grouting can provide the depth and breadth of protection obtained with a cutoff, as well as the confined and targeted protection of a collection element. Grouting is also uniquely suited to dealing with voids and flow paths in rock.

CERM Sec. 67.14 through Sec. 67.16 provides design properties of grout, including modulus of elasticity, slump, and compressive strength, along with a detailed discussion of these properties. *Handbook* table Physical Properties of Chemical Grouts gives values for unconfined compressive strength based on the class of grout used. *Handbook* figure Rough Guide for Grouting Pressure gives approximate pressures for grout installation at a depth of up to 150 ft below the ground surface as a function of rock type.

Particle Filtration

Handbook: Identification of Usual Primary Function Versus Type of Geosynthetics; Soil Retention Criteria for Geotextile Filter Design Using Steady-State Flow Conditions

CERM: Sec. 21.10

$$D_{\text{opening,filter}} \leq D_{85,\text{soil}} \quad \text{[screen filters]} \quad 21.21$$

$$[\text{filtering criterion}]^6 \; D_{15,\text{filter}} \leq 5 D_{85,\text{soil}} \quad 21.22$$
[filter beds]

$$[\text{permeability criterion}]^7 \; D_{15,\text{filter}} \leq 5 D_{15,\text{soil}} \quad 21.23$$
[filter beds]

CERM Eq. 21.21 provides a simplified means of evaluating opening requirements for geotextiles such as pipe socks and trench liners. CERM Eq. 21.22 (for migration) and Eq. 21.23 (for flow) provide a simplified means of evaluating the gradation requirements for filter-compatible earth material. Both equations require the diameter of the filter material to be less than five times the diameter of the soil material at the same diameter percentage.

Seepage-reduction efforts typically place materials of contrasting grain size distributions in contact (e.g., clay fill against a coarse sand drainage blanket). If the two materials are not filter compatible, particles from the finer-grained material may migrate into and potentially pass through or clog the coarser-grained material, defeating both intentions for the filter material.

The *Handbook* provides information related to the use of geosynthetics for filtration in section Identification of Usual Primary Function Versus Type of Geosynthetics. Geotextiles and geocomposites are shown to provide filtration as the primary function. *Handbook* figure Soil Retention Criteria for Geotextile Filter Design Using Steady-State Flow Conditions provides the tendency of the soil to favor retention or permeability.

Exam tip: Only uniformly graded materials ($C_u < 2.5$) and well-graded materials ($2.5 < C_u < 6$) are suitable for use as filters.

Cutoffs

Handbook: Grout Curtain or Cutoff Trench Around an Excavation

CERM: Sec. 30.23

Cutoffs prevent flow from critical exit points on or adjacent to vulnerable structures. They can be installed by driving, drilling, or excavating. Cutoffs have limited maintenance and performance concerns at the trade-off of mobilization and material costs, especially for long or deep cutoffs.

Cutoffs are an effective means of reducing seepage on a large scale. Variations and uncertainties in material consistency, permeability, and flow are dealt with by providing a consistent, comprehensive level of protection that extends from the base of an engineered structure to a low-permeability stratum at depth.

Cutoffs can reach substantial depths of 100 ft or more.

Handbook figure Grout Curtain or Cutoff Trench Around an Excavation provides a cross section of an excavation in which the initial water table is higher outside the excavation than inside. The grout curtain or cutoff trench is used to keep the water outside the excavation. CERM Sec. 30.23 describes how a slurry cutoff wall can be used to contain sludge in place of a landfill.

Exam tip: Cutoffs can be used to keep water, sludge, and other materials in or out of the site or area in question.

Seepage Berms

CERM: Sec. 80.20

Seepage berms provide an unsophisticated, earth-fill-based means of mitigating unfavorable exit gradients and improving seepage-based factors of safety. U.S. Army Corps of Engineers design principles typically specify a berm width equal to four times the height of the protected structure, which typically implies that a fair amount of property must be available for the placement of berm fill. Stability and performance can be augmented by employing berms with collection elements like pipes and blanket and trench drains.

CERM Sec. 80.20 describes how a temporary berm can be used to control runoff during construction. An isometric drawing of a site is provided in CERM Fig. 80.12, which shows how runoff is controlled by a temporary berm and directed into erosion control fabric.

23 Problematic Soil and Rock Conditions

Content in blue refers to the *NCEES Handbook*.

A. Karst; Collapsible, Expansive, and Sensitive Soils 23-1
B. Reactive/Corrosive Soils................................. 23-3
C. Frost Susceptibility 23-5

The knowledge area of Problematic Soil and Rock Conditions makes up between three and five questions out of the 80 questions on the PE Civil Geotechnical exam. The organization of this chapter follows the order of subtopics given by NCEES for this knowledge area. Each subtopic is covered in the following sections.

A. KARST; COLLAPSIBLE, EXPANSIVE, AND SENSITIVE SOILS

Key concepts: These key concepts are important for answering exam questions in subtopic 15.A, Karst; Collapsible, Expansive, and Sensitive Soils.

- Atterberg limits
- geophysical investigative methods
- liquid limit
- liquidity index

NCEES Handbook: To prepare for this subtopic, familiarize yourself with these sections in the *Handbook*.

- Soil Classification Chart
- Unified Soil Classification System (USCS)
- Plasticity Chart for Unified Soil Classification System
- Commonly Performed Laboratory Tests on Soils

PE Civil Reference Manual **(CERM):** Study these sections in CERM that either relate directly to this subtopic or provide background information.

- Section 35.14: Atterberg Limit Tests

Codes and standards: To prepare for this subtopic, familiarize yourself with these sections from codes and standards.

Geotechnical Site Characterization (FHWA NHI-16-072)

- Section 4.17.3: SBT Identification from Soil Behavior Type Index
- Section 4.17.6: Soil Identification from Dilatometer (DMT)
- Section 5.2.2: Indirect Identification of Collapsible Soils
- Section 5.3.1: Occurrence of Expansive Soils
- Section 5.3.2: Identification and Characterization of Expansive Soils Using Indirect Methods
- Section 5.11: Sensitive "Quick" Clays
- Section 5.11.3: Challenges for Subsurface Exploration and Testing in Sensitive Clays
- Section 12.2.1: Implications of Karst Hazard for Transportation Projects
- Section 12.2.2: Identification and Characterization of Karst Hazards

Soils and Foundations Reference Manual – Volume I (FHWA NHI-06-088)

- Section 5.7.2: Collapse Potential of Soils

The following equations, figures, tables, and concepts are relevant for subtopic 15.A, Karst; Collapsible, Expansive, and Sensitive Soils.

Collapsible Soil

Handbook: Commonly Performed Laboratory Tests on Soils

FHWA NHI-06-088: Sec. 5.7.2

FHWA NHI-16-072: Sec. 5.2.2

Collapsible soil is soil that decreases in volume with increased moisture content. The collapse potential can be correlated with the liquid limit and dry unit weight. With a known liquid limit and an estimate of a soil's dry unit weight, collapse potential can be assessed using a chart such as FHWA NHI-06-088 Fig. 5-28, which suggests that as dry density increases (meaning void ratio decreases) and plasticity increases (meaning liquid limit increases), the risk of collapse diminishes. The percent collapse can be calculated using FHWA NHI-06-088 Eq. 5-15.

$$\%C = \frac{100\Delta H_c}{H_o} \quad \text{[FHWA Eq. 5-15]}$$

When the entire sample is inundated, the magnitude of collapse can be calculated using FHWA NHI-06-088 Eq. 5-16.

$$s_{\text{collapse}} = H\left(\frac{\%C}{100}\right) \quad \text{[FHWA Eq. 5-16]}$$

The collapse potential (CP) is the percent collapse, $\%C$, of a sample that has a total load of 4 ksf applied. FHWA NHI-06-088 Table 5-12 provides a qualitative assessment of CP. A CP of zero indicates there is no collapse risk, and the severity increases as CP increases. A CP greater than 10% is the highest, most severe risk.

Handbook table Commonly Performed Laboratory Tests on Soils shows that ASTM D5333 is the standard to determine the collapse potential of soils. FHWA NHI-16-072 Table 5-2 provides four additional methods to determine whether a soil is collapsible.

Expansive Soil

Handbook: Plasticity Chart for Unified Soil Classification System

FHWA NHI-16-072: Sec. 4.17.3, Sec. 4.17.6, Sec. 5.3.1, Sec. 5.3.2

Expansive soil is soil that experiences an increase in volume with increased moisture content. Expansive soils cause damage to structures built on clays having a high affinity for water and highly active mineralogy.

A simple way to determine if the soils at a site are potentially expansive is to check where the liquid limit (LL) and plasticity index (PI) plot on *Handbook* figure Plasticity Chart for Unified Soil Classification System.

ASTM D4829 standardizes the estimation of expansion potential by specifying the load at which sample inundation and expansion occur. For perspective, an expansion index (EI) value of 10 corresponds to 1% expansion, which for a 10 ft thick clay layer amounts to just over 1 in of heave. An EI value of 100 corresponds to 10% expansion, which produces 1 ft of heave.

$$\text{EI} = 1000\left(\frac{\Delta H}{H_1}\right)$$

FHWA NHI-16-072 Fig. 5-4 shows the distribution of expansive soils in the United States. FHWA NHI-16-072 Table 5-4 lists the severity of swell as a function of the liquid limit of the soil. As the liquid limit increases, the potential for expansion increases. A liquid limit lower than 20 indicates a non-swelling soil, and a liquid limit higher than 90 indicates a soil with an extra high potential for expansion.

Exam tip: Expansive soils are often measured by their swell potential.

Sensitive Soils

Handbook: Unified Soil Classification System (USCS); Soil Classification Chart

CERM: Sec. 35.14

FHWA NHI-16-072: Sec. 4.17.3, Sec. 4.17.6, Sec. 5.11, Sec. 5.11.3

Sensitive clay is clay that may experience significant strength loss when disturbed. Sensitive clay potential can be assessed with Atterberg limit tests. The natural moisture content of sensitive clay will typically be close to or above the liquid limit (LL). The liquidity index (LI) will approach or exceed 1.0, which implies that the natural moisture content of the suspect soil is close to or above the LL. *Handbook* section Unified Soil Classification System (USCS) contains the liquidity index equation.

$$\text{LI} = \frac{w - \text{PL}}{\text{PI}}$$

The sensitivity of a soil can also be determined through in situ testing with a cone penetration test (CPT). FHWA NHI-16-072 Fig. 4-14, Fig. 4-15, Table 4-15, and Table 4-16 provide equations and graphs to correlate the readings from a CPT sounding to soil types and conditions. Table 4-18 in the same manual provides a chart based on the dilatometer testing (DMT) material index.

Handbook table Soil Classification Chart provides equations to determine the liquid limit, plastic limit, plasticity index, and liquidity index. These parameters, along with the shrinkage index, are discussed in detail in CERM Sec. 35.14. FHWA NHI-16-072 Table 5-17 provides a classification chart for sensitive clays as a function of the sensitivity measurement. A sensitivity less than 1.0 indicates an insensitive clay, and a sensitivity greater than 16 is defined as a quick clay.

Exam tip: Sensitive clay is also referred to as a quick clay.

Karst Topography

FHWA NHI-16-072: Sec. 12.2.1

Karst is a potentially devastating geologic hazard; the collapse of a large cavity can take lives in addition to destroying buildings and infrastructure. Even if known and judged to be only a minor threat, karst can unfavorably impact a project scope by introducing variables to the design and construction processes that may not be fully qualified or quantified until the project is built.

The design effort needs to consider the long-term effects of continued cavity dissolution and the possibility that not all karst features have been discovered. Construction needs to proceed with caution to give the project team the best chance at characterizing and mitigating unfavorable features as they are discovered.

Karst hazards are caused by dissolved rock. Carbonate rocks such as limestone and dolomite are susceptible to karst hazards, as are gypsum and salt rock. Karst can occur when the small amount of acid in groundwater dissolves the rock over time and creates sinkholes. FHWA NHI-16-072 Fig. 12-3 provides a map of karst and potential karst areas in the United States.

Karst Mitigation

FHWA NHI-16-072: Sec. 12.2.2

Foundations in karst terrain can span or penetrate unstable features. Cavities can be filled, and collapse structures can be partially cleared and backfilled. Additional work may be required, including grouting of joints and fractures and groundwater and surface drainage control.

Mitigating karst hazards can be difficult, particularly if the extent of the hazard (e.g., the size or number of cavities or collapse structures) is large relative to the size of the project or if depth is an issue. It may be difficult to drill through cavities and flowing groundwater down to sound bedrock for foundation support. Exposing cavities, joints, and fractures for filling is time consuming and expensive. In the long term, surface drainage must also be controlled to limit the impact that infiltrating water could have on any karst features only partially corrected.

The best mitigation for karst topography is avoidance. If karst topography is suspected on a site, an in-depth investigation should be performed. FHWA NHI-16-072 Table 12-1 lists the applicability of geophysical methods for identifying voids and sinkholes that are associated with karst hazards. FHWA NHI-16-072 Sec. 12.2.2 also includes descriptions of each geophysical method listed in the table.

B. REACTIVE/CORROSIVE SOILS

Key concepts: These key concepts are important for answering exam questions in subtopic 15.B, Reactive/Corrosive Soils.

- concrete exposure classes
- corrosion protection
- loss of thickness for buried steel
- soil testing

NCEES Handbook: To prepare for this subtopic, familiarize yourself with these sections in the *Handbook*.

- Concrete Exposure Categories and Classes
- Commonly Performed Laboratory Tests on Soils
- Effect of Resistivity on Corrosion
- Corrosion of Buried Steel
- Corrosion Protection Requirements
- Decision Tree for Selection of Corrosion Protection Level

PE Civil Reference Manual (CERM): Study these sections in CERM that either relate directly to this subtopic or provide background information.

- Section 22.37: Corrosion
- Section 48.2: Cementitious Materials
- Section 48.26: Electrical Protection of Rebar
- Section 48.27: Coated Rebar

Codes and standards: To prepare for this subtopic, familiarize yourself with these sections from codes and standards.

Design and Construction of Driven Pile Foundations – Volume I (FHWA NHI-16-009)

- Section 6.12.1.1: Corrosion in Non-Marine Environments
- Section 6.12.1.2: Corrosion in Marine Environments

Geotechnical Site Characterization (FHWA NHI-16-072)

- Section 5.12.2: Identification of High Sulfate Soils

The following equations, figures, and tables are relevant for subtopic 15.B, Reactive/Corrosive Soils.

Corrosion

Handbook: Commonly Performed Laboratory Tests on Soils; Effect of Resistivity on Corrosion

CERM: Sec. 22.37

Corrosion potential tests measure resistivity, pH, redox potential, and sulfide concentration. Moisture content is also taken into consideration. Fine-grained soils are generally more corrosive than coarse-grained soils. Corrosion impacts the longevity of ductile iron pipe and concrete placed in contact with soil, which includes utilities, foundations, slabs, below-grade walls, and grade-supported slabs.

Improperly specified piping or concrete can experience premature deterioration and loss of strength or functionality. At the least, this can lead to maintenance that is needed sooner and is more extensive than anticipated; at worst, it can cause damage to impacted and nearby structures. Corrosion potential is influenced by a number of factors that can occur alone or in combination, and so its qualification typically involves an array of tests.

The AASHTO and ASTM test numbers for the testing related to pH values, corrosion, sulfate content, and resistivity are provided in *Handbook* table Commonly Performed Laboratory Tests on Soils.

Handbook table Effect of Resistivity on Corrosion provides values of resistivity (ranging from less than 700 Ω·cm to over 10,000 Ω·cm) in relation to the level of aggressiveness of the soil (ranging from very corrosive to noncorrosive).

Corrosion Protection for Steel

Handbook: Corrosion of Buried Steel; Corrosion Protection Requirements; Decision Tree for Selection of Corrosion Protection Level

CERM: Sec. 22.37, Sec. 48.26, Sec. 48.27

FHWA NHI-16-009: Sec. 6.12.1.1, Sec. 6.12.1.2

Iron pipe is rated by a system of points to determine corrosion potential. Points are assigned to five categories: resistivity, pH, redox potential, sulfide concentration, and moisture content. Points are summed to rate corrosion potential. An overall score of 10 or more typically warrants protection.

Corrosion protection for iron pipe is generally provided through applied coatings, reactive coatings, anodization, and cathodic or anodic protection. CERM Sec. 22.37 provides a long list of various corrosion possibilities ranging from uniform attack corrosion to stress corrosion to intergranular corrosion. CERM Sec. 48.26 and Sec. 48.27 discuss two different methods of protecting rebar from corrosion; the former discusses electrical protection of rebar, and the latter discusses the use of epoxy coated rebar.

The equation in *Handbook* section Corrosion of Buried Steel estimates the loss of thickness at a given time.

$$x = Kt^n$$

Handbook table Corrosion Protection Requirements and figure Decision Tree for Selection of Corrosion Protection Level provide requirements for ground anchors used in earth structure design. FHWA NHI-16-009 Sec. 6.12.1 provides guidance for piles in aggressive subsurface environments, along with flowcharts and design considerations for both marine (Sec. 6.12.1.2) and non-marine (Sec. 6.12.1.1) conditions.

Loss of Thickness for Buried Steel

Handbook: Corrosion of Buried Steel

$$x = Kt^n$$

The *Handbook* equation estimates the loss of thickness at a given time. Note that the loss of thickness in the equation is given in microns, whereas you will often need to know the loss of thickness in units of mils. The *Handbook* provides the conversion in the same section.

$$1000 \text{ microns} = 1 \text{ mm} = 0.039 \text{ in} = 39 \text{ mils}$$

Sulfates in Soils

Handbook: Concrete Exposure Categories and Classes

CERM: Sec. 48.2

FHWA NHI-16-072: Sec. 5.12.2

Soils with sulfate concentrations may create problems when mixed with lime, fly ash, or cement to stabilize the soil. Heave in the soil can occur once mixed, which can cause disturbance or even cracking in structures or pavements. The calcium in the lime or cement reacts with the sulfate to expand the soil, increasing its volume. The high concentration of sulfate can also be corrosive to portland cement, leading to cracking or a loss of strength in concrete that is in contact with the soil.

FHWA NHI-16-072 Sec. 5.12.2 provides guidelines on how to identify soils that have a high concentration of sulfates. FHWA NHI-16-072 Table 5-18 lists the levels of risk as a function of soluble sulfate concentrations. *Handbook* table Concrete Exposure Categories and Classes also lists the concrete classes as a function of the water-soluble sulfate in the soil.

CERM Sec. 48.2 lists the different classifications of portland cement used in concrete production. Type I may be used when sulfate hazards are not present. Type II has moderate sulfate resistance. Type V is used when exposure to sulfate concentration is expected.

C. FROST SUSCEPTIBILITY

Key concepts: These key concepts are important for answering exam questions in subtopic 15.C, Frost Susceptibility.

- frost heave and expansion
- soil particle size/gradation
- subgrade preparation
- surface and subsurface drainage control
- water absorption

NCEES Handbook: To prepare for this subtopic, familiarize yourself with these sections in the *Handbook*.

- Frost Depth
- Frost Susceptibility Classification by Percentage of Mass

PE Civil Reference Manual **(CERM):** Study these sections in CERM that either relate directly to this subtopic or provide background information.

- Section 76.32: Damage from Frost and Freezing

Codes and standards: To prepare for this subtopic, familiarize yourself with these sections from codes and standards.

Geotechnical Aspects of Pavements (FHWA NHI-05-037)

- Section 5.5.1: 1993 AASHTO Guide

The following figures and concepts are relevant for subtopic 15.C, Frost Susceptibility.

Frost-Susceptible Soils

Handbook: Frost Depth; Frost Susceptibility Classification by Percentage of Mass

CERM: Sec. 76.32

Frost-susceptible soils are soils that experience volumetric expansion on freezing. The impact of frost heave on foundations, slabs, pavements, and utilities in cold regions is widespread and is especially pronounced where structures transition from frost-susceptible to non-frost-susceptible soils.

The heaving of structures in response to the expansion of frozen soils can cause structural distress, create tripping hazards, alter surface drainage patterns, and open joints and gaps between structures, which allow water to penetrate below the structures and magnify the problem. Water is a primary source of this phenomenon, and freezing temperatures and frost-susceptible soils are the other two critical contributing sources. When saturation of subgrade soils is prevented and sources of water eliminated, frost heave does not occur.

When soils thaw in the spring, they become weak and susceptible to disturbance. In cold regions, road restrictions are enacted so that the temporarily sensitive soils are not exposed to damaging wheel loads, and so that the pavements supported on those soils are not prematurely and unnecessarily damaged.

Handbook figure Frost Susceptibility Classification by Percentage of Mass provides classification of frost susceptibility based on soil type. This chart shows that the finer the particle size, the higher the rate of heave.

The figure in *Handbook* section Frost Depth shows contour lines of frost depth, measured in meters, across the United States. As can be expected, the frost depth increases as the latitude moves away from the equator.

Frost Susceptibility Mitigation

Handbook: Frost Depth

CERM: Sec. 76.32

Mitigating frost heave is mainly about removing or blocking sources of water that contribute to saturation and heaving. Proper surface drainage and conscientious landscaping go a long way toward limiting frost heave potential.

The next steps typically include partial or complete removal of frost-susceptible soils from below grade-supported structures and replacement with non-frost-susceptible backfill. Such work is typically supported with surface drainage collectors since excavations often terminate in soils that won't drain, causing the non-frost-susceptible backfill to act like a bathtub and potentially feed accumulating water to frost-susceptible soils along the excavation perimeter.

CERM Sec. 76.32 provides a list of techniques to prevent damage from frost and freezing to flexible pavement, including using stronger and thicker pavement, lowering the water table, and using waterproof or rigid foam sheets.

Another approach to preventing frost damage to a rigid foundation is to ensure the foundation is below the frost depth. The approximate frost depth for areas of the United States is shown in the figure in *Handbook* section Frost Depth. If the foundation of the structure is below the depth provided in the figure, the effects of the frost will not influence the structure's foundation.

Frost Heave Parameters

FHWA NHI-05-037: Sec. 5.5.1

FHWA NHI-05-037 Sec. 5.5.1 lists the requirements for the 1993 *AASHTO Guide for Design of Pavement Structures*. The input parameters for this design process include frost heave parameters, which consist of the frost heave rate, the maximum potential serviceability loss, and the frost heave probability. The frost heave rate depends on the gradation of the subgrade material.

The maximum potential serviceability loss is a function of the drainage and the frost depth. The frost heave probability is much more subjective and depends on the frost susceptibility of the subgrade, the amount of moisture in the system, the drainage, the number of freeze-thaw cycles per year, and the frost depth.

24 Earth Retaining Structures (ASD or LRFD)

Content in blue refers to the *NCEES Handbook*.

A. Lateral Earth Pressure 24-1
B. Load Distribution 24-5
C. Rigid Retaining Wall Stability Analysis 24-7
D. Flexible Retaining Wall Stability Analysis 24-9
E. Cofferdams... 24-11
F. Underpinning 24-12
G. Ground Anchors, Tie-Backs, Soil Nails, and
    Rock Anchors for Foundations and Slopes 24-13

The knowledge area of Earth Retaining Structures (ASD or LRFD) makes up between four and six questions out of the 80 questions on the PE Civil Geotechnical exam. The organization of this chapter follows the order of subtopics given by NCEES for this knowledge area. Each subtopic is covered in the following sections.

A. LATERAL EARTH PRESSURE

Key concepts: These key concepts are important for answering exam questions in subtopic 16.A, Lateral Earth Pressure.

- active, passive, and at-rest lateral earth pressure
- earth pressure coefficient
- Rankine and Coulomb theories
- stress-strain scenario
- vertical overburden pressure

NCEES Handbook: To prepare for this subtopic, familiarize yourself with these sections in the *Handbook*.

- At-Rest Coefficients
- Stress States on a Soil Element Subjected Only to Body Stresses
- Rankine Active and Passive Coefficients (Friction Only)
- Failure Surfaces, Pressure Distribution and Forces
- Rankine Active and Passive Coefficients (Friction and Cohesion)
- Wall Friction on Soil Wedges
- Coulomb Coefficients K_a and K_p for Sloping Wall with Wall Friction and Sloping Cohesionless Backfill
- Wall Friction and Adhesion for Dissimilar Materials
- Effects of Seismic Coefficients and Friction Angle on Seismic Active Pressure Coefficient

PE Civil Reference Manual (**CERM**): Study these sections in CERM that either relate directly to this subtopic or provide background information.

- Section 37.3: Earth Pressure
- Section 37.5: Active Earth Pressure
- Section 37.6: Passive Earth Pressure
- Section 37.7: At-Rest Soil Pressure
- Section 37.8: Determining Earth Pressure Coefficients
- Section 38.4: Theoretical Skin-Friction Capacity

The following equations and figures are relevant for subtopic 16.A, Lateral Earth Pressure.

Earth Pressure Coefficient

Handbook: At-Rest Coefficients; Stress States on a Soil Element Subjected Only to Body Stresses; Coulomb Coefficients K_a and K_p for Sloping Wall with Wall Friction and Sloping Cohesionless Backfill

CERM: Sec. 37.5

$$K_o = 1 - \sin \phi'$$

$$K_o = (1 - \sin \phi')(\text{OCR})^\Omega$$

Lateral earth pressure is derived from effective vertical overburden pressure multiplied by the earth pressure coefficient, represented in the *Handbook* by K. The value of the earth pressure coefficient can be anywhere between 0.2 and 5. The coefficient is highly dependent on the shear strength of the soil, the soil conditions, and various geometric and structural boundary conditions.

The earth pressure coefficient used depends on the condition of the soil. *Handbook* figure Stress States on a Soil Element Subjected Only to Body Stresses shows four different conditions of earth pressure. Part (a) and part (b) of the figure show the at-rest condition, in which the soil cannot move (part (a)) or is not allowed to move (part (b)). Part (c) shows the active condition, in which the soil expands or pushes on the wall. Part (d) shows the passive condition, in which the wall pushes on or contracts the soil. *Handbook* figure Coulomb Coefficients K_a and K_p for Sloping Wall with Wall Friction and Sloping Cohesionless Backfill shows the differences between the active and passive cases and all of the various geometric and structural "boundary conditions."

The at-rest earth pressure coefficient is a function of the friction angle and overconsolidation ratio, as shown in the *Handbook* equations.

> *Exam tip*: Because of how dependent the earth pressure coefficient is on the soil condition, make sure to select the correct soil condition—at-rest, active, or passive—before making any calculations. The at-rest coefficient value is greater than the active coefficient but less than the passive coefficient.

At-Rest Earth Pressure

Handbook: At-Rest Coefficients; Stress States on a Soil Element Subjected Only to Body Stresses

$$K_o = 1 - \sin\phi'$$

$$K_o = (1 - \sin\phi')(\text{OCR})^\Omega$$

CERM: Sec. 37.7

$$p_o = k_o p_v \qquad 37.20$$

$$R_o = \tfrac{1}{2} k_o \rho g H^2 \qquad [\text{SI}] \qquad 37.22(a)$$

$$R_o = \tfrac{1}{2} k_o \gamma H^2 \qquad [\text{U.S.}] \qquad 37.22(b)$$

CERM Eq. 37.20 is used to calculate the at-rest earth pressure, and CERM Eq. 37.22 is used to calculate the force on the soil when at rest. K_o in the *Handbook* equations and k_o in the CERM equations both represent the *at-rest earth pressure coefficient* (also known as the *at-rest coefficient*).

At-rest earth pressures are most commonly applied to earth retaining structures fixed at their tops to limit movement (e.g., below-grade building walls or walls bearing on/supporting stiff materials). At-rest earth pressure conditions are also assumed for flexible retaining structures whose performance tolerances limit movement, and for soil elements restrained from movement due to surrounding confining pressures, such as a buried pipe or soil element below the ground surface.

Handbook figure Stress States on a Soil Element Subjected Only to Body Stresses shows two at-rest conditions. Part (a) shows a soil element that is restrained from movement due to the confining soil surrounding it. Part (b) shows a soil element at a retaining wall that is restrained from movement. This type of wall could be a basement wall, a bridge abutment, a wall that is restrained at the top, or another wall that is not allowed to displace.

Active Earth Pressure

Handbook: Rankine Active and Passive Coefficients (Friction Only); Stress States on a Soil Element Subjected Only to Body Stresses; Failure Surfaces, Pressure Distribution and Forces

$$p_a = K_a p_o$$

CERM: Sec. 37.5

$$p_a = k_a p_v \quad [c = 0] \qquad 37.9$$

The *Handbook* equation uses p_o to represent overburden pressure, and CERM Eq. 37.9 uses p_v to represent vertical pressure, but these values are the same. Both the *Handbook* equation and CERM Eq. 37.9 assume vertical wall face and horizontal backfill.

Active earth pressure is the result of the soil expanding or pushing on the adjacent structure. The active earth pressure is the product of the coefficient of active earth pressure and the overburden pressure. The active earth pressure condition is shown in part (c) of *Handbook* figure Stress States on a Soil Element Subjected Only to Body Stresses and in part (a) of *Handbook* figure Failure Surfaces, Pressure Distribution and Forces.

Either Rankine earth pressure theory or Coulomb earth pressure theory may be used to determine the earth pressure coefficients. For more about the Rankine earth pressure coefficients, see the entries "Rankine Active and Passive Coefficients for Friction Only" and "Rankine Active and Passive Coefficients for Friction and Cohesion" in this chapter. For more about the Coulomb earth pressure coefficients, see the entry "Coulomb Coefficients for Sloping Wall" in this chapter.

> *Exam tip*: If multiple soil types are present in a problem, it is helpful to draw a quick sketch showing the soil layers and water conditions.

Passive Earth Pressure

Handbook: Rankine Active and Passive Coefficients (Friction Only); Failure Surfaces, Pressure Distribution and Forces

$$p_p = K_p p_o$$

CERM: Sec. 37.6

$$p_p = p_v k_p + 2c\sqrt{k_p} \qquad 37.13$$

The *Handbook* equation uses p_o to represent overburden pressure, and CERM Eq. 37.13 uses p_v to represent vertical pressure, but these values are the same.

Handbook figure Failure Surfaces, Pressure Distribution and Forces shows the passive pressure condition in part (b). The passive pressure condition (i.e., the movement of the wall) in part (b) is toward the backfill. Passive pressure is often thought of as a result of the retaining wall squeezing or compressing the soil.

In cases of rigid earth retaining structures with shallow foundation embedment, passive earth pressure is often discounted; the consistency and strength of the material providing passive resistance may be suspect for a number of reasons, including a lack of compaction oversight, no testing of the structural backfill, or possible susceptibility to freezing and associated strength loss.

Either Rankine earth pressure theory or Coulomb earth pressure theory may be used to determine the earth pressure coefficients. For more about the Rankine earth pressure coefficients, see the entries "Rankine Active and Passive Coefficients for Friction Only" and "Rankine Active and Passive Coefficients for Friction and Cohesion" in this chapter. For more about the Coulomb earth pressure coefficients, see the entry "Coulomb Coefficients for Sloping Wall" in this chapter.

Rankine Active and Passive Coefficients for Friction Only

Handbook: Rankine Active and Passive Coefficients (Friction Only)

$$K_a = \frac{1 - \sin\phi'}{1 + \sin\phi'} = \tan^2\left(45° - \frac{\phi'}{2}\right)$$

$$K_p = \frac{1 + \sin\phi'}{1 - \sin\phi'} = \tan^2\left(45° + \frac{\phi'}{2}\right)$$

CERM: Sec. 37.5, Sec. 37.6

$$k_a = \frac{1}{k_p} = \tan^2\left(45° - \frac{\phi}{2}\right) \qquad 37.7$$

$$= \frac{1 - \sin\phi}{1 + \sin\phi} \quad \begin{bmatrix} \text{Rankine: horizontal} \\ \text{backfill; vertical face} \end{bmatrix}$$

K in the *Handbook* equations is the same as k in CERM Eq. 37.7. All these equations assume a vertical wall face and horizontal backfill.

The coefficient of active earth pressure (active coefficient) and coefficient of passive earth pressure (passive coefficient) are functions of the angle of internal friction. The active pressure can also be found by taking the reciprocal of the passive coefficient, as shown in CERM Eq. 37.7. Because the coefficients are reciprocal to each other, the passive coefficient will be on the order of 4 to 25 times the value of the active coefficient, the gap being greatest for materials having the greatest shear strength.

In saturated clays, the angle of internal friction is 0, making either coefficient equal to 1, so long as tension cracks do not develop near the top of the retaining wall.

Rankine Active and Passive Coefficients for Friction and Cohesion

Handbook: Rankine Active and Passive Coefficients (Friction and Cohesion)

$$K_a = \tan^2\left(45° - \frac{\phi'}{2}\right) - \frac{2c'}{p_o'}\tan^2\left(45° - \frac{\phi'}{2}\right)$$

$$K_p = \tan^2\left(45° + \frac{\phi'}{2}\right) + \frac{2c'}{p_o'}\tan^2\left(45° + \frac{\phi'}{2}\right)$$

$$p_a' = K_a(\gamma z - u) - 2c'\sqrt{K_a}$$

$$p_p' = K_p(\gamma z - u) + 2c'\sqrt{K_p}$$

CERM: Sec. 37.5, Sec. 37.6

$$p_a = p_v k_a - 2c\sqrt{k_a} \qquad 37.4$$

$$p_p = p_v k_p + 2c\sqrt{k_p} \qquad 37.13$$

The *Handbook* equations for passive and active earth pressure and CERM Eq. 37.4 and Eq. 37.13 are similar. However, the *Handbook* equations account for the removal of hydrostatic pressure, u.

The Rankine active and passive earth pressure coefficients when both friction and cohesion are present are functions of the angle of internal friction, the at-rest earth pressure, and the cohesion of the soil. The active and passive earth pressures are functions of these coefficients, the unit weight of the soil, the height of the wall, and the hydrostatic pressure.

The figure in *Handbook* section Rankine Active and Passive Coefficients (Friction and Cohesion) shows both the active and passive pressures on a retaining wall. The passive pressure is below grade at the toe of the wall, and only comes into effect when the active pressure on the right side of the wall pushes the wall into the soil on the left side of the wall. The figure also shows the additional pressure from the presence of the water on both sides of the wall. The effect of cohesion is shown in part (b) of the figure.

> *Exam tip*: Do not multiply the hydrostatic pressure by the earth pressure coefficient.

Wall Friction on Soil Wedges

Handbook: Wall Friction on Soil Wedges

CERM: Sec. 37.3

Handbook figure Wall Friction on Soil Wedges and CERM Fig. 37.2 show what happens during an active earth pressure design condition. The active wedge on the right of the *Handbook* figure "activates" and pushes the wall to the left. In turn, the wall pushes into the passive wedge at the base of the wall and squeezes the soil into itself. The result of the pressure at the toe of the wall is shown as the passive wedge.

Coulomb earth pressure theory assumes a flat failure plane and includes the effect of wall friction. In situations with significant wall friction, Coulomb earth pressure theory can predict a lower active pressure than Rankine earth pressure theory.

> *Exam tip*: The active earth condition can be thought of as the condition in which the soil expands. The passive earth condition can be thought of as the condition in which the soil is compressed.

Coulomb Coefficients for Sloping Wall

Handbook: Coulomb Coefficients K_a and K_p for Sloping Wall with Wall Friction and Sloping Cohesionless Backfill; Wall Friction and Adhesion for Dissimilar Materials

$$K_a = \frac{\cos^2(\phi - \theta)}{\cos^2\theta \cos(\theta + \delta)\left(1 + \sqrt{\frac{\sin(\phi + \delta)\sin(\phi - \beta)}{\cos(\theta + \delta)\cos(\theta - \beta)}}\right)}$$

$$K_p = \frac{\cos^2(\theta + \phi)}{\cos^2\theta \cos(\theta - \delta)\left(1 - \sqrt{\frac{\sin(\phi + \delta)\sin(\phi + \beta)}{\cos(\theta - \delta)\cos(\theta - \beta)}}\right)}$$

CERM: Sec. 37.3, Sec. 37.5, Sec. 37.6

$$k_a = \frac{\sin^2(\lambda + \phi)}{\sin^2\lambda \sin(\lambda - \delta)\left(1 + \sqrt{\frac{\sin(\phi + \delta)\sin(\phi - \beta)}{\sin(\lambda - \delta)\sin(\lambda + \beta)}}\right)^2} \quad 37.5$$

[Coulomb]

$$k_p = \frac{\sin^2(\lambda - \phi)}{\sin^2\lambda \sin(\lambda + \delta)\left(1 - \sqrt{\frac{\sin(\phi + \delta)\sin(\phi + \beta)}{\sin(\lambda + \delta)\sin(\lambda + \beta)}}\right)^2} \quad 37.14$$

[Coulomb]

The *Handbook* equations correspond to CERM Eq. 37.5 and Eq. 37.14. The *Handbook* equations give the angle of the wall face, θ, while the CERM equations give the rake angle of the wall face, λ. θ is measured from the top of the wall, and λ is measured from the bottom of the wall.

Handbook figure Coulomb Coefficients K_a and K_p for Sloping Wall with Wall Friction and Sloping Cohesionless Backfill illustrates the variables used in the earth pressure coefficient calculations. It also illustrates measurements and variables used in the application of the active earth pressure. CERM Fig. 37.2 provides a similar illustration of this cross section. CERM Table 37.1 and *Handbook* table Wall Friction and Adhesion for Dissimilar Materials provide the value of the angle of external friction.

Exam tip: Make sure to choose the simplest possible equation for the earth pressure coefficient. All equations will result in the same answer, but if some terms are already eliminated from the equation by the conditions of the wall and soil (e.g., the backslope is flat), it will save time and reduce the potential for calculation errors.

Seismic Coefficients and Friction Angle

Handbook: Effects of Seismic Coefficients and Friction Angle on Seismic Active Pressure Coefficient

Handbook figure Effects of Seismic Coefficients and Friction Angle on Seismic Active Pressure Coefficient can be used to find the seismic active pressure coefficient for an earth retaining structure that meets the requirements given in the lower-right corner of each chart. As shown in the graphs, as the friction angle of the soil decreases, the value of the seismic active pressure coefficient increases. For a constant friction angle, the seismic active pressure coefficient also increases as the horizontal seismic coefficient increases.

Exam tip: If a problem does not meet the specific requirements of the graphs in the *Handbook*, the seismic pressure coefficient can be calculated using the equations listed in *LRFD Seismic Analysis and Design of Transportation Geotechnical Features and Structural Foundations Reference Manual* (FHWA NHI-11-032), but it is a much more rigorous process.

B. LOAD DISTRIBUTION

Key concepts: These key concepts are important for answering exam questions in subtopic 16.B, Load Distribution.

- concrete flexural design
- retaining wall design

NCEES Handbook: To prepare for this subtopic, familiarize yourself with these sections in the *Handbook*.

- Horizontal Pressure on Walls from Compaction Effort
- Forces Behind a Gravity Wall
- Beams—Flexure Strength

PE Civil Reference Manual (**CERM**): Study these sections in CERM that either relate directly to this subtopic or provide background information.

- Section 54.5: Design of Retaining Walls

The following equations and figures are relevant for subtopic 16.B, Load Distribution.

Horizontal Pressure on Walls from Compaction Effort

Handbook: Horizontal Pressure on Walls from Compaction Effort

$$Z_c = K_a \sqrt{\frac{2P}{\pi \gamma}}$$

$$d = \frac{1}{K_a} \sqrt{\frac{2P}{\pi \gamma}}$$

$$P(\text{roller load}) = \frac{\text{dead wt of roller} + \text{centrifugal force}}{\text{width of roller}}$$

For $Z_c \leq Z \leq d$,

$$\overline{\sigma}_h = \sqrt{\frac{2P\gamma}{\pi}} \left(\frac{L}{a+L} \right)$$

For $Z > d$,

$$\overline{\sigma}_h = K_a \gamma Z$$

In these equations,

a	distance of roller from wall	ft
d	depth below ground surface where earth pressure is greater than compaction pressure	ft
K_a	coefficient of active earth pressure	–
L	length of roller	ft
P	roller load	lbf/ft
Z	depth below ground surface	ft
γ	unit weight	lbf/ft^3
σ	stress	lbf/ft^2

Compaction effort above or behind a retaining wall can cause an increase in pressure due to the weight of the compactor and the dynamic force from the compaction equipment. *Handbook* figure Horizontal Pressure on Walls from Compaction Effort illustrates what the pressure on a retaining wall looks like when a compaction roller with load P is used at a distance a above the wall.

Exam tip: If you are calculating the total pressure resultant, it is critical to include the modification near the ground surface due to the compaction effort. However, if you are only interested in the pressure at a depth of $Z > d$, the isolated pressure value is no different from a typical calculation of earth pressure. Make sure you understand what value the problem statement is asking for and what depth the calculation is at.

Forces Behind a Gravity Wall

Handbook: Forces Behind a Gravity Wall

Handbook figure Forces Behind a Gravity Wall is a wall free-body diagram that illustrates the surface surcharge and the inertial forces of the active wedge for seismic analysis.

Heel Load Distribution

Handbook: Beams—Flexure Strength

CERM: Sec. 54.5

$$V_{u,\text{base}} = 1.2 V_{\text{soil}} + 1.2 V_{\text{heel weight}} \qquad 54.9$$
$$+ 1.6 V_{\text{surcharge}}$$

$$M_{u,\text{base}} = 1.2 M_{\text{soil}} + 1.2 M_{\text{heel weight}} \qquad 54.10$$
$$+ 1.6 M_{\text{surcharge}}$$

Handbook section Beams—Flexure Strength provides equations for concrete flexure design. CERM Eq. 54.9 can be used to find the ultimate shear in the base heel, and CERM Eq. 54.10 can be used to find the ultimate moment in the base heel.

Each portion of a concrete wall (stem, toe, heel, etc.) is essentially a cantilevered beam in flexure. The soil and any water or surface surcharges are the forces acting on the concrete, and each section of the wall should be designed to withstand those forces. For heel load distribution design, the critical section for shear checking is located at the face of the stem. CERM Fig. 54.3 shows heel details.

Guide to Design of Slabs-on-Ground (ACI 360R) and *Minimum Design Loads for Buildings and Other Structures* (ASCE/SEI 7) are both useful references for heel load distribution design.

Figure 54.3 *Heel Details (step 9)*

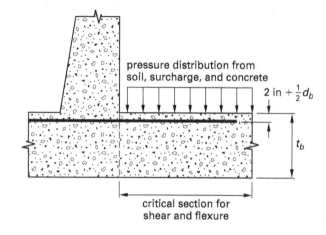

Exam tip: Upward pressure and effects from a key are disregarded in the heel design.

Toe Load Distribution

Handbook: Beams—Flexure Strength

CERM: Sec. 54.5

$$V_{u,\text{toe}} = 1.6 V_{\text{toe pressure}} - 1.2 V_{\text{toe weight}} \qquad 54.11$$

$$M_{u,\text{toe}} = 1.6 M_{\text{toe pressure}} - 1.2 M_{\text{toe weight}} \qquad 54.12$$

Handbook section Beams—Flexure Strength provides equations for concrete flexure design. CERM Eq. 54.11 can be used to find the ultimate shear in the toe, and CERM Eq. 54.12 can be used to find the ultimate moment in the toe.

Each portion of a concrete wall (stem, toe, heel, etc.) is essentially a cantilevered beam in flexure. The soil and any water or surface surcharges are the forces acting on the concrete, and each section of the wall should be designed to withstand those forces. For toe load distribution design, the toe loading is assumed to be caused by the toe self-weight and the upward soil pressure distribution. The passive soil loading is disregarded. For one-way shear, the critical section is located a distance, d, from the outer face of the stem. For flexure, the critical section is at the outer face of the stem. *Building Code Requirements for Structural Concrete* (ACI 318) Sec. 5.3 specifies a factor of 1.2 for concrete and soil dead loads and a factor of 1.6 for earth pressure. See CERM Fig. 54.4 for an example of toe details.

Figure 54.4 Toe Details (step 10)

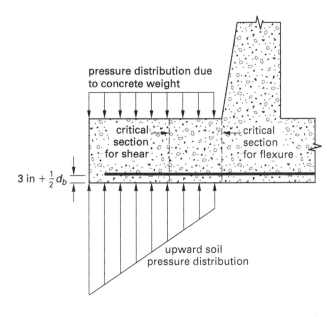

Exam tip: Upward pressure from the soil is used in the design of the toe.

C. RIGID RETAINING WALL STABILITY ANALYSIS

Key concepts: These key concepts are important for answering exam questions in subtopic 16.C, Rigid Retaining Wall Stability Analysis.

- bearing capacity
- overturning
- rotation
- translation

NCEES Handbook: To prepare for this subtopic, familiarize yourself with these sections in the *Handbook*.

- Pseudostatic Analysis and Earthquake Loads
- Forces Behind a Gravity Wall
- Effects of Seismic Coefficients and Friction Angle on Seismic Active Pressure Coefficient
- Bearing Capacity Equation for Concentrically Loaded Strip Footings
- Bearing Capacity Equation for Concentrically Loaded Square or Rectangular Footings
- Eccentric and Inclined Loaded Footings
- Eccentrically Loaded Footing
- Earth Pressure Concepts and Theory for Geotextile Wall Design
- Elements of Geogrid (or Geotextile) Reinforced Wall Design

PE Civil Reference Manual (**CERM**): Study these sections in CERM that either relate directly to this subtopic or provide background information.

- Section 37.1: Types of Retaining Wall Structures
- Section 37.12: Cantilever Retaining Walls: Analysis
- Section 40.14: Geotextiles

The following equations and figures are relevant for subtopic 16.C, Rigid Retaining Wall Stability Analysis.

Pseudostatic Analysis and Earthquake Loads

Handbook: Pseudostatic Analysis and Earthquake Loads; Forces Behind a Gravity Wall; Effects of Seismic Coefficients and Friction Angle on Seismic Active Pressure Coefficient

$$P_{AE} = \tfrac{1}{2} K_{AE} \gamma H^2 (1 - k_v)$$

$$P_{PE} = \tfrac{1}{2} K_{PE} \gamma H^2 (1 - k_v)$$

$$K_{AE} = \frac{\cos^2(\phi - \theta - \beta)}{\cos\theta (\cos^2 \beta) \cos(\beta + \delta + \theta) D}$$

$$D = \left(1 + \sqrt{\frac{\sin(\phi + \delta) \sin(\phi - \theta - i)}{\cos(\delta + \beta + \theta) \cos(i - \beta)}}\right)^2$$

$$K_{PE} = \frac{\cos^2(\phi - \theta + \beta)}{\cos\theta (\cos^2 \beta) \cos(\delta - \beta + \theta) D'}$$

$$D' = \left(1 + \sqrt{\frac{\sin(\phi + \delta) \sin(\phi + i - \theta)}{\cos(\delta - \beta + \theta) \cos(i - \beta)}}\right)^2$$

$$\theta = \tan^{-1} \frac{k_h}{1 - k_v}$$

K_{AE} is the seismic active earth pressure, and K_{PE} is the seismic passive earth pressure.

The difference between the pseudostatic analysis and static analysis for earthquake loads is the inclusion of the inertial load of the system itself in the calculations. The inertial force of the soil is shown as $k_h W_s$ (for the horizontal component) and $k_v W_s$ (for the vertical component) in *Handbook* figure Forces Behind a Gravity Wall.

Exam tip: Make sure to recalculate K_{AE} if a static and seismic calculation is performed. Note that K_a is not the same as K_{AE}.

Bearing Capacity Failure

Handbook: Bearing Capacity Equation for Concentrically Loaded Strip Footings; Bearing Capacity Equation for Concentrically Loaded Square or Rectangular Footings; Eccentric and Inclined Loaded Footings; Eccentrically Loaded Footing

$$q_{ult} = cN_c + qN_q + 0.5\gamma B_f N_\gamma$$

$$q = q_{appl} + \gamma_a D_f$$

$$N_q = e^{\pi \tan\phi} \tan^2\left(45° + \frac{\phi}{2}\right)$$

$$N_c = 2 + \pi = 5.14 \quad [\text{for } \phi = 0°]$$

$$N_c = (N_q - 1)\cot\phi = 5.14 \quad [\text{for } \phi > 0°]$$

$$N_\gamma = 2(N_q - 1)\tan\phi$$

For concentrically loaded square or rectangular footings,

$$q_{ult} = cN_c s_c + qN_q s_q + 0.5\gamma B_f N_\gamma s_\gamma$$

For eccentric or inclined loaded footings,

$$B_f' = B_f - 2e_B$$

$$L_f' = L_f - 2e_L$$

$$A' = B_f' - L_f'$$

CERM: Sec. 37.12

$$p_{v,max}, p_{v,min} = \left(\frac{\sum W_i + R_{a,v}}{B}\right) \cdot \left(1 \pm \frac{6\epsilon}{B}\right) \qquad 37.54$$

The maximum pressure on a base occurs at the toe, and the minimum pressure occurs at the heel. The bearing capacity analysis is fairly straightforward, as most of the needed parameters will have already been determined for the overturning analysis, as shown in CERM Fig. 36.6. However, the bearing capacity analysis requires that the applied pressure and allowable capacity both be determined so that it can be confirmed that the former is less than the latter.

Figure 36.6 *Footing with Overturning Moment*

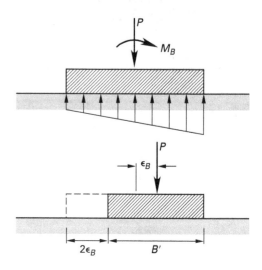

Equations for bearing capacity analysis vary depending on whether the footing is concentrically loaded or eccentrically loaded and on the type of footing employed. Bearing capacity equations are included in the *Handbook* under headings that indicate what situations and conditions the equations are appropriate for Bearing Capacity Equation for Concentrically Loaded Strip Footings, Bearing Capacity Equation for Concentrically Loaded Square or Rectangular Footings, and Eccentric and Inclined Loaded Footings. A footing is concentrically loaded when the load is applied to the center of the footing. A footing is eccentrically loaded when the load is applied at a distance away from the center of the footing.

CERM Eq. 37.54 can be used to find the maximum and minimum pressure on a base. If the vertical resultant, P, is not in the center of the footing, the equivalent footing width should be used to determine the applied pressure.

Handbook figure Eccentrically Loaded Footing illustrates how an eccentrically loaded footing can be converted to a concentrically loaded footing.

Exam tip: The bearing capacity is typically provided for comparison against the applied bearing pressure. If it is not provided in the problem statement, you will need to calculate it to ensure stability.

Reinforced Earth Retention

Handbook: Earth Pressure Concepts and Theory for Geotextile Wall Design

Failures in reinforced earth structures have revealed that successful projects specify permeable granular backfill for the reinforced zone. Geosynthetic or metallic reinforcing elements allow earth retention structures to be built using conventional construction equipment and techniques and more local, less expensive resources.

Handbook figure Earth Pressure Concepts and Theory for Geotextile Wall Design shows a wrapped face retaining wall cross section. The important design considerations specific to a geotextile-reinforced wall are the internal stability failure modes shown.

External stability (sliding, overturning, and bearing capacity) of a geotextile-reinforced retaining wall assumes the wall is rigid and the calculations are the same as any other retaining wall. The internal stability analysis of a geotextile-reinforced retaining wall is similar to the concrete and rebar design of a traditional concrete wall.

Exam tip: The applied forces on a retaining wall (earth pressure, water, surface surcharge, etc.) are the same regardless of wall type.

Overturning Failure

Handbook: Elements of Geogrid (or Geotextile) Reinforced Wall Design

CERM: Sec. 37.12

$$W_i = g\rho_i A_i \quad \text{[SI]} \quad 37.49(a)$$

$$W_i = \gamma_i A_i \quad \text{[U.S.]} \quad 37.49(b)$$

CERM Eq. 37.49 can be used to find the vertical forces acting at the base of a retaining wall.

Overturning of a retaining wall is a basic summation of moments (typically taken at the toe of the wall) to determine the ratio between the resisting and driving forces acting on the wall. *Handbook* figure Elements of Geogrid (or Geotextile) Reinforced Wall Design illustrates a potential overturning failure in a retaining wall. Applicable weights contributing to the resisting moment include those associated with all structural elements and overburden or backfill present above those elements, such as the concrete foundation and stem associated with a cast-in-place cantilever wall.

Exam tip: Drawing a quick, simple free-body diagram of the forces acting on the retaining wall is a good way to make sure everything is accounted for before beginning the analysis.

D. FLEXIBLE RETAINING WALL STABILITY ANALYSIS

Key concepts: These key concepts are important for answering exam questions in subtopic 16.D, Flexible Retaining Wall Stability Analysis.

- cantilever structures
- secant pile wall of drilled shafts
- sheet pile sections
- site considerations
- site constraints
- site geology
- soldier piles
- vertical and horizontal structural element selection

NCEES Handbook: To prepare for this subtopic, familiarize yourself with these sections in the *Handbook*.

- Maximum Allowable Stresses in Pile for Top-Driven Piles
- Components of a Ground Anchor
- Main Types of Grouted Ground Anchors
- Presumptive Ultimate Values of Load Transfer for Preliminary Design of Small-Diameter Straight-Shaft Gravity-Grouted Ground Anchors in Soil
- Presumptive Average Ultimate Bond Stress for Ground/Grout Interface Along Anchor Bond Zone
- Plotting Performance Test Data
- Plotting Elastic and Residual Movements for a Performance Test
- Plotting of Proof Test Data
- Steel Sheet Pile Properties
- Properties of NZ and PZ Hot-Rolled Steel

PE Civil Reference Manual (CERM): Study these sections in CERM that either relate directly to this subtopic or provide background information.

- Section 39.3: Braced Cuts in Sand
- Section 39.4: Braced Cuts in Stiff Clay
- Section 39.5: Braced Cuts in Soft Clay

- Section 39.12: Anchored Bulkheads
- Section 39.13: Analysis/Design of Anchored Bulkheads

Codes and standards: To prepare for this subtopic, familiarize yourself with these sections from codes and standards.

Foundations & Earth Structures Design Manual 7.02 (NAVFAC DM-7.02)

- Section 3.4: Design of Flexible Walls

The following figures, tables, and concepts are relevant for subtopic 16.D, Flexible Retaining Wall Stability Analysis.

Retention Systems

Handbook: Maximum Allowable Stresses in Pile for Top-Driven Piles; Steel Sheet Pile Properties; Properties of NZ and PZ Hot-Rolled Steel; Components of a Ground Anchor; Main Types of Grouted Ground Anchors

CERM: Sec. 39.1 through Sec. 39.13

Earth retention systems can be broken down into two different types: cantilevered support systems and anchored support systems.

Examples of cantilevered support systems include H-pile, secant pile, tangent pile, and sheet pile walls. These types of walls gain resistance and stability from the passive pressure in front of the wall and are designed similarly to a vertical cantilever beam.

Anchored support systems use (nearly) horizontal members that are anchored in the ground. Helical anchors, soil nails, and ground anchors are examples of anchored supported excavations.

Handbook table Maximum Allowable Stresses in Pile for Top-Driven Piles gives the maximum allowable stresses for different types of top-driven piles. *Handbook* figure Steel Sheet Pile Properties illustrates the properties of steel sheet piles, and *Handbook* table Properties of NZ and PZ Hot-Rolled Steel provides values for those properties and others. Anchored walls and horizontal elements are illustrated in *Handbook* figures Components of a Ground Anchor and Main Types of Grouted Ground Anchors.

Retention system elements must be selected to meet the project's service requirements. Ultimately, the project team—including the owners and contractors—may explore several alternatives and perform a cost- and risk-based analysis of each to facilitate the selection of a preferred system.

For more about anchored support systems, see Sec. G in this chapter.

Exam tip: Review the diagrams in CERM Sec. 39.3 through Sec. 39.12 to ensure you choose the appropriate earth pressure diagram for the specific soil type and earth retention system.

Anchored Retaining Structures

Handbook: Components of a Ground Anchor; Main Types of Grouted Ground Anchors; Presumptive Ultimate Values of Load Transfer for Preliminary Design of Small-Diameter Straight-Shaft Gravity-Grouted Ground Anchors in Soil; Presumptive Average Ultimate Bond Stress for Ground/Grout Interface Along Anchor Bond Zone; Plotting Performance Test Data; Plotting Elastic and Residual Movements for a Performance Test; Plotting of Proof Test Data

CERM: Sec. 39.12, Sec. 39.13

NAVFAC DM-7.02: Sec. 3.4

Anchored support systems use (nearly) horizontal members that are anchored in the ground. Helical anchors, soil nails, and ground anchors are examples of anchored supported excavations. *Handbook* figures Components of a Ground Anchor and Main Types of Grouted Ground Anchors illustrate ground anchors.

Any anchor resistance begins only after the point where the anchor passes through a theoretical Rankine failure plane. Full anchor resistance is only provided for portions of the anchor penetrating a plane extending up and away from the structure toe at the retained material's friction angle.

Handbook tables Presumptive Ultimate Values of Load Transfer for Preliminary Design of Small-Diameter Straight-Shaft Gravity-Grouted Ground Anchors in Soil and Presumptive Average Ultimate Bond Stress for Ground/Grout Interface Along Anchor Bond Zone provide estimated design bond based on the soil or rock type that the anchor is being installed into. These values are verified during construction through the load testing procedure. *Handbook* figures Plotting Performance Test Data, Plotting Elastic and Residual Movements for a Performance Test, and Plotting of Proof Test Data are examples of the plots produced by such load tests.

NAVFAC DM-7.02 Fig. 20 in Chap. 3 highlights important aspects of the assumptions behind a tied-back or anchored retaining structure design.

Flexible Retaining Wall Site Considerations

In addition to local practices, material availability, and cost, the selection of a practicable retention system depends on the project site's geologic and functional setting. For example, where retention elements have to be installed in a dense, cobbly medium, larger-diameter drilled shafts for secant wall construction are likely to be easier to install than driven H-piles or sheet piles. The overlapping concrete elements of a secant pile wall will also more effectively control seepage through a retained zone than an H-pile wall with lagging.

Even under favorable geologic conditions, retention system selection may be subject to functional constraints, such as utilities in line with tie-back anchors or restrictions on advancing anchors from the project site onto an adjacent site. Damage potential or restrictions on construction vibrations may also shift system selection from one involving driving to one involving drilling.

E. COFFERDAMS

Key concepts: These key concepts are important for answering exam questions in subtopic 16.E, Cofferdams.

- cofferdam arrangement
- cofferdam components
- fully enclosed flexible retaining structures
- seepage collection
- special considerations
- stabilization berms

NCEES Handbook: The *Handbook* does not contain any material relevant to this subtopic. Be sure to study the listed sections in CERM and from codes and standards.

PE Civil Reference Manual (**CERM**): Study these sections in CERM that either relate directly to this subtopic or provide background information.

- Section 39.14: Cofferdams

Codes and standards: To prepare for this subtopic, familiarize yourself with these sections from codes and standards.

Foundations & Earth Structures Design Manual 7.02 (NAVFAC DM-7.02)

- Section 3.5: Cofferdams

The following figures and concepts are relevant for subtopic 16.E, Cofferdams.

Cofferdam Overview

CERM: Sec. 39.14

A cofferdam is essentially a braced, cantilevered, or tied-back or anchored wall turned in upon itself and closed to create a three-dimensional retaining structure. A cofferdam isolates a work area in three dimensions from destabilizing earth, structure, and water loads; where water is concerned, it allows work to be performed "in the dry" at appreciable depths below water. Cofferdams are valuable in marine applications, as they can create a dry environment for construction and limit the extent of dewatering necessary (they have proven particularly valuable for bridge, pier, and dam construction). Cofferdams can be constructed individually or connected in series to create a retaining structure.

The *Handbook* does not specifically address cofferdams. However, a cofferdam is a specific application of an excavation, and there are numerous types of excavation support that can be used to create a cofferdam, such as H-piles, sheet piles, and so on. For more about different retention systems, see the entry "Retention Systems" in this chapter.

Exam tip: Once the specific type of cofferdam is identified (sheet pile, ground anchor, etc.), use the material related to that type of retaining system.

Cofferdam Components

CERM: Sec. 39.14

For the retention of earth materials and adjacent structures in conventional (urban, minimal, or no-seepage) environments, single-wall cofferdams may suffice. Bracing may be added to the cofferdam support to increase the factors of safety. Depending on the configuration of the cofferdam, the bracing may be anchored to the ground in front of the wall, or extend horizontally across to an opposing wall face, an anchor location at the ground surface, or even another structure.

Where it is necessary to retain water or control seepage, the complexity of the design increases. CERM Fig. 39.7 shows a double-wall cofferdam with a stabilization berm and seepage collection ditch.

Figure 39.7 Double-Wall Cofferdam

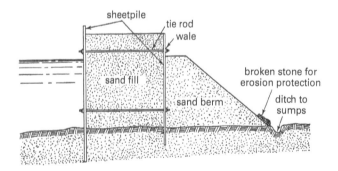

Exam tip: If bracing or anchors are used, make sure all cofferdam components are designed and accounted for.

Cofferdam Design

NAVFAC DM-7.02: Sec. 3.5

NAVFAC DM-7.02 Sec. 3.5 has a comprehensive design procedure for cellular cofferdams. The type of soil (rock, sand, or clay) determines the procedure to follow, and the presence of an exterior berm also affects the design. The width of the dam helps control the water seepage through the soil and resist overturning of the dam. The weight of the dam contributes to the sliding resistance of the system and also the overturning resistance. Once the exterior dimensions of the dam are determined, the individual components must be designed. The confining component design will be dependent upon the type of support system selected (H-pile, sheet pile, etc.).

Exam tip: While analyzing the soil inside the cofferdam, be sure to use the appropriate unit weight (effective, saturated, dry, etc.) based on the water level inside the cofferdam cell.

F. UNDERPINNING

Key concepts: These key concepts are important for answering exam questions in subtopic 16.F, Underpinning.

- augmentation of geotechnical capacity available to existing structures
- shaft friction
- transfer of existing structure loads from poor geotechnical materials to competent geotechnical materials
- vertical bearing

NCEES Handbook: To prepare for this subtopic, familiarize yourself with these sections in the *Handbook*.

- ASTM Standard Reinforcing Bars
- ASTM Standard Prestressing Strands
- ASTM Standard Plain Wire Reinforcement
- ASTM Standard Deformed Wire Reinforcement

PE Civil Reference Manual (**CERM**): Study these sections in CERM that either relate directly to this subtopic or provide background information.

- Section 38.10: Micropiles
- Section 38.11: Piers
- Section 39.14: Cofferdams
- Section 83.16: Temporary Structures

Codes and standards: To prepare for this subtopic, familiarize yourself with these sections from codes and standards.

Foundations & Earth Structures Design Manual 7.02 (NAVFAC DM-7.02)

- Section 5.2: Foundation Types and Design Criteria
- Section 5.3: Bearing Capacity and Settlement
- Section 5.4: Pile Installation and Load Tests
- Section 5.5: Distribution of Loads on Pile Groups
- Section 5.6: Deep Foundations on Rock
- Section 5.7: Lateral Load Capacity

The following tables and concepts are relevant for subtopic 16.F, Underpinning.

Cast-in-Place Concrete Piers

Handbook: ASTM Standard Reinforcing Bars; ASTM Standard Prestressing Strands; ASTM Standard Plain Wire Reinforcement; ASTM Standard Deformed Wire Reinforcement

CERM: Sec. 38.11

NAVFAC DM-7.02: Sec. 5.2 to Sec. 5.7

Cast-in-place concrete piers are generally employed on projects where shallow excavations are required adjacent to existing foundations, and where a more robust temporary excavation system has not already been installed. The piers are designed as concrete structures with geotechnical applications; they are commonly cylindrical in shape and are designed similarly to a concrete column. NAVFAC DM-7.02 Chap. 5 provides recommendations for concrete pier design. Section 5.2 provides the design criteria for deep foundations, Sec. 5.3 discusses bearing capacity and settlement, Sec. 5.4 discusses installation and testing procedures, Sec. 5.5 describes how to distribute loading on pile groups, Sec. 5.6 provides guidance for foundations on rock, and Sec. 5.7 includes lateral load capacity calculations. The concrete and rebar can be designed following the guidance in *Minimum Design Loads for Buildings and Other Structures* (ASCE/SEI 7).

Dimensions and weights for standard components of cast-in-place concrete piers are provided in *Handbook* tables ASTM Standard Reinforcing Bars, ASTM Standard Prestressing Strands, ASTM Standard Plain Wire Reinforcement, and ASTM Standard Deformed Wire Reinforcement.

Underpinning Overview

CERM: Sec. 83.16

Underpinning is a technique by which existing foundations are modified through the addition of structural elements that either increase the geotechnical capacity available to the foundations from their current bearing stratum or transfer the foundation loads to more competent or stable bearing strata.

The majority of underpinning applications fall into one of two categories: repair of distressed foundations impacted by unfavorable settlement or slope instability, or support of competent foundations at risk of distress from the removal of earth materials associated with pending adjacent excavations.

Screw Anchors and Micropiles

CERM: Sec. 38.10

Screw anchors and micropiles provide numerous underpinning options. Screw anchors often offer the most flexibility at the lowest cost, as they are generally available in small sections that are easily assembled and adjusted to gain the required capacity at varying depths. Since screw anchors have relatively slender shafts, they are vulnerable to bending, obstructions, and high torque in dense or stiff materials.

Micropiles consist of a slender steel rod and grout installed into a pre-drilled hole (like a vertical soil nail) for use as a deep foundation system or a retrofit or repair to an existing foundation system. They can be installed with lightweight, mobile construction equipment and can be used in almost any soil or rock condition. *Design and Construction of Driven Pile Foundations – Volume I* (FHWA NHI-16-009) can be referenced during the exam for limited micropile information.

Screw anchors and micropiles can be installed through (or attached to) haunches fixed in advance to the edges of existing foundations. Upon gaining required bearing, jacking may also be possible to raise or level the foundations. Screw anchors and micropiles can be installed at angles and also designed in either tension or compression. The design procedure and materials will be different for each case.

G. GROUND ANCHORS, TIE-BACKS, SOIL NAILS, AND ROCK ANCHORS FOR FOUNDATIONS AND SLOPES

Key concepts: These key concepts are important for answering exam questions in subtopic 16.G, Ground Anchors, Tie-Backs, Soil Nails, and Rock Anchors for Foundations and Slopes.

- foundation uplift resistance
- slope stabilization

NCEES Handbook: The *Handbook* does not contain any material relevant to this subtopic. Be sure to study the listed sections in CERM and from codes and standards.

PE Civil Reference Manual (**CERM**): Study these sections in CERM that either relate directly to this subtopic or provide background information.

- Section 15.19: Buoyancy of Submerged Pipelines
- Section 21.16: Hydrostatic Pressure Along Flow Path
- Section 36.13: General Considerations for Rafts
- Section 40.10: Slope Stability in Saturated Clay
- Section 40.15: Soil Nailing

Codes and standards: To prepare for this subtopic, familiarize yourself with these sections from codes and standards.

Drilled Shafts: Construction Procedures and Design Methods (FHWA NHI-18-024)

- Section 10.4: Design for Uplift

Foundations & Earth Structures Design Manual 7.02 (NAVFAC DM-7.02)

- Section 4.8: Uplift Resistance

The following figures and concepts are relevant for subtopic 16.G, Ground Anchors, Tie-Backs, Soil Nails, and Rock Anchors for Foundations and Slopes.

Foundation Elements

NAVFAC DM-7.02: Sec. 4.8

A foundation element subject to uplift will mobilize a block of soil (in the case of a column or pedestal footing) or cone of soil (in the case of a pier or screw anchor) whose geometry is governed by the shape and area of the foundation's base and the properties of the surrounding soil. NAVFAC DM-7.02 Fig. 17 in Chap. 4 applies to all types of foundations but shows the assumed failure geometry for the transient case. If this were a sustained-loading situation, the dashed "wedge" lines would be drawn vertically up from the foundation base.

Foundation Uplift Resistance in Rock

CERM: Sec. 36.13

FHWA NHI-18-024: Sec. 10.4

Evaluating uplift resistance in rock is trickier than evaluating uplift resistance in soil, as rock quality and (especially) rock strength will vary along the length of the anticipated anchor penetration.

The typical retaining element in uplift resistance in rock is grouted steel bar. The adhesion between the grout and the rock is what keeps the steel bar in place. In massive (i.e., not fractured) rock, the uplift capacity is the lesser of the rock compressive strength and the grout compressive strength. In fractured rock, the uplift capacity considers the strength of a 30° cone of rock extending up and out from the base of the bar. In the absence of a rock core (e.g., if the type of rock is unknown), values should be computed for both massive and fractured rock.

FHWA NHI-18-024 Sec. 10.4 provides guidance for foundation uplift.

> *Exam tip*: Pure uplift is a much simpler design condition than if shear and/or moment are included.

Buoyancy Resistance

CERM: Sec. 15.19, Sec. 21.16

Buoyancy is uplift below the ground surface due to water conditions. For buried structures more than 2 ft to 3 ft in diameter or height that may become submerged, and whose earth cover is no thicker than the diameter or height of the structure, uplift potential should be evaluated. In the case of a buried pipe, only the earth above the pipe midline should be assumed to be contributing to resistance.

> *Exam tip*: Don't forget to use the saturated or submerged soil unit weight when performing these calculations.

Slope Stabilization with Soil Nails and Screw Anchors

CERM: Sec. 40.15

A variety of screw-in and driven slope stabilization anchors are available for projects requiring only strengthening of a steep, unprotected surface, as well as for those requiring the stabilization of a slope failure. The design of surface protection projects is often governed by shear stress associated with surface runoff, whereas the design of slope failure projects is governed by the orientation of the known or assumed failure plane. In each case, anchor resistance is determined using the same procedures.

Slope Stabilization with Piles

CERM: Sec. 40.10

The design of structural elements intended to stabilize slopes generally proceeds under the same principles as for retaining structures. In the case of slope failures, however, the determination of the lateral earth load to be retained differs, as the force exerted on the retention system will be governed primarily by the shear strength along the failure plane rather than the shear strength within the failure mass (the former generally being much less than the latter).

During the analysis phase, if an unacceptable factor of safety is calculated, external materials such as piles may be used to increase the factor of safety to a desirable value. The piles must extend across the failure plane and be embedded deep enough to not pull out and become part of the failure itself.

Slope stabilization is a complicated technique, as multiple design codes are required. Guidance in USACE *Engineering and Design: Slope Stability* (EM 1110-2-1902) can help determine the stability analysis of the slope system. FHWA *Design and Construction of Driven Pile Foundations – Volume I* (FHWA NHI-16-009) and FHWA *Design and Construction of Driven Pile Foundations – Volume II* (FHWA NHI-16-010) give guidance for the design of driven piles. FHWA *Drilled Shafts: Construction Procedures and Design Methods* (FHWA NHI-18-024) is the design manual to use for drilled piles.

25 Shallow Foundations (ASD or LRFD)

Content in blue refers to the *NCEES Handbook*.

A. Bearing Capacity 25-1
B. Settlement, Including Vertical Stress
   Distribution 25-4

The knowledge area of Shallow Foundations (ASD or LRFD) makes up between four and six questions out of the 80 questions on the PE Civil Geotechnical exam. The organization of this chapter follows the order of subtopics given by NCEES for this knowledge area. Each subtopic is covered in the following sections.

A. BEARING CAPACITY

Key concepts: These key concepts are important for answering exam questions in subtopic 17.A, Bearing Capacity.

- allowable bearing capacity
- allowable strength design (ASD) method
- friction angle
- groundwater effects
- load and resistance factor design (LRFD) method
- net bearing capacity
- ultimate bearing capacity

NCEES Handbook: To prepare for this subtopic, familiarize yourself with these sections in the *Handbook*.

- Bearing Capacity Equation for Concentrically Loaded Strip Footings
- Bearing Capacity Equation for Concentrically Loaded Square or Rectangular Footings
- Shape Correction Factors
- Bearing Capacity Factors
- Eccentric and Inclined Loaded Footings

PE Civil Reference Manual **(CERM):** Study these sections in CERM that either relate directly to this subtopic or provide background information.

- Section 36.4: Allowable Bearing Capacity
- Section 36.5: General Bearing Capacity Equation
- Section 36.6: Selecting a Bearing Capacity Theory
- Section 36.7: Bearing Capacity of Clay
- Section 36.8: Bearing Capacity of Sand
- Section 36.9: Shallow Water Table Correction
- Section 36.10: Bearing Capacity of Rock
- Section 36.12: Eccentric Loads on Rectangular Footings
- Section 58.7: Methods of Design
- Section 65.5: Available Bearing Strength

The following equations, figures, and concepts are relevant for subtopic 17.A, Bearing Capacity.

Bearing Capacity for Concentrically Loaded Strip Footings

Handbook: Bearing Capacity Equation for Concentrically Loaded Strip Footings; Bearing Capacity Factors

$$q_{ult} = cN_c + qN_q + 0.5\gamma B_f N_\gamma$$

$$q = q_{appl} + \gamma_a D_f$$

CERM: Sec. 36.5

$$q_{ult} = \tfrac{1}{2}\rho g B N_\gamma + cN_c + (p_q + \rho g D_f)N_q \quad \text{[SI]} \qquad 36.1(a)$$

$$q_{ult} = \tfrac{1}{2}\gamma B N_\gamma + cN_c + (p_q + \gamma D_f)N_q \quad \text{[U.S.]} \qquad 36.1(b)$$

The ultimate bearing capacity, q_{ult}, is determined using the Terzaghi-Meyerhof equation, which is referred to in the *Handbook* only as the "bearing capacity equation." The equation varies depending on the footing used.

The nomenclature in the *Handbook* equations and the CERM equations is the same, except that the *Handbook* shows the soil unit weight as γ, and CERM shows the unit weight as the product of the soil density, ρ, and the gravitational acceleration, g.

These equations use bearing capacity factors based on surcharge, cohesion, and soil weight. These factors can be found in *Handbook* table Bearing Capacity Factors.

$$N_q = e^{\pi \tan \phi} \tan^2\left(45° + \frac{\phi}{2}\right)$$

$$N_c = 2 + \pi = 5.14 \quad [\phi = 0°]$$

$$N_c = (N_q - 1) \cot \phi > 5.14 \quad [\phi > 0°]$$

$$N_\gamma = 2(N_q - 1) \tan \phi$$

The equations shown in Bearing Capacity Equation for Concentrically Loaded Strip Footings should be used only with an allowable strength design (ASD) approach. For more about ASD, see the entry "Allowable Strength Design Method" in this chapter.

Exam tip: The ultimate bearing capacity found by these equations is not factored (i.e., is not the allowable bearing capacity). The allowable bearing capacity has a factor of safety applied to it (usually between 2 and 3 for the ASD method). The *Handbook* provides no factors of safety, so whenever a factor of safety is needed on the exam, it is likely to be provided in the question.

Bearing Capacity for Concentrically Loaded Square or Rectangular Footings

Handbook: Bearing Capacity Equation for Concentrically Loaded Square or Rectangular Footings; Shape Correction Factors; Bearing Capacity Factors

$$q_{\text{ult}} = cN_c s_c + qN_q s_q + 0.5\gamma B_f N_\gamma s_\gamma$$

The bearing capacity can be determined using the Terzaghi-Meyerhof equation, which is referred to in the *Handbook* only as the bearing capacity equation. The equation varies depending on the footing used. The subscript ult indicates that this equation is used to find the ultimate bearing capacity, a term used in the allowable strength design (ASD) method; however, this equation can be used with either the ASD method or the load resistance and factor design (LRFD) method.

These equations apply bearing capacity factors based on surcharge, cohesion, and soil weight. These factors can be found in *Handbook* table Bearing Capacity Factors.

These equations also apply shape correction factors based on the cohesion, surcharge, and unit weight. The equations for these factors are found in *Handbook* table Shape Correction Factors.

$$s_c = 1 + \frac{B_f}{5L_f} \quad [\phi = 0°]$$

$$s_c = 1 + \left(\frac{B_f}{L_f}\right)\left(\frac{N_q}{N_c}\right) \quad [\phi > 0°]$$

$$s_\gamma = 1 - 0.4\left(\frac{B_f}{L_f}\right) \quad [\phi > 0°]$$

$$s_q = 1 + \frac{B_f}{L_f} \tan \phi \quad [\phi > 0°]$$

The equations for the shape correction factors are based on the dimensions of the footing, the soil friction angle, and the bearing capacity factors for cohesion and surcharge. When the friction angle, ϕ, equals zero, s_γ and s_q are equal to 1.0.

Exam tip: The ultimate bearing capacity found by these equations is not factored (i.e., is not the allowable bearing capacity). The allowable bearing capacity has a factor of safety applied to it. The *Handbook* provides no factors of safety, so whenever a factor of safety is needed on the exam, it is likely to be provided in the question.

Moment Effect on Bearing Capacity

Handbook: Eccentric and Inclined Loaded Footings

$$B_f' = B_f - 2e_B$$

$$L_f' = L_f - 2e_L$$

$$A' = B_f' - L_f'$$

CERM: Sec. 36.12

$$\epsilon_B = \frac{M_B}{P}; \epsilon_L = \frac{M_L}{P} \qquad 36.18$$

$$L' = L - 2\epsilon_L; B' = B - 2\epsilon_B \qquad 36.19$$

$$A' = L'B' \qquad 36.20$$

$$p_{\max}, p_{\min} = \frac{P}{BL}\left(1 \pm \frac{6\epsilon}{B}\right) \quad 36.21$$

The nomenclature for the *Handbook* and CERM equations is the same, except that the *Handbook* uses e for eccentricity and CERM uses ϵ.

If a rectangular footing carries a moment in addition to its vertical load, an eccentric loading situation is created. Under these conditions, the footing bearing capacity should be analyzed with the area reduced by twice the eccentricity, as shown in the equations. This area is often referred to as an *equivalent area*, A'.

CERM Fig. 36.6 shows a footing that is eccentrically loaded due to the applied moment. The equivalent footing width, B', is shown in the lower section view. As shown, this loading situation means that the footing is effectively loaded in the center.

Figure 36.6 Footing with Overturning Moment

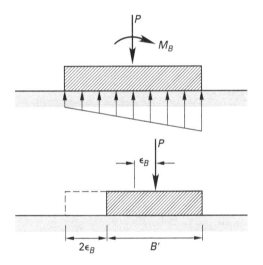

Where moments are introduced, eccentricity must be determined first so that the foundation dimensions can be adjusted appropriately to account for the moment effect. Bearing capacity is then determined based on the smaller equivalent dimensions. The bearing capacity accounting for eccentricity should be less than the bearing capacity based on actual foundation dimensions.

Exam tip: The equivalent footing dimensions may be used to determine both the ultimate bearing capacity and the shape factors.

Allowable Strength Design Method

CERM: Sec. 36.5, Sec. 58.7

The allowable strength design (ASD) method, also known as the allowable stress design method, is one of two methods of design permitted for use by the American Institute of Steel Construction, the other being the load and resistance factor design (LRFD) method. ASD is based on the premise that structural members remain elastic when subject to applied loads. According to this method, a structural member is designed so that its computed strength under service or working loads does not exceed available strength. The available strengths are prescribed by the building codes or specifications to provide a factor of safety against attaining some limiting strength, such as that defined by yielding or buckling.

The capacities found using the ASD method are referred to as *ultimate bearing capacity* and *allowable bearing capacity*. The equations for ultimate bearing capacity can be found in the entries "Bearing Capacity for Concentrically Loaded Strip Footings" and "Bearing Capacity for Concentrically Loaded Square or Rectangular Footings" in this chapter.

The ASD method is less rigorous than the LRFD method, as the LRFD method considers all the design elements, not just the geotechnical component. For more about the LRFD method, see the entry "Load and Resistance Factor Design Method" in this chapter.

Load and Resistance Factor Design Method

CERM: Sec. 58.7

The load and resistance factor design (LRFD) method, also referred to as limit states design, is one of two methods of design permitted for use by the American Institute of Steel Construction, the other being the allowable strength design (ASD) method. The LRFD method is the predominant design method for concrete structures. (For more about the ASD method, see the entry "Allowable Strength Design Method" in this chapter.)

LRFD methodology is more rigorous than the ASD method, as all design elements are assigned separate resistance factors based on their individual and combined reliabilities. The LRFD method also considers more service and limit state conditions than the ASD method.

The bearing capacities found using the LRFD method are referred to as nominal bearing capacity and factored bearing capacity. These correspond to the ASD method's ultimate bearing capacity and allowable bearing capacity, respectively.

A *limit state* is a condition at which a structure ceases to fulfill its intended function. There are two categories of limit states: strength and serviceability. Strength limit states include plastic strength, fracture, buckling, and fatigue. Serviceability limit states include excessive deflection, drift, vibration, and cracking.

B. SETTLEMENT, INCLUDING VERTICAL STRESS DISTRIBUTION

Key concepts: These key concepts are important for answering exam questions in subtopic 17.B, Settlement, Including Vertical Stress Distribution.

- foundation settlement
- stress dissipation with depth
- stress distribution principles

NCEES Handbook: To prepare for this subtopic, familiarize yourself with these sections in the *Handbook*.

- Vertical Stress Contours (Isobars) Based on Boussinesq's Theory for Continuous and Square Footings
- Distribution of Vertical Stress by the 2:1 Method
- Settlement (Elastic Method)
- Shape and Rigidity Factors, C_d, for Calculating Settlements of Points on Loaded Areas at the Surface of a Semi-Infinite, Elastic Half Space
- Elastic Constants of Various Soils
- Settlement (Schmertmann's Method)

PE Civil Reference Manual **(CERM):** Study these sections in CERM that either relate directly to this subtopic or provide background information.

- Section 36.1: Shallow Foundations
- Section 36.6: Selecting a Bearing Capacity Theory
- Section 38.6: Tensile Capacity
- Section 38.8: Settlement of Piles and Pile Groups
- Section 40.1: Pressure from Applied Loads: Boussinesq's Equation
- Section 40.2: Pressure from Applied Loads: Zone of Influence
- Section 40.4: Settling
- Appendix 40.A: Boussinesq Stress Contour Chart (infinitely long and square footings)
- Appendix 40.B: Boussinesq Stress Contour Chart (uniformly loaded circular footings)

The following equations, figures, and concepts are relevant for subtopic 17.B, Settlement, Including Vertical Stress Distribution.

Boussinesq Stress Contour Chart

Handbook: Vertical Stress Contours (Isobars) Based on Boussinesq's Theory for Continuous and Square Footings

CERM: Sec. 40.2, App. 40.A, App. 40.B

In the *Handbook* figure,

| B | footing width | ft |
| q_o | initial uniform load | lbf/ft |

The Boussinesq stress contour chart confirms not only how applied stresses propagate vertically and horizontally, but also that the rate at which they dissipate is not as uniform as predicted by simplified influence zone diagrams like CERM Fig. 40.2.

Stress contour charts like those in CERM App. 40.A and CERM App. 40.B relate foundation or structure dimensions to vertical depths and horizontal offsets of interest, and they assign influence factors that dictate the extent to which the applied stresses have dissipated.

Stresses applied at foundation subgrade elevation (or to a grade-supported structure like a levee or tank) spread out and dissipate with depth, which means that if a compressible clay layer is buried sufficiently far below the foundation, the reduction in stress may be enough to preclude a settlement issue.

Distribution of Vertical Stress by the 2:1 (60°) Method

Handbook: Distribution of Vertical Stress by the 2:1 Method

CERM: Sec. 40.2

In the *Handbook* figure,

B	footing width	ft
L	length	ft
Δp	change in uniform pressure	lbf/ft
P	pressure	lbf
$q_{applied}$	applied uniform pressure	lbf/ft
z	depth	ft
Z	depth	ft

The *Handbook* figure showing the 2:1 method (also called the 60° method) is helpful for determining where the stress below a rectangular footing or rectangular slab is distributed; it is also helpful for determining whether another structure or subsurface location is affected by the loading at the ground surface. The figure also provides equations that can be used to determine the increase in pressure at a given depth based on the surcharge at the ground surface. The surcharge at the surface is redistributed over the new area at the depth in question.

Elastic Method

Handbook: Settlement (Elastic Method); Elastic Constants of Various Soils

$$\delta_v = \frac{C_d \Delta p B_f (1-v^2)}{E_m}$$

This equation finds the vertical settlement at the surface of a soil. The elastic constant, or Young's modulus, can be found from *Handbook* table Elastic Constants of Various Soils, and the shape and rigidity factor, C_d, can be found from *Handbook* table Shape and Rigidity Factors, C_d, for Calculating Settlements of Points on Loaded Areas at the Surface of a Semi-Infinite, Elastic Half Space (see the entry "Shape and Rigidity Factors" in this chapter).

Shape and Rigidity Factors

Handbook: Shape and Rigidity Factors, C_d, for Calculating Settlements of Points on Loaded Areas at the Surface of a Semi-Infinite, Elastic Half Space

This *Handbook* chart provides the shape and rigidity factors for use in calculating settlement at the ground surface. The factors vary depending on the shape of the footing. For circular or square footings, factors vary based on whether the footing is rigid. For rectangular footings, factors vary based on the ratio of the footing's length to its width (sometimes called the *aspect ratio*). If the footing is not rigid, the factors also vary depending on the specific location of the settlement (edge of footing, center of footing, corner of footing, etc.). If the footing is rigid, the factor does not vary with location; in other words, settlement is constant across a rigid structure.

Schmertmann's Method

Handbook: Settlement (Schmertmann's Method)

$$S_i = C_1 C_2 \Delta p \sum_{i=1}^{n} \Delta H_i$$

$$\Delta H_i = H_c \left(\frac{I_z}{X E_s} \right)$$

$$C_1 = 1 - 0.5 \left(\frac{p_o}{\Delta p} \right) \geq 0.5$$

$$C_2 = 1 + 0.2 \log_{10} \frac{t}{0.1}$$

Schmertmann's method can be used to estimate the immediate settlement of a spread footing in a normally loaded sand. The equation for Schmertmann's method is complex, involving several factors that must themselves be calculated from the properties of the footing and the soil, but it is more accurate than other methods.

Shallow Foundation Settlement

CERM: Sec. 36.1

General grade fills often have footprints so broad that their applied stresses dissipate little, if at all, with depth; these tend to be the loading examples that engineers use to perform introductory settlement computations. Stresses associated with more compact fills and other structures like levees, tanks, and building foundations, however, are likely to experience measurable dissipation as they propagate through the bearing materials that extend below them. In addition to the determination of appropriate material properties and settlement equations, foundations require a characterization of stress dissipation with depth and the determination of strata-specific reduced stresses.

Foundation Subgrade Improvement Principles

CERM: Sec. 36.6

Stress dissipation charts suggest that, while foundation stresses extend to depths many times greater than typical foundation widths, efforts to improve the bearing capacity or reduce the compressibility of those materials need only extend below subgrade elevations to a depth equivalent to between 50% and 100% of the foundation width. This limited amount of work can, for many foundation types and shapes, effectively mitigate bearing and settlement issues. However, this depends on the absence of troublesome subgrade conditions at greater depths, such as deep uncompacted fills, debris-laden fills, and organic materials.

26 Deep Foundations (ASD or LRFD)

Content in blue refers to the *NCEES Handbook*.

A. Single-Element Axial Capacity 26-1
B. Lateral Load and Deformation Analysis 26-3
C. Single-Element Settlement 26-4
D. Downdrag 26-5
E. Group Effects 26-6
F. Installation Methods/Hammer Selection 26-8
G. Pile Dynamics 26-9
H. Pile and Drilled-Shaft Load Testing 26-11
I. Integrity Testing Methods 26-12

The knowledge area of Deep Foundations (ASD or LRFD) makes up between four and six questions out of the 80 questions on the PE Civil Geotechnical exam. The organization of this chapter follows the order of subtopics given by NCEES for this knowledge area. Each subtopic is covered in the following sections.

A. SINGLE-ELEMENT AXIAL CAPACITY

Key concepts: These key concepts are important for answering exam questions in subtopic 18.A, Single-Element Axial Capacity.

- drilled elements
  - auger-cast piles
  - drilled shafts
  - helical screw piles
  - micropiles
- driven elements
  - H-piles
  - pipe piles
  - precast concrete
  - timber

NCEES Handbook: The *Handbook* does not contain any material relevant to this subtopic. Be sure to study the listed sections in CERM and from codes and standards.

PE Civil Reference Manual (**CERM**): Study these sections in CERM that either relate directly to this subtopic or provide background information.

- Section 38.3: Theoretical Point-Bearing Capacity
- Section 38.4: Theoretical Skin-Friction Capacity

Codes and standards: To prepare for this subtopic, familiarize yourself with these sections from codes and standards.

Design and Construction of Driven Pile Foundations – Volume I (FHWA NHI-16-009)

- Section 7.2.1.3.4: Effective Stress β-Method – Mixed Soil Profiles

Foundations & Earth Structures, Design Manual 7.02 (NAVFAC DM-7.02)

- Section 5.3: Bearing Capacity and Settlement

Soils and Foundations Reference Manual – Volume II (FHWA NHI-06-089)

- Section 9.5.2.1: Total Stress – α-method

The following equations, figures, and tables are relevant for subtopic 18.A, Single-Element Axial Capacity.

Mohr-Coulomb Method
CERM: Sec. 38.3

NAVFAC DM-7.02: Sec. 5.3

$$Q_p = A_p\left(\tfrac{1}{2}\rho g B N_\gamma + cN_c + \rho g D_f N_q\right) \quad \text{[SI]} \quad 38.8(a)$$

$$Q_p = A_p\left(\tfrac{1}{2}\gamma B N_\gamma + cN_c + \gamma D_f N_q\right) \quad \text{[U.S.]} \quad 38.8(b)$$

$$Q_p = A_p \rho g D N_q \quad \text{[cohesionless; } D \leq D_c\text{]} \quad \text{[SI]} \quad 38.9(a)$$

$$Q_p = A_p \gamma D N_q \quad \text{[cohesionless; } D \leq D_c\text{]} \quad \text{[U.S.]} \quad 38.9(b)$$

CERM Eq. 38.8 and Eq. 38.9 are variations of the Terzaghi-Meyerhof equation and reflect allowable strength design (ASD) methodology. However, N_q and N_c correlations for deep foundations are different from those for shallow foundations, and both N_q and N_c also vary for driven versus drilled foundations.

Dynamic testing suggests that point bearing in cohesionless materials may not be constrained by a limiting depth, D_c. However, in the absence of a preinstallation drivability analysis or field testing program, it is prudent to use a limiting depth based on known or estimated relative density. For relative densities between 30% and 70%, D_c requires interpolation.

Q_p is the theoretical point-bearing capacity of a single pile, similar to the bearing of a strip footing. The first term of CERM Eq. 38.8 and Eq. 38.9 is usually very small due to the relatively small value of B and is commonly omitted from calculations. The values of N_c can be taken from Fig. 2 in NAVFAC DM-7.02 Sec. 5.3; they can also be found in CERM Table 38.2. Values of N_q are a function of both the friction angle of the soil and the type of pile being installed. They are listed in Fig. 1 in NAVFAC DM-7.02 Sec. 5.3 and CERM Table 38.1.

Exam tip: Be sure to use the correct N-values based on the type of pile and the soil type.

Skin Friction Capacity: Mohr-Coulomb Method

CERM: Sec. 38.4

$$Q_f = A_s f_s = p f_s L_e \qquad 38.11$$
$$= p f_s (L - \text{seasonal variation})$$

The skin friction capacity, Q_f (also known as side resistance, skin resistance, and shaft capacity), is given by CERM Eq. 38.11. L_e is the effective pile length, which can be estimated as the pile length less the depth of the seasonal variation, if any.

The values of f_s will vary if the pile is installed into varying soil types. When that is the case, a summation of the areas times the side friction factors is necessary to determine the skin friction capacity. CERM Eq. 38.13 can be used to calculate the side friction factor as a function of the adhesion, overburden pressure, and external friction angle for frictional soils.

$$f_s = c_A + \sigma_h \tan \delta \qquad 38.13$$

Exam tip: The skin friction capacity and the point-bearing capacity are added together to provide the total pile capacity.

Skin Friction Capacity: α-Method

CERM: Sec. 38.4

FHWA NHI-06-089: Sec. 9.5.2.1

$$Q_f = A_s f_s = p f_s L_e \qquad 38.11$$
$$= p f_s (L - \text{seasonal variation})$$

The skin friction capacity is given by CERM Eq. 38.11. The α-method is used for a total stress analysis in fine-grained soils with an undrained shear strength or cohesion value. The skin friction coefficient for the α-method, f_s, is the product of the bearing material's cohesion, c, and the adhesion factor, α, as given by CERM Eq. 38.14. This method and FHWA NHI-06-089 Fig. 9-15 are applicable to piles driven into stiff clay through overlying sands or sandy gravels; they are also applicable to the case when piles are driven into stiff clays through overlying soft clays.

$$f_s = \alpha c \qquad 38.14$$

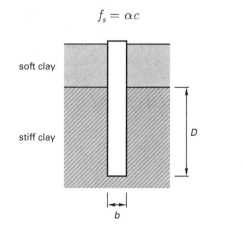

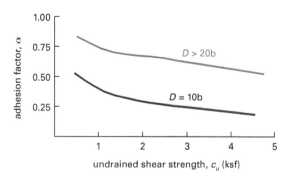

From FHWA NHI-06-089, *Soils and Foundations Reference Manual – Vol. II*, Fig. 9-15 (b) (Tomlinson, 1980), © 2006, U.S. Department of Transportation, Federal Highway Administration.

Selecting α from CERM Table 38.4 requires interpolation. α, like c as used with the Mohr-Coulomb method, is selected based on the average undrained shear strength (where it varies) along the length of the foundation.

FHWA NHI-06-089 Fig. 9-15 can be used to find α for more rigorous analyses, though the value of α is estimated here as well.

Skin Friction Capacity: β-Method

CERM: Sec. 38.4

FHWA NHI-16-009: Sec. 7.2.1.3.4

$$Q_f = p\beta\sigma' L \quad \text{[cohesive]} \quad 38.19$$

CERM Eq. 38.19 provides the friction capacity of a pile in cohesive clay. Typical values of β are provided in CERM Table 38.5. The value of β is a reduction factor applied to the average effective vertical stress.

The β-method applies a reduction factor to the average vertical effective stress along the length of the foundation, which, for a single layer, can be determined at the pile midpoint. For multiple layers, a stress profile is required. CERM Table 38.5 relates β to pile length.

The β-method for deep foundations in clays mimics the way skin friction is determined for cohesionless soils. (Refer to the effective stress portion of CERM Eq. 38.13.)

$$f_s = c_A + \sigma_h \tan\delta \quad 38.13$$

Research has shown that the β-method is best suited for the determination of drilled shaft skin friction in sand. The β-method is an effective stress analysis and should be used when soils have an effective friction angle. Table 7-9 in FHWA NHI-16-009 provides values of β as a function of soil type and friction angle.

B. LATERAL LOAD AND DEFORMATION ANALYSIS

Key concepts: These key concepts are important for answering exam questions in subtopic 18.B, Lateral Load and Deformation Analysis.

- NAVFAC design procedure
- pile stiffness

NCEES Handbook: The *Handbook* does not contain any material relevant to this subtopic. Be sure to study the listed sections in CERM and from codes and standards.

Codes and standards: To prepare for this subtopic, familiarize yourself with these sections from codes and standards.

Drilled Shafts: Construction Procedures and Design Methods (FHWA NHI-18-024)

- Section 9.3.3.1: Brief Description of the p-y Method

Foundations & Earth Structures Design Manual 7.02 (NAVFAC DM-7.02)

- Section 5.7.2: Deformation Analysis – Single Pile

Soils and Foundations Reference Manual – Volume II (FHWA NHI-06-089)

- Section 9.7: Design of Piles for Lateral Load

The following equations and figures are relevant for subtopic 18.B, Lateral Load and Deformation Analysis.

Lateral Load Design Process

FHWA NHI-06-089: Sec. 9.7

FHWA NHI-18-024: Sec. 9.3.3.1

NAVFAC DM-7.02: Sec. 5.7.2

Deep foundations may be designed to resist lateral and vertical loads or lateral loads only. Moment and deflection are important performance parameters in the lateral load design process; limiting deflection often governs design.

Computer programs are the primary medium through which deep foundations are designed and lateral load capacity and performance under lateral loads are qualified. The determination of ultimate and allowable lateral loads can be a complex geotechnical and structural process and is not covered in some geotechnical references (e.g., FHWA NHI-06-089).

FHWA NHI-18-024 Sec. 9.3 provides an in-depth discussion of the lateral load design process, but Sec. 9.3.3.1 acknowledges that the process is complicated and defers to computer software for modeling.

NAVFAC DM-7.02 Sec. 5.7.2 discusses lateral loading of piles. This section refers to the use of computer software for an accurate analysis and discusses approximate methods. The process is different for each type of soil. The design process for granular soil and normally to slightly overconsolidated cohesive soils is different from the process for heavily overconsolidated cohesive soils. Both processes are discussed in NAVFAC DM-7.02 Sec. 5.7.2.

Lateral Load Capacity

NAVFAC DM-7.02: Sec. 5.7.2

NAVFAC DM-7.02 Sec. 5.7 offers a simplified procedure for calculation of lateral deformation in free and fixed deep foundation elements in cohesionless materials and normally to slightly overconsolidated cohesive materials.

1. Determine the coefficient of variation of lateral subgrade reaction for soils.

2. Determine the stiffness of structural elements.

3. Select the appropriate design procedure graph.

4. Determine performance values graphically based on the length-to-stiffness or depth-to-stiffness ratio.

This design guidance can help geotechnical engineers to qualify proposed loading situations relatively simply in the absence of a computer program.

NAVFAC's guidance also helps illustrate the role of geotechnical and structural factors in lateral load evaluations, as well as load, deflection, and moment relationships.

Figure 10 in NAVFAC DM-7.02 Sec. 5.7.2 provides diagrams for the three principal loading conditions of a pile system. These diagrams work together with Fig. 11, Fig. 12, and Fig. 13 to determine the deflection, moment, and shear of the pile.

Relative Stiffness Factor

NAVFAC DM-7.02: Sec. 5.7.2

$$T = \left(\frac{EI}{f}\right)^{1/5}$$

The relative stiffness factor, T, is calculated based on the equation given in Fig. 10 in NAVFAC DM-7.02 Sec. 5.7.2. The relative stiffness factor is used to determine the L/T ratio and depth ratio for the graphs in NAVFAC DM-7.02 Fig. 11 (in Sec. 5.7.2) when determining the deflection, moment, and shear coefficients. These coefficients are then used with the equations given in Fig. 11 to determine the deflection, moment, and shear of a laterally loaded pile.

Single material values of modulus and moment of inertia can be obtained from general engineering references. Results can be evaluated for a range of foundation sizes, and the selection of those sizes can be determined through collaboration with the project structural engineer; alternatively, the structural engineer can choose a size that satisfies load demands. In case there is doubt about composite structural elements, you can confirm the modulus and moment of inertia values.

Initially, it is easiest to compute T in units of inches (only the coefficient of variation of subgrade modulus, f, needs to be converted). However, once the guidance is used to explore deflection, moment, shear, and load along the length of the structural elements being considered, it may be convenient to convert T into units of feet.

C. SINGLE-ELEMENT SETTLEMENT

Key concepts: These key concepts are important for answering exam questions in subtopic 18.C, Single-Element Settlement.

- axial compression
- axial deformation
- load transfer settlement
- pile-driving induced settlement

NCEES Handbook: The *Handbook* does not contain any material relevant to this subtopic. Be sure to study the listed sections in CERM and from codes and standards.

Codes and standards: To prepare for this subtopic, familiarize yourself with these sections from codes and standards.

Design and Construction of Driven Pile Foundations – Volume I (FHWA NHI-16-009)

- Section 3.5.3: Pile Driving Induced Vibrations
- Section 7.3.5.1: Elastic Compression of Piles

Foundations & Earth Structures, Design Manual 7.02 (NAVFAC DM-7.02)

- Section 5.3.4: Settlements of Pile Foundations

Soils and Foundations Reference Manual – Volume II (FHWA NHI-06-089)

- Section 9.4: Computation of Pile Capacity

The following equations and figures are relevant for subtopic 18.C, Single-Element Settlement.

Axial Compression

FHWA NHI-06-089: Sec. 9.4

FHWA NHI-16-009: Sec. 7.3.5.1

NAVFAC DM-7.02: Sec. 5.3.4

$$W_s = \frac{(Q_p + \alpha_s Q_s)L}{AE_p} \quad \text{[NAVFAC DM-7.02 Eq. 5.3.4.a(1)]}$$

Settlement due to axial deformation of a pile shaft can be determined using the preceding equation, given in NAVFAC DM-7.02 Sec. 5.3.4. Axial compression is not settlement—settlement is the result of bearing stratum compression—but it contributes to settlement, so these two concepts are often paired.

Q_p and Q_s equal the point load, R_t, and shaft or perimeter load, R_s, in FHWA NHI-06-089 Fig. 9-4. α is a coefficient that varies based on the density or stiffness of the bearing stratum per NAVFAC DM-7.02 Sec. 5.3.4.

Axial compression consists of a simple shortening of the element alone—regardless of whether the element moves downward beyond the depth to which it was installed—in proportion to its geometric and stiffness properties.

FHWA NHI-16-009 Eq. 7-48 addresses axial compression as the elastic compression of piles. FHWA NHI-16-009 Sec. 7.3.5.1 also provides the modulus of elasticity of steel and concrete for quick reference and notes that the compression of short piles is typically negligible and not included in most designs.

$$\Delta = \frac{QL}{AE} \quad \text{[FHWA NHI-16-009 Eq. 7-48]}$$

When a large amount of resistance is transferred to the soil via skin friction, this equation overestimates the compression of the pile.

Load Transfer Settlement

NAVFAC DM-7.02: Sec. 5.3.4

The point load settlement is given by NAVFAC DM-7.02 Eq. 5.3.4.a(2).

$$W_{\text{pp}} = \frac{C_p Q_p}{B q_o} \quad \text{[NAVFAC DM-7.02 Eq. 5.3.4.a(2)]}$$

The shaft or perimeter load transfer settlement is given by NAVFAC DM-7.02 Eq. 5.3.4.a(3).

$$W_{\text{ps}} = \frac{C_s Q_s}{D q_o} \quad \text{[NAVFAC DM-7.02 Eq. 5.3.4.a(3)]}$$

Of the two load transfer sources of settlement, the one corresponding to load transfer along the shaft or perimeter tends to be the smaller. However, settlement of an individual foundation element is generally not the subject of analysis, primarily because these types of structural elements are typically installed in groups; the load that an individual element might carry is not usually of concern.

The total pile settlement is a summation of the axial compression, the point load settlement, and the perimeter load transfer settlement. Table 5 in NAVFAC DM-7.02 Sec. 5.3.4 provides the values of C_p for use in NAVFAC DM-7.02 Eq. 5.3.4.a(2). The value of C_p varies between 0.02 to 0.18 depending on the soil and pile types. Q_p is the point load transmitted to the pile tip, B is the pile diameter, and q_o is the ultimate bearing capacity.

For calculation of the perimeter, the load transfer settlement, C_s, is defined in NAVFAC DM-7.02 Sec. 5.3.4, and D is the embedded pile length.

Exam tip: The methods discussed here are for a single pile; analysis of pile groups requires a different method.

Pile Driving Induced Settlement

FHWA NHI-16-009: Sec. 3.5.3

FHWA NHI-16-009 Sec. 3.5.3 discusses settlement resulting from pile driving or other vibrations due to construction activities. Soil densification and ensuing settlement can occur at velocities less than 2 in/sec, and even as low as 0.1 in/sec. This phenomenon is especially noticeable in clean sand where the coefficient of uniformity is 4 to 5 with a relative density of 50% to 55%. FHWA NHI-16-009 Fig. 3-9 shows the zone of influence that the settlement has on the pile itself. FHWA NHI-16-009 Eq. 3-1 and Eq. 3-2 provide, respectively, the maximum and average settlement.

$$S_{\max} = \alpha(D + 6b) \quad \text{[FHWA NHI-16-009 Eq. 3-1]}$$

$$S_{\text{avg}} = \frac{\alpha(D + 3b)}{3} \quad \text{[FHWA NHI-16-009 Eq. 3-2]}$$

D. DOWNDRAG

Key concepts: These key concepts are important for answering exam questions in subtopic 18.D, Downdrag.

- downdrag as a settlement condition
- downdrag mitigation
- drag loads (downdrag, or negative shaft resistance)
- soil compression

NCEES Handbook: The *Handbook* does not contain any material relevant to this subtopic. Be sure to study the listed sections in CERM and from codes and standards.

PE Civil Reference Manual **(CERM):** Study these sections in CERM that either relate directly to this subtopic or provide background information.

- Section 38.9: Downdrag and Adfreeze Forces

Codes and standards: To prepare for this subtopic, familiarize yourself with these sections from codes and standards.

Design and Construction of Driven Pile Foundations – Volume I (FHWA NHI-16-009)

- Section 7.3.6.1: Recommended Approach for Downdrag

Soils and Foundations Reference Manual – Volume II (FHWA NHI-06-089)

- Section 9.8: Downdrag or Negative Shaft Resistance

The following equations and concepts are relevant for subtopic 18.D, Downdrag.

Drag Loads (Downdrag)

CERM: Sec. 38.9

FHWA NHI-06-089: Sec. 9.8

Deep foundations are exposed to drag loads (*downdrag*) when the soils around them move downward relative to the structural elements. Downdrag can occur due to the compression of fill under its own weight, the compression of soft soils beneath existing fill, the introduction of new fill, nearby surcharges, or dewatering.

Historically, downdrag has been thought to have a potentially significant impact on deep foundations, and it is still standard practice in many areas to count it against the allowable capacity or geotechnical resistance of structural elements.

However, many professionals now believe that it counts less against geotechnical resistance than it does against structural resistance (since, at failure, deep foundations are still moving downward relative to the soils in contact with them).

FHWA NHI-06-089 Sec. 9.8 lists six cases when downdrag should be included in the design. They are as follows.

- a total settlement of the ground surface greater than 4 in
- a settlement of the ground surface after the piles are driven greater than 0.4 in
- an embankment height placed on the ground surface greater than 6.5 ft
- a thickness of a soft, compressible layer greater than 30 ft
- the water table being lowered by more than 13 ft
- piles longer than 80 ft

Exam tip: Downdrag is also referred to as *negative shaft resistance*.

Downdrag Mitigation

CERM: Sec. 38.9

FHWA NHI-06-089: Sec. 9.8

Mitigating downdrag is best accomplished by limiting exposure of structural elements to settlement. It may be necessary to design with structural elements that can accommodate higher stresses due to possible increased loading, but this adds cost to a project. Construction coordination is also important in mitigating downdrag. For example, stockpiling materials near installed foundations may cause distress to structural elements during construction.

FHWA NHI-06-089 Sec. 9.8 provides the following techniques to eliminate or minimize the impact of downdrag.

- reducing soil settlement (e.g., by preloading the soil surface)
- using lightweight fill material
- using friction reducers such as bitumen and plastic wrap
- increasing allowable pile stress
- preventing direct contact between the soil and the pile through the use of a pile sleeve

While friction reducers such as bitumen and plastic wrap are sometimes used, the practice is considered unreliable, as the material can be removed during installation.

Downdrag Designed as Settlement

FHWA NHI-06-089: Sec. 9.8

FHWA NHI-16-009: Sec. 7.3.6.1

FHWA NHI-06-089 Sec. 9.8 considers downdrag an additional force that is added to a pile during the design process. FHWA NHI-16-009 Sec. 7.3.6.1 provides a design approach that treats downdrag as a settlement condition that is addressed in the service limit state for the geotechnical design. The equation for settlement due to downdrag is given by FHWA NHI-16-009 Eq. 7-72 and is a function of the soil thickness, the increase in vertical stress, and the modulus of elasticity of the in situ soil.

$$S_{dd} = t_{soil}\left(\frac{\gamma_p \Delta \sigma}{E_s}\right) \quad \text{[FHWA NHI-16-009 Eq. 7-72]}$$

E. GROUP EFFECTS

Key concepts: These key concepts are important for answering exam questions in subtopic 18.E, Group Effects.

- block failure
- group capacity
- group efficiency
- lateral load capacity
- ultimate versus allowable capacity of pile groups

NCEES Handbook: The *Handbook* does not contain any material relevant to this subtopic. Be sure to study the listed sections in CERM and from codes and standards.

PE Civil Reference Manual (CERM): Study these sections in CERM that either relate directly to this subtopic or provide background information.

- Section 38.7: Capacity of Pile Groups
- Section 38.8: Settlement of Piles and Pile Groups

Codes and standards: To prepare for this subtopic, familiarize yourself with these sections from codes and standards.

Design and Construction of Driven Pile Foundations – Volume I (FHWA NHI-16-009)

- Section 7.3.7.6: Pile Groups

Foundations & Earth Structures, Design Manual 7.02 (NAVFAC DM-7.02)

- Section 5.3: Bearing Capacity and Settlement
- Section 5.7.6: Group Action

Soils and Foundations Reference Manual – Volume II (FHWA NHI-06-089)

- Section 9.6: Design of Pile Groups
- Section 9.6.1.3: Block Failure of Pile Groups

The following equations and figures are relevant for subtopic 18.E, Group Effects.

Group Efficiency

CERM: Sec. 38.7, Sec. 38.8

FHWA NHI-06-089: Sec. 9.6

NAVFAC DM-7.02: Sec. 5.3

Group efficiency (G_e per NAVFAC DM-7.02 Sec. 5.3, η_G in CERM Sec. 38.7, and η_g in FHWA NHI-06-089 Sec. 9.6) is the ratio of group capacity to the sum of individual capacities, as given by CERM Eq. 38.24.

$$\eta_G = \frac{Q_G}{\sum Q_i} \quad \quad 38.24$$

η_G is 0.7 for groups in cohesive materials with undrained shear strength less than 2000 lbf/ft² when center-to-center element spacing is greater than $3B$ (B = element diameter). η_G is 1.0 for cohesive materials with undrained shear strength greater than 2000 lbf/ft² when spacing is greater than $6B$. A minimum spacing of $3B$ is recommended.

Settlement of pile groups is typically greater than a summation of individual piles carrying an equivalent load per pile. FHWA NHI-06-089 Sec. 9.6 provides cross sections in Fig. 9-32 showing how the overlap of stresses from a pile group can result in increased settlement when compared to a single pile.

Exam tip: CERM Eq. 38.24 is for cohesive soils. Per CERM Sec. 38.7, for cohesionless (granular) soils, the capacity of a pile group is taken as the sum of the individual capacities.

Lateral Load Capacity

FHWA NHI-16-009: Sec. 7.3.7.6

NAVFAC DM-7.02: Sec. 5.7.6

NAVFAC DM-7.02 Sec. 5.7.6 considers groups when foundation element spacing in the direction of loading is less than $6B$ to $8B$. In this case, the recommended action is to reduce the coefficient of subgrade reaction, k_h (equivalent to reducing the coefficient of variation of subgrade reaction, f). The implied center-to-center spacing will be greater than $3B$.

The presumption is that proximity to other elements limits the extent to which any single element can fully engage the resistance available to it. The reduction for closely spaced elements is significant.

FHWA NHI-16-009 Sec. 7.3.7.6 discusses the application of modifications to the design required when pile groups are subjected to lateral loads. A p-multiplier (reduction) is used for the pile, depending on which row the pile is located in. For example, for a pile spacing of $3B$, the p-multiplier, P_m, is 0.8 for the first row, 0.4 for the second row, and 0.3 for the third and following rows.

Block Failure of Pile Groups

CERM: Sec. 38.7

FHWA NHI-06-089: Sec. 9.6.1.3

Block failure of a pile group is a design concern when the pile group is installed into cohesive soils or when the piles are installed into granular soils above a weak cohesive layer. CERM Eq. 38.20 through Eq. 38.23 provide the calculations to determine the ultimate and allowable capacity of the pile group.

$$Q_s = 2(b+w)L_e c_1 \quad \quad 38.20$$

$$Q_p = 9c_2 bw \quad \quad 38.21$$

$$Q_{\text{ult}} = Q_s + Q_p \quad \quad 38.22$$

$$Q_a = \frac{Q_{\text{ult}}}{F} \quad \quad 38.23$$

FHWA NHI-06-089 Sec. 9.6.1.3 provides similar equations as well as additional calculations for when the pile group is not a rectangular shape. The dimensions of the entire block of piles is considered for input into the calculations, and the overall pile group is considered as the unit under analysis.

F. INSTALLATION METHODS/HAMMER SELECTION

Key concepts: These key concepts are important for answering exam questions in subtopic 18.F, Installation Methods/Hammer Selection.

- actual versus rated energy transfer
- energy transfer
- pile capacity
- pile cushion changes on driving resistance
- pile drilling
- pile driving

NCEES Handbook: To prepare for this subtopic, familiarize yourself with these sections in the *Handbook*.

- Hammer-Pile-Soil System
- Modified Engineering-News Formula
- Gates Formula
- Maximum Allowable Stresses in Pile for Top-Driven Piles

PE Civil Reference Manual (**CERM**): Study these sections in CERM that either relate directly to this subtopic or provide background information.

- Section 38.2: Pile Capacity from Driving Data

Codes and standards: To prepare for this subtopic, familiarize yourself with these sections from codes and standards.

Soils and Foundations Reference Manual – Volume II (FHWA NHI-06-089)

- Section 9.9.2: Pile Drivability
- Section 9.9.3: Pile Driving Equipment and Operation

The following equations, figures, and tables are relevant for subtopic 18.F, Installation Methods/Hammer Selection.

Driven Foundations

Handbook: Hammer-Pile-Soil System; Maximum Allowable Stresses in Pile for Top-Driven Piles

CERM: Sec. 38.2

Bearing strata cannot be observed and tested directly. Driven foundation elements are only useful if they can be driven to intended depths without damage. Damage to such elements can result from unforeseen geotechnical or material issues and inappropriate driving equipment or operation.

Drivability is critical to the successful installation of driven deep foundation elements. The elements themselves can be qualified in advance of driving; some examples of this include

- making submittals demonstrating the yield strength of steel or the compressive strength of precast concrete
- measuring steel thickness
- examining welds

These measures offer some degree of certainty about the structural integrity of driven elements (although defects can still remain undiscovered). It is not until the driving of the elements that capacity is ultimately demonstrated and damage can occur.

CERM Eq. 38.3 through Eq. 38.7 provide calculations for the design capacity (also referred to as the maximum allowable vertical load per pile or the pile resistance) as a function of the type of hammer used to drive a pile into the ground. Equations for drop hammers, single-acting steam hammers, and double-acting steam hammers are provided.

$$Q_{a,\text{lbf}} = \frac{Q_{\text{ult}}}{FS}$$
$$= \frac{2W_{\text{hammer,lbf}} H_{\text{fall,ft}}}{S_{\text{in}} + 1} \qquad 38.3$$
[drop hammer]

$$Q_{a,\text{lbf}} = \frac{2W_{\text{hammer,lbf}} H_{\text{fall,ft}}}{S_{\text{in}} + 0.1} \qquad 38.4$$
$$\begin{bmatrix} \text{single-acting steam hammer;} \\ \text{driven weight } < \text{ striking weight} \end{bmatrix}$$

$$Q_{a,\text{lbf}} = \frac{2W_{\text{hammer,lbf}} H_{\text{fall,ft}}}{S_{\text{in}} + 0.1\left(\dfrac{W_{\text{driven}}}{W_{\text{hammer}}}\right)} \qquad 38.5$$
$$\begin{bmatrix} \text{single-acting steam hammer;} \\ \text{driven weight } > \text{ striking weight} \end{bmatrix}$$

$$Q_{a,\text{lbf}} = \frac{2E_{\text{ft-lbf}}}{S_{\text{in}} + 0.1} \qquad 38.6$$

$$\begin{bmatrix} \text{double-acting steam hammer;} \\ \text{driven weight} < \text{striking weight} \end{bmatrix}$$

$$Q_{a,\text{lbf}} = \frac{2E_{\text{ft-lbf}}}{S_{\text{in}} + 0.1\left(\dfrac{W_{\text{driven}}}{W_{\text{hammer}}}\right)} \qquad 38.7$$

$$\begin{bmatrix} \text{double-acting steam hammer;} \\ \text{driven weight} > \text{striking weight} \end{bmatrix}$$

Handbook table Maximum Allowable Stresses in Pile for Top-Driven Piles gives the maximum allowable stresses in piles as a function of the type of pile being driven. Values for steel H-piles, unfilled steel pipe piles, concrete-filled steel pipe piles, precast prestressed concrete piles, conventionally reinforced concrete piles, and timber piles are all provided in this chart.

The cross sections provided in *Handbook* figure Hammer-Pile-Soil System show how a pile is typically installed through a soft ground layer into a dense layer.

Drivability

Handbook: Modified Engineering-News Formula; Gates Formula

FHWA NHI-06-089: Sec. 9.9.2

Drivability is a two-part process, involving the following.

- design-phase work: predicting geotechnical-structural interaction and foundation element reaction to anticipated installation equipment and techniques in advance of construction

- construction-phase work: confirming those predictions in the field during driving

The *Handbook* provides two options for determining pile capacity at the time of installation. *Handbook* equation Modified Engineering-News Formula calculates the allowable pile load as a function of the driving energy, pile penetration per blow, and hammer type.

$$P_{\text{allow}} = \frac{R}{6} = \frac{2E_n}{S+k}$$

Handbook equation Gates Formula calculates the ultimate pile capacity as a function of the field-observed hammer energy and the number of blows per inch at final penetration.

$$R_u = 1.75\sqrt{E_r}(\log_{10} 10 N_b) - 100$$

FHWA NHI-06-089 Sec. 9.9.2 addresses pile drivability. During installation, a pile must be resilient against damage and stiff enough to overcome resistance from the soil. FHWA NHI-06-089 Table 9-9 lists the responsibilities of design and construction engineers for pile driving.

Equipment Selection

FHWA NHI-06-089: Sec. 9.9.3

Equipment selection is critical for driven deep foundations since equipment capacity and integrity are largely judged using information from the driving process.

Foundation element capacity is judged on the number of hammer blows required to achieve a specified penetration, and the blow-to-penetration relationship is based on energy transfer. The hammer and driving system components need to be matched to the types of structural elements being driven so that energy transfer can be optimized, the elements can be installed to their intended depths, and capacities can be confirmed without damage. Impact-induced driving stresses must match the types of elements being driven, whose design stresses vary.

Hammer efficiency must be known so that the capacities derived from the assumed imparted driving energy are not over- or underestimated. The quality of cushions for hammer and foundation elements (the latter for concrete piles) is particularly important, as excessive deterioration during driving will affect energy transfer.

FHWA NHI-06-089 Fig. 9-38 shows the typical components of a pile driving system. The critical elements of the system are the leads, the hammer cushion, the helmet, and the pile cushion (for concrete piles). The leads keep the pile in alignment with the hammer and the installation location. The helmet keeps the top of the pile in alignment with the hammer.

G. PILE DYNAMICS

Key concepts: These key concepts are important for answering exam questions in subtopic 18.G, Pile Dynamics.

- drivability assessment
- estimation of driving resistance
- high-strain dynamic testing
- interpretation of dynamic testing results
- static capacity

***NCEES Handbook*:** To prepare for this subtopic, familiarize yourself with these sections in the *Handbook*.

- Typical Wave Equation Models
- Wave Equation Bearing Graph

Codes and standards: To prepare for this subtopic, familiarize yourself with these sections from codes and standards.

Design and Construction of Driven Pile Foundations – Volume II (FHWA NHI-16-010)

- Section 10.6.6: Signal Matching

Soils and Foundations Reference Manual – Volume II (FHWA NHI-06-089)

- Section 9.9.6: Wave Equation Methodology
- Section 9.9.10: Dynamic Pile Monitoring

The following figures and concepts are relevant for subtopic 18.G, Pile Dynamics.

Wave Equation Analysis

Handbook: **Typical Wave Equation Models; Wave Equation Bearing Graph**

FHWA NHI-06-089: Sec. 9.9.6

Wave equation analysis is performed over extremely small, incremental time steps, as follows.

1. A soil model and foundation element capacity are assumed.

2. The geotechnical engineer estimates a static soil resistance profile consisting of point (or toe) and skin (or shaft) resistance for the singular or layered material profile through which the pile will penetrate.

3. The analysis begins with the selection of a hammer efficiency, which is used to compute the impact velocity of the hammer mass elements for an input hammer stroke.

4. For each time step, the model is used to calculate the acceleration, velocity, force, and displacement of each element. This process is repeated for all incremental time steps until the element point starts to rebound. Permanent element penetration is then calculated.

5. Penetration is plotted against capacity for one point on a wave equation bearing graph.

Wave equation analysis predicts capacity and other performance data, based on blow count, for elements of specific properties and lengths. Changes in any one of the model's input parameters warrants the analysis to be rerun to reflect and allow qualification of the changes.

FHWA NHI-06-089 Sec. 9.9.6 discusses the wave equation methodology and the concept behind *Handbook* figure **Typical Wave Equation Models**. The hammer and helmet, along with the hammer cushions and pile cushions, are displayed as springs. Static and dynamic components are used in the equations and act on each segment of the model. The dynamic soil resistance is a function of the damping factor, J. The graph at the bottom center of the *Handbook* figure shows the parameter q, known as the quake. q is used for the static portion of the problem, and J is used for the dynamic component.

Handbook figure **Wave Equation Bearing Graph** displays the results for a 50 ft long pile. This chart shows the predicted pile response based on the input of the wave equation analysis. The chart is used in the field and compared to the actual field measurements.

Dynamic Testing

FHWA NHI-06-089: Sec. 9.9.10

Dynamic testing not only confirms the results of static analyses performed in the project's design phase relative to ultimate capacity, but also provides data that can qualify foundation element/subsurface material profile interaction and installation equipment (hammer and driving system) performance.

For each hammer blow, the strain transducers and accelerometers obtain signals that the pile driving analyzer (PDA) converts into digital force and velocity data versus time. Given this data and the material properties of the foundation element, this process reveals the location and magnitude of the soil resistance forces along the element's length.

According to FHWA NHI-06-089 Sec. 9.9.10, a minimum of two diametrically opposite mounted strain transducers attached to the pile are required for an accurate test. Two accelerometers are also needed. These are commonly attached 2 to 3 diameters below the pile head. Results from dynamic testing are used to determine the static pile capacity, hammer and driving system performance, driving stresses, and pile integrity.

Signal Matching

FHWA NHI-16-010: Sec. 10.6.6

Signal matching creates a post-installation wave equation model of the driven foundation element. This model yields an ultimate element capacity estimate comparable to one determined through static load testing.

A signal matching program uses the continuous foundation element model, the recorded element's top velocity, and an assumed soil model consisting of the bearing stratum/strata resistance distribution, quakes, and damping characteristics particular to each bearing stratum segment and at the element toe. The program computes a force wave trace at the element's top, which is compared to the force wave trace measured in the field. During computational iterations, judgment is used to adjust the bearing material and field-measured force wave traces. The best match occurs when no further agreement can be obtained between the measured and computed wave traces.

FHWA NHI-16-010 Sec. 10.6.6 discusses the signal matching concept. The calculations are performed by computer models; an example of the output is shown in FHWA NHI-16-010 Fig. 10-17, which presents an overlay of the measured dynamic wave and the calculated dynamic wave. FHWA NHI-16-010 Fig. 10-18 shows a series of overlaid waves as the iterative process adjusts the calculated waves in order to match the measured wave.

H. PILE AND DRILLED-SHAFT LOAD TESTING

Key concepts: These key concepts are important for answering exam questions in subtopic 18.H, Pile and Drilled-Shaft Load Testing.

- elastic compression
- Osterberg cell compression test
- quick load compression test
- settlement
- static load compression test
- stress-strain relationships

NCEES Handbook: To prepare for this subtopic, familiarize yourself with these sections in the *Handbook*.

- Uniaxial Loading and Deformation

***PE Civil Reference Manual* (CERM):** Study these sections in CERM that either relate directly to this subtopic or provide background information.

- Section 44.3: Elastic Deformation

Codes and standards: To prepare for this subtopic, familiarize yourself with these sections from codes and standards.

Design and Construction of Driven Pile Foundations – Volume II (FHWA NHI-16-010)

- Section 9.2: Axial Compression Load Test
- Section 9.2.3: Recommended Axial Compression Test Loading Method
- Section 9.4.1: Lateral Load Test Equipment

Soils and Foundations Reference Manual – Volume II (FHWA NHI-06-089)

- Section 9.6.2.1: Elastic Compression of Piles
- Section 9.15.7.4: Plotting the Failure Criteria
- Section 9.15.8.1: The Osterberg Cell Method

The following equations, figures, and tables are relevant for subtopic 18.H, Pile and Drilled-Shaft Load Testing.

Elastic Compression

Handbook: Uniaxial Loading and Deformation

CERM: Sec. 44.3

FHWA NHI-06-089: Sec. 9.6.2.1, Sec. 9.15.7.4

Failure criteria is based on elastic compression and settlement, and performance criteria is dependent on allowable movement versus movement at failure load. FHWA NHI-06-089 Eq. 9-51 gives the equation for elastic deformation.

$$\Delta = \frac{QL}{AE} \quad \text{[FHWA NHI-06-089 Eq. 9-51]}$$

The allowable geotechnical load based on static load compression testing is typically determined by applying an ASD factor of safety as low as 2.0 or an LRFD load resistance factor of 0.75 to 0.80 for driven foundation elements and 0.70 for drilled foundation elements.

The elastic compression of a pile contributes to the overall movement of the pile itself and is added to the pile settlement to obtain the total drop in elevation of the top of the pile. Per FHWA NHI-06-089 Sec. 9.6.2.1, the modulus of elasticity for steel piles is 30,000 ksi, and the modulus of elasticity of concrete piles is around 4000 psi but will vary based on the compressive strength of the concrete.

The load input is frequently modified as shown in the equation from *Handbook* section Uniaxial Loading and Deformation, where the load is represented by *P*. The *Handbook* equation is equivalent to CERM Eq. 44.4, where the load is represented by *F*.

$$E = \frac{\sigma}{\varepsilon} = \frac{\frac{P}{A}}{\frac{\delta}{L}}$$

Exam tip: The elastic deformation equation given by FHWA NHI-06-089 Eq. 9-51 is the same for any material with a modulus of elasticity, *E*, cross-sectional area, *A*, and length, *L*.

Static Load Compression Testing

FHWA NHI-16-010: Sec. 9.2, Sec. 9.4.1

Static load testing is an accurate in situ means of determining bearing, tensile, and lateral load capacity. Static load testing replicates, in advance, application of the actual load to produce project-specific responses from which design details can be refined to reduce risk and cost. Static load tests are more often performed on drilled than driven foundations.

Static load testing is recommended when knowledge of local subsurface conditions is limited, conditions vary, knowledge regarding the magnitude of design loads is limited, loading conditions may vary over time, or discrepancies with dynamic testing exist.

Static load testing cannot predict long-term performance and is not representative of group action (unless the group can be tested). It may also be misleading for foundations on variable bearing materials.

FHWA NHI-16-010 Chap. 9 is devoted to static load tests and goes into detail on axial compression load tests, tension load tests, lateral load tests, and load transfer evaluations. This chapter also discusses practical issues and considerations concerning these tests, along with their advantages, disadvantages, and limitations. FHWA NHI-16-010 Sec. 9.2 is specific to compression testing.

An axial compression load diagram is shown in FHWA NHI-16-010 Fig. 9-1, and a static load test setup diagram is shown in Fig. 9-2. FHWA NHI-16-010 Sec. 9.2 also presents many images from actual axial load tests. Pictures of tension load tests in the field are shown in Fig. 9-9 and Fig. 9-10, and pictures of lateral load tests are shown in Fig. 9-12 through Fig. 9-16.

Quick Load Compression Test

FHWA NHI-16-010: Sec. 9.2.3

While several static load test methods are available, FHWA reports that the "quick" method is the most common. With this method, a load is applied to the top of the foundation element with a hydraulic jack. Resistance is provided by a beam anchored to or weighted independently of the foundation element.

Other points (known as *telltales*) along the length of the element may be established for monitoring. Loads are applied and increased in 2.5 min intervals, with the load increment equal to between 10% and 15% of the element's design load. Time, load, and movement are measured at the beginning and end of each increment.

The test is continued until continuous jacking is required to maintain the applied load or the capacity of the test system is reached, whereupon this final load is held for 5 min, and readings are taken at 2.5 min and 5 min. The results are plotted to determine a failure load and movement at that load.

The quick load test procedure is detailed in FHWA NHI-16-010 Sec. 9.2.3. This method is common because it only takes between 2 hr and 4 hr to perform. During this process, load increments are held between 4 min and 15 min and applied in increments of 5% of the anticipated resistance load. This approach provides around 20 data points for an accurate load movement curve to be analyzed.

Osterberg Cell Compression Test

FHWA NHI-06-089: Sec. 9.15.8.1

In the Osterberg cell method, the load is applied to the bottom of the foundation element (rather than the top) by a pressurized chamber between two plates that are fixed to the bottom of the element. A pressure-to-load calibration is made, and the element is then loaded to failure in a manner similar to the quick method. However, the Osterberg cell method determines only the failure load associated with skin friction developed along the element perimeter above the cell or with point resistance present below the cell.

The Osterberg cell method is discussed in FHWA NHI-06-089 Sec. 9.15.8.1. The Osterberg load cell is cast within the test shaft. FHWA NHI-06-089 Fig. 9-72 shows a comparison of the typical static test and the Osterberg cell test. The Osterberg cell is pressurized and expanded by means of a fluid, typically oil or water, that is pumped from the ground surface. Due to the size of the pistons, the Osterberg cell can apply loads up to 3000 tons with a standard model at a 32 in diameter. Osterberg cells can be used with drilled or driven piles. FHWA NHI-06-089 Table 9-12 and Table 9-13 provide sizes, capacities, and other important measurements for Osterberg cells used with drilled shafts and driven piles.

I. INTEGRITY TESTING METHODS

Key concepts: These key concepts are important for answering exam questions in subtopic 18.I, Integrity Testing Methods.

- coring
- crosshole sonic logging (CSL) test
- gamma density logging (GDL) test
- theoretical wave velocity of concrete
- thermal integrity profiling

NCEES Handbook: The *Handbook* does not contain any material relevant to this subtopic. Be sure to study the listed sections from codes and standards.

Codes and standards: To prepare for this subtopic, familiarize yourself with these sections from codes and standards.

Drilled Shafts: Construction Procedures and Design Methods (FHWA NHI-18-024)

- Section 16.2: Non-Destructive Integrity Tests
- Section 16.3: Drilling and Coring

Soils and Foundations Reference Manual – Volume II (FHWA NHI-06-089)

- Section 9.14.1: The Standard Crosshole Sonic Logging (CSL) Test
- Section 9.14.2: The Gamma Density Logging (GDL) Test

The following equations, figures, and concepts are relevant for subtopic 18.I, Integrity Testing Methods.

Integrity Testing

FHWA NHI-06-089: Sec. 9.14.1, Sec. 9.14.2

FHWA NHI-18-024: Sec. 16.2

Integrity testing is performed mainly on drilled deep foundations (auger-cast piles and drilled shafts) but can also be performed on driven foundations like precast concrete piles that are vulnerable to cracking or experience poor performance during driving due to material defects.

Most integrity testing is indirect, as it is difficult to observe structures in ground that has already been built on before testing. The indirect methods depend on signal continuity along, through, or between the foundation elements being tested.

The standard crosshole sonic logging (CSL) test and the gamma density logging (GDL) test are two common integrity testing methods; these tests are outlined in FHWA NHI-06-089 Sec. 9.14.1 and Sec. 9.14.2. FHWA NHI-18-024 Chap. 16 is dedicated to integrity testing. Nondestructive integrity testing is discussed in Sec. 16.2, and the concept, primary application, limitations, advantages, and variations are listed in Table 16-1 for CSL, thermal integrity profiling (TIP), and gamma-gamma testing.

Crosshole Sonic Logging (CSL) Test

FHWA NHI-06-089: Sec. 9.14.1

FHWA NHI-18-024: Sec. 16.2.1

In crosshole sonic logging (CSL) tests, a transmitter and receiver probes are lowered into access tubes. The travel time of ultrasonic waves between probes is recorded along with wave amplitude. The travel velocity is compared to the theoretical wave velocity of concrete. Lower average velocities are produced by anomalies.

CSL doesn't directly image anomalies and discontinuities in concrete, but it does provide a means of determining to what extent those features likely impact concrete strength and overall element capacity. When multiple pairs of probe holes are installed prior to concrete placement, the findings from one set of probes can be validated or refined with the findings from other probe sets.

FHWA NHI-06-089 Fig. 9-58 shows a CSL schematic.

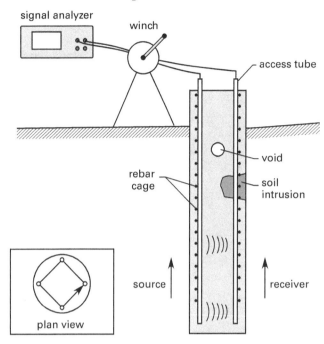

From FHWA NHI-06-089, *Soils and Foundations Reference Manual – Vol. II*, Fig. 9-58 (Samtani, et al.), © 2006, U.S. Department of Transportation, Federal Highway Administration.

FHWA NHI-06-089 Sec. 9.14.1 describes the CSL process and provides an example of the results of a CSL test. The time between the transmission and receiving of the signal is compared to the distance between each tube, and the calculated velocity is compared to the theoretical ultrasonic wave velocity of the concrete through which it is passing. The velocity reduction (VR) is then calculated. A VR between 0 and 10% is considered a good rating, a VR between 10% and 20% is questionable, and a VR greater than 20% is poor. FHWA NHI-18-024 Sec. 16.2.1 also discusses CSL. FHWA NHI-18-024 Eq. 16-1 through Eq. 16-3 provide calculations to determine the specific theoretical compressional wave velocity in the concrete.

$$V_c = \sqrt{\frac{\alpha E}{\rho}} \quad \text{[FHWA NHI-18-024 Eq. 16-1]}$$

$$\alpha = \frac{1-v}{(1+v)(1-2v)} \quad \text{[FHWA NHI-18-024 Eq. 16-2]}$$

$$\rho = \frac{\gamma}{g} \quad \text{[FHWA NHI-18-024 Eq. 16-3]}$$

Exam tip: New crosshole methods published by the Deep Foundations Institute are preliminary and are not covered in the current NCEES exam specifications.

Coring

FHWA NHI-18-024: Sec. 16.3

Coring is a direct method of evaluating material composition, consistency, and strength. Coring is often used as a follow-up to unfavorable indirect results. Cores can be sawed for petrographic inspection or subjected to compression testing. Cores are small and taken at discrete locations, so they must be targeted to suspected anomalies or discontinuities.

Coring provides physical and visual evidence of the composition, consistency, and strength of concrete. Cores are sometimes taken where steel reinforcement is (or is supposed to be) to verify reinforcement size and orientation.

In contrast to indirect methods, cores only provide information for the discrete sample that is obtained. For embedded portions of foundation elements, core location is often determined based on initial, indirect tests.

FHWA NHI-18-024 Fig. 16-15 provides four pictures of concrete cores of various sizes and qualities. FHWA NHI-18-024 Sec. 16.3 discusses the reason that each core was taken, the result of the coring process, and the action taken as a result of the inspection of the core.

Topic IV: Structural

Chapter

27. Analysis of Structures: Loads and Load Applications
28. Analysis of Structures: Forces and Load Effects
29. Design and Details of Structures: Concrete
30. Design and Details of Structures: Steel
31. Design and Details of Structures: Timber
32. Design and Details of Structures: Masonry
33. Codes, Standards, and Guidance Documents
34. Temporary Structures and Other Topics

27 Analysis of Structures: Loads and Load Applications

Content in blue refers to the *NCEES Handbook*.

A.1. Dead Loads 27-1
A.2. Live Loads 27-2
A.3. Construction Loads 27-3
A.4. Wind Loads 27-4
A.5. Seismic Loads 27-6
A.6. Moving Loads 27-8
A.7. Snow, Rain, Ice 27-9
A.8. Impact Loads 27-10
A.9. Earth Pressure and Surcharge Loads 27-11
A.10. Load Paths 27-15
A.11. Load Combinations 27-16
A.12. Tributary Areas 27-17

The knowledge area of Loads and Load Applications makes up between four and six questions out of the 80 questions on the PE Civil Structural exam. The organization of this chapter follows the order of subtopics given by NCEES for this knowledge area. Each subtopic is covered in the following sections.

A.1. DEAD LOADS

Key concepts: These key concepts are important for answering exam questions in subtopic 9.A.1, Dead Loads.

- dead loads in structural analysis
- distributed loads
- soil dead loads
- ultimate strength
- weights of fixed service equipment
- weights of materials and constructions

NCEES Handbook: The *Handbook* does not contain any material relevant to this subtopic. Be sure to study the listed sections in CERM and CEST and from codes and standards.

PE Civil Reference Manual (CERM): Study these sections in CERM that either relate directly to this subtopic or provide background information.

- Section 40.11: Loads on Buried Pipes
- Section 41.14: Distributed Loads
- Section 45.2: Ultimate Strength Design

Structural Depth Reference Manual (CEST): Study these sections in CEST that either relate directly to this subtopic or provide background information.

- Section 4.1: Plastic Design
- Section 5.2: Load Combinations
- Section 6.1: Design Principles

Codes and standards: To prepare for this subtopic, familiarize yourself with these sections from codes and standards.

Building Code Requirements for Structural Concrete and Commentary (ACI 318)

- Section 5.3: Load Factors and Combinations

Minimum Design Loads for Buildings and Other Structures (ASCE/SEI 7)

- Section 3.1.2: Weights of Materials and Constructions
- Section 3.1.3: Weight of Fixed Service Equipment
- Section 3.2: Soil Loads and Hydrostatic Pressure
- Chapter C3: Dead Loads, Soil Loads, and Hydrostatic Pressure

The following equations, figures, tables, and concepts are relevant for subtopic 9.A.1, Dead Loads.

Distributed Loads

CERM: Sec. 41.14

If an object is continuously loaded over a portion of its length, it is subject to a distributed load. Distributed loads result from dead load (i.e., self-weight), hydrostatic pressure, and materials distributed over the object. CERM Fig. 41.6 shows an example of a distributed load on a beam.

Figure 41.6 Distributed Loads on a Beam

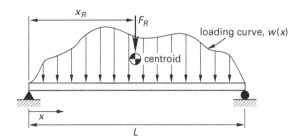

Ultimate Strength

CERM: Sec. 45.2

ACI 318: Sec. 5.3

The *ultimate strength* (i.e., the required strength) of a member is calculated from the actual service loads and multiplicative factors known as *overload factors* or *load factors*. Usually, a distinction is made between dead loads and live loads. For example, the required ultimate moment-carrying capacity in a concrete beam designed according to ACI 318 and CERM Eq. 45.5 is

$$M_u = 1.2 M_{\text{dead load}} + 1.6 M_{\text{live load}} \qquad 45.5$$

Dead Loads in Structural Analysis

CEST: Sec. 4.1, Sec. 5.2, Sec. 6.1

ASCE/SEI 7: Chap. C3

Dead loads consist of the weight of all materials of construction incorporated into the building, including walls, floors, roofs, ceilings, stairways, built-in partitions, finishes, cladding, and other similarly incorporated architectural and structural items and fixed service equipment (including cranes).

All structures have dead loads; ASCE/SEI 7 Sec. C3.1.2 contains useful information for determining what dead loads to apply in a structural analysis.

Soil Dead Loads

CERM: Sec. 40.11

ASCE/SEI 7: Sec. 3.2

In the design of structures below grade, provision must be made for the lateral pressure of adjacent soil. If soil loads are not given in a soil investigation report approved by the authority having jurisdiction, then the soil loads are as specified in ASCE/SEI 7 Table 3.2-1.

Weights of Materials and Constructions

ASCE/SEI 7: Sec. C3.1.2

ASCE/SEI 7 Table C3-1 includes a list of materials typically used in building construction and the weights of each material in pounds per square foot. ASCE/SEI 7 Table C3-2 gives minimum densities for design loads from materials.

Weights of Fixed Service Equipment

ASCE/SEI 7: Sec. 3.1.3

Fixed service equipment includes plumbing stacks and risers; electrical feeders; heating, ventilating, and air conditioning systems; and process equipment such as vessels, tanks, piping, and cable trays. The empty weight of the equipment and the maximum weight of the equipment contents are treated as dead loads.

A.2. LIVE LOADS

Key concepts: These key concepts are important for answering exam questions in subtopic 9.A.2, Live Loads.

- concentrated and uniformly distributed live loads
- live load reduction
- partition loads

NCEES Handbook: The *Handbook* does not contain any material relevant to this subtopic. Be sure to study the listed sections in CERM and from codes and standards.

PE Civil Reference Manual **(CERM):** Study these sections in CERM that either relate directly to this subtopic or provide background information.

- Section 50.5: Service Loads, Factored Loads, and Load Combinations

Codes and standards: To prepare for this subtopic, familiarize yourself with these sections from codes and standards.

Minimum Design Loads for Buildings and Other Structures (ASCE/SEI 7)

- Chapter 4: Live Loads

The following equations and concepts are relevant for subtopic 9.A.2, Live Loads.

Live Load Reduction

CERM: Sec. 50.5

ASCE/SEI 7: Chap. 4, Table 4-1, Table 4-2

The *live load* (the weight of objects not permanently supported by the structure) on a particular floor can be determined based on ASCE/SEI 7 Table 4-1. Live loads can be reduced in some circumstances. The reduction for all live loads other than roof live loads is calculated using ASCE/SEI 7 Eq. 4.7-1.

$$L = L_o \left(0.25 + \frac{15}{\sqrt{K_{\text{LL}} A_T}} \right)$$

In this equation,

A_T	tributary area	ft^2
K_{LL}	live load element factor (see ASCE/SEI 7 Table 4-2)	–
L	reduced design live load supported by the member; $\geq 0.50 L_o$ for members supporting one floor and $\geq 0.40 L_o$ for members supporting two or more floors	lbf/ft^2
L_o	unreduced design live load supported by the member (see ASCE/SEI 7 Table 4-1)	lbf/ft^2

The reduction in roof live loads can be calculated in accordance with ASCE/SEI 7 Eq. 4.8-1.

$$L_r = L_0 R_1 R_2 \quad \text{where} \quad 12 \leq L_r \leq 20 \quad [\text{U.S.}]$$
$$L_r = L_0 R_1 R_2 \quad \text{where} \quad 0.58 \leq L_r \leq 0.96 \quad [\text{SI}]$$

The reasoning behind this reduction is that the probability of the full nominal live load being distributed over the entire area decreases as the influence area—the floor area contributing load to a particular structural element (e.g., a beam or column)—increases. For large areas, it is excessively conservative to expect the entire area to be fully loaded.

ASCE/SEI 7 Sec. 4.7.2 through Sec. 4.7.6 give some limitations to using this reduction. Live load reductions may not be made for values of the influence area, $K_{LL} A_T$, less than 400 ft^2.

The maximum allowable reduction is 50% for members supporting one floor only and 60% for members (e.g., columns) supporting multiple floors.

Exam tip: When calculating loads, always check to see if live loads can be reduced.

Partition Loads

ASCE/SEI 7: Table 4-1

In buildings where partitions will be erected or rearranged, provisions for the partitions' weight must be made, whether or not partitions are shown on the plans. A partition load may not be taken as less than 15 psf, and it is considered a live load because the partition layout is often reworked over the lifetime of the structure.

Exam tip: Partition loads are not required when the floor live load exceeds 80 psf.

Concentrated and Uniformly Distributed Live Loads

ASCE/SEI 7: Sec. 4.4, Table 4-1

Floors, roofs, and other similar surfaces must be designed to safely support the uniformly distributed live loads or the concentrated load provided in ASCE/SEI 7 Table 4-1. The uniform and concentrated loads are not applied concurrently. Rather, the load that produces the greater load effect must be used in the design.

Access floor systems have both prescribed uniform and concentrated loads. These need not be considered to act simultaneously. Any load combination considering these live loads should therefore be checked twice: once with the uniform load and once with the concentrated load. This implicitly increases the total number of load combinations.

A.3. CONSTRUCTION LOADS

Key concepts: These key concepts are important for answering exam questions in subtopic 9.A.3, Construction Loads.

- design considerations
- shored construction
- stay-in-place formwork

NCEES Handbook: The *Handbook* does not contain any material relevant to this subtopic. Be sure to study the listed sections in CERM and from codes and standards.

PE Civil Reference Manual (CERM): Study these sections in CERM that either relate directly to this subtopic or provide background information.

- Section 51.13: Concrete Deck Systems in Bridges
- Section 57.7: Strength of Composite Sections

Codes and standards: To prepare for this subtopic, familiarize yourself with these sections from codes and standards.

AASHTO LRFD Bridge Design Specifications (AASHTO LRFD)

- Section 3.3: Notation

The following concepts are relevant for subtopic 9.A.3, Construction Loads.

Shored Construction

CERM: Sec. 57.7

Shored construction is used for a type of construction relating to composite steel beams. Concrete construction cannot be built without shoring. With shored construction, composite steel beams are supported by temporary falsework until the concrete cures, and the steel acts compositely with the concrete slab. All non-composite loads, including the steel beam itself, are supported by the falsework. Falsework supports the formwork, material, and construction loads in all forms of concrete construction.

Stay-in-Place Formwork

CERM: Sec. 51.13

AASHTO LRFD: Sec. 3.3

Stay-in-place formwork is often used in deck construction. The formwork is designed to behave elastically and prevent excessive sagging under construction loads. The formwork must be strong enough to carry the load of self-weight, the deck concrete weight, and an additional 50 psf. AASHTO LRFD Sec. 3.3 provides the limits of the stresses due to construction loads in the formwork to prevent failure of the formwork when vehicular loads are applied.

Design Considerations

Design should allow for construction loads and the possibility of damage from construction equipment. Construction loads consist of temporary loads applied during construction; stay-in-place formwork; construction machinery; weight of material; and formwork, falsework, scaffolding, and shoring. These loads are calculated in the same way as other loads. A construction live load is also required to account for the workers and miscellaneous equipment that are moved around during construction. Construction loads are typically applied as dead or live loads.

A.4. WIND LOADS

Key concepts: These key concepts are important for answering exam questions in subtopic 9.A.4, Wind Loads.

- basic wind speed
- surface roughness categories
- velocity pressure
- wind analysis
- wind exposure categories
- wind loads in load combinations

NCEES Handbook: The *Handbook* does not contain any material relevant to this subtopic. Be sure to study the listed sections in CEST and from codes and standards.

Structural Depth Reference Manual (**CEST**): Study these sections in CEST that either relate directly to this subtopic or provide background information.

- Section 4.1: Plastic Design
- Chapter 5: Design of Wood Structures

Codes and standards: To prepare for this subtopic, familiarize yourself with these sections from codes and standards.

International Building Code (IBC)

- Section 1605.2: Load Combinations Using Strength Design or Load and Resistance Factor Design
- Section 1605.3.1: Basic Load Combinations

Minimum Design Loads for Buildings and Other Structures (ASCE/SEI 7)

- Chapter 26: Wind Loads: General Requirements
- Chapter 27: Wind Loads on Buildings—MWFRS (Directional Procedure)
- Chapter 28: Wind Loads on Buildings—MWFRS (Envelope Procedure)
- Chapter 29: Wind Loads on Other Structures and Building Appurtenances—MWFRS
- Chapter 30: Wind Loads—Components and Cladding (C&C)

The following equations and figures are relevant for subtopic 9.A.4, Wind Loads.

Wind Loads in Load Combinations

CEST: Sec. 4.1, Sec. 5.2

IBC: Sec. 1605.2, Sec. 1605.3.1

The primary LRFD load combinations involving wind loads are given by IBC Eq. 16-3, Eq. 16-4, and Eq. 16-6.

$$\lambda W = 1.2(D+F) + 1.6(L_r \text{ or } S \text{ or } R) \\ + 1.6H + (f_1 L \text{ or } 0.5W) \quad \text{[IBC Eq. 16-3]}$$

$$\lambda W = 1.2(D+F) + 1.0W + f_1 L \\ + 1.6H + 0.5(L_r \text{ or } S \text{ or } R) \quad \text{[IBC Eq. 16-4]}$$

$$\lambda W = 0.9D + 1.0W + 1.6H \quad \text{[IBC Eq. 16-6]}$$

The load factor for wind loads is 0.6 in ASD load combinations, as given by IBC Eq. 16-12, Eq. 16-13, and Eq. 16-15.

$$\sum \gamma Q = D + H + F + (0.6W \text{ or } 0.7E) \quad \text{[IBC Eq. 16-12]}$$

$$\sum \gamma Q = D + H + F + 0.75(0.6W) \\ + 0.75L + 0.75(L_r \text{ or } S \text{ or } R) \quad \text{[IBC Eq. 16-13]}$$

$$\sum \gamma Q = 0.6D + 0.6W + H \quad \text{[IBC Eq. 16-15]}$$

Wind Analysis

ASCE/SEI 7: Chap. 26, Chap. 27, Chap. 28, Chap. 29, Chap. 30

Wind analysis is covered in ASCE/SEI 7 Chap. 26 through Chap. 30. Chapter 26 covers basic wind load parameters (e.g., basic wind speed) that are universally applicable to building and nonbuilding structures, main wind force resisting systems (MWFRSs), and components and cladding (C&C). ASCE/SEI 7 Chap. 27 through Chap. 30 cover wind analysis as it applies to various systems.

Exam tip: To calculate the wind loads on a building, it is necessary to determine a number of variables, including the exposure category of the site, the basic wind speed at the location of the structure, the velocity pressure exposure coefficient, the topography at the location of the building, the probable direction of the wind, and the type of building.

Basic Wind Speed

ASCE/SEI 7: Sec. 26.5

To determine basic wind speed, perform the following steps.

1. Locate the project site on the applicable wind speed map in ASCE/SEI 7.

    - Figure 26.5-1A for occupancy category II
    - Figure 26.5-1B for occupancy categories III and IV
    - Figure 26.5-1C for occupancy category I

2. Round up to the next highest wind contour or use linear interpolation.

The basic wind speed varies both by region and by structure risk category.

Exam tip: The basic wind speed is dependent on the occupancy category of the structure.

Surface Roughness Categories

ASCE/SEI 7: Sec. 26.7.2

Surface roughness category B has numerous obstructions and can be in urban, suburban, or wooded areas. Surface roughness category C has scattered obstructions but is usually open terrain. Surface roughness category D is unobstructed and is applicable to flat, unobstructed areas such as water surfaces. See ASCE/SEI 7 Sec. 26.7.2 for more complete definitions of roughness categories.

Exam tip: The more unobstructed the terrain is, the higher the wind load will be.

Wind Exposure Categories

ASCE/SEI 7: Sec. 26.7.3

For buildings 30 ft and shorter, wind exposure category B applies where prevailing upwinds are greater than 1500 ft; for buildings taller than 30 ft, this category applies where prevailing upwinds exceed the greater of 2600 ft or 20 times the building height.

Wind exposure category D is for prevailing upwinds that exceed the greater of 5000 ft or 20 times the building height, disregarding the greater of the first 600 ft or 20 times the building height.

Wind exposure category C covers all other cases. See ASCE/SEI 7 Sec. 26.7.3 for more complete definitions of exposure categories.

Exam tip: Exposure categories will typically correlate to surface roughness categories, although these are not directly equivalent.

Velocity Pressure

ASCE/SEI 7: Sec. 27.3

$$q_z = 0.00256 K_z K_{zt} K_d V^2 \quad (\text{lbf/ft}^2)$$

Velocity pressure, q_z, evaluated at height z is calculated using ASCE/SEI 7 Eq. 27.3-1.

The velocity pressure is a useful parameter to calculate when working with wind loads. It is dependent on the basic wind speed, importance factor, exposure, surface roughness, and height at which the pressure is being calculated.

Exam tip: As the height increases, the velocity pressure also increases.

A.5. SEISMIC LOADS

Key concepts: These key concepts are important for answering exam questions in subtopic 9.A.5, Seismic Loads.

- ASD required strength
- design acceleration response
- equivalent lateral forces
- load duration factors for wood structures
- site-specific seismic loads
- special reinforced shear wall reinforcement requirements
- spectral response parameter

NCEES Handbook: The *Handbook* does not contain any material relevant to this subtopic. Be sure to study the listed sections in CEST and from codes and standards.

Structural Depth Reference Manual (**CEST**): Study these sections in CEST that either relate directly to this subtopic or provide background information.

- Section 5.2: Load Combinations
- Section 5.6: Adjustment Factors
- Section 6.5: Design of Masonry Shear Walls

Codes and standards: To prepare for this subtopic, familiarize yourself with these sections from codes and standards.

Building Code Requirements for Masonry Structures (TMS 402)

- Section 7.3.2.6: Special Reinforced Masonry Shear Walls
- Commentary Section 8.1.1: Scope

International Building Code (IBC)

- Section 1605.3.1: Basic Load Combinations

Minimum Design Loads for Buildings and Other Structures (ASCE/SEI 7)

- Chapter 11: Seismic Design Criteria
- Chapter 12: Seismic Design Requirements for Building Structures
- Chapter 13: Seismic Design Requirements for Nonstructural Components
- Chapter 14: Material Specific Seismic Design and Detailing Requirements
- Chapter 15: Seismic Design Requirements for Nonbuilding Structures
- Section 20.3: Site Class Definitions

National Design Specification for Wood Construction with Commentary (NDS)

- Table 2.3.2: Frequently Used Load Duration Factors, C_D

The following equations, figures, tables, and concepts are relevant for subtopic 9.A.5, Seismic Loads.

ASD Required Strength

CEST: Sec. 5.2

IBC: Sec. 1605.3.1

TMS 402: Comm. Sec. 8.1.1

The seismic loads specified in the IBC are at the strength design level, in contrast to other loads that are at the service level. To reduce these loads to service-level values, the load factor for seismic loads is 0.7 in ASD load combinations. The load combinations, with uncommon load conditions (e.g., self-straining loads and fluid pressure) omitted, are given in IBC Eq. 16-12, Eq. 16-14, and Eq. 16-16.

$$\sum \gamma Q = D + H + F + (0.6W \text{ or } 0.7E) \quad [\text{IBC Eq. 16-12}]$$

$$\sum \gamma Q = D + H + F + 0.75(0.7E) \\ + 0.75L + 0.75S \quad [\text{IBC Eq. 16-14}]$$

$$\sum \gamma Q = 0.6(D + F) + 0.7E + H \quad [\text{IBC Eq. 16-16}]$$

For masonry structures, in accordance with TMS 402 Comm. Sec. 8.1.1, allowable stresses may not be increased by one-third for seismic load combinations.

Special Reinforced Shear Wall Reinforcement Requirements

CEST: Sec. 6.5

TMS 402: Sec. 7.3.2.6

Reinforcement is provided in special reinforced masonry shear walls in order to provide ductile behavior in the walls under seismic loads. To ensure this in seismic design categories D, E, and F, TMS 402 Sec. 7.3.2.6 requires walls to be reinforced with uniformly distributed vertical and horizontal reinforcement. Shear reinforcement must be anchored around vertical reinforcement with a standard 180° hook. The minimum required combined area of shear reinforcement and vertical reinforcement can be found in CEST Sec. 6.5.

$$A_{sh} + A_{sv} = 0.002 A_g$$

The requirements for horizontal shear reinforcement are shown in CEST Fig. 6.6.

Load Duration Factors for Wood Structures

CEST: Sec. 5.6

NDS: Table 2.3.2

The effects of transient loads, such as seismic loads, are allowed for by applying load duration factors to the reference design values. Values of the load duration factor for wood structures experiencing seismic loads are given in NDS Table 2.3.2 and summarized in CEST Table 5.3.

Site-Specific Seismic Loads

ASCE/SEI 7: Chap. 11, Chap. 12, Chap. 13, Chap. 14, Chap. 15

ASCE/SEI 7 Chap. 11 can be used to determine basic site-specific design criteria independent of the type of system being analyzed. ASCE/SEI 7 Chap. 12, Chap. 13, and Chap. 15 can be used to model seismic loads on building structures, nonstructural components, and nonbuilding structures, respectively. ASCE/SEI 7 Chap. 14 contains material-specific provisions for ensuring that the loads discussed in the other chapters are valid.

Exam tip: Exam problems are most likely to be based on ASCE/SEI 7 Chap. 11 and Chap. 12. The exam likely will not address concepts involving response history analysis, structures with isolation, or damping systems. However, code requirements for these items are given in ASCE/SEI 7 Chap. 16, Chap. 17, and Chap. 18, respectively.

Spectral Response Parameter

ASCE/SEI 7: Sec. 11.3, Sec. 11.4, Sec. 20.3

To find the spectral response parameter, S_{DS}, use the following steps.

1. Look up the site class in ASCE/SEI 7 Table 20.3-1.
2. Obtain the spectral response acceleration parameter, S_S, from the U.S. Geological Survey (USGS) or ASCE/SEI 7 Sec. 11.1.
3. Look up the site coefficient, F_a, in ASCE/SEI 7 Table 11.4-1.
4. Solve for S_{DS} using ASCE/SEI 7 Eq. 11.4-1 and Eq. 11.4-3.

$$S_{MS} = F_a S_S \quad \text{[ASCE/SEI 7 Eq. 11.4-1]}$$
$$S_{DS} = \tfrac{2}{3} S_{MS} \quad \text{[ASCE/SEI 7 Eq. 11.4-3]}$$

See ASCE/SEI 7 Sec. 11.3 for more thorough definitions of each factor. The seismic loads on a building are caused by inertial forces (i.e., the acceleration of mass in the building), not by direct pressures or weights. The spectral response parameter, S_{DS}, is needed to calculate seismic loads.

Design Acceleration Response

ASCE/SEI 7: Sec. 11.4, Sec. 12.8

A complete design acceleration response spectrum can be produced from S_{D1}, S_{DS}, and T_L using ASCE/SEI 7 Eq. 11.4-5, Eq. 11.4-6, and Eq. 11.4-7. From ASCE/SEI 7 Fig. 22-12, T_L is between 4 sec and 12 sec. The completed spectrum is shown in ASCE/SEI 7 Fig. 11.4-1.

$$S_a = S_{DS}\left(0.4 + 0.6\frac{T}{T_0}\right) \quad \text{[ASCE/SEI 7 Eq. 11.4-5]}$$
$$S_a = \frac{S_{D1}}{T} \quad \text{[ASCE/SEI 7 Eq. 11.4-6]}$$
$$S_a = \frac{S_{D1} T_L}{T^2} \quad \text{[ASCE/SEI 7 Eq. 11.4-7]}$$

If the equivalent lateral force method is used, the complete response spectrum does not need to be calculated (see ASCE/SEI 7 Sec. 12.8). However, students should have some conceptual understanding of what is happening in the equivalent force method.

The maximum instantaneous acceleration a structure will undergo in response to anticipated seismic ground motions can be determined as a function of the structure's fundamental dynamic period. As a result, only the structure's fundamental period and mass need to be known to determine the equivalent seismic force that the structure must resist.

Equivalent Lateral Forces

ASCE/SEI 7: Sec. 12.7.2, Sec. 12.8

ASCE/SEI 7 Sec. 12.8 covers how to determine equivalent lateral forces. The following table provides some of the variables used within ASCE/SEI 7.

variable	description	source
I_e	seismic importance factor	ASCE/SEI 7 Table 1.5-2
R	response modification coefficient	ASCE/SEI 7 Table 12.2-1
T_a	structure fundamental period	ASCE/SEI 7 Eq. 12.8-7
C_s	seismic response coefficient	ASCE/SEI 7 Eq. 12.8-2 through Eq. 12.8-6
W	effective seismic weight	ASCE/SEI 7 Sec. 12.7.2
V	seismic base shear	ASCE/SEI 7 Eq. 12.8-1

The seismic importance factor varies according to building risk category, as with snow and wind loads.

The structure's fundamental period (i.e., the period of the structure's first dynamic mode) is subject to certain limitations, as discussed in ASCE/SEI 7 Sec. 12.8.2.

The response modification coefficient is a reduction factor that allows for the building to be designed to resist reduced loads. The reduced loads are justified by an assumption of sufficient plastic ductility in the structure, characterized by the R-value. A response modification coefficient other than 1.0 assumes the structure will not remain elastic in a seismic event. Higher values of R mean stricter detailing requirements, per ASCE/SEI 7 Chap. 14. ASCE/SEI 7 Table 12.2-1 prescribes specific values for R based on the type of structural system being used to resist seismic loads.

The seismic response coefficient combines regional seismicity, site soil quality, building risk category, the type of structure used, and the structure's dynamic characteristics into one number that can be multiplied by the weight of the building to calculate the total seismic load.

The effective seismic weight consists of the structure's dead load, but it also must include other loads above the base of the structure, as described in ASCE/SEI 7 Sec. 12.7.2.

A.6. MOVING LOADS

Key concepts: These key concepts are important for answering exam questions in subtopic 9.A.6, Moving Loads.

- influence lines
- moving loads on beams
- vehicle loading
- vehicular live loading

NCEES Handbook: To prepare for this subtopic, familiarize yourself with these sections in the *Handbook*.

- Influence Lines for Beams and Trusses
- Moving Concentrated Load Sets

PE Civil Reference Manual (CERM): Study these sections in CERM that either relate directly to this subtopic or provide background information.

- Section 41.25: Influence Lines for Reactions
- Section 46.11: Moving Loads on Beams

Codes and standards: To prepare for this subtopic, familiarize yourself with these sections from codes and standards.

AASHTO LRFD Bridge Design Specifications (AASHTO LRFD)

- Section 3.4.2: Load Factors for Construction Loads

The following concepts are relevant for subtopic 9.A.6, Moving Loads.

Influence Lines

Handbook: Influence Lines for Beams and Trusses

CERM: Sec. 46.11

An *influence line* is a line on a graph showing the magnitude of a reaction as a function of load placement. The steps for using influence lines to determine the maximum force effects are provided in CERM Sec. 46.11. *Handbook* section Influence Lines for Beams and Trusses gives a definition for an influence line.

Exam tip: Influence lines are used in problems involving moving loads to quickly determine where the worst-case load effects occur.

Moving Loads on Beams

Handbook: Moving Concentrated Load Sets

CERM: Sec. 46.11

The absolute moment produced in a simply supported beam by a moving load occurs when the resultant of the load is placed at the center of the beam. The absolute shear occurs when the resultant is placed as close as possible to the end of the beam. *Handbook* section Moving Concentrated Load Sets includes a helpful figure for visualizing the moving loads on a beam.

Exam tip: Problems with moving loads are typically solved using influence lines.

Vehicle Loading

AASHTO LRFD: Sec. 3.6.1.2

Per AASHTO LRFD Sec. 3.6.1.2.2, a design truck has three axles consisting of an 8 kip lead axle and two 32 kip rear axles. The spacing between the two 32 kip axles varies from 14 ft to 30 ft, whatever produces the most critical effect. The wheels making up an axle have a transverse spacing of 6 ft, and there is one design truck per design lane.

A design tandem has two 25 kip axles separated 4 ft apart. There is a 6 ft transverse spacing of wheels and one design tandem per design lane.

Vehicular Live Loading

AASHTO LRFD: Sec. 3.6.1.2.1

Vehicular live loading for bridges is found under HL-93 in AASHTO. HL-93 is a type of theoretical vehicular loading proposed by AASHTO. HL-93 consists of the most critical combination of the

- design lane load plus design truck
- design lane load plus design tandem

A.7. SNOW, RAIN, ICE

Key concepts: These key concepts are important for answering exam questions in subtopic 9.A.7, Snow, Rain, Ice.

- flat roof snow load
- ice loads
- overview of snow loads
- rain loads
- sloped roof snow load
- snow drift

NCEES Handbook: The *Handbook* does not contain any material relevant to this subtopic. Be sure to study the listed sections from codes and standards.

Codes and standards: To prepare for this subtopic, familiarize yourself with these sections from codes and standards.

Minimum Design Loads for Buildings and Other Structures (ASCE/SEI 7)

- Section 7.3: Flat Roof Snow Loads, p_f
- Section 7.4: Sloped Roof Snow Loads, p_s
- Section 7.7: Drifts on Lower Roofs (Aerodynamic Shade)
- Section 7.8: Roof Projections and Parapets
- Section 8.3: Design Rain Loads
- Section 10.4: Ice Loads Due to Freezing Rain

The following equations, figures, and concepts are relevant for subtopic 9.A.7, Snow, Rain, Ice.

Overview of Snow Loads

ASCE/SEI 7: Sec. 7.3, Sec. 7.4

Snow load is the weight of accumulated snow on a structure. The demand depends on

- heating and insulation in the structure
- region
- roof slope
- surrounding terrain
- structure risk category

Snow loads do not need to be considered to act simultaneously with other kinds of roof live loads, but they should be combined with other live and lateral loads in the structure. This is considered in the required load combinations (see subtopic A.11, Load Combinations, in this chapter).

Exam tip: Refer to ASCE/SEI 7 Chap. 7 for the various conditions for snow loading.

Flat Roof Snow Load

ASCE/SEI 7: Sec. 7.3

$$p_f = 0.7 C_e C_t I_s p_g \quad \text{[ASCE/SEI 7 Eq. 7.3-1]}$$

In this equation,

variable	description	units	source
C_e	exposure factor	–	ASCE/SEI 7 Table 7-2
C_t	thermal factor	–	ASCE/SEI 7 Table 7-3
I_s	snow importance factor	–	ASCE/SEI 7 Table 1.5-2
p_f	flat roof snow load	lbf/ft²	ASCE/SEI 7 Eq. 7.3-1
p_g	ground snow load	lbf/ft²	ASCE/SEI 7 Fig. 7-1

Snow load is the weight of accumulated snow on a structure. The sloped roof snow load can be calculated using ASCE/SEI 7 Eq. 7.3-1. The exposure factor accounts for surrounding terrain, the thermal factor accounts for heating and insulation, and the snow importance factor accounts for the structure risk category. The ground snow load is the maximum considered weight of snow on the ground surrounding the structure and depends only on region.

Exam tip: In order to determine the load effects from snow, the flat roof snow load must be calculated first.

Sloped Roof Snow Load

ASCE/SEI 7: Sec. 7.3, Sec. 7.4

$$p_s = C_s p_f \quad \text{[ASCE/SEI 7 Eq. 7.4-1]}$$

Snow load is the weight of accumulated snow on a structure. The sloped roof snow load can be calculated with ASCE/SEI 7 Eq. 7.4-1.

C_s is the roof slope factor, as determined by ASCE/SEI 7 Fig. 7-2. The flat roof snow load, p_f, is determined from ASCE/SEI 7 Eq. 7.3-1.

Exam tip: The sloped roof snow load is a reduced load from the flat roof snow load. The amount that the sloped roof snow load is reduced is dependent on the slope of the roof, along with other factors.

Rain Loads

ASCE/SEI 7: Sec. 8.3

$$R = 5.2(d_s + d_h) \quad \text{[ASCE/SEI 7 Eq. 8.3-1]}$$

In this equation,

R	rain load	lbf/ft^2
d_s	depth of water up to inlet of secondary drainage system	in
d_h	depth of water above inlet of secondary drainage system	in

Rain loads are caused by improper drainage and blockages of primary drainage systems. The design rain load can be computed using ASCE/SEI 7 Eq. 8.3-1. In some cases, the design rain load can be greater than a roof live load or snow load.

In this equation, depths d_s and d_h are in inches, while the design rain load is in lbf/ft^2. The factor of 5.2 converts the head in inches to a usable load in lbf/ft^2 by using principles of fluid mechanics.

Ice Loads

ASCE/SEI 7: Sec. 10.4

$$t_d = 2.0 t I_i f_z (K_{zt})^{0.35} \quad \text{[ASCE/SEI 7 Eq. 10.4-5]}$$

Ice loads are caused by freezing rain, snow, and in-cloud icing. They are dependent on the following variables.

variable	description	units	source
f_z	height factor	–	ASCE/SEI 7 Eq. 10.4-4
I_i	ice importance factor	–	ASCE/SEI 7 Table 1.5-2
K_{zt}	topographic factor	–	ASCE/SEI 7 Sec. 10.4.5
t_d	design ice thickness	in (mm)	ASCE/SEI 7 Fig. 10-2

The design ice load can be computed using ASCE/SEI 7 Eq. 10.4-5.

Snow Drift

ASCE/SEI 7: Sec. 7.7, Sec. 7.8

Snow drifts need to be considered when there are roof areas that are higher and lower than each other, since wind can cause the snow from an adjacent roof to fall onto a lower roof, and the snow from a lower roof can accumulate against an exterior wall or parapet that sits higher than the roof. The amount of additional snow load can make an impact on the design of the members in these areas.

The geometry of the roof and the direction of the wind are the two factors that lead to snow drifts. The two directions of wind that cause snow drifts are windward and leeward. *Windward* snow drifts occur when wind blows snow from a lower-elevation roof toward the wall of an adjacent, higher-elevation roof. *Leeward* snow drifts occur when wind blows snow off a higher-elevation roof down onto an adjacent lower-elevation roof. Snow drifts are calculated in accordance with ASCE/SEI 7 Sec. 7.7 and Sec. 7.8.

Exam tip: Snow drifts need to be considered on all roof projections greater than 15 ft long.

A.8. IMPACT LOADS

Key concepts: These key concepts are important for answering exam questions in subtopic 9.A.8, Impact Loads.

- dynamic load allowance
- special machinery impact loads: elevators
- special machinery impact loads: machinery

NCEES Handbook: The *Handbook* does not contain any material relevant to this subtopic. Be sure to study the listed sections in CERM and from codes and standards.

ANALYSIS OF STRUCTURES: LOADS AND LOAD APPLICATIONS

PE Civil Reference Manual (CERM): Study these sections in CERM that either relate directly to this subtopic or provide background information.

- Section 44.10: Impact Loading
- Section 74.4: Load and Resistance Factor Rating (LRFR) Method

Codes and standards: To prepare for this subtopic, familiarize yourself with these sections from codes and standards.

AASHTO LRFD Bridge Design Specifications (AASHTO LRFD)

- Section 3.6.1.1.2: Multiple Presence of Live Load

International Building Code (IBC)

- Section 1607.7.4.1: Impact and Fatigue
- Section 1607.10: Impact Loads

Minimum Design Loads for Buildings and Other Structures (ASCE/SEI 7)

- Section 4.3: Uniformly Distributed Live Loads
- Section 4.4: Concentrated Live Loads
- Section 4.5: Loads on Handrail, Guardrail, Grab Bar, Vehicle Barrier Systems, and Fixed Ladders
- Section 4.6.2: Elevators
- Section 4.6.3: Machinery

The following equations and concepts are relevant for subtopic 9.A.8, Impact Loads.

Dynamic Load Allowance

CERM: Sec. 74.4

AAHSTO LRFD: Sec. 3.6.1.1.2

Dynamic load allowance is a factor applied to static loads that accounts for the effects of short impact loads. It is not applied to any other load. The following factors are from AASHTO LRFD Sec. 3.6.2.

$I = 1.15$ [fatigue and fracture limit states of all other components]

$I = 1.33$ [all other limit states for all other components]

Impact loads are more difficult to resist, but static loads are easier to analyze. The dynamic load allowance artificially increases the static load to estimate the potential danger of an impact load.

Exam tip: The dynamic load allowance should be used only with design truck loads and design tandem loads.

Special Machinery Impact Loads: Elevators

CERM: Sec. 44.10

ASCE/SEI 7: Sec. 4.3, Sec. 4.4, Sec. 4.5, Sec. 4.6.2

IBC: Sec. 1607.7.4.1, Sec. 1607.10

For structures that support special machinery, such as certain construction equipment, elevators, and cranes, the effects of impact loading must be considered.

According to ASCE/SEI 7 Sec. 4.6.2, elements subject to dynamic loads from elevators should be designed for impact loads and deflection limits prescribed by ASME A17.1.

Special considerations must be taken on structures subject to impact. See IBC Sec. 1607.7.4.1 and Sec. 1607.10 for factors to use when impact loads must be considered in the design. The live loads in ASCE/SEI 7 Sec. 4.3 through Sec. 4.5 include adequate allowance for ordinary impact conditions.

Special Machinery Impact Loads: Machinery

CERM: Sec. 44.10

ASCE/SEI 7: Sec. 4.3, Sec. 4.4, Sec. 4.5, Sec. 4.6.3

IBC: Sec. 1607.7.4.1, Sec. 1607.10

For structures that support special machinery, such as certain construction equipment, elevators, and cranes, the effects of impact loading must be considered.

According to ASCE/SEI 7 Sec. 4.6.3, for light machinery (including shaft- and motor-driven machinery), the weight and moving loads should be increased by 20% to allow for impact. For reciprocating machinery and power-driven units, the weight and moving loads should be increased by 50% to allow for impact. All percentages should be increased where specified by the manufacturer.

Special considerations must be taken on structures subject to impact. See IBC Sec. 1607.7.4.1 and Sec. 1607.10 for factors to use when impact loads must be considered in the design. The live loads in ASCE/SEI 7 Sec. 4.3 through Sec. 4.5 include adequate allowance for ordinary impact conditions.

A.9. EARTH PRESSURE AND SURCHARGE LOADS

Key concepts: These key concepts are important for answering exam questions in subtopic 9.A.9, Earth Pressure and Surcharge Loads.

- at-rest coefficients
- failure surfaces, pressure distribution, and forces
- lateral pressure due to line load surcharges
- lateral pressure due to point load surcharges

- lateral pressure due to strip load surcharges
- lateral pressure due to uniform load surcharges
- surcharge loading

NCEES Handbook: To prepare for this subtopic, familiarize yourself with these sections in the *Handbook*.

- Stress States on a Soil Element Subjected Only to Body Stresses
- At-Rest Coefficients
- Failure Surfaces, Pressure Distribution and Forces
- Lateral Pressure Due to Surcharge Loadings

PE Civil Reference Manual (CERM): Study these sections in CERM that either relate directly to this subtopic or provide background information.

- Section 37.4: Vertical Soil Pressure
- Section 37.5: Active Earth Pressure
- Section 37.6: Passive Earth Pressure
- Section 37.7: At-Rest Soil Pressure
- Section 37.8: Determining Earth Pressure Coefficients
- Section 37.10: Surcharge Loading

Codes and standards: To prepare for this subtopic, familiarize yourself with these sections from codes and standards.

Minimum Design Loads for Buildings and Other Structures (ASCE/SEI 7)

- Table 3.2-1: Design Lateral Soil Load

The following equations, figures, tables, and concepts are relevant for subtopic 9.A.9, Earth Pressure and Surcharge Loads.

Failure Surfaces, Pressure Distribution, and Forces

Handbook: Failure Surfaces, Pressure Distribution and Forces

Active pressure at depth z is

$$p_a = K_a \gamma z$$

Active force within depth z is

$$P_a = \frac{K_a \gamma z^2}{2}$$

Passive pressure at depth z is

$$p_p = K_p \gamma z$$

Passive force within depth z is

$$P_p = \frac{K_p \gamma z^2}{2}$$

CERM: Sec. 37.4, Sec. 37.5, Sec. 37.6

$$p_a = p_v k_a - 2c\sqrt{k_a} \qquad 37.4$$

$$p_p = p_v k_p + 2c\sqrt{k_p} \qquad 37.13$$

ASCE/SEI 7: Table 3.2-1

Vertical soil pressure is caused by the soil's own weight and is calculated in the same manner as a fluid column. There are notable differences between the *Handbook* equations and the CERM equations. CERM Eq. 37.3(a) and (b) refer to depth as H, while the *Handbook* refers to it as z. The pressure equations in CERM include a value for cohesion, c, while the *Handbook* assumes $c = 0$, meaning that the soil is assumed to be granular. In CERM Eq. 37.3, p_v is calculated as follows.

$$p_v = \rho g H \qquad \text{[SI]} \qquad 37.3(a)$$

$$p_v = \gamma H \qquad \text{[U.S.]} \qquad 37.3(b)$$

Handbook figure **Failure Surfaces, Pressure Distribution and Forces** graphically shows pressure distributions and failure surfaces behind a vertical wall. Because wall friction is neglected in the Rankine theory, the triangular resultant pressure acts at $H/3$ from the base and is normal to the wall.

Earth force can be calculated using CERM Eq. 37.10.

$$R_a = \tfrac{1}{2} p_a H \qquad \text{[SI]} \qquad 37.10(a)$$
$$= \tfrac{1}{2} k_a \rho g H^2$$

$$R_a = \tfrac{1}{2} p_a H \qquad \text{[U.S.]} \qquad 37.10(b)$$
$$= \tfrac{1}{2} k_a \gamma H^2$$

The equivalent fluid pressure for soils can be found in ASCE/SEI 7 Table 3.2-1. *Handbook* figure Stress States on a Soil Element Subjected Only to Body Stresses is useful for determining the forces acting on retaining walls needed in retaining wall design, and for determining the forces and pressures acting on soil at depth. See Chap. 3 in this book for entries on soil mechanics that apply to breadth concepts.

Exam tip: For granular soils, the cohesion, c, equals zero.

At-Rest Coefficients

Handbook: At-Rest Coefficients

CERM: Sec. 37.7

For normally consolidated soils,

$$K_o = 1 - \sin\phi'$$

For overconsolidated soils,

$$K_o = (1 - \sin\phi')\text{OCR}^\Omega$$
$$\Omega = \sin\phi'$$

The *Handbook* includes an additional equation for overconsolidated soils that is not included in CERM. *At rest* is defined as describing soil that is completely confined, cannot move, and is appropriate to use next to bridge abutments, basement walls, walls restrained at their tops, walls bearing on rock, and walls with soft-clay backfill, as well as for sand deposits of infinite depth and extent.

The horizontal pressure at rest is related to the coefficient of earth pressure at rest, K_o. CERM Table 37.2 contains typical ranges of earth pressure coefficients, which vary from 0.4 to 0.5 for untamped sand, 0.5 to 0.7 for normally consolidated clays, and 1.0 and up for overconsolidated clays.

Table 37.2 Typical Range of Earth Pressure Coefficients

condition	granular soil	cohesive soil
active	0.20–0.33	0.25–0.5
passive	3–5	2–4
at rest	0.4–0.6	0.4–0.8

Lateral Pressure Due to Point Load Surcharges

Handbook: Lateral Pressure Due to Surcharge Loadings

$$p_h\left(\frac{H^2}{Q_p}\right) = \frac{1.77\overline{m}^2\overline{n}^2}{(\overline{m}^2 + \overline{n}^2)^3} \quad [\text{for } \overline{m} > 0.4]$$

$$p_h\left(\frac{H^2}{Q_p}\right) = \frac{0.28\overline{n}^2}{(0.16 + \overline{n}^2)^3} \quad [\text{for } \overline{m} \leq 0.4]$$

CERM: Sec. 37.10

$$p_q = \frac{1.77 V_q m^2 n^2}{H^2(m^2 + n^2)^3} \quad [m > 0.4] \qquad 37.33$$

$$p_q = \frac{0.28 V_q n^2}{H^2(0.16 + n^2)^3} \quad [m \leq 0.4] \qquad 37.34$$

In these equations and the *Handbook* figure,

H	wall height	ft
Q_p, V_q	point load	lbf
x	horizontal distance to load from inner face of wall	ft
y, z	vertical distance to resultant horizontal force from bottom of wall	ft

CERM Eq. 37.38 and Eq. 37.39 assume elastic soil and a Poisson's ratio of 0.5.

$$m = \frac{x}{H} \qquad 37.38$$

$$n = \frac{y}{H} \qquad 37.39$$

CERM Eq. 37.35, Eq. 37.36, and Eq. 37.37 give the resultant, or total force applied, from the point load surcharge.

$$R_q \approx \frac{0.78 V_q}{H} \quad [m = 0.4] \qquad 37.35$$

$$R_q \approx \frac{0.60 V_q}{H} \quad [m = 0.5] \qquad 37.36$$

$$R_q \approx \frac{0.46 V_q}{H} \quad [m = 0.6] \qquad 37.37$$

For horizontal pressures caused by point load Q_p at different depths and distances from the inner face of the wall, see *Handbook* figure **Lateral Pressure Due to Surcharge Loadings**.

Lateral Pressure Due to Line Load Surcharges

Handbook: Lateral Pressure Due to Surcharge Loadings

$$p_h \left(\frac{H}{Q_l} \right) = \frac{1.28 \overline{m}^2 \overline{n}}{(\overline{m}^2 + \overline{n}^2)^2} \quad [\text{for } \overline{m} > 0.4]$$

$$p_h \left(\frac{H}{Q_l} \right) = \frac{0.20 \overline{n}^2}{(0.16 + \overline{n}^2)^2} \quad [\overline{m} \leq 0.4]$$

CERM: Sec. 37.10

$$p_q = \frac{4 L_q m^2 n}{\pi H (m^2 + n^2)^2} \quad [m > 0.4] \qquad 37.40$$

$$p_q = \frac{0.203 L_q n}{H(0.16 + n^2)^2} \quad [m \leq 0.4] \qquad 37.42$$

In these equations and the *Handbook* figure,

H	wall height	ft
Q_p	point load	lbf
x	horizontal distance to load from inner face of wall	ft
y, z	vertical distance to resultant horizontal force from bottom of wall	ft

CERM Eq. 37.38 and Eq. 37.39 assume elastic soil and a Poisson's ratio of 0.5.

$$m = \frac{x}{H} \qquad 37.38$$

$$n = \frac{y}{H} \qquad 37.39$$

From *Handbook* figure **Lateral Pressure Due to Surcharge Loadings**, the resultants, or total pressure, due to line load surcharges are

$$p_h = \frac{0.64 Q_l}{(\overline{m}^2 + 1)} \quad [\text{for } \overline{m} > 0.4]$$

$$p_h = 0.55 Q_l \quad [\text{for } \overline{m} \leq 0.4]$$

For horizontal pressures caused by the line load (Q_l in the *Handbook* and L_q in CERM) at different depths and distances from the inner face of the wall, see *Handbook* figure **Lateral Pressure Due to Surcharge Loadings**.

Lateral Pressure Due to Strip Load Surcharges

Handbook: Lateral Pressure Due to Surcharge Loadings

$$p_h = \frac{2q}{\pi}(\beta - \sin\beta \cos 2\alpha)$$

In this equation,

q	strip load	plf
α	angle between center of strip and vertical wall face	radians
β	angle between front and back boundaries of strip load	radians

Handbook figure **Lateral Pressure Due to Surcharge Loadings** shows the applied pressure due to a strip load at a given location along the wall, p_h.

Lateral Pressure Due to Uniform Load Surcharges

Handbook: Lateral Pressure Due to Surcharge Loadings

$$p_h(\text{due to } q) = qK$$

CERM: Sec. 37.10

$$p_q = k_a q \qquad 37.31$$

The appropriate coefficient, whether passive, active, or at rest, is indicated by K in the *Handbook* equation and k in the CERM equation. *Handbook* figure **Lateral Pressure Due to Surcharge Loadings** shows the horizontal pressures caused by the uniform surcharge load, q.

The resultant, or total force applied, from the uniform surcharge across the distance, H, is given by CERM Eq. 37.32.

$$R_q = k_a q H \qquad 37.32$$

Surcharge Loading

CERM: Sec. 37.10

A *surcharge* is an additional force applied at the exposed upper surface of the restrained soil. See CERM Fig. 37.3 for an illustrative example.

Figure 37.3 Surcharges

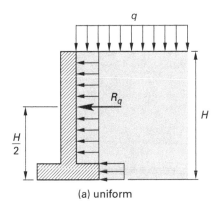

(a) uniform

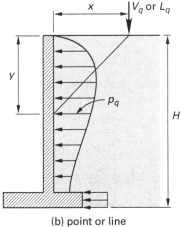

(b) point or line

A surcharge can result from a uniform load, point load, line load, or strip load. Sidewalks, railways, and roadways parallel to a retaining wall are examples of strip surcharges.

With a uniform load surcharge of q, at the surface, there will be an additional active force, R_q, that acts horizontally at $H/2$ above the base. This surcharge resultant is in addition to the backfill active force that acts at $H/3$ above the base.

Exam tip: If a load is applied on the soil that is being retained, a surcharge load needs to be considered.

A.10. LOAD PATHS

Key concepts: These key concepts are important for answering exam questions in subtopic 9.A.10, Load Paths.

- load paths

NCEES Handbook: The *Handbook* does not contain any material relevant to this subtopic. Be sure to study the listed sections from codes and standards.

Codes and standards: To prepare for this subtopic, familiarize yourself with these sections from codes and standards.

Minimum Design Loads for Buildings and Other Structures (ASCE/SEI 7)

- Section 1.3.5: Counteracting Structural Actions
- Section 1.4.1: Load Combinations of Integrity Loads
- Section 1.4.2: Load Path Connections

The following concepts are relevant for subtopic 9.A.10, Load Paths.

Load Path Overview

ASCE/SEI 7: Sec. 1.3.5, Sec. 1.4.1, Sec. 1.4.2

A *load path* is the series of structural elements and connections through which a load is transmitted, starting at the point where the force is applied and ending at the ground. Loads in a building must travel through the building, down to the foundation, and into the ground. ASCE/SEI 7 Sec. 1.3.5 states that a continuous load path must be provided for transmitting forces due to earthquakes and wind to the foundation. A continuous load path through a structure follows the requirements given in ASCE/SEI 7 Sec. 1.4.1. A complete lateral force-resisting system with adequate strength is given in ASCE/SEI 7 Sec. 1.4.2.

An understanding of the correct load path is needed so that connections and members do not bear unexpected loads. Each element and connection in a load path must be designed to resist the portion of the load it must carry. Failing to consider the entire load path is one of the most common and serious errors that inexperienced engineers can make.

Load paths from both vertical forces (e.g., dead, live, and snow loads) and horizontal forces (e.g., wind and seismic loads) must be considered. Gravity load paths are the most common that an engineer will work with from day to day.

Exam tip: When checking a load path, determine all the strength checks that must be completed.

A.11. LOAD COMBINATIONS

Key concepts: These key concepts are important for answering exam questions in subtopic 9.A.11, Load Combinations.

- load combination considerations
- load combinations for concrete, steel, and masonry
- load combinations for timber
- wind and seismic loads

NCEES Handbook: The *Handbook* does not contain any material relevant to this subtopic. Be sure to study the listed sections in CERM and CEST and from codes and standards.

PE Civil Reference Manual (**CERM**): Study these sections in CERM that either relate directly to this subtopic or provide background information.

- Section 50.5: Service Loads, Factored Loads, and Load Combinations
- Section 58.8: Loads

Structural Depth Reference Manual (**CEST**): Study these sections in CEST that either relate directly to this subtopic or provide background information.

- Section 4.1: Plastic Design
- Section 5.2: Load Combinations

Codes and standards: To prepare for this subtopic, familiarize yourself with these sections from codes and standards.

International Building Code (IBC)

- Section 1605: Load Combinations

Minimum Design Loads for Buildings and Other Structures (ASCE/SEI 7)

- Section 2.3: Combining Factored Loads Using Strength Design
- Section 2.4: Combining Nominal Loads Using Allowable Stress Design
- Section 12.8.4: Horizontal Distribution of Forces

National Design Specification for Wood Construction Supplement (NDS Supplement)

- Section 1.4.4: Load Combinations
- Table 2.3.2: Frequently Used Load Duration Factors, C_D

The following equations, tables, and concepts are relevant for subtopic 9.A.11, Load Combinations.

Load Combinations for Concrete, Steel, and Masonry

CERM: Sec. 50.5, Sec. 58.8

CEST: Sec. 4.1

ASCE/SEI 7: Sec. 2.3, Sec. 2.4

IBC: Sec. 1605

ASCE/SEI 7 Chap. 2 gives two sets of combinations, ASD (service loads) and LRFD (ultimate loads). The primary LRFD combinations are given as IBC Eq. 16-1 through Eq. 16-7. Equivalent equations can be found in CEST Sec. 4.1.

$$1.4(D+F) \text{ [IBC Eq. 16-1]}$$

$$1.2(D+F) + 1.6(L+H)$$
$$+ 0.5(L_r \text{ or } S \text{ or } R) \text{ [IBC Eq. 16-2]}$$

$$1.2(D+F) + 1.6(L_r \text{ or } S \text{ or } R)$$
$$+ 1.6H + (f_1 L \text{ or } 0.5W) \text{ [IBC Eq. 16-3]}$$

$$1.2(D+F) + 1.0W + f_1 L$$
$$+ 1.6H + 0.5(L_r \text{ or } S \text{ or } R) \text{ [IBC Eq. 16-4]}$$

$$1.2(D+F) + 1.0E + f_1 L + 1.6H + f_2 S \text{ [IBC Eq. 16-5]}$$

$$0.9D + 1.0W + 1.6H \text{ [IBC Eq. 16-6]}$$

$$0.9(D+F) + 1.0E + 1.6H \text{ [IBC Eq. 16-7]}$$

In IBC Eq. 16-3, Eq. 16-4, and Eq. 16-5, the load factor on L may be reduced by as much as 50% for occupancies in which L_o (minimum uniformly distributed live load) is ≤ 100 lbf/ft^2, excluding garages and places of public assembly.

ASCE/SEI 7 Sec. 2.4 shows the primary ASD combinations. Not all load combinations are given in ASCE/SEI 7 Chap. 2. For example, ASCE/SEI 7 Sec. 12.4 gives several load combinations, and the commentary in App. C suggests a few more load combinations for checking serviceability criteria.

The same load combinations are provided in IBC Sec. 1605. Either ASCE/SEI 7 or the IBC can be used for load combinations.

Exam tip: On the exam, use LRFD for concrete structures, use either ASD or LRFD for steel structures, and use ASD for wood and most masonry structures. Masonry walls with out-of-plane loads may be designed according to LRFD.

Load Combinations for Timber

CEST: 5.2

IBC: Sec. 1605

NDS: Sec. 1.4.4, Table 2.3.2

Examinees will use only the allowable stress design (ASD) method for wood design. In accordance with NDS Sec. 1.4.4, load combinations must be as specified in the applicable building code. The primary ASD combinations are given by IBC Eq. 16-8 through Eq. 16-16. Equivalent equations can be found in CEST Sec. 5.2.

$$D + F \quad \text{[IBC Eq. 16-8]}$$

$$D + H + F + L \quad \text{[IBC Eq. 16-9]}$$

$$D + H + F + (L_r \text{ or } S \text{ or } R) \quad \text{[IBC Eq. 16-10]}$$

$$D + H + F + 0.75(L) + 0.75(L_r \text{ or } S \text{ or } R) \quad \text{[IBC Eq. 16-11]}$$

$$D + H + F + (0.6W \text{ or } 0.7E) \quad \text{[IBC Eq. 16-12]}$$

$$D + H + F + 0.75(0.6W) \\ + 0.75L + 0.75(L_r \text{ or } S \text{ or } R) \quad \text{[IBC Eq. 16-13]}$$

$$D + H + F + 0.75(0.7E) + 0.75L + 0.75S \quad \text{[IBC Eq. 16-14]}$$

$$0.6D + 0.6W + H \quad \text{[IBC Eq. 16-15]}$$

$$0.6(D + F) + 0.7E + H \quad \text{[IBC Eq. 16-16]}$$

Normal load duration is equivalent to applying the maximum allowable load to a member for a period of 10 years. For loads of shorter duration, a member has the capacity to sustain higher loads, and the basic design values are multiplied by the load duration factor, C_D. Values of the load duration factor are given in NDS Table 2.3.2 and summarized in CEST Table 5.3.

Table 5.3 *Load Duration Factors*

design load	load duration	C_D
dead load	permanent	0.9
occupancy live load	10 years	1.0
snow load	2 months	1.15
construction load	7 days	1.25
wind or earthquake load	10 minutes	1.6
impact load	impact	2.0

Wind and Seismic Loads

CERM: Sec. 50.5, Sec. 58.8

CEST: Sec. 4.1

ASCE/SEI 7: Sec. 12.8.4

IBC: Sec. 1605

Wind and seismic loads are often reversible and have multiple subcases that must be considered. See the eccentric loading requirements in ASCE/SEI 7 Sec. 12.8.4. Maximum uplift is often given by LRFD combinations 6 and 7 (IBC Eq. 16-6 and Eq. 16-7).

$$\lambda W = 0.9D + 1.0W + 1.6H \quad \text{[IBC Eq. 16-6]}$$

$$\lambda W = 0.9(D + F) + 1.0E + 1.6H \quad \text{[IBC Eq. 16-7]}$$

Because wind and seismic loads are reversible, they should be considered to act with the same magnitude in either a positive or negative direction.

Load Combination Considerations

CERM: Sec. 58.8

For each element or connection, the most unfavorable effects from any considered combination govern the design. Generally, each combination is considered for each structural element. Engineering judgment may be used to decide, by inspection, which combination will govern the design.

Different load combinations may govern for different members or connections within the same structural system.

Exam tip: It is important to check the different load combinations to obtain the worst-case load effects on a member.

A.12. TRIBUTARY AREAS

Key concepts: These key concepts are important for answering exam questions in subtopic 9.A.12, Tributary Areas.

- influence area
- live load reduction
- tributary area

NCEES Handbook: The *Handbook* does not contain any material relevant to this subtopic. Be sure to study the listed sections from codes and standards.

Codes and standards: To prepare for this subtopic, familiarize yourself with these sections from codes and standards.

International Building Code (IBC)

- Section 1607.12: Roof Loads

Minimum Design Loads for Buildings and Other Structures (ASCE/SEI 7)

- Section 4.7.1: General
- Section 4.8: Reduction in Roof Live Loads
- Section C4.7.1: General

The following concepts are relevant for subtopic 9.A.12, Tributary Areas.

Influence Area

ASCE/SEI 7: Sec. C4.7.1

The *influence area* is the product of the tributary area and the live load element factor. From the ASCE/SEI 7 C4.7.1 commentary, the influence area is the floor area over which the influence surface for structural effects is significantly different from zero. This area is typically the total area supported by all the structural elements supported by the element under consideration.

ASCE/SEI 7 Fig. C4-1 illustrates the relationships between tributary and influence areas for various common structural elements in a framing plan. The ratio of influence area to tributary area is known as the live load element factor and is given in ASCE/SEI 7 Table 4-2.

Live Load Reduction

ASCE/SEI 7: Sec. 4.7.1, Sec. 4.8

IBC: Sec. 1607.12

Live loads may be reduced in some circumstances. The reasoning behind most reductions is that the probability of the full nominal live load being distributed over the entire area decreases as influence area (floor area contributing load to a particular structural element; e.g., a beam or column) increases. For large areas, it is excessively conservative to expect the entire area to be fully loaded.

The reduction for uniform live loads is calculated using ASCE/SEI 7 Eq. 4.7-1. ASCE/SEI 7 Sec. 4.8 contains equations for reduction in roof live loads. IBC Sec. 1607.12 provides reduction factors for roof tributary area and roof slope.

Exam tip: Use the tributary area, not the tributary width, in the live load reduction equation.

Tributary Area

The *tributary area* is the size of the area that influences a structural member. There are no set equations for determining tributary area because it is dependent on the geometry of the structure. The general steps to approximate a tributary area are as follows.

1. Identify the neighboring column in each direction.
2. Measure the distance between the pair.
3. Divide the distance by two.
4. Draw a perpendicular line at this point.
5. Repeat for all neighboring gridlines until a bounding polygon appears.

28 Analysis of Structures: Forces and Load Effects

Content in blue refers to the *NCEES Handbook*.

B.1. Diagrams.. 28-1
B.2. Axial.. 28-6
B.3. Shear 28-9
B.4. Flexure 28-10
B.5. Deflection..28-13
B.6. Special Topics 28-15

The knowledge area of Forces and Load Effects makes up between 9 and 14 questions out of the 80 questions on the PE Civil Structural exam. The organization of this chapter follows the order of subtopics given by NCEES for this knowledge area. Each subtopic is covered in the following sections.

B.1. DIAGRAMS

Key concepts: These key concepts are important for answering exam questions in subtopic 9.B.1, Diagrams.

- beams
- constructing bending moment diagrams
- constructing shear diagrams
- frames
- moment, shear, and deflection diagrams
  - beams fixed at both ends
  - beams fixed at one end
  - beams overhanging one support
  - beams with variable end moments
  - cantilevered beams
  - continuous beams
  - simple beams
- nomenclature in moment, shear, and deflection diagrams
- shear and moment

NCEES Handbook: To prepare for this subtopic, familiarize yourself with these sections in the *Handbook*.

- Systems of Forces
- Shearing Force and Bending Moment Sign Conventions
- Moment, Shear, and Deflection Diagrams

PE Civil Reference Manual **(CERM):** Study these sections in CERM that either relate directly to this subtopic or provide background information.

- Section 41.2: Internal and External Forces
- Section 44.11: Shear and Moment
- Section 44.12: Shear and Bending Moment Diagrams

Codes and standards: To prepare for this subtopic, familiarize yourself with these sections from codes and standards.

Steel Construction Manual (AISC)

- Part 3: Design of Flexural Members

The following equations, figures, tables, and concepts are relevant for subtopic 9.B.1, Diagrams.

Constructing Shear Diagrams

Handbook: Systems of Forces; Moment, Shear, and Deflection Diagrams

CERM: Sec. 41.2, Sec. 44.12

AISC: Part 3

The following guidelines and conventions should be observed when constructing a shear diagram.

- The shear at any point is equal to the sum of the loads and reaction from the point to either end.
- The magnitude of the shear at any point is equal to the slope of the moment diagram at that point. The magnitude of shear can be calculated using CERM Eq. 44.26.

$$V = \frac{dM}{dx} \qquad 44.26$$

- Loads and reactions acting upward are positive.
- The shear diagram is straight and sloping for uniformly distributed loads.
- The shear diagram is straight and horizontal between concentrated loads.
- The shear is a vertical line and is undefined at points of concentrated loads.

A beam is statically determinate if the number of equilibrium equations equals the number of unknowns. If this is not the case, the structure may be either unstable or indeterminate.

The equilibrium requirements can be found in *Handbook* section Systems of Forces.

$$\sum F_n = 0$$
$$\sum M_n = 0$$

For determining external reactions, use statics or a design reference such as in *Handbook* section Moment, Shear, and Deflection Diagrams or AISC Table 3-23. Once the external beam reactions are determined, cut the free-body diagram sections to determine the internal forces in the beam to finish creating the diagrams.

Exam tip: Using the applicable diagrams provided in *Handbook* section Moment, Shear, and Deflection Diagrams instead of drawing the diagrams from scratch will increase your speed during the exam.

Constructing Bending Moment Diagrams

Handbook: Systems of Forces; Shearing Force and Bending Moment Sign Conventions; Moment, Shear, and Deflection Diagrams

CERM: Sec. 44.12

AISC: Part 3

Observe the following guidelines and conventions when constructing a bending moment diagram.

- By convention, the moment diagram is drawn on the compression side of the beam.
- The moment at any point is equal to the sum of the moments and couples from the point to the left end.
- Clockwise moments about the point are positive.
- The magnitude of the moment at any point is equal to the area under the shear line up to that point.

This is equivalent to the integral of the shear function, as shown in CERM Eq. 44.27.

$$M = \int V dx \qquad 44.27$$

- The maximum moment occurs where the shear is zero.
- The moment diagram is straight and sloping between concentrated loads.
- The moment diagram is curved over uniformly distributed loads.

A beam is statically determinate if the number of equilibrium equations equals the number of unknowns.

The equilibrium requirements can be found in *Handbook* section Systems of Forces.

$$\sum F_n = 0$$
$$\sum M_n = 0$$

For determining external reactions, use statics or a design reference such as in *Handbook* section Moment, Shear, and Deflection Diagrams or AISC Table 3-23. Once the external beam reactions are determined, cut free-body diagram sections to determine the internal forces in the beam to finish creating the diagrams.

Exam tip: Using the applicable diagrams provided in *Handbook* section Moment, Shear, and Deflection Diagrams instead of drawing the diagrams from scratch will increase your speed during the exam. Moments at hinges are zero.

Nomenclature in Moment, Shear, and Deflection Diagrams

Handbook: Moment, Shear, and Deflection Diagrams

CERM: Chap. 44

The following nomenclature applies to all moment, shear, and deflection diagrams found in the *Handbook*. CERM nomenclature is included in parentheses where applicable.

l	total length of beam between reaction points	in
L (L)	total length of beam between reaction points	ft (in, m)
$M_{\max}$	maximum moment	in-kips
M_x	moment at distance x from end of beam	in-kips
M_1 (M)	maximum moment in left section of beam	in-kips (ft-lbf, N·m)
M_2	maximum moment in right section of beam	in-kips
M_3	maximum positive moment in beam with combined end moment conditions	in-kips
P (P)	concentrated load	kips (lbf, N)
P_1	concentrated load nearest left reaction	kips
P_2	concentrated load nearest right reaction, and of different magnitude than P_1	kips
R (R)	end beam reaction for any condition of symmetrical loading	kips (lbf, N)
R_1	left end beam reaction	kips
R_2	right end or intermediate beam reaction	kips
R_3	right end beam reaction	kips
V (V)	maximum vertical shear for any condition of symmetrical loading	kips (lbf, N)
V_x	vertical shear at distance x from end of beam	kips
V_1	maximum vertical shear in left section of beam	kips
V_2	vertical shear at right reaction point, or to left of intermediate reaction point of beam	kips
V_3	vertical shear at right reaction point, or to right of intermediate reaction point of beam	kips
w (w)	uniformly distributed load per unit of length	kip/in (lbf/in, N/m)
w_1	uniformly distributed load per unit of length nearest left reaction	kip/in
w_2	uniformly distributed load per unit of length nearest right reaction and of different magnitude than w_1	kip/in
W	total load on beam	kips
x (x)	any distance measured along beam from left reaction	in (in, m)
x_1	any distance measured along overhang section of beam from nearest reaction point	in
Δ_a (y)	deflection at point of load	in (in, m)
$\Delta_{\max}$ (x)	maximum deflection	in (in, m)
Δ_x	deflection at any point x distance from left reaction	in
Δ_{x1}	deflection of overhang section of beam at any distance from nearest reaction point	in

Moment, Shear, and Deflection Diagrams

Handbook: Moment, Shear, and Deflection Diagrams

CERM: Sec. 44.12

AISC: Part 3

The tables in *Handbook* section Moment, Shear, and Deflection Diagrams include shear, moment, and deflection diagrams for various support conditions and loading characteristics. See the entries "Constructing Shear Diagrams" and "Constructing Bending Moment Diagrams" in this chapter for the guidelines and conventions that should be observed when constructing a shear or bending moment diagram. See also CERM Fig. 44.9.

Simple Beams

The following *Handbook* tables include diagrams for simple beams with uniform loads.

- Shears, Moments, and Deflections: Simple Beam—Uniformly Distributed Load
- Shears, Moments, and Deflections: Simple Beam—Uniform Load Partially Distributed
- Shears, Moments, and Deflections: Simple Beam—Uniform Load Partially Distributed at One End
- Shears, Moments, and Deflections: Simple Beam—Uniform Load Partially Distributed at Each End

The following *Handbook* tables include diagrams for simple beams with loads increasing uniformly.

- Shears, Moments, and Deflections: Simple Beam—Load Increasing Uniformly to One End
- Shears, Moments, and Deflections: Simple Beam—Load Increasing Uniformly to Center
- Shears, Moments, and Deflections: Simple Beam—Load Increasing Uniformly from Center

The following *Handbook* tables include diagrams for simple beams with concentrated loads.

- Shears, Moments, and Deflections: Simple Beam—Concentrated Load at Center

- Shears, Moments, and Deflections: Simple Beam—Concentrated Load at Any Point
- Shears, Moments, and Deflections: Simple Beam—Two Equal Concentrated Loads Symmetrically Placed
- Shears, Moments, and Deflections: Simple Beam—Two Equal Concentrated Loads Unsymmetrically Placed
- Shears, Moments, and Deflections: Simple Beam—Two Unequal Concentrated Loads Unsymmetrically Placed
- Shears, Moments, and Deflections: Simple Beam—Concentrated Moment at End
- Shears, Moments, and Deflections: Simple Beam—Concentrated Moment at Any Point
- Shears, Moments, and Deflections: Simple Beam—One Concentrated Moving Load
- Shears, Moments, and Deflections: Simple Beam—Two Equal Concentrated Moving Loads
- Shears, Moments, and Deflections: Simple Beam—Two Unequal Concentrated Moving Loads
- Shears, Moments, and Deflections: General Rules for Simple Beams Carrying Moving Concentrated Loads

Beams Fixed at One End

The following *Handbook* tables include diagrams for beams fixed at one end.

- Shears, Moments, and Deflections: Beam Fixed at One End, Supported at Other—Uniformly Distributed Load
- Shears, Moments, and Deflections: Beam Fixed at One End, Supported at Other—Concentrated Load at Center
- Shears, Moments, and Deflections: Beam Fixed at One End, Supported at Other—Concentrated Load at Any Point
- Shears, Moments, and Deflections: Beam Fixed at One End, Free to Deflect Vertically But Not Rotate at Other—Uniformly Distributed
- Shears, Moments, and Deflections: Beam Fixed at One End, Free to Deflect Vertically But Not Rotate at Other—Concentrated Load at Deflected End

Beams Fixed at Both Ends

The following *Handbook* tables include diagrams for beams fixed at both ends.

- Shears, Moments, and Deflections: Beam Fixed at Both Ends—Uniformly Distributed Loads
- Shears, Moments, and Deflections: Beam Fixed at Both Ends—Concentrated Load at Center
- Shears, Moments, and Deflections: Beam Fixed at Both Ends—Concentrated Load at Any Point

Cantilevered Beams

The following *Handbook* tables include diagrams for cantilevered beams.

- Shears, Moments, and Deflections: Cantilevered Beam—Load Increasing Uniformly to Fixed End
- Shears, Moments, and Deflections: Cantilevered Beam—Uniformly Distributed Load
- Shears, Moments, and Deflections: Cantilevered Beam—Concentrated Load at Any Point
- Shears, Moments, and Deflections: Cantilevered Beam—Concentrated Load at Free End

Beams Overhanging One Support

The following *Handbook* tables include diagrams for beams overhanging one support.

- Shears, Moments, and Deflections: Beam Overhanging One Support—Uniformly Distributed Load
- Shears, Moments, and Deflections: Beam Overhanging One Support—Uniformly Distributed Load on Overhang
- Shears, Moments, and Deflections: Beam Overhanging One Support—Concentrated Load at End of Overhang
- Shears, Moments, and Deflections: Beam Overhanging One Support—Uniformly Distributed Load Between Supports
- Shears, Moments, and Deflections: Beam Overhanging One Support—Concentrated Load at Any Point Between Supports

Continuous Beams

The following *Handbook* tables include diagrams for continuous beams.

- Shears, Moments, and Deflections: Continuous Beam—Two Equal Spans—Uniform Load on One Span
- Shears, Moments, and Deflections: Continuous Beam—Two Equal Spans—Concentrated Load at Center of One Span
- Shears, Moments, and Deflections: Continuous Beam—Two Equal Spans—Concentrated Load at Any Point
- Shears, Moments, and Deflections: Continuous Beam—Three Equal Spans—One End Span Unloaded

- Shears, Moments, and Deflections: Continuous Beam—Three Equal Spans—End Spans Loaded
- Shears, Moments, and Deflections: Continuous Beam—Three Equal Spans—All Spans Loaded
- Shears, Moments, and Deflections: Continuous Beam—Four Equal Spans—Third Span Unloaded
- Shears, Moments, and Deflections: Continuous Beam—Four Equal Spans—First and Third Spans Unloaded
- Shears, Moments, and Deflections: Continuous Beam—Four Equal Spans—All Spans Loaded

Beams with Variable End Moments

The following *Handbook* tables include diagrams for beams with variable end moments.

- Shears, Moments, and Deflections: Beam—Uniformly Distributed Load and Variable End Moments
- Shears, Moments, and Deflections: Beam—Concentrated Load at Center and Variable End Moments

Beams

Handbook: Moment, Shear, and Deflection Diagrams

AISC: Part 3

Boundary conditions are important to understand when reading or creating shear and moment diagrams. Study reactions associated with boundary conditions and what symbol is used to represent these conditions.

The diagrams are read from left to right to determine the shears and moments in the beam. The shear diagram represents the slope of the beam loading, and the moment diagram represents the slope of the shear diagram. The area under the shear diagram represents the value of the moment.

To determine the magnitude of shear on a linear diagram at a location not labeled on a given diagram, create the equation of the sloped line in the slope intercept form and then solve for the slope. To find the corresponding moment, find the area under the curve to the location associated. If the loading condition matches a condition given in *Handbook* section Moment, Shear, and Deflection Diagrams or AISC Table 3-23, use the associated equations to determine the values.

For concentrated loads, there is no change in slope to the shear diagram. For distributed loads, there is a linear change in slope to the shear diagram. For varying distributed loads, there is a parabolic change in slope to the shear diagram.

Exam tip: When the shear is zero, the moment is maximum. When the change of slope is linearly decreasing, the moment diagram is concave. It is important to keep track of sign convention to determine the shape of the diagrams.

Shear and Moment

CERM: Sec. 44.11

Shear at a point is the sum of all vertical forces acting on an object. A typical application is shear at a point on a beam, V, defined as the sum of all vertical forces between the point and one of the ends.

Shear load acts transverse to the member cross section. The shear at a point on a beam can be calculated using CERM Eq. 44.24.

$$V = \sum_{\substack{\text{point to}\\ \text{one end}}} F_i \qquad 44.24$$

Shear is taken as positive when there is a net upward force to the left of a point and negative when there is a net downward force between the point and the left end of the beam.

Moment at a point is the total bending moment acting on an object. In the case of a beam, the moment, M, will be the algebraic sum of all moments and couples located between the investigation point and one of the beam ends. A beam moment can be calculated using CERM Eq. 44.25.

$$M = \sum_{\substack{\text{point to}\\ \text{one end}}} F_i d_i + \sum_{\substack{\text{point to}\\ \text{one end}}} C_i \qquad 44.25$$

Moment is taken as positive when the upper surface of the beam is in compression and the lower surface is in tension.

Frames

Shear and moment diagrams for frames are very similar to beams. Think of frames as multiple beams attached together. Use statics to find the external reactions of the frame. Once the external reactions have been determined, draw free-body diagrams of each frame member to determine the internal forces.

Frames are different from beams because they include vertical members as well as horizontal members. It is important to note that there could be axial loads on the vertical members. Sign convention is important for determining whether the vertical members are in tension or compression. Vertical members may have zero shear and moments.

B.2. AXIAL

Key concepts: These key concepts are important for answering exam questions in subtopic 9.B.2, Axial.

- axial, concentric, and eccentric loading
- axial deformation
- axial load in truss members
- design of axial load
  - concrete members
  - masonry members
  - steel members
  - wood members
- effective length factor
- Euler buckling

NCEES Handbook: To prepare for this subtopic, familiarize yourself with these sections in the *Handbook*.

- Uniaxial Loading and Deformation
- Columns

PE Civil Reference Manual (CERM): Study these sections in CERM that either relate directly to this subtopic or provide background information.

- Section 41.33: Zero-Force Members
- Section 41.34: Method of Joints
- Section 41.36: Method of Sections
- Section 44.1: Basic Concepts
- Section 44.2: Hooke's Law
- Section 44.3: Elastic Deformation
- Section 44.14: Bending Stress in Beams
- Section 44.16: Eccentric Loading of Axial Members
- Section 45.3: Slender Columns
- Section 46.4: Review of Elastic Deformation
- Chapter 52: Reinforced Concrete: Short Columns
- Chapter 53: Reinforced Concrete: Long Columns
- Chapter 60: Structural Steel: Tension Members
- Chapter 61: Structural Steel: Compression Members

Structural Depth Reference Manual (CEST): Study these sections in CEST that either relate directly to this subtopic or provide background information.

- Section 5.8: Design for Compression
- Section 5.9: Design for Tension
- Section 6.4: Design of Masonry Columns

Codes and standards: To prepare for this subtopic, familiarize yourself with these sections from codes and standards.

Building Code Requirements for Structural Concrete and Commentary (ACI 318)

- Section 6.2: General
- Section 6.7: Elastic Second-Order Analysis
- Section 22.4: Axial Strength or Combined Flexural and Axial Strength

Steel Construction Manual (AISC)

- Part 1: Dimensions and Properties
- Part 16 Chapter E: Design of Members for Compression
- Part 17: Miscellaneous Data and Mathematical Information
- Appendix 7: Alternative Methods of Design for Stability

Specification for Structural Steel Buildings (AISC 360) is included in Part 16 of the *Steel Construction Manual*.

The following equations, figures, and concepts are relevant for subtopic 9.B.2, Axial.

Axial, Concentric, and Eccentric Loading

Handbook: Uniaxial Loading and Deformation

$$\sigma = P/A$$

CERM: Sec. 44.14, Sec. 44.16

$$\sigma_{\text{max,min}} = \frac{F}{A} \pm \frac{Mc}{I_c} \quad \quad 44.42$$

$$\sigma_{\text{max,min}} = \frac{F}{A} \pm \frac{Fec}{I_c} \quad \quad 44.43$$

The equation for uniform axial stress on column cross sections is found in *Handbook* section Uniaxial Loading and Deformation.

The *Handbook* equation uses P for axial load, while CERM uses F. The variable e in CERM Eq. 44.43 represents the eccentricity of the load offset from the centroid. The axial load is in the longitudinal direction of the member.

If the load is not concentric, both the axial stress and bending stress are normal stresses oriented in the same direction, so simple addition can be used to combine them. Combined stress theory is not applicable. The axial stress can be calculated using CERM Eq. 44.42 and Eq. 44.43.

By convention, the axial load is negative if the force compresses the member, as shown in CERM Fig. 44.13.

Figure 44.13 Bending Stress Distribution in a Beam

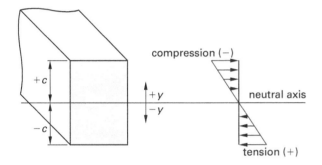

The actual length of a member under loading is given by CERM Eq. 44.5. The algebraic sign of the deformation must be observed.

$$L = L_o + \delta \qquad 44.5$$

The variables P and F are used interchangeably to stand for the loading. Thermal deformation is often lumped with axial deformation because changes in thermal loading cause axial forces and deformations. AISC Table 17-11 gives coefficient of thermal expansion values for common materials.

When an object with initial length L_o and coefficient of thermal expansion α experiences a temperature change of ΔT degrees, the deformation can be calculated using CERM Eq. 46.5.

$$\delta = \alpha L_o \Delta T \qquad 46.5$$

Exam tip: Hooke's law is also referred to as $\varepsilon = \delta/L$.

Axial Deformation

Handbook: Uniaxial Loading and Deformation

$$\varepsilon = \delta / L$$

CERM: Sec. 44.1, Sec. 44.2, Sec. 44.3, Sec. 46.4

$$\delta = L_o \epsilon = \frac{L_o \sigma}{E} = \frac{L_o F}{EA} \qquad 44.4$$

AISC: Part 17

Since stress is the load over the area, and the strain is the deflection over the initial length, Hooke's law can be rearranged to give the elongation of an axially loaded member with a uniform cross section experiencing normal stress, as shown in *Handbook* section Uniaxial Loading and Deformation and CERM Eq. 44.4. Tension loading is considered positive; compressive loading is negative.

In these equations,

A	cross-sectional area	in^2, m^2
F, P	loading	lbf, N
σ	stress on the cross section	lbf/in^2, MPa

Effective Length Factor

Handbook: Columns

CERM: Sec. 45.3, Sec. 53.3, Sec. 61.3

ACI 318: Sec. 6.7

AISC: Part 1, App. 7

The effective length is the product of the effective length factor and the length of the member. Effective length factor values for common boundary conditions are found in *Handbook* section Columns, as well as CERM Table 45.1 and CERM Table 61.1. Alternatively, the effective length factor can be determined using one of the following equations in conjunction with alignment charts (CERM Eq. 53.3 for concrete and CERM Eq. 61.4 for steel).

$$\Psi = \frac{\sum_{\text{columns}} \dfrac{EI}{l_c}}{\sum_{\text{beams}} \dfrac{EI}{l}} \qquad 53.3$$

$$G = \frac{\sum \left(\dfrac{I}{L}\right)_c}{\sum \left(\dfrac{I}{L}\right)_b} \qquad 61.4$$

Depending on the material, the equations will be depicted differently, but they are the same. For concrete design, the moment of inertia is permitted to be taken as

$$\text{modulus of elasticity of concrete} = 57{,}000\sqrt{f'_c}$$
$$\text{area} = A_g$$
$$\text{moments of inertia of columns} = 0.70 I_g$$
$$\text{moments of inertia of beams} = 0.35 I_g$$

For steel design, use the moment of inertia listed in AISC Table 1-1 through Table 1-20 for the given shape. If a built-up section is given, find the moment of inertia using the parallel axis theorem.

Exam tip: For ground-level columns, one of the column ends will not be framed to beams or other columns. In that case, G is 10 (theoretically, G is infinity) for pinned ends, and G is 1 (theoretically, G is zero) for rigid footing connections.

Euler Buckling

Handbook: Columns

$$P_{cr} = \frac{\pi^2 EI}{(Kl)^2}$$

CERM: Sec. 45.3, Sec. 61.2

$$P_e = \frac{\pi^2 EI}{L^2} = \frac{\pi^2 EA}{\left(\frac{L}{r}\right)^2} \quad 45.7$$

AISC: Part 16 Chap. E

The Euler load is the theoretical maximum load that an initially straight column can support without buckling. For columns with frictionless or pinned ends, this load is given by the *Handbook* equation, which is the same as CERM Eq. 45.7, with some minor differences in presentation. P_{cr} is the Euler buckling load. For steel design specifically, the stress is calculated and different equations are used for the design based on the calculated stress. The corresponding column stress is given by CERM Eq. 45.8. In order to use Euler's theory, this stress cannot exceed half of the compressive yield strength of the column material.

$$F_e = \frac{P_e}{A} = \frac{\pi^2 E}{\left(\frac{L}{r}\right)^2} \quad [\sigma_e < \tfrac{1}{2} S_y] \quad 45.8$$

The *Handbook* equation includes the effective length term, Kl, while the CERM equations denote only the length, L. CERM breaks out the equations for pin-pin where K is 1. For columns that do not have frictionless or pinned ends, use CERM Eq. 45.9.

$$L' = KL \quad 45.9$$

See *Handbook* section Columns for information on determining the value of the effective length factor, K. Euler buckling stresses are compared to varying limits (depending on material type) to determine if elements are slender or nonslender. This comparison will control how the element is designed.

Exam tip: Flexural buckling and Euler buckling are often used interchangeably.

Design of Axial Load in Concrete Members

Handbook: Columns

CERM: Chap. 52, Chap. 53

ACI 318: Sec. 6.2.5, Sec. 22.4

If the member is part of a braced structure, use CERM Eq. 52.1 to determine if a column is a short column.

$$\frac{k_b l_u}{r} \le 34 - 12\left(\frac{M_1}{M_2}\right) \le 40 \quad 52.1$$

If the member is part of an unbraced structure, use CERM Eq. 52.2 to determine if a column is a short column.

$$\frac{k_u l_u}{r} \le 22 \quad [k_u > 1] \quad 52.2$$

The ratio of M_1 to M_2 cannot be less than –0.5. In CERM Eq. 52.1, it is always conservative to take k_b as 1. Refer to ACI 318 Sec. 6.2.5 for more information on columns.

The radius of gyration is determined by the equation in *Handbook* section Columns.

$$r = \sqrt{\frac{I}{A}}$$

For a rectangular section, it can be simplified to

$$r = \frac{t}{\sqrt{12}}$$

If these equations are not satisfied, then the column is considered a long column.

See ACI 318 Sec. 22.4 for the design of concrete columns for axial load. Refer to Chap. 29 in this book for material properties and design components of concrete columns.

Exam tip: For circular columns, r is $d/4$. For rectangular columns, r is $0.289h$, but ACI 318 Sec. 6.2.5.1 permits r to be approximated $0.30h$.

Axial Load in Truss Members

CERM: Sec. 41.33, Sec. 41.34, Sec. 41.36

CEST: Sec. 5.8, Sec. 5.9

The method of joints can be used to find the internal forces in each truss member. This method is useful when most or all of the truss member forces are to be calculated.

The equations of equilibrium will be directly applied to each joint. Start with a joint where the reactions are known to minimize the unknowns in the equilibrium equations. Move from joint to joint, cutting free-body diagrams and using equilibrium to solve for the internal forces in each member.

The method of sections is a direct approach to finding forces in any truss member. This method is convenient when only a few truss member forces are unknown. Once the reactions are known, continue by cutting a section through the members of interest. Then use statics to determine the forces in the members.

Exam tip: Check for zero-force members prior to solving for forces. (See CERM Sec. 41.33.)

Design of Axial Load in Steel Members

CERM: Chap. 60, Chap. 61

AISC: Part 16 Chap. E

For steel columns, the slenderness is dependent on the makeup of the column and whether the flanges and webs are slender or nonslender. See AISC Part 16 Chap. E for design of steel compression members. See CERM Chap. 60 and Chap. 61 for the design of axial load in structural steel members.

Refer to Chap. 30 in this book for material properties and design components of steel columns.

Design of Axial Load in Wood Members

CEST: Sec. 5.8, Sec. 5.9

Refer to Chap. 31 in this book for material properties and design components of wood columns.

Design of Axial Load in Masonry Members

CEST: Sec. 6.4

Refer to Chap. 32 in this book for material properties and design components of masonry columns.

B.3. SHEAR

Key concepts: These key concepts are important for answering exam questions in subtopic 9.B.3, Shear.

- average shear stress
- design of masonry for shear
- design of reinforced concrete for shear
- design of wood for shear
- shear and moment diagrams
- shear modulus
- shear stress-strain relationship

NCEES Handbook: To prepare for this subtopic, familiarize yourself with these sections in the *Handbook*.

- Shear Stress-Strain

PE Civil Reference Manual (CERM): Study these sections in CERM that either relate directly to this subtopic or provide background information.

- Section 44.12: Shear and Bending Moment Diagrams
- Section 44.13: Shear Stress in Beams
- Section 45.16: Circular Shaft Design
- Chapter 50: Reinforced Concrete: Beams
- Section 85.23: Stress Measurements in Known Directions

Structural Depth Reference Manual (**CEST**): Study these sections in CEST that either relate directly to this subtopic or provide background information.

- Section 5.10: Design for Shear
- Section 6.3: Design for Shear

The following equations and concepts are relevant for subtopic 9.B.3, Shear.

Average Shear Stress

Handbook: Shear Stress-Strain

$$\gamma = \frac{\tau}{G}$$

CERM: Sec. 44.13

$$\tau = \frac{V}{A} \qquad 44.28$$

Average shear stress can be calculated using the shear stress equation in *Handbook* section Shear Stress-Strain or CERM Eq. 44.28. In CERM, V is the shear load and A is the cross-sectional area.

Shear load is a force that acts parallel to the cross section. The maximum shear stress a material can support without yielding is considered the shear strength.

Shear Stress-Strain Relationship

Handbook: Shear Stress-Strain

$$\gamma = \frac{\tau}{G}$$

CERM: Sec. 85.23

$$\tau = G\gamma = 2G\epsilon = \frac{2G\epsilon_t}{\text{BC}} \qquad 85.35$$

Similar to the relationship between stress and strain (Hooke's law) for axial loading, the relationship shown is Hooke's law for shear stress and shear strain. The shear modulus of the material relates the two, as does the modulus of elasticity with axial stresses.

Shear Modulus

Handbook: Shear Stress-Strain

CERM: Sec. 45.16

$$G = \frac{E}{2(1+v)}$$

The shear modulus is only valid for isotropic, homogeneous materials such as steel and aluminum. Poisson's ratio will vary depending on the material. E is the modulus of elasticity (also referred to as Young's modulus). Poisson's ratio, v, can be represented by the equation in *Handbook* section Shear Stress-Strain.

$$v = -\frac{\text{(lateral strain)}}{\text{(longitudinal strain)}}$$

Shear and Moment Diagrams

Refer to subtopic B.1, Diagrams, in this chapter for information on shear and moment diagrams.

Design of Reinforced Concrete for Shear

CERM: Chap. 50

Refer to Chap. 29 in this book for material properties and design components of concrete columns.

Design of Wood for Shear

CEST: Sec. 5.10

Refer to Chap. 31 in this book for material properties and design components of wood.

Design of Masonry for Shear

CEST: Sec. 6.3

Refer to Chap. 32 in this book for material properties and design components of masonry.

B.4. FLEXURE

Key concepts: These key concepts are important for answering exam questions in subtopic 9.B.4, Flexure.

- design of concrete beams for flexure
- design of wood for flexure
- plastic section modulus and moment
- stresses in beams
- transformed section properties

NCEES Handbook: To prepare for this subtopic, familiarize yourself with these sections in the *Handbook*.

- Stresses in Beams
- Composite Sections

PE Civil Reference Manual (**CERM**): Study these sections in CERM that either relate directly to this subtopic or provide background information.

- Section 44.14: Bending Stress in Beams
- Section 44.16: Eccentric Loading of Axial Members

- Chapter 50: Reinforced Concrete: Beams
- Chapter 57: Composite Concrete and Steel Bridge Girders
- Chapter 59: Structural Steel: Beams
- Chapter 63: Structural Steel: Built-Up Sections
- Section 64.4: Available Flexural Strength
- Chapter 68: Masonry Walls

Structural Depth Reference Manual (CEST): Study these sections in CEST that either relate directly to this subtopic or provide background information.

- Section 4.1: Plastic Design
- Section 4.4: Composite Beams
- Section 5.7: Design for Flexure
- Section 6.2: Design for Flexure
- Section 6.6: Wall Design for Out-of-Plane Loads

Codes and standards: To prepare for this subtopic, familiarize yourself with these sections from codes and standards.

Building Code Requirements and Specification for Masonry Structures (TMS 402/602)

- Section 6.2.4: Effective Embedment Length for Headed Anchor Bolts
- Section 6.2.5: Effective Embedment Length of Bent-Bar Anchor Bolts

National Design Specification for Wood Construction with Commentary (NDS)

- Part 3: Design Provisions and Equations

Steel Construction Manual (AISC)

- Part 3: Design of Flexural Members
- Part 16 Chapter F: Design of Members for Flexure

Specification for Structural Steel Buildings (AISC 360) is included in Part 16 of the *Steel Construction Manual*.

The following equations, figures, and concepts are relevant for subtopic 9.B.4, Flexure.

Stresses in Beams

Handbook: Stresses in Beams

$$\sigma = \frac{-My}{I}$$

$$\sigma = \frac{\pm Mc}{I}$$

CERM: Sec. 44.14, Sec. 44.16

Bending stress varies with location (depth) within a beam. It is zero at the neutral axis and increases linearly with distance from the neutral axis, as given by the equation for normal stress in *Handbook* section Stresses in Beams and CERM Eq. 44.36. The negative sign in the equation, required by the convention that compression is negative, is commonly omitted. The equations for maximum normal stress in *Handbook* section Stresses in Beams and CERM Eq. 44.37 show that the maximum bending stress occurs where the moment along the length of the beam is maximum.

For a rectangular or square cross section, the neutral axis will be half the depth. Both the axial and bending stress are normal stresses oriented in the same direction, so simple addition can be used to combine them. Combined stress theory is not applicable. By convention, the load is negative if the force compresses the member. CERM Fig. 44.14, CERM Eq. 44.42, and CERM Eq. 44.43 show this.

Figure 44.14 Eccentric Loading of an Axial Member

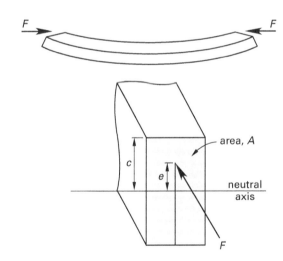

$$\sigma_{\text{max,min}} = \frac{F}{A} \pm \frac{Mc}{I_c} \qquad 44.42$$

$$\sigma_{\text{max,min}} = \frac{F}{A} \pm \frac{Fec}{I_c} \qquad 44.43$$

Tension exists when the Mc/I_c term in CERM Eq. 44.42 is larger than the F/A term. From CERM Eq. 44.43, the total stress is the sum of the direct axial tension and the bending stress.

Exam tip: Use the form of the equation that suits the variable given in the exam question.

Transformed Section Properties

Handbook: Composite Sections

$$\sigma_1 = \frac{-nMy}{I_T}$$

$$\sigma_2 = \frac{-My}{I_T}$$

CEST: Sec. 4.4

The bending stress in a beam of two dissimilar materials where the elastic modulus of material 1 is greater than the elastic modulus of material 2 is given by the *Handbook* equations.

The two materials can then be treated as a composite section composed of a single material separated by a modular ratio, n.

$$E_1 A = E_2 n A$$

The bending stress equations for the transformed section can be used after the transformed section centroid and moment of inertia are found.

Plastic Section Modulus and Moment

CERM: Sec. 59.13

CEST: Sec. 4.1

AISC: Part 3, Part 16 Chap. F

An increasing applied bending moment on a steel beam eventually causes the extreme fibers to reach the yield stress where they are no longer in the elastic range and have become plastic. The plastic moment is given by AISC Eq. F2-1.

$$M_p = F_y Z_x \quad \text{[AISC Eq. F2-1]}$$

As the moment on the section continues to increase, the yielding at the extreme fibers progresses toward the equal area axis until the whole of the section finally yields, as shown in CEST Fig. 4.2.

Figure 4.2 Shape Factor Determination

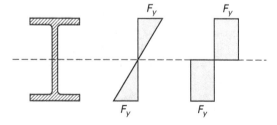

For compact sections with adequate lateral support, the nominal flexural strength is equal to the plastic moment strength. The M_p and M_n values differ by approximately 1.6, corresponding to the ratio Z/S, as shown in CERM Eq. 59.22.

$$M_p = F_y Z \approx 1.6 F_y S \qquad 59.22$$

For built-up shapes, the plastic section modulus equation can be found in AISC Part 16 Chap. F.

$$Z = \sum A_i D_i$$

The particular subsection is represented by i, and D is the distance from the centroid of subsection i to the plastic neutral axis (PNA) of the whole section. For the section to be in equilibrium, the tension force must be equal to the compression force, so the area above the PNA is equal to the area below the PNA, and the PNA location can be determined. Once the PNA location is determined, the plastic section modulus can be found. If the yield strength of the pieces within the built-up section varies, then the tension being equal to the compression is not as simple as the area above being equal to the area below. F_y will need to be accounted for.

Refer to Chap. 30 in this book for material properties and design components of steel.

Exam tip: Plastic design is applicable to steel design only.

Design of Concrete Beams for Flexure

CERM: Chap. 50

Refer to Chap. 29 in this book for concrete beam design components.

Design of Wood for Flexure

CEST: Sec. 5.7

NDS: Part 3

See Chap. 31 in this book for material properties and design components of wood.

Design of Masonry for Flexure

CERM: Chap. 68

CEST: Sec. 6.2, Sec. 6.6

TMS 402: Sec. 6.2.4., Sec. 6.2.5

See Chap. 32 in this book for material properties and design components of masonry.

B.5. DEFLECTION

Key concepts: These key concepts are important for answering exam questions in subtopic 9.B.5, Deflection.

- beam deflection
    - conjugate beam method
    - cracking moment
    - double integration method
    - moment area method
    - strain energy method
    - superposition
    - table lookup method
- frame deflection: unit load method
- truss deflection: unit load method

NCEES Handbook: To prepare for this subtopic, familiarize yourself with these sections in the *Handbook*.

- Truss Deflection by Unit Load Method
- Frame Deflection by Unit Load Method
- Moment, Shear, and Deflection Diagrams

PE Civil Reference Manual (**CERM**): Study these sections in CERM that either relate directly to this subtopic or provide background information.

- Section 44.15: Strain Energy Due to Bending Moment
- Section 44.17: Beam Deflection: Double Integration Method
- Section 44.18: Beam Deflection: Moment Area Method
- Section 44.19: Beam Deflection: Strain Energy Method
- Section 44.20: Beam Deflection: Conjugate Beam Method
- Section 44.21: Beam Deflection: Superposition
- Section 44.22: Beam Deflection: Table Lookup Method
- Section 44.24: Truss Deflection: Strain Energy Method
- Section 44.25: Truss Deflection: Virtual Work Method
- Section 47.6: Dummy Unit Load Method
- Section 47.9: Frame Deflections by the Dummy Unit Load Method
- Section 48.11: Splitting Tensile Strength
- Section 50.15: Serviceability: Deflections
- Appendix 44.A: Elastic Beam Deflection Equations

Codes and standards: To prepare for this subtopic, familiarize yourself with these sections from codes and standards.

Building Code Requirements for Structural Concrete and Commentary (ACI 318)

- Section 24.2.3: Calculation of Immediate Deflections

Steel Construction Manual (AISC)

- Part 3: Design of Flexural Members

Specification for Structural Steel Buildings (AISC 360) is included in Part 16 of the *Steel Construction Manual*.

The following equations, tables, and concepts are relevant for subtopic 9.B.5, Deflection.

Truss Deflection: Unit Load Method

Handbook: Truss Deflection by Unit Load Method; Frame Deflection by Unit Load Method

$$\Delta_{\text{joint}} = \sum_{i=1}^{\text{members}} f_i (\Delta L)_i$$

$$(\Delta L)_i = \left(\frac{FL}{AE}\right)_i$$

CERM: Sec. 44.25

$$f\delta = \sum \frac{SuL}{AE} \quad [f=1] \qquad 44.53$$

The unit load method is also known as the virtual work method. This method is an extension of the strain energy method. Use this method to determine the deflection at any location on a truss. This method requires using the equation in *Handbook* section Truss Deflection by Unit Load Method to solve for member forces twice, once with the actual loads and once with a unit load.

Exam tip: Use a table format to keep all the member loads accessible.

Frame Deflection: Unit Load Method

Handbook: Frame Deflection by Unit Load Method

$$\Delta = \sum_{i=1}^{\text{members}} \int_{x=0}^{x=L_i} \frac{m_i M_i}{EI_i} dx$$

CERM: Sec. 47.6, Sec. 47.9

$$W_m = \int \frac{m_Q m_P}{EI} ds \qquad 47.9$$

For typically proportioned frames, a good approximation of deflection can usually be obtained by considering only the flexural contribution. In rigid frames that resist loading primarily by flexure, using the unit dummy load method to calculate deflection is based on the equation in *Handbook* section Frame Deflection by Unit Load Method, which is equivalent to CERM Eq. 47.9. See CERM Sec. 47.9 for more details on frame deflection by the unit load method.

Exam tip: Use a table format to keep all the member loads accessible.

Beam Deflection: Table Lookup Method

Handbook: Moment, Shear, and Deflection Diagrams

CERM: Sec. 44.22

AISC: Part 3

Use the tables in the *Handbook* section Moment, Shear, and Deflection Diagrams (which are from AISC Table 3-23) to determine the deflection at the location of interest on a beam. Then use superposition to determine the total deflection if there are various loads on the beam. This will be the quickest way to determine the deflection on the exam.

Beam Deflection: Cracking Moment

CERM: Sec. 48.11, Sec. 50.15

ACI 318: Sec. 24.2.3

Cracking takes place whenever the bending moment at a cross section exceeds the cracking moment, M_{cr}. Since moments from the applied loads vary along the beam length, certain portions of the beam will be cracked while others will be uncracked. ACI 318 provides a simplified method for obtaining an effective moment of inertia that is assumed to be constant along the entire beam length. The effective moment of inertia is given by ACI 318 Eq. 24.2.3.5a and CERM Eq. 50.43.

$$I_e = \left(\frac{M_{cr}}{M_a}\right)^3 I_g + \left(1 - \left(\frac{M_{cr}}{M_a}\right)^3\right) I_{cr} \leq I_g \qquad 50.43$$

The cracking moment is calculated using ACI Eq. 24.2.3.5b (CERM Eq. 50.44).

$$M_{cr} = \frac{f_r I_g}{y_t} \qquad 50.44$$

In this equation, I_g is the gross moment of inertia, which is found by the following.

$$I_g = \frac{bh^3}{12}$$

The modulus of rupture (i.e., the stress at which the concrete is assumed to crack) is f_r, and y_t is the distance from the center of gravity of the gross section (neglecting reinforcement) to the extreme tension fiber. For a rectangular cross section, y_t is

$$y_t = \frac{h}{2}$$

For stone or gravel concrete (normalweight concrete), the average modulus of rupture used for deflection calculations is given in CERM Eq. 50.45.

$$f_r = 0.7\lambda\sqrt{f'_c} \qquad \text{[SI]} \qquad 50.45(a)$$

$$f_r = 7.5\lambda\sqrt{f'_c} \qquad \text{[U.S.]} \qquad 50.45(b)$$

The lightweight aggregate factor, λ, is 0.75 for all-lightweight concrete, 0.85 for sand-lightweight concrete, and 1.0 for normalweight concrete. (See CERM Table 48.2.)

Beam Deflection: Double Integration Method

CERM: 44.17

The deflection and slope of a loaded beam are related to moment and shear by way of integration. CERM Eq. 44.44, Eq. 44.45, Eq. 44.46, and Eq. 44.47 show this.

$$y = \text{deflection} \qquad 44.44$$

ANALYSIS OF STRUCTURES: FORCES AND LOAD EFFECTS

$$y' = \frac{dy}{dx} = \text{slope} \qquad 44.45$$

$$y'' = \frac{d^2y}{dx^2} = \frac{M(x)}{EI} \qquad 44.46$$

$$y''' = \frac{d^3y}{dx^3} = \frac{V(x)}{EI} \qquad 44.47$$

If the moment function, $M(x)$, is known for a section of the beam, the deflection at any point can be found from CERM Eq. 44.48.

$$y = \frac{1}{EI}\int\left(\int M(x)\,dx\right)dx \qquad 44.48$$

Beam Deflection: Moment Area Method

CERM: Sec. 44.18

This method is applicable when the slopes of the deflected shape are not too large; it is based on two theorems.

The angle between the tangents at any two points on the elastic line of a beam is equal to the area of the moment diagram between the two points divided by EI. CERM Eq. 44.49 can be used to calculate this.

$$\phi = \int \frac{M(x)\,dx}{EI} \qquad 44.49$$

One point's deflection away from the tangent of another point is equal to the statical moment of the bending moment between those two points divided by EI. CERM Eq. 44.50 can be used to calculate this.

$$y = \int \frac{xM(x)\,dx}{EI} \qquad 44.50$$

Beam Deflection: Strain Energy Method

CERM: Sec. 44.15, Sec. 44.19, Sec. 44.24

The strain energy method of beam deflection uses the work-energy principle and equates the external work to the total internal strain energy, as given by CERM Eq. 44.51.

$$\tfrac{1}{2}Fy = \sum U \qquad 44.51$$

The elastic strain energy due to a bending moment stored in a beam can be calculated using CERM Eq. 44.41.

$$U = \frac{1}{2EI}\int M^2(x)\,dx \qquad 44.41$$

Once the internal work is determined, equate the internal work to the external work to solve for deflection.

$$\sum U = W$$

This method can also be applied to trusses when all the member forces are known. (See CERM Sec. 44.24.)

Beam Deflection: Conjugate Beam Method

CERM: Sec. 44.20

The conjugate beam method has the advantage of being able to handle beams of varying cross sections (e.g., stepped beams) and materials. However, it cannot easily handle beams with two built-in ends. Follow the steps outlined in CERM Sec. 44.20 to use the conjugate beam method.

Beam Deflection: Superposition

CERM: Sec. 44.21

When multiple loads act simultaneously on a beam, all of the loads contribute to deflection. The principle of superposition permits the deflection at a point to be calculated as the sum of the deflections from each individual load acting singly. This principle is valid as long as none of the deflections are excessive and all stresses are less than the yield point of the beam material.

This method is applicable to all the deflection methods. Determine the deflection at a specific location along the length of the beam from each loading condition and sum them up for the total deflection.

B.6. SPECIAL TOPICS

Key concepts: These key concepts are important for answering exam questions in subtopic 9.B.6, Special Topics.

- bearing
- buckling
- fatigue
- thermal deformation
- torsional stress and strain

NCEES Handbook: To prepare for this subtopic, familiarize yourself with these sections in the *Handbook*.

- Shear Stress-Strain
- Thermal Deformations
- Torsion
- Torsional Strain

PE Civil Reference Manual (CERM): Study these sections in CERM that either relate directly to this subtopic or provide background information.

- Section 42.7: Polar Moment of Inertia
- Section 43.15: Torsion Test
- Section 43.17: Fatigue Testing
- Section 44.6: Thermal Deformation
- Section 45.16: Circular Shaft Design
- Section 45.19: Shear Center for Beams
- Section 58.9: Fatigue Loading
- Section 59.22: Beam Bearing Plates

Codes and standards: To prepare for this subtopic, familiarize yourself with these sections from codes and standards.

Building Code Requirements for Structural Concrete and Commentary (ACI 318)

- Section 21.2: Strength Reduction Factors for Structural Concrete Members and Connections
- Section 22.8: Bearing

Steel Construction Manual (AISC)

- Part 16 Chapter J: Design of Connections
- Appendix 3: Design for Fatigue

Specification for Structural Steel Buildings (AISC 360) is included in Part 16 of the *Steel Construction Manual*.

The following equations, figures, tables, and concepts are relevant for subtopic 9.B.6, Special Topics.

Torsional Stress and Strain

Handbook: Shear Stress-Strain; Torsion; Torsional Strain

$$\tau = \frac{Tr}{J}$$

$$\phi = \int_0^L \frac{T}{GJ}dz = \frac{TL}{GJ}$$

CERM: Sec. 42.7, Sec. 43.15, Sec. 45.16, Sec. 45.19

$$\tau = G\theta = \frac{Tr}{J} \quad \text{45.50}$$

$$\gamma = \frac{TL}{JG} = \frac{\tau L}{rG} \quad [\text{radians}] \quad \text{43.23}$$

Shear stress occurs when a shaft is placed in torsion. The shear stress at the outer surface of a bar of radius r, which is torsionally loaded by a torque, T, is given by the equation in *Handbook* section **Torsion** and CERM Eq. 45.50. Use the equation in *Handbook* section **Torsional Strain** or CERM Eq. 43.23 to determine the angle of twist (torsional strain) on a bar from torsional load.

The polar moment of inertia is derived from CERM Eq. 42.21.

$$J = \int (x^2 + y^2)dA \quad \text{42.21}$$

In the *Handbook* and CERM equations for torsional strain, G is the shear modulus. It can be calculated from the modulus of elasticity using the equation in *Handbook* section **Shear Stress-Strain** or CERM Eq. 45.55.

$$G = \frac{E}{2(1+v)}$$

CERM Fig. 45.12 represents a shear force that is eccentric to the shear center of a channel section.

Figure 45.12 Channel Beam in Pure Bending (shear resultant, V, directed through shear center, O)

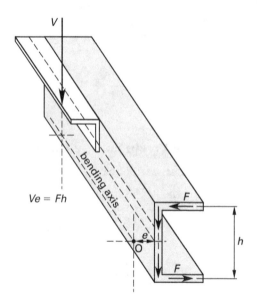

For beams with transverse loading, simple bending without torsion can only occur if the transverse load (shear resultant, V, or shear force of action) is directed through the shear center. Otherwise, a torsional moment calculated as the product of the shear resultant and the torsional eccentricity will cause the beam to twist. The torsional eccentricity is the distance between the line of action of the shear resultant and the shear center.

For design components of concrete for torsion, see Chap. 29 in this book.

Thermal Deformation

Handbook: Thermal Deformations

$$\delta_t = \alpha L(T - T_o)$$

CERM: Sec. 44.6

$$\Delta L = \alpha L_o(T_2 - T_1) \quad\quad 44.9$$

The *Handbook* equation and CERM Eq. 44.9 have different nomenclature, but they are the same. The coefficient of linear expansion will vary depending on the material in question. Refer to Subtopic B.2, Axial, in this chapter for more on thermal deformation.

Exam tip: This equation is often paired with the axial equation PL/EA to determine the force buildup in a rigid end beam.

Fatigue

CERM: Sec. 43.17, Sec. 58.9

AISC: App. 3

The behavior of a material under repeated loading is evaluated by a fatigue test. A specimen is loaded repeatedly to a specific stress amplitude, S, and the number of applications of that stress required to cause failure, N, is counted.

See AISC Table A-3.1 for many common loading applications. The table shows the stress category for these common scenarios and the C_f and F_{TH} constants to use in AISC Eq. A-3-1 and Eq. A-3-1M. The number of stress range fluctuations in design life is represented by n_{SR}.

$$F_{SR} = \left(\frac{C_f}{n_{SR}}\right)^{0.333} \geq F_{TH} \quad [\text{AISC Eq. A-3-1}]$$

$$F_{SR} = \left(\frac{C_f \cdot 329}{n_{SR}}\right)^{0.333} \geq F_{TH} \quad [\text{AISC Eq. A-3-1M}]$$

Exam tip: If cyclic application of live load is less than 20,000 cycles, fatigue need not be considered. Peak cyclic load should not exceed $0.66F_y$.

Bearing

CERM: Sec. 59.22

ACI 318: Sec. 21.2, Sec. 22.8

AISC: Part 16 Chap. J

It is important to check the plate strength for the induced bearing loads and the support bearing area for the bearing loads. See CERM Fig. 59.14 for nomenclature for beam bearing plates.

For a plate covering the full area of a concrete support, the nominal bearing strength can be calculated using AISC Eq. J8-1, or CERM Eq. 59.38.

$$P_p = 0.85 f'_c A_1 \quad [\text{AISC Eq. J8-1}] \quad\quad 59.38$$

For a plate covering less than the full area of a concrete support, the area of steel concentrically bearing on a concrete support is given by AISC Eq. J8-2, or CERM Eq. 59.40.

$$P_p = 0.85 f'_c A_1 \sqrt{\frac{A_2}{A_1}} \leq 1.7 f'_c A_1 \quad\quad 59.40$$

[AISC Eq. J8-2]

Compare the moment induced from the bearing to the moment capacity of the plate cross section.

For concrete, compare the load on the allowed area to the load allowed from ACI 318 Table 22.8.3.2. See ACI 318 Fig. R22.8.3.2 for an illustration of this application. Yield lines for A_1 will continue until area 2 is twice area 1 or an edge is encountered. In the event of an edge, all four sides are stopped at the same time. The loaded area strength reduction factor for bearing per ACI 318 Table 21.2.1 is 0.65.

Buckling

Refer to subtopic B.2, Axial, in this chapter for entries on buckling stress.

29 Design and Details of Structures: Concrete

Content in blue refers to the *NCEES Handbook*.

A.1. Concrete 29-1
B.1. Horizontal Members 29-3
B.2. Vertical Members 29-6
B.3. Systems.. 29-8
B.4. Connections.. 29-8
B.5. Foundations.. 29-9

The knowledge area of Design and Details of Structures makes up between 16 and 24 questions out of the 80 questions on the PE Civil Structural exam. Four to six questions are on materials and material properties, while 12 to 18 questions are on component design and detailing. This chapter focuses on these knowledge areas only in the context of concrete. The organization of this chapter follows the order of subtopics given by NCEES for this knowledge area, but includes only subtopics relevant to the design and details of structures for concrete.

A.1. CONCRETE

Key concepts: These key concepts are important for answering exam questions in subtopic 10.A.1, Concrete.

- concrete compressive strength
- concrete exposure and mix requirements
- concrete modulus of elasticity
- concrete testing
- reinforcing steel
- strength reduction factor

NCEES Handbook: To prepare for this subtopic, familiarize yourself with these sections in the *Handbook*.

- ASTM Standard Reinforcing Bars
- Design Provisions: Definitions

PE Civil Reference Manual (**CERM**): Study these sections in CERM that either relate directly to this subtopic or provide background information.

- Section 48.8: Compressive Strength
- Section 48.10: Modulus of Elasticity
- Section 48.23: Reinforcing Steel
- Section 48.24: Mechanical Properties of Steel
- Section 48.25: Exposure
- Section 49.1: Concrete Mix Design Considerations
- Section 49.2: Strength Acceptance Testing
- Section 50.6: Design Strength and Design Criteria

Codes and standards: To prepare for this subtopic, familiarize yourself with these sections from codes and standards.

Building Code Requirements for Structural Concrete and Commentary (ACI 318)

- Chapter 19: Concrete: Design and Durability Requirements
- Chapter 20: Steel Reinforcement Properties, Durability, and Embedments
- Chapter 21: Strength Reduction Factors
- Chapter 22: Sectional Strength
- Chapter 26: Construction Documents and Inspection
- Appendix A: Steel Reinforcement Information

The following equations and concepts are relevant for subtopic 10.A.1, Concrete.

Concrete Compressive Strength

CERM: Sec. 48.8

ACI 318: Sec. 19.2.1

The compressive strength of concrete, f'_c, is the maximum stress that a concrete specimen can sustain under compressive axial loading. It is based on the 28-day strength of the concrete and is measured in units of psi (U.S. customary) or MPa (SI).

Exam tip: Concrete material properties are discussed in ACI 318 Chap. 19.

Concrete Modulus of Elasticity

Handbook: Design Provisions: Definitions

$$E_c = 33w_c^{1.5}\sqrt{f_c'}$$

CERM: Sec. 48.10

ACI 318: Sec. 19.2.1

$$E_c = 4700\sqrt{f_c'} \quad [\text{SI}] \quad 48.3(a)$$

$$E_c = 57{,}000\sqrt{f_c'} \quad [\text{U.S.}] \quad 48.3(b)$$

The *Handbook* equation for E_c is valid for values of w_c (the density of normal weight concrete or the equilibrium density of lightweight concrete) between 90 lbf/ft³ and 160 lbf/ft³.

The units for f_c' are the same as for stress: psi (U.S. customary) or MPa (SI). By convention, when the square root of f_c' is taken, the units do not change; that is, $\sqrt{f_c'}$ is also in units of psi.

Exam tip: These equations are for buildings. Don't forget the units (psi).

Strength Reduction Factor

Handbook: Design Provisions: Definitions

CERM: Sec. 50.6

ACI 318: Sec. 21.2

The strength reduction factor, ϕ, is multiplied by the nominal capacity to obtain the design strength.

$$\text{design strength} = \phi(\text{nominal strength}) \quad 50.8$$

ϕ is dependent on the type of capacity (shear, flexural, etc.) and the tensile strain. For flexural capacity, ϕ is based on the tensile strength of the reinforcing steel. Refer to ACI 318 Sec. R21.2.2 and CERM Fig. 50.4.

ACI 318 uses the subscript n to indicate a nominal quantity. The nominal moment strength and nominal shear strength are designated M_n and V_n, respectively.

Exam tip: The equations for nominal strength capacities such as M_n are provided in the *Handbook*. The exam may test on design capacities, such as ϕM_n, that have been multiplied by ϕ. Practice using ACI 318 Table 21.2.1 to select the correct value of ϕ. Engineers will often try to limit the steel characteristics to ensure that a cross section is tension controlled; when this is the case, $\phi = 0.9$ can be used for flexure.

Concrete Testing

CERM: Sec. 49.2

ACI 318: Sec. 26.12.3.1(b)

According to ACI 318 Sec. 26.12, concrete must be evaluated and tested before it is considered acceptable. A single strength test is defined as the arithmetic average of the compressive strengths of at least two cylinders measuring 6 in × 12 in, or at least three cylinders measuring 4 in × 8 in.

Exam tip: The commentary in ACI 318 Chap. 26 covers testing requirements and lists ASTM standards used.

Concrete Exposure and Mix Requirements

CERM: Sec. 48.25, Sec. 49.1

ACI 318: Sec. 19.3, Sec. 26.4

Durability requirements for concrete are given in ACI 318 Chap. 19, and mix requirements based on exposure are given in ACI 318 Sec. 19.3.2.1. Additional ASTM mix proportion limits are provided in ACI 318 Sec. 26.4.

Exam tip: Answering exam questions based on these concepts requires knowing where to look in the code, rather than knowing an equation or a procedure. Material properties and ASTM standards are found in ACI 318 Chap. 19 and Chap. 26. Strength capacities of members are found in ACI 318 Chap. 22.

Reinforcing Steel

Handbook: ASTM Standard Reinforcing Bars

CERM: Sec. 48.23, Sec. 48.24

ACI 318: Sec. 20.2, App. A

In reinforced concrete, the concrete and rebar are intended to act compositely. As a result, the concrete grips the rebar as it is placed under stress and deforms. If there is slippage between the bar and the concrete, then the element loses strength. For this reason, reinforcing bars are often deformed to provide a better mechanical bond between the concrete and the reinforcing steel. ACI 318 Sec. 20.2 provides the material properties. ACI 318 App. A provides the area, diameter, and unit weight of typical rebar.

Exam tip: Typical exam questions on reinforced concrete will require using grade 60 reinforcing bar, which has a yield strength of 60 ksi.

B.1. HORIZONTAL MEMBERS

Key concepts: These key concepts are important for answering exam questions in subtopic 10.B.1, Horizontal Members.

- concrete beams
- concrete diaphragms
- concrete one-way shear
- concrete slabs
- concrete torsion
- concrete two-way shear
- moment strength of singly reinforced concrete sections

NCEES Handbook: To prepare for this subtopic, familiarize yourself with these sections in the *Handbook*.

- Beams—Flexure Strength

PE Civil Reference Manual (**CERM**): Study these chapters and sections in CERM that either relate directly to this subtopic or provide background information.

- Chapter 50: Reinforced Concrete: Beams
- Chapter 51: Reinforced Concrete: Slabs
- Section 55.3: Column Footings

Structural Depth Reference Manual (**CEST**): Study these sections in CEST that either relate directly to this subtopic or provide background information.

- Section 1.3: Design for Torsion

Codes and standards: To prepare for this subtopic, familiarize yourself with these chapters and sections from codes and standards.

Building Code Requirements for Structural Concrete and Commentary (ACI 318)

- Chapter 7: One-Way Slabs
- Chapter 8: Two-Way Slabs
- Chapter 9: Beams
- Chapter 12: Diaphragms
- Section 21.2: Strength Reduction Factors for Structural Concrete Members and Connections
- Section 22.3: Flexural Strength
- Section 22.5: One-Way Shear Strength
- Section 22.6: Two-Way Shear Strength
- Section 22.7: Torsional Strength

The following equations, figures, and concepts are relevant for subtopic 10.B.1, Horizontal Members.

Concrete Beams

CERM: Chap. 50

ACI 318: Chap. 9

Concrete beams can have different shapes, including rectangular, T-shaped, and L-shaped. Concrete beams are required to have flexural steel in the tension side of the beam and shear reinforcing stirrups along the beam.

Beams must be designed for flexure, shear, and deflection. Deflection does not need to be checked if the beam depth is greater than the depths listed in ACI 318 Table 9.3.1.1. Refer to the entry "Flexural Strength of Singly Reinforced Concrete Sections" in this chapter for flexural capacity equations. Refer to the entry "Concrete One-Way Shear" in this chapter for more on beam shear capacity.

Beam reinforcement limits are specified in ACI 318 Sec. 9.3.3.1 and Sec. 9.6. Required strength for shear and moment are specified in ACI 318 Sec. 9.4. For distributed loads, the critical shear is located at distance d from the face of the support. Beam shear and moment can be found using the approximation method in ACI 318 Sec. 6.5. ACI 318 Sec. 9.7 specifies reinforcing detailing requirements, including those for cover, spacing, and cutoff detailing.

Exam tip: The *Handbook* provides the flexural capacity for singly reinforced beams only. During the exam, refer to ACI 318 Chap. 9 for other locations in the code that contain general beam requirements.

Concrete Slabs

CERM: Chap. 51

ACI 318: Chap. 7, Chap. 8

Slabs are structural elements whose lengths and widths are large in comparison to their thicknesses. Slab thickness is typically governed by deflection criteria or fire rating requirements. Shear is generally carried by the concrete without the aid of shear reinforcement, which is difficult to place and anchor in shallow slabs. Longitudinal reinforcement is used to resist bending moments.

Slabs are one-way or two-way depending on the length and width of the slab supports (beams or walls). One-way slabs (shown in CERM Fig. 51.1) are covered in ACI 318 Chap. 7. Two-way slabs are covered in ACI 318 Chap. 8.

Figure 51.1 One-Way Slab

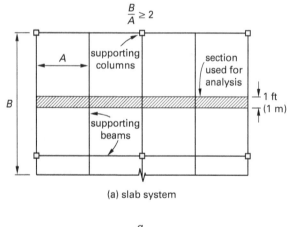

(a) slab system

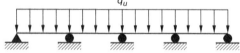

(b) model used to obtain moments and shears

There are two different types of shear failures. One-way shear, which is similar to beam shear, is covered in ACI 318 Sec. 22.5. Two-way shear, which is covered in ACI 318 Sec. 22.6, is sometimes referred to as *punching shear*, and it occurs when columns "punch" through the slab. Refer to the entry "Flexural Strength of Singly Reinforced Concrete Sections" in this chapter for flexural capacity equations. Refer to the entry "Concrete One-Way Shear" in this chapter for beam shear capacity. Refer to the entry "Concrete Two-Way Shear" in this chapter for two-way shear capacity.

Exam tip: The *Handbook* provides the flexural capacity for singly reinforced concrete only. Refer to ACI 318 Chap. 7 and Chap. 8 for other locations in the code for general slab requirements.

Concrete Diaphragms

ACI 318: Chap. 12

Diaphragms are horizontal structural members that transfer lateral loads (wind and seismic) to the vertical lateral force-resisting members (shear walls, moment frames, etc.). Concrete slabs act as rigid diaphragms, and they must be designed for shear and flexure. It is important to note that the loading for diaphragms is in the horizontal direction, not the vertical direction.

ACI 318 Chap. 12 provides shear capacity equations and limits (rather than referring to other chapters). ACI 318 Sec. 12.5.4 provides requirements for collector elements. ACI 318 Sec. 12.6 and Sec. 12.7 provide reinforcement limits and detailing requirements, including requirements for spacing.

Exam tip: If the exam asks a question about a building in seismic design category D, E, or F, you will need to refer to ACI 318 Sec. 18.12 for additional requirements.

Flexural Strength of Singly Reinforced Concrete Sections

Handbook: Beams—Flexure Strength

$$M_n = 0.85 f'_c ab \left(d - \frac{a}{2}\right) = A_s f_y \left(d - \frac{a}{2}\right)$$

CERM: Sec. 50.10

$$M_n = A_s f_y (d - \lambda) \quad \quad 50.27$$

The term λ in CERM Eq. 50.27 is the same as $a/2$ in the *Handbook* equation.

ACI 318 uses the subscript n to indicate a nominal quantity. A nominal value can be interpreted as being in accordance with theory for the specified dimensions and material properties. The nominal moment strength and nominal shear strength are designated M_n and V_n, respectively. The design strength is the result of multiplying the nominal strength by a strength reduction factor (also known as a capacity reduction factor), ϕ.

$$\text{design strength} = \phi(\text{nominal strength}) \quad 50.8$$

The moment capacity of a reinforced concrete cross section derives from the couple composed of the tensile force in the steel and the compressive force in the concrete. CERM Fig. 50.6 shows an example of the strain distribution and the equivalent systems used to determine the flexural capacity.

Figure 50.6 Conditions at Maximum Moment

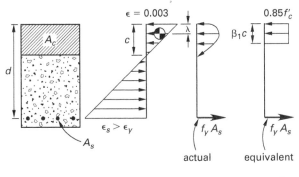

(a) strain distribution (b) compressive stress distribution (c) equivalent rectangular compressive stress block

Exam tip: The equations for nominal strength capacities (such as M_n) are provided in *Handbook*, such as section Beams—Flexure Strength. The exam may ask for design capacities, which are multiplied by ϕ. Be comfortable referring to ACI 318 Sec. 21.2 for selecting the correct value for ϕ.

Concrete One-Way Shear

CERM: Sec. 50.20 through Sec. 50.27

ACI 318: Sec. 22.5

$$\phi V_n = V_u \quad 50.50$$

$$V_n = V_c + V_s \quad 50.51$$

Beam shear is referred to in ACI 318 as *one-way shear*. The nominal shear strength of a concrete beam is the sum of contributions from the concrete strength and the shear reinforcement strength. This relationship is shown in ACI 318 Eq. 22.5.1.1 and CERM Eq. 50.51. Stirrups are used to resist shear in beams, and ties or circular hoops are used to resist shear in columns. Slabs and footings are designed to avoid the use of shear reinforcement.

The design strength is ϕV_n, and the value of ϕ for shear is 0.75. Cross-sectional dimensional restrictions for shear are provided in ACI 318 Sec. 22.5.1.2.

The shear resistance provided by the concrete is given in ACI 318 Sec. 22.5.5.1 for beams with no axial load, and in ACI 318 Sec. 22.5.6.1 for beams with axial compressive load. For no axial load, the equation is

$$V_c = 2\lambda\sqrt{f'_c}\, b_w d \quad [\text{U.S.}] \quad 50.53(b)$$

The shear resistance provided by steel stirrups is given in ACI 318 Sec. 22.5.10. The equation for the shear capacity of the reinforcement, ACI 318 Eq. 22.5.10.5.3, is the same as CERM Eq. 50.54.

$$V_s = \frac{A_v f_{yt} d}{s} \quad 50.54$$

CERM Fig. 50.13 illustrates anchorage requirements for shear reinforcement. Minimum shear reinforcement requirements for beams are provided in ACI 318 Sec. 9.6.3.

Figure 50.13 Anchorage Requirements for Shear Reinforcement (ACI 318 Sec. 25.7.1.3)

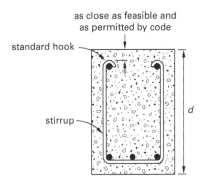

(a)
No. 5 and smaller, all f_{yt} values; No. 6, No. 7, No. 8 stirrups with $f_{yt} \leq 40{,}000$ psi

(b)
No. 6, No. 7, No. 8 stirrups with $f_{yt} > 40{,}000$ psi

Exam tip: In calculating A_v for stirrup reinforcing, each leg of the stirrup that crosses a potential crack can be counted. It is often easiest to start with A_v and then calculate backward to the required s.

Concrete Two-Way Shear

CERM: Sec. 55.3

ACI 318: Sec. 22.6

$$v_n = (2 + y)\lambda\sqrt{f'_c} \qquad 55.18$$

$$y = \min\{2,\ 4/\beta_c,\ 40d/b_o\} \qquad 55.19$$

$$\beta_c = \frac{\text{column long side}}{\text{column short side}} \qquad 55.20$$

$$b_o = \frac{A_p}{d} = 2(b_1 + b_2) \qquad 55.21$$

Two-way shear, also known as *punching shear*, occurs in footings and slabs where columns can punch through the member. The capacity of punching shear is based on the area around the column that the column can punch through. Refer to ACI 318 Fig. R22.6.4.3 for this critical section.

Engineers typically avoid adding shear reinforcement to resist punching shear, which means that all the shear force must be resisted by the concrete. When steel is needed to help resist two-way shear, then engineers may use stirrups or headed studs. The capacity for punching shear when there is no shear reinforcement is shown in ACI 318 Eq. 22.6.1.2.

$$v_n = v_c$$

The concrete capacity for two-way shear is provided in the equations in ACI 318 Table 22.6.5.2. CERM combines these equations with the use of the y term in CERM Eq. 55.18 and Eq. 55.19.

Although CERM Sec. 55.3 is about footings, it discusses two-way shear, which also applies to slabs.

Exam tip: Slabs and footings may have both one-way and two-way shear. Be sure to check both.

Concrete Torsional Strength

CEST: Sec. 1.3

ACI 318: Sec. 22.7

$$T_n = \frac{2A_o A_t f_{yt}}{s}\cot\theta \quad [\text{ACI 22.7.6.1a}]$$

$$T_n = \frac{2A_o A_l f_y}{p_h}\cot\theta \quad [\text{ACI 22.7.6.1b}]$$

Concrete torsional strength, shown in ACI 318 Eq. 22.7.6.1a and Eq. 22.7.6.1b, is provided by closed stirrups used with shear reinforcement and supplemental longitudinal reinforcement. Torsional requirements may be neglected if $T_u < \phi T_{th}$.

ACI 318 Table 22.7.4.1 provides the equations for T_{th} for different sections.

Exam tip: For exam problems involving concrete torsional strength, it is often easiest to select the stirrups and then calculate back to the required s.

B.2. VERTICAL MEMBERS

Key concepts: These key concepts are important for answering exam questions in subtopic 10.B.2, Vertical Members.

- concrete axial and combined flexural strength
- concrete columns
- concrete walls

NCEES Handbook: The *Handbook* does not contain any material relevant to this subtopic. Be sure to study the listed sections in CERM and from codes and standards.

***PE Civil Reference Manual* (CERM):** Study these chapters in CERM that either relate directly to this subtopic or provide background information.

- Chapter 52: Reinforced Concrete: Short Columns
- Chapter 53: Reinforced Concrete: Long Columns
- Chapter 54: Reinforced Concrete: Walls and Retaining Walls

Codes and standards: To prepare for this subtopic, familiarize yourself with these sections from codes and standards.

Building Code Requirements for Structural Concrete and Commentary (ACI 318)

- Chapter 6: Structural Analysis
- Chapter 10: Columns
- Chapter 11: Walls
- Section 18.10: Special Structural Walls
- Section 22.4: Axial Strength or Combined Flexural and Axial Strength

The following figures and concepts are relevant for subtopic 10.B.2, Vertical Members.

Concrete Columns

CERM: Chap. 52

ACI 318: Sec. 6.2.5, Chap. 10

Concrete columns are typically rectangular, square, or circular. There are two types of columns, slender and nonslender. A nonslender column is one that meets the requirements shown in ACI 318 Sec. 6.2.5. A slender column must be designed for second order effects due to combined flexure and axial. Figure R6.2.6 in ACI 318 provides a flowchart on how to analyze a slender column for these slenderness effects. Slender columns can be found in CERM Chap. 52. Refer to the entry "Concrete One-Way Shear" in this chapter for the shear capacity of columns.

Concrete Walls

CERM: Chap. 54

ACI 318: Chap. 6, Chap. 11, Sec. 18.10

Concrete walls must be designed for axial, in-plane, and out-of-plane loading. Axial loading comes from gravity loads on the floor and from the roof members that the wall is supporting. In-plane loading occurs when the wall acts as a shear wall to transfer wind and seismic loads to the foundation. Out-of-plane loading occurs when a wind or seismic event creates a load on the face of the wall. Refer to ACI 318 Fig. R11.4.1.3 for a general description of these loadings.

Required axial strength and moment strength are specified in ACI 318 Sec. 11.4. Walls are considered slender and should be analyzed per ACI 318 Sec. 6.6.4, Sec. 6.7, Sec. 6.8, or Sec. 11.8 for out-of-plane loading.

Walls that resist in-plane shear act as shear walls. The shear capacity for in-plane shear is found in ACI 318 Sec. 11.5.4. Minimum reinforcement requirements are described in ACI 318 Sec. 11.6.

Exam tip: Be sure to verify the seismic design category (SDC) before evaluating a shear wall. If the wall is in SDC D, E, or F, then refer to ACI 318 Sec. 18.10; if not, then refer to ACI 318 Chap. 11.

Concrete Axial and Combined Axial and Flexural Strength

CERM: Sec. 52.4, Sec. 52.5

ACI 318: Sec. 10.4, Sec. 22.4

Axial and moment forces must be analyzed simultaneously using an interaction diagram. Refer to ACI 318 Fig. R10.4.2.1 or CERM Fig. 52.5 for a general representation of an interaction diagram.

Figure 52.5 Relationship Between Nominal and Design Interaction Diagrams (per ACI 318 Chap. 21)

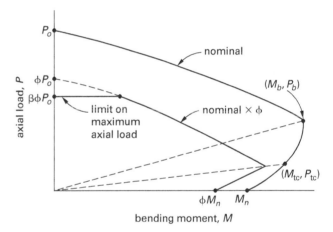

The ϕ factor can be found in ACI 318 Sec. 21.2 and needs to be determined for each location along the diagram. The values for ϕP_n and ϕM_n can be found by calculating the internal forces inside the member with different strain conditions. CERM Sec. 52.5 provides steps to create a general interaction diagram. Moment and axial combinations that follow inside the influence diagram are considered acceptable.

P_o is the nominal axial strength with zero eccentricity (see the definition in ACI 318 Sec. 22.4), which is pure axial strength. When a member is under pure axial load, P_o must be multiplied by 0.8 or 0.85, per ACI 318 Table 22.4.2.1, to account for accidental eccentricity.

Exam tip: When dealing with pure axial load, remember to multiply P_o by either 0.8 or 0.85 and ϕ. Interaction diagrams are not provided in the *Handbook* and take more than six minutes to create.

B.3. SYSTEMS

Concrete systems are made up of beams and frames. Refer to entries in Chap. 28 of this book for an analysis of these systems.

B.4. CONNECTIONS

Key concepts: These key concepts are important for answering exam questions in subtopic 10.B.4, Connections.

- anchor bolt connections in concrete
- concrete corbels
- development length of steel
- reinforcement splices

NCEES Handbook: The *Handbook* does not contain any material relevant to this subtopic. Be sure to study the listed sections in CERM and CEST and from codes and standards.

PE Civil Reference Manual (**CERM**): Study these sections in CERM that either relate directly to this subtopic or provide background information.

- Section 55.5: Development Length of Flexural Reinforcement
- Section 55.6: Transfer of Force at Column Base

Structural Depth Reference Manual (**CEST**): Study these sections in CEST that either relate directly to this subtopic or provide background information.

- Section 1.2: Corbels

Codes and standards: To prepare for this subtopic, familiarize yourself with these sections from codes and standards.

Building Code Requirements for Structural Concrete and Commentary (ACI 318)

- Section 16.5: Brackets and Corbels
- Chapter 17: Anchoring to Concrete
- Section 25.4: Development of Reinforcement
- Section 25.5: Splices

The following equations, figures, and concepts are relevant for subtopic 10.B.4, Connections.

Development Length of Steel

CERM: Sec. 55.5

ACI 318: Sec. 25.4.2

$$l_d = \frac{d_b f_y \psi_t \psi_e}{25 \lambda \sqrt{f_c'}} \geq 12 \text{ in} \quad [\text{no. 6 bars and smaller}] \quad 55.28$$

$$l_d = \frac{d_b f_y \psi_t \psi_e}{20 \lambda \sqrt{f_c'}} \geq 12 \text{ in} \quad [\text{no. 7 bars and larger}] \quad 55.29$$

$$l_d = \frac{3 d_b f_y \psi_t \psi_e}{50 \lambda \sqrt{f_c'}} \geq 12 \text{ in} \quad \begin{bmatrix}\text{no. 6 bars and smaller;} \\ \text{Eq. 55.27 requirements} \\ \text{not met}\end{bmatrix} \quad 55.30$$

$$l_d = \frac{3 d_b f_y \psi_t \psi_e}{40 \lambda \sqrt{f_c'}} \geq 12 \text{ in} \quad \begin{bmatrix}\text{no. 7 bars and larger;} \\ \text{Eq. 55.28 requirements} \\ \text{not met}\end{bmatrix} \quad 55.31$$

The development length is the length of embedded reinforcement required to develop the design strength of reinforcement at a critical section. ACI 318 Table 25.4.2.2 provides equations for development length.

The adjustment factors ψ and λ are defined in ACI 318 Table 25.4.2.4. The equations for l_d in Table 25.4.2.2 include adjustments (the ψ factors) for the development lengths of epoxy-coated bars, small bars, and bars that have 12 in or more fresh concrete cast below them. For normal weight concrete, $\lambda = 1.0$.

Exam tip: According to ACI 318 Sec. 25.4.1.4, the value $\sqrt{f_c'}$ used in these equations should not exceed 100 psi.

Reinforcement Splices

CERM: Sec. 55.6

ACI 318: Sec. 25.5

Concrete splices are used to transfer the force in one bar to another bar. Transfer of force is accomplished by

- attaching bars to each other using mechanical or welded connections
- lapping bars next to each other, the spacing being no more than 6 in or one-fifth the required lap splice length, whichever is smaller (see ACI 318 Sec. 25.5.1.3), and allowing the load to transfer through the concrete

Bars spliced mechanically or by weld must be strong enough to develop $1.25f_y$. Minimum lengths are provided for two classes of lap splice in ACI 318 Table 25.5.2.1. Lap splice lengths for bars in tension can be found in ACI 318 Sec. 25.5.4. Lap splice lengths for bars in compression can be found in ACI 318 Sec. 25.5.5.

For tension splices, the lap splice length must be the greater of 12 in and either $1.0l_d$ or $1.3l_d$, depending on the ratio of the area of the steel provided to the area of the steel required for the load at the splice. Refer to ACI 318 Table 25.5.2.1 for the different conditions.

Exam tip: When determining your splice length for tension splices, refer to ACI 318 Table 25.5.2.1 first to determine the case.

Anchor Bolt Connections in Concrete

ACI 318: Chap. 17

There are five failure modes for concrete-embedded anchors loaded in tension and three failure modes for concrete-embedded anchors in shear. ACI 318 Table 17.3.1.1 identifies each failure mode and where the check is found in the code. Figure R17.3.1 shows the different failure modes for anchor bolts in tension and shear. Each tension failure mode is discussed in a different subsection of ACI 318 Sec. 17.4, and each shear failure mode is discussed in a different subsection of ACI 318 Sec. 17.5. Each limit state must be checked to ensure the connection is adequate. Strength reduction factors are provided in ACI 318 Sec. 17.3.3.

Exam tip: Be sure to read the entire question, as it may require you to check only one failure mode.

Concrete Corbels

CEST: Sec. 1.2

ACI 318: Sec. 16.5

A *corbel* is a cantilever bracket supporting a load-bearing member, as shown in CEST Fig. 1.5. A similar image is found in ACI 318 Fig. R16.5.1b.

The dimensional requirement for a corbel is found in ACI 318 Sec. 16.5.2. The required strength is found in ACI 318 Sec. 16.5.3, and the provided strength is found in ACI 318 Sec. 16.5.4. Reinforcing limits and detailing are given in ACI 318 Sec. 16.5.5 and Sec. 16.5.6, respectively.

Exam tip: Horizontal tensile forces on corbels are treated like live loads, per ACI 318 Sec. 16.5.3.4.

Figure 1.5 Corbel Details

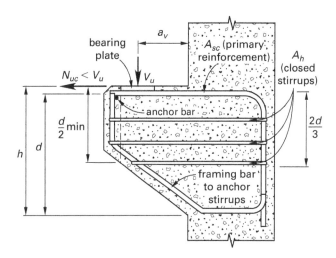

B.5. FOUNDATIONS

Key concepts: These key concepts are important for answering exam questions in subtopic 10.B.5, Foundations.

- retaining walls
- shallow foundations

NCEES Handbook: To prepare for this subtopic, familiarize yourself with these sections in the *Handbook*.

- Stress States on a Soil Element Subjected Only to Body Stresses
- Eccentrically Loaded Footing

PE Civil Reference Manual (CERM): Study these chapters in CERM that either relate directly to this subtopic or provide background information.

- Chapter 37: Rigid Retaining Walls
- Chapter 54: Reinforced Concrete: Walls and Retaining Walls
- Chapter 55: Reinforced Concrete: Footings

Structural Depth Reference Manual (CEST): Study these sections in CEST that either relate directly to this subtopic or provide background information.

- Chapter 2: Foundations

Codes and standards: To prepare for this subtopic, familiarize yourself with these sections from codes and standards.

Building Code Requirements for Structural Concrete and Commentary (ACI 318)

- Chapter 11: Walls
- Chapter 13: Foundations

The following equations and figures are relevant for subtopic 10.B.5, Foundations.

Shallow Foundations

Handbook: Eccentrically Loaded Footing

CERM: Chap. 55

CEST: Chap. 2

ACI 318: Sec. 13.3

There are multiple different types of shallow foundations. These include strip footings (footings under a wall), isolated footings (footings under a column), and combined footings, which support multiple columns. ACI 318 Fig. R13.1.1 shows several types of foundations. CERM Fig. 55.1 shows an example of an isolated footing, and CEST Fig. 2.9 shows an example of a combined footing.

Figure 55.1 General Footing Loading

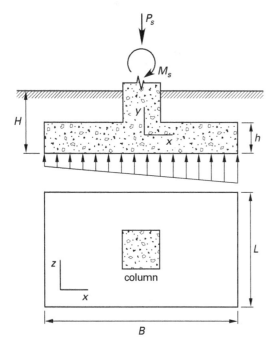

Figure 2.9 Soil Pressure for Service Loads

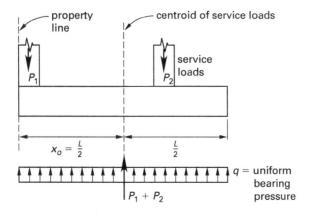

A shallow foundation needs to be designed for both the concrete capacity and the soil capacities. *Handbook* figure Eccentrically Loaded Footing illustrates the soil loading distribution for calculating the pressures in the soil. ACI 318 Sec. 13.3 provides the concrete requirements for shallow foundations.

The concrete in a shallow footing needs to be designed for flexure, one-way shear, and two-way shear. Two-way shear does not apply under a wall. Refer to the entries "Concrete One-Way Shear" and "Concrete Two-Way Shear" in this chapter.

Exam tip: Shallow footings are typically designed so that they do not require any shear reinforcement. The thickness of the footing is increased to avoid the need for shear reinforcement for one-way and two-way shear.

Retaining Walls

Handbook: Stress States on a Soil Element Subjected Only to Body Stresses

CERM: Sec. 37.1, Sec. 37.12, Sec. 37.13, Sec. 55.4

ACI 318: Chap. 11

There are four main types of retaining walls: gravity, buttress, counterfort, and cantilever. A gravity wall is a high-bulk structure that relies on self-weight and the weight of the earth over the heel to resist overturning. Retaining walls must be designed for both the capacity of the soil and the capacity of the concrete.

Soil capacities include overturning, sliding, and bearing pressure. Different cases of lateral soil loading acting on the wall are shown in *Handbook* figure Stress States on a Soil Element Subjected Only to Body Stresses. CERM Sec. 37.12 describes the steps for the soil analysis of a cantilevered retaining wall.

Exam tip: The cantilever retaining wall is the most likely type to appear on the exam.

30 Design and Details of Structures: Steel

Content in blue refers to the *NCEES Handbook*.

A.2. Steel 30-1
B.1. Horizontal Members 30-3
B.2. Vertical Members 30-6
B.3. Systems 30-9
B.4. Connections 30-9

The knowledge area of Design and Details of Structures makes up between 16 and 24 questions out of the 80 questions on the PE Civil Structural exam. Four to six questions are on materials and material properties, while 12 to 18 questions are on component design and detailing. This chapter focuses on these knowledge areas only in the context of steel. The organization of this chapter follows the order of subtopics given by NCEES for this knowledge area, but includes only subtopics relevant to the design and details of structures for steel.

A.2. STEEL

Key concepts: These key concepts are important for answering exam questions in subtopic 10.A.2, Steel.

- compact, noncompact, and slender elements
- material properties of structural steel
- slender and nonslender elements
- structural steel section properties

***NCEES Handbook*:** To prepare for this subtopic, familiarize yourself with these sections in the *Handbook*.

- Stress-Strain Curve for Mild Steel

***PE Civil Reference Manual* (CERM):** Study these sections in CERM that either relate directly to this subtopic or provide background information.

- Section 58.2: Types of Structural Steel and Connecting Elements
- Section 58.3: Steel Properties
- Section 58.4: Structural Shapes
- Section 59.4: Compact Sections
- Section 61.10: Local Buckling

Codes and standards: To prepare for this subtopic, familiarize yourself with these sections from codes and standards.

Steel Construction Manual (AISC)

- Part 1: Dimensions and Properties
- Part 2: General Design Considerations
- Part 4: Design of Compression Members
- Part 16 Section B4: Member Properties
- Part 16 Section E7: Members with Slender Elements
- Part 16 Chapter F: Design of Members for Flexure
- Part 16.1 Glossary

The following equations, figures, and concepts are relevant for subtopic 10.A.2, Steel.

Material Properties of Structural Steel

Handbook: Stress-Strain Curve for Mild Steel

CERM: Sec. 58.2, Sec. 58.3

AISC: Part 2, Part 16.1 Glossary

Steel can be classified into two categories: hot-rolled and cold-formed steel. *Hot-rolled steel*, also called structural steel, is steel rolled at a high temperature above the steel recrystallization temperature.

Cold-formed steel is a steel sheet that is rolled or pressed at relatively low temperatures.

Typical stress-strain curves for structural steels are found in CERM Fig. 58.1. Key properties from these curves are yield strength, ultimate tensile strength, and modulus of elasticity.

Yield strength, F_y, is the stress beyond which strain increases without a significant increase in stress. This is the point between elastic behavior and plastic behavior. Some steels have well-defined yield points, but for other steels, the yield strength is defined according to an offset (e.g., 0.2%) from ideal elastic behavior or as a stress at a specified strain. *Handbook* figure Stress-Strain Curve for Mild Steel shows an example of this 0.2% offset for mild steel.

Tensile strength, F_u, also called ultimate tensile strength, is the largest stress that a material can withstand in a tension test.

The yield strength and tensile strength for various types of structural steel are shown in AISC Table 2-4. See the glossary in AISC Part 16.1 for definitions.

Figure 58.1 Typical Stress-Strain Curves for Structural Steels

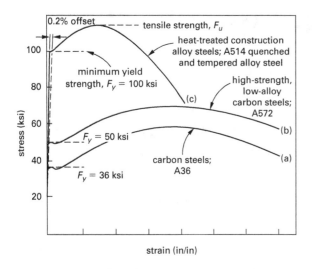

Figure 58.2 Structural Shapes

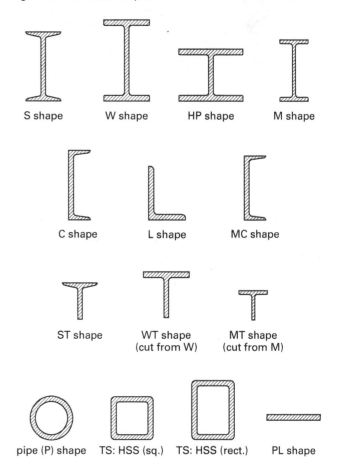

The modulus of elasticity, E, is the slope of the initial straight-line section of the stress-strain curve (the elastic region).

Exam tip: The modulus of elasticity for steel is approximately 29,000 ksi. It does not change significantly with steel type.

Structural Steel Section Properties

CERM: Sec. 58.4

AISC: Part 1

Structural steel comes in different typical shapes, as shown in CERM Fig. 58.2. These shapes are used for different member types. For example, many structural steel beams and columns are made of W shapes (wide-flange shapes). Hollow structural section (HSS) members can be used for beams, columns, and braces in brace frames.

Typical section properties for structural steel used in design are area, A; moment of inertia, I; elastic section modulus, S; plastic section modulus, Z; element dimensions; and nominal weight. These values are given in the tables of AISC Part 1 for typical hot-rolled shapes.

Exam tip: If the member in the exam question is a standard section, go to the AISC tables first before attempting to calculate the section properties.

Compact, Noncompact, and Slender Elements

CERM: Sec. 59.4

AISC: Part 1, Part 16 Sec. B4, Part 16 Chap. F

$$\frac{b_f}{2t_f} \leq 0.38 \sqrt{\frac{E}{F_y}} \quad \begin{bmatrix} \text{flanges in flexural} \\ \text{compression only} \end{bmatrix} \quad 59.4$$

$$\frac{h}{t_w} \leq 3.76 \sqrt{\frac{E}{F_y}} \quad \begin{bmatrix} \text{webs in flexural} \\ \text{compression only} \end{bmatrix} \quad 59.5$$

Flanges and webs in structural steel beams for flexure may be described as compact, noncompact, or slender based on their width-to-thickness ratios. AISC Table B4.1b identifies the width-to-thickness ratio limits for each case.

For a wide-flange member, the width-to-thickness ratio, λ, for flanges is

$$\lambda = \frac{b_f}{2t_f}$$

For a wide-flange member, the width-to-thickness ratio, λ, for webs is

$$\lambda = \frac{h}{t_w}$$

These values are provided in AISC Table 1-1 for typical rolled shapes. The limit between compact and noncompact elements is designated as λ_p, and the limit between noncompact and slender elements is λ_r. These are given in AISC Table B4.1b for various stiffened and unstiffened elements. CERM Eq. 59.4 and Eq. 59.5 demonstrate the check to determine whether the element is compact.

If a member has noncompact or slender elements, then the flexural strength will need additional checks for the reduced strength. AISC Sec. F3, Sec. F4, and Sec. F5 cover these additional checks.

Sections with a superscripted f next to the size in AISC Table 1-1 have noncompact flanges for flexure. The tables in AISC Part 3 take these reduced capacities into consideration.

Exam tip: If the member is a typical shape, be sure to check if it has a superscript f next to the name in the tables before trying to calculate the width-to-thickness ratio.

Slender and Nonslender Elements

CERM: Sec. 61.10

AISC: Part 1, Part 4, Part 16 Sec. B4.1, Part 16 Sec. E7

Flanges and webs in structural steel columns in pure axial compression may be described as slender and nonslender based on the width-to-thickness ratio. Nonslender elements have smaller width-to-thickness ratios and can hold more load than slender elements. AISC Table B4.1a identifies the width-to-thickness ratio limits for each case.

For a wide-flange member, the width-to-thickness ratio, λ, for flanges is

$$\lambda = \frac{b_f}{2t_f}$$

For a wide-flange member, the width-to-thickness ratio, λ, for webs is

$$\lambda = \frac{h}{t_w}$$

These values are provided in AISC Table 1-1 for typical rolled shapes. The limit between nonslender and slender is designated as λ_r. These are given in AISC Table B4.1a for axial members and CERM Table 61.2.

If a member has slender elements, then the axial compressive strength will need additional checks for the reduced strength due to local buckling. AISC Part 16 Sec. E7 covers these additional checks.

Sections with a superscript c next to the size in AISC Table 1-1 have slender webs for axial compression. The tables in AISC Part 4 take these reduced capacities into consideration.

Exam tip: If the member is a typical shape, be sure to check whether it has a superscript c next to the name in the tables before trying to calculate the width-to-thickness ratio.

B.1. HORIZONTAL MEMBERS

Key concepts: These key concepts are important for answering exam questions in subtopic 10.B.1, Horizontal Members.

- composite steel beams
- steel beam shear capacity
- steel beams
- steel flexural capacity

NCEES Handbook: To prepare for this subtopic, familiarize yourself with these sections in the *Handbook*.

- Moment, Shear, and Deflection Diagrams

PE Civil Reference Manual (CERM): Study these sections in CERM that either relate directly to this subtopic or provide background information.

- Section 59.1: Types of Beams
- Section 59.5: Lateral Bracing
- Section 59.10: Shear Strength in Steel Beams
- Section 59.12: Analysis of Steel Beams
- Section 59.13: Design of Steel Beams
- Chapter 64: Structural Steel: Composite Beams

Structural Depth Reference Manual (CEST): Study these sections in CEST that either relate directly to this subtopic or provide background information.

- Section 4.4: Composite Beams

Codes and standards: To prepare for this subtopic, familiarize yourself with these sections from codes and standards.

International Building Code (IBC)

- Section 1604: General Design Requirements

Steel Construction Manual (AISC)

- Part 3: Design of Flexural Members
- Part 16 Chapter F: Design of Members for Flexure
- Part 16 Chapter G: Design of Members for Shear
- Part 16 Section I3: Flexure
- Part 16 Section I8: Steel Anchors

The following equations, figures, tables, and concepts are relevant for subtopic 10.B.1, Horizontal Members.

Steel Beams

Handbook: Moment, Shear, and Deflection Diagrams

CERM: Sec. 59.1, Sec. 59.12, Sec. 59.13

AISC: Part 3, Part 16 Chap. F, Part 16 Chap. G

IBC: Sec. 1604

Steel beams primarily support transverse loads. They are subjected primarily to flexure (bending). Although some axial loading is unavoidable in any structural member, the effect of axial loads is generally negligible, and the member can be treated strictly as a beam. If an axial compressive load of substantial magnitude is also present with transverse loads, the member is called a *beam-column*.

Typical steel beams shapes are wide flanges, channels, and hollow structural section (HSS) members. Beam shear, flexure, and deflection can be found per the shear and bending moment diagrams in the tables included in *Handbook* section Moment, Shear, and Deflection Diagrams, as well as AISC Table 3-23. Allowable deflection limits are found in IBC Table 1604.3.

Flexural capacity for beams can be found in AISC Part 16 Chap. F. AISC Table 3-1, Table 3-2, and Table 3-10 can also be used to determine the flexural capacity. Refer to the entry "Steel Flexural Capacity" in this chapter for information on these tables. Shear capacity for beams can be found in AISC Part 16 Chap. G. CERM Fig. 59.6 and Fig. 59.7 provide flowcharts for the checks required for steel beam analysis and design.

Exam tip: Be familiar with the AISC design aid tables and know which ones you can use. Often when solving exam problems, you can either pull a value from the table or do a few short calculations to obtain the value you need.

Steel Flexural Capacity

CERM: Sec. 59.5

AISC: Part 3, Part 16 Chap. F

There are three types of flexural failure: plastic, inelastic, and elastic lateral torsional buckling. The nominal flexural capacity for each of these failure modes is provided in AISC Eq. F2-1, Eq. F2-2, and Eq. F2-3 and is dependent on the unbraced length, L_b, of the compression flange. In no case may the flexural capacity exceed the plastic capacity, M_p, provided in AISC Eq. F2-1. The design strength is the nominal strength multiplied by ϕ or divided by Ω_o, depending on the use of LRFD or ASD methodology.

CERM Fig. 59.5 is similar to AISC Fig. 3.1 and demonstrates how the unbraced length affects the overall flexural capacity. The equations for the unbraced length limits, L_p and L_r, are given by AISC Eq. F2-5 and Eq. F2-6 and CERM Eq. 59.6 and Eq. 59.7.

Figure 59.5 Available Moment Versus Unbraced Length

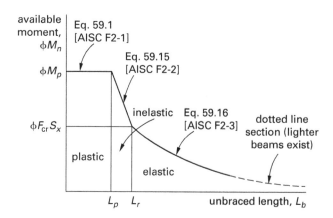

Plastic failure occurs when the entire cross section yields. Elastic lateral torsional buckling occurs when the beam buckles laterally due to flexural compressive forces, as shown in CERM Fig. 59.3.

Figure 59.3 Lateral Buckling in a Beam

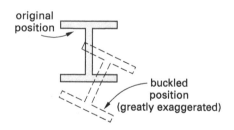

Inelastic lateral torsional buckling, shown in the following illustration, occurs when some of the cross section yields and the cross section is long enough to have some lateral torsional buckling.

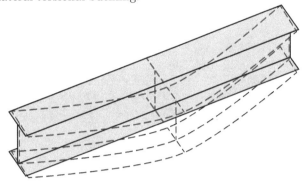

Inelastic and elastic lateral torsional buckling capacities have the lateral torsional buckling modification factor for nonuniform moment diagrams factor, C_b, included in the equation to account for a nonuniform moment diagram. C_b can be calculated using AISC Eq. F1-1 or found for typical loading conditions in AISC Table 3-1. C_b will never be less than 1.0.

M_p, L_p, L_r, and the flexural strength at $L_b = L_r$ are provided in AISC Table 3-2. The design aid tables (AISC Table 3-2 and Table 3-10) assume that C_b is 1.0. When using these tables in cases where C_b is not 1.0, do not multiply the plastic moment by C_b, because it only applies to inelastic and elastic lateral torsional buckling.

AISC Table 3-2 can be used to determine the flexural capacity for a beam with $L_b \leq L_r$. AISC Eq. 3-4 may be used with AISC Table 3-2 for calculating the capacity of a beam when $L_p < L_b \leq L_r$. AISC Table 3-10 can be used to determine the flexural capacity of a beam for all cases based on the unbraced length.

For beams that have noncompact flanges (indicated with a superscript f in the tables), the plastic capacity and L_p value must be adjusted for local buckling of the compression flange. When using AISC Table 3-2 for a noncompact member, use the adjusted ϕM_p, M_p/Ω_o, and L_p values from the table when applying AISC Eq. F2-2. Refer to the entry "Compact, Noncompact, and Slender Elements" in this chapter for discussion on noncompact flanges.

Exam tip: When using the tables and L_p is less than L_b, be sure to multiply the capacity by C_b, but check it against the plastic capacity, ϕM_p (or M_p/Ω_o when using ASD).

Steel Beam Shear Capacity

CERM: Sec. 59.10

AISC: Part 3, Part 16 Chap. G

$$V_n = 0.6 F_y A_w C_v \quad \text{[AISC Eq. G2-1]}$$

For steel I-beams, the shear strength is provided by the web. For built-up members, transverse stiffeners may need to be added to help strengthen the web. The nominal shear strength for I-shaped members and channels is provided in AISC Eq. G2-1 (and CERM Eq. 59.18). The design strength is the nominal strength multiplied by the resistance factor, ϕ, or divided by the safety factor, Ω_o, depending on the use of LRFD or ASD methodology.

For built-up members, $\phi = 0.9$, and the web shear coefficient, C_v, must be calculated. C_v accounts for the web buckling due to the shear load based on the web h/t_w ratio.

For most standard W shape members, $\phi = 1.0$ and $C_v = 1.0$. When there is a superscript v next to the shape in AISC Table 1-1, the web does not meet the criteria of AISC Sec. G2.1(a) for A992 steel and $\phi = 0.9$. The design shear capacity of typical wide flanges is provided in AISC Table 3-2.

Exam tip: Use AISC Table 3-2 whenever possible before spending time calculating the shear capacity.

Composite Steel Beams

CERM: Chap. 64

CEST: Sec. 4.4

AISC: Part 3, Part 16 Sec. I3, Part 16 Sec. I8

In steel construction, it is often more economical to design the steel beam and concrete deck to act compositely as a single member. This composite design often reduces the size of the steel beam because it no longer has to resist all the load.

Composite beams must be designed for both precomposite and composite capacity. Prior to the hardening of the concrete, the steel beam must be able to support the initial dead load and construction live load. After the concrete has hardened and reached strength, the concrete slab and steel beam act compositely to support the loads on the member. Composite sections must be designed for flexure, deflection, and shear transfer between the concrete and the steel. CEST Fig. 4.21 shows an example of a composite section.

The web of a steel beam must be designed for shear as well. AISC Part 16 Sec. I3 covers the flexural capacity of composite beams, including calculating the transfer force between concrete and steel. AISC Table 3-19 provides a design aid to calculate the flexural capacity of a composite beam. CERM Fig. 64.3 provides a general description of the methodology and modeling used in the composite flexural strength.

Figure 64.3 Strength Design Models for Composite Beams

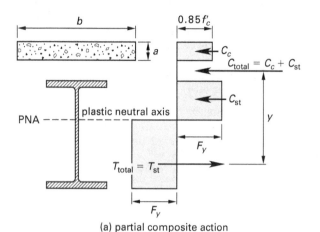

(a) partial composite action

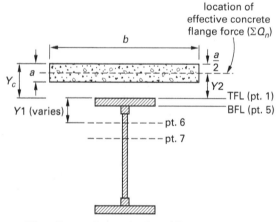

(b) steel beam dimensions

$$\Sigma Q_n(@ \text{ pt. } 6) = \frac{\Sigma Q_n(@ \text{ pt. } 5) + \Sigma Q_n(@ \text{ pt. } 7)}{2}$$

$$\Sigma Q_n(@ \text{ pt. } 7) = 0.25 F_y A_s$$

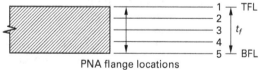

PNA flange locations

(c) concrete flange dimensions

*From *AISC Manual* Fig. 3-3.

Copyright © American Institute of Steel Construction, Inc. Reprinted with permission. All rights reserved.

AISC Part 16 Sec. I8 provides detailing requirements for the anchor studs used in composite beams. AISC Table 3-20 may be used to help calculate the deflection of the composite beam. AISC Table 3-21 may be used to calculate the shear transfer force capacity for each anchor stud.

Exam tip: Using the design aid will help you solve the problem faster. The total number of studs in AISC Eq. 3-7 is the number of studs to provide transfer on the side of the maximum moment. For a uniformly loaded beam, this will be half the studs in the beam.

B.2. VERTICAL MEMBERS

Key concepts: These key concepts are important for answering exam questions in subtopic 10.B.2, Vertical Members.

- effective length factor
- steel columns
- steel members with combined axial and flexural loading
- tensile capacity of steel members

NCEES Handbook: The *Handbook* does not contain any material relevant to this subtopic. Be sure to study the listed sections in CERM and from codes and standards.

PE Civil Reference Manual **(CERM):** Study these sections in CERM that either relate directly to this subtopic or provide background information.

- Section 60.2: Axial Tensile Strength
- Section 60.8: Analysis of Tension Members
- Section 61.3: Effective Length
- Section 61.7: Design Compressive Strength
- Section 61.8: Analysis of Columns
- Section 62.2: Flexural/Axial Compression
- Section 62.4: Analysis of Beam-Columns

Codes and standards: To prepare for this subtopic, familiarize yourself with these sections from codes and standards.

Steel Construction Manual (AISC)

- Part 4: Design of Compression Members
- Part 16 Section B4.3: Gross and Net Area Determination
- Part 16 Section C1.1: Direct Analysis Method of Design

- Part 16 Chapter D: Design of Members for Tension
- Part 16 Chapter E: Design of Members for Compression
- Part 16 Chapter H: Design of Members for Combined Forces and Torsion
- Appendix 7: Alternative Methods of Design for Stability
- Appendix 8: Approximate Second-Order Analysis

The following equations, figures, and concepts are relevant for subtopic 10.B.2, Vertical Members.

Tensile Capacity of Steel Members

CERM: Sec. 60.2, Sec. 60.8

AISC: Part 16 Sec. B4.3, Part 16 Chap. D

$$P_n = F_y A_g \quad \begin{bmatrix} \text{yielding criterion;} \\ \text{AISC Specification Eq. D2-1} \end{bmatrix} \quad 60.2$$

$$P_n = F_u A_e = A_n U F_u \quad \begin{bmatrix} \text{fracture criterion;} \\ \text{AISC Specification} \\ \text{Eq. D2-2 and Eq. D3-1} \end{bmatrix} \quad 60.3$$

Tension members must be checked for tension yielding and tension rupture. The nominal tensile strength for structural steel members in yielding is given in AISC Eq. D2-1 and CERM Eq. 60.2. The nominal tensile strength for tensile rupture is given in AISC Eq. D2-2 and CERM Eq. 60.3. The design strength is the nominal strength multiplied by the resistance factor, ϕ, or divided by the safety factor, Ω_o, depending on the use of LRFD or ASD methodology. ϕ is 0.9 for yield and 0.75 for rupture.

A_e is the effective area and is equal to U multiplied by the net area, A_n. The shear lag reduction coefficient, U, accounts for uneven tensile stress at the connecting elements and can be found in AISC Table D3.1. AISC Sec. B4.3 describes the adjustments made for staggered bolts and the net area of bolt holes. When calculating A_n in the capacity of tension and shear with bolts, an additional 1/16 in should be included in the diameter of the hole, per AISC Sec. B4.3b.

At the ends of tension members, additional considerations are block shear and bolt group action due to bolt shear and bolt hole bearing and tearing. Refer to the entries "Bolts" and "Block Shear" in this chapter for further discussion on these failure modes. Block shear can occur in both welded and bolted connections.

CERM Fig. 60.6 provides a flowchart for the analysis of tensile members.

Exam tip: Because of the design assumptions for the effective area used in the AISC tables, it is recommended to calculate the tensile rupture capacity using the equations from AISC Chap. D. When calculating the net area of bolt holes, do not forget to add 1/16 in for the bolt hole diameter.

Effective Length Factor, K

CERM: Sec. 61.3

AISC: Part 16 Sec. C1.1, App. 7 and commentary

The effective length factor, K, is used to modify the unbraced length of compression members for buckling behavior based on their end conditions.

When analyzing and modeling steel structures using the direct analysis method (refer to AISC Part 16 Sec. C1.1), K is taken as 1.0 for all members.

When analyzing and modeling steel structures using the effective length method (refer to AISC App. 7), K must be calculated. There are multiple methods for calculating K. These include

- using AISC Table C-A-7.1
- using the alignment chart in AISC Fig. C-A-7.1 for columns in braced frames (also shown in CERM Fig. 61.2(a))
- using the alignment chart in AISC Fig. C-A-7.2 for columns in moment frames (also shown in CERM Fig. 61.2(b))

The values of K in AISC Table C-A-7.1 do not require prior knowledge of the column size or shape. However, if the columns and beams of an existing design are known, the alignment chart can be used to obtain a more accurate effective length factor.

Exam tip: Be sure to read the problem statement and the provided information before selecting the appropriate K-value. Some compressive strength problems may just ask for the K-value, and some may provide the effective length, KL.

Steel Columns

CERM: Sec. 61.7, Sec. 61.8

AISC: Part 4, Part 16 Chap. E

$$P_n = F_{cr} A_g \quad [\text{AISC Specification Eq. E3-1}] \quad 61.8$$

The nominal compressive strength for I-shaped members and channels is provided in AISC Eq. E3-1 and CERM Eq. 61.8. The design strength is the nominal strength times the resistance factor, ϕ (for LRFD), or divided by the safety factor, Ω_o (for ASD), for the allowable strength. The critical stress of the cross section, F_{cr}, is dependent on the effective length ratio, KL/r. Refer to the entry "Effective Length Factor, K" in this chapter for an explanation on calculating the K-factor.

CERM Fig. 61.4 provides a visual of the available compressive stress versus the slenderness ratio (KL/r). The two regions in the plot represent inelastic buckling and elastic buckling; the regions are separated by the limit given in AISC Eq. E3-2.

Figure 61.4 Available Compressive Stress Versus Slenderness Ratio

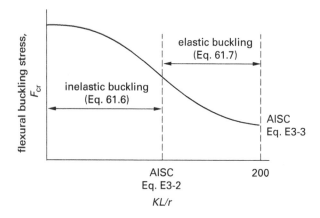

AISC Eq. E3-2 and CERM Eq. 61.6 give the design stress for intermediate columns that fail by inelastic buckling and account for the inelastic behavior due to residual stresses and initial crookedness in a real column.

$$F_{cr} = (0.658^{F_y/F_e})F_y \qquad 61.6$$

AISC Eq. E3-3 (CERM Eq. 61.7) must be used for long columns that fail by elastic buckling.

$$F_{cr} = 0.877 F_e \qquad 61.7$$

The capacity must be checked for both the strong and weak axes. The controlling effective length ratio is the larger of the two ratios and will provide the smallest capacity. Refer to the entry "Slender and Nonslender Elements" in this chapter for the capacity of members with slender elements.

The tables in AISC Part 4 provide the compressive strength of typical steel members based on the effective length, KL, of the weak axis. To account for members with longer effective lengths in the strong axis, use an equivalent length for the strong axis capacity per AISC Eq. 4-1.

Exam tip: Use the tables as much as possible to save time during the exam. AISC Table 4-1 is based on weak axis effective length, which typically controls. If the effective length ratio for the strong axis is larger than the effective length ratio for the weak axis due to a longer effective length, then use the equivalent effective length per AISC Eq. 4-1.

Steel Members with Combined Axial and Flexural Loading

CERM: Sec. 62.2, Sec. 62.4

AISC: Part 16 Chap. H, App. 8

$$\frac{P_r}{P_c} + \left(\frac{8}{9}\right)\left(\frac{M_{rx}}{M_{cx}} + \frac{M_{ry}}{M_{cy}}\right) \leq 1.0 \qquad 62.3$$

$$\frac{P_r}{2P_c} + \frac{M_{rx}}{M_{cx}} + \frac{M_{ry}}{M_{cy}} \leq 1.0 \qquad 62.4$$

Steel members that are loaded axially and in flexure must be analyzed per AISC Eq. H1-1a (CERM Eq. 62.3) and Eq. H1-1b (CERM Eq. 62.4). When the axial force is in compression, the member is considered a beam-column and must account for second-order effects. Second-order effects are the additional moments and forces caused by the horizontal displacements of vertical loads. AISC App. 8 provides an approximate method for calculating second-order effects on beam-columns. AISC Sec. H2 provides additional information for steel members loaded in axial tension and flexure.

P_r is the required axial strength. For beam-columns, this is the amplified axial load in the column due to second-order effects.

P_c is the design axial strength of the member. For LRFD, this is ϕP_n.

M_r is the required flexural strength. For beam-columns this is the amplified moment due to second-order effects.

M_c is the design flexural strength of the member. For LRFD, this is ϕM_n.

CERM Sec. 62.4 provides the steps for analysis of beam-columns.

Exam tip: When checking beam-columns, be sure to use the AISC tables to find the axial and flexural strengths to plug into the AISC Chap. H equations.

B.3. SYSTEMS

Key concepts: These key concepts are important for answering exam questions in subtopic 10.B.3, Systems.

- axial loads in truss members
- frame deflection
- frames
- tension and compression in steel members
- truss deflection
- trusses

Systems consist of trusses and frames. Trusses have only tension and compression members. Typical steel frames are brace frames and moment frames. Frame members have axial, shear, and moment. The analysis of trusses and frames, including internal forces and deflections, is discussed further in Chap. 28 of this book.

B.4. CONNECTIONS

Key concepts: These key concepts are important for answering exam questions in subtopic 10.B.4, Connections.

- block shear
- bolts
- column base plates
- eccentric bolt groups
- eccentrically loaded welds
- simple shear connections
- welds

NCEES Handbook: To prepare for this subtopic, familiarize yourself with these sections in the *Handbook*.

- Fastener Groups in Shear
- Basic Welding Symbols

PE Civil Reference Manual (CERM): Study these sections in CERM that either relate directly to this subtopic or provide background information.

- Section 60.6: Block Shear Strength
- Section 61.12: Column Base Plates
- Chapter 65: Structural Steel: Connectors
- Section 66.2: Types of Welds and Joints
- Section 66.3: Fillet Welds
- Section 66.4: Concentric Tension Connections
- Section 66.6: Eccentrically Loaded Welded Connections

Structural Depth Reference Manual (CEST): Study these sections in CEST that either relate directly to this subtopic or provide background information.

- Section 4.2: Eccentrically Loaded Bolt Groups
- Section 4.3: Eccentrically Loaded Weld Groups

Codes and standards: To prepare for this subtopic, familiarize yourself with these sections from codes and standards.

Steel Construction Manual (AISC)

- Part 7: Design Considerations for Bolts
- Part 8: Design Considerations for Welds
- Part 10: Design of Simple Shear Connections
- Part 14: Design of Beam Bearing Plates, Column Base Plates, Anchor Rods, and Column Splices
- Part 16 Section J2: Welds
- Part 16 Section J3: Bolts and Threaded Parts
- Part 16 Section J4: Affected Elements of Members and Connecting Elements
- Part 16 Section J8: Column Bases and Bearing on Concrete

The following equations, figures, and concepts are relevant for subtopic 10.B.4, Connections.

Eccentric Bolt Group

Handbook: Fastener Groups in Shear

CERM: Sec. 65.7

CEST: Sec. 4.2

AISC: Part 7

Bolt groups can be loaded eccentrically in the plane of the faying surface, as shown in CEST Fig. 4.9, or normal to the faying surface, as shown in the figure in CEST Ex. 4.12.

Figure 4.9 Eccentrically Loaded Bolt Group, LRFD Method

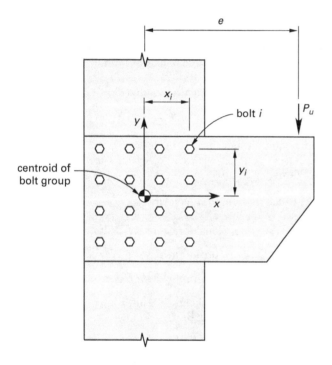

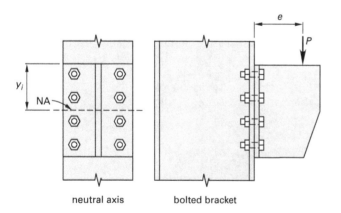

Bolts loaded eccentrically in the plane of the faying surface will be loaded in shear. Bolts loaded eccentrically normal to the faying surface will have tension and shear.

There are two methods provided in AISC for analyzing bolts loaded eccentrically in the plane of the faying surface: the elastic method and the instantaneous center of rotation method. The *elastic method* is covered in *Handbook* section Fastener Groups in Shear and in AISC Part 7 (AISC Eq. 7-2 through Eq. 7-8). The *instantaneous center of rotation method* uses AISC Table 7-6 through Table 7-13. Note that interpolation between the tables is not allowed. In both methods, the individual bolt effective strength is related to the bolt group capacity. Refer to the entry "Bolts" in this chapter for more information.

AISC Part 7 provides two cases for analyzing bolts loaded eccentrically normal to plane.

> *Exam tip*: Be sure to read the problem statement carefully. Test problems may require the use of the elastic method, which may also be used for non-typical configurations.

Welds

Handbook: Basic Welding Symbols

CERM: Sec. 66.2, Sec. 66.3, Sec. 66.4

AISC: Part 8, Part 16 Sec. J2

Welds types include groove welds, fillet welds, and plug and slot welds. CERM Fig. 66.1 illustrates types of welds. AISC Table 8-2 and *Handbook* section Basic Welding Symbols provide the typical symbols used for specifying welds.

The base material and the weld must be checked for capacity. The weld capacity is dependent on the effective area of the weld and the filler metal classification strength (F_{EXX}).

Weld detailing and capacities are found in AISC Part 16 Sec. J.2 and in Part 8. The effective area and limitations for groove welds are provided in AISC Part 16 Sec. J2.1. The effective area and limitations, including minimum and maximum sizes for fillet welds, are provided in AISC Part 16 Sec. J2.2. AISC Table J2.5 identifies the sections and equations for determining the available weld strength for each type of weld.

DESIGN AND DETAILS OF STRUCTURES: STEEL

The weld strength for shorter fillets weld with $F_{EXX} = 70$ ksi (typical) is provided in AISC Eq. 8-2 (CERM Eq. 66.5 and 66.6).

$$\phi R_n = 1.392 Dl \quad \text{[LRFD; E70 electrode]} \quad 66.5$$

$$\frac{R_n}{\Omega} = 0.928 Dl \quad \text{[ASD; E70 electrode]} \quad 66.6$$

In these equations, D is the size of the weld in sixteenths of an inch, and L is the length of the weld. The length can be taken as 1 in to obtain a unit capacity when considering eccentrically loaded weld groups.

Exam tip: For shorter fillet welds, use AISC Eq. 8-2a (for LRFD) or Eq. 8-2b (for ASD). It may be beneficial to memorize the equations.

Block Shear

CERM: Sec. 60.6

AISC: Part 16 Sec. J4

$$\begin{aligned} R_n &= 0.6 F_u A_{nv} + U_{bs} F_u A_{nt} \\ &\leq 0.6 F_y A_{gv} + U_{bs} F_u A_{nt} \end{aligned} \quad 60.10$$

$$\begin{bmatrix} \text{block shear criterion;} \\ \text{AISC Specification Eq. J4-5} \end{bmatrix}$$

Depending on the connection at the end of a tension member, *block shear failure* can occur in either the tension member itself or in the member to which it is attached (e.g., a gusset plate). Block shear failure represents a combination of two failures—a shear failure along a plane through the bolt holes or welds and a tension failure along a perpendicular plane—occurring simultaneously.

Block shear strength, R_n, can be computed from AISC Eq. J4-5 or CERM Eq. 60.10. A_{gv} is the gross area subject to shear, A_{nt} is the net area subject to tension, and A_{nv} is the net area subject to shear. Where the tension stress is uniform, U_{bs} is 1.0, and where the tension stress is nonuniform, U_{bs} is 0.5.

For tension members, another possible failure mode exists in addition to yielding and fracture. In this mode, a segment or "block" of material at the end of the member tears out, as shown in CERM Fig. 60.5.

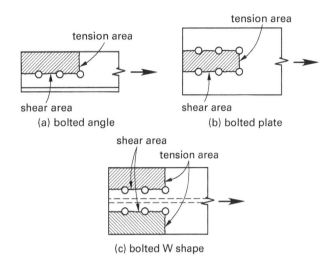

Figure 60.5 Block Shear Failures

The nominal strength must be multiplied by ϕ or divided by Ω_o to determine the design, depending on the use of LRFD or ASD methodology.

Exam tip: Don't forget to include an additional $1/16$ in to the bolt hole diameter, per AISC Sec. B4.3b, when considering block shear for bolts.

Column Base Plates

CERM: Sec. 61.12

AISC: Part 14, Part 16 Sec. J8

Column base plates must be designed for concrete bearing and base plate minimum thickness. Concrete bearing strength is found in AISC Chap. J8, which also references corresponding ACI equations. The minimum thickness is based on the cantilever flexural capacity of the plate. The minimum plate thickness can be found using AISC Eq. 14-7a (CERM Eq. 61.32(a)) for LRFD and Eq. 14-7b (CERM Eq. 61.32(b)) for ASD.

$$t_{\min} = l \sqrt{\frac{2 P_u}{0.9 F_y BN}} \quad \text{[LRFD]} \quad 61.32(a)$$

$$t_{\min} = l \sqrt{\frac{3.33 P_a}{F_y BN}} \quad \text{[ASD]} \quad 61.32(b)$$

In these equations, l is the largest of m, n, and $\lambda n'$, which are provided in AISC Eq. 14-2 through Eq. 14-5 and shown in AISC Fig. 14-3 and CERM Fig. 61.7. B and N are the base plate dimensions, P_u and P_a are the axial load in the column, and F_y is the yield strength of the plate.

Figure 61.7 Column Base Plate

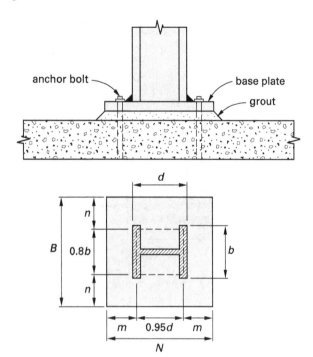

Exam tip: The factor λ is conservatively taken as 1.0. If n' is less than m and n, then skip the step on the calculation of λ and $\lambda n'$.

Bolts

CERM: Chap. 65

AISC: Part 7, Part 16 Sec. J3

Common bolts are classified by ASTM as A307 bolts. The two basic types of *high-strength bolts* have ASTM designations of A325 and A490. Bolts can be loaded in shear, tension, or combined shear and tension. Bolt groups can be in a bearing-type connection or a slip critical connection. A *bearing connection* relies on the shearing resistance of the fasteners to resist loading.

Bolt detailing and capacities are found AISC Part 7 and Part 16 Sec. J3. Detailing requirements such as hole size, edge distance, and bolt spacing are provided in AISC Part 16 Sec. J3.1 through Sec. J3.5.

Per the user note at the end of AISC Part 16 Sec. J3.6, the effective strength of an individual fastener in the bearing connection is the lesser of the bolt shear strength (AISC Part 16 Sec. J3.6) and the bearing/tearing strength (AISC Part 16 Sec. J3.10). The capacity of a bolt group is the sum of the effective strengths of the individual fasteners.

AISC Table 7-1 through Table 7-5 can be used to calculate individual bolt strengths. AISC Table 7-4 and Table 7-5 provide the bearing and tearing capacity at the holes per thickness of the element. AISC Table 7-5 is for edge bolts in the group. Interpolation is not allowed for these tables.

Exam tip: Use the tables when possible for typical bolt spacing. Do not forget to multiply by the plate thickness when using AISC Table 7-4 and Table 7-5.

Eccentrically Loaded Welds

CERM: Sec. 66.6

CEST: Sec. 4.3

AISC: Part 8, Part 16 Sec. J2.4

Welds may be loaded eccentrically in the plane of the faying surface or normal to the plane of the faying surface (refer to CERM Fig. 66.6 and Fig. 66.7). There are two methods for determining the capacity of eccentrically loaded weld groups: the elastic method and the instantaneous center of rotation method.

Figure 66.6 Welded Connection in Combined Shear and Torsion (eccentricity in plane of faying surfaces)

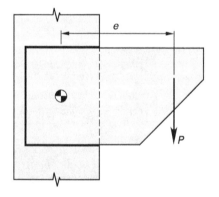

Figure 66.7 Welded Connection in Combined Shear and Bending (eccentricity normal to plane of faying surfaces)

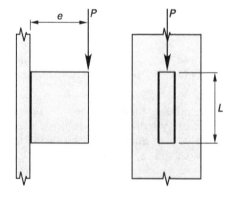

The *elastic method* is discussed in AISC Part 8 and compares a unit load calculated by an applied moment to the group and force to the unit strength of the weld. Refer to the entry "Welds" in this chapter for determining the unit strength of a weld.

The *instantaneous center of rotation method* uses AISC Table 8-4 through Table 8-11 for fillet welds. Interpolation between the tables is not appropriate.

Exam tip: The tables for the instantaneous center of rotation method apply only to fillet welds. If an exam problem concerns a groove weld, use the elastic method. If the problem concerns a fillet weld, be sure to read the problem statement carefully before using the tables.

Simple Shear Connections

AISC: Part 10

For simple shear connections, multiple failure modes should be checked. These include but are not limited to bolt shear, weld capacity, bolt group capacity, thickness requirements, local shear and tension capacity of the plate, and block shear.

AISC Part 10 provides tables for the capacity of different typical shear connections. A discussion on the limitations and assumptions for each table is provided before each table.

Exam tip: The checks necessary for a simple shear connection may take longer than six minutes. Use the design aid tables for the capacity of simple shear connections.

31 Design and Details of Structures: Timber

Content in blue refers to the *NCEES Handbook*.

A.3. Timber.. 31-1
B.1. Horizontal Members 31-2
B.2. Vertical Members 31-5
B.3. Systems.. 31-8
B.4. Connections................................. 31-8

The knowledge area of Design and Details of Structures makes up between 16 and 24 questions out of the 80 questions on the PE Civil Structural exam. Four to six questions are on materials and material properties, while 12 to 18 questions are on component design and detailing. This chapter focuses on these knowledge areas only in the context of timber. The organization of this chapter follows the order of subtopics given by NCEES for this knowledge area, but includes only subtopics relevant to the design and details of structures for timber.

A.3. TIMBER

Key concepts: These key concepts are important for answering exam questions in subtopic 10.A.3, Timber.

- timber design method
- wood adjustment factors
- wood shape and species properties

NCEES Handbook: The *Handbook* does not contain any material relevant to this subtopic. Be sure to study the listed sections in CEST and from codes and standards.

Structural Depth Reference Manual **(CEST):** Study these sections in CEST that either relate directly to this subtopic or provide background information.

- Chapter 5: Design of Wood Structures

Codes and standards: To prepare for this subtopic, familiarize yourself with these sections from codes and standards.

International Building Code (IBC)

- Section 1605.3.1: Basic Load Combinations

Minimum Design Loads for Buildings and Other Structures (ASCE/SEI 7)

- Section 2.4: Combining Nominal Loads Using Allowable Stress Design

National Design Specification for Wood Construction with Commentary (NDS)

- Section 1.4: Design Procedure
- Part 3: Design Provisions and Equations
- Part 4: Sawn Lumber
- Part 5: Structural Glued Laminated Timber

National Design Specification for Wood Construction Supplement (NDS Supplement)

- Chapter 3: Section Properties
- Chapter 4: Reference Design Values

The following equations, tables, and concepts are relevant for subtopic 10.A.3, Timber.

Wood Shape and Species Properties

CEST: Sec. 5.3

NDS Supplement: Chap. 3, Chap. 4

The NDS Supplement provides shape properties and unadjusted material properties. Shape properties such as area, section modulus, and moment of inertia for various wood members are found in NDS Supplement Chap. 3. Material properties such as tensile and compression strength parallel and perpendicular to grain are found in NDS Supplement Chap. 4.

Dressed sawn lumber is categorized into boards, dimensional lumber, and timbers. *Boards* have a nominal dimension of 0.75 in to 1.5 in. *Dimensional lumber* consists of members with 2 in to 4 in nominal thickness and 2 in or more nominal width. *Timbers* are members with a nominal size of 5 in × 5 in and larger. Section properties for dressed sawn lumber are found in NDS Supplement Table 1A and Table 1B. The unadjusted material capacities for sawn lumber are found in NDS Supplement Table 4A through Table 4F. The majority of timber problems on the exam will be with visually graded lumber (2 in and 4 in thick), which is covered in NDS Supplement Table 4A (all species except southern pine) and Table 4B (southern pine), and visually graded timber (5 in × 5 in and larger), which is covered in NDS Supplement Table 4D.

Structural glued laminated timber members, or *glulams*, are built up from wood laminations bonded together with adhesives. The grain of all laminations is parallel to the length of the beam, and the laminations are typically 1.5 in thick. Section properties can be found in NDS Supplement Table 1C and Table 1D. Unadjusted material capacities for glulams can be found in NDS Supplement Table 5A through Table 5D.

Exam tip: Be sure to determine whether your member is timber or sawn lumber before referencing the tables. The unadjusted capacities for 2 in and 4 in thick members are different from those for 5 in thick members and are found in a different table.

Wood Adjustment Factors

CEST: Sec. 5.6

NDS: Part 4, Part 5

Wood design capacities are determined using referenced material-specific properties and adjustment values. Adjusted design values are obtained by multiplying reference design values by the applicable adjustment factors. For example, the adjusted design strength for sawn lumber in bending using allowing strength design (ASD) is provided in the following equation.

$$F'_b = F_b C_F C_r C_i C_D C_M C_t C_{fu} C_L$$

F'_b is the adjusted bending capacity in psi. F_b is the unadjusted material-specific bending capacity found in the NDS Supplement tables. The other variables are adjustment factors. These values and equations are found in NDS Table 4.3.1 (sawn lumber) and Table 5.3.1 (glulams). The adjusted capacity is compared with the internal stress (in psi) in the member.

Wood may be designed using either the ASD method or the load and resistance factor design (LRFD) method. When referring to NDS Table 4.3.1 and Table 5.3.1, the adjustment factor C_D is used only for ASD design. The factors K_F, f, and l are used only for LRFD design. All other indicated factors in the tables are used for both design methods.

CEST Table 5.1 shows the applicability of various adjustment factors. Each column on the right side of this table corresponds to a reference design value (e.g., bending strength, tensile strength) in psi. Each row corresponds to an adjustment factor. The allowable stress for a given load application is found by multiplying the reference design value by all applicable adjustment factors. Some adjustment factors apply to only sawn lumber or glulam, while some apply to both.

NDS Sec. 4.3 and Sec. 5.3 give the adjustment values or specify where in the code they can be found. Adjustment factors consider the use of the member, the moisture content in the member, the size of the member, and other considerations that can affect the capacity of the wood.

Exam tip: NDS Table 4.3.1 and Table 5.3.1 will direct you to which factors apply for each case. Some factors are different for glulam and sawn lumber. Be sure to go to the correct table first.

Timber Design Method

CEST: Chap. 5

ASCE/SEI 7: Sec. 2.4

IBC: Sec. 1605.3.1

NDS: Sec. 1.4

NDS permits wood design using both the allowable strength design (ASD) method and the load resistance factor design (LRFD) method. ASD is the traditional wood design method and is the only method covered in CEST Chap. 5.

The ASD method calculates the stresses produced by factored working loads, where working loads (also known as nominal or service loads) are applied to a member (see the load combinations in IBC Sec. 1605.3.1 or ASCE/SEI 7 Sec. 2.4). These loads are factored using the ASD load combinations. The designer must ensure that the stresses in the member do not exceed the allowable stresses. The allowable stresses are found by multiplying the reference design values by the appropriate adjustment factors. See the entry "Wood Adjustment Factors" in this chapter for more information.

Exam tip: When comparing capacities, be sure to use the ASD load combinations when using ASD adjustment factors and code values. Wind and seismic forces are strength-level forces and are adjusted in the ASD load combinations. When consulting the shear wall and diaphragm tables, be sure to compare the ASD capacity with $0.6W$ or $0.7E$.

B.1. HORIZONTAL MEMBERS

Key concepts: These key concepts are important for answering exam questions in subtopic 10.B.1, Horizontal Members.

- notched beams
- wood beams
- wood diaphragms
- wood flexural strength
- wood shear strength

DESIGN AND DETAILS OF STRUCTURES: TIMBER

NCEES Handbook: The *Handbook* does not contain any material relevant to this subtopic. Be sure to study the listed sections in CEST and from codes and standards.

***Structural Depth Reference Manual* (CEST):** Study these sections in CEST that either relate directly to this subtopic or provide background information.

- Chapter 5: Design of Wood Structures

Codes and standards: To prepare for this subtopic, familiarize yourself with these sections from codes and standards.

National Design Specification for Wood Construction with Commentary (NDS)

- Part 3: Design Provisions and Equations
- Part 4: Sawn Lumber
- Part 5: Structural Glued Laminated Timber

Special Design Provisions for Wind and Seismic with Commentary (SDPWS)

- Section 4.2: Wood-Frame Diaphragms

The following equations, figures, and concepts are relevant for subtopic 10.B.1, Horizontal Members.

Wood Beams

CEST: Chap. 5

NDS: Part 3, Part 4, Part 5

There are multiple types of wood beams. The exam will cover sawn lumber beams and glulam beams.

Beams must be designed for shear, bending, and deflection. Consideration must also be made for compression perpendicular to grain, depending on bearing conditions.

The capacities of wood are provided as stresses (in psi) that are compared with the internal stresses in the members. The internal stresses for shear and flexure are found in NDS Part 3. The capacity equations for wood are different for sawn lumber and glued laminated members. The equations and adjustment factors for the adjusted design capacities for sawn lumber are found in NDS Sec. 4.3, and those for glued laminated members are found in NDS Section 5.3.

> *Exam tip*: Be sure to consult NDS Table 4.3.1 for sawn lumber and Table 5.3.1 for glued laminated members.

Wood Diaphragms

SDPWS: Sec. 4.2

Diaphragms are horizontal structural members that transfer lateral loads (wind and seismic) to the vertical lateral force resisting members (shear walls). Wood sheathing act as flexible diaphragms. The sheathing must be designed for unit shear. Chords and collectors are wood members such as top plates or beams and must be designed for tension and compression.

Wood frame diaphragms are covered in SDPWS Sec. 4.2, and the nominal shear capacities are given in Table 4.2A (blocked diaphragms) and Table 4.2C (unblocked diaphragms). When using allowable strength design (ASD) methodology, the nominal table values must be divided by two per SDPWS Sec. 4.2.3 and adjusted based on the specific gravity per SDPWS Table 4.2A and Table 4.2C, note 2.

The unit shear capacity of a wood diaphragm is dependent on the sheathing size and the nailing. Wood diaphragms can either be blocked or unblocked. A *blocked diaphragm* has blocking at the edges of each panel, with each panel being nailed on all four sides at the job site. Refer to the figures shown for clarification on the difference between unblocked (first figure) and blocked (second figure) diaphragms.

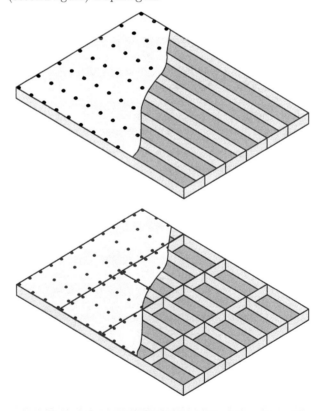

> *Exam tip*: When using ASD, be sure to use the appropriate load combinations for unit shear (0.6W or 0.7E) before comparing them to the capacities.

Wood Flexural Strength

CEST: Sec. 5.7

NDS: Sec. 3.3, Sec. 4.3, Sec. 5.3

For sawn lumber,

$$F'_b = F_b C_F C_r C_i C_D C_M C_t C_{\text{fu}} C_L$$

For glued laminated timber,

$$F'_b = F_b C_c C_I C_D C_M C_t C_{\text{fu}} (C_L \text{ or } C_V)$$

The capacities of wood are provided as stresses (in psi) that are compared to the internal stresses in the members. The internal stress for flexure is found in NDS Sec. 3.3. The capacity equations for wood are different for sawn lumber and glued laminated members. The equations and adjustment factors for the adjusted design capacities for sawn lumber are found in NDS Sec. 4.3, and those for glued laminated members are found in NDS Sec. 5.3. F'_b is the adjusted bending capacity in psi. F_b is the unadjusted material-specific bending capacity found in the NDS Supplement tables. See the entry "Wood Adjustment Factors" in this chapter for details on the wood factors. The equations provided in this entry are for allowable strength design (ASD).

The beam stability factor, C_L, is applicable to the reference bending design value for sawn lumber and glued laminated members. In accordance with NDS Sec. 3.3.3, C_L is 1.0 when the depth of a bending member does not exceed its breadth, or when the compression edge of a bending member is provided with continuous lateral restraint and the ends are restrained against rotation.

For glued laminated members, C_L is not applied simultaneously with the volume factor, C_V, and the lesser of these two factors is applicable.

C_L is dependent on effective length, l_e. NDS Table 3.3.3 provides the equation for different l_e values. CEST Fig. 5.1 provides typical l_e conditions.

Exam tip: When designing glued laminated members, be sure to use the smaller of C_V and C_L. If the member is continuously laterally braced, and the ends are restrained from rotation, then C_V will control because neither can be greater than 1.0.

Wood Shear Strength

CEST: Sec. 5.10

NDS: Sec. 3.4, Sec. 4.3, Sec. 5.3

For sawn lumber,

$$F'_v = F_v C_i C_D C_M C_t$$

For glued laminated timber,

$$F'_v = F_v C_{\text{vr}} C_D C_M C_t$$

The capacities of wood are provided as stresses (in psi) that are compared to the internal stresses in the members. NDS Sec. 3.4 provides the equation for the internal stress in a member due to shear and the critical sections near the end of beams for shear. The capacity equations for wood are different for sawn lumber and glued laminated members. The equations and adjustment factors for the adjusted design capacities for sawn lumber are found in NDS Sec. 4.3, and those for glued laminated members are found in NDS Sec. 5.3. F'_v is the adjusted shear capacity in psi. F_v is the unadjusted material-specific bending capacity found in the NDS Supplement tables. See the entry "Wood Adjustment Factors" in this chapter for details on the wood factors. The equations provided in this entry are for allowable strength design (ASD).

Exam tip: When calculating shear, if a beam has only a uniform load, you can neglect the load up to a distance d from the support.

Notched Beams

CEST: Sec. 5.10

NDS: Sec. 4.4.3, Sec. 5.4.5

Notches reduce the shear capacity of a beam. NDS Sec. 4.4.3 imposes restrictions on their size and location in sawn lumber members, as shown in CEST Fig. 5.6. For glued laminated members (shown in CEST Fig. 5.7), similar restrictions are specified in NDS Sec. 5.4.5.

Figure 5.6 Notches in Sawn Lumber Beams

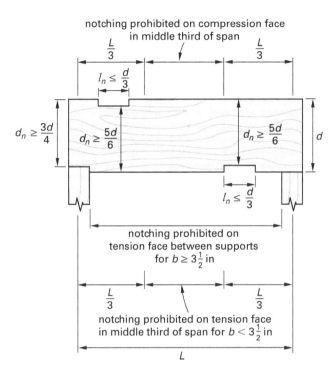

Figure 5.7 Notches in Glued Laminated Beams

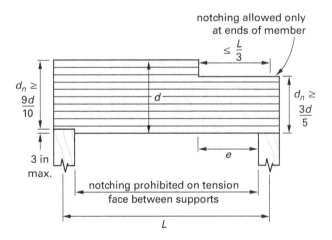

Exam tip: When calculating shear, if a beam has only a uniform load, you can neglect the load up to a distance d from the support.

B.2. VERTICAL MEMBERS

Key concepts: These key concepts are important for answering exam questions in subtopic 10.B.2, Vertical Members.

- wood compressive capacity
- wood posts
- wood shear walls
- wood stud walls
- wood tensile capacity

NCEES Handbook: The *Handbook* does not contain any material relevant to this subtopic. Be sure to study the listed sections in CEST and from codes and standards.

***Structural Depth Reference Manual* (CEST):** Study these sections in CEST that either relate directly to this subtopic or provide background information.

- Chapter 5: Design of Wood Structures

Codes and standards: To prepare for this subtopic, familiarize yourself with these sections from codes and standards.

National Design Specification for Wood Construction with Commentary (NDS)

- Section 3.6: Compression Members—General
- Section 3.7: Solid Columns
- Section 3.9: Combined Bending and Axial Loading
- Section 4.3: Adjustment of Reference Design Values
- Section 5.3: Adjustment of Reference Design Values

Special Design Provisions for Wind and Seismic with Commentary (SDPWS)

- Section 4.3: Wood-Frame Shear Walls

The following equations, figures, and concepts are relevant for subtopic 10.B.2, Vertical Members.

Wood Posts

CEST: Sec. 5.8

NDS: Sec. 3.6, Sec. 3.7, Sec. 3.9, Sec. 4.3, Sec. 5.3

Wood posts are typically 4×4 or larger members. They can be solid members or built-up members, such as two 2×4 members nailed together. Depending on their locations, wood posts must be designed for either axial loading or combined axial and flexural loading. The exam will most likely require wood posts to be designed with axial loading only.

Exam tip: Wood members that are 5 × 5 or larger (timbers) are shown in a different table than members less than 5 × 5 for unfactored capacity. Be sure to double-check what the member is before looking for the material-specific unfactored capacity. On the exam, columns will usually be axially controlled, unless the problem mentions a horizontal load such as a wind load from components and cladding.

Wood Stud Walls

Stud walls consist of 2 in thick studs typically spaced at 16 in on center with a double top plate and a bottom plate. Window and doors openings are framed with a header, trimmer, and king studs.

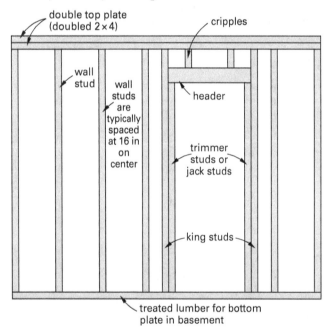

Studs are designed for axial and combined axial and flexure loading. Headers are beams and are designed for shear, flexure, and deflection. See the entry "Wood Beams" in this chapter for more information. Trimmers act as posts supporting the header and are designed only for axial loading. King studs span from the bottom plate to the top plate and are designed for flexure and combined axial and flexure loading due to wind. Bottom plates are designed for compression perpendicular to grain.

Exam tip: Designing a whole wall would take too long for an exam problem. It is reasonable, however, to expect to be asked for a single element in the wall.

Wood Shear Walls

SDPWS: Sec. 4.3

Wood shear walls are used to transfer lateral loads (wind and seismic) to the ground. The design and requirements for wood shear walls are found in SDPWS Sec. 4.3. There are three different types of shear walls provided in NDS: segmented, perforated, and force transfer. The three different types can be seen in SDPWS Fig. 4C, Fig. 4D, and Fig. 4E.

All three wall types need to be designed for unit shear of the sheathing and tension, as well as for compression chord forces. The bottom plates of shear walls need to be designed to transfer the forces. Unit shear capacities are provided in SDPWS Table 4.3A. Unit shear values are dependent on nail spacing, sheathing, and specific gravity of the studs. Unit shear values must be divided by two for ASD load combinations and adjusted for wood-specific gravity per SDPWS Table 4.3A, note 3. Per note 2, when the spacing of the studs is 16 in on center or less, the unit shear values may be increased to those for 15/32 in sheathing with the same nailing pattern.

The capacities and forces of perforated shear walls are adjusted using SDPWS Eq. 4.3-8 and Eq. 4.3-9. Adjusted unit shear values should be compared with the unit shear values per the applicable load combinations.

Exam tip: The exam will not likely contain problems involving force transfer, since the calculations would take too long, and no equations for force transfer are provided in the *Handbook*.

Wood Tensile Capacity

CEST: Sec. 5.9

NDS: Sec. 3.8, Sec. 4.3, Sec. 5.3

For sawn lumber,

$$F'_t = F_t C_F C_i C_D C_M C_t$$

For glued laminated timber,

$$F'_t = F_t C_D C_M C_t$$

The capacities of wood are provided as stresses (in psi) that are compared to the internal stresses in the members. NDS Sec. 3.8 provides the requirements for tension. The capacity equations for wood are different for sawn lumber and glued laminated members. The equations and adjustment factors for the adjusted design capacities for sawn lumber are found in NDS Sec. 4.3, and those for glued laminated members are found in NDS Sec. 5.3. F'_t is the adjusted tensile capacity parallel to grain in psi. F_t is the unadjusted material-specific tensile capacity parallel to grain found in the NDS Supplement tables. See the entry "Wood Adjustment Factors" in this chapter for details on the wood factors. The equations provided in this entry are for allowable strength design (ASD).

Exam tip: The area used for calculating the tensile stress is the net area.

Wood Compressive Capacity

CEST: Sec. 5.8

NDS: Sec. 3.6, Sec. 3.7, Sec. 4.3, Sec. 5.3

For sawn lumber,

$$F'_c = F_c C_F C_i C_D C_M C_t C_P$$

For glued laminated timber,

$$F'_c = F_c C_D C_M C_t C_P$$

The capacities of wood are provided as stresses (in psi) that are compared to the internal stresses in the members. NDS Sec. 3.6 provides the requirements for compression members, including when to use the gross and net areas. The capacity equations for wood are different for sawn lumber and glued laminated members. The equations and adjustment factors for the adjusted design capacities for sawn lumber are found in NDS Sec. 4.3, and those for glued laminated members are found in NDS Sec. 5.3. F'_c is the adjusted compressive strength parallel to grain in psi. F_c is the unadjusted material-specific compressive strength parallel to grain found in the NDS Supplement tables. See the entry "Wood Adjustment Factors" in this chapter for details on the wood factors. The equations provided in this entry are for allowable strength design (ASD).

The column stability factor is applicable to the reference compression design values parallel to the grain for sawn lumber and glued laminated members. In accordance with NDS Sec. 3.7.1, C_P is 1.0 when a compression member is provided with continuous lateral restraint in all directions. For other conditions, the column stability factor is specified by NDS Sec. 3.7.1. The stability factor is dependent on the slenderness ratio (l_e/d) of the column and should be checked for the maximum ratio. Refer to CEST Fig. 5.3 or NDS Fig. 3F for details on the dimensions used for the slenderness ratio.

Figure 5.3 Slenderness Ratios for a Rectangular Column

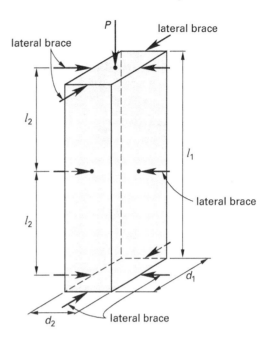

Exam tip: Before calculating C_p, determine the worst-case slenderness ratio. If the member is continuously braced in both directions, then C_p is 1.0.

Wood Combined Axial and Flexural Capacity

CEST: Sec. 5.8, Sec. 5.9

NDS: Sec. 3.9

Members subjected to combined tension and flexural stresses caused by axial and transverse loading must satisfy the two expressions given in NDS Sec. 3.9.1. Members subjected to combined compression and flexural stresses from axial and transverse loading must satisfy the interaction equations given in NDS Sec. 3.9.2. Refer to NDS Sec. 4.3 and Sec. 5.3 for adjusted capacities used in these equations.

Exam tip: When solving problems involving wood, be sure to use the correct material tables based on member size (2 in thick versus 5 × 5 or larger) and type (sawn lumber versus glued laminated timber).

B.3. SYSTEMS

Wood systems consist of trusses and walls. Trusses have only tension and compression members. Refer to entries in Chap. 28 of this book for an analysis of trusses, including internal forces and deflections. Refer to the entries "Wood Tensile Capacity" and "Wood Compressive Capacity" in this chapter for the capacity of wood members in trusses. Refer to the entries "Wood Stud Walls" and "Wood Shear Walls" in this chapter for the capacity of wood walls.

B.4. CONNECTIONS

Key concepts: These key concepts are important for answering exam questions in subtopic 10.B.4, Connections.

- wood fasteners

NCEES Handbook: The *Handbook* does not contain any material relevant to this subtopic. Be sure to study the listed sections in CEST and from codes and standards.

Structural Depth Reference Manual **(CEST):** Study these sections in CEST that either relate directly to this subtopic or provide background information.

- Section 5.11: Design of Connections

Codes and standards: To prepare for this subtopic, familiarize yourself with these sections from codes and standards.

National Design Specification for Wood Construction with Commentary (NDS)

- Part 11: Mechanical Connections
- Part 12: Dowel-Type Fasteners
- Part 13: Split Ring and Shear Plate Connectors
- Part 14: Timber Rivets

The following concepts are relevant for subtopic 10.B.4, Connections.

Wood Fasteners

CEST: Sec. 5.11

NDS: Part 11, Part 12, Part 13, Part 14

Common wood fasteners include bolts, lag screws, screws, and nails. The allowable design values for wood fasteners depend on the type of fastener and the service conditions. The reference design values, tabulated in NDS Part 11 through Part 14, are applicable to normal conditions of use and normal load duration. To determine the relevant design values for other conditions of service, the reference design values are multiplied by adjustment factors specified in NDS Sec. 11.3. A summary of adjustment factors follows. The applicability of each to the basic design values is given in NDS Table 11.3.1; CEST Table 5.6 provides a summary.

Bolts, lag screws, screws, and nails must be designed for lateral loads, such as dowel bearings. Lag screws, screws, and nails must also be designed for withdrawal, and combined withdrawal and lateral loads. General requirements are found in NDS Sec. 12.1. Fastener withdrawal capacity is covered in NDS Sec. 12.2. NDS Table 12.2A through Table 12.2D provide typical unfactored withdrawal capacities as a pound per inch of penetration into the wood. Lateral design values such as dowel bearing strength are provided in NDS Sec. 12.3. Design of fasteners with combined withdrawal and lateral loads is covered in NDS Sec. 12.4. Typical unadjusted bearing capacities are provided in NDS Table 12A through Table 12T.

Required adjustment factors for each fastener type and condition are identified in NDS Table 11.3.1. Fastener-specific adjustment factors are given in NDS Sec. 12.5.

Exam tip: NDS Table 12A through Table 12T are for typical lateral conditions. For any nontraditional cases, Z must be calculated per NDS Table 12.3.1A. The exam will not likely require you to check every case with the equations in NDS Table 12.3.1A; if such calculations are required on the exam, they will likely be for a specific yield mode.

32 Design and Details of Structures: Masonry

Content in blue refers to the *NCEES Handbook*.

A.4. Masonry... 32-1
B.1. Horizontal Members 32-3
B.2. Vertical Members............................. 32-3
B.3. Systems.. 32-8
B.4. Connections...................................... 32-8

The knowledge area of Design and Details of Structures makes up between 16 and 24 questions out of the 80 questions on the PE Civil Structural exam. Four to six questions are on materials and material properties, while 12 to 18 questions are on component design and detailing. This chapter focuses on these knowledge areas only in the context of masonry. The organization of this chapter follows the order of subtopics given by NCEES for this knowledge area, but includes only subtopics relevant to the design and details of structures for masonry.

A.4. MASONRY

Key concepts: These key concepts are important for answering exam questions in subtopic 10.A.4, Masonry.

- masonry compressive strength
- masonry construction components
- masonry design methods
- masonry modulus of elasticity

NCEES Handbook: The *Handbook* does not contain any material relevant to this subtopic. Be sure to study the listed sections in CERM and CEST and from codes and standards.

PE Civil Reference Manual (**CERM**): Study these sections in CERM that either relate directly to this subtopic or provide background information.

- Chapter 67: Properties of Masonry
- Section 68.1: Methods of Design
- Section 68.3: Allowable Stress Wall Design
- Section 68.4: Strength Design

Structural Depth Reference Manual (**CEST**): Study these sections in CEST that either relate directly to this subtopic or provide background information.

- Section 6.1: Design Principles

Codes and standards: To prepare for this subtopic, familiarize yourself with these sections from codes and standards.

Building Code Requirements for Masonry Structures (TMS 402)

- Chapter 2: Notation and Definitions
- Section 4.2.2.2: Clay and Concrete Masonry
- Section 6.1.4: Protection of Reinforcement and Metal Accessories
- Chapter 8: Allowable Stress Design of Masonry
- Chapter 9: Strength Design of Masonry

Specification for Masonry Structures (TMS 602)

- Section 1.4 B: Compressive Strength Determination
- Section 3.4 B: Reinforcement

The following equations, figures, and concepts are relevant for subtopic 10.A.4, Masonry.

Masonry Construction Components

CERM: Chap. 67

TMS 402: Sec. 6.1.4

TMS 602: Sec. 3.4 B

Structural masonry walls consist of masonry units, mortar, grout, and reinforcement. The most common horizontal reinforcements for a masonry wall are rebar (shown in the top portion of the following illustration) and galvanized steel wire ladder (shown in the bottom portion of the illustration) laid in the mortar between courses of brick or block.

Masonry units used in the United States include concrete, clay (both brick and structural clay tile), glass block, and stone. Concrete and clay masonry are used for the majority of structural masonry construction.

The relevant structural properties of clay and concrete masonry are similar, although test methods and requirements can vary significantly between the two materials. Masonry unit dimensions are listed in the order of width × height × length. Nominal dimensions are used for planning and layout; actual dimensions of masonry units are typically 3/8 in (9.5 mm) less than nominal dimensions to account for the mortar thickness used in the wall.

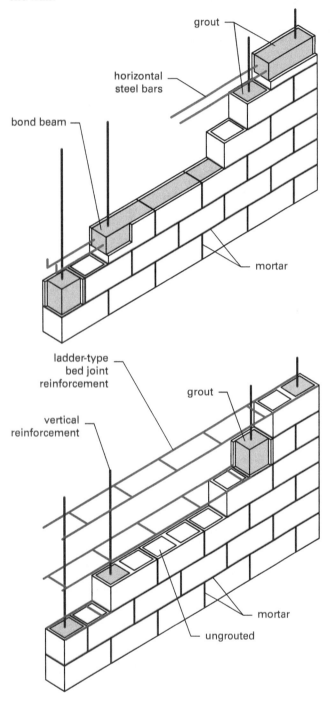

Mortar structurally bonds units together, seals joints against air and moisture penetration, accommodates small wall movements, and bonds to joint reinforcement, ties, and anchors. Mortar joints are 3/8 in (9.5 mm) thick. Mortar types are defined in *Standard Specification for Mortar for Unit Masonry* (ASTM C270). Restrictions are placed on the use of mortar types based on criteria such as seismic design category.

Grout is a fluid cementitious mixture used to fill masonry cores or cavities in order to increase structural performance. Grout is most commonly used in reinforced construction to bond steel reinforcement to the masonry.

Two principal types of reinforcement used in masonry are reinforcing bars and cold-drawn wire products. Reinforcing bars are placed vertically and/or horizontally in the masonry and are grouted into position. TMS 602 Sec. 3.4 B covers reinforcement installation requirements, and TMS 402 Sec. 6.1.4 dictates a minimum masonry and grout cover around reinforcing bars.

Exam tip: Any masonry problems on the exam will likely relate to concrete masonry units (CMU).

Masonry Design Methods

CERM: Sec. 68.1, Sec. 68.3, Sec. 68.4

CEST: Sec. 6.1

TMS 402: Chap. 8, Chap. 9

Masonry structures are designed using one of several methods, including allowable stress design (ASD) and strength design (SD). ASD is the most widely used method, although SD is gaining popularity. The ASD method is covered in TMS 402 Chap. 8, and the SD method is covered in TMS 402 Chap. 9.

Exam tip: Per NCEES, only the ASD method will be used on the exam, with the exception of problems on out-of-plane loads on walls, which may use the SD method.

Masonry Compressive Strength

CERM: Sec. 67.6

TMS 402: Chap. 2

TMS 602: Sec. 1.4 B

The compressive strength of a masonry assembly is dependent on unit properties and mortar type. Compressive strength is typically 28-day strength. TMS 602 Table 1 and Table 2 provide typical compressive strengths based on mortar type and unit strength. CERM Table 67.2 and Table 67.3 are equivalent tables.

Exam tip: The majority of exam problems on this topic will include the compressive strength in the problem statement.

Masonry Modulus of Elasticity

CERM: Sec. 67.8

TMS 402: Sec. 4.2.2.2.1

$$E_m = 700 f'_m \quad \text{[clay masonry]} \qquad 67.1$$

$$E_m = 900 f'_m \quad \text{[concrete masonry]} \qquad 67.2$$

The modulus of elasticity for masonry is dependent on the masonry compressive strength and can be estimated using the equations in TMS 402 Sec. 4.2.2.2.1; CERM Eq. 67.1 and Eq. 67.2 are equivalent equations. The compressive strength and modulus of elasticity in these equations are of the masonry assembly (masonry units and mortar).

Exam tip: The majority of the exam problems on this topic will involve concrete masonry, but always read the question thoroughly to be sure.

B.1. HORIZONTAL MEMBERS

Key concepts: These key concepts are important for answering exam questions in subtopic 10.B.1, Horizontal Members.

- beams

NCEES Handbook: The *Handbook* does not contain any material relevant to this subtopic. Be sure to study the listed sections in CEST and from codes and standards.

Structural Depth Reference Manual **(CEST):** Study these sections in CEST that either relate directly to this subtopic or provide background information.

- Section 6.1: Design Principles
- Section 6.2: Design for Flexure

Codes and standards: To prepare for this subtopic, familiarize yourself with these sections from codes and standards.

Building Code Requirements for Masonry Structures (TMS 402)

- Section 5.2: Beams

The following figures and concepts are relevant for subtopic 10.B.1, Horizontal Members.

Masonry Beams

CEST: Sec. 6.1, Sec. 6.2

TMS 402: Sec. 5.2

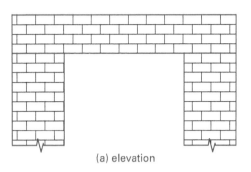

(a) elevation

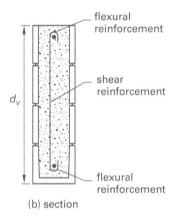

(b) section

Masonry beams are built within masonry walls. All other horizontal members in masonry structures are made of other materials. Masonry beams must be designed for flexure, shear, and deflection.

TMS 402 Sec. 5.2 provides requirements for the properties of masonry beams.

Exam tip: In exam problems on masonry beams, the beams will be designed and analyzed using allowable stress design.

B.2. VERTICAL MEMBERS

Key concepts: These key concepts are important for answering exam questions in subtopic 10.B.2, Vertical Members.

- allowable stress design (ASD)
- masonry columns
- masonry shear walls
- masonry walls
- reinforced masonry
  - allowable axial compressive stress
  - allowable combined axial compressive and flexural stresses

- allowable flexural stress
- allowable shear stress
- axial compressive strength
- combined axial compressive and flexural strength
- flexural strength
- shear strength

- strength design (SD)

NCEES Handbook: To prepare for this subtopic, familiarize yourself with these sections in the *Handbook*.

- Beams—Flexure Service

***PE Civil Reference Manual* (CERM):** Study these sections in CERM that either relate directly to this subtopic or provide background information.

- Chapter 68: Masonry Walls
- Chapter 69: Masonry Columns

***Structural Depth Reference Manual* (CEST):** Study these sections in CEST that either relate directly to this subtopic or provide background information.

- Section 6.3: Design for Shear
- Section 6.4: Design of Masonry Columns
- Section 6.5: Design of Masonry Shear Walls
- Section 6.6: Wall Design for Out-of-Plane Loads

Codes and standards: To prepare for this subtopic, familiarize yourself with these sections from codes and standards.

Building Code Requirements for Masonry Structures (TMS 402)

- Section 2.2: Definitions
- Chapter 5: Structural Elements
- Section 6.1: Details of Reinforcement and Metal Accessories
- Chapter 7: Seismic Design Requirements
- Chapter 8: Allowable Stress Design of Masonry
- Section 8.3.3: Steel Reinforcement—Allowable Stresses
- Section 8.3.4: Axial Compression and Flexure
- Section 8.3.5: Shear
- Chapter 9: Strength Design of Masonry

The following equations and figures are relevant for subtopic 10.B.2, Vertical Members.

Reinforced Masonry Allowable Flexural Stress: Allowable Stress Design

Handbook: Beams—Flexure Service

$$f_s = \frac{M_s}{A_s j d}$$

$$f_c = \frac{2M_s}{k b j d^2}$$

CERM: Sec. 68.8

TMS 402: Sec. 8.3.3, Sec. 8.3.4.2.2

Actual stresses are determined using the model shown in the figure in *Handbook* section Beams—Flexure Service, which uses the same variables for the beam properties as the *Handbook* equations shown. These equations are provided for concrete beams, but the calculations are the same for designing masonry beams using allowable stress design (ASD).

The flexural stresses in tension in the steel, f_s, are compared to the allowable tensile strength of the reinforcement. The allowable tensile stress is 20,000 lbf/in² for grades 40 and 50 steel and 32,000 lbf/in² for grade 60 steel, per TMS 402 Sec. 8.3.3.

The allowable flexural stress in compression in the masonry (see *Handbook* section Beams—Flexure Service) is equivalent to 0.45 times the actual compressive strength of masonry in flexure, per the equation in TMS 402 Sec. 8.3.4.2.2. CERM Eq. 68.1 is an equivalent equation.

$$F_b = 0.45 f'_m \qquad \text{68.1}$$

Masonry Walls

CERM: Chap. 68

TMS 402: Sec. 6.1, Chap. 7

Masonry walls must be designed for axial, in-plane, and out-of-plane loading. *Axial loading* comes from gravity loads on the floor and roof members that the wall is supporting. *In-plane loading* occurs when the wall acts as a shear wall to transfer wind and seismic loads to the foundation. *Out-of-plane loading* occurs when a wind or seismic event creates a load on the face of the wall. Reinforcement requirements in walls are provided in TMS 402 Sec. 6.1.

Walls that resist in-plane shear act as shear walls. Seismic requirements for walls, including shear walls, are provided in TMS 402 Chap. 7.

Exam tip: Out-of-plane walls may be designed using allowable stress design or strength design. In-plane walls may only be designed using allowable stress design.

Masonry Shear Walls

CERM: Sec. 68.13 through Sec. 68.17

CEST: Sec. 6.5

TMS 402: Sec. 2.2, Sec. 7.3.2

Shear walls transfer in-plane seismic and wind loads from diaphragms to the foundation through shear and flexure. There are several types of masonry shear walls, which are classified in TMS 402 Sec. 2.2. The type of shear wall permitted for a given structure is dependent on the seismic design category of the structure.

CEST Fig. 6.6 shows reinforcement details for special reinforced masonry shear walls.

Figure 6.6 Reinforcement Details for Special Reinforced Masonry Shear Walls Laid in Running Bond

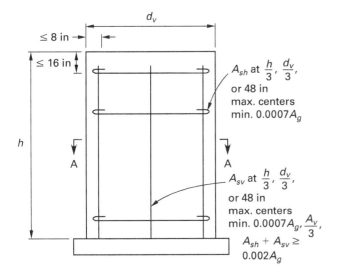

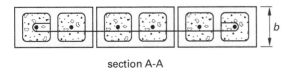

section A-A

Shear walls must comply with the requirements given in TMS 402 Table CC-7.3.2-1.

Masonry Columns

CERM: Chap. 69

CEST: Sec. 6.4

TMS 402: Sec. 5.3

Columns are isolated structural elements subject to axial loads and flexure. To distinguish columns from short walls, TMS 402 defines a column as an isolated vertical member not built integrally into a wall and designed primarily to resist compressive loads parallel to its longitudinal axis. CEST Fig. 6.5 shows an illustration of column details. TMS 402 Sec. 5.3 provides requirements for masonry columns.

Exam tip: Masonry columns must be designed using allowable stress design.

Reinforced Masonry Flexural Strength: Strength Design

CEST: Sec. 6.6

TMS 402: Sec. 9.3.5

$$M_u \leq \phi M_n$$

$$M_n = A_{se} f_y \left(d - \frac{a}{2} \right)$$

$$a = \frac{\dfrac{P_u}{\phi} + A_s f_y}{0.80 f'_m L_w}$$

The design flexural capacity of a masonry wall with out-of-plane loading is dependent on the equivalent stress block for masonry stress and rebar tensile forces. The commentary for TMS 402 Sec. 9.3.5.2 provides the equation for the nominal flexural strength of the masonry wall; the equations from CEST Sec. 6.6 are equivalent equations derived from the TMS 402 equations.

CEST Fig. 6.7 provides an illustration of the stress block used for the equations for pure flexure.

The flexural design capacity of a reinforced masonry wall is the product of nominal flexural capacity, M_n, multiplied by a factor, ϕ, which is equal to 0.9 for reinforced masonry in flexure.

Figure 6.7 Nominal Moment for Out-of-Plane Loading of Concrete Masonry

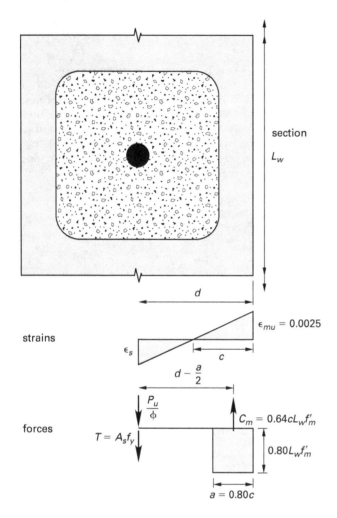

Exam tip: Do not confuse the stress block equations for concrete and masonry.

Reinforced Masonry Allowable Axial Compressive Stress: Allowable Stress Design

CERM: Sec. 68.11, Sec. 69.5

CEST: Sec. 6.4

TMS 402: Sec. 8.3.4

$$P_a = (0.25 f'_m A_n + 0.65 A_{st} F_s)\left(1 - \left(\frac{h}{140r}\right)^2\right) \quad 69.1$$
[when $h/r \leq 99$]

$$P_a = (0.25 f'_m A_n + 0.65 A_{st} F_s)\left(\frac{70r}{h}\right)^2 \quad 69.2$$
[when $h/r > 99$]

When a compression member is in pure compression, the axial compressive load must be compared to the allowable compressive force. TMS 402 Eq. 8-21 and Eq. 8-22 are used to determine the allowable compressive force for reinforced masonry members, depending on the ratio of the member's effective height and radius of gyration. CERM Eq. 69.1 and Eq. 69.2 are equivalent equations.

Reinforced Masonry Axial Compressive Strength: Strength Design

CERM: Sec. 69.6

TMS 402: Sec. 9.3.4.1.1

$$P_n = 0.80 \left(0.80 f'_m (A_n - A_s) + f_y A_s\right) \cdot \left(1 - \left(\frac{h}{140r}\right)^2\right) \quad 69.3$$
[when $h/r \leq 99$]

$$P_n = 0.80 \left(0.80 f'_m (A_n - A_s) + f_y A_s\right)\left(\frac{70r}{h}\right)^2 \quad 69.4$$
[when $h/r > 99$]

$$P_u \leq \phi P_n \quad 69.5$$

The axial capacity for pure compression for reinforced masonry is dependent on the compressive strength of the masonry and the steel reinforcement, as shown in TMS 402 Eq. 9-19 and Eq. 9-20. CERM Eq. 69.3 and Eq. 69.4 are equivalent equations. Which equation is used depends on the ratio of the member's effective height and radius of gyration. The design capacity is the nominal axial capacity multiplied by a factor, ϕ, which is equal to 0.9 for reinforced masonry in axial compression.

Exam tip: Per NCEES, only out-of-plane walls may require strength design.

Reinforced Masonry Allowable Combined Axial Compressive and Flexural Stresses: Allowable Stress Design

CERM: Sec. 68.11, Sec. 69.7

TMS 402: Sec. 8.3.4

$$F_b = 0.45 f'_m \quad\quad 68.26$$

$$P_a = (0.25 f'_m A_n + 0.65 A_s F_s)\left[1 - \left(\frac{h}{140r}\right)^2\right] \quad\quad 68.27$$

[when $h/r \leq 99$]

$$P_a = (0.25 f'_m A_n + 0.65 A_s F_s)\left(\frac{70r}{h}\right)^2 \quad\quad 68.28$$

[when $h/r > 99$]

The axial compression, flexural tension, and combined compression on reinforced masonry members must be checked separately when comparing the combined effects of axial and flexural forces on masonry members. The axial compressive load must not exceed the allowable compressive force, and the actual combined axial compressive and flexural compressive stresses must not exceed the allowable compressive strength of masonry in flexure, which is 0.45 times the specified compressive strength, per TMS 402 Sec. 8.3.4.2.2. CERM Eq. 68.26 is an equivalent equation. TMS 402 Eq. 8-21 and Eq. 8-22 are used to find the allowable compressive force. CERM Eq. 68.27 and Eq. 68.28 are equivalent equations.

Reinforced Masonry Combined Axial Compressive and Flexure Strength: Strength Design

CERM: Sec. 68.12, Sec. 69.8

TMS 402: Sec. 9.3.5.2, Sec. 9.3.5.4

Masonry walls designed for out-of-plane loading using the strength design (SD) method must take slenderness effects into consideration. Slenderness effects can be accounted for by increasing the factored moment for second-order effects using the procedures set forth in TMS 402 Sec. 9.3.5.4.2 and Sec. 9.3.5.4.3.

Per TMS 402 Sec. 9.3.5.2, the increased factored moment must be compared to the design flexural capacity, and the factored axial load must be compared to the design axial capacity. The design capacity is the nominal capacity multiplied by a factor, ϕ, of 0.9 for reinforced masonry in axial compression.

Exam tip: Per NCEES, only out-of-plane walls may require SD.

Reinforced Masonry Allowable Shear Stress: Allowable Stress Design

CERM: Sec. 68.13, Sec. 68.16

CEST: Sec. 6.3, Sec. 6.5

TMS 402: Sec. 8.3.5

$$f_v = \frac{V}{bd} = \frac{V}{A_{nv}} \quad\quad 68.46$$

The shear stress in masonry is found using TMS 402 Eq. 8-24, which is equivalent to CERM Eq. 68.46 and the equation given in CEST Sec. 6.3. The allowable shear stress in reinforced masonry is a function of the allowable shear stress of the masonry and the allowable shear stress provided by the reinforcing steel; it can be found using TMS 402 Eq. 8-25, Eq. 8-26, or Eq. 8-27, depending on the loading conditions. The allowable shear stress of the masonry can be found using TMS 402 Eq. 8-28 for special reinforced masonry shear walls or TMS 402 Eq. 8-29 for other walls. The allowable shear stress provided by the steel can be found using TMS 402 Eq. 8-30.

Masonry walls for out-of-plane loading do not have shear reinforcement and rely solely on the masonry to resist shear.

Reinforced Masonry Shear Strength: Strength Design

CERM: Sec. 68.17

TMS 402: Sec. 9.3.4.1.2

$$V_n = (V_{nm} + V_{ns})\gamma_g \quad [MSJC \text{ Sec. } 9.3.4.1.2] \quad\quad 68.51$$

$$\leq \left(6 A_n \sqrt{f'_m}\right)\gamma_g \quad [\text{when } M_u/V_u d_v < 0.25] \quad\quad 68.52$$

$$\leq \left(4 A_n \sqrt{f'_m}\right)\gamma_g \quad [\text{when } M_u/V_u d_v > 1.0] \quad\quad 68.53$$

The nominal shear capacity of masonry members is dependent on the masonry shear strength, the strength of the shear reinforcing steel, and a factor, γ_g, which is equal to 0.8 for reinforced masonry in shear. The nominal shear capacity can be found using TMS 402 Eq. 9-21 and cannot exceed the values given by TMS 402 Eq. 9-22 or Eq. 9-23, depending on loading conditions. CERM Eq. 68.51 through Eq. 68.53 are equivalent to TMS 402 Eq. 9-21 through Eq. 9-23. The design shear capacity is equal to the nominal shear capacity, V_n, multiplied by the factor ϕ. The factored shear load must be less than the design shear capacity.

The masonry shear strength can be found using TMS 402 Eq. 9-24. CERM Eq. 68.54 is an equivalent equation. The strength of the shear reinforcing steel can be found using TMS 402 Eq. 9-25.

$$V_{nm} = \left(4 - 1.75\frac{M_u}{V_u d_v}\right)A_n\sqrt{f'_m} + 0.25P_u \quad\quad 68.54$$

If $M_u/V_u d_v$ is conservatively taken as its maximum value of 1.0 and the axial force is minimal, CERM Eq. 68.54 becomes CERM Eq. 68.55.

$$V_{nm} = 2.25 A_n \sqrt{f'_m} \quad\quad 68.55$$

CERM Eq. 68.56 is an equivalent equation.

$$V_{ns} = 0.5\left(\frac{A_v}{s}\right)f_y d_v \quad\quad 68.56$$

Masonry walls for out-of-plane loading do not have shear reinforcement and rely solely on the masonry to resist shear.

Exam tip: Flexure for out-of-plane loading typically controls and should be checked first.

B.3. SYSTEMS

Masonry systems are made up of masonry walls; all material pertaining to masonry systems is covered in the entries in Sec. B.2 in this chapter.

B.4. CONNECTIONS

Key concepts: These key concepts are important for answering exam questions in subtopic 10.B.4, Connections.

- masonry anchor bolts
- reinforcement development length

NCEES Handbook: The *Handbook* does not contain any material relevant to this subtopic. Be sure to study the listed sections in CERM and CEST and from codes and standards.

PE Civil Reference Manual (CERM): Study these sections in CERM that either relate directly to this subtopic or provide background information.

- Section 67.18: Development Length

Structural Depth Reference Manual (CEST): Study these sections in CEST that either relate directly to this subtopic or provide background information.

- Section 6.2: Design for Flexure
- Section 6.7: Design of Anchor Bolts

Codes and standards: To prepare for this subtopic, familiarize yourself with these sections from codes and standards.

Building Code Requirements for Masonry Structures (TMS 402)

- Section 6.2: Anchor Bolts
- Section 8.1.3: Anchor Bolts Embedded in Grout
- Section 8.1.6: Development of Reinforcement Embedded in Grout
- Section 9.3.3: Reinforcement Requirements and Details

The following equations and figures are relevant for subtopic 10.B.4, Connections.

Masonry Anchor Bolts

CEST: Sec. 6.7

TMS 402: Sec. 6.2, Sec. 8.1.3

Anchor bolts are embedded into the grouted portion of masonry units and are loaded in tension and/or shear. CEST Fig. 6.10 illustrates both headed and bent-bar anchor bolts.

Figure 6.10 Anchor Bolts in Concrete Masonry

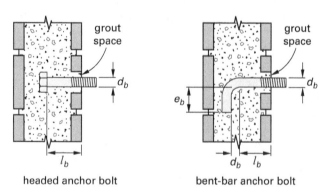

Anchor bolts may fail under tensile forces by tensile breakout, shear breakout, shear crushing, anchor bolt shear pryout, or tensile yielding of the steel anchor. Bent-bar anchors may also fail due to straightening of the hook followed by pullout from the masonry.

Details of headed and bent-bar anchor bolts, including placement and projected area requirements, are given in TMS 402 Sec. 6.2. TMS 402 Sec. 8.1.3 provides the capacity equations for anchor bolts in masonry.

Exam tip: Some exam problems may require you to check only one failure mode, so be sure to read the problem thoroughly.

Reinforcement Development Length

CERM: Sec. 67.18

CEST: Sec. 6.2

TMS 402: Sec. 8.1.6, Sec. 9.3.3

$$l_d = \frac{0.13 d_b^2 f_y \gamma}{K\sqrt{f'_m}} \quad \text{[TMS 402 Eq. 8-12]}$$
$$\geq 12 \text{ in}$$

Development length (also known as *anchorage* or *embedment length*) ensures that forces can be transferred in the reinforced masonry. Reinforcing bars can be anchored by embedment length, hooks, or mechanical devices. Reinforcing bars anchored by embedment length rely on interlock of the bar deformations with grout, and on the masonry cover being sufficient to prevent splitting from the reinforcing bar to the free surface.

The required development length of reinforcing bars is determined using TMS 402 Eq. 8-12 for allowable stress design (ASD), or TMS 402 Eq. 9-16 for strength design (SD). The development length is dependent on the bar size, bar cover and spacing, and masonry compressive strength.

33 Codes, Standards, and Guidance Documents

Content in blue refers to the *NCEES Handbook*.

A.1. International Building Code (IBC) 33-1
A.2. American Concrete Institute (ACI 318, 530) 33-3
A.3. Precast/Prestressed Concrete Institute (PCI Design Handbook) 33-8
A.4. Steel Construction Manual (AISC) 33-11
A.5. National Design Specification for Wood Construction (NDS) 33-15
A.6. LRFD Bridge Design Specifications (AASHTO) 33-18
A.7. Minimum Design Loads for Buildings and Other Structures (ASCE/SEI 7) 33-20
A.8. American Welding Society (AWS D1.1, D1.2, and D1.4) 33-22
A.9. OSHA 1910 General Industry and OSHA 1926 Construction Safety Standards 33-23

The knowledge area of Codes, Standards, and Guidance Documents makes up between four and six questions out of the 80 questions on the PE Civil Structural exam. The organization of this chapter follows the order of subtopics given by NCEES for this knowledge area. Each subtopic is covered in the following sections.

A.1. INTERNATIONAL BUILDING CODE (IBC)

Key concepts: These key concepts are important for answering exam questions in subtopic 11.A.1, International Building Code (IBC).

- codes and standards referenced in the IBC
- navigating the IBC
- structure of the IBC

NCEES Handbook: The *Handbook* does not contain any material relevant to this subtopic.

The following concepts are relevant for subtopic 11.A.1, International Building Code (IBC).

Overview of the IBC

The IBC is used as a model code in most of the United States. It covers inspections, building occupancies, building heights and areas, interior and exterior finishes, fire safety, elevators and escalators, provisions of means of egress, construction materials, and existing structures.

Some jurisdictions have their own building codes, which may be designed to align with the IBC.

Exam tip: Exam answers are based on the particular edition of the IBC that is listed in the exam specifications. This is usually *not* the most recent edition. Be sure you are using the correct edition when studying.

Organization of the IBC

The IBC consists of 35 chapters and 13 appendices. In theory, any part of the IBC can be the subject of a question on the PE exam. In practice, however, the chapters most likely to be referenced on the exam are IBC Chap. 16, Chap. 19, and Chap. 21 through Chap. 23, which provide load- and material-specific standards.

You should also be familiar with IBC Chap. 17 and Chap. 18, as these are applicable to everyday structural engineering practice. Chapter 20, however, covers aluminum structures; aluminum is not part of the exam specifications, so this chapter is unlikely to come up during the exam.

IBC Chapter 16: Structural Design

IBC Chap. 16 covers loads, including dead, live, snow, wind, and earthquake loads. This chapter includes specific information on load combinations, deflection limits, strength requirements, lateral force resisting system detailing requirements, and structural integrity. Chapter 16 references *Minimum Design Loads for Buildings and Other Structures* (ASCE/SEI 7).

Sections and tables to study in this chapter include

- Table 1604.3: Deflection Limits
- Table 1604.5: Risk Category of Buildings and Other Structures
- Section 1605: Load Combinations
- Section 1607: Live Loads
- Section 1609: Wind Loads
- Section 1610: Soil Lateral Loads
- Section 1613: Earthquake Loads

IBC Chapter 17: Special Inspections and Tests

IBC Chap. 17 covers inspection and testing requirements for different materials of construction.

Sections and tables to study in this chapter include

- Section 1704: Special Inspections and Tests, Contractor Responsibility and Structural Observation
- Section 1705: Required Special Inspections and Tests

Exam tip: A nonquantitative problem related to this chapter may appear on the exam.

IBC Chapter 18: Soils and Foundations

IBC Chap. 18 includes information that can be used for foundation design. Sections and tables to study in this chapter include

- Table 1806.2: Presumptive Load-Bearing Values
- Section 1807: Foundation Walls, Retaining Walls and Embedded Posts and Poles
- Table 1808.8.1: Minimum Specified Compressive Strength f'_c of Concrete or Grout
- Section 1809: Shallow Foundations
- Section 1810: Deep Foundations

Exam tip: Quantitative exam questions will most likely reference codes related to the foundation element's respective material, not the IBC, but be familiar with the organization of this chapter in case the exam does reference the IBC.

IBC Chapter 19: Concrete

IBC Chap. 19 covers the materials, quality control, design, and construction of concrete.

The most notable section in this chapter is Sec. 1905, which contains guidelines that may override *Building Code Requirements for Structural Concrete* (ACI 318). IBC Sec. 1908 is also notable because it covers shotcrete, which is not discussed in ACI 318.

Sections and tables to study in this chapter include

- Section 1905: Modifications to ACI 318
- Section 1908: Shotcrete

IBC Chapter 21: Masonry

IBC Chap. 21 covers the materials, design, construction, and quality of masonry. Design standards in IBC Chap. 21 may override standards given in *Building Code Requirements for Masonry Structures* (TMS 402, also known as ACI 530) or *Specification for Masonry Structures* (TMS 602, also known as ACI 530.1). Sections in this chapter that may modify TMS 402 and TMS 602 include

- Section 2101: General
- Section 2103: Masonry Construction Materials
- Section 2104: Construction
- Section 2105: Quality Assurance
- Section 2107: Allowable Stress Design
- Section 2108: Strength Design of Masonry
- Section 2109: Empirical Design of Masonry
- Section 2110: Glass Unit Masonry

The following sections cover areas that are not covered in TMS 402 or TMS 602.

- Section 2111: Masonry Fireplaces
- Section 2112: Masonry Heaters
- Section 2113: Masonry Chimneys

IBC Chapter 22: Steel

IBC Chap. 22 governs the quality, design, fabrication, and erection of steel structures. Sections and tables to study in this chapter include

- Section 2207: Steel Joists
- Section 2208: Steel Cable Structures
- Section 2209: Steel Storage Racks
- Section 2210: Cold-Formed Steel
- Section 2211: Cold-Formed Steel Light-Frame Construction

Exam tip: Exam problems referencing IBC Chap. 22 will most likely be nonquantitative.

IBC Chapter 23: Wood

IBC Chap. 23 covers the design of wood structures. Sections and tables to study in this chapter include

- Section 2304: General Construction Requirements
- Section 2305: General Design Requirements for Lateral Force-Resisting Systems
- Section 2308: Conventional Light-Frame Construction

Exam tip: For problems relating to wood, the exam is more likely to reference the *National Design Specification for Wood Construction* (NDS) than the IBC. Students should be prepared, however, to solve at least one problem using the prescriptive design tables in IBC Chap. 23.

A.2. AMERICAN CONCRETE INSTITUTE (ACI 318, 530)

Key concepts: These key concepts are important for answering exam questions in subtopic 11.A.2, American Concrete Institute (ACI 318, 530).

- locations of exam-relevant sections and tables
- overview and organization of ACI 318, *Building Code Requirements for Structural Concrete*
- overview and organization of ACI 530 (TMS 402), *Building Code Requirements for Masonry Structures*

NCEES Handbook: To prepare for this subtopic, familiarize yourself with these sections in the *Handbook*.

- ASTM Standard Reinforcing Bars
- ASTM Standard Prestressing Strands
- ASTM Standard Plain Wire Reinforcement
- ASTM Standard Deformed Wire Reinforcement
- Design Provisions: Definitions

The following concepts are relevant for subtopic 11.A.2, American Concrete Institute (ACI 318, 530).

ACI 318 Overview

The design of reinforced concrete beams is governed by the provisions of ACI 318. The topics covered by ACI 318 include, but are not limited to, material requirements, mixing and placement, evaluation of existing structures, strength and serviceability, rebar protection, rebar development, two-way slabs, walls, footings, precast and prestressed concrete, unreinforced concrete, and anchors.

The code and commentary are presented side by side. The commentary explains in detail on code requirements, gives references to outside literature, and provides helpful illustrations.

Many of the code equations are empirical; in other words, the units of the result cannot be derived from those of the input values.

Exam tip: ACI 318 should be the first resource checked for any problems relating to concrete design. Don't ignore the commentary; it may contain information needed for solving a problem.

ACI 318 Organization

ACI 318 is informally organized into nine parts.

- Part 1 is for general knowledge and contains Chap. 1 through Chap. 4.
- Part 2 is for loads and analysis and contains Chap. 5 and Chap. 6.
- Part 3 is for concrete members and contains Chap. 7 through Chap. 14.
- Part 4 is for joints, connections, and anchors and contains Chap. 15 through Chap. 17.
- Part 5 is for earthquake resistance and contains Chap. 18.
- Part 6 is for materials and durability and contains Chap. 19 and Chap. 20.
- Part 7 is for strength and serviceability and contains Chap. 21 through Chap. 24.
- Part 8 is for reinforcement and contains Chap. 25.
- Part 9 is for structural evaluation and construction documentation and contains Chap. 26 and Chap. 27.

ACI 318 Part 1: Chapters 1–4

Handbook: Design Provisions: Definitions

ACI 318 Chap. 1 contains general information about ACI 318.

Chapter 2 contains definitions for notations and terms.

Chapter 3 lists standards referenced in ACI 318.

Chapter 4 describes different design requirements, structural systems, and types of construction.

Exam tip: A quick way to determine which portion of ACI 318 is needed to solve a problem is to look up the relevant variables in Chap. 2.

ACI 318 Part 2: Chapters 5–6

The most important task for a designer when beginning a project is to determine the loads on the structure. After these are calculated, the capacity of the sections is designed.

Load factors and combinations for design can be found in ACI 318 Chap. 5. The following section and table from Chap. 5 should be reviewed for the exam.

- Section 5.3: Load factors and combinations
- Table 5.3.1: Load combinations

Chapter 6 describes different methods of analysis. The following sections from Chap. 6 should be reviewed for the exam.

- Section 6.5: Simplified method of analysis for non-prestressed continuous beams and one-way slabs
- Section 6.6: First-order analysis
- Section 6.7: Elastic second-order analysis

Exam tip: For concrete design, the load combinations required by ACI 318 take precedence over those provided by *Minimum Design Loads for Buildings and Other Structures* (ASCE/SEI 7).

ACI 318 Part 3: Chapters 7–14

ACI 318 Chap. 7 through Chap. 14 each covers a specific type of concrete element. Each chapter follows the same format, first presenting the scope of the chapter and then discussing the design strengths and limits for that element type. The chapters in Part 3 follow the load path for a reinforced concrete structure and therefore follow the design process.

Chapter 7 gives the design requirements for one-way slabs. The following sections and tables from Chap. 7 should be reviewed for the exam.

- Table 7.3.1.1: Minimum thickness of solid nonprestressed one-way slabs
- Section 7.5: Design strength
- Section 7.6: Reinforcement limits
- Section 7.7: Reinforcement detailing

Chapter 8 gives the design requirements for two-way slabs. The following sections from Chap. 8 should be reviewed for the exam.

- Section 8.3: Design limits
- Section 8.4: Required strength
- Section 8.5: Design strength
- Section 8.6: Reinforcement limits
- Section 8.7: Reinforcement detailing
- Section 8.10: Direct design method
- Section 8.11: Equivalent frame method

Chapter 9 gives the design requirements for beams. The following sections and tables from Chap. 9 should be reviewed for the exam.

- Table 9.3.1.1: Minimum depth of nonprestressed beams
- Section 9.5: Design strength
- Section 9.6: Reinforcement limits
- Section 9.7: Reinforcement detailing

Chapter 10 gives the design requirements for columns. The following sections from Chap. 10 should be reviewed for the exam.

- Section 10.5: Design strength
- Section 10.6: Reinforcement limits
- Section 10.7: Reinforcement detailing

Chapter 11 gives the design requirements for walls. The following sections and tables from Chap. 11 should be reviewed for the exam.

- Table 11.3.1.1: Minimum wall thickness h
- Section 11.5: Design strength
- Section 11.6: Reinforcement limits
- Section 11.7: Reinforcement detailing

Chapter 12 gives the design requirements for diaphragms. The following sections and figures from Chap. 12 should be reviewed for the exam.

- Figure R12.1.1: Typical diaphragm actions
- Section 12.5: Design strength
- Section 12.6: Reinforcement limits
- Section 12.7: Reinforcement detailing

Chapter 13 gives the design requirements for foundations. The following sections from Chap. 13 should be reviewed for the exam.

- Section 13.3: Shallow foundations
- Section 13.4: Deep foundations

Chapter 14 gives the design requirements for plain concrete. The following sections from Chap. 14 should be reviewed for the exam.

- Section 14.3: Design limits
- Section 14.4: Required strength
- Section 14.5: Design strength
- Section 14.6: Reinforcement detailing

ACI 318 Part 4: Chapters 15–17

ACI 318 Chap. 15 through Chap. 17 discuss connecting elements of a reinforced concrete structure.

Chapter 15 provides design and detailing requirements for beam-column and slab-column joints. The following sections from Chap. 15 should be reviewed for the exam.

- Section 15.3: Transfer of column axial force through the floor system
- Section 15.4: Detailing of joints

Chapter 16 provides design and detailing requirements for connections between members. The following sections from Chap. 16 should be reviewed for the exam.

- Section 16.3: Connections to foundations
- Section 16.5: Brackets and corbels

Chapter 17 provides design and detailing requirements for anchoring to concrete. This chapter is amended by *International Building Code* (IBC) Sec. 1905.1.8. The following sections from Chap. 17 should be reviewed for the exam.

- Section 17.3: General requirements for strength of anchors
- Section 17.4: Design requirements for tensile loading
- Section 17.5: Design requirements for shear loading

Exam tip: The exam will most likely cover only cast-in-place anchors, not post-installed anchors.

ACI 318 Part 5: Chapter 18

ACI 318 Chap. 18 provides design and detailing requirements for earthquake-resistant structures. Which structures Chap. 18 applies to depends on the structure's seismic design category, as explained in Sec. R18.1. The following sections from Chap. 18 should be reviewed for the exam.

- Section 18.3: Ordinary moment frames
- Section 18.4: Intermediate moment frames
- Section 18.6: Beams of special moment frames
- Section 18.7: Columns of special moment frames
- Section 18.8: Joints of special moment frames
- Section 18.10: Special structural walls
- Section 18.12: Diaphragms and trusses
- Section 18.13: Foundations

ACI 318 Part 6: Chapters 19–20

ACI 318 Chap. 19 provides information about the properties of concrete and its design and durability requirements. The durability requirements are prescriptive in order to ensure concrete will resist loads without deteriorating. The following sections from Chap. 19 should be reviewed for the exam.

- Section 19.2: Concrete design properties
- Section 19.3: Concrete durability requirements

Chapter 20 provides information about the properties of steel reinforcement and embedments. The following sections from Chap. 20 should be reviewed for the exam.

- Section 20.2: Nonprestressed bars and wires
- Section 20.7: Embedments

Exam tip: Exam problems regarding durability can be answered by looking up the durability requirements in Chap. 19.

ACI 318 Part 7: Chapters 21–24

ACI 318 Chap. 21 covers the resistance factors for concrete. Section 21.2.2 defines the assumed strain limit of concrete, and Table 21.2.1 lists the strength reduction factors. Both of these should be studied for the exam.

Chapter 22 provides data needed to calculate the sectional strengths of reinforced concrete elements. Chapter 22 also provides the equations used to determine members' flexural strength, axial strength, shear strength, torsional strength, and bearing strength. The chapter does not give explicit equations for nominal moment or nominal axial capacity; the nominal moment must be derived from the general design principles. The following sections from Chap. 22 should be reviewed for the exam.

- Section 22.3: Flexural strength
- Section 22.5.6: V_c for nonprestressed members with axial compression
- Section 22.5.10: One-way shear reinforcement

Chapter 23 applies when strut-and-tie models are required to analyze and design a member. The following sections from Chap. 23 should be reviewed for the exam.

- Section 23.4: Strength of struts
- Section 23.7: Strength of ties
- Section 23.9: Strength of nodal zones

Chapter 24 provides information on serviceability requirements. The following sections and tables from Chap. 24 should be reviewed for the exam.

- Table 24.2.2: Maximum permissible calculated deflections
- Section 24.4: Shrinkage and temperature reinforcement
- Section 24.5: Permissible stress in prestressed concrete flexural members

ACI 318 Part 8: Chapter 25

ACI 318 Chap. 25 provides reinforcement detailing requirements. The following sections from Chap. 25 should be reviewed for the exam.

- Section 25.4.3: Development of standard hooks in tension
- Section 25.4.8: Development of pretensioned seven-wire strands in tension
- Section 25.5.5: Lap splice lengths of deformed bars in compression

ACI 318 Part 9: Chapters 26–27

ACI 318 Chap. 26 provides information on construction documents and inspections. The following sections from Chap. 26 should be reviewed for the exam.

- Section 26.4: Concrete materials and mixture requirements
- Section 26.5: Concrete production and construction
- Section 26.6: Reinforcement materials and construction requirements
- Section 26.12: Concrete evaluation and acceptance
- Section 26.13: Inspection

Chapter 27 provides information on strength evaluation of existing structures. The following sections from Chap. 27 should be reviewed for the exam.

- Section 27.3: Analytical strength evaluation
- Section 27.4: Strength evaluation by load test
- Section 27.5: Reduced load rating

Exam tip: Questions on construction documents often show up on the exam as multiple-choice questions.

ACI 318 Appendices

ACI 318 has two appendices. Appendix A, Steel Reinforcement Information, provides tabulated data from the American Society of Testing and Materials (ASTM) and Wire Reinforcement Institute (WRI). Appendix B, Equivalence between SI-Metric, MKS-Metric, and U.S. Customary Units of Nonhomogeneous Equations in the Code, provides tabulated equivalencies between the three systems of units.

Exam tip: The tables in App. A include data on diameters and areas required for concrete reinforcement and often need to be referenced to solve exam problems.

ACI 530 Overview

The design of masonry is governed by *Building Code Requirements for Masonry Structures* (ACI 530, also known as TMS 402 and ASCE 5) or *Specification for Masonry Structures* (ACI 530.1, also known as TMS 602 and ASCE 6).

ACI 530 is a two-document book, where ACI 530 is the code, and ACI 530.1 is the specification based on that code. Topics in ACI 530 include material requirements, strength and serviceability, and requirements for walls, columns, pilasters, beams, and lintels. Topics in ACI 530.1 include material quality assurance and placement, bonding, and anchoring of units. Commentary is presented side by side with the code standards.

ACI 530 should be the first resource for any problems relating to masonry design. Some ACI 530 standards for concrete construction are incorporated into *International Building Code* (IBC) Chap. 21.

Organization of ACI 530

ACI 530 is organized into 14 chapters grouped into five parts.

Part 1, General (Chap. 1 through Chap. 3), contains general information about ACI 530, notation, and information on quality assurance programs.

Part 2, Design Requirements (Chap. 4 through Chap. 7), covers general analysis and design considerations; seismic design requirements; and design requirements for structural elements, reinforcements, metal accessories, and anchor bolts.

Part 3, Engineered Design Methods (Chap. 8 through Chap. 11), covers allowable stress design and strength design of masonry and prestressed masonry, and strength design of autoclaved aerated concrete.

Part 4, Prescriptive Design Methods (Chap. 12 through Chap. 14), covers design methods for veneer, glass unit masonry, and masonry partition walls.

Part 5, Appendices, follows the main chapters, and contains references for equation conversions, unit conversions, and prefixes. ACI 530 also includes a References for the Code Commentary section.

The structure of ACI 530.1 follows the structure of ACI 530. ACI 530.1 has three parts, followed by the sections Foreword to Specification Checklists, Optional Requirements Checklist, and References for the Specification Commentary.

Exam tip: Students should be prepared to solve one prestressed masonry problem on the exam.

ACI 530 Chapter 1: General Requirements

Chapter 1 contains general information about ACI 530. It is unlikely that any concrete-related exam questions will reference this chapter.

ACI 530 Chapter 2: Notation and Definitions

Chapter 2 contains definitions for notations and terms.

ACI 530 Chapter 3: Quality and Construction

Chapter 3 contains quality assurance and construction considerations. The following sections from Chap. 3 should be reviewed for the exam.

- Section 3.1.1: Level A Quality Assurance
- Section 3.1.2: Level B Quality Assurance
- Section 3.1.3: Level C Quality Assurance

Exam tip: Masonry-related exam questions referencing this chapter will most likely be qualitative, not quantitative.

ACI 530 Chapter 4: General Analysis and Design Considerations

Chapter 4 contains equations for analysis and design. The following sections from Chap. 4 should be reviewed for the exam.

- Section 4.2: Material properties
- Section 4.3.4: Bearing area
- Section 4.5: Masonry not laid in running bond

ACI 530 Chapter 5: Structural Elements

Each section of Chap. 5 focuses on a specific masonry element. The following sections from Chap. 5 should be reviewed for the exam.

- Section 5.2: Beams
- Section 5.3: Columns
- Section 5.4: Pilasters
- Section 5.5: Corbels

ACI 530 Chapter 6: Reinforcement, Metal Accessories, and Anchor Bolts

Chapter 6 contains design requirements and equations for reinforcement, metal accessories, and anchor bolts. The following sections from Chap. 6 should be reviewed for the exam.

- Section 6.1.5: Standard hooks
- Section 6.1.6: Minimum bend diameter for reinforcing bars
- Section 6.2: Anchor bolts

ACI 530 Chapter 7: Seismic Design Requirements

Chapter 7 covers design requirements for seismic design. The seismic design requirements vary based on the seismic design category of the structure. Minimum amounts of reinforcement, required types of anchorage ties, and material requirements are given. The following sections and tables from Chap. 7 should be reviewed for the exam.

- Section 7.3: Element classification
- Table CC-7.3.2-1: Requirements for Masonry Shear Walls Based on Shear Wall Designation

ACI 530 Chapter 8: Allowable Stress Design of Masonry

Chapter 8 provides requirements for allowable stress design (ASD). Only the ASD method may be used on the exam, with the exception of problems on walls with out-of-plane loads, for which the strength design (SD) method can be used.

The following sections from Chap. 8 should be reviewed for the exam.

- Section 8.1: General
- Section 8.2: Unreinforced masonry
- Section 8.3: Reinforced masonry

Equations in Sec. 8.1 sometimes reference Sec. 6.2. For example, Eq. 8-3 defines the allowable breakout tensile load on an anchor bolt, but this equation depends on the projected tension area in the masonry, which is defined in Sec. 6.2.2.

ACI 530 Chapter 9: Strength Design of Masonry

The only section of Chap. 9 that is applicable to the exam is Sec. 9.3.5, which focuses on wall design for out-of-plane loads.

ACI 530 Chapter 10: Prestressed Masonry

Chapter 10 provides design requirements for prestressed masonry. The following sections from Chap. 10 should be reviewed for the exam.

- Section 10.3: Permissible stresses in prestressing tendons
- Section 10.4: Axial compression and flexure
- Section 10.6: Shear

Shear design, discussed in Sec. 10.6, has the same requirements as reinforced masonry, except that the normal force includes the prestress.

ACI 530 Chapters 11–14

ACI 530 Chap. 11 through Chap. 14 provide requirements for design and details for different masonry product types under engineered design methods and prescriptive design methods. Chapter 11 covers strength design of autoclaved aerated concrete (AAC) masonry. Chapter 12 discusses veneers. Chapter 13 discusses glass unit masonry. Chapter 14 discusses masonry partition walls.

Exam tip: Of these topics, the only one likely to appear on the exam is masonry partition walls.

ACI 530 Appendices

Appendix A discusses empirical design. Appendix B includes provisions for both participating and non-participating infills. Appendix C covers the limit design method.

Exam tip: Masonry-related exam questions are not likely to reference the appendices.

A.3. PRECAST/PRESTRESSED CONCRETE INSTITUTE (PCI DESIGN HANDBOOK)

Key concepts: These key concepts are important for answering exam questions in subtopic 11.A.3, Precast/Prestressed Concrete Institute (PCI Design Handbook).

- important tables and figures in PCI
- navigating PCI
- organization of PCI

NCEES Handbook: To prepare for this subtopic, familiarize yourself with these sections in the *Handbook*.

- Moment, Shear, and Deflection Diagrams

The following concepts are relevant for subtopic 11.A.3, Precast/Prestressed Concrete Institute (PCI Design Handbook).

PCI Overview

The PCI Design Handbook: Precast and Prestressed Concrete (PCI) is the guide to designing precast and prestressed concrete structures, components, and connections. It is based on and reproduces some information from older versions of some exam design standards, such as *Building Code Requirements for Structural Concrete* (ACI 318), *Minimum Design Loads for Buildings and Other Structures* (ACSE/SEI 7), and the *International Building Code* (IBC).

Exam tip: Avoid consulting PCI for any exam problem that can be solved only with reference to other codes. Some of the information from the older versions of those codes may be outdated and may lead to incorrect answers.

Organization of PCI

PCI contains many helpful design examples and aids. It is broken up into 15 chapters and an appendix.

Chapter 1 contains general information on precast and prestressed concrete and precast applications in the built environment, and Chap. 2 provides definitions for notations and terms.

Chapter 3 and Chap. 4 include design tables for preliminary component sizing and design aids that contain several factors crucial to design and analysis.

Chapter 5 and Chap. 6 cover the procedures used in the design of precast and prestressed concrete components and connections.

Chapter 7 and Chap. 8 cover the design of architectural precast concrete and design guidelines for precast and prestressed components for temporary loads.

Chapter 9 covers descriptions of types and uses of concrete and its constituent materials.

Chapter 10 through Chap. 13 contain information on miscellaneous topics.

Chapter 14 covers specifications and standard practices in the construction industry.

Chapter 15 includes design aids covering various topics.

The Appendix covers the impact of *Building Code Requirements for Structural Concrete* (ACI 318) regulations on PCI.

Exam tip: An exam problem that requires reference to PCI is likely to focus on whether you can find and use a particular design aid. A precast or prestressed structure problem that cannot be answered using these design aids alone can be solved by determining whether the problem relates to a precast or prestressed component, connection, or structure, and then searching the table of contents of the appropriate PCI chapter.

PCI Chapter 3: Preliminary Design of Precast/Prestressed Concrete Structures

Chapter 3 covers gravity and lateral force systems and includes design tables for preliminary component sizing. Important items to study in Chap. 3 include

- Section 3.4: Double-Tee Load Tables
- Design Aid 3.12.7: Recommended Height-to-Thickness Ratio Limits for Precast, Reinforced Solid Non-Load Bearing Wall Panels

PCI Chapter 4: Analysis and Design of Precast/Prestressed Concrete Structures

Chapter 4 provides definitions of various loads, load effects, load combinations, structural modeling, and gravity systems. Chapter 4 also includes design aids that contain importance factors; snow, wind, and seismic maps; tables from *Minimum Design Loads for Buildings and Other Structures* (ASCE/SEI 7) used for calculating volume change effects; shear wall deflections; lateral load resisting systems; diaphragms; and seismic considerations for intermediate prestressed concrete shear walls and moment-resisting frames. Important sections, figures, and examples to study in Chap. 4 include

- Section 4.2.6.1: Load Factors for Diaphragms
- Figure 4.5.3: Effective width of wall perpendicular to shear walls
- Example 4.5.7.1: Design of Unsymmetrical Shear Walls
- Section 4.8: Diaphragm Design

Exam tip: For yielding elements of seismic connections, refer to *Building Code Requirements for Structural Concrete* (ACI 318) and the *International Building Code* (IBC), as they will be more up to date than PCI.

PCI Chapter 5: Design of Precast and Prestressed Concrete Components

Chapter 5 covers the procedures used in the design of precast and prestressed concrete components. Design procedures are based on the provisions in *Building Code Requirements for Structural Concrete* (ACI 318), but PCI also includes information not in ACI 318, such as a discussion of torsion of solid precast sections.

Chapter 5 contains worked examples, the most important being Ex. 5.7.1, which covers equations for prestress losses using a combination of elastic shortening loss, creep loss, shrinkage loss, and relaxation loss. In addition, Chap. 5 includes several useful design aids cited in the working examples. Important sections, tables, and examples to study in Chap. 5 include

- Example 5.3.5.1: Horizontal Shear Design for Composite Beam
- Example 5.4.2: Shear and Torsion Design of a Prestressed Concrete Component
- Example 5.7.1: Loss of Prestress
- Table 5.7.1: Values of K_{re} and J
- Table 5.7.2: Values of C
- Section 5.7.3: Estimating Prestress Loss
- Example 5.10.1: Shear-Wall Analysis
- Section 5.14: Design Aids

Exam tip: Chapter 5 will be the chapter most often needed during the exam. Be sure to review the sections on prestress losses.

PCI Chapter 6: Design of Connections

Chapter 6 covers connections between components and includes discussions of the component details in precast construction. This chapter includes 26 worked examples. In addition, Sec. 6.15 includes several useful design aids cited in the working examples. Important sections and examples to study in Chap. 6 include

- Table 6.2.1: Diaphragm over-strength factors
- Section 6.4: Connection Hardware and Load-Transfer Devices

- Example 6.7.4.1: Strength Analysis of Weld Group
- Figure 6.11.1: Typical column base-plate detail
- Figure 6.13.1: Typical precast concrete connections
- Example 6.13.6: Wall-to-Wall Shear Connection with Combined Loading
- Section 6.15: Design Aids

PCI Chapter 7: Structural Considerations for Architectural Precast Concrete

Chapter 7 covers the design of architectural precast concrete and the production and erection procedures for precast concrete. This chapter includes discussion of non-load-bearing wall panels, spandrels, and column covers; load-bearing wall panels and spandrels; structural and connection considerations of components and cladding; wind loads; and seismic design considerations. Important sections and examples to study in Chap. 7 include

- Section 7.3: Structural Design Considerations
- Section 7.5.1: Wind Loads, ASCE 7-05 Method 1
- Section 7.5.2: Wind Loads, ASCE 7-05 Method 2
- Section 7.5.3: Seismic Loads

PCI Chapter 8: Component Handling and Erection Bracing

Chapter 8 covers design guidelines regarding temporary loads on precast and prestressed components. This chapter includes discussions on stripping, storage, transportation, erection, and stability of the overall structure and components. Important sections and examples to study in Chap. 8 include

- Example 8.3.1: Design of Wall Panel for Stripping
- Section 8.3.4.2: Factors of Safety
- Section 8.7: Erection Bracing

PCI Chapter 9: Precast and Prestressed Concrete: Materials

Chapter 9 covers types and uses of concrete and its constituent materials. This chapter includes discussions of aggregate durability, admixtures, fresh concrete properties (e.g., slump, air content, workability, and curing), and hardened concrete properties (e.g., elastic modulus, shrinkage, creep, and deterioration mechanisms). Also included is a discussion of welding of structural bolts, anchor bolts, nuts, and washers. Important sections to study in Chap. 9 include

- Section 9.2.1.3: Aggregates
- Section 9.2.2.3: Tensile Strength
- Section 9.6: Reinforcement

PCI Chapter 10: Design for Fire Resistance of Precast and Prestressed Concrete

Chapter 10 covers design of precast and prestressed concrete for heat transmission, fire endurance, and fire resistance. This chapter also includes miscellaneous considerations such as code and economic concerns.

Exam tip: This chapter is not likely to be referenced during the exam.

PCI Chapter 11: Thermal and Acoustical Properties of Precast Concrete

Chapter 11 covers thermal properties, transmittance, thermal storage effects, condensation control, and acoustical properties of precast concrete. Important sections and examples to study in Chap. 11 include

- Example 11.1.4.1: Thermal Resistance of Wall
- Section 11.2.4: Sound Transmission Loss
- Section 11.2.5: Impact Noise Reduction

PCI Chapter 12: Vibration Design of Precast/Prestressed Concrete Floor Systems

Chapter 12 covers types of floor vibrations, the fundamental frequency of vibration, damping devices, vibrations for mixed-use structures, and vibration isolation for mechanical equipment. Much of the information in this chapter is also covered in AISC Steel Design Guide 11, *Vibrations of Steel-Framed Structural Systems Due to Human Activity*.

Exam tip: This chapter is not likely to be referenced during the exam.

PCI Chapter 13: Tolerances for Precast and Prestressed Concrete

Chapter 13 covers product tolerances, erection tolerances, clearances, and interfacing tolerances.

Exam tip: This chapter is not likely to be referenced during the exam.

PCI Chapter 14: Specifications and Standard Practices

Chapter 14 covers standard design practices for precast concrete. The practices discussed in this chapter reference *Building Code Requirements for Structural*

Concrete (ACI 318) standards and provide additional guidance on composite concrete flexural members, footings, two-way slab systems, reinforcement, shear and torsion design values for prestressed members, flexural and axial loads, strength and serviceability requirements, concrete quality, and durability requirements. It also covers definitions and terminology of project participants and includes a discussion of typical contract requirements, such as shop drawings, tests and inspections, delivery, erection responsibilities, and warranty.

PCI Chapter 15: General Design Information

Handbook: Moment, Shear, and Deflection Diagrams

Chapter 15 includes 33 design aids covering design information (loads, equations, and diagrams); material properties of concrete, prestressing steel, reinforcing bars, and structural welded-wire reinforcement; standard bolts, nuts, and washers; welding; section properties; and metric conversion. Several of these design aids are available in other codes.

- Design Aid 15.1.1 is sourced from *Minimum Design Loads for Buildings and Other Structures* (ASCE/SEI 7).
- Design Aid 15.1.3 is sourced from the *Steel Construction Manual* (AISC) and is also available in *NCEES Handbook* section Moment, Shear, and Deflection Diagrams.
- Information in Design Aid 15.2.1 is also available in *Building Code Requirements for Structural Concrete* (ACI 318).

Important design aids to study in Chap. 15 include

- Design Aid 15.1.1: Dead Weights of Floors, Ceilings, Roofs, and Walls
- Design Aid 15.1.3: Beam Design Equations and Diagrams
- Design Aid 15.2.1: Concrete Modulus of Elasticity as Affected by Concrete Density and Strength
- Design Aid 15.3.3: Typical Design Stress-Strain Curve, 7-Wire Low-Relaxation Prestressing Strand
- Design Aid 15.4.1: Reinforcing Bar Data
- Design Aid 15.7.2: Typical Welded Joints in Precast Concrete Construction
- Design Aid 15.8.1: Properties of Geometric Sections
- Design Aid 15.8.2: Plastic Section Moduli and Shape Factors

A.4. STEEL CONSTRUCTION MANUAL (AISC)

Key concepts: These key concepts are important for answering exam questions in subtopic 11.A.4, Steel Construction Manual (AISC).

- AISC organization
- AISC overview
- important sections, tables, and figures in AISC

NCEES Handbook: To prepare for this subtopic, familiarize yourself with these sections in the *Handbook*.

- Conic Sections
- Area & Centroid (Table)
- Area Moment of Inertia (Table)
- Radius of Gyration (Table)
- Product of Inertia (Table)
- Simply Supported Beam Slopes and Deflections
- Moment, Shear, and Deflection Diagrams
- Fastener Groups in Shear
- Basic Welding Symbols

The following concepts are relevant for subtopic 11.A.4, Steel Construction Manual (AISC).

AISC Overview

Of the manuals, design guides, and specifications published by the American Institute of Steel Construction, AISC is the most widely used for steel design. AISC is based on two standards, *Specification for Structural Steel Buildings* (AISC 360) and *Code of Standard Practice for Steel Buildings and Bridges* (AISC 313). Both of these standards are included in Part 16 of AISC.

Exam tip: AISC should be your first resource for problems relating to steel design.

AISC Organization

AISC is organized into 17 parts. Parts 1 through 15 give standard shape geometries and materials, and contain design aids based on the specification. Some of the design aids are approximate, intended to help designers develop a good preliminary design. Parts 1 through 15 also contain additional information relating to certain aspects of steel design that are not contained in *Specification for Structural Steel Buildings* (AISC 360).

Parts 1 and 2 include information on dimensions, properties, and general design considerations.

Parts 3 through 6 include information on the design of flexural members, compression members, tension members, and members subject to combined forces.

Parts 7 and 8 cover design considerations for bolts and welds.

Parts 9 and 10 include information on connecting elements and simple shear connections.

Parts 11 and 12 include information on moment connections. Information from these two parts probably won't be referenced during the exam.

Parts 13 through 15 cover information on a variety of additional connecting elements, such as bracing, trusses, plates, rods, splices, and hanger connections.

Part 16, Specifications and Codes, contains the following.

- Part 16.1: *Specification for Structural Steel Buildings* (AISC 360)
- Part 16.2: *Specification for Structural Joints Using High-Strength Bolts*
- Part 16.3: *Code of Standard Practice for Steel Buildings and Bridges*

Part 17 provides miscellaneous data and mathematical information.

AISC Part 1: Dimensions and Properties

Part 1 gives the defining dimensions and calculated section properties for standard steel shapes.

Exam tip: Using the already calculated section properties in the tables will reduce the time needed to solve steel problems on the exam.

AISC Part 2: General Design Considerations

Part 2 covers material specifications for steel products, as well as general specification requirements and other design considerations. Part 2 includes load combinations that should be used for steel design, and material strengths for the different standard shape families.

AISC Part 3: Design of Flexural Members

Handbook: Simply Supported Beam Slopes and Deflections; Moment, Shear, and Deflection Diagrams

Part 3 gives requirements for the design of flexural members subject to uniaxial flexure without axial forces or torsion. Important tables to study in Part 3 include

- Table 3-2: W-Shapes—Selection by Z_x
- Table 3-10: W-Shapes
- Table 3-19: Composite W-Shapes
- Table 3-23: Shears, Moments and Deflections

The information in AISC Table 2-23 can also be found in *Handbook* table Simply Supported Beam Slopes and Deflections and section Moment, Shear, and Deflection Diagrams.

AISC Part 4: Design of Compression Members

Requirements for the design of members subject to axial compression are covered in Part 4. Important tables to study in Part 4 include

- Table 4-1: W-Shapes in Axial Compression
- Table 4-21: Stiffness Reduction Factor τ_b
- Table 4-22: Available Critical Stress for Compression Members

Exam tip: Using the tables and design aids provided in Part 4 can help you solve exam problems faster.

AISC Part 5: Design of Tension Members

Requirements for the design of members subject to static axial tension are covered in Part 5. Important figures and tables to study in Part 5 include Table 5-1: W-Shapes.

AISC Part 6: Design of Members Subject to Combined Forces

Requirements for the design of members subject to combined forces are covered in Part 6. The most important table to study in Part 6 is Table 6-1, Combined Flexure and Axial Force, W-Shapes.

AISC Part 7: Design Considerations for Bolts

Handbook: Fastener Groups in Shear

Part 7 covers the requirements for the design of bolts in steel-to-steel structural connections. Important tables to study in Part 7 include

- Table 7-1: Available Shear Strength of Bolts
- Table 7-2: Available Tensile Strength of Bolts
- Table 7-4: Available Bearing Strength at Bolt Holes Based on Bolt Spacing
- Table 7-5: Available Bearing Strength at Bolt Holes Based on Edge Distance

AISC Part 8: Design Consideration for Welds

Handbook: Basic Welding Symbols

Part 8 covers requirements for the design of welded joints. Important sections and tables to study in Part 8 include

- Section: Proper Specification of Joint Type
- Table 8-2: Prequalified Welded Joints
- Table 8-3: Electrode Strength Coefficient, C_1

Handbook table Basic Welding Symbols is based on AISC Table 8-2.

AISC Part 9: Design of Connecting Elements

Part 9 discusses beam cope buckling, prying on bolts, gusset strength, and other topics related to the design of connecting elements. Important tables to study in Part 9 include

- Table 9-1: Reduction in Area for Holes
- Table 9-3: Block Shear Rupture
- Table 9-4: Beam Bearing Constants

AISC Part 10: Design of Simple Shear Connections

Part 10 discusses double-angle and single-plate beam end connections, among other topics. Important tables to study in Part 10 include

- Table 10-1: All-Bolted Double-Angle Connections
- Table 10-2: Available Weld Strength of Bolted/Welded Double-Angle Connections
- Table 10-3: Available Weld Strength of All-Welded Double-Angle Connections

AISC Part 13 and Part 14

Part 13, "Design of Bracing Connections and Truss Connections," and Part 14, "Design of Beam Bearing Plates, Column Base Plates, Anchor Rods, and Column Splices," contain analysis methods and equations for solving steel-related problems. Important sections and tables to study in Part 14 include

- Section: Column Base Plates for Axial Compression
- Section: Design Considerations for HSS Cap Plates
- Table 14-3: Typical Column Splices

AISC Part 15: Design of Hanger Connections, Bracket Plates, and Crane-Rail Connections

Requirements for the design of hanger connections, bracket plates, and crane-rail connections are covered in Part 15. Important sections, figures, and tables to study in Part 15 include

- Table 15-2: Preliminary Hanger Connection Selection Table
- Table 15-3: Net Plastic Section Modulus, Z_{net}

AISC Part 16.1: Specification for Structural Steel Buildings

Part 16.1, *Specification for Structural Steel Buildings* (AISC 360), includes Chap. A through Chap. N, eight appendices, and commentaries at the end that correspond with each chapter and appendix.

AISC Part 16.1 Chapter A: General Provisions

Chapter A covers the scope of AISC 360; the referenced specifications, codes, and standards; and requirements for material and structural design documents.

Exam tip: The exam will not likely contain questions on the material in Chap. A.

AISC Part 16.1 Chapter B: Design Requirements

The general requirements for the analysis and design of steel structures are covered in Chap. B. Important tables to study in this chapter include

- Table B4.1a: Width-to-Thickness Ratios: Compression Elements Members Subject to Axial Compression
- Table B4.1b: Width-to-Thickness Ratios: Compression Elements Members Subject to Flexure

AISC Part 16.1 Chapter C: Design for Stability

Chapter C includes analysis methods and equations referenced by other chapters in AISC. Important sections to study in this chapter include

- Section C1: General Stability Requirements
- Section C2: Calculation of Required Strengths
- Section C3: Calculation of Available Strengths

AISC Part 16.1 Chapter D: Design of Members for Tension

Members subject to axial tension are covered in Chap. D. Important sections and tables to study in this chapter include

- Section D2: Tensile Strength
- Section D3: Effective Net Area
- Table D3.1: Shear Lag Factors for Connections to Tension Members

AISC Part 16.1 Chapter E: Design of Member for Compression

Members subject to axial compression are covered in Chap. E. Important sections and tables to study in this chapter include

- Table User Note E1.1: Selection Table for the Application of Chapter E Sections
- Section E3: Flexural Buckling of Members Without Slender Elements
- Section E7: Members with Slender Elements

Exam tip: This chapter is the most likely reference required for exam problems related to steel columns.

AISC Part 16.1 Chapter F: Design of Member for Flexure

Members subject to bending are covered in Chap. F. Important sections and tables to study in this chapter include

- Table User Note F1.1: Selection Table for the Application of Chapter F Sections
- Section F1: General Provisions
- Section F2: Doubly Symmetric Compact I-Shaped Members and Channels Bent About Their Major Axis

Exam tip: This chapter is the most likely reference required for exam problems related to steel beams.

AISC Part 16.1 Chapter G: Design of Members for Shear

Chapter G discusses the webs of singly and doubly symmetric members subject to shear in the plane of the web, single angles, hollow structural sections (HSSs), and shear in the weak direction of singly and doubly symmetric shapes. Important sections to study in this chapter include

- Section G2: Members with Unstiffened or Stiffened Webs
- Section G3: Tension Field Action
- Section G6: Round HSS

AISC Part 16.1 Chapter H: Design of Members for Combined Forces and Torsion

Chapter H discusses members subject to axial force and flexure about one or both axes, with or without torsion, as well as members subject to torsion only. Important sections to study in this chapter include

- Section H1: Doubly and Singly Symmetric Members Subject to Flexure and Axial Force
- Section H2: Unsymmetric and Other Members Subject to Flexure and Axial Force
- Section H3: Members Subject to Torsion and Combined Torsion, Flexure, Shear and/or Axial Force

AISC Part 16.1 Chapter I: Design of Composite Members

Composite members are covered in Chap. I. Important sections to study in this chapter include

- Section I3: Flexure
- Section I6: Load Transfer
- Section I8: Steel Anchors

AISC Part 16.1 Chapter J: Design of Connections

Chapter J discusses connecting elements, connectors, and the affected elements of connected members not subject to fatigue loads. Important tables to study in this chapter include

- Table J2.4: Minimum Size of Fillet Welds
- Table J2.5: Available Strength of Welded Joints
- Table J3.2: Nominal Strength of Fasteners and Threaded Parts
- Table J3.3: Nominal Hole Dimensions

AISC Part 16.1 Chapter K: Design of HSS and Box Member Connections

Chapter K discusses the design of hollow structural section (HSS) connections. Important tables to study in this chapter include

- Table K1.1: Available Strengths of Plate-to-Round HSS Connections
- Table K1.2: Available Strengths of Plate-to-Rectangular HSS Connections
- Table K2.1: Available Strengths of Round HSS-to-HSS Truss Connections
- Table K2.2: Available Strengths of Rectangular HSS-to-HSS Truss Connections

AISC Part 16.1 Appendices

The eight appendices address additional topics related to steel design and expand upon the provisions discussed in Chap. A through Chap. N. Important appendices to study include

- Appendix 3: Design for Fatigue
- Appendix 5: Evaluation of Existing Structures
- Appendix 6: Stability Bracing for Columns and Beams

AISC Part 17: Miscellaneous Data and Mathematical Information

Handbook: Conic Sections; Area & Centroid (Table); Area Moment of Inertia (Table); Radius of Gyration (Table); Product of Inertia (Table)

Part 17 includes tables for conversion factors, geometric section properties, trigonometric formulas, and other helpful references for the exam. Much of the information in Part 17 can also be found in the *Handbook* section and tables listed previously.

A.5. NATIONAL DESIGN SPECIFICATION FOR WOOD CONSTRUCTION (NDS)

Key concepts: These key concepts are important for answering exam questions in subtopic 11.A.5, National Design Specification for Wood Construction (NDS).

- important sections and tables in the NDS
- NDS scope
- organization of NDS documents

NCEES Handbook: The *Handbook* does not contain any material relevant to this subtopic.

The following concepts are relevant for subtopic 11.A.5, National Design Specification for Wood Construction (NDS).

NDS Overview and Scope

The NDS consists of three documents: the *National Design Specification for Wood Construction with Commentary*, which contains design guidance; the *National Design Specification Supplement*, which contains standard section and material properties; and *Special Design Provisions for Wind & Seismic with Commentary* (SDPWS). The topics covered include sawn lumber, glued laminated timber (i.e., glulam), composite lumber, connections, and fire design.

The intent of the NDS is to provide a national standard of practice. The NDS should be used in tandem with sound engineering judgment and design, proper fabrication, and sufficient construction supervision. The provisions in the specification are based on the most reliable data available from laboratory tests, as well as analysis and evaluation of structures already in service.

Some NDS requirements are incorporated into *International Building Code* (IBC) Chap. 23.

Exam tip: The NDS should be your first resource for any problems relating to wood design. The NDS allows for both allowable strength design (ASD) and load and resistance factor design (LRFD), but the exam tests only on ASD.

Organization of the NDS

The NDS consists of 16 chapters.

Chapter 1 covers scope and notation.

Chapter 2 covers design adjustment factors that are not specific to a material.

Chapter 3 covers general design procedures for all wood products.

Chapter 4 through Chap. 10 cover specific wood products. Chapter 4 covers sawn lumber. Chapter 5 is about structural glued laminated timber (glulam). Chapter 6 is about round timber poles and piles. Chapter 7 is about prefabricated wood I-joists. Chapter 8 is about structural composite lumber. Chapter 9 is about wood structural panels. Chapter 10 is about cross-laminated timber.

The structures of Chap. 4 through Chap. 10 each follow a similar pattern: the first section (e.g., Sec. 4.1) covers general considerations, the second section provides reference design values, the third specifies adjustment factors, and the fourth covers special design considerations. The adjustment factors provided in these chapters are used to convert an allowable stress for a given material to an allowable stress that accounts for the material strength, the geometry of the member, and the condition in which it will be used.

Chapter 11 through Chap. 14 cover connection design provisions. Chapter 11 is about mechanical connections. Chapter 12 is about dowel-type fasteners. Chapter 13 is about split ring and shear plate connectors. Chapter 14 is about timber rivets.

Chapter 15 covers special loading conditions.

Chapter 16 includes equations related to fire design.

Following the chapters are an appendix section and a commentary section.

Exam tip: Chapter 3 is very important to review and contains many useful design equations. Chapter 4 and Chap. 5 are likely to be required during the exam and are important to review. Chapter 6 through Chap. 10, Chap. 15, and Chap. 16 are less likely to be required during the exam.

NDS Chapter 1: General Requirements for Structural Design

This chapter includes the scope and notation for symbols used throughout the document.

Exam tip: The descriptions of symbols in NDS Chap. 1 do not provide references to the relevant portions of the code. It is important to review the notation defined in this chapter, but this chapter will not be of use in determining which chapter to consult while solving an exam problem.

NDS Chapter 2: Design Values for Structural Members

Chapter 2 contains adjustment factors for load duration and temperature effects, as well as load and resistance factor design (LRFD) criteria. The following sections and tables from Chap. 2 should be thoroughly reviewed for the exam.

- Table 2.3.2: Frequently Used Load Duration Factors, C_D
- Table 2.3.3: Temperature Factor, C_t
- Section 2.3.4: Fire Retardant Treatment

NDS App. B and App. C contain additional, nonmandatory provisions and explanations for the load duration and temperature factors.

Exam tip: Carefully reading and applying relevant footnotes can make the difference in selecting the correct answer choice.

NDS Chapter 3: Design Provisions and Equations

This is the go-to chapter for design equations and other adjustment factors. The design procedures given in Chap. 3 are meant to be generic and applicable to various member types, with subsequent chapters giving adjustments based on member type. The following sections and tables are important to review for the exam.

- Section 3.3.2: Flexural Design Equations
- Table 3.3.3: Effective Length, l_e, for Bending Members
- Section 3.4.2: Shear Design Equations
- Section 3.5.1: Deflection Calculations
- Section 3.7.1: Column Stability Factor, C_P
- Section 3.9.1: Bending and Axial Tension
- Section 3.9.2: Bending and Axial Compression
- Table 3.10.4: Bearing Area Factors, C_b

NDS Chapter 4: Sawn Lumber

Chapter 4 provides standards specific to sawn lumber. The following sections and tables are important to review for the exam.

- Section 4.2: Reference Design Values
- Section 4.3: Adjustment of Reference Design Values
- Table 4.3.1: Applicability of Adjustment Factors for Sawn Lumber
- Section 4.4: Special Design Considerations

Exam tip: One of the most vital portions of the NDS for examinees is Table 4.3.1 because it provides in one place all the capacity formulas for sawn lumber members, which are the type of wood element most likely to be used in exam problems.

NDS Chapter 5: Structural Glued Laminated Timber

Chapter 5 provides standards specific to structural glued laminated timber (glulam). The following sections and tables from Chap. 5 should be thoroughly reviewed for the exam.

- Section 5.2: Reference Design Values
- Section 5.3: Adjustment of Reference Design Values
- Table 5.3.1: Applicability of Adjustment Factors for Structural Glued Laminated Timber
- Section 5.4: Special Design Considerations

Because glulams may be oriented with the laminations parallel to either axis, and because glulam layups may be balanced or unbalanced (i.e., the grade of lumber is placed either symmetrically or unsymmetrically about the neutral axis), glulam reference values are more complex than those for sawn lumber.

NDS Chapters 11–14

Chapter 11 includes general design provisions that apply to all fastener types covered in Chap. 12 through Chap. 14, while Chap. 12 through Chap. 14 contain reference design values that are specific to one type of connection. The following sections and tables from Chap. 11 through Chap. 14 should be reviewed for the exam.

- Table 11.3.1: Applicability of Adjustment Factors for Connections
- Table 11.3.4: Temperature Factors, C_t, for Connections
- Table 11.3.6A: Group Action Factors, C_g, for Bolt or Lag Screw Connections with Wood Side Members
- Table 11.3.6B: Group Action Factors, C_g, for 4" Split Ring or Shear Plate Connectors with Wood Side Members
- Table 11.3.6C: Group Action Factors, C_g, for Bolt or Lag Screw Connections with Steel Side Plates
- Table 11.3.6D: Group Action Factors, C_g, for 4" Shear Plate Connectors with Steel Side Plates
- Table 12.2A: Lag Screw Reference Withdrawal Design Values
- Table 13.2.3: Penetration Depth Factors, C_d, for Split Ring and Shear Plate Connectors Used with Lag Screws
- Section 13.2.4: Metal Side Plate Factor, C_{st}
- Section 13.3.2: Geometry Factor, C_Δ, for Split Ring and Shear Plate Connectors in Side Grain
- Section 13.3.3: Geometry Factor, C_Δ, for Split Ring and Shear Plate Connectors in End Grain
- Table 14.2.3: Metal Side Plate Factor, C_{st}, for Timber Rivet Connections

NDS Chapter 15: Special Loading Conditions

This chapter gives design equations and adjustment factors for special loading conditions.

Exam tip: The content in this chapter is not likely to be referenced during the exam.

NDS Chapter 16: Fire Design of Wood Members

This chapter contains equations for calculating member strength and adjustment factors for fire design.

Exam tip: The content in this chapter is not likely to be referenced during the exam.

National Design Specification Supplement

The *National Design Specification Supplement* (NDS Supplement) is a collection of references for wood design. The NDS Supplement has four chapters.

- Chapter 1: Sawn Lumber Grading Agencies
- Chapter 2: Species Combinations
- Chapter 3: Section Properties
- Chapter 4: Reference Design Values

These chapters contain information for lumber, structural glued laminated timber, and round timber piles and poles. Reference design values are given in NDS Supplement Table 4A through Table 6B for various types, species, and dimensions of visually graded dimensional lumber, mechanically graded dimensional lumber, visually graded decking, visually graded timber, structural glued laminated timber, timber piles, and timber poles. The NDS Supplement does not provide reference design values for I-joists or composite lumber.

Exam tip: Carefully reading and applying relevant footnotes can make the difference in selecting the correct answer option.

Special Design Provisions for Wind and Seismic

There are four chapters in *Special Design Provisions for Wind and Seismic* (SDPWS).

Chapter 1 contains a designer flowchart. Chapter 2 contains terminology and notations. Chapter 3 contains some design information for wall framing and sheathing.

Chapter 4 contains information for the exam for diaphragms and shear walls. Section 4.1 contains general information on design requirements. Section 4.2 contains information related to timber diaphragms. Section 4.3 contains information for shear walls. Section 4.4 contains information for a combination of shear and tension loads.

Limiting aspect ratios for diaphragms are given in Table 4.2.4. It is unlikely that the exam will test on shear walls with openings. Similar to diaphragms, maximum aspect ratios for shear walls are given in Table 4.3.4.

Exam tip: SDPWS Chap. 3 is the least likely to be referenced on the exam, and Chap. 4 is the most likely to be referenced, with the exception of Sec. 4.4, which is unlikely to be needed. The most important information for shear walls is in Table 4.3A through Table 4.3D.

A.6. LRFD BRIDGE DESIGN SPECIFICATIONS (AASHTO)

Key concepts: These key concepts are important for answering exam questions in subtopic 11.A.6, LRFD Bridge Design Specifications (AASHTO).

- AASHTO scope
- important sections in AASHTO
- organization of AASHTO

NCEES Handbook: The *Handbook* does not contain any material relevant to this subtopic.

The following concepts are relevant for subtopic 11.A.6, LRFD Bridge Design Specifications (AASHTO).

AASHTO Overview

AASHTO LRFD Bridge Design Specifications (AASHTO) is a set of specifications for bridges. The specifications are intended to be used for the design, evaluation, and rehabilitation of bridges in the United States. AASHTO must be used for bridges that receive federal funding.

The topics in AASHTO include, but are not limited to, loading, analysis, steel bridge member design, concrete bridge member design, and foundations. AASHTO uses different load combinations, equations, and factors for bridge design than the code books used for buildings.

Exam tip: The PE exam may contain a problem based on AASHTO. However, familiarity with AASHTO is not crucial for success on the exam. If you encounter a problem relating to AASHTO and you are not familiar with using the AASHTO standard, you should skip the problem and come back to it at the end.

Organization of AASHTO

AASHTO LRFD Bridge Design Specifications (AASHTO) consists of 15 sections.

Section 1 and Sec. 2 cover basic limit states and general design and location features.

Section 3 covers loads and load factors.

Section 4 covers structural analysis and evaluation.

Section 5 through Sec. 8 cover structural design of concrete, steel, aluminum, and wood structures, respectively.

Section 9 covers decks and deck system general requirements.

Section 10 covers foundation design.

Section 11 covers walls, abutments, and piers.

Section 12 through Sec. 15 cover buried structures, tunnel liners, railings, joints, bearings, and the design of sound barriers.

Exam tip: The most important sections for the exam are Sec. 3 through Sec. 6, Sec. 10, and Sec. 11. Section 7, Sec. 8, and Sec. 12 through Sec. 15 are unlikely to be referenced on the exam.

AASHTO Section 1: Introduction

Section 1 gives an introduction and covers the basic limit states. An important portion of this section gives the values for η, which are used with the load factors.

Important items to study in this section include

- Section 1.3.2: Limit States
- Section 1.3.3: Ductility
- Section 1.3.4: Redundancy
- Section 1.3.5: Operational Importance

AASHTO Section 2: General Design and Location Features

Section 2 covers general design and location features, including safety and serviceability considerations. Deflection requirements are provided in this section.

Important items to study in this section include

- Section 2.5.2.6.2: Criteria for Deflection
- Table 2.5.2.6.3-1: Traditional Minimum Depths for Constant Depth Superstructures

AASHTO Section 3: Loads and Load Factors

Section 3 covers loads and load factors. All bridge loading is found in this section, including but not limited to live, permanent, ice, wind, and seismic loads. The load factors and combinations in this section should be used along with the η-factors from Sec. 1.

Important items to study in this section include

- Table 3.4.1-1: Load Combinations and Load Factors
- Figure 3.6.1.2.2-1: Characteristics of the Design Truck
- Appendix A3: Seismic Design Flowcharts

AASHTO Section 4: Structural Analysis and Evaluation

Section 4 covers structural analysis and evaluation. Analysis methods discussed in this section include the distribution of loads to girders to determine maximum shear and moments within the members. Adjustments provided in this section are made based on the number of design lanes. Appendix A4, found at the end of this section, provides a deck slab design table for maximum live load moment per unit width. Important items to study in this section include

- Table 4.6.2.2.2a-1: Distribution of Live Load per Lane for Moment and Shear in Interior Beams with Wood Decks
- Table 4.6.2.2.2b-1: Distribution of Live Loads per Lane for Moment in Interior Beams
- Appendix A4: Deck Slab Design Table

AASHTO Section 5: Concrete Structures

Section 5 covers concrete structure design, including capacities and concrete-specific requirements (e.g., cover). Prestressing concrete is covered in this section.

Important items to study in this section include

- Figure C5.5.4.2.1-1: Variation of ϕ with Net Tensile Strain ε_t for Nonpresstressed Reinforcement and for Prestressing Steel
- Table 5.9.3-1: Stress Limits for Prestressing Tendons
- Table 5.9.4.2.2-1: Tensile Stress Limits in Prestressed Concrete at Service Limit State after Losses

AASHTO Section 6: Steel Structures

Section 6 covers steel structure design, including capacities and steel-specific considerations (e.g., fatigue). Composite design is discussed in Sec. 6, along with the other design methods.

Appendix C6 and App. D6 are both noteworthy parts of this section. Appendix C6 provides steps and flowcharts for the basic design of steel bridge superstructures. Appendix D6 provides a table of fundamental equations used to find the flexural capacity of composite girders.

Important items to study in this section include

- Table 6.9.4.1.1-1: Selection Table for Determination of Nominal Compressive Resistance, P_n
- Figure C6.4.5: Flowchart for LRFD Article 6.10.7—Composite Sections in Positive Flexure
- Appendix D6: Fundamental Calculations for Flexural Members

AASHTO Section 9: Decks and Deck Systems

Section 9 covers decks and deck system general requirements, including the minimum reinforcement area required in a concrete slab. Many requirements in this section refer to other sections in AASHTO.

Exam tip: Section 9 is unlikely to be referenced during the exam.

AASHTO Section 10: Foundations

Section 10 covers foundation design. Included in this section are soil properties and methods for the design of different types of foundations, including spread footings, driven piles, and drilled shafts. Important tables and figures to study in this section include

- Table 10.5.5.2.2-1: Resistance Factors for Geotechnical Resistance of Shallow Foundations at the Strength Limit State
- Figure C10.6.1.3-1: Reduced Footing Dimensions

AASHTO Section 11: Abutments, Piers and Walls

Section 11 covers walls, abutments, and piers. This section includes methods for the design of conventional retaining walls, piers, nongravity cantilevered walls, anchored walls, and mechanically stabilized earth walls. This section provides several figures depicting the loading for these different wall types.

Important items to study in this section include

- Figure 11.6.3.2-1: Bearing Stress Criteria for Conventional Wall Foundations on Soil
- Figure 11.10.5.2-1: External Stability for Wall with Horizontal Backslope and Traffic Surcharge

A.7. MINIMUM DESIGN LOADS FOR BUILDINGS AND OTHER STRUCTURES (ASCE/SEI 7)

Key concepts: These key concepts are important for answering exam questions in subtopic 11.A.7, Minimum Design Loads for Buildings and Other Structures (ASCE/SEI 7).

- frequently referenced chapters and sections in ASCE/SEI 7
- locating important equations and tables in ASCE/SEI 7
- structure of ASCE/SEI 7

NCEES Handbook: The *Handbook* does not contain any material relevant to this subtopic.

The following concepts are relevant for subtopic 11.A.7, Minimum Design Loads for Buildings and Other Structures (ASCE/SEI 7).

ASCE/SEI 7 Overview

Minimum Design Loads for Buildings and Other Structures (ASCE/SEI 7) gives the minimum load requirements for building and nonbuilding structures. It covers both strength design (SD) and allowable stress design (ASD). The topics include load combinations, dead loads, live loads, snow loads, wind loads, seismic loads, and special detailing requirements. The exam often uses the older designation ASCE7 instead of ASCE/SEI 7.

> *Exam tip*: ASCE/SEI 7 deals primarily with prescribing loads. It only rarely specifies how a structure must perform under prescribed loading.

Organization of ASCE/SEI 7

Generally, ASCE/SEI 7 chapters are organized by load type, with specific chapters dedicated to specific load types. Seismic and wind loads have multiple chapters dedicated to them, with 13 chapters for seismic loads and 6 chapters for wind loads. Four chapters (Chap. 6, Chap. 9, Chap. 24, and Chap. 25) are empty, reserved for future provisions. There is also an appendix with four sections. Commentary is included, numbered to indicate the section it concerns, with a C in front of the number to indicate commentary.

ASCE/SEI 7 Chapter 1: General

Chapter 1 contains definitions, notations, and general requirements for the calculation of design loads for buildings and other structures. Two important tables to study in this chapter are

- Table 1.5-1: Risk Category of Buildings and Other Structures for Flood, Wind, Snow, Earthquake, and Ice Loads
- Table 1.5-2: Importance Factors by Risk Category of Buildings and Other Structures for Snow, Ice, and Earthquake Loads

ASCE/SEI 7 Chapter 2: Combinations of Loads

Load combinations are given in Chap. 2, including both load and resistance factor design (LRFD) and allowable strength design (ASD) load factors. Design standards typically permit both LRFD and ASD design, but steel and concrete design are usually performed with LRFD loads, and masonry and wood design are usually performed with ASD loads. For all materials, serviceability checks, including deflection checks, are typically performed using ASD or modified ASD loading.

ASCE/SEI 7 Chapter 3: Dead Loads, Soil Loads, and Hydrostatic Pressure

The most important concept to understand from Chap. 3 is dead load because different load types are multiplied by different factors when calculating load combinations. Table 3.2-1 specifies the design lateral soil load for different backfill materials and is useful for calculations where soil loads are not already known.

ASCE/SEI 7 Chapter 4: Live Loads

Live load values are provided in the following sections and tables in Chap. 4.

- Section 4.3.2: Provision for Partitions
- Section 4.5.1: Loads on Handrail and Guardrail Systems
- Section 4.5.2: Loads on Grab Bar Systems

- Section 4.5.3: Loads on Vehicle Barrier Systems
- Section 4.5.4: Loads on Fixed Ladders
- Table 4-1: Minimum Uniformly Distributed Live Loads, L_o, and Minimum Concentrated Live Loads

The following sections describe how different live loads should be treated, in addition to load factors from load combinations.

- Section 4.6: Impact Loads
  - Section 4.6.3: Machinery
- Section 4.9: Crane Loads
  - Section 4.9.3: Vertical Impact Force
  - Section 4.9.4: Lateral Force
  - Section 4.9.5: Longitudinal Force

The following sections include equations for calculating live load reduction and roof live load reduction.

- Section 4.7.2: Reduction in Uniform Live Loads
- Section 4.8.2: Flat, Pitched, and Curved Roofs

ASCE/SEI 7 Chapter 5: Flood Loads

Chapter 5 covers information on calculating flood loads. The different possible flood loads are each discussed in the following sections.

- Section 5.4.2: Hydrostatic Loads
- Section 5.4.3: Hydrodynamic Loads
- Section 5.4.4: Wave Loads
- Section 5.4.5: Impact Loads

Section 5.4.4 is further split into subsections because the load equations differ depending on how the waves break.

ASCE/SEI 7 Chapter 7: Snow Loads

Chapter 7 covers information on calculating snow loads.

Exam tip: Every section of Chap. 7 has the potential to be used in an exam problem. Know how to calculate the values found in each section and how to read the tables and figures included in the chapter.

ASCE/SEI 7 Chapter 8: Rain Loads

Chapter 8 covers information on calculating rain loads. The most important section to review is Sec. 8.3, "Design Rain Loads," which includes Eq. 8.3-1, the equation for calculating rain load on an undeflected roof.

ASCE/SEI 7 Chapter 10: Ice Loads—Atmospheric Icing

Chapter 10 covers information on calculating atmospheric ice loads. A helpful way to navigate and understand this chapter is by following Sec. 10.8, "Design Procedure," which outlines how to calculate ice loads. Each step references the section and/or equation used for that step of the calculation.

ASCE/SEI 7 Chapters 11–23: Seismic Loads

Chapter 11 through Chap. 23 cover topics related to seismic loads. The most important chapters to review for the exam are

- Chapter 11: Seismic Design Criteria
- Chapter 12: Seismic Design Requirements for Building Structures
- Chapter 20: Site Classification Procedure for Seismic Design

The seismic design category (SDC) of a structure is determined using the methods in Chap. 11.

ASCE/SEI 7 Chapters 26–31: Wind Loads

Chapter 26 through Chap. 31 cover topics related to wind loads. A helpful way to navigate and understand these chapters is to follow the tables that outline the steps to determine the wind loads for different building types. Each step references a table, figure, or equation that is relevant for the calculation. The relevant tables are

- Table 27.2-1: Steps to Determine MWFRS Wind Loads for Enclosed, Partially Enclosed, and Open Buildings of All Heights
- Table 27.5-1: Steps to Determine MWFRS Wind Loads Enclosed Simple Diaphragm Buildings ($h \leq$ 160 ft (48.8 m))
- Table 28.2-1: Steps to Determine Wind Loads on MWFRS Low-Rise Buildings
- Table 28.5-1: Steps to Determine Wind Loads on MWFRS Simple Diaphragm Low-Rise Buildings
- Table 29.1-1: Steps to Determine Wind Loads on MWFRS Rooftop Equipment and Other Structures
- Table 30.4-1: Steps to Determine C&C Wind Loads Enclosed and Partially Enclosed Low-Rise Buildings

- Table 30.5-1: Steps to Determine C&C Wind Loads Enclosed Low-Rise Buildings (Simplified Method)
- Table 30.6-1: Steps to Determine C&C Wind Loads Enclosed or Partially Enclosed Building with $h > 60$ ft
- Table 30.7-1: Steps to Determine C&C Wind Loads Enclosed Building with $h \leq 160$ ft
- Table 30.8-1: Steps to Determine C&C Wind Loads Open Buildings
- Table 30.9-1: Steps to Determine C&C Wind Loads Parapets
- Table 30.10-1: Steps to Determine C&C Wind Loads Roof Overhangs

In addition, Fig. 26.1-1 shows how the wind load chapters are organized, outlines the steps for determining wind loads, and references a section or figure for the calculation.

A.8. AMERICAN WELDING SOCIETY (AWS D1.1, D1.2, AND D1.4)

Key concepts: These key concepts are important for answering exam questions in subtopic 11.A.8, American Welding Society (AWS D1.1, D1.2, and D1.4).

- scope of AWS
- where to find AWS materials in
  - *Building Code Requirements and Specification for Masonry Structures* (TMS 402/602)
  - *Building Code Requirements for Structural Concrete and Commentary* (ACI 318)
  - *PCI Design Handbook: Precast and Prestressed Concrete* (PCI)
  - *Steel Construction Manual* (AISC)

NCEES Handbook: To prepare for this subtopic, familiarize yourself with these sections in the *Handbook*.

- Basic Welding Symbols

The following concepts are relevant for subtopic 11.A.8, American Welding Society (AWS D1.1, D1.2, and D1.4).

Scope of AWS D1.1, AWS D1.2, and AWS D1.4

The American Welding Society *Structural Welding Code* (AWS) contains material and performance specifications for studs (both stud-connectors and headed studs), as well as guidance for documentation, inspection, quality control, installation, and testing.

AWS D1.1 is the governing standard for structural welding. The *Steel Construction Manual* (AISC) references it several times, with some minor differences.

AWS D1.2 covers the welding requirements for any structure made from aluminum structural alloys, except for aluminum pressure vessels and pressure piping.

AWS D1.4 covers the design, workmanship, technique, qualification, and inspection requirements for welding reinforcing steel in most reinforced concrete.

Exam tip: AWS will not be provided during the exam. You will be able to reference its content only by way of sections, figures, and tables in AISC and the *PCI Design Handbook: Precast and Prestressed Concrete* (PCI), as cited in this section.

AWS D1.1

Handbook: Basic Welding Symbols

Welding requirements for hollow structural sections (HSSs) from AWS D1.1 can be found in *Steel Construction Manual* (AISC) Chap. 8.

Details for prequalified welded joints given in AWS D1.1 can be found in AISC Table 8-2. An excerpt of this table is provided in *Handbook* figure Basic Welding Symbols. Joint geometries, such as root openings, angles, and clearances (like those shown in AISC Fig. 8-21 and Fig. 8-22), are also provided in AISC Table 8-2.

PCI Design Handbook: Precast and Prestressed Concrete (PCI) Table 6.5.1, adapted from AWS D1.1-08 Table 7.1, shows the current minimum tensile and yield strengths for headed studs.

AWS D1.2

AWS D1.2 covers the welding requirements for any structure made from aluminum structural alloys, except for aluminum pressure vessels and pressure piping.

Exam tip: Any details from this standard will be provided in the problem statement.

AWS D1.4

AWS D1.4 covers the design, skill, technique, qualification, and inspection requirements for welding reinforcing steel in most reinforced concrete.

Welding of reinforcement should be in accordance with AWS D1.4, which is detailed in *Building Code Requirements for Structural Concrete* (ACI 318) Sec. 26.6.4. ACI Sec. 26.6.4 (R26.6.4) gives more details on AWS D1.4 standards.

The commentary in *Building Code Requirements for Masonry Structures* (TMS 402) Sec. 8.1.6.7.2 contains information on AWS D1.4 requirements for welded splices.

A.9. OSHA 1910 GENERAL INDUSTRY AND OSHA 1926 CONSTRUCTION SAFETY STANDARDS

Key concepts: These key concepts are important for answering exam questions in subtopic 11.A.9, OSHA 1910 General Industry and OSHA 1926 Construction Safety Standards.

- important safety standards in OSHA
- navigating OSHA
- organization of OSHA

NCEES Handbook: The *Handbook* does not contain any material relevant to this subtopic.

The following concepts are relevant for subtopic 11.A.9, OSHA 1910 General Industry and OSHA 1926 Construction Safety Standards.

Overview of OSHA 1910 and 1926

The Occupational Safety and Health Administration (OSHA) is a large regulatory agency of the U.S. Department of Labor, created in 1970 by the Occupational Safety and Health Act (also abbreviated OSHA). OSHA safety standards are designed to ensure safe working conditions. These standards are contained in Part 1910 and Part 1926 of Title 29 of the *Code of Federal Regulations* (CFR). The standards in 29 CFR 1910 (often referred to as OSHA 1910) are for general industry safety, while the standards in 29 CFR 1926 (OSHA 1926) focus on the construction industry. Together, they are vital documents that outline in detail the regulations and standards for a safe workplace. They apply to most work sites and establish safety protocols for employers, private sector workers, state and local government workers, and federal government workers.

Organization of OSHA 1910 and 1926

OSHA 1910 contains 26 subparts, lettered A through Z. The three most important subparts to study are

- Subpart D: Walking-Working Surfaces
- Subpart F: Powered Platforms, Manlifts, and Vehicle-Mounted Work Platforms
- Subpart I Section 140: Personal Fall Protection Systems

OSHA 1926 contains 29 subparts, lettered A through CC. The five most important subparts to study are

- Subpart E: Personal Protective and Life Saving Equipment
- Subpart L: Scaffolds
- Subpart M: Fall Protection
- Subpart Q: Concrete and Masonry Construction
- Subpart R: Steel Erection

OSHA 1910 Subpart D

OSHA 1910 Subpart D covers all general industry workplaces. There are 10 sections in Subpart D. The first two sections cover definitions and general requirements for all walking and working surfaces, and are considered the "horizontal" part of this standard. The next five sections cover vertical surfaces such as ladders, stairways, scaffolding, and ropes. The final three sections cover fall protection equipment, systems, and training requirements. The most important sections of Subpart D to study are

- Section 1910.28: Duty to have fall protection and falling object protection
- Section 1910.29: Fall protection systems and falling object protection—criteria and practices
- Section 1910.30: Training requirements

Important standards to note are

- Standard 1910.28(b)(1)(i): Unprotected sides or edges that are 4 ft or more above a lower level must be protected from falling by a guardrail, a safety net, or a personal fall protection system.
- Standard 1910.28(b)(6)(i): Employees less than 4 ft above dangerous equipment should be protected from falling into or onto the dangerous equipment by a guardrail system or a travel restraint system.
- Standard 1910.28(b)(13)(i): When work is performed less than 6 ft from the roof edge, the employer must ensure each employee is protected from falling by a guardrail system, safety net system, travel restraint system, or personal fall arrest system.
- Standard 1910.29(b)(1): The top edge height of top rails, or equivalent guardrail system members, are 42 in plus or minus 3 in above the walking-working surface.
- Standard 1910.29(b)(3): Guardrail systems must be capable of withstanding, without failure, a force of at least 200 lbf.
- Standard 1910.29(g)(4): Platforms used with fixed ladders should provide a horizontal surface of at least 24 in × 30 in.
- Standard 1910.30(a)(3)(iii): Employers must ensure that the employee is trained in the correct procedures for installing, inspecting, operating, maintaining, and disassembling their personal fall protection systems.

OSHA 1910 Subpart F

Subpart F covers powered platforms, manlifts, and vehicle-mounted work platforms. The most important sections to study are

- Section 1910.66: Powered platforms for building maintenance

- Section 1910.66 App. A: Guidelines (Advisory)

- Section 1910.66 App. B: Exhibits (Advisory)

- Section 1910.66 App. D: Existing Installations (Mandatory)

- Section 1910.68: Manlifts

Important standards to note are

- Standard 1910.66(e)(2): The exterior of each building must be provided with tie-in guides, or must have one of the following systems: an intermittent stabilization system, a button guide stabilization system, or a system using angulated roping and building face rollers.

- Standard 1910.66(f)(5)(i)(A): Each suspended unit component, except suspension ropes and guardrail systems, must be capable of supporting at least four times the maximum live load intended for or transmitted to that component.

- Standard 1910.68(b)(6)(v): For manlifts, one or more emergency landings must be provided every 25 ft or less of manlift travel where there is a travel of 50 ft or more between floor landings.

- Standard 1910.68(c)(5)(i): For manlifts, two separate automatic stop devices are to be provided to cut power and apply the brake when a loaded step passes the upper terminal landing.

OSHA 1910 Subpart I Section 1910.140

Subpart I Sec. 1910.140 covers the performance, care, and use of personal fall protection systems. The most important standards from this section are

- Standard 1910.140(c)(4): Lanyards and vertical lifelines must have a minimum breaking strength of 5000 lbf.

- Standard 1910.140(c)(5): Self-retracting lifelines and lanyards that automatically limit free fall distance to 2 ft or less must have components capable of sustaining a minimum tensile load of 3000 lbf.

- Standard 1910.140(c)(7): D-rings, snaphooks, and carabiners must be capable of sustaining a minimum tensile load of 5000 lbf.

- Standard 1910.140(c)(11)(ii): The employer must ensure that each horizontal lifeline is part of a complete personal fall arrest system that maintains a safety factor of at least two.

OSHA 1926 Subpart E Standard 1926.104

OSHA 1926 Subpart E Standard 1926.104 covers safety belts, lifelines, and lanyards. The most important standards from this section are

- Standard 1926.104(a): Any lifeline, safety belt, or lanyard that has been subjected to a load, apart from static load testing, should be immediately removed from service and not used again.

- Standard 1926.104(b): Lifelines should be capable of supporting a minimum dead weight of 5400 lbf.

- Standard 1926.104(f): All safety belt and lanyard hardware, except rivets, should be capable of supporting a tensile load of 4000 lbf.

OSHA 1926 Subpart L Appendix A

OSHA 1926 Subpart L App. A contains guidelines for scaffold specifications. This appendix contains tables that can be used as a starting point for designing scaffold systems. Important tables in this section include the table in App. A.1(c), which shows rated load capacity versus intended loads for fabricated planks and platforms, and the table in App. A.2(b), which provides minimum sizes of members used in tube and coupler scaffolds.

OSHA 1926 Subpart M

OSHA 1926 Subpart M covers standards for fall protection. The most important sections to study in Subpart M are

- Section 1926.500: Scope, application, and definitions applicable to this subpart

- Section 1926.501: Duty to have fall protection

- Section 1926.502: Fall protection systems criteria and practices

- Section 1926.503: Training requirements

Important standards to note for the exam are

- Standard 1926.501(b)(5): Employees on the face of formwork or reinforcing steel are to be protected from falling 6 ft or more to lower levels by personal fall arrest systems, safety net systems, or positioning device systems.
- Standard 1926.502(c)(8): Each safety net should have a border rope for webbing with a minimum breaking strength of 5000 lbf.
- Standard 1926.502(d)(3): D-rings and snaphooks should have a minimum tensile strength of 5000 lbf.
- Appendix D.I(2): Fixed anchorage for a positioning device system should be rigid and not have a deflection greater than 0.04 in when a force of 2250 lbf is applied.

OSHA 1926 Subpart Q

OSHA 1926 Subpart Q covers standards for concrete and masonry construction. Important sections of the subpart to study are

- Section 1926.703: Requirements for cast-in-place concrete
- Section 1926.704: Requirements for precast concrete
- Section 1926.705: Requirements for lift-slab construction operations
- Section 1926.706: Requirements for masonry construction

Important standards to note for the exam are

- Standard 1926.703(e)(1): Forms and shores can only be removed once the concrete has gained enough strength to support its weight and superimposed loads.
- Standard 1926.704(b): Lifting inserts attached to precast concrete members should be capable of supporting at least two times the maximum intended or transmitted load.
- Standard 1926.705(g): During lifting, all points of a slab are to be kept within 1/2 in of a level position.
- Standard 1926.706(b): Masonry walls over 8 ft tall should be braced to prevent overturning and collapse.

OSHA 1926 Subpart R

OSHA 1926 Subpart R covers standards for erecting steel buildings. Important sections of the subpart to study are

- Section 1926.752: Site layout, site-specific erection plan and construction sequence
- Section 1926.754: Structural steel assembly
- Section 1926.755: Column anchorage
- Section 1926.756: Beams and columns
- Section 1926.757: Open web steel joists
- Section 1926.758: Systems-engineered metal buildings

Important standards to note for the exam are

- Standard 1926.752(a)(1): Concrete in footings, piers, and walls and mortar in masonry piers and walls must have attained 75% of the minimum compressive design strength or be strong enough to support the loads imposed during steel erection.
- Standard 1926.754(b)(2): There should never be more than four floors or 48 ft, whichever is less, of unfinished bolting or welding above the foundation or uppermost secured floor.
- Standard 1926.755(a)(2): Each column anchor rod assembly should be able to resist a minimum gravity load of 300 lbf located 18 in from the outer face of the column in each direction at the top of the shaft.
- Standard 1926.756(d): Each column splice must be designed to resist a minimum gravity load of 300 lbf located 18 in from the outer face of the column in each direction at the top of the shaft.
- Standard 1926.757(a)(3): Where steel joists at or near columns span 60 ft or less, the joist should be designed with enough strength to allow one employee to release the hoisting cable without erection bridging.
- Standard 1926.758(c): For rigid frames, 50% of their bolts or the number of bolts specified by the manufacturer (whichever is greater) should be installed and tightened on both sides of the web adjacent to each flange before hoisting.

34 Temporary Structures and Other Topics

Content in blue refers to the *NCEES Handbook*.

B.1. Special Inspections 34-1
B.2. Submittals 34-2
B.3. Formwork 34-3
B.4. Falsework and Scaffolding 34-5
B.5. Shoring and Reshoring 34-5
B.6. Concrete Maturity and Early Strength
    Evaluation 34-6
B.7. Bracing 34-8
B.8. Anchorage 34-9
B.9. OSHA Regulations 34-10
B.10. Safety Management 34-11

The knowledge area of Temporary Structures and Other Topics makes up between two and four questions out of the 80 questions on the PE Civil Structural exam. The organization of this chapter follows the order of subtopics given by NCEES for this knowledge area. Each subtopic is covered in the following sections.

B.1. SPECIAL INSPECTIONS

Key concepts: These key concepts are important for answering exam questions in subtopic 11.B.1, Special Inspections.

- concrete construction
- masonry construction
- seismic and wind resistance
- steel construction
- substructure construction
- wood construction

NCEES Handbook: The *Handbook* does not contain any material relevant to this subtopic. Be sure to study the listed sections from codes and standards.

Codes and standards: To prepare for this subtopic, familiarize yourself with these sections from codes and standards.

Building Code Requirements for Masonry Structures (TMS 402)

- Section 3.1: Quality Assurance Program
- Section 7.3.2.6: Special Reinforced Masonry Shear Walls

Building Code Requirements for Structural Concrete and Commentary (ACI 318)

- Section 26.13: Inspection

International Building Code (IBC)

- Chapter 2: Definitions
- Chapter 17: Special Inspections and Tests
- Section 110: Inspections

Steel Construction Manual (AISC)

- Section N5: Minimum Requirements for Inspection of Structural Steel Buildings

The following equations and tables are relevant for subtopic 11.B.1, Special Inspections.

Concrete Construction

ACI 318: Sec. 26.13

IBC: Chap. 2, Sec. 1705.3

IBC Table 1705.3 displays the required special inspections for various types of concrete work. Definitions for *continuous special inspection* and *periodic special inspection* are provided in IBC Chap. 2.

ACI 318 Sec. 26.13.1.1 refers to the general building code (IBC or local building code) for inspections.

Steel Construction

AISC: Sec. N5

IBC: Chap. 2, Sec. 1705.2

IBC Table 1705.2.3 provides the required special inspections for various types of steel work. Definitions for *continuous special inspection* and *periodic special inspection* are provided in IBC Chap. 2.

The following tables and sections are relevant to special inspections for steel construction.

- AISC Table N5.4-1 through Table N5.4-3
- AISC Table N5.6-1 through Table N5.6-3
- IBC Sec. 1705.2.1
- IBC Sec. 1705.2.3

Substructure Construction

IBC: Chap. 2, Sec. 1705

IBC Table 1705.6, Table 1705.7, and Table 1705.8 display the required special inspections for various types of substructure work. Definitions for *continuous special inspection* and *periodic special inspection* are provided in IBC Chap. 2.

The following sections are also relevant to special inspections for substructure construction.

- IBC Sec. 1705.6
- IBC Sec. 1705.7
- IBC Sec. 1705.8
- IBC Sec. 1705.9

Masonry Construction

IBC: Chap. 2, Sec. 1705.4

TMS 402: Sec. 3.1

TMS 402 Table 3.1.1, Table 3.1.2, and Table 3.1.3 include the required special inspections for various types of masonry work. Definitions for *continuous special inspection* and *periodic special inspection* are provided in IBC Chap. 2.

Wood Construction

IBC: Sec. 1705.5, Sec. 1705.12.2

Fewer special inspections are required for wood construction than for the other materials. IBC does not provide tables for special inspection of wood. IBC Sec. 1705.5 and Sec. 1705.12.2 provide more information on wood construction.

Special Inspection Requirements in the *International Building Code*

IBC: Chap. 2, Sec. 110, Chap. 17

All work in which a permit is required must be inspected per IBC Sec. 110.1. Any such work must be accessible and exposed until it is approved by the building official. This is classified as a general inspection; however, certain types of construction require special expertise. IBC Chap. 17 discusses special inspections and requirements. See IBC Sec. 1705 for the types of construction that require special inspections and IBC Sec. 1704.2 for when special inspections can be waived. Definitions for special inspections are provided in IBC Chap. 2.

Exam tip: IBC Chap. 17 does not cover all construction materials. Refer to specific material codes for further special inspection requirements.

Seismic Resistance and Wind Resistance

IBC: Sec. 1705.11, Sec. 1705.12

IBC Sec. 1705.11 and Sec. 1705.12 provide further requirements for special inspections in high-wind and high-seismic constructions.

B.2. SUBMITTALS

Key concepts: These key concepts are important for answering exam questions in subtopic 11.B.2, Submittals.

- project samples
- shop drawings
- submittal documents
- submittal requirements

NCEES Handbook: The *Handbook* does not contain any material relevant to this subtopic. Be sure to study the listed sections from codes and standards.

Codes and standards: To prepare for this subtopic, familiarize yourself with these sections from codes and standards.

International Building Code (IBC)

- Section 107: Submittal Documents
- Chapter 2: Definitions
- Section 1603: Construction Documents

The following concepts are relevant for subtopic 11.B.2, Submittals.

Submittal Requirements

IBC: Sec. 107

After a work contract is awarded, the contractor is responsible for providing the submittals called for in the contract documents. These include shop drawings, samples, product data, and documents related to sustainability issues, where applicable.

Submittal requirements are listed in each specification section. The requested information may vary depending on the type of material under consideration. The submittals are sometimes prepared by the contractor, but most often they are prepared by subcontractors, vendors, and material suppliers and reviewed by the general contractor for coordination before they are submitted to the architect or engineer. Although submittals show in detail how much of the work is going to be built and installed, they are not contract documents. The contractor should wait until the submittal is approved and returned before completing the associated work. If the contractor proceeds with work prior to approval, they are operating at risk and may have to rectify any work or materials that were not approved.

A collection of samples may also be submitted so that the owner or architect can choose a color or finish when the choice has not been specified. Samples become standards of appearance and workmanship by which the final work will be judged.

Construction Documents

IBC: Sec. 107, Chap. 2

IBC Chap. 2 defines *construction documents* as "written, graphic, and pictorial documents prepared or assembled for describing the design, location, and physical characteristics of the elements of a project necessary for obtaining a building permit." See IBC Sec. 107 for an explanation concerning the documents required to obtain a permit, including fire protection system shop drawings, means of egress, and a description of the exterior wall envelope.

Shop Drawings

IBC: Sec. 1603

Shop drawings are drawings, diagrams, schedules, and other data prepared to show how a subcontractor or supplier proposes to supply and install work to conform to the requirements of the contract documents for a specific project. Shop drawings are usually very detailed, showing how portions of the work will be constructed, and may include drawings showing how the product or assembly will fit into the building. IBC Sec. 1603 discusses the items that are required to be included in the construction documents.

The contractor should wait until the shop drawings are approved and returned before completing the associated work. If the contractor proceeds with work prior to approval, they are operating at risk and may have to rectify any work that was not approved.

B.3. FORMWORK

Key concepts: These key concepts are important for answering exam questions in subtopic 11.B.3, Formwork.

- concrete and masonry construction
- lateral pressure on formwork
- special formwork
- stay-in-place formwork
- types of formwork

NCEES Handbook: The *Handbook* does not contain any material relevant to this subtopic. Be sure to study the listed sections in CERM and from codes and standards.

PE Civil Reference Manual (**CERM**): Study these sections in CERM that either relate directly to this subtopic or provide background information.

- Section 49.10: Placing
- Section 49.15: Formwork
- Section 49.16: Lateral Pressure on Formwork
- Section 51.13: Concrete Deck Systems in Bridges

Codes and standards: To prepare for this subtopic, familiarize yourself with these sections from codes and standards.

AASHTO LRFD Bridge Design Specifications (AASHTO LRFD)

- Section 9.7.4: Stay-in-Place Formwork

Building Code Requirements for Structural Concrete and Commentary (ACI 318)

- Section 26.11: Formwork

International Building Code (IBC)

- Section 3304: Site Work
- Section 1808.8.5: Forming of Concrete

Occupational Safety and Health Act (OSHA)

- Section 1926.703: Requirements for Cast-in-Place Concrete
- Section 1926.704: Requirements for Precast Concrete
- Section 1926.705: Requirements for Lift-Slab Construction Operations
- Section 1926.706: Requirements for Masonry Construction
- Subpart Q Appendix A

The following concepts are relevant for subtopic 11.B.3, Formwork.

Types of Formwork

CERM: Sec. 49.10, Sec. 49.15, Sec. 49.16

ACI: Sec. 26.11

IBC: Sec. 1808.8.5

Formworks are constructed out of a variety of materials. Unless the concrete is finished in some way, the shape and pattern of the formwork will affect the appearance of the final product. Wood grain, knotholes, joints, and other imperfections in the form will show in their negative image when the form is removed. Plywood is the most common forming material. It is usually coated on one side with oil, water-resistant glue, or plastic to prevent water from penetrating the wood and to increase the reusability of the form.

Prefabricated steel forms are often used because of their strength and reusability.

Other types of forms include glass-fiber reinforced plastic, hardboard, and various kinds of proprietary systems. Plastic forms can be manufactured with a variety of patterns embedded into them, and special form liners can be used to impart a deeply embossed pattern.

IBC Sec. 1808.8.5 allows earth forming (concrete cast directly against the earth) where the building official deems that formwork is not required. See ACI Sec. 26.11 for formwork design and removal requirements for cast-in-place concrete.

Exam tip: For cast-in-place concrete construction, one of the greatest expenses is formwork. It is possible that providing similar design thickness even when not required is more economical for formwork because it can be reusable.

Stay-in-Place Formwork

CERM: Sec. 51.13

AASHTO LRFD: Sec. 9.7.4

ACI: Sec. 26.11

IBC: Sec. 3304

Stay-in-place formwork is often used in deck construction. The formwork is designed to behave elastically and to prevent excessive sagging under construction loads. The formwork must be strong enough to carry the load of self-weight, the deck concrete weight, and an additional 50 psf. AASHTO LRFD Sec. 9.7.4.1 limits the stresses due to construction loads in the formwork to prevent failure of the formwork when vehicular loads are applied. In addition, deflection limits are specified for the formwork to ensure adequate cover for reinforcing steel.

It is not permitted to leave most concrete formwork in place. ACI Sec. 26.11 describes the removal of formwork, and IBC Sec. 3304 explains that formwork should be removed. Formwork that is to stay in place should be designed specifically for this purpose.

Stay-in-place formwork is often used in tight places where formwork is hard to remove, as well as in discrete items such as pile caps, columns, and bent caps in bridges. Stay-in-place forms are often metal and can be prefabricated, which can save on materials and time. Since these forms do not require removal, construction can proceed without this step.

Concrete and Masonry Construction

OSHA: Sec. 1926.703 through Sec. 1926.706, Subpart Q App. A

OSHA 1926 explains formwork requirements on job sites. OSHA Sec. 1926.703(a) states that formwork should be designed to resist all anticipated vertical and lateral loads and that drawings and plans for the formwork should be available at the job site.

OSHA Sec. 1926.703(c) describes vertical slip forms and their requirements.

OSHA Sec. 1926.703(e) explains that forms may not be removed until the concrete has reached adequate strength. Determination of adequate strength is made either by the construction documents or through testing.

Special Formwork

CERM: Sec. 49.15

Most formwork is designed and constructed to remain in place until the concrete cures sufficiently to stand on its own. However, with *slip forming*, the formwork moves as the concrete cures. Slip forming is used to form continuous surfaces such as curbs and gutters, open channels, tunnels, and high-rise building cores.

Flying forms are large fabricated sections of framework that are removed, once the concrete has cured, to be reused in forming identical adjacent sections. They are often used in buildings with repetitive units, such as hotels and apartments.

Lateral Pressure on Formwork

CERM: Sec. 49.16

As concrete is being placed, it is in a liquid state. This causes a hydraulic loading on the formwork. The hydraulic load is greatest immediately after pouring. As the concrete sets up, it begins to support itself, and the lateral force against the formwork is reduced.

Lateral pressure is calculated with CERM Eq. 49.2.

$$p = \rho g h \quad \text{[SI]} \quad 49.2(a)$$

$$p = \gamma h = \frac{\rho g h}{g_c} \quad \text{[U.S.]} \quad 49.2(b)$$

CERM Eq. 49.2 implies that the pressure, p, can be increased to any level by increasing the height of the pour, h. In practice, however, the pressure is limited by pour and setup (i.e., curing) rates.

B.4. FALSEWORK AND SCAFFOLDING

Key concepts: These key concepts are important for answering exam questions in subtopic 11.B.4, Falsework and Scaffolding.

- falsework
- navigating and interpreting exam codes (OSHA)
- scaffolding
- types of construction loads

NCEES Handbook: The *Handbook* does not contain any material relevant to this subtopic. Be sure to study the listed sections in CERM and from codes and standards.

PE Civil Reference Manual (**CERM**): Study these sections in CERM that either relate directly to this subtopic or provide background information.

- Section 57.7: Strength of Composite Sections
- Section 83.15: Scaffolds
- Section 83.16: Temporary Structures

Codes and standards: To prepare for this subtopic, familiarize yourself with these sections from codes and standards.

Occupational Safety and Health Act (OSHA)

- Section 1926.451: General Requirements

The following concepts are relevant for subtopic 11.B.4, Falsework and Scaffolding.

Scaffolding

CERM: Sec. 83.15

OSHA: Sec. 1926.451

A *scaffold* is any temporary elevated platform and its supporting structure used for supporting employees, materials, or both. The construction and use of scaffolds are regulated by OSHA Sec. 1926.451.

Fall protection must be provided for each employee on a scaffold more than 10 ft above a lower level. This differs from the 6 ft threshold for fall protection for other walking/working surfaces in construction because scaffolds are temporary structures that provide a work platform for employees who are constructing or demolishing other structures.

Falsework

CERM: Sec. 57.7

Falsework is any temporary structure used to support spanning or arched structures or concrete formwork. It is meant to hold a component in place until the construction has advanced enough to support itself.

With shored construction, steel beams are supported by temporary falsework until the concrete cures. All noncomposite loads, including the steel beam itself, are supported by the falsework. After the concrete cures, the falsework is removed, and the section acts compositely to resist all non-composite and composite loads. Shored construction results in a reduction in the service load stresses.

Temporary Structures

CERM: Sec. 83.16

Scaffolding and falsework are forms of temporary structures. Temporary structures may or may not require separate permits and/or drawings. Most temporary structures need ongoing inspection during their existence, since connections loosen and members get moved, weaken, deform, and/or fail. Temporary structures are often erected by a separate subcontractor, complicating the inspection tasks and responsibilities.

Temporary structures need to be designed for construction loading. This may include earth loads, the weight of concrete before self-sufficiency, machinery, materials, personnel, and wind and earthquake loading during construction. Temporary structures usually remain in place for less than 180 days and may be designed for lesser loads than the permanent structures they support.

B.5. SHORING AND RESHORING

Key concepts: These key concepts are important for answering exam questions in subtopic 11.B.5, Shoring and Reshoring.

- allowable compressive load
- Euler buckling formula
- reshoring
- shoring

NCEES Handbook: To prepare for this subtopic, familiarize yourself with these sections in the *Handbook*.

- Columns

PE Civil Reference Manual (**CERM**): Study these sections in CERM that either relate directly to this subtopic or provide background information.

- Section 57.7: Strength of Composite Sections
- Section 61.2: Euler's Column Buckling Theory
- Section 64.2: Effective Width of Concrete Slab

The following equations and concepts are relevant for subtopic 11.B.5, Shoring and Reshoring.

Allowable Compressive Load

Handbook: Columns

$$P_{cr} = \frac{\pi^2 EI}{(Kl)^2}$$

CERM: Sec. 61.2

$$P_e = \frac{\pi^2 EI}{L^2} \quad \text{61.1}$$

The Euler buckling formula calculates the allowable compressive load for shoring. The *Handbook* form of the equation includes both the effective length factor, K, and the unbraced column length, l. The product, Kl, is known as the *effective length* of the column. The effective length approximates the length over which a column actually buckles. Effective length factors can be found in *Handbook* section Columns. Note that CERM Eq. 61.1 does not include the effective length factor for the column.

Self-weight must be considered for the compressible load. Field testing is usually used to determine compressive strength before shoring can be performed. Shoring elements should be designed similarly to other compression and flexural members. Refer to subtopic B.2, Axial, and subtopic B.4, Flexure, in Chap. 28 of this book for design of axial and flexural members, respectively.

Shoring

CERM: Sec. 57.7, Sec. 64.2

Shoring supports formwork, workers, and fresh concrete at the top level. It distributes load from the form to the slab below (the top of the reshoring system). Shoring may be wood, aluminum, or steel. Shoring may be expensive and may not be feasible based on the construction sequence. It may be desirable to design a structure as unshored in a composite system where the supporting members carry the wet concrete prior to composite action.

There are three common types of shoring: vertical shoring, trench shoring, and lateral shoring to support a wall.

See CERM Sec. 57.7 for more information on unshored and shored construction.

Reshoring

Reshoring is a common practice in multistory concrete construction. The placement and loading of shores from story to story are different from the load path used in the design of a permanent structure. Once one level has cured and forms have been stripped, the level can be reshored to help carry the load from the shores above. The entire level is stripped of formwork before reshoring. This means the slab and beams will start to carry their own weight and will undergo immediate deflection. Then the level is reshored, and the reshores will carry an imposed load from construction.

B.6. CONCRETE MATURITY AND EARLY STRENGTH EVALUATION

Key concepts: These key concepts are important for answering exam questions in subtopic 11.B.6, Concrete Maturity and Early Strength Evaluation.

- concrete compressive strength
- concrete durability
- concrete maturity
- concrete strength testing

NCEES Handbook: To prepare for this subtopic, familiarize yourself with these sections in the *Handbook*.

- Design Provisions: Definitions

PE Civil Reference Manual (**CERM**): Study these sections in CERM that either relate directly to this subtopic or provide background information.

- Section 48.2: Cementitious Materials
- Section 48.4: Water
- Section 49.4: Water-Cement Ratio and Cement Content
- Section 49.12: Curing

Codes and standards: To prepare for this subtopic, familiarize yourself with these sections from codes and standards.

Building Code Requirements for Structural Concrete and Commentary (ACI 318)

- Section 19.3.1: Exposure Categories and Classes
- Section 26.11: Formwork
- Section 26.12: Concrete Evaluation and Acceptance
- Section 26.4.1.5: Steel Fiber Reinforcement
- Section 26.5.7: Construction of Concrete Members

The following equations, tables, and concepts are relevant for subtopic 11.B.6, Concrete Maturity and Early Strength Evaluation.

Concrete Strength Testing

Handbook: Design Provisions: Definitions

ACI 318: Sec. 26.11, Sec. 26.12

In-plane concrete strength can be estimated using field-cured concrete cylinders or another method outlined in ACI Sec. R26.11.2.1(e). If field-cured cylinders are chosen, they need not be approved by the design professional but by the building official when requested.

The concrete should be tested at 28 days or at the test age for which the concrete compressive stress, f'_c, is specified. After 28 days, concrete will plateau in strength.

At a minimum, samples are provided once a day, once for each 150 yd^3 of concrete, and once for each 5000 ft^2 of surface area for slabs or walls. *Strength test* is a precisely defined technical term in ACI 318; it refers to an average of tests (e.g., the average of two 6 × 12 cylinder strengths or three 4 × 8 cylinder strengths), not just one test. Concrete strength testing is governed by ACI 318 Sec. 26.12, which states that concrete must be evaluated and tested before it is judged to be acceptable.

According to ACI 318 Sec. 26.12.3.1(b), the average of any three consecutive strength tests must equal or exceed f'_c. No strength test may fall below f'_c by more than 500 psi or 0.1 f'_c, whichever is greater. Additional requirements apply if the concrete compressive strength falls short.

Concrete Compressive Strength

Handbook: Design Provisions: Definitions

CERM: Sec. 48.2, Sec. 49.4, Sec. 49.12

ACI 318: Sec. 26.12

Concrete strength is inversely proportional to the *water-cement ratio*, which is the ratio of the amount of water to the amount of cement in a mixture; this is usually stated as a decimal by weight. ACI 318 uses the more general term *water-cementitious materials ratio*. Typical ratios are approximately 0.45 to 0.60 by weight.

The strength that concrete achieves depends on the curing process. During curing, the temperature and humidity must be controlled. Concrete strength increases with age as long as moisture and favorable temperatures are present for hydration of the cement. Concrete should be kept saturated or nearly saturated until the chemical reaction between water and cement is completed. Loss of water will reduce the hydration process and cause the concrete to shrink and crack.

In cold weather or fast-paced construction schedules, high-early-strength concrete might be desired. Type III portland cement provides high concrete strength earlier. Type III cement has higher percentages of tricalcium aluminate, which contributes to early heat of hydration and, in turn, concrete strength.

Concrete should not be subjected to external loads until it is shown that it has sufficient strength. *The modulus of elasticity* is a measurement of *concrete* stiffness, which is an indicator of strength. *Handbook* section Design Provisions: Definitions includes an equation for the modulus of elasticity.

$$E_c = 33 w_c^{1.5} \sqrt{f'_c} \quad [90 \text{ lbf/ft}^3 \leq w_c \leq 160 \text{ lbf/ft}^3]$$

Concrete Durability

CERM: Sec. 48.2, Sec. 48.4

ACI 318: Sec. 19.3.1, Sec. 26.4.1.5, Sec. 26.5.7

A concrete structure's exposure must be classified for four separate categories per ACI 318 Sec. 19.3.1.1. ACI 318 Table 19.3.1.1 defines the exposure classes for each category. The commentary elaborates on the meaning of each exposure class. Exposure classes are assigned according to the severity of the anticipated exposure of structural concrete members.

- freezing and thawing (F)
- sulfate (S)
- in contact with water (W)
- corrosion protection of reinforcement (C)

When a concrete member is assigned multiple exposure classes, the most restrictive requirements from ACI 318 Table 19.3.2.1 apply.

The durability of concrete can be increased by adding different materials (pozzolans) to the concrete mixture and by using different types of cement. Other methods to increase durability include using admixtures and air entrainment. Discontinuous steel fibers are routinely used in concrete, as permitted in accordance with ACI 318 Sec. 26.4.1.5 and Sec. 26.5.7.1. Discrete synthetic fibers, including carbon and alkali-resistant glass, can be added to create fiber-reinforced concrete.

Concrete Maturity

When concrete maturity is understood, concrete strength can be estimated. This becomes useful when trying to determine when it is appropriate to complete certain construction tasks, such as removing formwork. Concrete maturity is a product of the relationship between temperature, time, and strength gain. Below a certain temperature (known as the *datum temperature*), concrete cannot gain any strength. As time passes, concrete gains more strength. Temperature and time aren't the only two factors that affect concrete strength, but they are major factors, especially when it comes to maturity.

To understand concrete maturity, the maturity index must be calculated. There are multiple methods to determine concrete maturity, but they all require the maturity index. One of the most popular methods to determine the index is the Nurse-Saul function (temperature-time factor). It is suggested that the datum temperature be taken as 32°F (0°C). Using the maturity index, the concrete strength can be estimated.

B.7. BRACING

Key concepts: These key concepts are important for answering exam questions in subtopic 11.B.7, Bracing.

- braced versus unbraced frames
- concrete braced frames
- construction bracing
- lateral bracing

NCEES Handbook: The *Handbook* does not contain any material relevant to this subtopic. Be sure to study the listed sections in CERM and from codes and standards.

PE Civil Reference Manual **(CERM):** Study these sections in CERM that either relate directly to this subtopic or provide background information.

- Section 49.15: Formwork
- Section 53.2: Braced and Unbraced Columns
- Section 59.5: Lateral Bracing

Codes and standards: To prepare for this subtopic, familiarize yourself with these sections from codes and standards.

Building Code Requirements for Structural Concrete and Commentary (ACI 318)

- Section 6.2.5: Slenderness Effect
- Section 6.6.4: Slenderness Effects, Moment Magnification Method

Steel Construction Manual (AISC)

- Part 3: Design of Flexural Members

The following equations and figures are relevant for subtopic 11.B.7, Bracing.

Concrete Braced Frames

CERM: Sec. 53.2

ACI 318: Sec. 6.2.5, Sec. 6.6.4

A column is considered braced if its buckling mode shape does not involve translation of the endpoints. CERM Fig. 53.2 shows an example of both a braced (non-sway) and unbraced (sway) frame.

Figure 53.2 Braced and Unbraced Frames

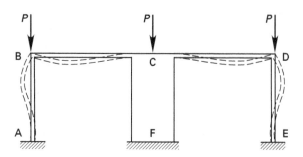

(a) frame braced by wall

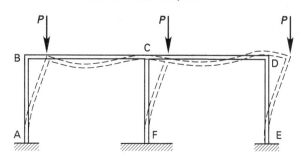

(b) unbraced frame

There are instances where inspection is not sufficient to distinguish between braced and unbraced structures. For example, as the size of the wall CF in CERM Fig. 53.2 is reduced, the condition changes from braced to unbraced. It is difficult to determine when the transition occurs. ACI 318 provides three quantitative methods to distinguish between braced and unbraced conditions. In particular, columns in any given level of a building can be treated as braced when one of the following criteria is satisfied.

1. The ratio of the moment from a second-order analysis to the first-order moment is less than or equal to 1.05. Review ACI 318 Sec. 6.6.4.3(a) for more information.

2. The stability coefficient or index, Q, is less than or equal to 0.05. Q is given by CERM Eq. 53.1. Review ACI 318 Sec. 6.6.4.3(b) for more information.

$$Q = \frac{\sum P_u \Delta_o}{V_{us} l_c} \leq 0.05 \qquad 53.1$$

3. The sum of the lateral stiffness of the bracing elements in that story exceeds 12 times the gross lateral stiffness of all columns within that story. Review ACI 318 Sec. 6.2.5 for more information.

Beam Lateral Bracing

CERM: Sec. 59.5

AISC: Part 3

To determine the failure mode of the beam and subsequently the moment strength, the beam properties L_p and L_r must be calculated using AISC Eq. F2-5 (CERM Eq. 59.6) and AISC Eq. F2-6 (CERM Eq. 59.7). The failure mode depends on the relationship of the lateral bracing, L_b, to these properties. Failure by yielding (material failure) occurs when $L_b < L_p$. Failure by inelastic buckling occurs when $L_p < L_b < L_r$. Failure by elastic buckling occurs when $L_b > L_r$.

$$L_p = 1.76 r_y \sqrt{\frac{E}{F_y}} \quad \text{[AISC Eq. F2-5]} \quad 59.6$$

$$L_r = 1.95 r_{ts} \left(\frac{E}{0.7 F_y}\right) \sqrt{\frac{Jc}{S_x h_o}}$$
$$\cdot \sqrt{1 + \sqrt{1 + 6.76\left(\frac{0.7 F_y S_x h_o}{E J c}\right)^2}} \quad 59.7$$

[AISC Eq. F2-6]

The effective radius of gyration, r_{ts}, can be calculated from AISC Eq. F2-7 (CERM Eq. 59.9), in which both the elastic section modulus, S_x, and the warping constant, C_w, are tabulated in the AISC shape tables.

$$r_{ts}^2 = \frac{\sqrt{I_y C_w}}{S_x} \quad 59.9$$

For doubly symmetrical I-shapes, $c = 1$ (AISC Eq. F2-8a). For a channel, the torsional constant, c, is calculated using AISC Eq. F2-8b (CERM Eq. 59.12).

$$c = \frac{h_o}{2}\sqrt{\frac{I_y}{C_w}} \quad 59.12$$

To prevent lateral torsional buckling (illustrated in CERM Fig. 59.3), a beam's compression flange must be supported at frequent intervals. Complete support is achieved when a beam is fully encased in concrete or has its flange welded or bolted along its full length. In many designs, however, lateral support is provided only at regularly spaced intervals. The actual spacing between points of lateral bracing is designated as L_b.

Construction Bracing

CERM: Sec. 49.15

During construction, a structure may need temporary external bracing to help carry loads until enough construction has taken place to ensure the structure can carry loads as intended. Bracing can occur in all types of structures and materials. Walls and frames need to be braced until the walls have cured or the frames have been completely constructed. The mechanics and principles of designing temporary bracing should follow standard design practices for permanent construction; however, the loading may be different because the bracing is temporary.

B.8. ANCHORAGE

Key concepts: These key concepts are important for answering exam questions in subtopic 11.B.8, Anchorage.

- breakout strength
- column anchorage
- fall protection anchor point
- steel anchor strength
- tensile loading

NCEES Handbook: The *Handbook* does not contain any material relevant to this subtopic. Be sure to study the listed sections from codes and standards.

Codes and standards: To prepare for this subtopic, familiarize yourself with these sections from codes and standards.

Building Code Requirements for Structural Concrete and Commentary (ACI 318)

- Section 17.4: Design Requirements for Tensile Loading

Occupational Safety and Health Act (OSHA)

- Section 1926.104: Safety Belts, Lifelines, and Lanyards
- Section 1926.755: Column Anchorage

The following equations, figures, and concepts are relevant for subtopic 11.B.8, Anchorage.

Steel Anchor Strength in Concrete

ACI 318: Sec. 17.4

$$N_{sa} = A_{se,N} f_{uta} \quad \text{[ACI 318 Eq. 17.4.1.2]}$$

Steel strength is increased by increasing the yield strength or providing a larger-diameter anchor. The steel force is limited to the product of the steel area times the steel yield strength.

$A_{se,N}$	effective cross-sectional area	in²
f_{uta}	lesser of $1.9 f_{ya}$ and 125,000 psi	psi
f_{ya}	yield strength of anchor	psi

Steel Anchor Breakout Strength in Concrete

ACI 318: Sec. 17.4

Breakout strength is governed mainly by the embedment, h_{ef}, of the anchor. The deeper the embedment, the greater the concrete breakout strength will be. When anchors are placed near edges, the concrete breakout failure plane is reduced. Edges should be avoided when possible.

There are three modification factors for single anchors and four modification factors for groups of anchors. These factors should be calculated first and then used in the N_{cb} equations.

For a single anchor, the breakout strength in tension is given by ACI 318 Eq. 17.4.2.1a.

$$N_{cb} = \left(\frac{A_{Nc}}{A_{Nco}} \right) \psi_{ed,N} \psi_{c,N} \psi_{cp,N} N_b \quad \text{[ACI 318 Eq. 17.4.2.1a]}$$

For a group of anchors, the breakout strength in tension is given by ACI 318 Eq. 17.4.2.1b.

$$N_{cbg} = \left(\frac{A_{Nc}}{A_{Nco}} \right) \psi_{ec,N} \psi_{ed,N} \psi_{c,N} \psi_{cp,N} N_b$$
$$\text{[ACI 318 Eq. 17.4.2.1b]}$$

ACI 318 Fig. R17.3.1(a)(iii) shows how anchors are pulled by force N.

Minimum Column Anchorage

OSHA: Sec. 1926.755

Column anchorage minimums are in place to help with column stability during construction. Columns are anchored with a minimum of four anchor bolts. Each assembly is able to resist 300 lbf at an eccentricity of 18 in plus half the column depth. If the column is set on a floor that is not level, then leveling nuts and pre-grouted leveling plates can be used. Adequately designed shim packs are acceptable as well. The erection engineer should determine if the columns need to be braced during construction. The project structural engineer of record is responsible for approving the repair, replacement, or modification of the anchor bolts.

Fall Protection Anchor Point

OSHA: Sec. 1926.104

Anchorage means a secure point of attachment for lifelines, lanyards, or deceleration devices. Joists and girders should not be used for anchorage. Purlins and girts should not be used unless allowed by a qualified person.

Anchorage to support or suspend platforms cannot be used as anchorage for fall protection. Anchorage must be capable of supporting 5000 lbf per user or two times the potential impact of the user's fall, whichever is greater. Unless specifically designed as fall protection anchor points, many structural members are not prepared for sudden impact from a person falling.

Many times, anchorage is removed once construction is complete. However, it may be designed to stay in place.

B.9. OSHA REGULATIONS

Key concepts: These key concepts are important for answering exam questions in subtopic 11.B.9, OSHA Regulations.

- construction sequence regulations
- personal protective and life saving equipment
- site layout regulations
- site-specific erection plan
- steel erection regulations
- structural steel assembly regulations

NCEES Handbook: The *Handbook* does not contain any material relevant to this subtopic. Be sure to study the listed sections from codes and standards.

Codes and standards: To prepare for this subtopic, familiarize yourself with these sections from codes and standards.

Occupational Safety and Health Act (OSHA)

- Section 1926.104: Safety Belts, Lifelines, and Lanyards
- Section 1926.752: Site Layout, Site-Specific Erection Plan and Construction Sequence
- Section 1926.754: Structural Steel Assembly

The following concepts are relevant for subtopic 11.B.9, OSHA Regulations.

Personal Protective and Life Saving Equipment

OSHA: Sec. 1926.104

For employee safeguarding, the use of lifelines, safety belts, and lanyards is required. If equipment is subject to in-service loading, it must be immediately removed from service. Lifelines must be anchored to a structural member that can resist 5400 lbf of dead weight. In addition to the structural member, lifeline rope must also be able to resist 5400 lbf and be 3/4 in manila or equivalent. If the rope is subject to cutting or abrasion, it must be 7/8 in manila or equivalent. Safety belts must be 1/2 in nylon or equivalent, and their length must not exceed a 6 ft fall. Hardware must be cadmium-plated drop-forged steel or pressed steel.

Steel Erection: Site Layout, Site-Specific Erection Plan, and Construction Sequence

OSHA: Sec. 1926.752

Before steel erection can begin, the contractor must provide the steel erector with the following written notifications.

1. Concrete and mortar have attained 75% designed strength or enough to support construction loads.
2. Repairs, replacements, and modifications to anchor bolts have occurred.

The contractor must ensure that adequate access roads have been provided to the site. Such roads must include a plan for vehicular traffic and safe delivery and movement of equipment. The contractor must also ensure that the site is properly drained and graded, and that adequate space for safe storage of equipment and materials has been provided. The contractor must preplan hoisting operations and have a site-specific erection plan.

Steel Erection: Structural Steel Assembly

OSHA: Sec. 1926.754

During the erection process, a structure must maintain stability. Installation of permanent floors occurs as the structural erection progresses. There must not be more than four stories erected before bolting and welding are completed. Beam flanges must remain clear until surfaces have been installed. When plumbing-up equipment is used, it must be installed with the steel erection process and must be in place before loads are applied.

Bundles of metal decking must not be hoisted by their packaging and strapping. Holes in metal decking must not be cut until immediately before use. If an opening must be in place prior to use, it must be temporarily covered and capable of supporting two times the weight of employees, equipment, and materials. Covers must be marked and secured. Gaps between columns and decking must contain mesh or plywood. Metal decking must be placed to ensure full support, and once in place, it must be secured immediately.

B.10. SAFETY MANAGEMENT

Key concepts: These key concepts are important for answering exam questions in subtopic 11.B.10, Safety Management.

- crane use and safety
- safeguards during construction

NCEES Handbook: To prepare for this subtopic, familiarize yourself with these sections in the *Handbook*.

- Safety Incidence Rate

PE Civil Reference Manual (**CERM**): Study these sections in CERM that either relate directly to this subtopic or provide background information.

- Chapter 83: Construction and Job Site Safety

Codes and standards: To prepare for this subtopic, familiarize yourself with these sections from codes and standards.

International Building Code (IBC)

- Chapter 33: Safeguards During Construction

The following equations, tables, and concepts are relevant for subtopic 11.B.10, Safety Management.

Safety Management

Handbook: Safety Incidence Rate

$$\text{IR} = \frac{N \cdot 200{,}000}{T}$$

CERM: Chap. 83

Safety management refers to having processes and procedures in place to reduce risk, especially of injury on a job site. OSHA has regulations in place to help reduce this risk. The equation in *Handbook* section Safety Incidence Rate and CERM Eq. 83.1 can be used to calculate the total injury/illness rate.

OSHA 1910 applies to general industry, and OSHA 1926 applies to the construction industry. CERM Table 83.1 lists subjects for general industry, and CERM Table 83.2 lists subjects for the construction industry.

Implementing safety management programs reduces risk and saves money due to injury, time lost, and rework. Safety management programs should include steps to identify, assess, and mitigate risk. It is helpful to have plans and workflows in place to ensure that processes are streamlined. Safety management programs are also useful for educating employees and ensuring that contractors are acting in accordance with OSHA regulations.

Safeguards During Construction

IBC: Chap. 33

IBC Chap. 33 covers new construction, alterations, repairs, additions, and demolition. Contractors must store equipment and materials in a manner that does not endanger anyone for the duration of the project. Egress paths and fire protection must remain intact during alterations, repairs, and additions. Sanitary waste should be disposed of in a manner that does not endanger workers or the public.

IBC Sec. 3306 covers protections for pedestrians during construction. See IBC Table 3306.1 for types of protection required for different heights of construction.

Crane Use and Safety

CERM: Sec. 83.18

Crane operation must be performed only by qualified and trained personnel. The ground underneath cranes must be firm, stable, and within 1% of level, especially when using truck cranes.

Crane loading is typically governed by stability issues (i.e., overturning of the crane during service). Crane stability can be affected by many issues, including the weight distribution of a load, the angle of the crane boom, and soil conditions under the crane. Engineers need to take all these factors into consideration when dealing with cranes.

Topic V: Transportation

Chapter

35. Traffic Engineering
36. Horizontal Design
37. Vertical Design
38. Intersection Geometry
39. Roadside and Cross-Section Design
40. Signal Design
41. Traffic Control Design
42. Geotechnical and Pavement
43. Drainage
44. Alternatives Analysis

35 Traffic Engineering

Content in blue refers to the *NCEES Handbook*.

A. Uninterrupted Flow 35-1
B. Street Segment Interrupted Flow 35-5
C. Intersection Capacity 35-6
D. Traffic Analysis 35-8
E. Trip Generation and Traffic Impact Studies 35-10
F. Accident Analysis 35-11
G. Nonmotorized Facilities 35-12
H. Traffic Forecast 35-14
I. Highway Safety Analysis 35-15

The knowledge area of Traffic Engineering makes up between 10 and 15 questions out of the 80 questions on the PE Civil Transportation exam. The organization of this chapter follows the order of subtopics given by NCEES for this knowledge area. Each subtopic is covered in the following sections.

A. UNINTERRUPTED FLOW

Key concepts: These key concepts are important for answering exam questions in subtopic 9.A, Uninterrupted Flow.

- flow rate
- free-flow travel speed
- levels of service (LOS)
- LOS criteria
- peak hour factor

NCEES Handbook: To prepare for this subtopic, familiarize yourself with these sections in the *Handbook*.

- Traffic Flow, Density, Headway, and Speed Relationships
- Headway
- Space Mean Speed
- Lane Occupancy Used in Freeway Surveillance
- Greenshields Maximum Flow Rate Relationship

PE Civil Reference Manual (**CERM**): Study these sections in CERM that either relate directly to this subtopic or provide background information.

- Section 73.3: Facilities Terminology
- Section 73.4: Design Vehicles
- Section 73.5: Levels of Service
- Section 73.10: Speed, Flow, and Density Relationships
- Section 73.13: Freeways
- Section 73.14: Multilane Highways
- Section 73.15: Two-Lane Highways

Transportation Depth Reference Manual (**CETR**): Study these sections in CETR that either relate directly to this subtopic or provide background information.

- Section 2.1: Travel Time and Delay Studies
- Section 2.3: Capacity Analysis for Uninterrupted Flow

Codes and standards: To prepare for this subtopic, familiarize yourself with these sections from codes and standards.

A Policy on Geometric Design of Highways and Streets (AASHTO *Green Book*)

- Section 1.4.3: Functional System Characteristics

Highway Capacity Manual (HCM)

- Section 4.2: Motorized Vehicle Mode
- Section 5.3: Level of Service
- Section 12.2: Concepts
- Section 12.3: Motorized Vehicle Core Methodology
- Section 14.3: Core Methodology
- Section 15.2: Concepts

The following equations, tables, and concepts are relevant for subtopic 9.A, Uninterrupted Flow.

Traffic Flow, Density, Headway, and Speed Relationships

Handbook: Traffic Flow, Density, Headway, and Speed Relationships

$$q = \frac{n}{t}$$

$$q = \frac{n(3600)}{t}$$

$$k = \frac{n}{L}$$

CERM: Sec. 73.10

$$v = SD = \frac{3600 \frac{\text{sec}}{\text{hr}}}{\text{headway}_{\text{sec/veh}}} \qquad 73.10$$

CETR: Sec. 2.3

$$D_{(\text{veh/mi})} = \frac{\text{flow rate (veh/hr)}}{\text{average travel speed (mi/hr)}} \qquad 2.12$$
$$= \frac{v_{\text{vph}}}{S} \quad [\text{HCM Eq. 4-4}]$$

HCM: Sec. 4.2

In the *Handbook* equations,

k	traffic density	veh/ft	veh/m
L	length of roadway	ft	m
n	number of vehicles passing some designated point during time t	–	–
q	traffic flow	veh/sec	veh/s
t	time	sec	s

In the CERM equation,

D	traffic density	vpm/lane	vpk/lane
S	speed	mi/hr	km/h
v	rate of flow	vph/lane	vph/lane

In the CETR equation,

D	density	veh/mi
S	travel speed	mi/hr
v	rate of flow	veh/hr

Traffic flow is defined in the *Handbook* as the equivalent hourly rate, q, at which vehicles pass a point on a highway during a period of less than one hour. CERM Eq. 73.10 finds the flow rate as the number of vehicles crossing a point per hour per lane. Flow rate is closely related to density and speed, as shown in CETR Eq. 2.12, and knowing any two values can determine the third.

Exam tip: Remember that traffic flow is over time at a point in space while density is over space at a point in time.

Headway

Handbook: Headway

$$t = \sum_{i=1}^{n} h_i$$

$$q = \frac{n}{\sum_{i=1}^{n} h_i}$$

$$q = \frac{1}{\bar{h}}$$

CERM: Sec. 73.10

$$\text{headway}_{\text{s/veh}} = \frac{\text{spacing}_{\text{m/veh}}}{\text{space mean speed}_{\text{m/s}}} \quad [\text{SI}] \qquad 73.12(a)$$

$$\text{headway}_{\text{sec/veh}} = \frac{\text{spacing}_{\text{ft/veh}}}{\text{space mean speed}_{\text{ft/sec}}} \quad [\text{U.S.}] \qquad 73.12(b)$$

Headway is the time between successive vehicles, calculated as the average of the headways of a stream of vehicles measured at a designated point. The average hourly flow can be found from the same values. Headway can also be calculated as the ratio of the spacing between vehicles and the mean speed of the vehicles, as shown in CERM Eq. 73.12.

A concept related to headway is *gap*, or the time interval between vehicles in a traffic stream. Gap is important for pedestrians trying to cross a traffic flow.

Space Mean Speed

Handbook: Space Mean Speed

$$u_s = \frac{q}{k}$$

$$u_s = \frac{nL}{\sum_{i=1}^{n} t_i}$$

CETR: Sec. 2.1

The *space mean speed* (also known as the *mean travel speed* or *mean running speed*) is the average speed of vehicles traveling along a given segment of roadway, and is the ratio of the traffic flow on that segment to the traffic density along that segment. If these properties are not known, the space mean speed can also be found using the number of vehicles on the roadway, the length of the segment, and the total time it takes all vehicles to travel across the segment. This second method can also be used for a single vehicle taking several trips.

Exam tip: Know the difference between space mean speed and time mean speed. Time mean speed is the average speed of vehicles that travel an equal amount of time, while space mean speed is the average speed of vehicles that travel an equal amount of distance. Space mean speed is usually less than time mean speed.

Lane Occupancy Used in Freeway Surveillance

Handbook: Lane Occupancy Used in Freeway Surveillance

$$R = \frac{\sum L_i}{D}$$

Lane occupancy, R, is the ratio of the lengths of all vehicles on a given roadway section at a given time, divided by the length of the roadway section. The lane occupancy can be divided by the average length of a vehicle to get an estimate of density, k.

Greenshields Maximum Flow Rate Relationship

Handbook: Greenshields Maximum Flow Rate Relationship

$$q_{\max} = \frac{k_j u_f}{4}$$

Greenshields maximum flow rate relationship is one of the most commonly adopted mathematical models of the relationship between traffic speed, traffic flow, and traffic density. The relationship can be used to determine how a given roadway or facility is performing.

Road Designation

CERM: Sec. 73.3

AASHTO *Green Book*: Sec. 1.4.3

Although common usage does not always distinguish among highways, freeways, and other types of roadways, the major transportation engineering references are more specific, as shown in CERM Table 73.2.

Table 73.2 General Functional Classifications of Roadways*

road designation	ADT (vpd)
local road	2000 or less
collector road	2000–12,000
arterial/urban road	12,000–40,000
freeway	30,000 and above

*Classifications can also be established based on percentages of total length and travel volume.

Free-Flow Speed

CERM: Sec. 73.13

CETR: Sec. 2.3

HCM: Sec. 12.2, Sec. 12.3, Sec. 14.3, Sec. 15.2

Free-flow speed, FFS, is the mean speed of passenger cars measured during flows of less than 1000 cars per hour per lane (pcphpl); it assumes ideal conditions, such as 12 ft lanes and adequate lateral clearance.

Free-flow speed is determined from the base free-flow speed, BFFS, with adjustments for factors such as median type, lane width, total lateral clearance, and density of access points. Adjustment factors for free-flow speed are provided in HCM Exh. 12-21 through 12-24. Speed variations at lower flow rates (500–800 pcphpl) are not drastically affected by these adjustments, so long as the attributes in question are consistent throughout the segment.

HCM Eq. 12-2 finds the free-flow speed for freeways, HCM Eq. 12-3 finds the free-flow speed for multilane highways, and HCM Eq. 15-1 and 15-2 find the free-flow speeds for two-lane highways.

$$\text{FFS} = \text{BFFS} - f_{\text{LW}} - f_{\text{RLC}} - 3.22 \cdot \text{TRD}^{0.84}$$
[HCM Eq. 12-2]

$$\text{FFS} = \text{BFFS} - f_{\text{LW}} - f_{\text{TLC}} - f_M - f_A \quad \text{[HCM Eq. 12-3]}$$

$$\text{FFS} = S_{\text{FM}} + 0.00776 \frac{v}{f_{\text{HV,ATS}}} \quad \text{[HCM Eq. 15-1]}$$

$$\text{FFS} = \text{BFFS} - f_{\text{LS}} - f_A \quad \text{[HCM Eq. 15-2]}$$

The average freeway speed as a function of free-flow speed is found from HCM Eq. 15-6.

$$\text{ATS}_d = \text{FFS} - 0.0076(v_{d,\text{ATS}} + v_{o,\text{ATS}}) - f_{\text{np,ATS}}$$
[HCM Eq. 15-6]

Levels of Service

CERM: Sec. 73.5

HCM: Sec. 5.3

Level of service (LOS) is a user's quality of service through or over a specific facility, such as a highway, an intersection, or a crosswalk. The designations run from A, unimpeded flow, to F, gridlock. A lower LOS (where A is highest and F is lowest) typically indicates less safety afforded by the roadway.

The parameters used to define the level of service vary with the type of facility. CERM Table 73.4 provides a summary of the primary parameters used to measure LOS in several different facilities.

Table 73.4 Primary Measures of Level of Service

type of facility	level of service parameter
freeways	
basic freeway segment	density (pc/mi/ln or pc/km/ln)
weaving areas	density (pc/mi/ln or pc/km/ln)
ramp junctions	density (pc/mi/ln or pc/km/ln)
multilane highways	density (pc/mi/ln or pc/km/ln)
two-lane rural highways	speed or percent time spent following
signalized intersections	average control delay (sec/veh or s/veh)
unsignalized intersections	average control delay (sec/veh or s/veh)
urban streets	average travel speed (mph or km/h)
mass transit	various (pers/seat, veh/hr, people/hr, p/seat, veh/h, p/h)
pedestrians	space per pedestrian or delay (ft^2/ped or m^2/p)
bicycles	event, delay (sec/veh or s/veh)

Compiled from *Highway Capacity Manual*, 6th Edition: A Guide for Multimodal Mobility Analysis, 2016, by the Transportation Research Board of the National Academies of Sciences, Engineering, and Medicine, Washington DC. DOI: 10.17226/24798

Economic considerations favor an LOS with higher traffic volumes. However, political considerations favor an LOS with less obstruction.

> *Exam tip*: Pedestrian LOS is a consideration of space per pedestrian.

Levels of Service for Two-Lane Highways

HCM: Sec. 15.2

The performance criteria of Class I, two-lane highways in non-mountainous terrain with no traffic signals, include average travel speed and percent time spent following (PTSF). Performance of Class II highways is determined only by PTSF. Performance of Class III highways is determined only by percent free-flow speed (PFFS), as these highways are generally limited in length and have lower posted speed limits.

Levels of Service for Multilane Highways

CERM: Sec. 73.14

HCM: Sec. 12.3

$$\text{FFS} = \text{BFFS} - f_{\text{LW}} - f_{\text{TLC}} - f_M - f_A \quad \text{[HCM Eq. 12-3]}$$

The level of service of a multilane highway is based on the free-flow speed, FFS, found from HCM Eq. 12-3, and the traffic density, as shown in HCM Exh. 12-16 and 12-17. CERM Table 73.10 is an equivalent table.

The 15 min passenger car equivalent flow rate in passenger cars per hour per lane (pcphpl) is given by HCM Eq. 12-9.

$$v_p = \frac{V}{(\text{PHF})Nf_{\text{HV}}} \quad \text{[HCM Eq. 12-9]}$$

In this equation, V is the volume of vehicles passing a point each hour, f_{HV} is the same heavy vehicle factor used in freeway analysis, N is the number of lanes, and PHF is the peak hour factor. HCM Sec. 12.3 recommends that where specific local data is not available, a PHF from 0.75 to 0.95 should be used as a reasonable estimate; for congested conditions, a PHF of 0.95 should be used. Lower PHF values are typical of low-volume rural conditions, and higher values are typical of urban and suburban peak-hour conditions.

Levels of Service for Freeways

CETR: Sec. 2.3

HCM: Sec. 12.3

$$D = \frac{v_p}{S} \quad \text{[HCM Eq. 12-11]}$$

The process of determining the level of service (LOS) of a freeway section generally involves determining the vehicular density, D. Maximum flow rate (pc/hr/ln) can also be used for LOS designations on basic freeway sections.

B. STREET SEGMENT INTERRUPTED FLOW

Key concepts: These key concepts are important for answering exam questions in subtopic 9.B, Street Segment Interrupted Flow.

- average speed
- flow-density model
- running time
- speed-density model
- speed-flow model
- time mean speed

NCEES Handbook: To prepare for this subtopic, familiarize yourself with these sections in the *Handbook*.

- Speed-Density Model
- Flow-Density Model
- Speed-Flow Model
- Time Mean Speed
- Average Speed (Mean Speed)
- Segment Running Time Simplified

PE Civil Reference Manual (**CERM**): Study these sections in CERM that either relate directly to this subtopic or provide background information.

- Section 73.6: Speed Parameters
- Section 73.10: Speed, Flow, and Density Relationships

Transportation Depth Reference Manual (**CETR**): Study these sections in CETR that either relate directly to this subtopic or provide background information.

- Section 2.1: Travel Time and Delay Studies

Codes and standards: To prepare for this subtopic, familiarize yourself with these sections from codes and standards.

Highway Capacity Manual (HCM)

- Section 4.2: Motorized Vehicle Mode
- Section 12.3: Motorized Vehicle Core Methodology
- Section 18.3: Motorized Vehicle Methodology
- Section 19.2: Concepts
- Section 19.3: Core Motorized Vehicle Methodology

The following equations and figures are relevant for subtopic 9.B, Street Segment Interrupted Flow.

Speed-Density Model

Handbook: Speed-Density Model

$$u_s = u_f\left(1 - \frac{k}{k_j}\right)$$

CERM: Sec. 73.10

$$S = S_f\left(1 - \frac{D}{D_j}\right) \qquad 73.9$$

HCM: Sec. 12.3

The *Handbook* equation and CERM Eq. 73.9 are the same equation with different nomenclature: S is the same as u_s, S_f is the same as u_f, and D is the same as k.

As traffic density increases, traffic can become interrupted, and the speed on the roadway will no longer be the free-flow speed. The density of the roadway eventually increases until it reaches the *jam density*, the density at which the vehicles or pedestrians are all at a standstill. Knowing the current density of the roadway, the jam density, and the free-flow speed allow for calculation of the speed of the roadway when traffic is not in the free-flow condition. The speed-density relationship is shown graphically in CERM Fig. 73.1.

Flow-Density Model

Handbook: Flow-Density Model

$$q = u_f\left(k - \frac{k^2}{k_j}\right)$$

$$q_{\text{cap}} = u_f\left(\frac{k_j}{4}\right)$$

HCM: Sec. 4.2

Flow-density relationship refers to the relationship between the flow rate, the free-flow speed, and the traffic density. This relationship is used to calculate the maximum traffic volume at the peak of a curve, and it can be mathematically described using the *Handbook* equations shown.

Speed-Flow Model

Handbook: Speed-Flow Model

$$k = k_j \left(1 - \frac{u_s}{u_f}\right)$$

$$q = k_j \left(u_s - \frac{u_s^2}{u_f}\right)$$

CERM: Sec. 73.10

HCM: Sec. 4.2

The speed-flow relationship can be used to find the critical speed of the system, as shown in the *Handbook* equations. CERM Fig. 73.2 illustrates the speed-flow relationship.

Time Mean Speed

Handbook: Time Mean Speed

$$\overline{u}_t = \frac{\sum_{i}^{t} u_i}{n}$$

HCM: Sec. 4.2

Time mean speed (or *average spot speed*) is the arithmetic mean of the instantaneous speed of all cars at a particular point.

> *Exam tip*: Know the difference between space mean speed and time mean speed. Time mean speed is the average speed of vehicles that travel an equal amount of time, while space mean speed is the average speed of vehicles that travel the same distance. Space mean speed is usually less than time mean speed.

Average Speed

Handbook: Average Speed (Mean Speed)

$$\overline{x} = \frac{\sum n_i S_i}{N}$$

HCM: Sec. 4.2

The average speed of vehicles on a given segment is measured under low-volume conditions when drivers are free to drive at their desired speed and are not constrained by the presence of other vehicles or downstream traffic control devices (e.g., traffic signals, roundabouts, and stop signs).

Running Time

Handbook: Segment Running Time Simplified

$$\text{average running time} = \frac{\text{segment length}}{\text{average travel speed}}$$

CETR: Sec. 2.1

HCM: Sec. 18.3

Running time is the time that a vehicle is actually in motion; it is a function of the length of a given roadway segment and the average travel speed along that segment, including the dwell times at all stations encountered. HCM Eq. 18-7 also considers traffic control at boundary intersections, free-flow speed, vehicle proximity, and various midsegment delay sources, but can be simplified to the *Handbook* equation shown if all lost time and boundary control delays are accounted for or given as part of the average travel speed.

C. INTERSECTION CAPACITY

Key concepts: These key concepts are important for answering exam questions in subtopic 9.C, Intersection Capacity.

- elements of an intersection
- intersection capacity
- intersection delay
- saturation flow rate

***NCEES Handbook*:** The *Handbook* does not contain any material relevant to this subtopic. Be sure to study the listed sections in CERM and CETR and from codes and standards.

***PE Civil Reference Manual* (CERM):** Study these sections in CERM that either relate directly to this subtopic or provide background information.

- Section 73.6: Speed Parameters
- Section 73.17: Signalized Intersections

***Transportation Depth Reference Manual* (CETR):** Study these sections in CETR that either relate directly to this subtopic or provide background information.

- Section 2.5: Traffic Signals

Codes and standards: To prepare for this subtopic, familiarize yourself with these sections from codes and standards.

Highway Capacity Manual (HCM)

- Section 19.2: Concepts
- Section 19.3: Core Motorized Vehicle Methodology

The following equations, figures, and tables are relevant for subtopic 9.C, Intersection Capacity.

Elements of an Intersection

CERM: Sec. 73.17

CERM Fig. 73.5 shows typical elements of an intersection. Auxiliary lanes are also known as *turning lanes* and are a form of acceleration/deceleration lane.

Figure 73.5 Elements of an Intersection

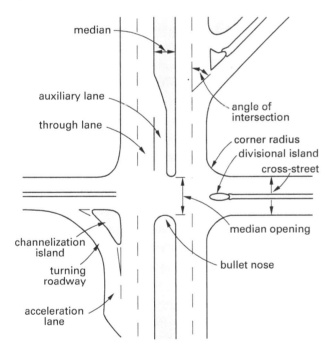

Intersection Levels of Service

CETR: Sec. 2.5

HCM: Sec. 19.2

Intersection level of service (LOS) is determined by the average control delay per vehicle. The LOS designations are found in HCM Exh. 19-8. CETR Table 2.42 is an equivalent table.

Table 2.42 LOS Designation for Automobile Mode Based on Signal Delay and V/C

Control Delay (s/veh)	LOS by Volume-to-Capacity Ratio*	
	≤1.0	>1.0
≤10	A	F
>10-20	B	F
>20-35	C	F
>35-55	D	F
>55-80	E	F
>80	F	F

*For approach-based and Intersectionwide assessments, LOS is defined society by control delay.

Used with permission from *Highway Capacity Manual*, 6th Edition: A Guide for Multimodal Mobility Analysis, 2016, Exhibit 19-8, by the Transportation Research Board of the National Academies of Sciences, Engineering, and Medicine, Washington DC. DOI: 10.17226/24798

Intersection Delay

CETR: Sec. 2.5

HCM: Sec. 19.3

The *intersection delay* (or *average control delay*) experienced by all vehicles that arrive in the analysis period determines the level of service for the intersection. HCM Eq. 19-18 is used to determine the average control delay.

$$d = d_1 + d_2 + d_3 \quad \text{[HCM Eq. 19-18]}$$

d_1 is the uniform delay, determined assuming uniform arrivals, stable flow, and no initial queue at the beginning of the green time.

d_2 is the incremental delay, which accounts for the effect of random arrivals and oversaturation queues and is adjusted for the duration of the analysis period and the type of signal control. The incremental delay is determined assuming that there is an initial queue for the lane group at the start of the analysis period.

d_3 is the initial queue delay, which accounts for the delay to all vehicles in the analysis period due to the initial queue at the start of the analysis period.

Aggregate Delay

CETR: Sec. 2.5

HCM: Sec. 19.3

Lane groups for an approach can be aggregated to provide an estimation of control delay for the entire approach. This aggregation computes weighted averages, where the lane group delays are weighted by the adjusted flows in the lane groups. The aggregate delay can be found using HCM Eq. 19-28. CETR Eq. 2.61 is an equivalent equation.

$$d_{A,j} = \frac{\sum_{i=1}^{m_j} d_i v_i}{\sum_{i=1}^{m_j} v_i} \quad \text{[HCM Eq. 19-28]}$$

Saturation Flow Rate

CETR: Sec. 2.5

HCM: Sec. 19.3

The *saturation flow rate* is the total maximum flow rate on an approach or group of lanes that can pass through an intersection under prevailing traffic and roadway conditions when 100% of the effective green time is available.

HCM Eq. 19-8 is the base equation used for interrupted flow through a signalized intersection. CETR Eq. 2.34 is an equivalent equation.

$$s = s_o f_w f_{HVg} f_p f_{bb} f_a f_{LU} f_{LT} f_{RT} f_{Lpb} f_{Rpb} f_{wz} f_{ms} f_{sp}$$
$$\text{[HCM Eq. 19-8]}$$

HCM Eq. 19-9 gives the heavy vehicle adjustment factor for a downgrade approach. CETR Eq. 2.35 is an equivalent equation.

$$f_{HVg} = \frac{100 - 0.79 P_{HV} - 2.07 P_g}{100} \quad \text{[HCM Eq. 19-9]}$$

HCM Eq. 19-10 gives the heavy vehicle adjustment factor for a zero approach or an upgrade approach. CETR Eq. 2.36 is an equivalent equation.

$$f_{HVg} = \frac{100 - 0.78 P_{HV} - 0.31 P_g^2}{100} \quad \text{[HCM Eq. 19-10]}$$

HCM Eq. 19-11 gives the parking adjustment factor. CETR Eq. 2.37 is an equivalent equation.

$$f_p = \frac{N - 0.1 - \frac{18 N_m}{3600}}{N} \geq 0.050 \quad \text{[HCM Eq. 19-11]}$$

HCM Eq. 19-12 gives the bus blocking factor. CETR Eq. 2.38 is an equivalent equation.

$$f_{bb} = \frac{N - \frac{14.4 N_b}{3600}}{N} \geq 0.050 \quad \text{[HCM Eq. 19-12]}$$

HCM Eq. 19-13 gives the right-turn adjustment factor, and HCM Eq. 19-14 gives the left-turn adjustment factor. CETR Eq. 2.39(a) is equivalent to HCM Eq. 19-14, and CETR Eq. 2.39(b) is equivalent to HCM Eq. 19-13.

$$f_{RT} = \frac{1}{E_R} \quad \text{[HCM Eq. 19-13]}$$

$$f_{LT} = \frac{1}{E_L} \quad \text{[HCM Eq. 19-14]}$$

HCM Exh. 19-15 gives the lane utilization adjustment factors; the lane width adjustment factor can be found using HCM Exh. 19-20. CETR Table 2.36 is an equivalent table to HCM Exh. 19-20.

Intersection Capacity

CETR: Sec. 2.5

HCM: Sec. 19.3

HCM Eq. 19-16 gives the capacity of an intersection, calculated from the saturation flow rate, green time, and cycle length.

$$c = N s \frac{g}{C} \quad \text{[HCM Eq. 19-16]}$$

HCM Eq. 19-17 gives the degree of saturation of the intersection.

$$X = \frac{v}{c} \quad \text{[HCM Eq. 19-17]}$$

D. TRAFFIC ANALYSIS

Key concepts: These key concepts are important for answering exam questions in subtopic 9.D, Traffic Analysis.

- flow-density model
- methods of collecting traffic data
- relationships between traffic parameters
- speed-density model
- speed-flow model
- test vehicle methods
- time mean speed
- traffic studies
- traffic volume and density
- travel time

TRAFFIC ENGINEERING 35-9

NCEES Handbook: To prepare for this subtopic, familiarize yourself with these sections in the *Handbook*.

- Traffic Flow, Density, Headway, and Speed Relationships
- Peak-Hour Factor

PE Civil Reference Manual (**CERM**): Study these sections in CERM that either relate directly to this subtopic or provide background information.

- Section 73.10: Speed, Flow, and Density Relationships

Transportation Depth Reference Manual (**CETR**): Study these sections in CETR that either relate directly to this subtopic or provide background information.

- Section 2.1: Travel Time and Delay Studies
- Section 2.3: Capacity Analysis for Uninterrupted Flow

Codes and standards: To prepare for this subtopic, familiarize yourself with these sections from codes and standards.

Highway Capacity Manual (HCM)

- Section 4.2: Motorized Vehicle Mode

The following equations and concepts are relevant for subtopic 9.D, Traffic Analysis.

Density

Handbook: Traffic Flow, Density, Headway, and Speed Relationships

$$k = \frac{n}{L}$$

CETR: Sec. 2.3
HCM: Sec. 4.2

Traffic density is the ratio of the number of vehicles crossing over a segment of roadway to the length of that segment. The *critical density*, also called the *optimum density*, is the traffic density at a point where the traffic volume is at the maximum value for the roadway.

Peak Hour Factor

Handbook: Peak-Hour Factor

$$v = \frac{V}{\text{PHF}}$$

$$\text{PHF} = \frac{V}{V_{15} \cdot 4}$$

CETR: Sec. 2.3
HCM: Sec. 4.2

The *peak hour factor* (PHF) describes the traffic flow rate during the peak hour for a given roadway. The *Handbook* equation for PHF is equivalent to HCM Eq. 4-2, and the *Handbook* equation for the rate of flow for a 15 min period is equivalent to HCM Eq. 4-3.

Running Speed

CETR: Sec. 2.1

$$S = \frac{L}{t} \qquad 2.3$$

CETR Eq. 2.3 is used to calculate the running speed, S, for a single test run. L is the study segment length, and t is the travel time.

Exam tip: This formula assumes the speed of each vehicle remains the same between the points where data is collected.

Hourly Volume Method

CETR: Sec. 2.1

The *hourly volume* is the number of vehicles that will pass the beginning of a study segment in the time it takes a test vehicle to make a round-trip through the study segment. The adjusted average travel time is determined from the average of all test run travel times, adjusted by the number of vehicles passed and the number of overtaking vehicles, as shown in CETR Eq. 2.8 and CETR Eq. 2.9. The space mean speed for the moving vehicle method is calculated using CETR Eq. 2.10 or CETR Eq. 2.11.

$$\bar{t}_s = t_s - \frac{O_s - P_s}{V_s} \quad \text{[southbound]} \qquad 2.8$$

$$\bar{t}_n = t_n - \frac{O_n - P_n}{V_n} \quad \text{[northbound]} \qquad 2.9$$

$$S_s = \frac{d}{\bar{t}_s} \quad \text{[southbound]} \qquad 2.10$$

$$S_n = \frac{d}{t_n} \quad \text{[northbound]} \qquad 2.11$$

E. TRIP GENERATION AND TRAFFIC IMPACT STUDIES

Key concepts: These key concepts are important for answering exam questions in subtopic 9.E, Trip Generation and Traffic Impact Studies.

- regression equations
- trip generation studies

NCEES Handbook: The *Handbook* does not contain any material relevant to this subtopic. Be sure to study the listed sections in CERM and CETR.

***PE Civil Reference Manual* (CERM):** Study these sections in CERM that either relate directly to this subtopic or provide background information.

- Section 73.9: Trip Generation

***Transportation Depth Reference Manual* (CETR):** Study these sections in CETR that either relate directly to this subtopic or provide background information.

- Section 1.3: Trip Generation and Attraction

The following equations and figures are relevant for subtopic 9.E, Trip Generation and Traffic Impact Studies.

Trip Generation Models

CETR: Sec. 1.3

$$T = a + bX \quad \text{[linear model]} \qquad 1.1$$

$$\ln T = a + b \ln X \quad \text{[nonlinear model]} \qquad 1.2$$

The regression equations shown are derived from data obtained from traffic studies. Once independent variables are selected, and the necessary data is collected, the total number of trips, or total trip ends, can be calculated using either a linear model, CETR Eq. 1.1, or a nonlinear model, CETR Eq. 1.2. T is the total number of trips, bX and $b \ln X$ are the slopes of the plotted curve, and a is the curve's vertical intercept. The units of X depend on the variable being used. For example, if X were the gross leasable area, the units would be square feet.

Trip Generation Curve

CETR: Sec. 1.3

The simple form of a trip has an origin, a destination, and a link between the two. The trip destination, commonly called the *trip generator*, determines significant characteristics of the trip. Trip generation studies are used to determine the predominant trip generators in the study region, which are then used to estimate the number of trips generated.

The Institute of Transportation Engineers (ITE) publishes the *Trip Generation Manual*, which defines standard methodologies and analysis methods for trip generation. CETR Fig. 1.1 provides an example of data from the *Trip Generation Manual* for an office park and shows the data given for each land use.

Figure 1.1 *Office Park Trip Generation Curve*

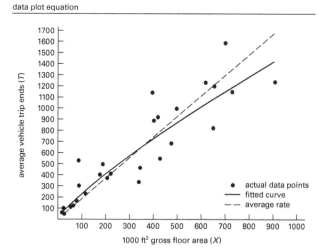

(Multiply ft² by 0.929 to obtain m².)

From *Trip Generation Manual*, © 2000, Institute of Transportation Engineers. Used by permission.

F. ACCIDENT ANALYSIS

Key concepts: These key concepts are important for answering exam questions in subtopic 9.F, Accident Analysis.

- crash analysis
- crash factors
- crash mitigation effectiveness
- crash rates
- sight distance
- vehicle exposure

NCEES Handbook: To prepare for this subtopic, familiarize yourself with these sections in the *Handbook*.

- Accident Analysis: Acceleration
- Vehicle Acceleration Rates

PE Civil Reference Manual (**CERM**): Study these sections in CERM that either relate directly to this subtopic or provide background information.

- Section 75.11: Factors Contributing to Crashes
- Section 75.12: Roadway Segment Crash Factors
- Section 75.13: Intersection Crash Factors

Transportation Depth Reference Manual (**CETR**): Study these sections in CETR that either relate directly to this subtopic or provide background information.

- Section 6.4: Crash Analysis

Codes and standards: To prepare for this subtopic, familiarize yourself with these sections from codes and standards.

Highway Safety Manual (AASHTO)

- Section 3.2.4: Factors Contributing to a Crash
- Section 4.4.2.2: Crash Rate
- Section 9A.1: Computational Procedure for Implementing the EB Before/After Safety Effectiveness Evaluation Method

The following equations and concepts are relevant for subtopic 9.F, Accident Analysis.

Crash Analysis

CETR: Sec. 6.4

AASHTO: Sec. 9A.1

A crash analysis should cover the direct effects of pertinent site features, such as sight distance limits, signal timing, and traffic patterns, to establish a cause-and-effect relationship. If necessary, statistical analysis can be weighted for the severity of crashes and to account for observed near-miss situations.

The analysis should focus on the effects of possible engineering improvements and the prediction of crash reduction due to these improvements. The effectiveness of the improvements is found using AASHTO Eq. 9A.1-10 and is a function of the odds ratio calculated using AASHTO Eq. 9A.1-7 and AASHTO Eq. 9A.1-8.

$$OR' = \frac{\sum_{\text{all sites}} N_{\text{observed},A}}{\sum_{\text{all sites}} N_{\text{expected},A}} \quad [\text{AASHTO Eq. 9A.1-7}]$$

$$OR = \frac{OR'}{1 + \dfrac{\text{var}\left(\sum_{\text{all sites}} N_{\text{expected},A}\right)}{\left(\sum_{\text{all sites}} N_{\text{expected},A}\right)^2}} \quad [\text{AASHTO Eq. 9A.1-8}]$$

$$\text{safety effectiveness} = 100 \cdot (1 - OR) \quad [\text{AASHTO Eq. 9A.1-10}]$$

CETR Eq. 6.5 is an equivalent equation to AASHTO Eq. 9A.1-10 that incorporates the calculations for the odds ratio directly into the equation.

$$\text{effectiveness} = \frac{N_{\text{cr},b} - N_{\text{cr},a}}{N_{\text{cr},b}} \cdot 100\% \qquad 6.5$$

Crash Factors

AASHTO: Sec. 3.2.4

AASHTO Fig. 3-3 shows contributing factors for vehicle crashes. As it can be seen from the figure, human factors contribute far more to crashes than the other factors.

Crash Rates

CETR: Sec. 6.4

AASHTO: Sec. 3.4.1, Sec. 4.4.2.2

The *crash rate* is the ratio of the number of crashes to the exposure. *Exposure* is the number of vehicles that travel over a roadway in a period of time and/or that travel a specific distance. AASHTO Eq. 3-2 is the general equation for crash rate; it is equivalent to CETR Eq. 6.1.

$$\text{crash rate} = \frac{\text{average crash frequency in a period}}{\text{exposure in same period}} \quad [\text{AASHTO Eq. 3-2}]$$

$$R = \frac{N_{cr}}{\text{exposure}} \quad 6.1$$

AASHTO Eq. 4-2 finds the number of exposures per million entering vehicles. AASHTO Eq. 4-3 finds the crash rate.

$$\text{MEV} = \left(\frac{\text{TEV}}{1{,}000{,}000}\right) n(365) \quad [\text{AASHTO Eq. 4-2}]$$

$$R_i = \frac{N_{\text{observed},i(\text{total})}}{\text{MEV}_i} \quad [\text{AASHTO Eq. 4-3}]$$

G. NONMOTORIZED FACILITIES

Key concepts: These key concepts are important for answering exam questions in subtopic 9.G, Nonmotorized Facilities.

- bicycle facilities
- effective walkway width
- levels of service for shared-use paths
- pedestrian facilities
- pedestrian levels of service
- pedestrian platoons
- pedestrian unit flow rates
- pedestrian walking space requirements
- shared-use facilities
- uninterrupted and interrupted flow pedestrian facilities
- walking speed

NCEES Handbook: The *Handbook* does not contain any material relevant to this subtopic. Be sure to study the listed sections in CERM and CETR and from codes and standards.

PE Civil Reference Manual (**CERM**): Study these sections in CERM that either relate directly to this subtopic or provide background information.

- Section 73.24: Pedestrians and Walkways

Transportation Depth Reference Manual (**CETR**): Study these sections in CETR that either relate directly to this subtopic or provide background information.

- Section 3.1: Non-Motorized Facilities: Pedestrian

Codes and standards: To prepare for this subtopic, familiarize yourself with these sections from codes and standards.

Highway Capacity Manual (HCM)

- Section 4.3: Pedestrian Mode
- Section 18.4: Pedestrian Methodology
- Section 19.5: Pedestrian Methodology
- Section 24.2: Concepts
- Section 24.3: Core Methodologies

The following equations, figures, tables, and concepts are relevant for subtopic 9.G, Nonmotorized Facilities.

Pedestrian Walking Speed

CETR: Sec. 3.1

HCM: Sec. 4.3, Sec. 18.4

HCM Exh. 4-13 shows that younger people can walk faster than older people at an average of 1 ft/sec; CETR Fig. 3.1 is an equivalent figure. As such, per HCM Sec. 18.4, the average pedestrian walking speed for a given pedestrian flow is assumed to be 4.4 ft/sec if 20% or less of the pedestrians are elderly, and 3.3 ft/sec if more than 20% of the pedestrians are elderly.

The average walking speed is greater in terminals than in crosswalks or on street sidewalks. Terminal average walking speeds can be increased by 0.5 ft/sec (0.2 m/s) over similar conditions on general sidewalk connections.

HCM Exh. 4-14 shows the relationship between pedestrian speed and pedestrian density. CETR Fig. 3.5 is an equivalent figure.

Pedestrian Walking Space Requirements

CETR: Sec. 3.1

HCM: Sec. 4.3

For simplicity, a pedestrian is assumed to occupy an oval-shaped body ellipse that represents the minimum space required for the pedestrian to stand and move around without impediment. The elliptical space occupied by a pedestrian standing upright is assumed to be 24 in (0.6 m) wide at the shoulders and 20 in (0.5 m) deep front to back.

Figure 3.5 Relationship Between Pedestrian Speed and Density

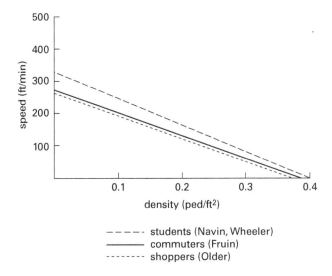

(Multiply ped/min-ft by 0.305 to obtain ped/min·m.)
(Multiply ft²/ped by 0.093 to obtain m²/ped.)

Used with permission from *Highway Capacity Manual*, 6th Edition: A Guide for Multimodal Mobility Analysis, 2016, Exhibit 4-14, by the Transportation Research Board of the National Academies of Sciences, Engineering, and Medicine, Washington DC. DOI: 10.17226/24798

HCM Exh. 4-12 shows an example of the body ellipse and pedestrian walking space requirements. CETR Fig. 3.2 is an equivalent figure. HCM Exh. 4-17 shows the relationship between pedestrian space and pedestrian speed.

Pedestrian Unit Flow Rate

CETR: Sec. 3.1

HCM: Sec. 4.3

The pedestrian unit flow rate is the number of pedestrians passing a point per unit of time, typically expressed as pedestrians per minute. The pedestrian unit flow rate is given by HCM Eq. 4.11; CETR Eq. 3.1 is an equivalent equation.

$$v_{\text{ped}} = S_{\text{ped}} D_{\text{ped}} \quad \text{[HCM Eq. 4-11]}$$

As the equation shows, the pedestrian unit flow rate in a given area is equal to the product of the average walking speed of pedestrians in that area and the pedestrian density in that same area.

Effective Walkway Width

CETR: Sec. 3.1

$$W_E = W_t - W_o \qquad 3.4$$

HCM: Sec. 18.4, Sec. 24.2

$$W_E = W_T - W_{O,i} - W_{O,o} - W_{s,i} - W_{s,o} \geq 0.0$$
[HCM Eq. 18-23]

HCM Eq. 18-23 is used to calculate the effective walkway width. As the equation shows, the effective width of a walkway is equal to the total width of the walkway minus the shy distances of any obstructions in the walkway. CETR Eq. 3.4 is an equivalent equation, with the various shy widths summed into one variable, W_o.

HCM Exh. 18-18 illustrates typical shy distances caused by objects placed in the total walkway width, and HCM Exh. 24-9 provides a list of effective widths of typical fixed objects. CETR Fig. 3.7 is an equivalent figure to HCM Exh. 18-18.

Exam tip: HCM does not provide a clear definition of which side of a walkway is the outside and which is the inside, but it can be inferred from the text that the curbside edge of the walkway is the outside and the edge farthest from the curb is the inside.

Figure 3.7 Width Adjustments for Fixed Objects

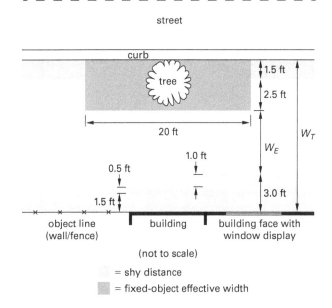

Used with permission from *Highway Capacity Manual*, 6th Edition: A Guide for Multimodal Mobility Analysis, 2016, Exhibit 18-18, by the Transportation Research Board of the National Academies of Sciences, Engineering, and Medicine, Washington DC. DOI: 10.17226/24798

Pedestrian Level of Service

CERM: Sec. 73.24

CETR: Sec. 3.1

HCM: Sec. 24.2

The level of service (LOS) for pedestrians in walkways, sidewalks, and queuing areas is categorized in much the same way as for freeway and highway vehicles. The primary criterion for determining pedestrian level of service is space per pedestrian (i.e., the inverse of density). This, in turn, affects the speed at which pedestrians can walk. Mean speed and flow rate are supplementary criteria. HCM Exh. 24-4 gives criteria for LOS on shared-use paths. CETR Table 3.13 is an equivalent table.

HCM Exh. 24-1 relates important parameters to the LOS. CERM Table 73.16 is an equivalent table.

Table 73.16 Pedestrian Level of Service on Walkways and Sidewalks[a]

LOS	pedestrian space (ft^2/ped)	average speed (ft/sec)	flow rate (ped/min-ft)[b]	volume-capacity (v/c) ratio[c]
A	> 60	> 4.25	≤ 5	≤ 0.21
B	> 40–60	> 4.17–4.25	> 5–7	> 0.21–0.31
C	> 24–40	> 4.00–4.17	> 7–10	> 0.31–0.44
D	> 15–24	> 3.75–4.0	> 10–15	> 0.44–0.65
E	> 8–15[d]	> 2.50–3.75	> 15–23	> 0.65–1.0
F	≤ 8[d]	≤ 2.50	variable	variable

(Multiply ft^2/ped by 0.0929 to obtain m^2/ped.)
(Multiply ft/sec by 0.3048 to obtain m/s.)
(Multiply ped/min-ft by 3.28 to obtain ped/min·m.)

[a]This table does not apply to walkways with steep grades (for example, grades >5%).
[b]Pedestrians per minute per foot width of walkway.
[c]v/c ratio = flow rate/23. LOS is based on average space per pedestrian.
[d]In cross-flow situations, the LOS E-F threshold is 13 ft^2/ped.
Used with permission from *Highway Capacity Manual*, 6th Edition: A Guide for Multimodal Mobility Analysis, 2016, Exhibit 24-1, by the Transportation Research Board of the National Academies of Sciences, Engineering, and Medicine, Washington DC. DOI: 10.17226/24798

Exam tip: HCM Exh. 24-1 and Exh. 24-2 should not be confused; they both provide LOS criteria for walkways, but Exh. 24-1 is based on random flow and Exh. 24-2 is platoon adjusted.

Pedestrian Platoons

CETR: Sec. 3.1

HCM: Sec. 24.2

Platoon flow occurs on narrow passageways of transportation terminals when large groups of people arrive on an airplane, train, or ferry and discharge along a boarding ramp or passageway. Faster walkers have less opportunity to pass slower walkers, and so form queues along the passageway. These queues eventually dissipate after the queue flow moves to a wider concourse or walkway.

The LOS for platoon flow is shown in HCM Exh. 24-2. CETR Table 3.5 is an equivalent table.

Exam tip: HCM Exh. 24-1 and Exh. 24-2 should not be confused; they both provide LOS criteria for walkways, but Exh. 24-1 is based on random flow and Exh. 24-2 is adjusted for platoon flow.

Uninterrupted Flow Pedestrian Facilities

CETR: Sec. 3.1

HCM: Sec. 24.3

Field observations of space are made by counting the number of pedestrians in a given area per unit of time. When calculating the pedestrian unit flow rate, a 15 min unit of time is often used to compensate for normal flow variations, as shown in HCM Eq. 24-3. CETR Eq. 3.5 is an equivalent equation.

$$v_p = \frac{v_{15}}{15 W_E} \quad \text{[HCM Eq. 24-3]}$$

Interrupted Flow Pedestrian Facilities

CETR: Sec. 3.1

HCM: Sec. 19.5

$$d_p = \frac{(C - g_{\text{walk,mi}})^2}{2C} \quad \text{[HCM Eq. 19-70]}$$

Pedestrians experience interrupted flow at signalized intersections. The delay experienced at the intersection is calculated from HCM Eq. 19-70. CETR Eq. 3.6 is an equivalent equation. C is the cycle length in seconds, and g is the effective green time in seconds. Pedestrians use both the WALK interval and the first few seconds of the flashing DON'T WALK (FDW) interval to enter the intersection, so for delay calculations, the effective green time interval is equal to the walk interval plus the first 4 sec of the FDW interval.

H. TRAFFIC FORECAST

Key concepts: These key concepts are important for answering exam questions in subtopic 9.H, Traffic Forecast.

- traffic forecasting

NCEES Handbook: To prepare for this subtopic, familiarize yourself with these sections in the *Handbook*.

- Traffic Forecast

The following equations are relevant for subtopic 9.H, Traffic Forecast.

Gravity Model

Handbook: Traffic Forecast

$$T'_{ab} = T'_a \frac{A_b f_{ab} K_{ab}}{\sum_{\forall b} A_b f_{ab} K_{ab}}$$

The *gravity model* is one of the models used to calculate the total number of trips from one location to another. The gravity model assumes that the trips beginning at a starting location and attracted to a specific destination are directly proportional to the total trips taken from that starting location and the total attractions at the destination.

I. HIGHWAY SAFETY ANALYSIS

Key concepts: These key concepts are important for answering exam questions in subtopic 9.I, Highway Safety Analysis.

- highway safety standards
- intersection safety
- railroad crossing safety
- roadway elements for safe design
- safe design practices
- traffic safety

NCEES Handbook: To prepare for this subtopic, familiarize yourself with these sections in the *Handbook*.

- Deflection Angles on a Simple Circular Curve

***PE Civil Reference Manual* (CERM):** Study these sections in CERM that either relate directly to this subtopic or provide background information.

- Section 79.1: Horizontal Curves

Transportation Depth Reference Manual (CETR): Study these sections in CETR that either relate directly to this subtopic or provide background information.

- Section 6.1: Traffic Safety and Roadway Design Analysis
- Section 6.3: Roadway Elements for Safe Design

Codes and standards: To prepare for this subtopic, familiarize yourself with these sections from codes and standards.

A Policy on Geometric Design of Highways and Streets (AASHTO *Green Book*)

- Section 3.2.2: Stopping Sight Distance
- Section 3.2.3: Decision Sight Distance
- Section 3.2.4: Passing Sight Distance for Two-Lane Highways
- Section 3.3.3.3: Minimum Radius
- Section 3.3.10.2: Application of Widening on Curves
- Section 3.3.11.2: Design Values
- Section 3.4.6.4: Sight Distance at Undercrossings
- Section 3.4.6.5: General Controls for Vertical Alignment
- Section 4.2.2: Cross Slope
- Section 4.3: Lane Widths
- Section 4.4: Shoulders
- Section 4.6.1: Clear Zones
- Section 4.11: Medians
- Section 4.20: On-Street Parking
- Section 5.2.2: Cross-Sectional Elements
- Section 6.2.6: Railroad-Highway Grade Crossings
- Section 9.12.4: Sight Distance
- Section 10.8.4: Underpass Roadways
- Section 10.8.5: Overpass Roadways
- Section 10.9.2: Three-Leg Designs
- Section 10.9.3: Four-Leg Designs
- Section 10.9.5.10: Auxiliary Lanes
- Section 10.9.6: Ramps

The following equations and concepts are relevant for subtopic 9.I, Highway Safety Analysis.

Curve Radius and Superelevation

Handbook: Deflection Angles on a Simple Circular Curve

$$R = \frac{5729.6}{D}$$

$$R = \frac{50}{\sin \frac{D}{2}}$$

CERM: Sec. 79.1

$$R = \frac{5729.578 \frac{\text{ft}}{\text{deg}}}{D} \quad \text{[U.S.; arc definition]} \qquad 79.1$$

$$R = \frac{50 \text{ ft}}{\sin \frac{D}{2}} \quad \text{[U.S.; chord definition]} \qquad 79.2$$

CETR: Sec. 6.3

AASHTO *Green Book*: Sec. 3.3.3.3

The radius of a curve is a function of the deflection angle, as shown in the equations in *Handbook* section **Deflection Angles on a Simple Circular Curve**. CERM Eq. 79.1 and CERM Eq. 79.2 are equivalent equations, with the primary difference being that the CERM equations include units for the constants.

The superelevation is the difference in heights of the inside and outside edges of a curve.

Roadway Elements for Safe Design

CETR: Sec. 6.3

AASHTO *Green Book*: Sec. 3.3.3.3, Sec. 3.3.10.2, Sec. 3.4.6.5, Sec. 5.2.2, Sec. 6.2.6, Sec. 10.9.5.10

The main roadway elements of concern when designing safe roadways are the curve radius, superelevation, sight distance, cross section elements, intersections, ramps, weaving and maneuvering sections, and railroad crossings. For more about each of these elements, see the entries "Curve Radius and Superelevation," "Curve Widening," "Sight Distance," "Cross Section Elements," "Ramps and Intersections," "Weaving and Maneuvering Sections," and "Railroad Crossing Safety" in this chapter.

All roadway features are selected based on the design speed, which is typically set for the 85th percentile driver. Safe design incorporates both design speed considerations and consistent design measures, such as roadway signage color, size, and shape.

The driver's eye is considered to be 3.5 ft from the existing grade elevation for profile applications. The minimum object height a driver is expected to see is 2.0 ft above ground.

Curve Widening

CETR: Sec. 6.3

AASHTO *Green Book*: Sec. 3.3.10.2

Curve widening, especially for two-lane roads, is necessary to account for off-tracking of larger vehicles, to minimize unnecessary slowing of traffic around curves, and to avoid collisions.

Sight Distance

AASHTO *Green Book*: Sec. 3.2.2, Sec. 3.2.3, Sec. 3.2.4, Sec. 3.4.6.4, Sec. 5.2.2.9

Sight distance must be maintained for the design speed, whether the minimum for stopping or a longer distance for passing and making complex decisions.

Cross Section Elements

CETR: Sec. 6.3

AASHTO *Green Book*: Sec. 4.2.2, Sec. 4.3, Sec. 4.4, Sec. 4.6.1, Sec. 4.11, Sec. 4.20

Cross section elements must be matched to the volume, design speed, and vehicle mix. The number of lanes, provisions for curbs and shoulders, and the width of the roadway must match at least the minimum requirements for the desired service level. Examples of cross section elements include roadside barricades, roadway signage, and guide rails. AASHTO *Green Book* Chap. 4 provides geometric design specifics.

Ramps and Intersections

CETR: Sec. 6.3

AASHTO *Green Book*: Sec. 10.8.4, Sec. 10.8.5, Sec. 10.9.2, Sec. 10.9.3, Sec. 10.9.6

Intersections and ramps must provide adequate sight distance for the design speed, provide sufficient room for vehicles to maneuver, and allow the expected volume of traffic to flow in a normal fashion without requiring drivers to make extraordinary moves. AASHTO *Green Book* Chap. 10 provides geometric design specifics.

Weaving and Maneuvering Sections

CETR: Sec. 6.3

Weaving and maneuvering sections are areas where one-way traffic streams cross at contiguous access points. These sections must be long enough to accommodate the expected flow volumes, especially on freeways and expressways. These sections experience higher traffic accident frequency than other roadway elements due to weaving maneuvers of vehicles traveling at different speeds.

Railroad Crossing Safety

AASHTO *Green Book*: Sec. 9.12.4

Sight distance is a primary consideration at crossings without train-activated warning devices. AASHTO *Green Book* Fig. 9-67 and Fig. 9-68 illustrate these scenarios.

36 Horizontal Design

Content in blue refers to the *NCEES Handbook*.

A. Basic Curve Elements 36-1
B. Sight Distance Considerations 36-3
C. Superelevation 36-6
D. Special Horizontal Curves 36-8

The knowledge area of Horizontal Design makes up between three and five questions out of the 80 questions on the PE Civil Transportation exam. The organization of this chapter follows the order of subtopics given by NCEES for this knowledge area. Each subtopic is covered in the following sections.

A. BASIC CURVE ELEMENTS

Key concepts: These key concepts are important for answering exam questions in subtopic 10.A, Basic Curve Elements.

- bearing, angles, and azimuth
- degree of curve
- parts of a circular curve (terminology)
- stationing on a horizontal curve

NCEES Handbook: To prepare for this subtopic, familiarize yourself with these sections in the *Handbook*.

- Basic Curve Elements
- Parts of a Circular Curve
- Arc and Chord Definitions for a Circular Curve
- Deflection Angles on a Simple Circular Curve
- Example Bearings
- Example Directions for Lines in the Four Quadrants

PE Civil Reference Manual (**CERM**): Study these sections in CERM that either relate directly to this subtopic or provide background information.

- Section 78.10: Units
- Section 78.23: Direction Specification
- Section 79.1: Horizontal Curves
- Section 79.2: Degree of Curve
- Section 79.3: Stationing on a Horizontal Curve

Transportation Depth Reference Manual (**CETR**): Study these sections in CETR that either relate directly to this subtopic or provide background information.

- Section 4.3: Specifying Line Direction
- Section 4.4: Horizontal Curves
- Section 4.5: Degree of Curve
- Section 4.6: Stationing on a Horizontal Curve

The following equations, figures, and tables are relevant for subtopic 10.A, Basic Curve Elements.

Elements of Horizontal Curves

Handbook: Basic Curve Elements; Parts of a Circular Curve

CERM: Sec. 79.1

CETR: Sec. 4.4

Horizontal curves are used to change from one direction to another. A horizontal circular curve (or simple curve) is a circular arc between two straight lines known as tangents. When traveling in a particular direction, the first tangent encountered is the back tangent (approach tangent), and the second tangent encountered is the forward tangent (departure or ahead tangent). These elements of a curve are illustrated in *Handbook* figure Parts of a Circular Curve, CERM Fig. 79.1, and CETR Fig. 4.3. Refer to *Handbook* section Basic Curve Elements or CERM Table 79.1 for the definitions of the variables. The major difference between the *Handbook* nomenclature and the nomenclature in CERM and CETR is the use of Δ instead of I to represent the intersection angle.

Figure 79.1 Horizontal Curve Elements

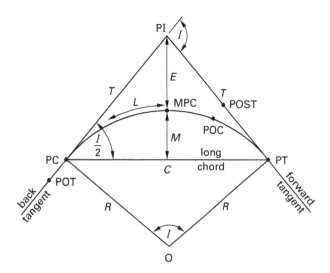

Figure 4.4 Degree of Curve Definition

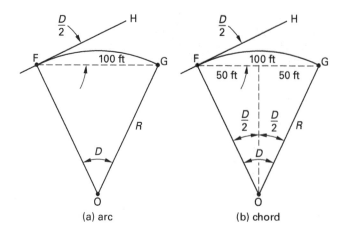

The middle ordinate and the external distance are used in calculating offsets from obstructions and are discussed in the entry "Middle Ordinate Distance" in Chap. 37 of this book.

CERM Eq. 79.8 can be used to calculate the degree of curve from the radius of the curve for the arc definition, and CERM Eq. 79.9 can be used to calculate the degree of curve from the radius for the chord definition. Both equations require that the calculation be made using customary U.S. units.

Degree of Curve

Handbook: Arc and Chord Definitions for a Circular Curve; Deflection Angles on a Simple Circular Curve

CERM: Sec. 79.2

CETR: Sec. 4.5

$$D = \frac{(360°)(100 \text{ ft})}{2\pi R} = \frac{5729.578 \text{ ft}}{R} \quad \text{[arc basis]} \quad 79.8$$

$$\sin \frac{D}{2} = \frac{50}{R} \quad \text{[chord basis]} \quad 79.9$$

In the United States, the curvature of a roadway can be specified by either the radius or the degree of curve. The degree of curve is defined by either the arc or the chord of a curve. Using the arc definition, the degree of curve is equal to the intersection angle divided by 100 ft along the arc of the curve; using the chord definition, the degree of curve is the intersection angle divided by 100 ft along the chord of the curve. The arc definition is commonly used with highways and streets, while the chord definition is used with railroads and was used on pre-interstate highways. These concepts are illustrated in *Handbook* figures Arc and Chord Definitions for a Circular Curve and Deflection Angles on a Simple Circular Curve, as well as CETR Fig. 4.4.

Bearing, Angles, and Azimuth

Handbook: Example Bearings; Example Directions for Lines in the Four Quadrants

CERM: Sec. 78.10, Sec. 78.23

CETR: Sec. 4.3

The direction of any line can be specified by an angle from a principal meridian. The azimuth of a line is given as a clockwise angle from the reference direction, either from the north or from the south. An azimuth never exceeds 360°.

The bearing of a line is referenced to the quadrant in which the line falls and the angle that line makes with the meridian in that quadrant. The bearing will specify the two cardinal directions (i.e., north or south and east or west) that define the quadrant in which the line is found; the north or south direction is specified first (e.g., N 45°E). The angle of a bearing never exceeds 90°. These concepts are illustrated in *Handbook* figure Example Bearings. *Handbook* table Example Directions for Lines in the Four Quadrants provides formulas for calculating the bearing of a line from the azimuth, depending on which quadrant the line is in.

Exam tip: When measuring the angle between two tangent lines given in bearings, make sure to pay attention to which quadrant the bearings are in (northwest, northeast, southwest, or southeast).

Stationing on a Horizontal Curve

CERM: Sec. 79.3

CETR: Sec. 4.6

$$\text{sta PT} = \text{sta PC} + L \quad \quad 79.11$$

$$\text{sta PC} = \text{sta PI} - T \quad \quad 79.12$$

In route surveying, stationing is carried ahead continuously from a starting point designated as station 0+00. The term *station* is applied to each subsequent 100 ft (or 100 m) length. Thus, sta 8+33.20 is a unique point 833.2 ft (or 833.2 m) from the starting point as measured along the survey line. Moving or looking toward increasing stations is called *ahead stationing*. Moving or looking toward decreasing stations is called *back stationing*. Offsets from the centerline are either left or right looking ahead on stationing.

The first point of the curve is the *point of curvature* (PC) or *beginning of curvature* (BC), which can be found using CERM Eq. 79.12 or CETR Eq. 4.25. Stationing is carried ahead along the arc of the curve to the endpoint of the curve, called the *point of tangency* (PT) or *end of curvature* (EC), which can be found from CERM Eq. 79.11 or CETR Eq. 4.24. PI is the *point of intersection*.

Exam tip: When given a problem regarding horizontal curvature, first solve for any missing information (radius, tangent length, curve length, etc.), and then use that information to solve for PC, PT, and PI as needed.

B. SIGHT DISTANCE CONSIDERATIONS

Key concepts: These key concepts are important for answering exam questions in subtopic 10.B, Sight Distance Considerations.

- braking distance
- decision sight distance
- horizontal sightline offsets
- intersection sight distance triangles
- passing sight distance
- perception-reaction distance
- stopping sight distance

NCEES Handbook: The *Handbook* does not contain any material relevant to this subtopic. Be sure to study the listed sections in CERM and CETR and from codes and standards.

PE Civil Reference Manual (CERM): Study these sections in CERM that either relate directly to this subtopic or provide background information.

- Section 79.14: Stopping Sight Distance
- Section 79.15: Passing Sight Distance
- Section 79.16: Minimum Horizontal Curve Length for Stopping Distance

Transportation Depth Reference Manual (CETR): Study these sections in CETR that either relate directly to this subtopic or provide background information.

- Section 4.11: Sight Distance
- Section 4.13: Vertical and Horizontal Clearances
- Section 4.15: Intersections and Interchanges

Codes and standards: To prepare for this subtopic, familiarize yourself with these sections from codes and standards.

A Policy on Geometric Design of Highways and Streets (AASHTO *Green Book*)

- Section 3.2.2.1: Brake Reaction Time
- Section 3.2.2.2: Braking Distance
- Section 3.2.2.3: Design Values
- Section 3.2.3: Decision Sight Distance
- Section 3.2.4.2: Design Values
- Section 3.3.12.1: Stopping Sight Distance
- Section 9.5.3: Intersection Control

The following equations, figures, tables, and concepts are relevant for subtopic 10.B, Sight Distance Considerations.

Stopping Sight Distance

CERM: Sec. 79.14

CETR: Sec. 4.11

AASHTO *Green Book*: Sec. 3.2.2.1, Sec. 3.3.12.1

Sight distance is the length of roadway a driver can see ahead of the vehicle. *Stopping sight distance* is the total distance required for a driver traveling at a design speed to stop a vehicle before reaching an object in its path. The stopping sight distance comprises two distances: the *perception-reaction distance* (the distance traveled at a constant approach speed while the driver's brain is reacting to and processing the obstruction) and the *braking distance* (the distance traveled after the driver has begun to apply an avoidance or stopping maneuver due to vehicle and sight conditions). Refer to AASHTO

Green Book Table 3-1 for stopping sight distances on level roadways and Table 3-2 for stopping site distances on grades.

> *Exam tip*: The AASHTO *Green Book* uses an average perception-reaction time of 2.5 sec, a driver's eye height of 3.5 ft, and an obstruction height of 2.0 ft for stopping sight distance calculations. If you are given no other information, these are the values you should use.

Braking Distance

CETR: Sec. 4.11

$$d_{\text{m}} = 0.039\left(\frac{v_{\text{kph}}^2}{a_{\text{m/s}^2}}\right) \quad \text{[SI]} \quad 4.52(a)$$

$$d_{\text{ft}} = 1.075\left(\frac{v_{\text{mph}}^2}{a_{\text{ft/sec}^2}}\right) \quad \text{[U.S.]} \quad 4.52(b)$$

$$d_{\text{m}} = \frac{v_{\text{kph}}^2}{254\left(\left(\frac{a_{\text{m/s}^2}}{g}\right) \pm G_{\%/100\,\text{m}}\right)} \quad \text{[SI]} \quad 4.53(a)$$

$$= \frac{v_{\text{kph}}^2}{254(f \pm G_{\%/100\,\text{m}})}$$

$$d_{\text{ft}} = \frac{v_{\text{mph}}^2}{30\left(\left(\frac{a_{\text{ft/sec}^2}}{g}\right) \pm G_{\%/100\,\text{ft}}\right)} \quad \text{[U.S.]} \quad 4.53(b)$$

$$= \frac{v_{\text{mph}}^2}{30(f \pm G_{\%/100\,\text{ft}})}$$

AASHTO *Green Book*: Sec. 3.2.2.2, Sec. 3.2.2.3

$$d_B = 0.039\left(\frac{V^2}{a}\right) \quad \text{[AASHTO Eq. 3-1, SI]}$$

$$d_B = 1.075\left(\frac{V^2}{a}\right) \quad \text{[AASHTO Eq. 3-1, U.S.]}$$

$$d_B = \frac{V^2}{254\left(\left(\frac{a}{9.81}\right) \pm G\right)} \quad \text{[AASHTO Eq. 3-3, SI]}$$

$$d_B = \frac{V^2}{30\left(\left(\frac{a}{32.2}\right) \pm G\right)} \quad \text{[AASHTO Eq. 3-3, U.S.]}$$

The braking distance required to bring a vehicle to a complete stop is determined from the design speed, the deceleration rate, and the grade of the roadway. The general braking distance can be calculated using AASHTO *Green Book* Eq. 3-1, and the braking distance for roadways on a grade can be calculated using AASHTO *Green Book* Eq. 3-3. CETR Eq. 4.52 and Eq. 4.53 are equivalent equations.

For braking distance (in ft or m) and design speed (in mi/hr or km/h), respectively, the AASHTO equations use d_B and V, and the CETR equations use d_{m} or d_{ft} and v. a is deceleration (in ft/sec^2 or m/s^2), G is grade (in ft/ft or m/m), and g is gravitational acceleration (32.2 ft/sec^2 or 9.81 m/s^2). The coefficient of friction, f, is equal to a/g and can be substituted for it, as shown in the CETR equations.

Per AASHTO *Green Book* Sec. 3.2.2.2, the maximum comfortable deceleration rate is 11.2 ft/sec^2 (3.4 m/s^2).

Total Stopping Sight Distance

CERM: Sec. 79.14

CETR: Sec. 4.11

$$S_{\text{m}} = \left(0.278\,\frac{\frac{\text{m}}{\text{s}}}{\frac{\text{km}}{\text{h}}}\right)v_{\text{kph}}t_{p,\text{s}} + 0.039\left(\frac{v_{\text{kph}}^2}{a_{\text{m/s}^2}}\right) \quad \text{[SI]} \quad 4.54(a)$$

$$S_{\text{ft}} = \left(1.47\,\frac{\frac{\text{ft}}{\text{sec}}}{\frac{\text{mi}}{\text{hr}}}\right)v_{\text{mph}}t_{p,\text{sec}} + 1.075\left(\frac{v_{\text{mph}}^2}{a_{\text{ft/sec}^2}}\right) \quad \text{[U.S.]} \quad 4.54(b)$$

AASHTO *Green Book*: Sec. 3.2.2.3

$$\text{SSD} = 0.278\,Vt + 0.039\left(\frac{V^2}{a}\right)$$

[AASHTO Eq. 3-2, SI]

$$\text{SSD} = 1.47\,Vt + 1.075\left(\frac{V^2}{a}\right)$$

[AASHTO Eq. 3-2, U.S.]

As explained in AASHTO *Green Book* Sec. 3.2.2.3 and shown in AASHTO *Green Book* Eq. 3-2 and CETR Eq. 4.54, the total stopping sight distance is the sum of the distance traveled during the perception-reaction time and the braking distance.

For stopping sight distance (in feet or meters) and design speed (in mi/hr or km/h), respectively, the AASHTO equations use SSD and V, and the CETR equations use S and v. t is brake reaction time (2.5 sec), and a is deceleration rate (in ft/sec^2 or m/s^2).

Decision Sight Distance

CERM: Sec. 79.16

CETR: Sec. 4.11

$$d_m = \left(0.278 \, \frac{\frac{m}{s}}{\frac{km}{h}}\right) v_{kph} t_{p,s} + 0.039 \left(\frac{v_{kph}^2}{a_{m/s^2}}\right) \quad [SI] \quad 4.55(a)$$

$$d_{ft} = \left(1.47 \, \frac{\frac{ft}{sec}}{\frac{mi}{hr}}\right) v_{mph} t_{p,sec} + 1.075 \left(\frac{v_{mph}^2}{a_{ft/sec^2}}\right) \quad [U.S.] \quad 4.55(b)$$

$$d_m = 0.278 v_{kph} t_{t,s} \quad [SI] \quad 4.56(a)$$

$$d_{ft} = 1.47 v_{mph} t_{t,sec} \quad [U.S.] \quad 4.56(b)$$

AASHTO *Green Book*: Sec. 3.2.3

$$DSD = 0.278 Vt + 0.039 \left(\frac{V^2}{a}\right)$$
[AASHTO Eq. 3-4, SI]

$$DSD = 1.47 Vt + 1.075 \left(\frac{V^2}{a}\right)$$
[AASHTO Eq. 3-4, U.S.]

Decision sight distance should be considered where hazards exist that require drivers to make decisions regarding maneuvers other than a stop, such as lane changes or exit ramp selections. Avoidance maneuvers are classified as maneuvers A through E, as defined in AASHTO *Green Book* Table 3-3, and decision sight distances for those maneuvers can often be found from that table. CETR Table 4.5 is an equivalent table.

If a decision sight distance cannot be found from the table, AASHTO *Green Book* Eq. 3-4 can be used to find the distance for avoidance maneuvers A and B, and AASHTO *Green Book* Eq. 3-5 can be used to find the distance for avoidance maneuvers C, D, and E. CETR Eq. 4.55 is equivalent to AASHTO *Green Book* Eq. 3-4, and CETR Eq. 4.56 is equivalent to AASHTO *Green Book* Eq. 3-5.

For decision sight distance (in ft or m) and design speed (in mi/hr or km/h), respectively, the AASHTO equations use DSD and V, and the CETR equations use d and v. t is pre-maneuver time (in seconds), and a is deceleration rate (in ft/sec^2 or m/s^2).

Passing Sight Distance

CERM: Sec. 79.15

CETR: Sec. 4.11

AASHTO *Green Book*: Sec. 3.2.4.2

Passing sight distance is applicable on two-lane highways where there are sufficient gaps in opposing flows to allow passing maneuvers to occur, as well as few access points with sufficient sight distance to observe traffic in both lanes and both directions. Passing sight distances for the design of two-lane highways can be found from AASHTO *Green Book* Table 3-4. Passing sight distance is not applicable to multilane highways.

Clearances on Horizontal Curves

CERM: Sec. 79.16

CETR: Sec. 4.13

AASHTO *Green Book*: Sec. 3.3.12.1

$$HSO = R\left(1 - \cos\left(\frac{28.65 S}{R}\right)\right) \quad [AASHTO \text{ Eq. 3-37}]$$

For horizontal curves, the sight distance typically refers to the driver's line of sight around the horizontal curve without obstructions such as trees, signs, cut slopes, or buildings. Sight distances on the inside of a curve can be equated to the chord of a circular arc centered at the point of obstruction closest to the edge of the traveled lane. For analysis, the radius of the centerline of the extreme right lane must be used. CETR Fig. 4.31 illustrates these concepts.

Figure 4.31 Horizontal Sight Distance Components on the Inside of a Curve

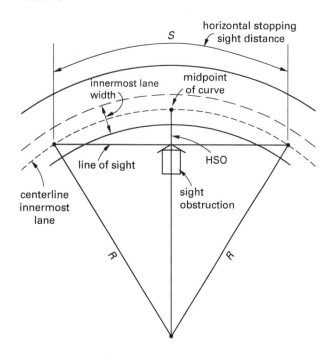

The horizontal sightline offset (HSO) can be found from AASHTO *Green Book* Eq. 3-37. CERM Eq. 79.45 and CETR Eq. 4.85 are equivalent equations.

Intersection Sight Distance

CETR: Sec. 4.15

AASHTO *Green Book*: Sec. 9.5.3

$$\text{ISD} = 0.278 V_{\text{major}} t_g \quad [\text{AASHTO Eq. 9-1, SI}]$$
$$\text{ISD} = 1.47 V_{\text{major}} t_g \quad [\text{AASHTO Eq. 9-1, U.S.}]$$

Intersection and driveway approaches to a traveled roadway should be clear of obstructions so that approaching drivers are able to see oncoming traffic. The sightline needed, referred to as the *intersection sight distance* (ISD), is determined by traffic control and approach speeds on each leg of the intersection; it can be found from AASHTO *Green Book* Eq. 9-1. CETR Eq. 4.86 is an equivalent equation.

The length of sight triangles for an intersection varies with the traffic control measures used. The AASHTO *Green Book* classifies intersection layouts and controls using letters A through G.

- Case A is for intersections with no control; it is covered in Sec. 9.5.3.1.
- Case B is for intersections with stop control on the minor road; it is covered in Sec. 9.5.3.2.
- Case C is for intersections with yield control on the minor road; it is covered in Sec. 9.5.3.3.
- Case D is for intersections with traffic signal control; it is covered in Sec. 9.5.3.4.
- Case E is for intersections with all-way stop control; it is covered in Sec. 9.5.3.5.
- Case F is for left turns from the major road; it is covered in Sec. 9.5.3.6.
- Case G is for roundabouts; it is covered in Sec. 9.5.3.7.

C. SUPERELEVATION

Key concepts: These key concepts are important for answering exam questions in subtopic 10.C, Superelevation.

- average running speed
- design speed
- distribution of superelevation and side friction for curve design
- lateral forces on a vehicle
- minimum runoff length
- minimum tangent runout length
- rate of superelevation
- side friction factor
- superelevation axes of rotation
- superelevation design for low-speed urban streets
- superelevation transitions

NCEES Handbook: The *Handbook* does not contain any material relevant to this subtopic. Be sure to study the listed sections in CERM and CETR and from codes and standards.

PE Civil Reference Manual **(CERM):** Study these sections in CERM that either relate directly to this subtopic or provide background information.

- Section 79.10: Superelevation
- Section 79.12: Transitions to Superelevation

Transportation Depth Reference Manual **(CETR):** Study these sections in CETR that either relate directly to this subtopic or provide background information.

- Section 4.12: Superelevation

Codes and standards: To prepare for this subtopic, familiarize yourself with these sections from codes and standards.

A Policy on Geometric Design of Highways and Streets (AASHTO *Green Book*)

- Section 3.3.1: Theoretical Considerations
- Section 3.3.2.2: Side Friction Factor
- Section 3.3.8.2.1: Minimum Length of Superelevation Runoff
- Section 3.3.8.2.2: Minimum Length of Tangent Runout

The following equations, figures, and concepts are relevant for subtopic 10.C, Superelevation.

Superelevation Design

CERM: Sec. 79.10

CETR: Sec. 4.12

$$e = \tan\phi = \frac{v^2}{gR} \quad\quad 4.61$$

AASHTO *Green Book*: Sec. 3.3.1

$$\frac{0.01e + f}{1 - 0.01ef} = \frac{v^2}{gR} \quad \text{[AASHTO Eq. 3-6]}$$

The difference between a curve's inside and outside elevations is its *superelevation*. A roadway is superelevated to resist the centrifugal force acting on a vehicle as it rounds a curve, which allows the driver to comfortably maneuver the curve at a higher speed.

As shown by CETR Eq. 4.61, the superelevation is related to the angle of the slope (also called the *curve banking* or *cross slope*), the design speed, and the curve radius. This equation is equivalent to AASHTO *Green Book* Eq. 3-6 when the superelevation and the design speed are balanced so that the friction factor, f, equals zero. In these equations, design speed (v in CETR and v in the *Green Book*) must be in ft/sec or m/s.

CETR Fig. 4.23 shows the forces acting on a vehicle while rounding a curve and illustrates the parameters that must be considered when determining superelevation.

Figure 4.23 Forces Acting on a Vehicle While Rounding a Curve

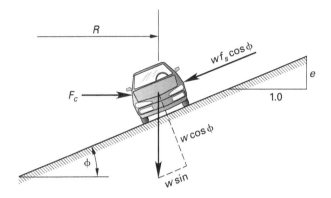

w = weight of vehicle
f_s = coefficient of side friction
F_c = centrifugal force
R = radius of curve
ϕ = angle of cross slope
e = rate of superelevation, $\tan\phi$

The proper curve radius for a given vehicle speed, superelevation, and side friction factor can be found from AASHTO *Green Book* Eq. 3-10.

Side Friction

CETR: Sec. 4.12

AASHTO *Green Book*: Sec. 3.3.2.2

The side friction factor is dependent on the lateral traction available where a vehicle's tires meet the roadway. The friction factor reduces as speed increases, and varies with tire tread design, tire wear, and roadway surface conditions. Graphs of side friction factors for a range of design speeds can be found in AASHTO *Green Book* Fig. 3-3. CETR Table 4.10 tabulates these side friction factors, as well.

Minimum Runoff Length

CERM: Sec. 79.12

CETR: Sec. 4.12

AASHTO *Green Book*: Sec. 3.3.8.2.1

$$L_r = \left(\frac{(wn_1)e_d}{\Delta}\right)b_w \quad \text{[AASHTO Eq. 3-23]}$$

The distance required to rotate from the point of no adverse crown to full superelevation is called the *superelevation transition length* or the *length of runoff*. This length is determined by the design speed of the roadway and the change in slope required to achieve the full superelevation rate. AASHTO *Green Book* Eq. 3-23 can be used to determine the minimum runoff length. CETR Eq. 4.67 is an equivalent equation.

Minimum Tangent Runout Length

CETR: Sec. 4.12

AASHTO *Green Book*: Sec. 3.3.8.2.2

$$L_t = \left(\frac{e_{\text{NC}}}{e_d}\right) L_r \quad \text{[AASHTO Eq. 3-24]}$$

Tangent runout is the section of a roadway over which the adverse crown is removed. In order to achieve a smooth transition, the rate at which the adverse crown is removed should equal the maximum relative gradient used to determine the minimum runoff length. The minimum length of tangent runout can be found using AASHTO *Green Book* Eq. 3-24. CETR Eq. 4.70 is an equivalent equation.

D. SPECIAL HORIZONTAL CURVES

Key concepts: These key concepts are important for answering exam questions in subtopic 10.D, Special Horizontal Curves.

- spiral curves
- three-centered compound curves
- two-centered compound curves

NCEES Handbook: To prepare for this subtopic, familiarize yourself with these sections in the *Handbook*.

- Two-Centered Compound Curve
- Three-Centered Compound Curve

PE Civil Reference Manual **(CERM):** Study these sections in CERM that either relate directly to this subtopic or provide background information.

- Section 79.9: Compound Horizontal Curves
- Section 79.26: Spiral Curves

Transportation Depth Reference Manual **(CETR):** Study these sections in CETR that either relate directly to this subtopic or provide background information.

- Section 4.8: Spiral Curves

Codes and standards: To prepare for this subtopic, familiarize yourself with these sections from codes and standards.

A Policy on Geometric Design of Highways and Streets (AASHTO *Green Book*)

- Section 3.3.8.4: Length of Spiral

The following equations and figures are relevant for subtopic 10.D, Special Horizontal Curves.

Compound Horizontal Curves

Handbook: Two-Centered Compound Curve; Three-Centered Compound Curve

CERM: Sec. 79.9

A compound curve comprises two or more curves of different radii that share a common tangent point, with their centers on the same side of the common tangent. The point of tangent (PT) for the first curve and the point of curve (PC) for the second curve coincide. This is referred to as the point of continuing curve (PCC). Compound curves are typically reserved for applications where design constraints, such as topography or high land cost, prevent the use of a circular or spiral curve. *Handbook* figures Two-Centered Compound Curve and Three-Centered Compound Curve and CERM Fig. 79.7 illustrate compound curves.

Figure 79.7 Compound Circular Curve

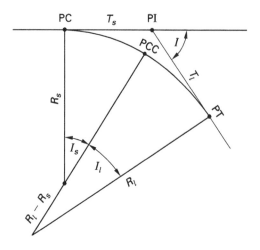

Spiral Curves

CERM: Sec. 79.26

CETR: Sec. 4.8

AASHTO *Green Book*: Sec. 3.3.8.4

$$L_{s,\min} = \sqrt{24 p_{\min} R} \quad \text{[AASHTO Eq. 3-27]}$$

$$L_{s,\min} = \frac{0.0214 V^3}{RC} \quad \text{[AASHTO Eq. 3-28, SI]}$$

$$L_{s,\min} = \frac{3.15 V^3}{RC} \quad \text{[AASHTO Eq. 3-28, U.S.]}$$

$$L_{s,\max} = \sqrt{24 p_{\max} R} \quad \text{[AASHTO Eq. 3-29]}$$

Spiral curves are introduced at the ends of a circular curve (i.e., at the point of curvature and/or point of tangent) in order to provide a transition, or easement, between the straight tangents and the circular parts of the curve. The degree of curve of the spiral increases linearly along the spiral transition from zero at the tangent section to the degree of curve of the circular curve. CETR Fig. 4.15 illustrates a spiral curve.

Figure 4.15 Fully Spiraled Circular Curve Layout

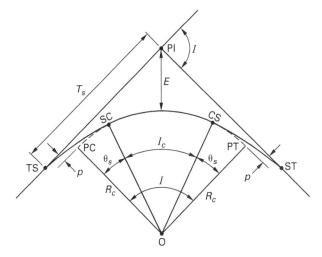

The minimum length of a spiral curve is based on driver comfort, so the minimum length of a spiral should be the larger of the values found using AASHTO *Green Book* Eq. 3-27 and Eq. 3-28. C is the rate of increase of lateral (centripetal) acceleration and is selected by judgment; transportation agencies use values ranging from 1 ft/sec^3 to 4 ft/sec^3. CERM Eq. 79.62(b) is an equivalent equation.

Spiral curves must also not be too long in relation to the length of the circular curve. The maximum length of a spiral curve can be found from AASHTO *Green Book* Eq. 3-29. CETR Eq. 4.35 is an equivalent equation.

AASHTO *Green Book* Sec. 3.3.8.4.4 recommends a maximum lateral offset of 3.3 ft. AASHTO *Green Book* Table 3-19 gives desirable lengths of spiral curve transitions.

37 Vertical Design

Content in blue refers to the *NCEES Handbook*.

A. Vertical Curve Geometry.................................. 37-1
B. Stopping and Passing Sight Distance 37-3
C. Vertical Clearance... 37-5

The knowledge area of Vertical Design makes up between three and five questions out of the 80 questions on the PE Civil Transportation exam. The organization of this chapter follows the order of subtopics given by NCEES for this knowledge area. Each subtopic is covered in the following sections.

A. VERTICAL CURVE GEOMETRY

Key concepts: These key concepts are important for answering exam questions in subtopic 11.A, Vertical Curve Geometry.

- crest curves
- location along vertical curves
- nomenclature for points along a vertical curve
- sag curves
- symmetrical and unsymmetrical vertical curves
- vertical curve formulas

NCEES Handbook: To prepare for this subtopic, familiarize yourself with these sections in the *Handbook*.

- Symmetrical Vertical Curve Formula
- Vertical Curve Formulas

PE Civil Reference Manual (**CERM**): Study these sections in CERM that either relate directly to this subtopic or provide background information.

- Section 79.17: Vertical Curves
- Section 79.18: Vertical Curves Through Points
- Section 79.19: Vertical Curve to Pass Through Turning Point
- Section 79.25: Unequal Tangent (Unsymmetrical) Vertical Curves

Transportation Depth Reference Manual (**CETR**): Study these sections in CETR that either relate directly to this subtopic or provide background information.

- Section 4.9: Vertical Curves

The following equations and figures are relevant for subtopic 11.A, Vertical Curve Geometry.

Vertical Curves, Sag Curves, and Crest Curves

Handbook: Symmetrical Vertical Curve Formula; Vertical Curve Formulas

CETR: Sec. 4.9

Vertical curves are used to connect two vertical tangents in order to change the grade of a highway. Vertical curves may be sag curves or crest curves.

Sag curves have a concave upward shape. They are most often used to transition from negative to positive vertical grades, but they can also be used to transition from positive to positive or from negative to negative, as shown in CETR Fig. 4.17.

Figure 4.17 Three Conditions of Sag Vertical Curves

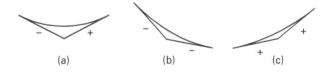

Crest curves have a concave downward shape. They are most often used to transition from positive to negative vertical grades, but they can also be used to transition from positive to positive or from negative to negative, as shown in CETR Fig. 4.18.

Figure 4.18 Three Conditions of Crest Vertical Curves

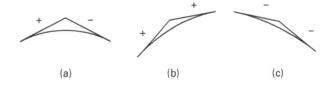

Handbook section Symmetrical Vertical Curve Formula provides a list of common nomenclature terms and definitions for vertical curve formulas. *Handbook* figure Vertical Curve Formulas illustrates the locations of several of these labels on a vertical curve.

Exam tip: The biggest pitfall in solving vertical curve problems is not being familiar with the nomenclature.

Rate of Grade Change per Station

Handbook: Symmetrical Vertical Curve Formula

CERM: Sec. 79.17

CETR: Sec. 4.9

$$r = \frac{g_2 - g_1}{L}$$

A vertical parabolic curve is defined by two grades and a curve length. The rate of grade change per station, r, is a measure of how quickly the curve changes grade, and it can be calculated from the same data, as shown by the *Handbook* equation. r has units of %/sta, which are equivalent to ft/sta^2.

Curve Elevation at Any Point on a Vertical Curve

Handbook: Symmetrical Vertical Curve Formula; Vertical Curve Formulas

CERM: Sec. 79.17

CETR: Sec. 4.9

$$\begin{aligned} \text{curve elevation} &= Y_{\text{PVC}} + g_1 x + a x^2 \\ &= Y_{\text{PVC}} + g_1 x + x^2 \left(\frac{g_2 - g_1}{2L} \right) \end{aligned}$$

The elevation at any point along a curve can be found if the PVC elevation, incoming grade, and rate of change of the curve are known. The *Handbook* equation is the quadratic equation for the parabola of the curve. CERM Eq. 79.47 is similar.

The first term, Y_{PVC}, is the starting elevation. The second term represents a projection along the incoming grade to a point off the curve. The third term is the vertical adjustment back onto the curve. x is the distance to any point on the curve, measured in stations beyond the beginning of the vertical curve (BVC).

Elevation is measured in feet. If measuring the grade as a decimal (e.g., 0.03 rather than 3%), then measure the distance, x, in feet instead of stations. Use the same datum (or reference point) to measure all elevations.

Handbook section Symmetrical Vertical Curve Formula provides a list of common nomenclature terms and definitions for vertical curve formulas. *Handbook* figure Vertical Curve Formulas illustrates the locations of several of these labels on a vertical curve.

Turning Point

Handbook: Symmetrical Vertical Curve Formula; Vertical Curve Formulas

CERM: Sec. 79.17

CETR: Sec. 4.9

$$x_m = -\frac{g_1}{2a} = \frac{g_1 L}{g_1 - g_2}$$

The maximum or minimum elevation along a curve occurs where the slope is equal to zero. This turning point can be found using the *Handbook* equation. CETR Eq. 4.51 is similar.

$$x_{\text{turning point,sta}} = \frac{-G_{1,\%}}{R_{\%/\text{sta}}} \quad 4.51$$

x_m or $x_{\text{turning point}}$ is the horizontal distance from the point of vertical curvature (PVC). The elevation of the turning point can be found using either the rate of grade change or the proportional offset method.

As shown in the following illustration, the turning point of the curve does not necessarily line up with the intersection point between the incoming and outgoing grades. The turning point is the x-coordinate where the curve changes direction.

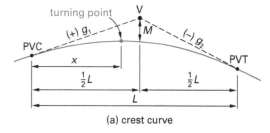

(a) crest curve

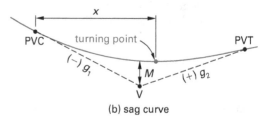

(b) sag curve

The equation for the turning point is particularly useful for sag vertical curves because the turning point is the lowest point on a sag curve, as well as the location at which catch basins should be installed for drainage.

Handbook section Symmetrical Vertical Curve Formula provides a list of common nomenclature terms and definitions for vertical curve formulas. *Handbook* figure Vertical Curve Formulas illustrates the locations of several of these labels on a vertical curve.

Middle Ordinate Distance

Handbook: Symmetrical Vertical Curve Formula

$$E = \frac{AL}{800}$$

CERM: Sec. 79.17

$$M_{\text{ft}} = \frac{AL_{\text{sta}}}{8} \qquad 79.49$$

The *Handbook* equation and CERM Eq. 79.49 are fundamentally the same, with only minor differences in nomenclature. The *Handbook* equation measures curve length, L, in feet, while the CERM equation measures L in stations. The conversion from length units of stations to feet is 100 ft = 1 sta. The middle ordinate distance occurs at the station where the two tangents intersect. The middle ordinate distance is not the same as the turning point; the turning point occurs at the middle ordinate station only if the approaching and departing grades are the same (i.e., $g_1 = g_2$).

B. STOPPING AND PASSING SIGHT DISTANCE

Key concepts: These key concepts are important for answering exam questions in subtopic 11.B, Stopping and Passing Sight Distance.

- height of drivers and obstructions
- K-value method
- passing sight distance
- stopping sight distance
- vertical curve length criteria
- vertical curve length for both sag and crest curves

NCEES Handbook: To prepare for this subtopic, familiarize yourself with these sections in the *Handbook*.

- Symmetrical Vertical Curve Formula

PE Civil Reference Manual (**CERM**): Study these sections in CERM that either relate directly to this subtopic or provide background information.

- Section 79.14: Stopping Sight Distance
- Section 79.15: Passing Sight Distance

- Section 79.20: Minimum Vertical Curve Length for Sight Distances (Crest Curves)
- Section 79.21: Design of Crest Curves Using K-Value
- Section 79.22: Minimum Vertical Curve Length for Headlight Sight Distance: Sag Curves
- Section 79.23: Minimum Vertical Curve Length for Comfort: Sag Curves
- Section 79.24: Design of Sag Curves Using K-Value

Transportation Depth Reference Manual (**CETR**): Study these sections in CETR that either relate directly to this subtopic or provide background information.

- Section 4.11: Sight Distance

Codes and standards: To prepare for this subtopic, familiarize yourself with these sections from codes and standards.

A Policy on Geometric Design of Highways and Streets (AASHTO *Green Book*)

- Section 3.2.2: Stopping Sight Distance
- Section 3.2.4: Passing Sight Distance for Two-Lane Highways
- Section 3.2.6.1: Height of Driver's Eye
- Section 3.4.5: Emergency Escape Ramps
- Section 3.4.6.2: Crest Vertical Curves
- Section 3.4.6.3: Sag Vertical Curves

The following equations, tables, and concepts are relevant for subtopic 11.B, Stopping and Passing Sight Distance.

Stopping Sight Distance

CETR: Sec. 4.11

AASHTO *Green Book*: Sec. 3.2.2

Stopping sight distance is the total distance required for a driver traveling at design speed to stop a vehicle before reaching an object in its path. Stopping sight distance calculations are determined using the driver's eye height set at 3.5 ft (1080 mm) and the object height at 2.0 ft (600 mm), which is equivalent to the height of a passenger car's taillight.

Typical stopping sight distance values for various design speeds on level roadways and on grades are given in AASHTO *Green Book* Table 3-1 and Table 3-2, respectively. CETR Table 4.3 and Table 4.4 are equivalent tables.

When the needed design speed cannot be found in these tables, use AASHTO *Green Book* Eq. 3-2 to find the total stopping sight distance needed to bring the vehicle to a complete stop on a level roadway, given the initial design speed. In this equation, the first term is the *pre-braking* distance (the distance traveled as the driver is reacting to the perceived need to stop but has not yet fully applied the brakes to the vehicle in motion), and the second term (the term with V^2/a) is the *braking distance* (the distance traveled after braking has been applied). CETR Eq. 4.54 is a similar equation using slightly different nomenclature.

$$\text{SSD} = 0.278\,Vt + 0.039\left(\frac{V^2}{a}\right)$$

[AASHTO Eq. 3-2, SI]

$$\text{SSD} = 1.47\,Vt + 1.075\left(\frac{V^2}{a}\right)$$

[AASHTO Eq. 3-2, U.S.]

When calculating the total stopping sight distance on a grade, replace the second term (the braking distance on a level roadway) with the result of AASHTO *Green Book* Eq. 3-3, which gives the braking distance on a grade.

$$d_B = \frac{V^2}{254\left(\left(\frac{a}{9.81}\right) \pm G\right)} \quad \text{[AASHTO Eq. 3-3, SI]}$$

$$d_B = \frac{V^2}{30\left(\left(\frac{a}{32.2}\right) \pm G\right)} \quad \text{[AASHTO Eq. 3-3, U.S.]}$$

The applied braking may be due to initial design speed, friction factor due to road conditions, and up- or downhill grade. An uphill grade will shorten the stopping distance due to gravity, whereas a downhill grade will lengthen the stopping distance. Similarly, a low friction factor, such as in wet road conditions or poor vehicle tire quality, will lengthen the stopping distance.

If solving for the decision sight distance, only the first term is needed. If solving for the braking distance, only the second term is needed. If solving for the total stopping sight distance, use the entire equation shown.

Exam tip: Note the nonlinear increase in the required stopping sight distance as design speed increases.

Passing Sight Distance

CETR: Sec. 4.11

AASHTO *Green Book*: Sec. 3.2.4

Passing sight distance is applicable on two-lane highways when there are enough gaps in opposing traffic flows to allow passing maneuvers and when there are few access points to the highway, with only occasional traffic entering the highway and with enough sight distance to observe traffic in both lanes and both directions.

Typical passing sight distance values for two-lane highways are given in AASHTO *Green Book* Table 3-4 and CETR Table 4.6.

Determining Crest Curve Length

CERM: Sec. 79.20

AASHTO *Green Book*: Sec. 3.4.6.2

The sight distance over the crest of a vertical curve is given by CERM Eq. 79.53 and Eq. 79.54. These equations use the terms h_1 and h_2 to represent the eye height of the driver and the height of the object above the roadway surface, respectively.

$$L = \frac{AS^2}{100(\sqrt{2h_1} + \sqrt{2h_2})^2} \quad [S < L] \qquad 79.53$$

$$L = 2S - \frac{200(\sqrt{h_1} + \sqrt{h_2})^2}{A} \quad [S > L] \qquad 79.54$$

CERM Table 79.4 compiles equations from AASHTO *Green Book* Chap. 3. These equations assume that the driver's eye height is 3.5 ft (1080 mm) and the object height is 2.0 ft (600 mm), which is the height of a typical passenger car's taillight.

Exam tip: An exam problem may ask you to determine the curve length, L, when the stopping sight distance, S, is known or vice versa. To do so, calculate the unknown quantity twice, using the equations for both $S < L$ and $S > L$, and then compare S and L to see which equation is correct.

Table 79.4 AASHTO Required Lengths of Curves on Grades[a]

	stopping sight distance[b] (crest curves)	passing sight distance[c] (crest curves)	stopping sight distance (sag curves)
SI units			
$S < L$	$L = \dfrac{AS^2}{658}$	$L = \dfrac{AS^2}{864}$	$L = \dfrac{AS^2}{120 + 3.5S}$
$S > L$	$L = 2S - \dfrac{658}{A}$	$L = 2S - \dfrac{864}{A}$	$L = 2S - \dfrac{120 + 3.5S}{A}$
U.S. units			
$S < L$	$L = \dfrac{AS^2}{2158}$	$L = \dfrac{AS^2}{2800}$	$L = \dfrac{AS^2}{400 + 3.5S}$
$S > L$	$L = 2S - \dfrac{2158}{A}$	$L = 2S - \dfrac{2800}{A}$	$L = 2S - \dfrac{400 + 3.5S}{A}$

[a] $A = |G_2 - G_1|$, absolute value of the algebraic difference in grades, in percent.
[b] The driver's eye is 3.5 ft (1080 mm) above road surface, viewing an object 2.0 ft (600 mm) high.
[c] The driver's eye is 3.5 ft (1080 mm) above road surface, viewing an object 3.5 ft (1080 mm) high.

Compiled from *A Policy on Geometric Design of Highways and Streets*, Chap. 3, copyright © 2018 by the American Association of State Highway and Transportation Officials, Washington, D.C.

K-Value Method for Crest Vertical Curves

Handbook: Symmetrical Vertical Curve Formula

CERM: Sec. 79.21

AASHTO *Green Book*: Sec. 3.4.6.2

$$K = \frac{L}{A}$$

$$K = \frac{L}{A} = \frac{L}{|G_2 - G_1|} \quad \text{[always positive]} \quad 79.57$$

The K-value method of analysis is a simplified, conservative method of choosing a stopping sight distance for a crest vertical curve. AASHTO *Green Book* Fig. 3-36 shows the crest vertical curves obtained using the K-method. Similar tables are shown in CERM Fig. 79.12 and 79.13. The length of the vertical curve per percent grade difference, K, is the ratio of the curve length, L, to the grade difference, A.

Exam tip: SI and U.S. units are given in separate tables, so be sure to use the correct one for your calculations.

K-Value Method for Sag Vertical Curves

Handbook: Symmetrical Vertical Curve Formula

AASHTO *Green Book*: Sec. 3.4.6.3

$$K = \frac{L}{A}$$

The procedure for determining sag curve length based on K-value AASHTO design control criteria is to select one of the curves based on the speed, or the K-value from *Handbook* section Symmetrical Vertical Curve Formula, and read the curve length that corresponds to the grade difference, A.

Sag Vertical Curve Length for Comfort

CERM: Sec. 79.22

AASHTO *Green Book* Sec. 3.4.6.3

$$L = \frac{AV^2}{395} \quad \text{[AASHTO Eq. 3-52, SI]}$$

$$L = \frac{AV^2}{46.5} \quad \text{[AASHTO Eq. 3-52, U.S.]}$$

In a sag curve, gravitational and centrifugal forces combine to act simultaneously on the driver and passengers, and comfort becomes the design control. Use AASHTO *Green Book* Eq. 3-52 for calculating the minimum length of curve to keep the centripetal acceleration on passengers below 1 ft/sec² (0.3 m/s²). CERM Eq. 79.58 is equivalent.

C. VERTICAL CLEARANCE

Key concepts: These key concepts are important for answering exam questions in subtopic 11.C, Vertical Clearance.

- critical clearance distance
- curve length
- external distance on vertical curves
- overhead obstructions
- sight distance

NCEES Handbook: To prepare for this subtopic, familiarize yourself with these sections in the *Handbook*.

- Vertical Curve Formulas

PE Civil Reference Manual (CERM): Study these sections in CERM that either relate directly to this subtopic or provide background information.

- Section 79.18: Vertical Curves Through Points
- Section 79.19: Vertical Curve to Pass Through Turning Point
- Section 79.21: Design of Crest Curves Using K-Value
- Section 79.24: Design of Sag Curves Using K-Value

Transportation Depth Reference Manual (CETR): Study these sections in CETR that either relate directly to this subtopic or provide background information.

- Section 4.13: Vertical and Horizontal Clearances

Codes and standards: To prepare for this subtopic, familiarize yourself with these sections from codes and standards.

A Policy on Geometric Design of Highways and Streets (AASHTO *Green Book*)

- Section 3.4.6.4: Sight Distance at Undercrossings
- Section 8.2.9: Vertical Clearance

Manual on Uniform Traffic Control Devices for Streets and Highways (MUTCD)

- Section 2A.16: Standardization of Location

The following equations and figures are relevant for subtopic 11.C, Vertical Clearance.

Clearance on Sag Vertical Curves

CERM: Sec. 79.18

AASHTO *Green Book*: Sec. 3.4.6.4

$$L = 2S - \frac{800\left(C - \frac{h_1 + h_2}{2}\right)}{A}$$
[AASHTO Eq. 3-53, $S > L$]

$$L = \frac{AS^2}{800\left(C - \frac{h_1 + h_2}{2}\right)}$$
[AASHTO Eq. 3-54, $S < L$]

AASHTO *Green Book* Eq. 3-53 and Eq. 3-54 show sag vertical curve length for $S > L$ and for $S < L$, respectively. C is the clearance between the top of the roadway and the bottom of an obstruction, such as a low bridge or sign elevation. A is the algebraic difference in grades expressed as a percentage. The same equation is used with both U.S. and SI units. For U.S. units, all distances must be in feet, and the vertical curve length, L, is in feet. For SI units, all distances are in meters, and L is in meters.

Vertical clearance on a sag curve is important when designing bridges and overpasses. The overhead obstruction is not necessarily above the low point of the bridge. It is important to be able to determine the elevation along the curve at any given location so that you can check for adequate clearance. A diagram of a vertical curve with an obstruction is shown in CERM Fig. 79.11.

Figure 79.11 Vertical Curve with an Obstruction

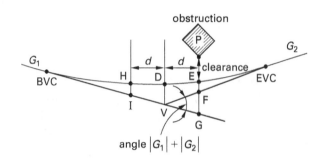

Line of Sight on Sag Vertical Curves

Handbook: Vertical Curve Formulas

CETR: Sec. 4.13

AASHTO *Green Book*: Chap. 3

$$E = \frac{AL}{800}$$

The *Handbook* equation shown assumes that both the external length, E, and the length of curve, L, are measured in feet. If the length of the curve is given in stations, convert the length to feet (1 sta = 100 ft). CETR Eq. 4.73 is equivalent.

CETR Fig. 4.29 shows the line of sight below an underpass when $S > L$. The stopping sight distance length, S, extends beyond PVT and PVC on the curve—that is, S is longer than the curve length, L.

Figure 4.29 Line of Sight Below Underpass When S > L

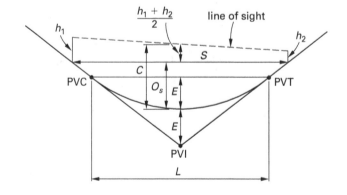

Figure 4.30 Line of Sight Below Underpass When S < L

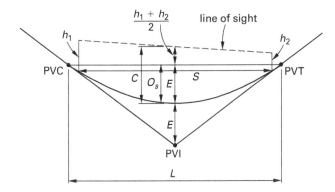

CETR Fig. 4.30 shows the line of sight below an underpass when $S < L$. When this is the case, the parabolic curve between PVC and PVT resembles a half circle.

Per AASHTO *Green Book* Chap. 3, if given no other information, the standard eye height of the driver above the road surface, h_1, is 3.5 ft, and the standard height of the obstruction above the road surface, h_2, is 3.5 ft.

Line of Sight for Trucks on Sag Vertical Curves

CETR: Sec. 4.13

AASHTO *Green Book*: Sec. 3.4.6.4

When $S > L$,

$$L = 2S - \frac{800(C-5)}{A}$$
[AASHTO Eq. 3-55, U.S.]

$$L = 2S - \frac{800(C-1.5)}{A}$$
[AASHTO Eq. 3-55, SI]

When $S < L$,

$$L = \frac{AS^2}{800(C-5)}$$
[AASHTO Eq. 3-56, U.S.]

$$L = \frac{AS^2}{800(C-1.5)}$$
[AASHTO Eq. 3-56, SI]

When solving for truck stopping sight distance on a sag vertical curve, the default AASHTO value in U.S. units for average truck driver eye height, h_1, is 8.0 ft, and the average automobile taillight height, h_2, is 2.0 ft. Thus, when using U.S. units, $(h_1 + h_2)/2 = 5$ ft.

In SI units, these default values are $h_1 = 2.4$ m and $h_2 = 0.6$ m. Thus, when using SI units, $(h_1 + h_2)/2 = 1.5$ m.

Substituting these values into AASHTO *Green Book* Eq. 3-53 and Eq. 3-54 gives Eq. 3-55 and Eq. 3-56.

38 Intersection Geometry

Content in blue refers to the *NCEES Handbook*.

A. Intersection Sight Distance 38-1
B. Interchanges 38-2
C. At-Grade Intersection Layout 38-4

The knowledge area of Intersection Geometry makes up between three and five questions out of the 80 questions on the PE Civil Transportation exam. The organization of this chapter follows the order of subtopics given by NCEES for this knowledge area. Each subtopic is covered in the following sections.

A. INTERSECTION SIGHT DISTANCE

Key concepts: These key concepts are important for answering exam questions in subtopic 12.A, Intersection Sight Distance.

- intersection control cases
- obstructions
- perception reaction
- sight triangles
- time gap

***NCEES Handbook*:** The *Handbook* does not contain any material relevant to this subtopic. Be sure to study the listed sections in CETR and from codes and standards.

Transportation Depth Reference Manual **(CETR):** Study these sections in CETR that either relate directly to this subtopic or provide background information.

- Section 4.15: Intersections and Interchanges

Codes and standards: To prepare for this subtopic, familiarize yourself with these sections from codes and standards.

A Policy on Geometric Design of Highways and Streets (AASHTO *Green Book*)

- Section 9.3: Types and Examples of Intersections
- Section 9.5.2: Sight Triangles
- Section 9.5.3: Intersection Control
- Section 9.10.1.1: Size and Space Needs

The following equations and concepts are relevant for subtopic 12.A, Intersection Sight Distance.

Intersection Sight Triangles

CETR: Sec. 4.15

AASHTO *Green Book*: Sec. 9.5.2.1, Sec. 9.5.2.2, Sec. 9.5.3.2.1

$$\text{ISD} = 1.47 V_{\text{major}} t_g \quad \text{[AASHTO Eq. 9-1]}$$

CETR Eq. 4.86 and AASHTO *Green Book* Eq. 9-1 are identical calculations for intersection sight distance (ISD).

ISD is the length of the leg of the sight triangle along the major road. AASHTO *Green Book* Fig. 9-16 and Fig. 9-17 illustrate the approach and departure sight triangles for an intersection, respectively.

In the equation for ISD, V_{major} is the design speed of the major road (in mph), and t_g is the time gap for the minor road vehicle to completely enter the major road (in seconds). The time gap is dependent on the turning movement. For example, a left turn from a minor to a major road involves crossing lanes of oncoming traffic and may have a larger time gap. AASHTO *Green Book* Table 9-5 provides adjustment factors for ISD based on the approach grade.

The AASHTO *Green Book* divides intersections into 10 cases according to their traffic control. This ISD equation can be used to find the ISD regardless of the intersection control case.

Intersection Control Cases

CETR: Sec. 4.15

AASHTO *Green Book*: Sec. 9.5.3.1

The AASHTO *Green Book* divides intersections into 10 cases according to their traffic control.

Case A is an intersection with no traffic control. CETR Fig. 4.32 illustrates the sight triangles for case A. AASHTO *Green Book* Table 9-4 can be used to find the length of the sight triangle leg for case A based on the design speed.

Case B is an intersection with stop control on the minor road. Case B comprises three subcases.

Case B1 is an intersection with a left turn from a minor road. CETR Fig. 4.33 illustrates the sight triangles for a case B1 intersection. AASHTO *Green Book* Table 9-6 can be used to find the time gap for case B1 based on the design vehicle, and AASHTO *Green Book* Table 9-7 can be used to find the stopping sight distance or intersection sight distance for case B1 based on the design speed.

Case B2 is an intersection with a right turn from the minor road. CETR Fig. 4.33 illustrates the sight triangles for case B2. AASHTO *Green Book* Table 9-8 can be used to find the time gap for case B2 based on the design vehicle, and AASHTO *Green Book* Table 9-9 can be used to find the intersection sight distance or stopping sight distance for case B2 based on the design speed.

Case B3 is an intersection with a crossing maneuver from the minor road. CETR Fig. 4.33 illustrates the sight triangles for case B3. AASHTO *Green Book* Table 9-10 can be used to find the time gap for case B3 based on the design vehicle. AASHTO *Green Book* Table 9-11 can be used to find the stopping sight distance or intersection sight distance for case B3 based on the design speed.

Case C is an intersection with yield control on the minor road. Case C comprises two subcases.

Case C1 is a crossing maneuver from the minor road. AASHTO *Green Book* Eq. 9-2 is used to find the time gap for case C1. V_{minor} is the design speed of the minor road in mi/hr, w is the width of the intersection to be crossed in feet, and L_a is the length of the design vehicle in feet. t_g is the total travel time in seconds required to reach and clear the major road. ta is the travel time required for a vehicle that does not stop to reach the major road from the decision point. t_a and t_g can be found from AASHTO *Green Book* Table 9-12 based on the design speed. CETR Eq. 4.87 and AASHTO *Green Book* Eq. 9-2 are identical.

$$t_g = t_a + \frac{w + L_a}{0.88 V_{\text{minor}}} \quad \text{[AASHTO Eq. 9-2]}$$

CETR Fig. 4.32 illustrates the sight triangles for case C1. AASHTO *Green Book* Table 9-13 can be used to find the length of the sight triangle leg for case C1 based on the design speed.

Case C2 is a left or right turn from the minor road. CETR Fig. 4.32 illustrates the sight triangles for case C2. AASHTO *Green Book* Table 9-14 can be used to find the time gap for case C2. AASHTO *Green Book* Table 9-15 can be used to find the stopping sight distance or intersection sight distance for case C2 based on the design speed.

Case D is for use at intersections controlled by traffic signals.

Case E is for use at intersections with all-way stop control.

Case F is a left turn from the major road. AASHTO *Green Book* Table 9-16 can be used to find the time gap for case F, and AASHTO *Green Book* Table 9-17 can be used to find the stopping sight distance or intersection sight distance for case F based on the design speed.

Case G is for use on roundabouts.

B. INTERCHANGES

Key concepts: These key concepts are important for answering exam questions in subtopic 12.B, Interchanges.

- acceleration length and deceleration length for interchange ramps
- auxiliary lanes
- interchange types
- interchange warrants
- major forks
- turning roadway lane widths
- weaving sections

NCEES Handbook: The *Handbook* does not contain any material relevant to this subtopic. Be sure to study the listed sections in CERM and CETR and from codes and standards.

PE Civil Reference Manual (**CERM**): Study these sections in CERM that either relate directly to this subtopic or provide background information.

- Section 73.27: Highway Interchanges
- Section 73.28: Weaving Areas

Transportation Depth Reference Manual (**CETR**): Study these sections in CETR that either relate directly to this subtopic or provide background information.

- Section 4.14: Acceleration and Deceleration
- Section 4.15: Intersections and Interchanges

Codes and standards: To prepare for this subtopic, familiarize yourself with these sections from codes and standards.

A Policy on Geometric Design of Highways and Streets (AASHTO *Green Book*)

- Section 2.8.3: Vehicle Performance
- Section 3.3.11: Widths for Turning Roadways at Intersections
- Section 10.1: Introduction and General Types of Interchanges

- Section 10.9.5: General Design Considerations
- Section 10.9.6: Ramps

The following figures and concepts are relevant for subtopic 12.B, Interchanges.

Types of Interchanges

CERM: Sec. 73.27

CETR: Sec. 4.15

AASHTO *Green Book*: Sec. 10.1

Interchanges vary from single ramps connecting local streets to complex and comprehensive layouts involving two or more highways. The basic interchange configurations are shown in AASHTO *Green Book* Fig. 10-1, CERM Fig. 73.11, and CETR Fig. 4.34.

Acceleration and Deceleration Length

CETR: Sec. 4.14

AASHTO *Green Book*: Sec. 2.8.3, Sec. 10.9.6.5.1

Average traffic accelerates at a rate of approximately 4.4 ft/sec^2 (about 3 mph/sec) at speeds up to 30 mph. The acceleration rate decreases to 2.8 ft/sec^2 as speeds approach 70 mph. AASHTO *Green Book* Fig. 2-33 and Fig. 2-34 illustrate acceleration and deceleration distances, respectively. CETR App. 4.E and App. 4.F are equivalent illustrations.

Highway design must consider acceleration and deceleration rates, as well as the distances necessary to safely achieve these rates. The distance required to reach a certain speed will increase for acceleration up a grade and decrease for acceleration down a grade. AASHTO *Green Book* Table 10-4 gives the minimum acceleration lane lengths for flat grades, and Table 10-5 gives the adjustment factors for both acceleration and deceleration lanes to account for steeper grades.

Exam tip: Deceleration rates are usually dictated by the comfort of the driver and passengers, not the braking ability of the vehicles.

Weaving Segments

CERM: Sec. 73.28

Weaving is the crossing of at least two traffic streams traveling in the same general direction along a length of highway without traffic control. Interchanges without weaving are favored over interchanges with weaving. CERM Fig. 73.13 illustrates a standard freeway weaving segment.

Figure 73.13 Freeway Weaving Segment

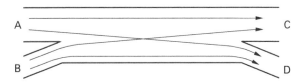

Used with permission from *Highway Capacity Manual*, 6th Edition: A Guide for Multimodal Mobility Analysis, 2016, Exhibit 13-1, by the Transportation Research Board of the National Academies of Sciences, Engineering, and Medicine, Washington DC. DOI: 10.17226/24798

One-sided ramp weaving segments have the entry and the exit from the weaving segment on the same side of the freeway. A major weaving segment is a segment where three or more entry or exit legs have multiple lanes. CERM Fig. 73.15 illustrates these segments.

Figure 73.15 One-Sided Weaving Segments

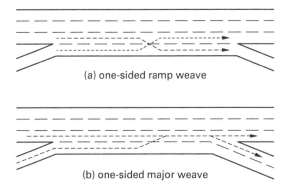

Used with permission from *Highway Capacity Manual*, 6th Edition: A Guide for Multimodal Mobility Analysis, 2016, Exhibit 13-3, by the Transportation Research Board of the National Academies of Sciences, Engineering, and Medicine, Washington DC. DOI: 10.17226/24798

Two-sided weaving segments have an on-ramp on one side of the freeway closely followed by an off-ramp on the opposite side of the freeway, as shown in CERM Fig. 73.16.

Figure 73.16 Two-Sided Weaving Segments

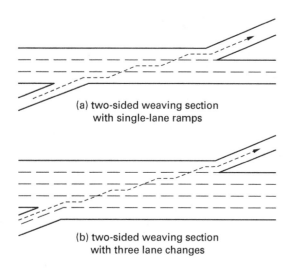

(a) two-sided weaving section with single-lane ramps

(b) two-sided weaving section with three lane changes

Used with permission from *Highway Capacity Manual*, 6th Edition: A Guide for Multimodal Mobility Analysis, 2016, Exhibit 13-4, by the Transportation Research Board of the National Academies of Sciences, Engineering, and Medicine, Washington DC. DOI: 10.17226/24798

C. AT-GRADE INTERSECTION LAYOUT

Key concepts: These key concepts are important for answering exam questions in subtopic 12.C, At-Grade Intersection Layout.

- four-leg intersections
- intersection conflict points
- intersection design elements
- roundabouts
- three-leg intersections

NCEES Handbook: The *Handbook* does not contain any material relevant to this subtopic. Be sure to study the listed sections in CERM and CETR and from codes and standards.

PE Civil Reference Manual **(CERM):** Study these sections in CERM that either relate directly to this subtopic or provide background information.

- Section 73.17: Signalized Intersections

Transportation Depth Reference Manual **(CETR):** Study these sections in CETR that either relate directly to this subtopic or provide background information.

- Section 4.15: Intersections and Interchanges
- Section 6.11: Conflict Analysis

Codes and standards: To prepare for this subtopic, familiarize yourself with these sections from codes and standards.

A Policy on Geometric Design of Highways and Streets (AASHTO *Green Book*)

- Section 9.3: Types and Examples of Intersections
- Section 9.3.1: Three-Leg Intersections
- Section 9.3.2: Four-Leg Intersections
- Section 9.3.4: Roundabouts

The following figures and concepts are relevant for subtopic 12.C, At-Grade Intersection Layout.

General Types of Intersections

AASHTO *Green Book*: Sec. 9.3

The basic types of at-grade intersections are three-leg intersections, four-leg intersections, multi-leg intersections, and roundabouts. Further classification includes whether an intersection is unchannelized, flared, or channelized; whether an intersection is offset; and whether an intersection is indirect. AASHTO *Green Book* Fig. 9-4 illustrates the general types of intersections.

Three-Leg Intersections

AASHTO *Green Book*: Sec. 9.3.1.1, Sec. 9.3.1.2

A three-leg intersection is an intersection where two roads meet and one of them ends, as in a T-type intersection. The most common type of three-leg intersection has the normal pavement width of both highways maintained, except for the paved corner radii where widening is needed to accommodate the selected design vehicle. AASHTO *Green Book* Fig. 9-5 and Fig. 9-6 illustrate unchannelized and channelized three-leg intersections, respectively.

Four-Leg Intersections

AASHTO *Green Book*: Sec. 9.3.2.1, Sec. 9.3.2.2

A four-leg intersection is a crossroads where two roads meet. AASHTO *Green Book* Fig. 9-7, Fig. 9-8, Fig. 9-9, and Fig. 9-10 illustrate various types of four-leg intersections. The same overall design principles for three-leg intersections also apply to four-leg intersections.

Roundabouts

AASHTO *Green Book*: Sec. 9.3.4

In modern roundabout design, all entering traffic is required to yield to the circulating traffic. Modern roundabouts operate at a higher level of efficiency and a lower crash frequency than other intersection configurations.

Mini-roundabouts are small roundabouts used in low-speed urban environments with average operating speeds of 30 mph or less. The central island of a mini-roundabout is small and fully mountable so that larger vehicles may cross over it if needed. AASHTO *Green Book* Fig. 9-12 shows a typical mini-roundabout.

Single-lane roundabouts have a single entry lane at all legs and one circulatory lane. The central island of a single-lane roundabout is nonmountable and larger in diameter than that of a mini-roundabout. AASHTO *Green Book* Fig. 9-13 shows a typical single-lane roundabout.

Multilane roundabouts include all roundabouts that have at least one entry with two or more lanes. AASHTO *Green Book* Fig. 9-14 shows a typical multilane roundabout.

Intersection Conflict Points

CETR: Sec. 6.11

Conflicts in traffic streams occur at any location where two traffic flows cross, merge, or diverge. Conflicts can be within the same type of traffic, such as automobile traffic, or between different types of traffic, such as between auto traffic and pedestrian traffic. CETR Fig. 6.13 illustrates conflict points at an intersection.

39 Roadside and Cross-Section Design

Content in blue refers to the *NCEES Handbook*.

A. Forgiving Roadside Concepts 39-1
B. Barrier Design 39-2
C. Cross Section Elements 39-4
D. Americans with Disabilities Act (ADA) Design Considerations 39-5

The knowledge area of Roadside and Cross-Section Design makes up between three and five questions out of the 80 questions on the PE Civil Transportation exam. The organization of this chapter follows the order of subtopics given by NCEES for this knowledge area. Each subtopic is covered in the following sections.

A. FORGIVING ROADSIDE CONCEPTS

Key concepts: These key concepts are important for answering exam questions in subtopic 13.A, Forgiving Roadside Concepts.

- clear zone
- clear zone distance
- recoverable slopes
- roadside obstacles

NCEES Handbook: The *Handbook* does not contain any material relevant to this subtopic. Be sure to study the listed sections in CERM and CETR and from codes and standards.

PE Civil Reference Manual (**CERM**): Study these sections in CERM that either relate directly to this subtopic or provide background information.

- Section 75.17: Road Safety Features

Transportation Depth Reference Manual (**CETR**): Study these sections in CETR that either relate directly to this subtopic or provide background information.

- Section 6.6: Roadside Clearance Analysis

Codes and standards: To prepare for this subtopic, familiarize yourself with these sections from codes and standards.

Roadside Design Guide (AASHTO RSDG)

- Section 3.1: The Clear-Zone Concept
- Section 3.2.1: Foreslopes

The following figures, tables, and concepts are relevant for subtopic 13.A, Forgiving Roadside Concepts.

Clear Zones

CETR: Sec. 6.6

AASHTO RSDG: Sec. 3.1

A *clear zone* is a total roadside area that begins at the edge of the roadway and allows drivers to stop safely if they leave the roadway. The clear zone may consist of a shoulder, a recoverable slope, a nonrecoverable slope, and/or a clear runout area. The clear zone includes the area past the edge of the traveled way to the shoulder and beyond.

AASHTO RSDG Table 3-1 gives suggested clear-zone distances based on traffic volumes, speeds, and roadside geometry. For more about specific types of clear zones, see the entries "Recoverable Slopes," "Nonrecoverable Slopes," and "Clear Runout Areas" in this chapter.

Recoverable Slopes

CETR: Sec. 6.6

AASHTO RSDG: Sec. 3.2.1

A *recoverable slope* is a slope flatter than 1V:4H. A recoverable slope generally allows a motorist to regain control of a vehicle by slowing or stopping. It is assumed that a vehicle's ability to return up a steep slope decreases as the slope distance increases. Recoverable slopes provide safer roadways than nonrecoverable slopes; however, recoverable slopes have high right-of-way costs, and in some cases, it is not possible to build recoverable slopes.

AASHTO RSDG Fig. 3-2 illustrates a recoverable slope as part of a larger variable-slope design. CETR Fig. 6.4 is an equivalent figure.

Nonrecoverable Slopes

CETR: Sec. 6.6

AASHTO RSDG: Sec. 3.2.1

Nonrecoverable slopes are slopes steeper than 1V:4H. Nonrecoverable slopes are considered traversable, but it is assumed the motorist will not be able to regain control and the vehicle will continue to the bottom of the slope, so nonrecoverable slopes should be evaluated for the need of a guardrail system.

AASHTO RSDG Fig. 3-2 illustrates a nonrecoverable slope as part of a larger variable-slope design. CETR Fig. 6.4 is an equivalent figure.

Clear Runout Areas

CETR: Sec. 6.6

AASHTO RSDG: Sec. 3.2.1

A *clear runout area* is part of a clear zone and is defined as the area at the toe of a nonrecoverable slope that is available for a vehicle's safe use. A clear runout area is typically part of a variable slope design, in which a recoverable slope is placed adjacent to the paved shoulder for a certain distance, followed by a nonrecoverable slope with a clear runout area at the bottom. A variable slope design with a clear runout area is illustrated in AASHTO RSDG Fig. 3-2. CETR Fig. 6.4 is an equivalent figure.

B. BARRIER DESIGN

Key concepts: These key concepts are important for answering exam questions in subtopic 13.B, Barrier Design.

- barrier design criteria
- barrier installation criteria
- barrier types
- crash cushions
- flaring
- guardrail end treatments
- length of need
- runout length
- shy lines

NCEES Handbook: The *Handbook* does not contain any material relevant to this subtopic. Be sure to study the listed sections in CERM and CETR and from codes and standards.

***PE Civil Reference Manual* (CERM):** Study these sections in CERM that either relate directly to this subtopic or provide background information.

- Section 75.18: Roadside Safety Railings

***Transportation Depth Reference Manual* (CETR):** Study these sections in CETR that either relate directly to this subtopic or provide background information.

- Section 6.3: Roadway Elements for Safe Design
- Section 6.5: Encroachment Crash Analysis
- Section 6.6: Roadside Clearance Analysis

Codes and standards: To prepare for this subtopic, familiarize yourself with these sections from codes and standards.

Roadside Design Guide (AASHTO RSDG)

- Section 5.1: Performance Requirements
- Section 5.5.2: Barrier Deflection Characteristics
- Section 5.6: Placement Recommendations
- Section 6.4.1.2: Low-Tension Cable Barrier
- Section 8.2: Anchorage Design Concepts
- Section 8.3: Terminal Design Concepts
- Section 8.4.2.2: Reusable Crash Cushions

The following figures, tables, and concepts are relevant for subtopic 13.B, Barrier Design.

Rigid and Semirigid Barriers

CETR: Sec. 6.6

Barriers are used to prevent damage to travelers. Rigid barriers are intended to protect those behind the barrier, such as vehicles on the other side of the road or workers in a designated work zone. Semirigid barriers are intended to protect the driver from striking an object within the clear zone.

Per AASHTO RSDG Sec. 8.2, every flexible and semirigid barrier needs to be terminated with an anchor system at both ends.

Exam tip: All barriers are designed for the same purpose, but different kinds of barriers will act in different ways. Barriers should be selected based on the intended work and the cost.

Cable Barriers

AASHTO RSDG: Sec. 6.4.1.2

Cable barriers typically deflect more during impacts than other types of barriers. Per AASHTO RSDG, cables are effective when placed on a 1V:6H side slope. The maintenance of cable guardrail systems is critical to their safe function because all the cables must be intact to properly decelerate a vehicle. Single-cable systems have been linked to safety issues.

Barrier Design Criteria

AASHTO RSDG: Sec. 5.1

Barriers and crash attenuators are tested using the criteria from either National Cooperative Highway Research Program (NCHRP) Report 350 or AASHTO *Manual for Assessing Safety Hardware* (MASH). AASHTO RSDG Table 5-1(a) shows the test conditions for longitudinal barriers per MASH criteria. AASHTO RSDG Table 5-1(b) shows the test conditions for longitudinal barriers per NCHRP Report 350 criteria. The test conditions vary depending on the vehicle type, vehicle weight, speed, and slope angle.

Barrier Installation Criteria

AASHTO RSDG: Sec. 5.6

The three barrier installation criteria are the length of need, the lateral extent of the area of concern, and the runout length. The *length of need* is the total length of the longitudinal barrier needed to shield a fixed object. The *lateral extent of the area of concern* is the distance from the edge of the traveled way to the far edge of the obstruction. The *runout length* is the horizontal distance from the obstruction to the location where the vehicle is last expected to leave the road and strike the far edge of the obstruction. These factors are used to determine the appropriate location for the upstream end of a barrier in relation to an obstruction in the clear zone.

AASHTO RSDG Fig. 5-32 shows an example guardrail and embankment layout sheet.

Barrier Flare Rate

CETR: Sec. 6.6

AASHTO RSDG: Sec. 5.6.3

A roadside barrier is considered to be *flared* when it is not parallel to the edge of the traveled way. Flare is used for a variety of purposes, such as to allow a barrier terminal to be located farther from the roadway than if the barrier were not flared, to minimize a driver's reaction to an obstacle near the road by gradually introducing a parallel barrier installation, or to transition a roadside barrier to an obstacle nearer the roadway.

AASHTO RSDG Table 5-9 shows recommended maximum flare rates for semirigid and rigid barriers. CETR Table 6.8 is an equivalent table.

Table 6.8 Suggested Barrier Flare Rates

design speed		flare rate for barrier inside shy line	flare rate for barrier at or beyond shy line	
kph	[mph]		A	B
110	[70]	30:1	20:1	15:1
100	[60]	26:1	18:1	14:1
90	[55]	24:1	16:1	12:1
80	[50]	21:1	14:1	11:1
70	[45]	18:1	12:1	10:1
60	[40]	16:1	10:1	8:1
50	[30]	13:1	8:1	7:1

A = suggested maximum flare rate for rigid barrier system
B = suggested maximum flare rate for semi-rigid barrier system

The MGS has been tested in accordance with the NCHRP Report 350 TL-3 at 5:1 flare.

Flatter flare rates for the MGS installations also are acceptable. The MGS should be installed using the flare rates shown or flatter for semi-rigid barriers beyond the shy line when installed in rock formations.

Reprinted with permission from the American Association of State Highway and Transportation Officials, *Roadside Design Guide*, Table 5-9, copyright © 2011.

Shy Lines

CETR: Sec. 6.6

AASHTO RSDG: Sec. 5.6.1

A *shy line* is the distance from the edge of a roadway at which a roadside object will not be perceived as an obstacle by a driver (i.e., the distance at which a driver will not change speed or position in the roadway in reaction to the object). This distance varies with speed. Suggested shy-line offsets are shown in AASHTO RSDG Table 5-7. CETR Table 6.7 is an equivalent table.

Table 6.7 Suggested Shy Distances from RSDG

design speed		shy-line offset, L_s	
kph	[mph]	m	[ft]
130	[80]	3.7	[12]
120	[75]	3.2	[10]
110	[70]	2.8	[9]
100	[60]	2.4	[8]
90	[55]	2.2	[7]
80	[50]	2.0	[6.5]
70	[45]	1.7	[6]
60	[40]	1.4	[5]
50	[30]	1.1	[4]

Reprinted with permission from the American Association of State Highway and Transportation Officials, *Roadside Design Guide*, Table 5-7, copyright © 2011.

Barrier Runout Length

CETR: Sec. 6.6

AASHTO RSDG: Sec. 5.6.4

The *runout length* (length of need) of a barrier is the distance from a fixed object to the closest point that an errant vehicle is likely to leave the roadway and strike that object. This distance has been found to vary with the design speed and the average daily traffic (ADT).

AASHTO RSDG Fig. 5-39 shows the variables that should be considered in determining a runout length. AASHTO RSDG Table 5-10 gives suggested runout lengths based on the design speed and traffic volume. CETR Table 6.6 is an equivalent table.

Table 6.6 Suggested Runout Lengths, L_R for Barrier Design

	SI units			
	runout length, L_R, given traffic volume, ADT, (m)			
design speed (kph)	over 10,000 veh/day	5000 to 10,000 veh/day	1000 to 5000 veh/day	under 1000 veh/day
130	143	131	116	101
110	110	101	88	76
100	91	76	64	61
80	70	58	49	46
60	49	40	34	30
50	34	27	24	21
	customary U.S. units			
	runout length, L_R, given traffic volume, ADT, (ft)			
design speed (mph)	over 10,000 veh/day	5000 to 10,000 veh/day	1000 to 5000 veh/day	under 1000 veh/day
80	470	430	380	330
70	360	330	290	250
60	300	250	210	200
50	230	190	160	150
40	160	130	110	100
30	110	90	80	70

Reprinted with permission from the American Association of State Highway and Transportation Officials, *Roadside Design Guide*, Tables 5-10a and 5-10b, copyright © 2011.

Guardrail End Treatment

AASHTO RSDG: Sec. 8.2, Sec. 8.3

A variety of commercial end treatments exist, ranging from compression devices to water-filled bladders, to help decelerate the driver during impact.

Crash Cushions

AASHTO RSDG: Sec. 8.4.2.2, Sec. 8.4.2.2.1

There are three different types of crash cushions: sand-filled barrels, sliding-rail weight systems, and collapsible systems. AASHTO RSDG Fig. 8-27 shows a standard QuadGuard® crash cushion.

C. CROSS SECTION ELEMENTS

Key concepts: These key concepts are important for answering exam questions in subtopic 13.C, Cross Section Elements.

- bicycle lanes
- cross slopes
- lane widths
- shoulder widths

NCEES Handbook: The *Handbook* does not contain any material relevant to this subtopic. Be sure to study the listed sections in CERM and CETR and from codes and standards.

***PE Civil Reference Manual* (CERM):** Study these sections in CERM that either relate directly to this subtopic or provide background information.

- Section 79.12: Transitions to Superelevation

***Transportation Depth Reference Manual* (CETR):** Study these sections in CETR that either relate directly to this subtopic or provide background information.

- Section 4.11: Sight Distance

Codes and standards: To prepare for this subtopic, familiarize yourself with these sections from codes and standards.

A Policy on Geometric Design of Highways and Streets (AASHTO *Green Book*)

- Section 4.3: Lane Widths
- Section 4.4.2: Shoulder Width
- Section 4.9.1: Normal Crown Sections
- Section 7.3.9.1: Bicycle Facilities

The following figures and concepts are relevant for subtopic 13.C, Cross Section Elements.

Cross Slopes

AASHTO *Green Book*: Sec. 4.9.1

A *cross slope* is a slope that is transverse relative to the horizon. Cross slopes are often used on low-speed, undivided roads to drain the rainwater away from the roads. A cross slope typically is crowned at the centerline (referred to as a *normal crown*) and has a slope of 1.6% to 2%. AASHTO *Green Book* Fig. 4-1A shows a typical cross section of a modern highway with a normal crown.

Lane Widths

AASHTO *Green Book*: Sec. 4.3

AASHTO *Green Book* Sec. 4.3 indicates that most lanes should be between 10 ft and 12 ft wide. On high-speed facilities, it can be more economical to use 12 ft lanes. Drivers at high speeds are better able to stay within the confines of a wider lane, leading to less deterioration of the shoulder and less maintenance needed.

Shoulders

AASHTO *Green Book*: Sec. 4.4.2

The main functions of shoulders are to

- prevent movement off the traveled pavement section
- extend the pavement edge beyond the expected wheel loading area
- prevent stormwater from encroaching into the traveled way
- provide additional horizontal offset for sight distance

AASHTO *Green Book* Sec. 4.4.2 suggests a minimum of 2 ft of paved shoulder width and a shoulder slope of 2–6%, depending on the superelevation.

Bicycle Lanes

AASHTO *Green Book*: Sec. 7.3.9.1

A bicycle lane that is integrated into a road should be from 3 ft to 6 ft wide; a bicycle lane adjacent to a road for two-way pedestrian traffic should be 10 ft wide. Bicycle lane design is covered more in the AASHTO *Guide for the Development of Bicycle Facilities*.

D. AMERICANS WITH DISABILITIES ACT (ADA) DESIGN CONSIDERATIONS

Key concepts: These key concepts are important for answering exam questions in subtopic 13.D, Americans with Disabilities Act (ADA) Design Considerations.

- auditory indicators
- corner ramps
- cross slopes
- curb ramps
- mobility impairments
- signage
- truncated domes
- visual impairments
- walkway slopes

NCEES Handbook: The *Handbook* does not contain any material relevant to this subtopic. Be sure to study the listed sections in CETR and from codes and standards.

***Transportation Depth Reference Manual* (CETR):** Study these sections in CETR that either relate directly to this subtopic or provide background information.

- Section 3.1: Non-Motorized Facilities: Pedestrian

Codes and standards: To prepare for this subtopic, familiarize yourself with these sections from codes and standards.

Manual on Uniform Traffic Control Devices (MUTCD)

- Section 3B.18: Crosswalk Markings
- Section 3I.06: Pedestrian Islands and Medians
- Section 4E.08: Pedestrian Detectors
- Section 4E.09: Accessible Pedestrian Signals and Detectors—General
- Section 4E.10: Accessible Pedestrian Signals and Detectors—Location
- Section 6F.03: Sign Placement
- Section 8D.04: Stop Lines, Edge Lines, and Detectable Warnings
- Section 8D.05: Passive Devices for Pathway Grade Crossings

The *2010 ADA Standards for Accessible Design* will not be available for reference during the exam, but it can be useful during study.

The following figures and concepts are relevant for subtopic 13.D, Americans with Disabilities Act (ADA) Design Considerations.

Walkway Slopes

CETR: Sec. 3.1

For purposes of accessibility, sidewalks should be graded no more than 2% and should have a smooth surface. Section 405.3 of the *2010 ADA Standards* states that the cross slope of a ramp run must not be greater than 1:48, or 2%.

Exam tip: The codes and standards available during the exam do not give calculations for the minimum length of a walkway slope, so you should know how to perform this calculation.

Visual Impairments

MUTCD: Sec. 3B.18, Sec. 8D.04, Sec. 8D.05

Pedestrians who are visually impaired need detectable warnings or sounds to assist them in interacting with other pedestrians and drivers. *2010 ADA Standards* Sec. 705 defines a detectable warning as a standardized surface feature built in or applied to walking surfaces or other elements to warn people who are visually impaired of hazards on a circulation path. MUTCD Sec. 3B.18 describes detectable warning surfaces in detail. For detectable warnings related to pathway grade crossings, see MUTCD Sec. 8D.04 and Sec. 8D.05.

Pedestrian Islands and Medians

MUTCD: Sec. 3I.06

MUTCD Sec. 3I.06 discusses minimum widths for accessible refuge islands and for the design and placement of detectable warning surfaces.

Pedestrian Detectors

MUTCD: Sec. 4E.08, Sec. 4E.09, Sec. 4E.10

Accessible pedestrian signals and detectors provide information in nonvisual formats. MUTCD Sec. 4E.08 discusses pedestrian detectors. MUTCD Sec. 4E.09 and Sec. 4E.10 discuss the specifications and location of braille or raised print for traffic control devices.

Curb Ramps

CETR: Sec. 3.1

A wheelchair encountering a sharp drop-off can easily overturn. For this reason, a pathway with wheelchair traffic should not be constructed on a longitudinal grade greater than 5%. The flares of a curb ramp should be limited to 10%, and the center path limited to 8.33%. *2010 ADA Standards* Fig. 406.3 illustrates appropriate slopes for a curb ramp. See CETR Sec. 3.1 and *2010 ADA Standards* Sec. 406 for more details.

Providing 3.5 ft (1.1 m) for each lane of pedestrian traffic, not including handrails or other obstructions, gives room for a wheelchair user to reach over the wheels and permits wheelchair traffic in both directions.

> *Exam tip*: The codes and standards available during the exam do not give calculations for the minimum length of a curb ramp, so you should know how to perform this calculation.

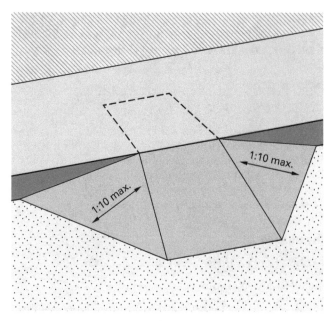

curb ramp

Corner Ramps

Corner ramps may be installed at each pedestrian street crossing. *2010 ADA Standards* Sec. 406.6 dictates that the bottoms of diagonal curb ramps have a minimum clear space of 48 in outside active traffic lanes of the roadway, as shown in *ADA Standards* Fig. 406.6.

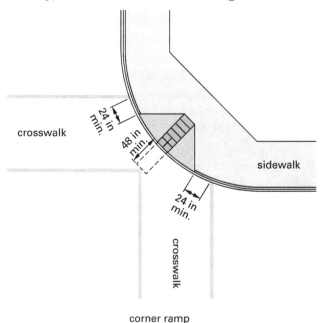

corner ramp

ADA accessibility guidelines suggest a maximum cross slope of 5% for pedestrian access routes within crossings without yield or stop control.

Exam tip: The codes and standards available during the exam do not provide calculations for the minimum length of a corner ramp, so you should know how to perform this calculation.

Temporary Traffic Control Provisions

MUTCD: Sec. 6F.03

ADA provisions about the placement of temporary traffic signs are covered in MUTCD Sec. 6F.03.

40 Signal Design

Content in blue refers to the *NCEES Handbook*.

A. Signal Timing.. 40-1
B. Signal Warrants... 40-3

The knowledge area of Signal Design makes up between three and five questions out of the 80 questions on the PE Civil Transportation exam. The organization of this chapter follows the order of subtopics given by NCEES for this knowledge area. Each subtopic is covered in the following sections.

A. SIGNAL TIMING

Key concepts: These key concepts are important for answering exam questions in subtopic 14.A, Signal Timing.

- adjustment factors
- critical lane group
- estimated cycle length
- pedestrian crossing factors
- pedestrian crossing time
- platoon ratio
- saturation flow rate
- signalized intersections
- volume-capacity ratio
- Webster's equation
- yellow time, red time, green time, and effective green time

NCEES Handbook: To prepare for this subtopic, familiarize yourself with these sections in the *Handbook*.

- Dilemma Zones
- Dilemma Zone Formation
- Offsets

PE Civil Reference Manual (**CERM**): Study these sections in CERM that either relate directly to this subtopic or provide background information.

- Section 73.18: Cycle Length: Webster's Equation

Transportation Depth Reference Manual (**CETR**): Study these sections in CETR that either relate directly to this subtopic or provide background information.

- Section 2.5: Traffic Signals
- Section 3.1: Non-Motorized Facilities: Pedestrian

Codes and standards: To prepare for this subtopic, familiarize yourself with these sections from codes and standards.

Highway Capacity Manual (HCM)

- Section 19.2: Concepts
- Section 19.3: Core Motorized Vehicle Methodology
- Section 19.4: Extensions to the Motorized Vehicle Methodology
- Section 19.5: Pedestrian Methodology
- Section 31.2: Capacity and Phase Duration

Manual on Uniform Traffic Control Devices for Streets and Highways (MUTCD)

- Section 4D.26: Yellow Change and Red Clearance Intervals

The following equations, figures, and concepts are relevant for subtopic 14.A, Signal Timing.

Volume-Capacity Ratio

Handbook: Dilemma Zones; Dilemma Zone Formation

$$X_c = 1.47 V t_{stop} + 1.075 \left(\frac{V^2}{a_2} \right)$$

$$X_0 = 1.47 V Y - W + \frac{1}{2} a_1 (Y - t_{passing})^2$$

CETR: Sec. 2.5

$$X_i = \left(\frac{v}{c} \right) = \frac{v}{s\left(\frac{g}{C} \right)} = \frac{vC}{sg} \qquad 2.47$$

HCM: Sec. 19.3

$$X = \frac{v}{c} \quad \text{[HCM Eq. 19-17]}$$

The volume-capacity ratio of a lane group is the ratio of the hourly demand volume to the hourly capacity of the lane group, as shown in HCM Eq. 19-17. CETR Eq. 2.47 is an equivalent equation, differing in the addition of a subscript to the variable X and two alternative ways to calculate the volume-capacity ratio as a function of the effective green time, cycle length, and saturation flow rate.

Volume-capacity ratios greater than 1.0 indicate an excess of demand over capacity. Excess demand over capacity may result in the formation of a *dilemma zone*, a zone within which a driver can neither bring the vehicle to a stop safely nor go through a signal-controlled intersection before the signal turns red. The formation of a dilemma zone is depicted in *Handbook* figure Dilemma Zone Formation. X_c is the minimum safe stopping distance, and X_0 is the maximum yellow passing distance, found from the equations in *Handbook* section Dilemma Zones. If X_c is greater than X_0, the difference between the two is the length of the dilemma zone, as shown in the figure.

Critical Volume-Capacity Ratio

CETR: Sec. 2.5

HCM: Sec. 19.4

$$X_c = \frac{C}{C-L} \sum_{i \in \text{ci}} y_{c,i} \quad \text{[HCM Eq. 19-30]}$$

$$L = \sum_{i \in \text{ci}} l_{t,i} \quad \text{[HCM Eq. 19-31]}$$

For signalized intersections without overlapping phases, the intersection can be evaluated by considering critical lane groups for each phase. The critical volume-capacity ratio can be calculated using HCM Eq. 19-30; the lost cycle time required for this calculation can be found using HCM Eq. 19-31. CETR Eq. 2.48 and Eq. 2.49 are equivalent equations.

For peak periods, evaluating the volume-capacity and critical volume-capacity ratios of a signalized intersection can help in assessing timing problems. For instance, if all volume-capacity ratios are greater than 1.0 and all critical volume-capacity ratios are 1.0 or less, then the cycle length is adequate, but the green time allocation is incorrect. If several volume-capacity ratios and critical volume-capacity ratios are greater than 1.0, a different phasing may yield improvement.

Exam tip: The equations in this entry and the ones referenced in the entry "Volume-Capacity Ratio" in this chapter look similar, but the critical volume-capacity ratio equations are for intersections only, not segments.

Capacity (Signalized Intersection)

CETR: Sec. 3.1

HCM: Sec. 19.3

$$c = Ns\left(\frac{g}{C}\right) \quad \text{[HCM Eq. 19-16]}$$

The capacity of any lane group with no shared lanes or permitted left-turn lane group can be found from HCM Eq. 19-16. CETR Eq. 2.46 is an equivalent equation.

Saturation Flow Rate

CETR: Sec. 2.5

HCM: Sec. 19.3

$$s = s_o f_w f_{\text{HV}g} f_p f_{\text{bb}} f_a f_{\text{LU}} f_{\text{LT}} f_{\text{RT}}$$
$$\cdot f_{\text{Lpb}} f_{\text{Rpb}} f_{wz} f_{ms} f_{sp} \quad \text{[HCM Eq. 19-8]}$$

The base equation for interrupted flow through a lane in a signalized intersection is HCM Eq. 19-8. CETR Eq. 2.34 is an equivalent equation.

A base saturation flow rate of 1900 pc/hr-ln indicates normal conditions (i.e., the best conditions). This value will need to be adjusted downward to suit the conditions of the intersections being evaluated.

Exam tip: Study the nomenclature for this equation ahead of time to reduce confusion and save time during the exam.

Pedestrian Crossing Time

CETR: Sec. 3.1

HCM: Sec. 19.5

For a crosswalk more than 10 ft wide,

$$t_{\text{ps,do}} = 3.2 + \frac{L_d}{S_p} + 2.7\left(\frac{N_{\text{ped,do}}}{W_d}\right) \quad \text{[HCM Eq. 19-64]}$$

For a crosswalk 10 ft wide or less,

$$t_{\text{ps,do}} = 3.2 + \frac{L_d}{S_p} + 2.7 N_{\text{ped,do}} \quad \text{[HCM Eq. 19-65]}$$

The *pedestrian crossing time* is the time it takes for a crosswalk to clear once pedestrians enter it. The pedestrian crossing time can be found from HCM Eq. 19-64 if the crosswalk width is greater than 10 ft, or HCM Eq. 19-65 if the crosswalk width is 10 ft or less. CETR Eq. 3.17 is equivalent to both equations.

For more information on pedestrian crossings, see the entry "Pedestrian Crossing Factors" in this chapter.

Pedestrian Crossing Factors

CETR: Sec. 2.5, Sec. 3.1

HCM: Sec. 19.2, Sec. 19.5, Sec. 31.2

HCM Eq. 19-66 can be used to calculate the number of pedestrians crossing during a single cycle, which is used to calculate the pedestrian crossing time. CETR Eq. 3.16 is an equivalent equation. For more about the pedestrian crossing time, see the entry "Pedestrian Crossing Time" in this chapter.

$$N_{\text{ped,do}} = N_{\text{do}}\left(\frac{C - g_{\text{walk,mi}}}{C}\right) \quad \text{[HCM Eq. 19-66]}$$

The signal phase duration is calculated using HCM Eq. 19-2. CETR Eq. 2.44 is an equivalent equation.

$$D_p = l_1 + g_s + g_e + Y + R_c \quad \text{[HCM Eq. 19-2]}$$

For intersections with pedestrian crossings, the minimum green interval may be calculated using HCM Eq. 31-70. CETR Eq. 2.50 is an equivalent equation.

$$G_{p,\min} = t_{\text{pr}} + \frac{L_{cc}}{S_p} - Y - R_c \quad \text{[HCM Eq. 31-70]}$$

Effective Green Interval

CETR: Sec. 2.5

HCM: Sec. 19.2

$$\begin{aligned} g &= D_p - l_1 - l_2 \\ g &= g_s + g_e + e \end{aligned} \quad \text{[HCM Eq. 19-3]}$$

The effective green interval is the time during which a given traffic movement or set of movements may proceed at saturation flow rate. The effective green interval can be found from HCM Eq. 19-3. CETR Eq. 2.45 is an equivalent equation.

Exam tip: The extension of the effective green interval, e, is typically 2 sec.

Estimated Cycle Length

CERM: Sec. 73.18

$$C = \frac{1.5L + 5}{1 - \sum_{\substack{\text{critical phases}}} Y_i} = \frac{1.5L + 5}{1 - \sum_{\substack{\text{critical phases}}} \left(\frac{v}{c}\right)_i} \quad 73.34$$

Webster's equation for minimum delay, as shown in CERM Eq. 73.34, determines cycle lengths for an isolated, pre-timed location. The equation becomes unstable when used in situations with high levels of saturation and should not be used for locations where demand approaches capacity.

Yellow Change Interval

HCM: Sec. 19.2

MUTCD: Sec. 4D.26

The yellow change interval includes the reaction time, a deceleration element, and an intersection clearing time. Yellow change intervals longer than 5 sec can encourage drivers to treat the yellow as part of the green, and short yellow change intervals can result in an increase in rear-end accidents. The relationship between the yellow change interval and the other variables contributing to the signal phase duration is illustrated in HCM Exh. 19-7.

Red Clearance Interval

HCM: Sec. 19.2

MUTCD: Sec. 4D.26

The red clearance interval is the time that follows the yellow change interval and precedes the next conflicting green interval. Per MUTCD Sec. 4D.26, this interval should not be less than 3 sec nor greater than 6 sec. The relationship between the red clearance interval and the other variables contributing to the signal phase duration is illustrated in HCM Exh. 19-7.

B. SIGNAL WARRANTS

Key concepts: These key concepts are important for answering exam questions in subtopic 14.B, Signal Warrants.

- signal warrant 1, eight-hour vehicular volume
- signal warrant 2, four-hour vehicular volume
- signal warrant 3, peak hour

- signal warrant 4, pedestrian volume
- signal warrant 5, school crossing
- signal warrant 6, coordinated signal system
- signal warrant 7, crash experience
- signal warrant 8, roadway network
- signal warrant 9, intersection near a grade crossing

NCEES Handbook: The *Handbook* does not contain any material relevant to this subtopic. Be sure to study the listed sections in CERM and CETR and from codes and standards.

***PE Civil Reference Manual* (CERM):** Study these sections in CERM that either relate directly to this subtopic or provide background information.

- Section 73.20: Warrants for Intersection Signaling

***Transportation Depth Reference Manual* (CETR):** Study these sections in CETR that either relate directly to this subtopic or provide background information.

- Section 2.5: Traffic Signals
- Appendix 2.B: Warrant 1, Eight-Hour Vehicular Volume
- Appendix 2.C: Warrant 2
- Appendix 2.D: Warrant 3
- Appendix 2.E: Warrant 4
- Appendix 2.F: Warrant 9

Codes and standards: To prepare for this subtopic, familiarize yourself with these sections from codes and standards.

Manual on Uniform Traffic Control Devices for Streets and Highways (MUTCD)

- Section 4C.02: Warrant 1, Eight-Hour Vehicular Volume
- Section 4C.03: Warrant 2, Four-Hour Vehicular Volume
- Section 4C.04: Warrant 3, Peak Hour
- Section 4C.05: Warrant 4, Pedestrian Volume
- Section 4C.06: Warrant 5, School Crossing
- Section 4C.07: Warrant 6, Coordinated Signal System
- Section 4C.08: Warrant 7, Crash Experience
- Section 4C.09: Warrant 8, Roadway Network
- Section 4C.10: Warrant 9, Intersection Near a Grade Crossing

The following figures, tables, and concepts are relevant for subtopic 14.B, Signal Warrants.

Warrant 1: Eight-Hour Vehicular Volume

CERM: Sec. 73.20

CETR: Sec. 2.5, App. 2.B

MUTCD: Sec. 4C.02

The need for a traffic signal is determined by a traffic engineering study showing that the intersection satisfies one or more of the various prescribed traffic signal warrants, described in detail in MUTCD. Signal warrant 1 is the eight-hour vehicular volume, and it is satisfied by either condition A or condition B or a combination of both.

- Condition A specifies a minimum hourly volume of traffic on both the major street and minor street during any eight-hour period of the day.
- Condition B specifies that traffic on the major street (of a certain minimum volume) prevents traffic on the minor street (also of a certain minimum volume) from entering the intersection without significant delay during any eight-hour period of the day.

These minimums are given in MUTCD Table 4C-1, along with adjustments for intersections that satisfy both conditions. CETR App. 2.B is an equivalent table.

Warrant 2: Four-Hour Vehicular Volume

CERM: Sec. 73.20

CETR: Sec. 2.5, App. 2.C

MUTCD: Sec. 4C.03

The need for a traffic signal is determined by a traffic engineering study showing that the intersection satisfies one or more of the various prescribed traffic signal warrants, described in detail in MUTCD. Signal warrant 2 is the four-hour vehicular volume, which is satisfied if, during any four-hour period of the day, the plot of the traffic volumes on the major street and minor street falls above the curve shown in MUTCD Fig. 4C-1. If the posted or statutory speed limit is above 40 mph, or the intersection is in an area with a population of less than 10,000 people, MUTCD Fig. 4C-2 should be consulted instead of Fig. 4C-1. The highest volumes for the major street and the highest volumes for the minor street do not have to occur during the same four-hour period.

CETR App. 2.C presents an equivalent figure to MUTCD Fig. 4C-2.

Warrant 3: Peak Hour

CERM: Sec. 73.20

CETR: Sec. 2.5, App. 2.D

MUTCD: Sec. 4C.04

The need for a traffic signal is determined by a traffic engineering study showing that the intersection satisfies one or more of the various prescribed traffic signal warrants, described in detail in MUTCD. Signal warrant 3 is the peak hour, which is satisfied if traffic on the minor approach experiences undue delay for at least one hour during the average day.

Three conditions must be met during the same one-hour period.

- The vehicle delay on a minor approach is a minimum of four vehicle-hours for one-lane approaches and five vehicle-hours for two-lane approaches.
- The four-hour volume is at least 100 vph for one-lane approaches and 150 vph for two-lane approaches.
- The total volume is at least 650 vph for a three-approach intersection or 800 vph for intersections with four or more approaches.

Alternatively, warrant 3 is satisfied if the plot of the traffic volumes on the major street and minor street falls above the curve shown in MUTCD Fig. 4C-3, or Fig. 4C-4 if the posted or statutory speed limit is above 40 mph or the intersection is in an area with a population of less than 10,000 people. CETR App. 2.D provides equivalent figures to the MUTCD figures.

Warrant 4: Pedestrian Volume

CERM: Sec. 73.20

CETR: Sec. 2.5, App. 2.E

MUTCD: Sec. 4C.05

The need for a traffic signal is determined by a traffic engineering study showing that the intersection satisfies one or more of the various prescribed traffic signal warrants, described in detail in MUTCD. Signal warrant 4 is the pedestrian volume, which requires that two conditions be satisfied.

- The nearest existing signal must be at least 300 ft away from the intersection.
- The plot of the traffic volume on the major street and the corresponding pedestrian volume crossing the major street must fall above the curve shown in MUTCD Fig. 4C-5 for any four-hour period of an average day, or above the curve shown in MUTCD Fig. 4C-7 for any one-hour period of an average day. If the posted or statutory speed limit is above 40 mph, or the intersection is in an area with a population of less than 10,000 people, the plot is compared to MUTCD Fig. 4C-6 for a four-hour period or MUTCD Fig. 4C-8 for a one-hour period. CETR App. 2.E provides equivalent figures to the MUTCD figures.

Warrant 5: School Crossing

CERM: Sec. 73.20

CETR: Sec. 2.5

MUTCD: Sec. 4C.06

The need for a traffic signal is determined by a traffic engineering study showing that the intersection satisfies one or more of the various prescribed traffic signal warrants, described in detail in MUTCD. Signal warrant 5 is for a traffic signal at a school crossing, which requires that two conditions be satisfied.

- The nearest existing signal must be at least 300 ft away from the intersection, unless the new signal will not restrict the progressive movement of traffic.
- During the time period that children are using the crossing, there must be fewer adequate crossing gaps in the traffic stream than the minutes required for the children to cross, and there must be at least 20 students crossing during the highest-volume crossing hour.

Warrant 6: Coordinated Signal System

CERM: Sec. 73.20

CETR: Sec. 2.5

MUTCD: Sec. 4C.07

The need for a traffic signal is determined by a traffic engineering study showing that the intersection satisfies one or more of the various prescribed traffic signal warrants, described in detail in MUTCD. Signal warrant 6 is the coordinated signal system warrant, which is satisfied if either of the following conditions are met.

- The street is a one-way street or a street with traffic that is predominantly in one direction, and the existing signals are too far apart to maintain efficient platooning.
- The street is a two-way street, existing adjacent signals do not provide efficient platooning, and the new signals will collectively provide a progressive operation.

In either case, to satisfy this warrant, the new signals should be spaced at least 1000 ft from existing signals.

Warrant 7: Crash Experience

CERM: Sec. 73.20

CETR: Sec. 2.5

MUTCD: Sec. 4C.08

The need for a traffic signal is determined by a traffic engineering study showing that the intersection satisfies one or more of the various prescribed traffic signal warrants, described in detail in MUTCD. Signal warrant 7 is the crash experience warrant, which requires three conditions all be met.

- Trial periods of lesser measures have not reduced crash frequency.

- Five or more reportable crashes that were susceptible to correction by a traffic control signal, and that involved property damage or personal injury, have occurred at the intersection within a 12-month period.

- The traffic volume at the intersection is 80% of the traffic volume indicated in MUTCD Table 4C-1.

Warrant 8: Roadway Network

CERM: Sec. 73.20

CETR: Sec. 2.5

MUTCD: Sec. 4C.09

The need for a traffic signal is determined by a traffic engineering study showing that the intersection satisfies one or more of the various prescribed traffic signal warrants, described in detail in MUTCD. Signal warrant 8 is the roadway network warrant. This warrant is satisfied if the traffic volume in the intersection is at least 1000 vph during the peak hour of a typical workday or is projected to reach that volume within the next five years, and any one of the following conditions is met.

- The intersection is part of a principal network for through traffic.

- The intersection includes rural or suburban highways near or through a city.

- The intersection appears as a major route on an official transportation plan.

Warrant 9: Intersection Near a Grade Crossing

CERM: Sec. 73.20

CETR: Sec. 2.5, App. 2.F

MUTCD: Sec. 4C.10

The need for a traffic signal is determined by a traffic engineering study showing that the intersection satisfies one or more of the various prescribed traffic signal warrants, described in detail in MUTCD. Signal warrant 9 is intended for use in the case of an intersection that meets none of the other eight warrants, but is near a grade crossing that is not currently signalized. The warrant is satisfied if all three of the following conditions are met.

- The grade crossing is on an approach controlled by a stop or yield sign.

- The nearest track is within 140 ft of the stop line or yield line on the approach.

- The plot of the traffic volumes on the major street and the minor street approach that crosses the track is above the curve in the appropriate MUTCD figure. If there is only one lane approaching the intersection at the track crossing location, the plot is compared to MUTCD Fig. 4C-9. If there are two or more lanes approaching the intersection, the plot is compared to MUTCD Fig. 4C-10.

41 Traffic Control Design

Content in blue refers to the *NCEES Handbook*.

A. Signs and Pavement Markings.......................... 41-1
B. Temporary Traffic Control.............................. 41-2

The knowledge area of Traffic Control Design makes up between three and five questions out of the 80 questions on the PE Civil Transportation exam. The organization of this chapter follows the order of subtopics given by NCEES for this knowledge area. Each subtopic is covered in the following sections.

A. SIGNS AND PAVEMENT MARKINGS

Key concepts: These key concepts are important for answering exam questions in subtopic 15.A, Signs and Pavement Markings.

- guide signs
- pavement markings
- regulatory sign types
- traffic sign placement
- traffic sign shapes
- traffic sign types
- warning signs

NCEES Handbook: The *Handbook* does not contain any material relevant to this subtopic. Be sure to study the listed sections in CERM and CETR and from codes and standards.

PE Civil Reference Manual **(CERM):** Study these sections in CERM that either relate directly to this subtopic or provide background information.

- Section 73.37: Temporary Traffic Control Zones

Transportation Depth Reference Manual **(CETR):** Study these sections in CETR that either relate directly to this subtopic or provide background information.

- Section 6.8: Driver Behavior and Performance

Codes and standards: To prepare for this subtopic, familiarize yourself with these sections from codes and standards.

Manual on Uniform Traffic Control Devices (MUTCD)

- Section 2A.09: Shapes
- Section 2A.16: Standardization of Location
- Section 2B.13: Speed Limit Sign (R2-1)
- Section 2B.37: DO NOT ENTER Sign (R5-1)
- Section 2C.05: Placement of Warning Signs
- Section 2C.06: Horizontal Alignment Warning Signs
- Section 2D.03: Color, Retroreflection, and Illumination
- Section 2E.36: Exit Direction Signs
- Chapter 3B: Pavement and Curb Markings

The following concepts are relevant for subtopic 15.A, Signs and Pavement Markings.

Traffic Sign Shapes

CETR: Sec. 6.8

MUTCD: Sec. 2A.09, Sec. 2A.16

Traffic signs are designed and placed according to the guidelines in MUTCD and any additional local requirements. MUTCD Table 2A-4 describes the shapes of common traffic signs. Standards for a sign's height, color, letter spacing, message content, panels, layout, and location are discussed in detail in MUTCD Sec. 2A.09. Some examples of sign usages are given in MUTCD Fig. 2A-4.

Traffic Sign Placement

MUTCD: Sec. 2A.09, Sec. 2A.16

Traffic signs and critical road markings should be placed so that they are within the established normal central cone of vision in order to be clearly read by drivers. Some examples of heights and lateral locations of sign installations are given in MUTCD Fig. 2A-2, Fig. 2A-3, and Fig. 2A-4.

Traffic Sign Coloration

MUTCD: Sec. 2A.09

The colors prescribed by MUTCD for traffic control devices and informational signs are meant to minimize the effects of color blindness on a sign's readability. The prescribed background and letter coloring provide enough contrast that lettering does not disappear when viewed by someone with color-impaired vision. MUTCD Table 2A-5 gives examples of common usage of sign colors.

> *Exam tip*: Exam questions may require looking up information in MUTCD Table 2A-5. To find the table quickly in an exam setting, search for "Common Uses of Sign Color."

Regulatory Signs

MUTCD: Sec. 2B.13, Sec. 2B.37

Regulatory signs include speed limit, photo enforcement, exclusion, and wrong-way signs. Speed limit signs indicate the speed limits required by law at points of change between speed limits. Photo enforcement signs indicate that traffic laws are enforced using photographic equipment in a certain area. MUTCD Fig. 2B-3 shows examples of speed limit and photo enforcement signs.

Exclusion signs indicate that certain types of traffic are forbidden from using some or all of a roadway. MUTCD Fig. 2B-11 shows examples of exclusion signs.

Wrong-way signage indicates that a driver reading the sign is driving their vehicle the wrong way. MUTCD Fig. 2B-12 shows examples of locations for wrong-way signage.

Warning Signs

CETR: Sec. 6.8

MUTCD: Chap. 3B

Warning signs need to be placed on roadways such that they provide motorists with an adequate perception-reaction time, but not so far ahead of a hazard that drivers may forget the warning. MUTCD Table 2C-4 shows the guidelines for advance placement of warning signs.

Horizontal alignment warning signs are used to inform motorists of a change in the roadway. MUTCD Fig. 2C-1 shows examples of various horizontal alignment signs and plaques. MUTCD Table 2C-5 describes horizontal alignment sign selection, including whether a horizontal alignment warning is required, recommended, or even allowed.

Guide Signs

MUTCD: Sec. 2D.03

Road guide signs are uniform in design. MUTCD Fig. 2D-1 shows examples of color-coded road guide signs.

Pavement Markings

MUTCD: Chap. 3B

Pavement markings are discussed in MUTCD Chap. 3. The placement, shape, coloration, and other parameters of pavement markings are specified in MUTCD Chap. 3B. Critical road markings should be placed so that they can be clearly read by drivers. MUTCD Fig. 3B-8 and Fig. 3B-10 show examples of this.

B. TEMPORARY TRAFFIC CONTROL

Key concepts: These key concepts are important for answering exam questions in subtopic 15.B, Temporary Traffic Control.

- advance warning sign placement
- channelizing devices
- detour signs
- one-way traffic control
- taper length criteria
- temporary traffic barriers
- temporary traffic control scenarios
- temporary traffic control tapers
- types of lane tapering
- typical applications

NCEES Handbook: The *Handbook* does not contain any material relevant to this subtopic. Be sure to study the listed sections in CERM and CETR and from codes and standards.

PE Civil Reference Manual (**CERM**): Study these sections in CERM that either relate directly to this subtopic or provide background information.

- Section 73.37: Temporary Traffic Control Zones

Transportation Depth Reference Manual (**CETR**): Study these sections in CETR that either relate directly to this subtopic or provide background information.

- Section 2.5: Traffic Signals
- Section 6.10: Temporary Traffic Control Zones

Codes and standards: To prepare for this subtopic, familiarize yourself with these sections from codes and standards.

Manual on Uniform Traffic Control Devices (MUTCD)

- Section 3H.01: Channelizing Devices
- Section 6C.08: Tapers
- Section 6F.19: DETOUR Sign (W20-2)
- Section 6F.63: Channelizing Devices
- Section 6F.70: Temporary Traffic Barriers as Channelizing Devices
- Section 6F.85: Temporary Traffic Barriers
- Section 6G.01: Typical Applications
- Section 6G.06: Work Outside of the Shoulder
- Section 6G.07: Work on the Shoulder with No Encroachment
- Section 6H.01: Typical Applications

The following figures, tables, and concepts are relevant for subtopic 15.B, Temporary Traffic Control.

Temporary Traffic Control Tapers

MUTCD: Sec. 6C.08

Tapers are created by using a series of channelizing devices and/or pavement markings to move traffic into or out of a given pathway. Types of tapers are illustrated in MUTCD Fig. 6C-2.

Taper Length Criteria

MUTCD: Sec. 6C.08

The required length of a taper depends on the type of taper, the speed of traffic, the posted speed limit, and the width of the offset. MUTCD Table 6C-4 provides formulas for calculating the taper length, and MUTCD Table 6C-3 provides modifications to that value based on the type of taper.

Temporary Traffic Barriers

MUTCD: Sec. 6F.63, Sec. 6F.70

Temporary traffic barriers are barriers designed to prevent vehicles from entering certain areas (e.g., work areas) on or near a roadway. Temporary traffic barriers and other channelizing devices are shown in MUTCD Fig. 6F-7. All such devices must be designed to withstand crashes and changes in weather conditions.

Detour Signs

MUTCD: Sec. 6F.19, Sec. 6H.01

Detour signs are used to indicate that a road user must detour onto a different street or route. Requirements for detour signs are discussed in MUTCD Sec. 6F-19. MUTCD Fig. 6H-20 shows examples of how detour signs can be applied for a closed street.

Other Temporary Traffic Control Scenarios

MUTCD: Sec. 6G.06, Sec. 6G.07, Sec. 6H.01

Instead of hazard lights, vehicles should utilize rotating, flashing, oscillating, and strobe lights if they are available.

MUTCD Fig. 6H-1 and Fig. 6H-3 show signage that applies to roadwork on or beyond the shoulder. If 8 ft or more of a shoulder is closed, at least one warning sign must be used.

42 Geotechnical and Pavement

Content in blue refers to the *NCEES Handbook*.

A. Sampling and Testing.. 42-1
B. Soil Stabilization Techniques............................ 42-2
C. Design Traffic Analysis and Pavement Design
   Procedures.. 42-5
D. Pavement Evaluation and Maintenance
   Measures... 42-8

The knowledge area of Geotechnical and Pavement makes up between four and six questions out of the 80 questions on the PE Civil Transportation exam. The organization of this chapter follows the order of subtopics given by NCEES for this knowledge area. Each subtopic is covered in the following sections.

A. SAMPLING AND TESTING

Key concepts: These key concepts are important for answering exam questions in subtopic 16.A, Sampling and Testing.

- California bearing ratio (CBR) tests and values
- soil testing methods
- subgrade modulus

NCEES Handbook: To prepare for this subtopic, familiarize yourself with these sections in the *Handbook*.

- Hammer Efficiency
- Soil Properties Correlated with Standard Penetration Test Values
- Commonly Performed Laboratory Tests on Soils

PE Civil Reference Manual (**CERM**): Study these sections in CERM that either relate directly to this subtopic or provide background information.

- Section 35.8: Standardized Soil Testing Procedures
- Section 35.9: Standard Penetration Test
- Section 35.22: California Bearing Ratio Test
- Section 35.23: Plate Bearing Value Test

Codes and standards: To prepare for this subtopic, familiarize yourself with these sections from codes and standards.

Guide for Design of Pavement Structures (AASHTO GDPS)

- Part I Section 1.5: Roadbed Soil
- Part II Section 2.2.1: Serviceability

The following equations, tables, and concepts are relevant for subtopic 16.A, Sampling and Testing.

Standardized Soil Testing Procedures

Handbook: Hammer Efficiency; Soil Properties Correlated with Standard Penetration Test Values

$$N_{60} = \left(\frac{E_{\text{eff}}}{60}\right)N_{\text{meas}}$$

CERM: Sec. 35.8, Sec. 35.9

The most commonly used soil test is the *standard penetration test* (SPT). The SPT measures resistance to the penetration of a standard split-spoon sampler that is driven by a 140 lbm (63.5 kg) hammer dropped from a height of 30 in (0.76 m). The number of blows required to drive the sampler a distance of 12 in (0.305 m) after an initial penetration of 6 in (0.15 m) is referred to as the *N*-value or *standard penetration resistance*, measured in blows per foot. The adjusted *N*-value can be found from the measured *N*-value and the hammer efficiency using the equation in *Handbook* section Hammer Efficiency. Adjusted *N*-values for various densities and consistencies of soil can be found in *Handbook* table Soil Properties Correlated with Standard Penetration Test Values.

Exam tip: CERM Chap. 35 lists all of the available soil testing procedures. Understanding when each of the procedures is used and what advantages and disadvantages they each have may be useful for the exam.

California Bearing Ratio Test

Handbook: Commonly Performed Laboratory Tests on Soils

CERM: Sec. 35.22

$$\text{CBR} = \frac{\text{actual load}}{\text{standard load}} \cdot 100 \qquad 35.47$$

The *California bearing ratio* (CBR) test is used to determine the suitability of a soil for use as a subbase in pavement sections. The test measures the relative load required to cause a standard 3 in^2 (19.3 cm^2) plunger to penetrate a water-saturated soil specimen at a specific rate to a specific depth. The resulting data is measured in a ratio of inches of penetration to pounds-force of load. The CBR test is most reliable when used to test cohesive soils. CERM Eq. 35.47 can be used to find the CBR value for a given load.

Plate Bearing Value Test

CERM: Sec. 35.23

The *plate bearing value test* provides an indication of the shear strength of pavement components. A standard diameter round steel plate is set over the soil on a bedding of fine sand or plaster of paris. After the plate is seated by a quick but temporary load, it is loaded to a deflection of about 0.04 in (1 mm). This load is maintained until the deflection rate decreases to 0.001 in/min (0.03 mm/min). The load is then released. The deflection prior to loading, the final deflection, and the deflection at each minute are recorded, and the test is repeated 10 times. The bearing value is the interpolated load that would produce a deflection of 0.5 in (12 mm).

B. SOIL STABILIZATION TECHNIQUES

Key concepts: These key concepts are important for answering exam questions in subtopic 16.B, Soil Stabilization Techniques.

- average end area method
- compaction and settlement
- cut and fill volumes
- erosion control methods
- excavation and embankment
- mass balance
- soil classification systems
- swell and shrinkage

NCEES Handbook: To prepare for this subtopic, familiarize yourself with these sections in the *Handbook*.

- Cross-Section Methods
- Construction of Field Virgin Consolidation Relationships
- Typical Consolidation Curve for Normally Consolidated Soil
- Typical Consolidation Curve for Overconsolidated Soil
- Soil Classification Chart
- AASHTO Soil Classification System
- A Unit of Soil Mass and Its Idealization
- Weight-Volume Relationships
- Volume and Weight Relationships
- Slope and Shield Configurations
- Allowable Slopes
- Slope Configurations: Excavations in Layered Soils
- Excavations Made in Type A Soil
- Excavations Made in Type B Soil
- Shrinkage of Soil Mass
- Soil Compaction

PE Civil Reference Manual **(CERM):** Study these sections in CERM that either relate directly to this subtopic or provide background information.

- Section 35.2: Soil Classification
- Section 35.3: AASHTO Soil Classification
- Section 35.4: Unified Soil Classification
- Section 35.11: Proctor Test
- Section 40.4: Settling
- Section 40.6: Consolidation Parameters
- Section 40.7: Primary Consolidation
- Section 40.8: Primary Consolidation Rate
- Section 40.9: Secondary Consolidation
- Section 80.2: Unit of Measure
- Section 80.3: Swell and Shrinkage
- Section 80.4: Classification of Materials
- Section 80.5: Cut and Fill
- Section 80.7: Cross Sections
- Section 80.8: Original and Final Cross Sections

- Section 80.13: Earthwork Volumes
- Section 80.14: Average End Area Method
- Section 80.15: Prismoidal Formula Method
- Section 80.16: Borrow Pit Geometry
- Section 80.17: Mass Diagrams
- Section 80.20: Slope and Erosion Control Features

Transportation Depth Reference Manual (CETR): Study these sections in CETR that either relate directly to this subtopic or provide background information.

- Section 5.2: Excavation and Embankment
- Section 5.3: Mass Diagrams

The following equations, figures, and tables are relevant for subtopic 16.B, Soil Stabilization Techniques.

Earthwork Volumes

Handbook: Cross-Section Methods

For the average end area method,

$$V = L\left(\frac{A_1 + A_2}{2}\right)$$

For the prismoidal formula method,

$$V = L\left(\frac{A_1 + 4A_m + A_2}{6}\right)$$

CERM: Sec. 80.14, Sec. 80.15, Sec. 80.16

$$V = \frac{L(A_1 + A_2)}{2} \qquad 80.8$$

$$V_{\text{pyramid}} = \frac{L A_{\text{base}}}{3} \qquad 80.9$$

$$V = \left(\frac{L}{6}\right)(A_1 + 4A_m + A_2) \qquad 80.10$$

There are two methods of calculating the earthwork volume between cross sections. The first is the *average end area method*, which averages the two cross-sectional areas and multiplies them by the length between cross sections to get a total volume, as shown in the *Handbook* equation and CERM Eq. 80.8.

The precision obtained from the average end area method is generally sufficient unless one of the end areas is very small or zero. In that case, the volume should be computed as a truncated pyramid, as shown in CERM Eq. 80.9.

The second method of calculating earthwork volume is the *prismoidal formula*, which is preferred when the two end areas differ greatly or when the ground surface is irregular. The prismoidal formula uses the mean area midway between the two end sections by averaging similar dimensions of the start and end areas, as shown in the *Handbook* equation and CERM Eq. 80.10. CERM Fig. 80.6 illustrates these dimensions.

Figure 80.6 Soil Prismoid

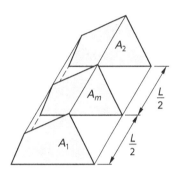

Exam tip: The averaging of similar dimensions of the start and end areas is not the same as averaging the two end areas themselves.

Compaction and Settlement

Handbook: Construction of Field Virgin Consolidation Relationships; Typical Consolidation Curve for Normally Consolidated Soil; Typical Consolidation Curve for Overconsolidated Soil; Soil Compaction

CERM: Sec. 35.11, Sec. 40.4, Sec. 40.6, Sec. 40.7, Sec. 40.8, Sec. 40.9

Settlement is generally caused by consolidation, a decrease in the void fraction of the supporting soil. In order to reduce the effects of settlement, soils are intentionally compacted during construction. This is usually accomplished by placing the soil in lifts (i.e., layers) and then mechanically compacting the lifts. Compaction equipment, as shown in CERM Fig. 35.5, can densify the soil using static loading, impact, vibration, and/or kneading actions.

Soil Classification Systems

Handbook: Soil Classification Chart; AASHTO Soil Classification System

CERM: Sec. 35.2, Sec. 35.3, Sec. 35.4

The classification of a soil depends mostly on the percentages of gravel, sand, silt, and clay. Organic matter can also be present in a soil. There are two major soil classification systems, the American Association of State Highway Officials (AASHTO) classification system and the Unified Soil Classification System (USCS).

The AASHTO system is based on sieve analysis and the liquid limit and plasticity index of the soil. The best soils for use as roadway subgrades are classified as A-1. Highly organic soils not suitable for roadway subgrades are classified as A-8. See *Handbook* table **AASHTO Soil Classification System** for data on how to classify soils using the AASHTO soil classification system. CERM Table 35.4 is an equivalent table.

The USCS is based on the grain size distribution, liquid limit, and plasticity index of the soil. Soils are classified into groups that contain two letters: the first represents the most significant particle size fraction (e.g., *G*, which is for gravelly soils), and the second is a descriptive modifier (e.g., *W*, which is for well-graded, fairly clean soils). See *Handbook* table **Soil Classification Chart** for data on how to classify soils using the USCS. CERM Table 35.5 is an equivalent table.

Shrinkage and Swell

Handbook: A Unit of Soil Mass and Its Idealization; Weight-Volume Relationships; Volume and Weight Relationships; Shrinkage of Soil Mass

CERM: Sec. 80.2, Sec. 80.3

CETR: Sec. 5.2

The volume and density of earth changes under natural conditions and during excavation, hauling, and placing. The volume of a loose pile of excavated earth will be greater than the original, in-place natural volume, and if the earth is compacted after it is placed, the volume may be less than its original, in-place natural volume.

The change in volume of earth from its natural state to its loose state is known as *swell*, expressed as a percentage of the natural volume. The decrease in volume of earth from its natural state to its compacted state is known as *shrinkage*, expressed as a percent decrease from the natural state. Equations showing the relationships between volumes and weights of soils can be found in *Handbook* tables **Weight-Volume Relationships** and **Volume and Weight Relationships**, and the different volumes and weights in soil are illustrated in *Handbook* figure **A Unit of Soil Mass and Its Idealization**.

Excavation and Embankment

Handbook: Slope and Shield Configurations; Allowable Slopes; Slope Configurations: Excavations in Layered Soils; Excavations Made in Type A Soil; Excavations Made in Type B Soil

CERM: Sec. 80.4, Sec. 80.7, Sec. 80.8, Sec. 80.13

CETR: Sec. 5.2

Earthwork that is to be excavated is known as *cut*. Earthwork that is placed in an embankment is known as *fill*. These concepts are illustrated in CETR Fig. 5.2.

Figure 5.2 *Cut and Fill Sections*

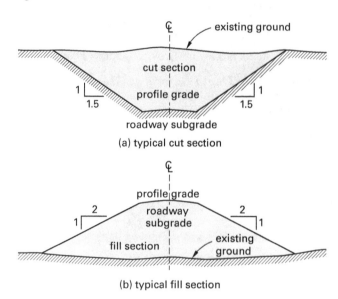

Depending on the depth of the excavation and the type of soil, excavation and embankment for a roadway project may need additional support to prevent collapse and ensure worker safety. *Handbook* figure **Slope and Shield Configurations** illustrates the allowable depths and height-depth ratios of various possible supporting configurations, and *Handbook* table **Allowable Slopes** provides guidelines for allowable height-depth ratios and slope angles for different soil types.

Cross Sections

CERM: Sec. 80.7

Cross sections are profiles of the earth taken at right angles to the centerline of an engineering project. To obtain a total volume measurement, cross sections are cut at a regular interval along the project centerline, with area shapes showing where cut and fill will be taking place at the cross section. These areas can then be interpolated in order to calculate approximate earthwork volumes. CERM Fig. 80.1 illustrates a typical highway cross section.

Figure 80.1 *Typical Highway Cross Section*

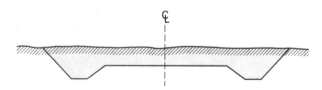

Mass Diagrams

CERM: Sec. 80.17

CETR: Sec. 5.3

A *profile diagram* is a cross section of the existing ground elevation along a route alignment. A *mass diagram* is a record of the cumulative earthwork volume moved along an alignment, plotted above and below the finished roadway profile. CERM Fig. 80.8 illustrates profile and mass diagrams.

Figure 80.8 Profile and Mass Diagrams

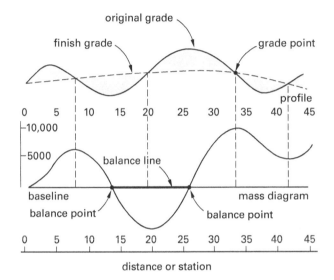

Erosion Control Methods

CERM: Sec. 80.20

Almost all excavations require temporary measures to control erosion and water pollution of the construction job site and surrounding areas. These measures include berms, dikes, dams, sediment basins, and fiber mats. See CERM Fig. 80.12 for illustrations of these measures and CERM Sec. 80.20 for in-depth descriptions of them all.

C. DESIGN TRAFFIC ANALYSIS AND PAVEMENT DESIGN PROCEDURES

Key concepts: These key concepts are important for answering exam questions in subtopic 16.C, Design Traffic Analysis and Pavement Design Procedures.

- asphalt concrete design methods
- asphalt concrete pavement
- concrete design
- design traffic and trucks
- pavement preservation
- pavement structural numbers
- paving equipment
- rigid pavements

NCEES Handbook: To prepare for this subtopic, familiarize yourself with these sections in the *Handbook*.

- Predicting Truck Traffic Volumes
- Monthly Adjustment Factor
- Traffic Growth Rate

PE Civil Reference Manual (**CERM**): Study these sections in CERM that either relate directly to this subtopic or provide background information.

- Section 49.1: Concrete Mix Design Considerations
- Section 49.6: Absolute Volume Method
- Section 76.1: Asphalt Concrete Pavement
- Section 76.9: Weight-Volume Relationships
- Section 76.10: Placement and Paving Equipment
- Section 76.11: Rolling Equipment
- Section 76.12: Characteristics of Asphalt Concrete
- Section 76.13: Hot Mix Asphalt Concrete Mix Design Methods
- Section 76.14: Marshall Mix Design
- Section 76.16: Hveem Mix Design
- Section 76.17: Superpave
- Section 76.18: Flexible Pavement Structural Design Methods
- Section 76.19: Traffic
- Section 76.20: Truck Factors
- Section 76.21: Design Traffic
- Section 76.26: Pavement Structural Number
- Section 76.27: Asphalt Institute Method of Full-Depth Flexible Pavement Design
- Section 77.1: Rigid Pavement
- Section 77.2: Mixture Proportioning
- Section 77.3: AASHTO Method of Rigid Pavement Design
- Section 77.4: Layer Material Strengths
- Section 77.6: Load Transfer and Dowels
- Section 77.7: Pavement Design Methodology
- Section 77.8: Steel Reinforcing
- Section 77.10: Pavement Joints

Transportation Depth Reference Manual (CETR): Study these sections in CETR that either relate directly to this subtopic or provide background information.

- Section 5.4: Pavement Design
- Section 5.5: Hot Mix Asphalt
- Section 5.6: Marshall Mix Design
- Section 5.7: Hveem Mix Design
- Section 5.8: Superpave Mix Design Procedures

Codes and standards: To prepare for this subtopic, familiarize yourself with these sections from codes and standards.

Guide for Design of Pavement Structures (AASHTO GDPS)

- Part I Section 1.2: Design Considerations
- Part II Section 2.1.2: Traffic
- Part II Section 2.3.3: Pavement Layer Materials Characterization
- Part II Section 2.3.4: PCC Modulus of Rupture
- Part II Section 2.3.5: Layer Coefficients
- Part II Section 3.1.1: Determine Required Structural Number
- Part II Section 3.1.4: Selection of Layer Thickness
- Part II Section 3.1.5: Layered Design Analysis
- Part II Section 3.2.1: Develop Effective Modulus of Subgrade Reaction
- Part II Section 3.2.2: Determine Required Slab Thickness
- Appendix D: Conversion of Mixed Traffic to Equivalent Single Axle Loads for Pavement Design

The following equations and figures are relevant for subtopic 16.C, Design Traffic Analysis and Pavement Design Procedures.

Design Traffic and Trucks

Handbook: Predicting Truck Traffic Volumes; Monthly Adjustment Factor; Traffic Growth Rate

CERM: Sec. 76.19, Sec. 76.20, Sec. 76.21

AASHTO GDPS: Part II Sec. 2.1.2, App. D

In order to use the AASHTO pavement design method, all traffic must be converted into *equivalent single-axle loads* (ESALs), the equivalent number of 18,000 lbf single axles (with dual tires). This can be calculated using CERM Eq. 76.23. AASHTO GDPS App. D gives load equivalency factors (LEFs) for flexible pavements for various terminal serviceability indices.

$$\text{ESALs} = (\text{no. of axles})(\text{LEF}) \qquad 76.23$$

Truck factors (TFs) are the average LEFs for a given class of vehicle, calculated by the total ESALs for all axles divided by the number of trucks. Truck factors can be calculated using CERM Eq. 76.24.

$$\text{TF} = \frac{\text{ESALs}}{\text{no. of trucks}} \qquad 76.24$$

Asphalt Concrete

CERM: Sec. 76.1, Sec. 76.9, Sec. 76.12

CETR: Sec. 5.4, Sec. 5.5

Hot mix asphalt, commonly referred to as *asphalt concrete*, is a mixture of asphalt cement and well-graded, high-quality aggregate, heated and compacted by a pavement machine into a uniform dense mass. Asphalt pavement is easily repaired, additional thicknesses can be placed at any time, and its nonskid properties do not significantly deteriorate over time. The mineral aggregate component of the asphalt mixture comprises 90–95% of the weight and 75–85% of the volume.

Asphalt Concrete Design Methods

CERM: Sec. 76.12, Sec. 76.13, Sec. 76.14, Sec. 76.16, Sec. 76.17

CETR: Sec. 5.6, Sec. 5.7, Sec. 5.8

There are several methods of designing the mix of asphalt concrete.

The *Marshall mix design method* is focused on determination of the optimum asphalt content of a mix. It is a fairly simple procedure that does not require complex equipment, and the needed materials and equipment are portable.

The *Hveem mix design method* expands upon the Marshall mix method by additionally considering resistance to shear and asphalt absorption by aggregates. This leads to the disadvantage of requiring more specialized and nonportable equipment for mixing, compaction, and testing.

The *Superpave* (Superior Performing Asphalt Pavement) *mixture design method* is a product of the U.S. Federal Highway Administration; it has rapidly become the standard design methodology for hot mix asphalt highway pavements in North America. Superpave is separated into three levels of design, with level 1 design requiring the least arduous specifications and level 3 the most. Level 1 Superpave design is used by most highway agencies and in most new highway construction.

Concrete Design

CERM: Sec. 49.1, Sec. 49.6

Concrete can be designed for compressive strength or durability in its hardened state. In its wet state, concrete should have good workability. Mixes are often described by the number of sacks of cement needed to produce 1 yd^3 of concrete. Concrete strength is inversely proportional to the water-cement ratio by weight.

Rigid Pavements

CERM: Sec. 77.1, Sec. 77.3

AASHTO GDPS: Part II Sec. 2.3.4, Sec. 3.2.1, Sec. 3.2.2

Portland cement concrete pavement (PCCP) is the most common form of rigid pavement because of its excellent durability and long service life. Its raw materials are readily available and reasonably inexpensive, it is easily formed, it withstands exposure to water without deterioration, and it is recyclable. The primary disadvantages of PCCP are that it can lose its nonskid surface over time, it must be used with an even subgrade where uniform settling is expected, and it may fault at transverse joints.

Pavement Structural Numbers

CERM: Sec. 76.26

AASHTO GDPS: Part II Sec. 2.3.3, Sec. 2.3.5, Sec. 3.1.1, Sec. 3.1.4, Sec. 3.1.5

The goal of flexible pavement structural design is to specify the thicknesses of all structural layers in a pavement. The design of a hot mix asphalt pavement requires knowledge of climate, traffic, subgrade soil support, and drainage.

Prior to designing a flexible pavement, the strengths of the pavement layers and the underlying soil must be determined or assumed. The American Association of State Highway and Transportation Officials (AASHTO) combines pavement layer properties and thicknesses into one variable called the *design structural number*, SN, as shown in CERM Eq. 76.33.

$$\text{SN} = D_1 a_1 + D_2 a_2 m_2 + D_3 a_3 m_3 \qquad 76.33$$

D_1, D_2, and D_3 represent the actual thicknesses, in inches, of the surface, base, and subbase courses, respectively, as shown in CERM Fig. 76.11. If a subbase layer is not used, the third term is omitted. The a terms are the layer coefficients based on material, which can be found using CERM Table 76.13, and the m_i terms are drainage coefficients, typically ranging from 0.40 to 1.40, as recommended in AASHTO GDPS Table 2.4.

Figure 76.11 Procedure for Determining Thicknesses of Layers Using a Layered Analysis Approach*

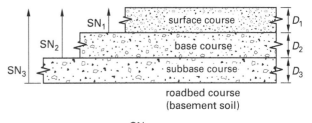

$$D_1^* \geq \frac{\text{SN}_1}{a_1}$$

$$\text{SN}_1^* = a_1 D_1^* \geq \text{SN}_1$$

$$D_2^* \geq \frac{\text{SN}_2 - \text{SN}_1^*}{a_2 m_2}$$

$$\text{SN}_1^* + \text{SN}_2^* \geq \text{SN}_2$$

$$D_3^* \geq \frac{\text{SN}_3 - \left(\text{SN}_1^* + \text{SN}_2^*\right)}{a_3 m_3}$$

*Asterisks indicate values actually used, which must be equal to or greater than the required values.

Paving Equipment

CERM: Sec. 76.10, Sec. 76.11

Large asphalt paving machines can place a layer, or lift, of asphalt with a thickness between 1 in and 10 in over a width of 6 in to 32 ft, at a forward speed of 10 ft/min to 70 ft/min.

Most specifications for pavement now require 92–97% density, which means rolling is necessary even with high-density screeds. Production efficiency is higher with vibratory rollers than static rollers, but consistency, noise, and control are sometimes problems with vibratory rollers. Rubber-tired pneumatic rollers provide a kneading action in the finish roll. When static wheel rollers are used, the maximum lift thickness that can be compacted is about 3 in. When pneumatic or vibratory rollers are used, the maximum lift thickness that can be compacted is almost unlimited, but is generally between 6 in and 8 in.

Pavement Preservation

CERM: Sec. 77.8, Sec. 77.10

The many types of reinforcement used to preserve concrete rigid pavements include dowels, aggregate interlock, and tied shoulders. Reinforcing steel, such as reinforcing bar, smooth wire mesh, or wire fabric, can also be used. The purpose of the steel is not to add structural strength but to hold cracks tightly together and restrain their growth.

CERM Fig. 77.5 shows the three standard forms of joints used on a concrete pavement. *Contraction joints* relieve tensile stresses in the pavement. *Construction joints* are used transversely between construction periods and longitudinally between pavement lanes, and are usually spaced to coincide with pavement lane markings. *Isolation joints* (expansion joints) are used to relieve compressive stresses in the pavement where it adjoins another structure (e.g., bridges or intersections).

Figure 77.5 Types of Concrete Pavement Joints

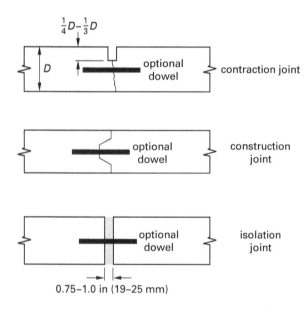

Grooving is a method of increasing skid resistance and reducing hydroplaning on all types of pavements. Grooves permit water to escape under tires and prevent water buildup on the surface. Grooving should only be used with structurally adequate pavements.

D. PAVEMENT EVALUATION AND MAINTENANCE MEASURES

Key concepts: These key concepts are important for answering exam questions in subtopic 16.D, Pavement Evaluation and Maintenance Measures.

- damage from frost and freezing
- pavement problems and defects
- pavement properties

NCEES Handbook: The *Handbook* does not contain any material relevant to this subtopic. Be sure to study the listed sections in CERM.

PE Civil Reference Manual (**CERM**): Study these sections in CERM that either relate directly to this subtopic or provide background information.

- Section 76.5: Pavement Properties
- Section 76.6: Problems and Defects
- Section 76.32: Damage from Frost and Freezing

The following concepts are relevant for subtopic 16.D, Pavement Evaluation and Maintenance Measures.

Properties of Pavement Quality

CERM: Sec. 76.5

Pavement quality is rated on several factors. *Stability* measures how well a pavement can resist permanent deformation over a long time period. *Durability* measures how well a pavement can resist disintegration by weathering. *Flexibility* measures how well a pavement can conform to gradual changes such as temperature and settlement. *Fatigue resistance* measures how well a pavement can withstand repeated wheel loads. *Skid resistance* measures how well a pavement can resist tire slipping or skidding. *Impermeability* measures how well a pavement can resist the passage of water and air. *Workability* measures how easily a pavement can be placed and compacted.

Pavement Problems and Defects

CERM: Sec. 76.6

There are several common types of problems and defects associated with mixing, placing, and functionality of pavements. *Alligator cracks* are interconnected cracks forming a series of small blocks. *Corrugations* are plastic deformations characterized by ripples across the pavement. *Faulting* is a difference in the elevations of the edges of two adjacent pavement slabs. *Plastic instability* is an excessive displacement of pavement under traffic. A *pothole* (or *blowout*) is a bowl-shaped hole in the pavement. *Pumping* is a bellows-like movement of the pavement, causing trapped water to be forced or ejected through cracks and joints. *Reflective cracking* is cracking in asphalt overlays that follows the crack or joint pattern of layers underneath. *Rutting* is a channelized depression that occurs in the normal paths of wheel travel. *Shoving* is bulging where the pavement abuts an immobile edge, caused by the pavement being pushed around during heavy loading. *Spalling* is the breaking or chipping of pavement at joints, cracks, and edges. *Upheaving* is the local upward displacement of pavement due to the swelling of layers below.

Pavement Damage from Frost and Freezing

CERM: Sec. 76.32

Frost heaving and reduced subgrade strength, along with the accompanying pumping during spring thaw, can quickly destroy a pavement. Several techniques can be used to reduce damage to a flexible pavement in areas susceptible to frost, such as constructing stronger and thicker pavement sections, lowering the water table by use of subdrains and drainage ditches, and using layers of coarse sands or waterproof sheets beneath the pavement surface to reduce capillary action.

43 Drainage

Content in blue refers to the *NCEES Handbook*.

A. Hydrology 43-1
B. Hydraulics 43-5

The knowledge area of Drainage makes up between two and four questions out of the 80 questions on the PE Civil Transportation exam. The organization of this chapter follows the order of subtopics given by NCEES for this knowledge area. Each subtopic is covered in the following sections.

A. HYDROLOGY

Key concepts: These key concepts are important for answering exam questions in subtopic 17.A, Hydrology.

- hydrographs
- Natural Resources Conservation Service curve numbers
- NRCS method
- rainfall gauging
- rational method
- runoff coefficients
- runoff retention and detention basins
- storm/flood frequency probabilities
- time of concentration
  - peak flow
  - sheet flow

NCEES Handbook: To prepare for this subtopic, familiarize yourself with these sections in the *Handbook*.

- Probability of Single Occurrence in a Given Storm Year
- Risk (Probability of Exceedance)
- Runoff Analysis: Rational Formula Method
- Runoff Coefficients for Rational Formula
- SCS Hydrologic Soils Group Type Classifications
- Recommended Manning's Roughness Coefficients for Overland Flow
- Time of Concentration: SCS Lag Formula
- Time of Concentration: Sheet Flow Formula
- Time of Concentration: Inlet Flow Formula
- Unit Hydrograph
- Rainfall Gauging Stations: Precipitation Gauge Analysis
- Detention and Retention: Rational Method

PE Civil Reference Manual **(CERM):** Study these sections in CERM that either relate directly to this subtopic or provide background information.

- Section 20.2: Storm Characteristics
- Section 20.3: Precipitation Data
- Section 20.4: Estimating Unknown Precipitation
- Section 20.5: Time of Concentration
- Section 20.7: Floods
- Section 20.8: Total Surface Runoff from Stream Hydrograph
- Section 20.9: Hydrograph Separation
- Section 20.10: Unit Hydrograph
- Section 20.14: Hydrograph Synthesis
- Section 20.15: Peak Runoff from the Rational Method
- Section 20.16: NRCS Curve Number
- Section 20.17: NRCS Graphical Peak Discharge Method
- Section 20.18: Reservoir Sizing: Modified Rational Method

The following equations, figures, and tables are relevant for subtopic 17.A, Hydrology.

Storm/Flood Frequency Probabilities

Handbook: Probability of Single Occurrence in a Given Storm Year; Risk (Probability of Exceedance)

$$P(x \geq x_T) = p = \frac{1}{T}$$

$$P(x \geq x_T \text{ at least once in } n \text{ years}) = 1 - \left(1 - \frac{1}{T}\right)^n$$

CERM: Sec. 20.7

The equation given in *Handbook* section Probability of Single Occurrence in a Given Storm Year is equivalent to CERM Eq. 20.19, but CERM uses F instead of T for the recurrence interval frequency. These equations describe the probability, P, of a flood event, such as a 100-year storm or flood, as the ratio of 1 divided by the return period of that storm.

The equation given in *Handbook* section Risk (Probability of Exceedance) is equivalent to CERM Eq. 20.20, but CERM again uses F rather than T for the recurrence interval frequency.

A *flood* occurs when more water arrives than can be drained away. Runoff will occur only when the rain falls on a very wet watershed that is unable to absorb additional water, or when a very large amount of rain falls on a dry watershed faster than it can be absorbed.

The same storm may be described as either a *100-year storm* or a *1% storm*. (These two terms refer to the same event.) Note that a 100-year storm is one whose probability of happening in a given year is 1/100, not a storm that will happen exactly once every 100 years. Design to accommodate 10- and 50-year storms is common for low-volume roads, while design to accommodate 100-year storms is common on freeways and other areas where standing water is particularly risky.

Rainfall Gauging

Handbook: Rainfall Gauging Stations: Precipitation Gauge Analysis

$$\overline{P} = \sum_{i=1}^{n} \left(\frac{A_i}{A}\right) P_i$$

CERM: Sec. 20.3

Precipitation data on rainfall can be collected using an open precipitation rain gauge. If rain gauge stations are uniformly distributed over a flat site, their precipitation depth in inches can be averaged. The equation given in *Handbook* section Rainfall Gauging Stations: Precipitation Gauge Analysis showcases this relationship, indicating that the average precipitation depth at a site, $\overline{P}$, is equal to the sum of the individual station precipitation depths, weighted by the ratio of the individual station drainage area to the total drainage area of the site.

Another method, the *Thiessen method*, calculates the average precipitation by weighting the precipitation gauge measurements by the size of their assumed drainage watersheds. These watersheds are found by drawing dotted lines between all stations and then bisecting the dotted lines with solid lines; the solid lines form the boundaries of the drainage areas.

Time of Concentration

Handbook: Time of Concentration: SCS Lag Formula

$$t_L = 0.000526 \, L^{0.8} \left(\frac{1000}{\text{CN}} - 9\right)^{0.7} S^{-0.5}$$

CERM: Sec. 20.4, Sec. 20.5

$$t_c = t_{\text{sheet}} + t_{\text{shallow}} + t_{\text{channel}} \quad\quad 20.5$$

CERM Eq. 20.5 shows that *time of concentration*, t_c, is the time of travel from the hydraulically most remote point (timewise) in the watershed to the watershed outlet or another design point. For points along storm drains being fed from a watershed, time of concentration is taken as the largest combination of overland flow time (sheet flow), swale or ditch flow (shallow concentrated flow), and storm drain, culvert, or channel time.

The Natural Resources Conservation Service (NRCS) lag equation given in *Handbook* section Time of Concentration: SCS Lag Formula and by CERM Eq. 20.11 was developed for use where overland flow paths are poorly defined and channel flow is absent. L is the length of flow in feet, CN is the NRCS curve number, and S is the average flow path slope as a percentage. It is unusual for time of concentration to be less than 0.1 hr (6 min) when using the NRCS method (previously known as the *Soil Conservation Service* (SCS) *method*) or less than 10 min when using the rational method.

Time of Concentration for Sheet Flow

Handbook: Recommended Manning's Roughness Coefficients for Overland Flow; Time of Concentration: Sheet Flow Formula

$$T_{\text{ti}} = \left(\frac{K_u}{I^{0.4}}\right)\left(\frac{nL}{\sqrt{S}}\right)^{0.6}$$

CERM: Sec. 20.5

$$t_{\text{sheet flow}} = \frac{0.007(nL_o)^{0.8}}{\sqrt{P_2}\, S_{\text{decimal}}^{0.4}} \qquad 20.6$$

The equation given in *Handbook* section Time of Concentration: Sheet Flow Formula can be used to calculate sheet flow travel time, T_{ti}, based on flow length, L, surface slope, S, and rainfall intensity, I. K_u is a constant equal to 6.92 for SI units and 0.933 for U.S. units, and n is the roughness coefficient. CERM Eq. 20.6 is the NRCS form of this equation, using the Manning kinematic equation for calculating sheet flow travel time over distances less than 100 ft in developed areas and 300 ft in undeveloped areas. It is slightly different from the *Handbook* equation. The *Handbook* equation should be used on the exam.

Time of Concentration for Inlet Flow

Handbook: Time of Concentration: Inlet Flow Formula

$$t = t_i + t_s$$
$$t_i = C\left(\frac{Li^2}{S}\right)^{1/3}$$

The time of concentration for a particular inlet to a storm sewer is defined by the inlet time, t_i, which is the time it takes water to enter an inlet based on the capacity and grate configuration. t_s is the time of sewer flow, which is the time that flow in the trunk line from upstream takes to arrive at the inlet.

Rational Method

Handbook: Runoff Analysis: Rational Formula Method; Runoff Coefficients for Rational Formula

CERM: Sec. 20.15

$$Q = CIA$$

The rational formula for peak discharge, given in *Handbook* section Runoff Analysis: Rational Formula Method and by CERM Eq. 20.36, is applicable to small areas (less than several hundred acres or so). Typical values of C are found in *Handbook* table Runoff Coefficients for Rational Formula. The intensity, I, depends on the time of concentration and the design storm frequency. When more than one area contributes to the runoff, the coefficient, C_w, is weighted by the areas, as given by the following equation.

$$C_w = \frac{A_1 C_1 + A_2 C_2 + \cdots + A_n C_n}{A_1 + A_2 + \cdots + A_n}$$

The rational method runoff coefficient, C, is unitless, while rainfall intensity, I, is in in/hr, and rainfall area, A, is in acres. The unit conversion from ac-in/hr to ft^3/sec is generally approximated to 1.0, so the resultant peak discharge, Q, is already in units of cubic feet per second.

Hydrographs

Handbook: Unit Hydrograph

CERM: Sec. 20.2, Sec. 20.8, Sec. 20.9, Sec. 20.10

After a rain, runoff and groundwater increase stream flow. A plot of the stream discharge versus time is known as a *hydrograph*. Hydrograph periods may be very short (e.g., hours) or very long (e.g., days, weeks, or months). A typical hydrograph is shown in CERM Fig. 20.7. The time base is the length of time that the stream flow exceeds the original base flow. The flow rate increases on the rising limb (concentration curve) and decreases on the falling limb (recession curve).

Figure 20.7 Stream Hydrograph

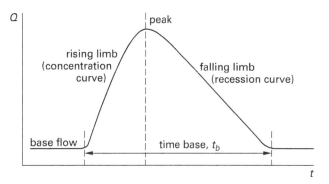

A stream discharge consists of both surface runoff and subsurface groundwater flow. There are several methods of separating groundwater from runoff. In the *straight-line method*, shown in CERM Fig. 20.8, a horizontal line is drawn from the start of the rising limb to the falling limb. All flow under the horizontal line is considered base groundwater flow and can be subtracted from the hydrograph for the purpose of peak surface runoff analysis.

The overland flow hydrograph can be used to find the total runoff volume, V, from the storm as the area under the curve. The hydrograph can be approximated with a histogram, as shown in CERM Fig. 20.12, by summing the areas of the rectangles.

Figure 20.8 Straight-Line Method

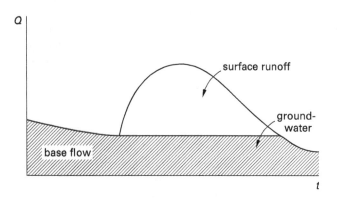

Figure 20.12 Hydrograph Histogram

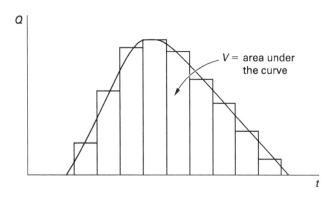

When the area of the watershed is known, the average depth of the excess precipitation, $P_{ave,excess}$, can be calculated, as shown in CERM Eq. 20.21.

$$V = A_d P_{ave,excess} \quad\quad 20.21$$

Unit Hydrographs

Handbook: Unit Hydrograph

CERM: Sec. 20.10

A *unit hydrograph* is developed by dividing every point on the overland flow hydrograph by the average excess precipitation, $P_{ave,excess}$. This is the hydrograph of a storm dropping 1 in of runoff evenly on the entire watershed. Units of the unit hydrograph are in/in. Once a unit hydrograph has been developed from historical data of a particular storm volume, it can be used for other storm volumes with the same duration. CERM Sec. 20.10 contains two example problems regarding hydrographs.

Hydrograph Synthesis

Handbook: Unit Hydrograph

CERM: Sec. 20.14

When a storm's duration is not the same as the unit hydrograph duration, the unit hydrograph cannot be used directly to predict runoff. However, the technique of hydrograph synthesis can be used to construct the hydrograph of the longer storm from the unit hydrograph. See CERM Sec. 20.14 for figures illustrating hydrograph synthesis.

The *lagging storm method*, shown in CERM Fig. 20.17, can be used to construct the unit hydrograph of a storm whose duration is a whole multiple of the original unit hydrograph duration. Let the whole multiple be n. Draw n original unit hydrographs in order, offset by the duration of the original unit hydrograph, t_R, and then add the ordinates to obtain a scaled hydrograph. To reduce to a unit hydrograph with the new storm duration, divide the new ordinates by the whole multiple n.

Figure 20.17 Lagging Storm Method

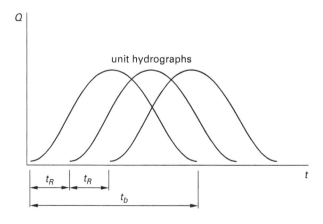

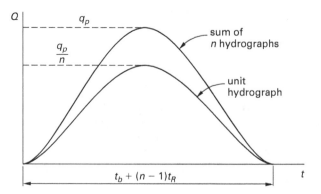

The *S-curve method*, shown in CERM Fig. 20.18, can be used to construct the unit hydrograph of a storm whose duration is *not* a whole multiple of the original unit hydrograph duration. Draw multiple original unit hydrographs in order, offset by the duration of the original unit hydrograph, t_R, and then add the ordinates to obtain a scaled hydrograph. When the summed total of the unit hydrographs starts to level off, you can stop adding them. This cumulative S-curve is offset from itself by the new storm duration, t'_R. The differences between the two curves can be plotted and scaled to a unit hydrograph by multiplying by the ratio of t_R/t'_R.

Figure 20.18 Constructing the S-Curve

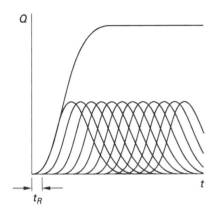

NRCS Curve Number

Handbook: SCS Hydrologic Soils Group Type Classifications

CERM: Sec. 20.16, Sec. 20.17

When a watershed is unmonitored such that no historical records are available to produce a unit hydrograph, the NRCS method can be used for watersheds up to 5000 ac to develop a synthetic unit hydrograph based on *curve number*, CN. The NRCS method specifies using the Manning kinematic equation for sheet flow surface runoff. The curve number is based on land use and soil type. See CERM Table 20.4 and Table 20.5 and *Handbook* section SCS Hydrologic Soils Group Type Classifications for more information on the hydrologic soil types used to determine the curve number for the NRCS method.

Runoff Detention and Retention Basins

Handbook: Detention and Retention: Rational Method

$$V_{\text{in}} = i \sum A C t$$
$$V_{\text{out}} = Q_0 t$$

CERM: Sec. 20.18

An effective method of preventing flooding is to store surface runoff temporarily. After the storm is over, the stored water can be gradually released. The volume of a reservoir needed to hold excess runoff from a storm is the total area of the hydrograph and can be determined with the equations in *Handbook* section Detention and Retention: Rational Method.

B. HYDRAULICS

Key concepts: These key concepts are important for answering exam questions in subtopic 17.B, Hydraulics.

- culverts and culvert design
- Manning equation
- open channel flow
- pipe capacity
- stormwater systems

NCEES Handbook: To prepare for this subtopic, familiarize yourself with these sections in the *Handbook*.

- Principles of One-Dimensional Fluid Flow: Continuity Equation
- Hydraulic Radius
- Manning's Equation
- Conveyance
- Approximate Values of Manning's Roughness Coefficient
- Chezy Equation
- Minimum Diameter of Circular Pipe Under Full Flow Conditions
- Geometric Elements of Channel Sections
- Channel Section Critical Depths
- Normal and Critical Flow
- Culvert Flow Types (USGS)

PE Civil Reference Manual (CERM): Study these sections in CERM that either relate directly to this subtopic or provide background information.

- Section 19.4: Velocity Distribution
- Section 19.5: Parameters Used in Open Channel Flow
- Section 19.6: Governing Equations for Uniform Flow
- Section 19.7: Variations in the Manning Constant
- Section 19.11: Sizing Trapezoidal and Rectangular Channels
- Section 19.12: Most Efficient Cross Section
- Section 19.25: Critical Flow and Critical Depth in Rectangular Channels
- Section 19.26: Critical Flow and Critical Depth in Nonrectangular Channels
- Section 19.27: Froude Number
- Section 19.36: Erodible Channels

- Section 19.37: Culverts
- Section 19.38: Determining Type of Culvert Flow
- Section 19.39: Culvert Design
- Section 28.8: Sewer Velocities
- Section 28.9: Sewer Sizing
- Section 28.10: Street Inlets
- Section 28.11: Manholes

Transportation Depth Reference Manual (CETR): Study these sections in CETR that either relate directly to this subtopic or provide background information.

- Section 6.6: Roadside Clearance Analysis

The following equations, figures, tables, and concepts are relevant for subtopic 17.B, Hydraulics.

Open Channel Flow

Handbook: Principles of One-Dimensional Fluid Flow: Continuity Equation; Hydraulic Radius; Geometric Elements of Channel Sections

CERM: Sec. 19.4, Sec. 19.5

$$Q = A\text{v}$$

An open channel is a fluid passageway that allows part of the fluid to be exposed to the atmosphere. Open channels include natural waterways, culverts, and pipes flowing under the influence of gravity. The continuity equation given in *Handbook* section Principles of One-Dimensional Fluid Flow: Continuity Equation and by CERM Eq. 19.1 is a rewrite of the conservation of momentum equation for fluids. Recall that fluids are assumed incompressible, so the flow, Q, is equal to the cross-sectional area of the channel, A, multiplied by the average velocity of the flow, v.

The *hydraulic radius*, R_H, is the ratio of the area in flow to the wetted perimeter. The equation for hydraulic radius is given in *Handbook* section Hydraulic Radius.

$$R_H = \frac{\text{cross-sectional area}}{\text{wetted perimeter}}$$

CERM Eq. 19.2 is equivalent to the *Handbook* equation but uses variable A for the cross-sectional area and P for the wetted perimeter.

Handbook table Geometric Elements of Channel Sections gives hydraulic parameters of basic channel sections.

Manning Equation

Handbook: Manning's Equation; Conveyance; Approximate Values of Manning's Roughness Coefficient; Chezy Equation

CERM: Sec. 19.6, Sec. 19.7

The most common equation used to calculate the flow velocity in open channels is the Chezy equation.

$$\text{v} = C\sqrt{R_H S}$$

Combining the Chezy equation with the Manning formula for calculating the Chezy resistance coefficient and the incompressibility of water results in the Manning equation for flow.

$$Q = \left(\frac{1.486}{n}\right) A R_H^{2/3} S^{1/2}$$

All of the coefficients and constants in the Manning equation may be combined into a factor known as the *conveyance*, K.

$$K = \left(\frac{1.486}{n}\right) A R_H^{2/3}$$

Exam tip: *Handbook* table Approximate Values of Manning's Roughness Coefficient can be used to find values of n.

Open Channel Flow in a Rectangular or Trapezoidal Channel

Handbook: Geometric Elements of Channel Sections; Channel Section Critical Depths

CERM: Sec. 19.11, Sec. 19.12, Sec. 19.25, Sec. 19.26, Sec. 19.27, Sec. 19.36

CETR: Sec. 6.6

Trapezoidal and rectangular cross sections are the most common for surface channels. Equations for flow in a channel can be found in *Handbook* table Channel Section Critical Depths.

This table also provides cross sections of channels.

The cross section of a trapezoidal channel illustrates the variables used in the equation for flow rate: b is the bottom width of the channel, y is the depth of the channel, and z is the reciprocal of the slope of the channel.

$$Q = \frac{K_n S^{0.5}}{n} \left(\frac{(b+zy)^5 y^5}{\left(b + 2y\sqrt{1+z^2}\right)^2} \right)^{1/3}$$

The preceding equation, derived from the equation for uniform flow section factor given in *Handbook* table Channel Section Critical Depths, can be used to calculate the flow rate, Q, in a trapezoidal or rectangular channel. K is the conveyance of the channel, S is the slope of the channel bottom, and n is the Manning value for roughness. Note that the term being raised to the 1/3 power is the hydraulic radius. In the instance of a rectangular channel, the reciprocal of the slope of the channel, z, is 0, and the equation can be simplified as shown.

$$Q = \frac{K_n S^{0.5}}{n} \left(\frac{b^5 y^5}{(b+2y)^2} \right)^{1/3}$$

The most efficient open channel cross section will maximize the flow for a given Manning coefficient, slope, and flow area (i.e., the hydraulic radius will be at a maximum while the wetted perimeter is at a minimum). The cross section with the highest efficiency is a semicircle, although this shape can only be constructed with concrete, not with earthen ditches. In CERM Fig. 19.3, a semicircle is inscribed in rectangular and trapezoidal channel bottoms to illustrate the loss of efficiency in channel shapes other than semicircles.

Figure 19.3 Circles Inscribed in Efficient Channels

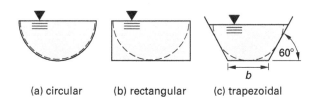

(a) circular   (b) rectangular   (c) trapezoidal

CERM Eq. 19.34 shows that the most efficient rectangle is one having a depth equal to one-half the width, while Eq. 19.38 shows that the most efficient trapezoid has a flow depth equal to twice the hydraulic radius.

$$d = \frac{w}{2} \quad \text{[most efficient rectangle]} \quad 19.34$$

$$d = 2R \quad \text{[most efficient trapezoid]} \quad 19.38$$

Stormwater Systems

Handbook: Normal and Critical Flow; Minimum Diameter of Circular Pipe Under Full Flow Conditions

$$D = \left(\frac{C_0 Q_n}{\sqrt{S}} \right)^{3/8}$$

CERM: Sec. 28.8 through Sec. 28.11

The equation given in *Handbook* section Minimum Diameter of Circular Pipe Under Full Flow Conditions can be used to calculate the pipe diameter needed for full flow conditions.

A storm sewer flow velocity of 2 ft/sec is commonly accepted as the minimum velocity for the pipe to be self-cleaning (i.e., preventing silt and other materials from clogging the bottom of the pipe). CERM Table 28.3 lists the minimum pipe slopes recommended to achieve self-cleaning velocity. Flow velocity greater than 10 ft/sec is usually avoided to prevent pipe erosion and hydraulic jump shocks to the network.

Table 28.3 Minimum Slopes and Capacities for Sewers

sewer diameter		minimum change in elevation[b] (ft/100 ft or m/100 m)	full flow discharge at minimum slope[c,d] (ft³/sec)
(in)	(mm)		
8	(200)	0.40	0.771
9	(230)	0.33	0.996
10	(250)	0.28	1.17
12	(300)	0.22	1.61
14	(360)	0.17	2.23
15	(380)	0.15	2.52
16	(410)	0.14	2.90
18	(460)	0.12	3.67
21	(530)	0.10	5.05
24	(610)	0.08	6.45
27	(690)	0.067	8.08
30	(760)	0.058	9.96
36	(910)	0.046	14.4

(Multiply in by 25.4 to obtain mm.)
(Multiply ft/100 ft by 1 to obtain m/100 m.)
(Multiply ft³/sec by 448.8 to obtain gal/min.)
(Multiply ft³/sec by 28.32 to obtain L/s.)
(Multiply gal/min by 0.0631 to obtain L/s.)

[a]to achieve a velocity of 2 ft/sec (0.6 m/s) when flowing full
[b]as specified in Sec. 24.31 of *Recommended Standards for Sewage Works* (RSSW), also known as "*Ten States' Standards*" (TSS), published by the Health Education Service, Inc. [TSS Sec. 33.41]
[c]$n = 0.013$ assumed
[d]For any diameter in inches and $n = 0.013$, calculate the full flow as $Q = 0.0472 d_{in}^{8/3} \sqrt{S}$

Street inlets are required at all low points where ponding could occur. A common practice is to install three inlets in a sag vertical curve, as a factor of safety against clogging. Inlets on grade should be placed no more than 600 ft apart, but a spacing closer to 300 ft is preferable where achievable.

Manholes along storm sewer lines should be provided at major changes in elevation, direction, size, diameter, and slope. If a sewer line is too small for a person to enter, manholes should be placed at every 400 ft to allow for cleaning.

Culverts and Culvert Design

Handbook: Culvert Flow Types (USGS)

CERM: Sec. 19.37, Sec. 19.38, Sec. 19.39

Handbook section Culvert Flow Types (USGS) includes figures that show different culvert flow control types and the respective equations for calculating flow.

A *culvert* is a pipe that carries water under a feature that would otherwise block the flow of water. Culverts connect two open channels directly, unlike closed storm sewer systems, which are connected by inlets and manholes. Culverts are classified by which of their ends controls the discharge capacity. When water can flow through and out of a culvert faster than it can enter, the culvert is under *inlet control*; otherwise, the culvert is under *outlet control*.

Culvert flow is classified into six different types based on the type of control, the steepness of the barrel, the relative tailwater and headwater heights, and in some cases the relationship between the critical depth and culvert size. CERM Fig. 19.25 illustrates the culvert flow classifications, which are described in detail in CERM Sec. 19.38.

Type 1 flow in the *Handbook* is referred to as type 4 flow in CERM. For this type, the backwater elevation is the controlling factor. Critical depth cannot occur, and the upstream water surface elevation is a function of the tailwater elevation. Discharge is independent of barrel slope. The culvert is submerged at both the headwater and the tailwater.

Type 2 flow in the *Handbook* is referred to as type 6 flow in CERM. This flow is considered a high-head flow. The culvert is full under pressure with a free outfall.

Type 3 flow in the *Handbook* is referred to as type 5 flow in CERM. It is classified as partially full flow under a high head, with rapid flow near the entrance. The barrel length, roughness, and bed slope must be sufficient to keep the velocity high throughout the culvert.

Type 4 flow in the *Handbook* is referred to as type 2 flow in CERM. Flow passes through the critical depth at the culvert outlet, and the barrel flows partially full. The tailwater elevation does not exceed the elevation of the water surface at the control section.

Type 5 flow in the *Handbook* is referred to as type 1 flow in CERM. Water passes through the critical depth near the culvert entrance, and the culvert flows partially full. The tailwater elevation is less than the elevation of the water surface at the control section.

Type 6 flow in the *Handbook* is referred to as type 3 flow in CERM. Backwater is the controlling factor in this type of culvert flow, and the critical depth cannot occur. The upstream water surface elevation is a function of the height of the tailwater. The flow is subcritical for the entire length of the culvert, with the flow being partial. The outlet is not submerged, but the tailwater elevation does not exceed the elevation of critical depth at the terminal section.

44 Alternatives Analysis

Content in blue refers to the *NCEES Handbook*.

Economics Analysis... 44-1

ECONOMICS ANALYSIS

The knowledge area of Alternatives Analysis makes up between one and three questions out of the 80 questions on the PE Civil Transportation exam. See Chap. 9, Engineering Economics, in this book for entries relevant to this knowledge area.

Topic VI: Water Resources and Environmental

Chapter
45. Analysis and Design
46. Hydraulics—Closed Conduit
47. Hydraulics—Open Channel
48. Hydrology
49. Groundwater and Wells
50. Wastewater Collection and Treatment
51. Water Quality
52. Drinking Water Distribution and Treatment
53. Engineering Economics Analysis

45 Analysis and Design

Content in blue refers to the *NCEES Handbook*.

A. Mass Balance 45-1
B. Hydraulic Loading 45-3
C. Solids Loading 45-5
D. Hydraulic Flow Measurement 45-8

The knowledge area of Analysis and Design makes up between four and six questions out of the 80 questions on the PE Civil Water Resources and Environmental exam. The organization of this chapter follows the order of subtopics given by NCEES for this knowledge area. Each subtopic is covered in the following sections.

A. MASS BALANCE

Key concepts: These key concepts are important for answering exam questions in subtopic 9.A, Mass Balance.

- compressible and incompressible fluids
- conservation of mass
- continuity equation and its applications
- density
- energy equation and its uses
- specific gravity
- specific volume
- specific weight
- steady-state mass balance
- volumetric flow rate

NCEES Handbook: To prepare for this subtopic, familiarize yourself with these sections in the *Handbook*.

- Density, Specific Volume, Specific Weight, and Specific Gravity
- Principles of One-Dimensional Fluid Flow: Continuity Equation
- Principles of One-Dimensional Fluid Flow: Energy Equation

PE Civil Reference Manual (CERM): Study these sections in CERM that either relate directly to this subtopic or provide background information.

- Section 14.1: Characteristics of a Fluid
- Section 14.4: Density
- Section 14.5: Specific Volume
- Section 14.6: Specific Gravity
- Section 14.7: Specific Weight
- Section 16.2: Kinetic Energy
- Section 16.3: Potential Energy
- Section 16.4: Pressure Energy
- Section 17.1: Hydraulics and Hydrodynamics
- Section 17.2 Conservation of Mass
- Section 17.3 Typical Velocities in Pipes

Water Resources and Environmental Depth Reference Manual (CEWE): Study these sections in CEWE that either relate directly to this subtopic or provide background information.

- Section 2.4: Continuity Equation
- Section 3.5: Open Channel Flow

The following equations and concepts are relevant for subtopic 9.A, Mass Balance.

Density, Specific Weight, and Specific Gravity

Handbook: Density, Specific Volume, Specific Weight, and Specific Gravity

$$SG = \frac{\gamma}{\gamma w} = \frac{\rho}{\rho w}$$

$$\rho = \frac{m}{V}$$

$$\gamma = \rho g$$

CERM: Sec. 14.4 through Sec. 14.7

$$\rho = \frac{p}{RT} \quad\quad 14.4$$

$$v = \frac{1}{\rho} \qquad 14.5$$

$$SG_{liquid} = \frac{\rho_{liquid}}{\rho_{water}} \qquad 14.6$$

$$SG_{gas} = \frac{\rho_{gas}}{\rho_{air}} \qquad 14.7$$

$$\gamma = g\rho \qquad [SI] \qquad 14.13(a)$$

$$\gamma = \rho \cdot \frac{g}{g_c} \qquad [U.S.] \qquad 14.13(b)$$

The nomenclature in the *Handbook* and CERM equations is identical, except that CERM Eq. 14.4 gives the density as a function of the pressure, p, specific gas constant, R, and temperature, T, rather than of mass and volume.

The *Handbook* defines key concepts related to the density, specific weight, and specific gravity of fluids. CERM Sec. 14.4 through Sec. 14.7 go into the details of these properties.

The density of an ideal gas can be found from its mass and volume, as shown in the *Handbook* equation for density. It can also be found from the specific gas constant and the ideal gas law, as shown in CERM Eq. 14.4.

Specific volume is the volume occupied by a unit mass of fluid, and is also the reciprocal of density. CERM Table 14.2 lists the approximate densities of common fluids.

Specific gravity is the ratio of a substance's density to the density of water or air, depending on whether the substance is a liquid or a gas.

Specific weight is the weight of a substance per unit volume, and is equal to the substance's density times the gravitational acceleration. Specific weight is often referred to as *density*, but these are different properties and not precisely interchangeable.

Continuity Equation

Handbook: Principles of One-Dimensional Fluid Flow: Continuity Equation

$$A_1 v_1 = A_2 v_2$$

$$Q = Av$$

$$\dot{m} = \rho Q = \rho A v$$

CERM: Sec. 17.2

$$\dot{V}_1 = \dot{V}_2 \qquad 17.4$$

The continuity equation applies the conservation of mass to fluid flow. The mass flow rate of a fluid must be the same at two different points within a system as given by CERM Eq. 17.4. This means that the product of a fluid's density, area, and velocity must be the same at any two points in the same system. The continuity equation can be used to determine fluid characteristics within a closed system under continuous, one-dimensional flow. For more about conservation of mass, see the entry "Conservation of Mass" in this chapter.

The continuity equation as written in the *Handbook* and CERM Eq. 17.3 assumes that the fluid is incompressible. Density is constant, so the product of area and velocity is the same at all points in the system. See also the entry "Compressible and Incompressible Fluids" in this chapter.

Exam tip: Before starting any calculations, check whether the fluid is compressible or incompressible. If density is not variable, the equation for the mass flow rate can be simplified. If friction can be ignored, then the equation for mass flow rate can be used to calculate the cross-sectional area or velocity at any point within a gravity-fed fluid dynamic system, based on the flow rate and density of the fluid.

Energy Equation

Handbook: Principles of One-Dimensional Fluid Flow: Energy Equation

$$\frac{p_1}{\gamma} + z_1 + \frac{v_1^2}{2g} = \frac{p_2}{\gamma} + z_2 + \frac{v_2^2}{2g} + h_f$$

$$\frac{p_1}{\rho g} + z_1 + \frac{v_1^2}{2g} = \frac{p_2}{\rho g} + z_2 + \frac{v_2^2}{2g} + h_f$$

CERM: Sec. 16.2, Sec. 16.3, Sec. 16.4

The energy equation is based on the continuity equation. It mathematically expresses the mass balance between two points in a one-dimensional, steady-state fluid flow. CERM Sec. 16.2 through Sec. 16.4 offer more qualitative descriptions of the kinetic, potential, and pressure energies within a fluid dynamic system. These sections do not address energy losses due to friction.

ANALYSIS AND DESIGN   **45-3**

Exam tip: The energy equation can be used to calculate various components of the energy in a system based on more detailed information in the system. For instance, if a specific amount of velocity is required at the downstream end of a pipe that discharges the contents of a storage tank, the velocity and other physical parameters at the downstream end can be used to calculate the amount of pressure head required to meet the velocity requirements.

Conservation of Mass

CERM: Sec. 17.2

$$\dot{m}_1 = \dot{m}_2 \qquad 17.1$$

CERM Eq. 17.1 is the basic conservation of mass equation as expressed for fluid dynamics. In a closed system, the mass of the fluid going into the system is equal to the mass of the fluid leaving it (i.e., the mass flow rate at point 1 must equal to the mass flow rate at point 2).

The *Handbook* uses the conservation of mass equation as the basis of the continuity equation, covered more in the entry "Continuity Equation" in this chapter.

Compressible and Incompressible Fluids

Compressible fluids respond to pressures within the system by changing density. *Incompressible fluids* do not change density. Incompressible fluids are liquids such as water, sewage, and graywater. Compressible fluids are gases such as steam, air, and carbon dioxide.

If a fluid is incompressible, then the density at any two points in the fluid will be equal. When dealing with liquids, which are so barely compressible as to be effectively incompressible, the continuity equation essentially simplifies to consider only the cross-sectional area and velocity of a fluid within a non-pressurized gravity-based system where friction is negligible. For more on this, see the entry "Continuity Equation" in this chapter.

Exam tip: Exam problems may include a conservation of mass problem using a gas instead of a liquid. The main distinction between a gas and a liquid is that a gas can be compressed, while a liquid cannot. Therefore, densities must be considered for gases.

B. HYDRAULIC LOADING

Key concepts: These key concepts are important for answering exam questions in subtopic 9.B, Hydraulic Loading.

- batch reactors, variable volume
- detention time
- flow reactors, steady state
- hydraulic loading rate

NCEES Handbook: To prepare for this subtopic, familiarize yourself with these sections in the *Handbook*.

- Batch Reactor, Variable Volume
- Flow Reactors, Steady State
- Clarifiers: Hydraulic Loading Rate

PE Civil Reference Manual (**CERM**): Study these sections in CERM that either relate directly to this subtopic or provide background information.

- Section 26.12: Sedimentation Tanks
- Section 29.11: Aerated Lagoons
- Section 29.17: Plain Sedimentation Basins/Clarifiers
- Section 29.18: Chemical Sedimentation Basins/Clarifiers
- Section 30.3: Aeration Staging Methods
- Section 30.8: Plug Flow and Stirred Tank Models
- Section 30.10: Aeration Tanks

Water Resources and Environmental Depth Reference Manual (**CEWE**): Study these sections in CEWE that either relate directly to this subtopic or provide background information.

- Section 5.4: Hydraulic Loading Rates and Detention Times
- Section 5.7: Sedimentation

The following equations and concepts are relevant for subtopic 9.B, Hydraulic Loading.

Batch Reactors, Variable Volume

Handbook: Batch Reactor, Variable Volume

$$V = V_{X,A=0}(1 + \varepsilon_A X_A)$$

$$\varepsilon_A = \frac{V_{X,A=1} - V_{X,A=0}}{V_{X,A=0}} = \frac{\Delta V}{V_{X,A=0}}$$

$$C_A = C_{A0}\left(\frac{1-X_A}{1+\varepsilon_A X_A}\right)$$

$$t = C_{A0}\int_0^{X_A} \frac{dX_A}{(1+\varepsilon_A X_A)(-r_A)}$$

$$kt = -\ln(1-X_A) = -\ln\left(1 - \frac{\Delta V}{\varepsilon_A X_{X,A=0}}\right)$$

C	concentration	mol/L
k	reaction rate constant	1/s
r_A	reaction rate of chemical constituent A	mol/L·s
t	time	s
V	volume	L
$V_{X,A=0}$	molar volume of chemical constituent A	L
X_A	fractional conversion of chemical constituent A	mol/mol
ε_A	variable volume coefficient of chemical constituent A	L/L

These equations can be used to find the properties of a reacting mass with a volume that varies at constant pressure.

Flow Reactors, Steady State

Handbook: Flow Reactors, Steady State

For a plug-flow reactor,

$$\tau = \frac{C_{A0}V_{\text{PFR}}}{F_{A0}} = C_{A0}\int_0^{X_A} \frac{dX_A}{-r_A}$$

For a continuous stirred tank reactor,

$$\frac{\tau}{C_{A0}} = \frac{V_{\text{CSTR}}}{F_{A0}} = \frac{X_A}{-r_A}$$

C	concentration	mol/L
F	molar flow rate	mol/h
r	rate of reaction	mol/L·h
V	volume	L
X	fractional conversion	mol/mol
τ	space-time	h

These equations are used for the proper sizing of a reactor. CERM Eq. 30.13 through Eq. 30.17 also cover kinetic models for plug-flow and continuous stirred tank reactors, but the equations in CERM focus on calculations specific to microorganism and biochemical oxygen demand (BOD) reduction via aeration processes.

Exam tip: Reactor sizing will depend on the influent BOD levels, the target effluent BOD levels, and the influent flow.

Hydraulic Loading Rate

Handbook: Clarifiers: Hydraulic Loading Rate

$$v_0 = \frac{Q}{A}$$

CERM: Sec. 29.17

$$v^* = \frac{Q}{A} \qquad 29.4$$

The hydraulic loading rate is based on the volumetric flow rate going into the clarifier and its cross-sectional area. The *Handbook* refers to this measurement as the *hydraulic loading rate*, and CERM refers to it as the *surface loading rate*, *overflow rate*, or *settling rate*. The *Handbook* equation and CERM Eq. 29.4 are essentially identical except for the variable used for the volumetric flow rate.

The hydraulic loading rate is a target design parameter—in other words, it is a value that must be achieved by the design. Once the desired hydraulic loading rate is identified, that value can be used to size the storage basin or set influent flow rates. The rate will also depend greatly on the nature of the system and the quality or quantity of influent and effluent parameters needed or available.

In treatment, the hydraulic loading rate is determined as a function of *retention time*, which is how long the water must remain in the tank to be treated. For more about this, see the entry "Hydraulic Retention Time" in this chapter.

CERM Table 29.10 includes a list of parameter values common for wastewater clarifiers. The hydraulic loading rate is a key design parameter for achieving these values.

Exam tip: The calculation for hydraulic loading rate appears to be solving for velocity, but think of it as a rate to measure how containers are filled or overflow.

Hydraulic Retention Time

Handbook: Clarifiers: Hydraulic Loading Rate

$$\theta = \frac{V}{Q}$$

CERM: Sec. 29.17

$$t_d = \frac{V}{Q} \qquad 29.5$$

Retention time (also referred to as *residence time* and *detention time*) is the amount of time water is retained within a storage system. In water resources engineering, retention time has a secondary meaning: the time it takes a fluid to clear a container (e.g., tank or basin).

To find the hydraulic retention time, the volume and flow rate must be known. The *Handbook* equation and CERM Eq. 29.5 are identical except for the variable used for hydraulic retention time. Note that the tank can be of any shape without affecting the equation.

In water quality applications, the retention time may need to match the contact time required for a specific chemical to improve the quality of the water; if the water moves too fast through the tank, then it will not be treated properly.

Exam tip: The hydraulic retention time can be used to determine the parameters of a settling basin. The smallest size of particles to be removed can be used to determine a terminal settling velocity, and then the equation can be applied to find the volume or retention time.

C. SOLIDS LOADING

Key concepts: These key concepts are important for answering exam questions in subtopic 9.C, Solids Loading.

- impulse-momentum principle
- permissible shear stress
- sedimentation and types of sediment loading
- sludge production
- solids loading rate
- Stokes's law
- stress, viscosity, and pressure
- types of solids loading

NCEES Handbook: To prepare for this subtopic, familiarize yourself with these sections in the *Handbook*.

- Stress, Pressure, and Viscosity
- Impulse-Momentum Principle
- Revised Universal Soil Loss Equation
- Computed K Values for Soils on Erosion Research Stations
- Values of the Topographic Factor, LS, for Specific Combinations of Slope Length and Steepness
- Factor C for Permanent Pasture, Range, and Idle Land
- Average Annual Values of the Rainfall Erosion Index
- Solids Loading Rate (Clarifiers)
- Clarifiers: Settling
- Clarifiers: Design Criteria for Sedimentation Basins
- Activated Sludge Treatment: Activated Sludge Process
- Stokes's Law
- Type 1 Settling - Discrete Particle Settling
- Design Criteria for Sedimentation Basins/Clarifiers

PE Civil Reference Manual (**CERM**): Study these sections in CERM that either relate directly to this subtopic or provide background information.

- Section 14.3: Fluid Pressure and Vacuum
- Section 14.9: Viscosity
- Section 17.38: Impulse-Momentum Principle
- Section 17.53: Drag on Spheres and Disks
- Section 17.54: Terminal Velocity
- Section 26.11: Sedimentation Physics
- Section 26.12: Sedimentation Tanks
- Section 30.2: Activated Sludge Process

Water Resources and Environmental Depth Reference Manual (**CEWE**): Study these sections in CEWE that either relate directly to this subtopic or provide background information.

- Section 1.13: Sedimentation
- Section 8.7: Primary and Secondary Treatment Processes
- Section 9.1: Sludge

- Section 9.8: Biomass Concentration
- Section 9.16: Sludge Mass and Volume

The following equations, figures, and concepts are relevant for subtopic 9.C, Solids Loading.

Stress, Pressure, and Viscosity

Handbook: Stress, Pressure, and Viscosity

$$\tau(1) = \lim_{\Delta A \to 0} \frac{\Delta F}{\Delta A}$$

CERM: Sec. 14.9, App. 14.C

$$\frac{F}{A} = \mu \frac{d\text{v}}{dy} \qquad 14.18$$

$$\tau = \mu \frac{d\text{v}}{dy} \qquad 14.19$$

Stress is the change in force over the change in surface area experienced by an object. The *Handbook* equation, equivalent to a combination of CERM Eq. 14.18 and Eq. 14.19, indicates the relationship between stress, pressure, and area. CERM Eq. 14.18 shows how pressure and area relate to velocity and distance.

Stress can be broken down into normal stress and tangential stress. *Normal stress* is measured in units of force per units of area, and *tangential stress* (also called *shear stress*) is a function of viscosity. CERM Eq. 14.19 shows how shear stress relates to absolute viscosity.

CERM Sec. 14.9 provides a good explanation of the differences between absolute viscosity and kinematic viscosity.

> *Exam tip*: *Handbook* tables Physical Properties of Water and Properties of Water (SI Metric Units) include the viscosity of water under a range of temperatures. These can be used to estimate viscosity when it is needed for an exam problem.

Impulse-Momentum Principle

Handbook: Impulse-Momentum Principle

$$\sum F = \sum Q_2 \rho_2 \text{v}_2 - \sum Q_1 \rho_1 \text{v}_1$$

CERM: Sec. 17.38

$$F = \dot{m}\Delta\text{v} \qquad \text{[SI]} \quad 17.171(a)$$

$$F = \frac{\dot{m}\Delta\text{v}}{g_c} \qquad \text{[U.S.]} \quad 17.171(b)$$

In the *Handbook* equation,

F	force	lbf
Q	volumetric flow rate	ft^3/hr
v	kinematic viscosity	ft^3/sec
ρ	density	lbf/ft^3

The impulse-momentum principle states that the force in a direction acting on a fluid equals the rate of change in momentum of the fluid. The *Handbook* equation multiplies the volumetric flow rate, density, and kinematic viscosity of the fluid to arrive at the impulse momentum (i.e., the sum of the forces acting on the fluid). CERM Eq. 17.171 is based on this same principle, but the mass flow rate is used and multiplied directly by the velocity of the water.

Revised Universal Soil Loss Equation

Handbook: Revised Universal Soil Loss Equation; Computed K Values for Soils on Erosion Research Stations; Values of the Topographic Factor, LS, for Specific Combinations of Slope Length and Steepness; Factor C for Permanent Pasture, Range, and Idle Land; Average Annual Values of the Rainfall Erosion Index

$$A = RKLSCP$$

The revised universal soil loss equation (RUSLE) is used to estimate the erosion of soils in a watershed. The various factors, indices, and constants used in this equation can be found in *Handbook* tables Computed K Values for Soils on Erosion Research Stations; Values of the Topographic Factor, LS, for Specific Combinations of Slope Length and Steepness; and Factor C for Permanent Pasture, Range, and Idle Land; as well as *Handbook* figure Average Annual Values of the Rainfall Erosion Index.

The RUSLE is not used to calculate instream erosive velocities.

Solids Loading Rate

Handbook: Solids Loading Rate (Clarifiers)

$$\text{SLR} = \frac{8.34QX}{A}$$

This equation is used to calculate the solids loading rate for clarifiers and activated sludge systems. The solids loading rate is directly related to the hydraulic loading rate and the concentration of solids within the influent. Different solid particle sizes will behave differently in treatment, requiring different tank sizes for clarification and sedimentation.

Exam tip: If the influent volumetric flow rate and the solids loading rate for a system are known, they can be used to determine the volume of the system or regularity of system maintenance required.

Settling Velocity

Handbook: Clarifiers: Settling

$$v_h = \frac{Q}{A_x}$$

CERM: Sec. 26.12

$$t_{\text{settling}} = \frac{h}{v_s} \qquad 26.5$$

$$t_d = \frac{V_{\text{tank}}}{Q} = \frac{Ah}{Q} \qquad 26.6$$

The *Handbook* equation and CERM Eq. 26.5 and Eq. 26.6 are rearrangements of the same relationships between the properties of the clarifier.

The *Handbook* equation finds the horizontal velocity of solids in a clarifier of a specified area. The horizontal velocity is a function of the flow rate and the area of the tank in question. CERM Fig. 26.3 illustrates the interaction between horizontal velocity, shown in the figure as v_f, and settling velocity, shown in the figure as v_t, in a rectangular sedimentation tank.

Figure 26.3 Sedimentation Basin

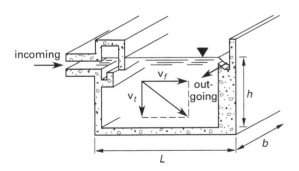

Activated Sludge

Handbook: Activated Sludge Treatment: Activated Sludge Process

$$X = \frac{\theta_c Y(S_0 - S_e)}{\theta(1 + k_d \theta_c)}$$

CERM: Sec. 30.2

CEWE: Sec. 9.1, Sec. 9.8

$$X = \frac{\left(\dfrac{\theta_c}{t_h}\right) Y(S_o - S)}{1 + k_d \theta_c} \qquad 9.9$$

The *Handbook* equation and CEWE Eq. 9.9 are nearly the same, with only slight changes to the presentation of variables. These equations calculate the concentration of suspended solids in a sludge mix.

There are several water treatment processes in which sludge is produced. *Handbook* section Activated Sludge Treatment: Activated Sludge Reactions includes the stoichiometry of a typical activated sludge chemical process.

Exam tip: The specific gravity of the sludge through all types of treatments is always very close to the specific gravity of water, which is 1.0.

Stokes's Law

Handbook: Stokes's Law

$$v_t = \frac{g(\rho_p - \rho_w)d^2}{18\mu} = \frac{g\rho_w(\text{SG} - 1)d^2}{18\mu}$$

CERM: Sec. 17.53, Sec. 17.54

$$F_D = 6\pi\mu v R = 3\pi\mu v D \quad \text{[laminar]} \qquad 17.220$$

$$F_D = mg - F_b \quad \begin{bmatrix}\text{at terminal} \\ \text{velocity}\end{bmatrix} \quad \text{[SI]} \qquad 17.221(a)$$

$$F_D = \frac{mg}{g_c} - F_b \quad \begin{bmatrix}\text{at terminal} \\ \text{velocity}\end{bmatrix} \quad \text{[U.S.]} \qquad 17.221(b)$$

$$v_{terminal} = \frac{D^2 g(\rho_{sphere} - \rho_{fluid})}{18\mu} \quad \text{[laminar]} \quad \text{[SI]} \quad 17.224(a)$$

$$v_{terminal} = \frac{D^2(\rho_{sphere} - \rho_{fluid})}{18\mu} \cdot \frac{g}{g_c} \quad \text{[laminar]} \quad \text{[U.S.]} \quad 17.224(b)$$

The Stokes's law equation calculates the terminal velocity of settling particles. The *Handbook* includes detailed calculations of the factors going into the Stokes's law equation. Stokes's law can also be expressed in terms of particle velocity, as in CERM Eq. 17.220, Eq. 17.221, and Eq. 17.224.

Exam tip: Stokes's law applies to laminar fluid flow, meaning that it is applicable to most open channel systems. One exception is anywhere in the system that induces turbulent flow, such as a critical depth downstream.

Sedimentation Parameters

Handbook: Type 1 Settling - Discrete Particle Settling

$$v_0 = \frac{Q}{A} = \frac{Q}{WL} = \frac{g(\rho_s - \rho)d^2}{18\mu}$$

$$H = v_0 t$$

$$t_0 = \frac{H}{v_t} = \frac{L}{v_H}$$

CERM: Sec. 26.12

$$t_d = \frac{V_{tank}}{Q} = \frac{Ah}{Q} \quad 26.6$$

The *Handbook* equations and CERM Eq. 26.6 can be used to estimate the design parameters for a settling basin. v_0 is the terminal settling velocity, which is used to find the height (i.e., depth) of the basin. For additional design criteria, see *Handbook* tables Clarifiers: Design Criteria for Sedimentation Basins and Design Criteria for Sedimentation Basins/Clarifiers.

Exam tip: The equations in this entry can be used in combination with those from other parts of this chapter, such as in the entries "Settling Velocity" and "Hydraulic Loading Rate," to size a settling basin.

Sediment Load

CEWE: Sec. 1.13

Bedload is the portion of sediment load that is transported along the bottom of a channel. *Suspended load* is the portion of sediment load that is carried in the body of the water flow. *Dissolved load* is the portion of the sediment load that is chemically carried in the water.

D. HYDRAULIC FLOW MEASUREMENT

Key concepts: These key concepts are important for answering exam questions in subtopic 9.D, Hydraulic Flow Measurement.

- flow measurement of compressible fluids
- hydraulic flow measurement with weirs
- orifice meters
- Parshall flumes
- pitot-static gauges
- pitot tubes
- types of weirs and flow conditions
- Venturi meters
- weir flow rates

NCEES Handbook: To prepare for this subtopic, familiarize yourself with these sections in the *Handbook*.

- Reynolds Number—Open Channel (Newtonian Fluid)
- Reynolds Number—Circular Pipes
- Pitot Tube
- Venturi Meters
- Parshall Flume
- Rectangular Weirs: Sharp-Crested Weirs
- Rectangular Weirs: Sharp-Crested Weirs with End Contractions (Francis Equation)
- Rectangular Weirs: Typical Rectangular Weir Proportions
- Rectangular Weirs: Horizontal Broad Crested Weirs
- Triangular (V-Notch) Weirs: General V-Notch
- Coefficient of Discharge vs. Head Over Discharge Chart
- Hazen-Williams Equation
- Values of Hazen-Williams Coefficient C
- Manning's Equation

- Approximate Values of Manning's Roughness Coefficient
- Chezy Equation

***PE Civil Reference Manual* (CERM):** Study these sections in CERM that either relate directly to this subtopic or provide background information.

- Section 16.7: Pitot Tube
- Section 16.10: Reynolds Number
- Section 17.33: Pitot-Static Gauge
- Section 17.34: Venturi Meter
- Section 17.37: Flow Measurements of Compressible Fluids
- Section 19.6: Governing Equations for Uniform Flow
- Section 19.8: Hazen-Williams Velocity
- Section 19.14: Flow Measurement with Weirs
- Section 19.15: Triangular Weirs
- Section 19.17: Broad-Crested Weirs and Spillways
- Section 19.19: Flow Measurement with Parshall Flumes
- Section 19.29: Occurences of Critical Flow

***Water Resources and Environmental Depth Reference Manual* (CEWE):** Study these sections in CEWE that either relate directly to this subtopic or provide background information.

- Section 2.15: Flow Measuring Devices
- Section 3.5: Open Channel Flow
- Section 3.12: Spillway Capacity
- Section 3.13: Flow Measurement Devices

The following equations and figures are relevant for subtopic 9.D, Hydraulic Flow Measurement.

Reynolds Number

Handbook: Reynolds Number—Open Channel (Newtonian Fluid); Reynolds Number—Circular Pipes

For an open channel,

$$\text{Re} = \frac{\text{v}R}{\nu}$$

For circular pipes,

$$\text{Re} = \frac{\text{v}D\rho}{\mu} = \frac{\text{v}D}{\nu}$$

CERM: Sec. 16.10

The *Reynolds number* is a dimensionless number that represents the ratio of inertial forces to viscous forces in a liquid. The Reynolds number is used in many fluid mechanics calculations. The value of the Reynolds number also indicates whether the state of flow is laminar, transitional, or turbulent. If the Reynolds number for an open channel is less than approximately 500, the flow is laminar; if between 500 and 2000, the flow is transitional; if 2000 or greater, the flow is fully turbulent. Based on this, the Reynolds number can be used to determine the right equations and tools to use for other calculations.

Exam tip: To determine the friction factor in pipes, use the Reynolds number and *Handbook* figure Moody, Darcy, or Stanton Friction Factor Diagram.

Pitot Tube

Handbook: Pitot Tube

$$\text{v} = \sqrt{\left(\frac{2}{\rho}\right)(p_0 - p_s)} = \sqrt{\frac{2g(p_0 - p_s)}{\gamma}}$$

g	gravitational acceleration	ft/sec²
p_0	stagnation pressure	lbf/ft²
p_s	static pressure	lbf/ft²
v	velocity	ft/sec
γ	unit weight	lbf/ft³
ρ	density	lbf/ft³

CERM: Sec. 16.7

A pitot tube measures the total energy within a channel or pipe, also referred to as the *energy grade line*. The *Handbook* equation can be used to calculate the upstream velocity if the static and stagnation pressures are known. If the pressure energy within the system is known at that point, the velocity head can be calculated, and the upstream velocity can be calculated if the static and stagnation pressures are known.

Exam tip: In non-pressurized systems, velocity head can be calculated as the additional depth of water above the surface water elevation within the pitot tube.

Venturi Meters

Handbook: Venturi Meters

$$Q = \frac{C_v A_2}{\sqrt{1 - \left(\frac{A_2}{A_1}\right)^2}} \sqrt{2g\left(\frac{p_1}{\gamma} + z_1 - \frac{p_2}{\gamma} - z_2\right)}$$

CERM: Sec. 17.34

$$v_2 = C_v v_{2,\text{ideal}}$$

$$= \left(\frac{C_v}{\sqrt{1 - \left(\frac{A_2}{A_1}\right)^2}}\right) \sqrt{\frac{2(p_1 - p_2)}{\rho}} \quad [\text{SI}] \quad 17.150(a)$$

$$v_2 = C_v v_{2,\text{ideal}}$$

$$= \left(\frac{C_v}{\sqrt{1 - \left(\frac{A_2}{A_1}\right)^2}}\right) \sqrt{\frac{2g_c(p_1 - p_2)}{\rho}} \quad [\text{U.S.}] \quad 17.150(b)$$

A Venturi meter measures the differences in pressure at two points in a system. The basic Venturi meter equation in CERM is slightly different from the one given in the *Handbook*. Both equations are derived from the Bernoulli equation, but the *Handbook* equation calculates the flow instead of the velocity, and the calculation includes a change in elevation, which means the equation assumes values are being calculated for a long Venturi meter or a pipe with a steep slope.

CERM Fig. 17.24 illustrates a simple Venturi meter.

Figure 17.24 Venturi Meter

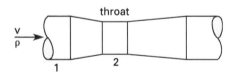

Exam tip: The coefficient C_v is usually close to 1 (0.98 or 0.99), so if the difference in pressures and the cross-sectional areas are known, the velocity through the meter can be approximated easily.

Parshall Flume

Handbook: Parshall Flume

For a width from 1 ft to 8 ft and a submergence ratio of 0.7 or less,

$$Q = 4WH_a^{1.522W^{0.026}}$$

For a width from 8 ft to 50 ft and a submergence ratio of 0.8 or less,

$$Q = (3.688W + 2.5)H_a^{1.6}$$

CERM: Sec. 19.19

$$Q = KbH_a^n \qquad 19.64$$

$$n = 1.522b^{0.026} \qquad 19.65$$

A Parshall flume is a fixed structure that measures flows. It is commonly used in wastewater measurements and can be used in systems where a large amount of sediment is being conveyed. The *Handbook* equations are broken down according to the width of the throat and the submergence ratio, H_b/H_a, which is the ratio of the depth of water downstream of the flume to the depth of water upstream of the flume. At a low submergence ratio, the downstream depth of water does not greatly affect the flow characteristics through the flume. As the ratio increases, the downstream depths have increasingly significant backwater effects on the flow through the flume, requiring different equations.

CERM Eq. 19.64 is essentially the same equation as the *Handbook* equation for a flume width of 1 ft to 8 ft and a submergence ratio of 0.7 or less. CERM Eq. 19.65 provides the numerical values used for the exponent of H that make CERM Eq. 19.64 resemble the *Handbook* equation.

Exam tip: The flow equations given by the *Handbook* differ substantially depending on the submergence ratio. Calculating that ratio first can provide information about the hydraulic nature of the system and ensure the right equation is used to calculate the flow.

Sharp-Crested Weirs

Handbook: Rectangular Weirs: Sharp-Crested Weirs; Rectangular Weirs: Typical Rectangular Weir Proportions

$$Q = CLH^{3/2}$$

ANALYSIS AND DESIGN 45-11

$$C = 3.27 + 0.4\left(\frac{H}{h}\right)$$

Rectangular weirs are organized into three categories: broad crested, short crested, and sharp crested. *Handbook* section Rectangular Weirs: Typical Rectangular Weir Proportions includes equations to assist in determining the classification of the weir.

Sharp-crested weirs are constructed with a sharp weir edge, as the name suggests, that is used for measurements. The *Handbook* equation for the flow, Q, is the basic equation used to calculate flows from rectangular sharp-crested weirs. C is the weir coefficient, H is the depth of water from the weir edge, and h is the height of the bottom of the weir with respect to the floor of the approach channel. Per *Handbook* section Rectangular Weirs: Sharp-Crested Weirs, the ratio of H to h must be less than 10.

Sharp-Crested Weirs with End Contractions (Francis Equation)

Handbook: Rectangular Weirs: Sharp-Crested Weirs with End Contractions (Francis Equation)

$$Q = C\big(L - n(0.1H)\big)H^{3/2}$$

Rectangular weirs are organized into three categories: broad crested, short crested, and sharp crested. *Handbook* section Rectangular Weirs: Typical Rectangular Weir Proportions includes equations to assist in determining the classification of the weir.

Sharp-crested weirs are constructed with a sharp weir edge, as the name suggests, that is used for measurements. The Francis equation finds the flow in a sharp-crested weir with end contractions. For calculations in customary U.S. units where the height of the bottom of the weir is negligible, the coefficient of discharge, C, is approximately 3.33.

Triangular (V-Notch) Weirs

Handbook: Triangular (V-Notch) Weirs; Coefficient of Discharge vs. Head Over Discharge Chart

$$Q = \tfrac{8}{15} C_d \sqrt{2g}\left(\tan\frac{\theta}{2}\right)H^{5/2}$$

CERM: Sec. 19.15

$$Q = C_2\left(\frac{8}{15}\tan\frac{\theta}{2}\right)\sqrt{2g}\,H^{5/2} \qquad 19.55$$

Triangular (V-notch) weirs should be used when small flow rates are to be measured. CERM Fig. 19.8 shows a triangular weir.

Figure 19.8 Triangular Weir

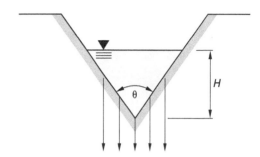

The coefficient of discharge (C_d in the *Handbook* equation and C_2 in CERM Eq. 19.55) can be found from *Handbook* figure Coefficient of Discharge vs. Head Over Discharge Chart, and generally varies from 0.58 to 0.61, depending on the notch angle.

Horizontal Broad-Crested Weirs

Handbook: Rectangular Weirs: Typical Rectangular Weir Proportions; Rectangular Weirs: Horizontal Broad Crested Weirs

$$Q = \tfrac{2}{3} C_v L_e \sqrt{2g}\, H_e^{3/2}$$

$$C_v = 0.602 + 0.075\left(\frac{H}{h}\right)$$

CERM: Sec. 19.14, Sec. 19.29

$$Q = \tfrac{2}{3} C_1 b \sqrt{2g}\, H^{3/2} \qquad 19.49$$

Rectangular weirs are organized into three categories: broad crested, short crested, and sharp crested. *Handbook* section Rectangular Weirs: Typical Rectangular Weir Proportions includes equations to assist in determining the classification of the weir.

Broad-crested weirs are designed for larger systems and can measure large river flow where momentum may overpower a sharp-crested weir. Broad-crested weirs allow for critical flow to occur at the discharge point, thus controlling flow. CERM Fig. 19.17 shows critical flow in a broad-crested weir. d_c and d_b are the critical depth and normal depth, respectively.

Figure 19.17 Broad-Crested Weir

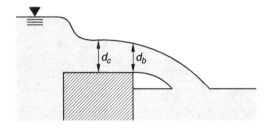

The equation used to calculate flow in a broad-crested weir is essentially the same as the standard weir equation, with the exception that the empirical velocity coefficient of discharge is calculated differently from the coefficient used for sharp-crested weirs, as shown in the *Handbook* equation for C_v.

The *Handbook* equation for flow in a broad-crested weir is similar to CERM Eq. 19.49, with some minor differences in nomenclature. The effective weir length is indicated by L_e in the *Handbook* equation, while b is used in CERM Eq. 19.49, and the velocity coefficient of discharge is represented by C_v in the *Handbook* equation and C_1 in the CERM equation.

Hazen-Williams Equation

Handbook: Hazen-Williams Equation; Values of Hazen-Williams Coefficient C

$$v = k_1 C R_H^{0.63} S^{0.54}$$

$$Q = k_1 C A R_H^{0.63} S^{0.54}$$

$$S = \frac{h_f}{L}$$

CERM: Sec. 19.8

$$v = 0.85 C R^{0.63} S_0^{0.54} \quad \text{[SI]} \quad 19.14(a)$$

$$v = 1.318 C R^{0.63} S_0^{0.54} \quad \text{[U.S.]} \quad 19.14(b)$$

In these equations,

A	cross-sectional area	ft²
C	Hazen-Williams coefficient	–
h_f	friction head	ft
k_1	coefficient equal to 0.849 (1.318)	–
L	length	ft
Q	discharge	ft³/sec
R_H	hydraulic radius	ft
S	friction slope	–
v	velocity	ft/sec

The Hazen-Williams equation can be used to find velocity or flow in an open channel. In the *Handbook* equation, k_1 is a coefficient based on whether the calculation is done in customary U.S. units or SI units. CERM Eq. 19.14 uses rounded values for the coefficient and includes them in the base equation, but is otherwise the same as the *Handbook* equations.

The Hazen-Williams coefficient, C, is used to characterize the roughness of the channel and does not depend on the fluid characteristics. *Handbook* table Values of Hazen-Williams Coefficient C includes values for common materials.

Exam tip: The Hazen-Williams equation is applicable to water flows at reasonably high Reynolds numbers in closed conduits.

Manning's Equation

Handbook: Manning's Equation; Approximate Values of Manning's Roughness Coefficient

$$Q = \left(\frac{1.486}{n}\right) A R_H^{2/3} S^{1/2}$$

$$v = \left(\frac{1.486}{n}\right) R_H^{2/3} S^{1/2}$$

CERM: Sec. 19.6

$$v = \left(\frac{1}{n}\right) R^{2/3} \sqrt{S} \quad \text{[SI]} \quad 19.12(a)$$

$$v = \left(\frac{1.49}{n}\right) R^{2/3} \sqrt{S} \quad \text{[U.S.]} \quad 19.12(b)$$

CEWE: Sec. 3.5

Manning's equation calculates the flow within an open channel system based on steady-state conditions. CERM Eq. 19.12 and CEWE Eq. 3.7 are nearly the same as the *Handbook* equation, with the only difference being that the conversion factor of 1.486 is rounded to 1.49. Manning's equation can be multiplied by the area to calculate the flow rate in an open channel.

The *Handbook* table Approximate Values of Manning's Roughness Coefficient contains values for different materials and channel types. CEWE Table 3.1 also provides these values.

Exam tip: Exam questions using Manning's equation will likely provide a material type, requiring that test takers estimate the Manning's roughness coefficient using the *Handbook* table Approximate Values of Manning's Roughness Coefficient.

Chezy Equation

Handbook: Chezy Equation; Approximate Values of Manning's Roughness Coefficient

$$v = C\sqrt{R_H S}$$

$$Q = CA\sqrt{R_H S}$$

$$C = \left(\frac{1}{n}\right) R_H^{1/6}$$

CERM: Sec. 19.6

$$v = C\sqrt{RS} \qquad 19.9$$

The Chezy equation is the most common equation used to calculate the flow velocity in open channels.

The Chezy coefficient, C, can be found using the *Handbook* equation or CERM Eq. 19.9 if the channel is large and the flow turbulent. The *Handbook* does not provide a calculation for the Chezy coefficient for other sizes of channel or types of flow. n is the Manning's roughness coefficient, which can be found from *Handbook* table Approximate Values of Manning's Roughness Coefficient.

46 Hydraulics—Closed Conduit

Content in blue refers to the *NCEES Handbook*.

A. Energy and/or Continuity Equation 46-1
B. Pressure Conduit 46-4
C. Pump Application and Analysis 46-8
D. Pipe Network Analysis 46-12

The knowledge area of Hydraulics—Closed Conduit makes up between four and six questions out of the 80 questions on the PE Civil Water Resources and Environmental exam. The organization of this chapter follows the order of subtopics given by NCEES for this knowledge area. Each subtopic is covered in the following sections.

A. ENERGY AND/OR CONTINUITY EQUATION

Key concepts: These key concepts are important for answering exam questions in subtopic 10.A, Energy and/or Continuity Equation.

- application of continuity equation
- Bernoulli equation
- momentum equation and its application

PE Civil Reference Manual (**CERM**): Study these sections in CERM that either relate directly to this subtopic or provide background information.

- Section 15.2: Manometers
- Section 15.4: Fluid Height Equivalent to Pressure
- Section 16.5: Bernoulli Equation
- Section 16.15: Energy Grade Line
- Section 17.2: Conservation of Mass
- Section 17.5: Head Loss Due to Friction
- Section 17.8: Energy Loss Due to Friction: Laminar Flow
- Section 17.9: Energy Loss Due to Friction: Turbulent Flow
- Section 17.15: Minor Losses
- Section 17.20: Energy and Hydraulic Grade Lines with Friction
- Section 17.38: Impulse-Momentum Principle

Water Resources and Environmental Depth Reference Manual (**CEWE**): Study these sections in CEWE that either relate directly to this subtopic or provide background information.

- Section 2.3: Energy Equation
- Section 2.4: Continuity Equation
- Section 2.5: Momentum Equation
- Section 2.6: The Darcy-Weisbach Equation

NCEES Handbook: To prepare for this subtopic, familiarize yourself with these sections in the *Handbook*.

- Pressure Field in a Static Liquid
- Manometers
- Barometers
- Principles of One-Dimensional Fluid Flow: Continuity Equation
- Principles of One-Dimensional Fluid Flow: Energy Equation
- Hydraulic Gradient (Grade Line)
- Energy Line (Bernoulli Equation)
- Impulse-Momentum Principle
- Darcy-Weisbach Equation (Head Loss)
- Momentum Equation

The following equations and figures are relevant for subtopic 10.A, Energy and/or Continuity Equation.

Pressure Relationships in a Static Liquid

Handbook: Pressure Field in a Static Liquid; Manometers; Barometers

CERM: Sec. 15.2, Sec. 15.4

The *Handbook* includes several equations demonstrating pressure relationships. Pressure varies linearly with depth. The relationship between pressure and depth for an incompressible fluid is given by CERM Eq. 15.6.

$$p = \rho g h \quad \text{[SI]} \quad 15.6(a)$$

$$p = \frac{\rho g h}{g_c} = \gamma h \quad \text{[U.S.]} \quad 15.6(b)$$

Since ρ and g are constants, CERM Eq. 15.6 shows that p and h are linearly related. Knowing one determines the other. For example, the height of a fluid column needed to produce a pressure is given by CERM Eq. 15.7.

$$h = \frac{p}{\rho g} \quad \text{[SI]} \quad 15.7(a)$$

$$h = \frac{p g_c}{\rho g} = \frac{p}{\gamma} \quad \text{[U.S.]} \quad 15.7(b)$$

In *Handbook* section **Pressure Field in a Static Liquid**, the difference in pressure between two different points can be determined by the following equation.

$$p_2 - p_1 = -\gamma(z_2 - z_1) = -\gamma h = -\rho g h$$

Handbook section **Manometers** provides the equation for the difference in pressure between two points and illustrates this with a manometer.

$$p_0 = p_2 + \gamma_2 h_2 - \gamma_1 h_1 = p_2 + g(\rho_2 h_2 - \rho_1 h_1)$$

When h_1, h_2, and h are equal,

$$p_0 = p_2 + (\gamma_2 - \gamma_1)h = p_2 + (\rho_2 - \rho_1)g h$$

The *Handbook* equations given in the **Manometers** section are similar to CERM Eq. 15.2 and Eq. 15.3, but differ slightly in that they solve for a specific pressure, assuming that one of the pressures is known. The *Handbook* equations also assume different fluids with different densities.

$$(p_2 - p_1)A = \rho_m g h A \quad 15.2$$

$$p_2 - p_1 = \rho_m g h \quad \text{[SI]} \quad 15.3(a)$$

$$p_2 - p_1 = \rho_m h \cdot \frac{g}{g_c} = \gamma_m h \quad \text{[U.S.]} \quad 15.3(b)$$

A barometer is another device that works on the same principle as the manometer.

$$p_{\text{atm}} = p_A = p_v + \gamma h = p_B + \gamma h = p_B + \rho g h$$

The equation given in *Handbook* section **Barometers** is similar to CERM Eq. 15.9 but differs in that it solves for atmospheric pressure.

$$p_a - p_v = \rho g h \quad \text{[SI]} \quad 15.9(a)$$

$$p_a - p_v = \frac{\rho g h}{g_c} = \gamma h \quad \text{[U.S.]} \quad 15.9(b)$$

Impulse-Momentum Principle

Handbook: Impulse-Momentum Principle

$$\sum F = \sum Q_2 \rho_2 v_2 - \sum Q_1 \rho_1 v_1$$

CERM: Sec. 17.38

$$F = \dot{m} \Delta v \quad \text{[SI]} \quad 17.171(a)$$

$$F = \frac{\dot{m} \Delta v}{g_c} \quad \text{[U.S.]} \quad 17.171(b)$$

CEWE: Sec. 2.5

The *Handbook* equation and CERM Eq. 17.171 are equivalent once you consider that the mass flow rate equals flow rate times density.

The *momentum* of a moving object is a vector quantity defined as the product of the object's mass and velocity. Force is equal to the product of mass and acceleration. In this case, the product of flow, density, and velocity produces the same results.

> *Exam tip*: The momentum equation calculates the force of flows to calculate the total force from flows. Assuming the flow vectors are meeting, the total force will indicate which direction the flow will go.

Bernoulli Equation

Handbook: Principles of One-Dimensional Fluid Flow: Energy Equation; Darcy-Weisbach Equation (Head Loss)

$$\frac{p_1}{\gamma} + z_1 + \frac{v_1^2}{2g} = \frac{p_2}{\gamma} + z_2 + \frac{v_2^2}{2g} + h_f$$

CERM: Sec. 16.5, Sec. 17.5, Sec. 17.8, Sec. 17.9, Sec. 17.15

CEWE: Sec. 2.3, Sec. 2.6

$$\frac{p_1}{\gamma} + \frac{v_1^2}{2g} + z_1 = \frac{p_2}{\gamma} + \frac{v_2^2}{2g} + z_2 + h_L \quad 2.2$$

The *Bernoulli equation*, also known as the *conservation of energy equation*, can be broken down into three energy/head loss components: pressure energy (pressure head), kinetic energy (velocity head), and potential energy (gravitational head). The *Handbook* equation and CEWE Eq. 2.2 are equivalent, with the *Handbook* equation using subscript f and CEWE Eq. 2.2 using subscript L to denote head loss.

The head loss has two components: major losses (fluid friction losses) and minor losses (turbulence or changes in flow direction).

Major friction losses can be calculated using the Darcy-Weisbach equation or the Hazen-Williams equation. The Darcy-Weisbach equation is dependent on the Reynolds number, is used for gases or complex fluids, and is useful for nonstandard water properties at high pressures. The Hazen-Williams equation is applicable for turbulent flow only, is used for standard water properties and pressures, and does not require the Reynolds number.

Minor friction losses can be calculated either with the friction loss coefficients or by using the equivalent pipe length.

Exam tip: The Bernoulli equation is an energy balance, meaning both sides must be equal. Unless otherwise noted, variables without a subscript number should be equal on both sides, leaving pressure, velocity, and water height as the only variables that would change.

Continuity Equation

Handbook: Principles of One-Dimensional Fluid Flow: Continuity Equation

$$A_1 v_1 = A_2 v_2$$
$$Q = A v$$

CERM: Sec. 17.2

CEWE: Sec. 2.4

Because mass cannot be created or destroyed, it is a constant throughout pipelines and open channels. Flow rates have many different units. Units of volumetric flow rate include gal/min, gal/day, and m^3/s. Flow rate simply measures the velocity of a moving volume.

A classic example of using the continuity equation involves evaluating two segments within the same pipe or channel to determine the velocity, friction loss, and other flow properties within the channel or pipeline. A more powerful application is the one shown in the illustration, for pipe distribution networks. The flow along the main line will equal the sum of the flows along the distribution lines (pipes in parallel).

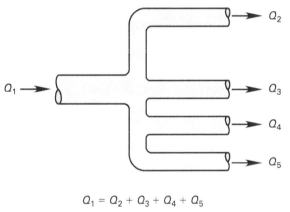

$Q_1 = Q_2 + Q_3 + Q_4 + Q_5$

Exam tip: Total flow in must equal total flow out, functioning the same as conservation of mass and momentum. However, velocity of the fluid can be different between the inlet and the discharge pipes.

Momentum Equation

Handbook: Momentum Equation

$$M = \frac{Q}{gA} + Ay$$

Momentum has a directional vector. The most common way to solve for it is to divide the vector into the two Cartesian coordinates of x and y. One of the main applications of the momentum equation is to design pipe restraints and thrust blocks. Two acting forces, the pressure and momentum forces, develop the reaction force.

Exam tip: The required bearing area of a thrust block is dependent upon the reaction force and the soil bearing pressure so that the soil is not "punched" in. The allowable soil bearing pressure (i.e., capacity) must be determined by geotechnical testing and will usually be provided in a problem.

Energy and Hydraulic Grade Lines

Handbook: Hydraulic Gradient (Grade Line); Energy Line (Bernoulli Equation)

$$H = z_a + \frac{p_a}{\gamma} + \frac{v_a^2}{2g}$$

CERM: Sec. 16.15, Sec. 17.20

$$\text{elevation of EGL} = h_p + h_v + h_z \quad 16.28$$

The *energy grade line* (EGL) is a graph of the total energy (total specific energy) along a length of pipe. In a frictionless pipe without pumps or turbines, the total specific energy is constant, and the EGL will be horizontal. (This is a restatement of the Bernoulli equation.) The *Handbook* equation and CERM Eq. 16.28 are equivalent, with CERM replacing the *Handbook* terms with simpler variables.

The *hydraulic grade line* (HGL) is the line connecting the sum of pressure and elevation heads at different points in conveyance systems. If a series of piezometers were placed at intervals along the pipe, the HGL would join the water levels in the piezometer water columns.

The HGL is the graph of the sum of the pressure and gravitational heads, plotted as a position along the pipeline. Since the pressure head can increase due to a pump or at the expense of the velocity head, the HGL can increase in elevation if the flow increases, if a pump or fitting adds pressure, or if the pipe size decreases, pressurizing the flow, as shown in CERM Eq. 16.29.

$$\text{elevation of HGL} = h_p + h_z \quad 16.29$$

The difference between the EGL and the HGL is the velocity head, h_v, of the fluid, as shown in CERM Eq. 16.30.

$$h_v = \text{elevation of EGL} - \text{elevation of HGL} \quad 16.30$$

CERM Sec. 16.15 provides a set of rules that apply to these grade lines in a frictionless environment, in a pipe flowing full (i.e., under pressure), without pumps or turbines. (See CERM Fig. 16.4.)

Figure 16.4 Energy and Hydraulic Grade Lines Without Friction

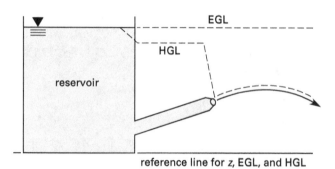

CERM Fig. 17.7 shows how EGL changes with different head losses.

Figure 17.7 Energy and Hydraulic Grade Lines

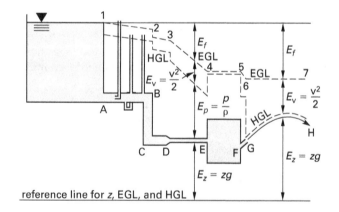

B. PRESSURE CONDUIT

Key concepts: These key concepts are important for answering exam questions in subtopic 10.B, Pressure Conduit.

- equivalent pipe length
- Hazen-Williams roughness coefficient
- major losses: Darcy-Weisbach equation
- major losses: Hazen-Williams equation
- minor losses
- Moody friction chart
- pressure drop in closed conduit
- Reynolds number

PE Civil Reference Manual (CERM): Study these sections in CERM that either relate directly to this subtopic or provide background information.

- Section 16.10: Reynolds Number
- Section 16.11: Laminar Flow
- Section 16.12: Turbulent Flow
- Section 17.3: Typical Velocities in Pipes
- Section 17.5: Head Loss Due to Friction
- Section 17.6: Relative Roughness
- Section 17.8: Energy Loss Due to Friction: Laminar Flow
- Section 17.9: Energy Loss Due to Friction: Turbulent Flow

Water Resources and Environmental Depth Reference Manual (CEWE): Study these sections in CEWE that either relate directly to this subtopic or provide background information.

- Section 2.2: Water Pressure
- Section 2.6: The Darcy-Weisbach Equation
- Section 2.7: The Hazen-Williams Equation
- Section 2.10: Minor Losses

NCEES Handbook: To prepare for this subtopic, familiarize yourself with these sections in the *Handbook*.

- Moody, Darcy, or Stanton Friction Factor Diagram
- Reynolds Number—Circular Pipes
- Head Loss Due to Flow (Darcy-Weisbach Equation)
- Pressure Drop for Laminar Flow
- Hazen-Williams Equation
- Circular Pipe Head Loss (as Feet)
- Circular Pipe Head Loss (as Pressure)
- Values of Hazen-Williams Coefficient *C*
- Darcy-Weisbach Equation (Head Loss)
- Minor Losses in Pipe Fittings, Contractions, and Expansions
- Minor Head Loss Coefficients

The following equations and tables are relevant for subtopic 10.B, Pressure Conduit.

Reynolds Number

Handbook: Reynolds Number—Circular Pipes

$$\mathrm{Re} = \frac{vD\rho}{\mu} = \frac{vD}{\nu}$$

CERM: Sec. 16.10, Sec. 16.11, Sec. 16.12, Sec. 17.3

CEWE: Sec. 2.2

The diameter used in the Reynolds equation is the *hydraulic diameter*, which in pressured circular pipes (flowing full) is the actual geometric diameter of the pipe. This is not the same as the hydraulic radius used in open channel flow. With hydraulic pressurized conduits, the pipe is always assumed to be flowing full.

The Reynolds number indicates the characteristics of flow that will guide how to solve certain equations. When it is important to know whether flow is laminar, turbulent, or transitional, calculate this first. Once this characteristic is determined, other equations, such as sedimentation, will use the Reynolds number in determining other constants or variables.

Laminar flow is typical with Reynolds numbers less than 2100, and turbulent flow is typical with Reynolds numbers greater than 4000. Reynolds numbers in between these values (i.e., the critical zone) indicate that the fluid is in transition between laminar and turbulent. It is difficult to design in this flow regime.

CERM Table 17.1 lists maximum fluid velocities in circular pipes.

Exam tip: The Reynolds number can be used to characterize flow, independent of other equations.

Moody Friction Chart

Handbook: Moody, Darcy, or Stanton Friction Factor Diagram

CERM: Sec. 17.6

CEWE: Sec. 2.6

The *Moody friction factor chart* can be used to determine the value of other variables used in calculating flow. A Moody friction chart is included in *Handbook* figure Moody, Darcy, or Stanton Friction Factor Diagram. CEWE Fig. 2.2 shows a similar chart.

To use the chart, start by determining the Reynolds number, which is shown at the bottom of the chart. (See the entry "Reynolds Number" in this chapter for information.) Next, note the right side of the chart, where the relative roughness value can be found. The specific roughness, *e*, is dictated by the pipe material and can be determined at the lower-left corner of the Moody chart. Divide the specific roughness by the pipe diameter to find the relative roughness. At the point where the Reynolds number intersects the relative roughness curve, the friction factor can be found on the left vertical axis.

> *Exam tip*: When using the Moody chart, the Reynolds number is important for determining the region of the chart being used. It can also be used in reverse if other factors on the table are known.

Energy Loss Due to Friction: Laminar Flow

Handbook: Head Loss Due to Flow (Darcy-Weisbach Equation) Pressure Drop for Laminar Flow Darcy-Weisbach Equation (Head Loss)

CERM: Sec. 17.5, Sec. 17.8

Two methods are available for calculating the frictional energy loss for fluids experiencing laminar flow. The most common is the *Darcy-Weisbach equation* (which is also known as the *Darcy equation* and the *Weisbach equation*). The Darcy-Weisbach equation can be used for both laminar and turbulent flow. One advantage of using the Darcy-Weisbach equation is that the assumption of laminar flow does not need to be confirmed if f is known.

Handbook section Head Loss Due to Flow (Darcy-Weisbach Equation) includes the Darcy-Weisbach equation for pipe flow. The *Handbook* equation and CERM Eq. 17.22 are identical.

$$h_f = f\left(\frac{L}{D}\right)\left(\frac{v^2}{2g}\right)$$

Note that the equation in *Handbook* section Darcy-Weisbach Equation (Head Loss) is arranged to be used to calculate head losses between two different locations in a pipe system.

$$h_f = f_a\left(\frac{L_a}{D_a}\right)\left(\frac{v_a^2}{2g}\right) = f_b\left(\frac{L_b}{D_b}\right)\left(\frac{v_b^2}{2g}\right)$$

Handbook section Pressure Drop for Laminar Flow expresses the Hagen-Poiseuille equation as a change in pressure. The relation is valid only for flow in the laminar region.

$$Q = \frac{\pi R^4 \Delta p_f}{8\mu L} = \frac{\pi D^4 \Delta p_f}{128\mu L}$$

Energy Loss Due to Friction: Turbulent Flow

Handbook: Head Loss Due to Flow (Darcy-Weisbach Equation); Hazen-Williams Equation; Circular Pipe Head Loss (as Feet); Circular Pipe Head Loss (as Pressure); Values of Hazen-Williams Coefficient C; Darcy-Weisbach Equation (Head Loss)

CERM: Sec. 17.9

The Darcy-Weisbach equation can be used to calculate friction losses for both turbulent and laminar flows. *Handbook* section Head Loss Due to Flow (Darcy-Weisbach Equation) presents the equation in a format nearly identical to CERM Eq. 17.27.

$$h_f = f\left(\frac{L}{D}\right)\left(\frac{v^2}{2g}\right)$$

Handbook section Darcy-Weisbach Equation (Head Loss) includes a variation on the same equation.

$$h_f = f_a\left(\frac{L_a}{D_a}\right)\left(\frac{v_a^2}{2g}\right) = f_b\left(\frac{L_b}{D_b}\right)\left(\frac{v_b^2}{2g}\right)$$

The *Handbook* also includes several different variations of the Hazen-Williams equation to solve for different variables. *Handbook* section Hazen-Williams Equation shows a variation that solves for the flow velocity.

$$v = k_1 C R_H^{0.63} S^{0.54}$$

The head loss equation shown in *Handbook* section Circular Pipe Head Loss (as Feet) most closely approximates the Hazen-Williams equation shown in CERM Eq. 17.29.

$$h_f = \left(\frac{4.73L}{C^{1.852}D^{4.87}}\right)Q^{1.852}$$

$$h_{f,\text{ft}} = \frac{3.022 v_{\text{ft/sec}}^{1.85} L_{\text{ft}}}{C^{1.85} D_{\text{ft}}^{1.17}} = \frac{10.44 L_{\text{ft}} Q_{\text{gpm}}^{1.85}}{C^{1.85} d_{\text{in}}^{4.87}} \quad [\text{U.S.}] \quad 17.29$$

The second part of CERM Eq. 17.29 provides the head loss equation in terms of flow in gallons per minute. This is convenient for most calculations related to drinking water conveyance, while flood and stormwater conveyance systems are generally communicated in cubic feet per second of flow. The Hazen-Williams equation should be used only for turbulent flow.

The *Handbook* also includes the table Values of Hazen-Williams Coefficient C, which provides different values for the Hazen-Williams coefficient based on the pipe material.

Handbook section Circular Pipe Head Loss (as Pressure) includes an equation for head loss in terms of pressure.

$$p = \frac{4.52\,Q^{1.85}}{C^{1.85} D^{1.87}}$$

Exam tip: For the *Handbook* equation, flow must be in cubic feet per second, and the diameter of the pipe must be in feet. Most pipe diameters are given in inches, so units will need to be converted.

Hazen-Williams Equation

Handbook: Circular Pipe Head Loss (as Feet); Values of Hazen-Williams Coefficient C

$$h_f = \left(\frac{4.73\,L}{C^{1.852} D^{4.87}}\right) Q^{1.852}$$

CEWE: Sec. 2.7

$$h_{L,\text{ft}} = \frac{10.44\,L_{\text{ft}}\,Q_{\text{gpm}}^{1.85}}{C^{1.85} D_{\text{in}}^{4.87}} \quad\quad 2.11$$

The Hazen-Williams equation is an empirical equation; it was determined based on laboratory testing and observation, and the units of the variables within the equation are not consistent. This is why it has a correction factor that varies according to the units being used for the different variables. It simplifies the calculations for head loss because it does not depend on the Reynolds number.

The *Handbook* equation for circular pipe head loss (as feet) uses different units for flow and pipe diameter than CEWE Eq. 2.11. In the *Handbook*, flow is measured in cubic feet per second (cfs), and pipe diameter is measured in feet. In CEWE, flow is measured in gallons per minute (gpm), and pipe diameter is measured in inches.

The values in *Handbook* table Values of Hazen-Williams Coefficient C should not be confused with Manning's roughness coefficient for open channel flow. Manning's roughness coefficient helps calculate the flow rate. The Hazen-Williams coefficient is used in calculating only the friction loss. Smoother pipes have higher C-values than rougher pipes. The Hazen-Williams equation is referenced also for calculating head loss. The slope of the energy grade line, S, is the friction head loss, h_f, divided by the distance, L. This can also be represented as drop in height per unit length.

Exam tip: When calculating the equation, units must be consistent. When choosing, the conversion factor k_1 must be selected based on using SI or U.S. units.

Darcy-Weisbach Equation

Handbook: Darcy-Weisbach Equation (Head Loss)

$$h_f = f_a\left(\frac{L_a}{D_a}\right)\left(\frac{\text{v}_a^2}{2g}\right) = f_b\left(\frac{L_b}{D_b}\right)\left(\frac{\text{v}_b^2}{2g}\right)$$

CERM: Sec. 17.8

$$h_f = \frac{f L \text{v}^2}{2Dg} \quad\quad 17.22$$

CEWE: Sec. 2.6

The Darcy-Weisbach equation is used for calculating friction head. This equation can be used with a single conduit for flow, or it can be used to compare similar conduits of flow. The *Handbook* equation and CERM Eq. 17.22 are the same, with only minor differences in formatting.

The Darcy-Weisbach equation is used with both laminar and turbulent flow. It is dependent on the Reynolds number, is used for gases or complex fluids, and is useful for nonstandard water properties at high pressures.

Exam tip: Exam problems will make it clear when to use the Darcy-Weisbach equation for friction loss. You may also need to use the Moody friction chart to determine the friction factor, f.

Friction Loss Coefficient

Handbook: Minor Losses in Pipe Fittings, Contractions, and Expansions; Minor Head Loss Coefficients

$$h_{f,\text{fitting}} = C\left(\frac{\text{v}^2}{2g}\right)$$

CEWE Sec. 2.10

$$h_L = \frac{K\text{v}^2}{2g} \qquad 2.20$$

The equation used to determine minor losses multiplies kinetic energy, $\text{v}^2/2g$, by a coefficient. This coefficient is C in the *Handbook* and K in CEWE. The friction head loss depends on the characteristics of the pipe, including the type of fitting used. Loss coefficients for a variety of common fittings can be found in *Handbook* table **Minor Head Loss Coefficients**.

For longer problems, this equation can be part of calculations for pipes and pump calculations, such as net positive suction head (NPSH) and total daily head (TDH).

Exam tip: On the exam, the loss coefficient for specific fittings will be given in the problem statement (per *Handbook* section **Minor Losses in Pipe Fittings, Contractions, and Expansions**).

C. PUMP APPLICATION AND ANALYSIS

Key concepts: These key concepts are important for answering exam questions in subtopic 10.C, Pump Application and Analysis.

- cavitation
- characteristics of pumps
- lift stations and wet wells
- pumps and turbines
- types of pumps

***PE Civil Reference Manual* (CERM):** Study these sections in CERM that either relate directly to this subtopic or provide background information.

- Section 17.18: Introduction to Pumps and Turbines
- Section 17.19: Extended Bernoulli Equation
- Section 18.1: Hydraulic Machines
- Section 18.2: Types of Pumps
- Section 18.7: Sewage Pumps
- Section 18.10: Pumping Power
- Section 18.11: Pumping Efficiency
- Section 18.16: Cavitation
- Section 18.17: Net Positive Suction Head
- Section 18.18: Preventing Cavitation
- Section 18.19: Cavitation Coefficient
- Section 18.24: Pumps in Parallel
- Section 18.25: Pumps in Series

***Water Resources and Environmental Depth Reference Manual* (CEWE):** Study these sections in CEWE that either relate directly to this subtopic or provide background information.

- Section 2.3: Energy Equation
- Section 2.11: Pump Application and Analysis
- Section 2.12: Cavitation

***NCEES Handbook*:** To prepare for this subtopic, familiarize yourself with these sections in the *Handbook*.

- Total Dynamic Pumping Head
- Net Positive Suction Head Available (NPSHA)
- Fluid Power Equations
- Pump Brake Horsepower
- Multiple Pumps in Series
- Multiple Pumps in Parallel
- Cavitation Parameter
- Lift Station Pumping and Wet Wells
- Single Pump System Cycle Time
- Ideal Minimum Wet Well Volume

The following equations, figures, and tables are relevant for subtopic 10.C, Pump Application and Analysis.

Total Dynamic Pumping Head

Handbook: Total Dynamic Pumping Head

$$\text{TDH} = H_L + H_F + H_V$$

CEWE: Sec. 2.3, Sec. 2.11

$$\text{TDH} = h_L + \Delta z + h_\text{v} \qquad 2.21$$

The *Handbook* equation and CEWE Eq. 2.21 are equivalent, but there are a few notable differences in the variables used for total static head and total friction loss.

- total dynamic pumping head: TDH

- sum of total static head (the vertical distance between fluid water levels from the pump centerline to the fluid water level): H_L in the *Handbook*, Δz in CEWE

- total friction loss: H_F in the *Handbook*, h_L in CEWE
- velocity head: H_V in the *Handbook*, h_v in CEWE

CEWE Fig. 2.5 shows a total dynamic head schematic.

Figure 2.5 Total Dynamic Head Schematic

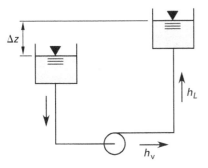

The total dynamic head is the head the pump has to work against.

Exam tip: Head losses can be calculated using different methods. Friction head loss can be calculated from either the Darcy-Weisbach or Hazen-Williams equation. Velocity head is $v_2/2g$. Head loss is the vertical difference the pump is working against.

Net Positive Suction Head

Handbook: Net Positive Suction Head Available (NPSHA)

$$\text{NPSH}_A = H_{pa} + H_s - \sum h_L - H_{vp} = \frac{p_{\text{inlet}}}{\rho g} + \frac{v_{\text{inlet}}^2}{2g} - \frac{p_{\text{vapor}}}{\rho g}$$

CERM: Sec. 18.17

$$\text{NPSHA} = h_{\text{atm}} + h_{z(s)} - h_{f(s)} - h_{vp} \quad \quad 18.30$$

The *Handbook* equation and CERM Eq. 18.30 are equivalent, with only minor differences in nomenclature.

Cavitation is a predictable occurrence. It will occur when the available head is less than the head required for satisfactory operation. The *net positive suction head required* is the minimum fluid energy required at the pump inlet for satisfactory operation. The *net positive suction head available* is the actual total fluid energy at the pump inlet.

The equation is calculating pump head using pressure and velocity. This equation shares many variables with the equation for total dynamic pumping head, TDH.

Exam tip: The equation depends on the density of water. If no temperature is provided, assume ambient temperature. Use the **Physical Properties of Water** table from *the Handbook*.

Fluid Power Equations

Handbook: Fluid Power Equations

CERM: Sec. 18.10

$$\dot{W}_{\text{fluid}} = \rho g H Q$$

$$\dot{W} = \frac{\rho g H Q}{\eta_{\text{pump}}}$$

$$\dot{W}_{\text{purchased}} = \frac{\dot{W}}{\eta_{\text{motor}}}$$

The fluid power equations in the *Handbook* are similar to the equations given in CERM Table 18.5 and Table 18.6. The variables H in the *Handbook* and W in CERM represent work being done on the fluid. H is the head increase provided by the pump, while W is work done.

CERM Eq. 18.10 can be used to calculate head added by the pump.

$$h_A = \frac{\eta_{\text{impeller}} v_{\text{impeller}} v_{\text{fluid}}}{g} \quad \quad 18.10$$

Head added can be thought of as energy added per unit mass. The total pumping power depends on the head added, h_A, and the mass flow rate, $\dot{m}$. The product $h_A \dot{m}$ has units of foot-pounds per second (in customary U.S. units), which can be easily converted to horsepower. Pump output power is known as *hydraulic power* or *water power*. Hydraulic power is the net power actually transferred to the fluid.

Horsepower is a common unit of power, which results in *hydraulic horsepower* and *water horsepower*, WHP, being used to designate the power that is transferred into the fluid. Various relationships for finding the hydraulic horsepower are given in CERM Table 18.5, and hydraulic kilowatt equations are given in CERM Table 18.6.

Table 18.5 Hydraulic Horsepower Equations

	Q (gal/min)	$\dot{m}$ (lbm/sec)		$\dot{V}$ (ft³/sec)
h_A in feet	$\dfrac{h_A Q(\text{SG})}{3956}$	$\dfrac{h_A \dot{m}}{550} \cdot \dfrac{g}{g_c}$		$\dfrac{h_A \dot{V}(\text{SG})}{8.814}$
Δp in psi	$\dfrac{\Delta p Q}{1714}$	$\dfrac{\Delta p \dot{m}}{(238.3)(\text{SG})} \cdot \dfrac{g}{g_c}$		$\dfrac{\Delta p \dot{V}}{3.819}$
Δp in psf[b]	$\dfrac{\Delta p Q}{2.468 \cdot 10^5}$	$\dfrac{\Delta p \dot{m}}{(34{,}320)(\text{SG})} \cdot \dfrac{g}{g_c}$		$\dfrac{\Delta p \dot{V}}{550}$
W in $\dfrac{\text{ft-lbf}}{\text{lbm}}$	$\dfrac{WQ(\text{SG})}{3956}$	$\dfrac{W\dot{m}}{550}$		$\dfrac{W\dot{V}(\text{SG})}{8.814}$

(Multiply horsepower by 0.7457 to obtain kilowatts.)
[a]based on $\rho_{\text{water}} = 62.4$ lbm/ft³ and $g = 32.2$ ft/sec²
[b]Velocity head changes must be included in Δp.

Table 18.6 Hydraulic Kilowatt Equations[a]

	Q (L/s)	$\dot{m}$ (kg/s)	$\dot{V}$ (m³/s)
h_A in meters	$\dfrac{9.81 h_A Q(\text{SG})}{1000}$	$\dfrac{9.81 h_A \dot{m}}{1000}$	$9.81 h_A \dot{V}(\text{SG})$
Δp in kPa[b]	$\dfrac{\Delta p Q}{1000}$	$\dfrac{\Delta p \dot{m}}{1000(\text{SG})}$	$\Delta p \dot{V}$
W in $\dfrac{\text{J}}{\text{kg}}$[b]	$\dfrac{WQ(\text{SG})}{1000}$	$\dfrac{W\dot{m}}{1000}$	$W\dot{V}(\text{SG})$

(Multiply kilowatts by 1.341 to obtain horsepower.)
[a]based on $\rho_{\text{water}} = 1000$ kg/m³ and $g = 9.81$ m/s²
[b]Velocity head changes must be included in Δp.

Exam tip: Be aware of units when using the fluid power equations. Metric units will require different conversion factors and constants than U.S. customary units. Remember that the pump is designed to counter head loss, so if a pump cannot move the fluid to its target location, the pump may run continuously if it is undersized or create a pressurized flow that can damage upstream pipes.

Pump Brake Horsepower

Handbook: Pump Brake Horsepower

$$\text{BHP} = \frac{\text{HP}}{\eta}$$

$$\text{BHP} = \frac{Q\gamma H}{\left(550 \dfrac{\text{ft-lbf}}{\text{hp-sec}}\right)\eta}$$

CERM: Sec. 18.11

$$\text{BHP} = \frac{\text{WHP}}{\eta_p} \qquad 18.11$$

The first *Handbook* equation can be used to determine pump brake horsepower when the hydraulic horsepower is known, whereas the second can be used when the flow rate and head added by the pump are known. CERM Eq. 18.11 is equivalent to the first *Handbook* equation, using WHP for the water (or hydraulic) horsepower rather than HP.

Hydraulic power is the net energy transferred to the fluid per unit time. The input power delivered by the motor to the pump is known as the *brake pump power*. The term *brake horsepower* (BHP) is commonly used to designate both the quantity and its units. Brake horsepower is calculated using the horsepower and efficiency of the pump and can be used to calculate the flow from the pump.

The difference between the brake and hydraulic powers is known as the *friction power* and is calculated using CERM Eq. 18.12.

$$\text{FHP} = \text{BHP} - \text{WHP} \qquad 18.12$$

Exam tip: Brake horsepower will be greater than the hydraulic power of the pump because of friction losses.

Cavitation

Handbook: Cavitation Parameter

$$\sigma = \frac{\text{NPSH}}{h_p}$$

CERM: Sec. 18.16, Sec. 18.18, Sec. 18.19

$$\sigma = \frac{2(p - p_{\text{vp}})}{\rho v^2} = \frac{\text{NPSHA}}{h_A} \qquad [\text{SI}] \quad 18.36(a)$$

$$\sigma = \frac{2g_c(p - p_{\text{vp}})}{\rho v^2} = \frac{\text{NPSHA}}{h_A} \qquad [\text{U.S.}] \quad 18.36(b)$$

CEWE: Sec. 2.12

The *Handbook* equation and CERM Eq. 18.36 are equivalent, with the *Handbook* equation using subscript p and CERM using subscript A to denote "added by the pump." CERM includes two forms that yield slightly different results. The first form is essentially the ratio of the net pressure for collapsing a vapor bubble to the velocity pressure creating the vapor. The second form is applicable to tests of production model pumps.

Cavitation is a spontaneous vaporization of the fluid inside the pump, resulting in a degradation of pump performance. Small vapor pockets form wherever the fluid pressure is less than the vapor pressure. Cavitation is undesirable, as vapor bubbles create friction and vibration that will damage the inside of a pump. The *net positive suction head required* (NPSHR) is provided by the pump manufacturer, while the *net positive suction head available* (NPSHA) is calculated. The vapor pressure is a property of the fluid and is dependent on the type of fluid and temperature.

Exam tip: Cavitation can be caused by excessive discharge from the pump or by high temperature or pressure. To determine whether cavitation is occurring, review the design specifications of the pump and compare these to actual data.

Multiple Pumps in Series

Handbook: Multiple Pumps in Series

CERM: Sec. 18.25

$$H = H_A = H_B = \cdots$$
$$Q = Q_A + Q_B + \cdots$$
$$\eta = \frac{H_A + H_B + \cdots}{\dfrac{H_A}{\eta_A} + \dfrac{H_B}{\eta_B} + \cdots}$$
$$\dot{W} = \frac{\gamma Q(H_A + H_B + \cdots)}{\eta}$$

Series operation is achieved by having one pump discharge into the suction of the next. This arrangement is used primarily to increase the discharge head, although a small increase in capacity also results. CERM Fig. 18.16 shows performance curves for a set of pumps in series.

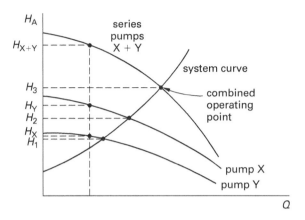

Figure 18.16 Pumps Operating in Series

Exam tip: Pumps in series follow the same principles as pumps in parallel, with the exception that the fluid is moved from one pump to the next. This means that the same fluid flows through all pumps while each pump is working to counter its own portion of head losses. The total work done by all the pumps together must offset the total head losses.

Multiple Pumps in Parallel

Handbook: Multiple Pumps in Parallel

CERM: Sec. 18.24

$$H = H_A = H_B = \cdots$$
$$Q = Q_A + Q_B + \cdots$$
$$\eta = \frac{Q_A + Q_B + \cdots}{\dfrac{Q_A}{\eta_A} + \dfrac{Q_B}{\eta_B} + \cdots}$$
$$\dot{W} = \frac{\gamma H(Q_A + Q_B + \cdots)}{\eta}$$

Parallel operation is obtained by having two pumps discharging into a common header. This type of connection is advantageous when the system demand varies greatly or when high reliability is required. With two pumps in parallel, one can be shut down during low demand.

CERM Fig. 18.15 illustrates that parallel operation increases the capacity of the system while maintaining the same total head.

Figure 18.15 Pumps Operating in Parallel

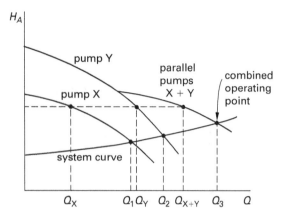

Exam tip: The *Handbook* equations for multiple pumps in parallel follow the same conservation of energy and mass principles as other flow calculations. The sum of the flow through pipes in parallel must equal the flow in. If the pipes are in parallel and they meet at the same termination point, energy losses due to pipe flow must be equal.

Lift Station and Wet Wells

Handbook: Lift Station Pumping and Wet Wells; Single Pump System Cycle Time; Ideal Minimum Wet Well Volume

CERM: Sec. 18.7

Lift stations (also known as pumping stations) are systems designed to move water by providing pumping power. CERM Fig. 18.5 shows a simplified pump installation.

Figure 18.5 Typical Wastewater Pump Installation (greatly simplified)

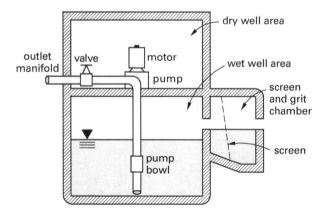

There are four main design requirements for a lift station.

- Determine the capacity and the amount of flow coming in.
- Determine the volume of the wet well needed to counteract the difference between inflows and outflows.
- Determine the design of the system curve or the energy required to pump the liquid.
- Determine the required design to counteract buoyancy from the groundwater table.

The time required to fill a wet well pump is determined using the equation in *Handbook* section Single Pump System Cycle Time.

$$T_{\min} = \frac{V_{\min}}{Q_{\text{in}}} + \frac{V_{\min}}{Q_{\text{out}} - Q_{\text{in}}}$$

The minimum storage volume of a wet well is found using the equation in *Handbook* section Ideal Minimum Wet Well Volume.

$$V_{\min} = \frac{T_{\min} Q_{\text{out}}}{4}$$

The volume of the wet well is determined by the fact that the critical inflow equals half the design outflow. The cycling time will be at a minimum when the inflow is equal to half the pumping capacity.

The weight of the wet well is the weight of the concrete and soil above the chamber or above the lip of the foundation slab. For the buoyancy check, the weight of the wet well components must be greater than the weight of the water that they displace. The factor of safety for buoyancy ranges between 1.2 and 1.5.

Exam tip: Lift station problems are lengthy, so it is unlikely that a full lift station design problem will be included on the exam. Unless calculating for pump power, expect to use wet well equations for determining how many pumps are needed to return a wet well to the minimum volume.

D. PIPE NETWORK ANALYSIS

Key concepts: These key concepts are important for answering exam questions in subtopic 10.D, Pipe Network Analysis.

- Hardy Cross method
- pipe network analysis

- pipes in parallel
- pipes in series

PE Civil Reference Manual (**CERM**): Study these sections in CERM that either relate directly to this subtopic or provide background information.

- Section 17.28: Series Pipe Systems
- Section 17.29: Parallel Pipe Systems
- Section 17.31: Pipe Networks

Water Resources and Environmental Depth Reference Manual (**CEWE**): Study these sections in CEWE that either relate directly to this subtopic or provide background information.

- Section 2.8: Parallel Piping
- Section 2.9: Branched Pipe Networks
- Section 2.13: Pipe Network Analysis

NCEES Handbook: To prepare for this subtopic, familiarize yourself with these sections in the *Handbook*.

- Hazen-Williams Equation
- Darcy-Weisbach Equation (Head Loss)
- Multiple Pipes in Parallel
- Pipe Network Analysis: Continuity
- Pipe Network Analysis: Single Path Adjustment (Hardy Cross)
- Pipe Network Analysis: Linear Method

The following equations and figures are relevant for subtopic 10.D, Pipe Network Analysis.

Pipes in Series

Handbook: Pipe Network Analysis: Single Path Adjustment (Hardy Cross)

CERM: Sec. 17.28

CEWE: Sec. 2.9, Sec. 2.13

Pipes in series are two or more pipes connected end to end, with changes in pipe diameter accomplished with reducers or increasers. Pipes can change direction and still be in series. Some of the most common parameters needed are pressure, head loss, velocity, and the pressure required to meet a capacity demand. If the flow rate or velocity in any part of the system is known, the friction loss can easily be found using CERM Eq. 17.89 as the sum of the friction losses in the individual section.

$$h_{f,t} = h_{f,a} + h_{f,b} \qquad 17.89$$

$$A_a \mathrm{v}_a = A_b \mathrm{v}_b \qquad 17.90$$

CERM Eq. 17.90 suggests that the flow in each section of pipe is the same. This equation assumes that there are no inflows or outflows related to stormwater inlets or drinking water laterals and that losses due to leaking in pipes are negligible.

Changes in materials, diameter, and other factors can impact velocity and head loss. However, the flow must be the same from input to discharge, as expressed by the flow equation in *Handbook* section Pipe Network Analysis: Single Path Adjustment (Hardy Cross).

$$Q_1 = Q_2 + Q_3 = Q_0$$

Exam tip: Determining the rate of flow is a key item in problem solving. Calculate the total friction and minor losses by summing the losses in each pipe segment. If the flow or velocity is not known, flag the problem as one that will require a long time to solve.

Pipes in Parallel

Handbook: Darcy-Weisbach Equation (Head Loss); Multiple Pipes in Parallel; Pipe Network Analysis: Continuity; Pipe Network Analysis: Single Path Adjustment (Hardy Cross); Pipe Network Analysis: Linear Method; Darcy-Weisbach Equation (Head Loss)

$$P = J + L + F - 1$$

$$\sum Q_{\mathrm{in}} - \sum Q_{\mathrm{out}} = Q_c l$$

CERM: Sec. 17.29

CEWE: Sec. 2.8, Sec. 2.13

Handbook section Multiple Pipes in Parallel provides an equation to determine the number of pipes in parallel based on the schematics of the system. *Handbook* section Pipe Network Analysis: Continuity includes an equation to determine the total flow rate in parallel pipes. An example of network analysis is shown in *Handbook* sections Pipe Network Analysis: Single Path Adjustment (Hardy Cross) and Pipe Network Analysis: Linear Method.

Parallel pipe systems such as the one shown in CERM Fig. 17.17 help increase the capacity of a system while also creating redundancy and forming loops. The most common parallel piping problems involve determining the individual flow and pressure levels through each pipe branch.

Figure 17.17 Parallel Pipe System

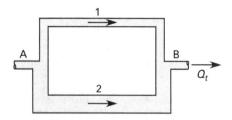

Three key principles govern the flow and energy going through the pipes. One is conservation of mass—the continuity equation. The flow coming into the parallel system is equal to the sum of the flows going through the parallel branches.

The other two key principles that govern the flow and energy going through pipes are two types of conservation of energy: one showing that the head loss is equal across all branches (see CEWE Eq. 2.14) and one showing that the head loss between the two main joints is equal to the head loss through any of the branches.

$$h_L = h_{L1} = h_{L2} = h_{L3} \qquad 2.14$$

Solving CEWE Eq. 2.14 for two branches requires determining that all head losses are equal and that the branch flows add up to the total flow. The only variable is head loss. Since head loss across the pipe network and inlet and exit are equal, use the equation in *Handbook* section **Darcy-Weisbach Equation (Head Loss)** to find the head loss, which can then be used to determine the total flow.

$$h_f = f_a \left(\frac{L_a}{D_a}\right)\left(\frac{v_a^2}{2g}\right) = f_b \left(\frac{L_b}{D_b}\right)\left(\frac{v_b^2}{2g}\right)$$

$$\left(\frac{\pi D^2}{4}\right)v = \left(\frac{\pi D_A^2}{4}\right)v_A + \left(\frac{\pi D_B^2}{4}\right)v_B$$

The Darcy-Weisbach equation in the *Handbook* is equivalent to CERM Eq. 17.98 and Eq. 17.100.

$$\frac{f_1 L_1 v_1^2}{2 D_1 g} = \frac{f_2 L_2 v_2^2}{2 D_2 g} \qquad 17.98$$

$$\frac{\pi}{4}(D_1^2 v_1 + D_2^2 v_2) = \dot{V}_t \qquad 17.100$$

Exam tip: CERM Eq. 17.100 shows flow through circular pipes, but this calculation can work for any shape.

Hardy Cross

Handbook: Pipe Network Analysis: Single Path Adjustment (Hardy Cross); Hazen-Williams Equation

$$Q_1 = Q_2 + Q_3 = Q_0$$

$$H_{f,\mathrm{BCD}} = H_{f,\mathrm{BFD}}$$

CERM: Sec. 17.31

CEWE: Sec. 2.8

$$Q = Q_1 + Q_2 + Q_3 \qquad 2.15$$

$$h_L = h_{L1} = h_{L2} = h_{L3} \qquad 2.14$$

The Hardy Cross method uses conservation of mass to help solve for flow through a complex series of pipes. Assuming a closed system, all flow in must equal all flow out. For two or more pipes in parallel, the energy equation (Bernoulli equation) is combined with the Hazen-Williams equation for head loss to form CEWE Eq. 2.14 and Eq. 2.15 (see *Handbook* section Hazen-Williams Equation).

A special case of a pipes-in-parallel problem can be solved using the equations in *Handbook* section **Pipe Network Analysis: Single Path Adjustment (Hardy Cross)**. This case is illustrated in the same *Handbook* section and is also shown in CEWE Fig. 2.3.

Figure 2.3 Piping System in Parallel

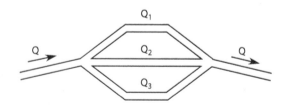

Exam tip: Problems involving three pipes in parallel are common; they take trial and error to solve but are straightforward. Use the principles of conservation of energy and conservation of mass. For water distribution, use the Hazen-Williams equation. Estimate the head loss through the system, and then solve for each branch flow and check whether the branch flows add up to the main flow. If not, adjust the head loss estimate and repeat. An error of ±5% is acceptable.

Multiloop Systems

CERM: Sec. 17.31

Multiloop systems have no closed-form equations. They can be solved using an iterative design using computers or using the Hardy Cross method. (See the entry "Hardy Cross Method" in this chapter for more information.) The Hardy Cross method uses the conservation of mass or the head loss in a closed loop.

Multiloop pipe systems require significant computation. In most day-to-day applications, computer programs are used to solve these types of problems. CERM Fig. 17.19 shows an example of a multiloop system.

Figure 17.19 Multiloop System

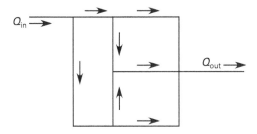

Exam tip: Large multiloop systems will not be included on the exam. However, a small system of up to three loops may be included. It typically takes about 8 to 15 minutes to solve each loop. Most likely, an exam problem will not require students to solve for all loops. If it seems that more than one loop needs to be solved, there may be some hidden variable within the problem statement that needs to be calculated before the loop can be solved. Consider marking problems of this type as difficult and saving them for the end of the exam.

47 Hydraulics—Open Channel

Content in blue refers to the *NCEES Handbook*.

A. Open-Channel Flow 47-1
B. Hydraulic Energy Dissipation 47-5
C. Stormwater Collection and Drainage 47-7
D. Sub- and Supercritical Flow............................. 47-11

The knowledge area of Hydraulics—Open Channel makes up between four and six questions out of the 80 questions on the PE Civil Water Resources and Environmental exam. The organization of this chapter follows the order of subtopics given by NCEES for this knowledge area. Each subtopic is covered in the following sections.

A. OPEN-CHANNEL FLOW

Key concepts: These key concepts are important for answering exam questions in subtopic 11.A, Open-Channel Flow.

- analysis of flow using the Bernoulli equation
- analysis of flow using the continuity equation
- channel configurations
- efficient hydraulic cross sections
- energy gradient
- hydraulic radius
- Manning equation
- Manning roughness coefficient
- nomographs
- open channel flow and its types
- types of open channel flow
  - nonuniform flow
  - uniform flow

PE Civil Reference Manual **(CERM):** Study these sections in CERM that either relate directly to this subtopic or provide background information.

- Section 19.2: Types of Flow
- Section 19.4: Velocity Distribution
- Section 19.5: Parameters Used in Open Channel Flow
- Section 19.6: Governing Equations for Uniform Flow
- Section 19.7: Variations in the Manning Constant
- Section 19.10: Energy and Friction Relationships
- Section 19.11: Sizing Trapezoidal and Rectangular Channels
- Section 19.12: Most Efficient Cross Section
- Section 19.13: Analysis of Natural Watercourses

Water Resources and Environmental Depth Reference Manual **(CEWE):** Study these sections in CEWE that either relate directly to this subtopic or provide background information.

- Section 2.3: Energy Equation
- Section 3.2: Bernoulli Energy Equation
- Section 3.3: Continuity in Open Channel Flow
- Section 3.4: Flow Regimes
- Section 3.5: Open Channel Flow

NCEES Handbook: To prepare for this subtopic, familiarize yourself with these sections in the *Handbook*.

- Principles of One-Dimensional Fluid Flow: Continuity Equation
- Energy Line (Bernoulli Equation)
- Hydraulic Radius
- Manning's Equation
- Approximate Values of Manning's Roughness Coefficient
- Geometric Elements of Channel Sections
- Gradually Varied Flow Profile Diagrams

The following equations, figures, and tables are relevant for subtopic 11.A, Open-Channel Flow.

Continuity Equation

Handbook: Principles of One-Dimensional Fluid Flow: Continuity Equation

$$A_1 v_1 = A_2 v_2$$
$$Q = A v$$

CERM: Sec. 19.4, Sec. 19.6, Sec. 19.13

CEWE: Sec. 3.3

The continuity equation given in *Handbook* section Principles of One-Dimensional Fluid Flow: Continuity Equation and by CERM Eq. 19.8 is a conservation of mass principle. Conservation of mass is universal and applies to all types of flow. Applying the continuity equation is an important part of analyzing nonuniform flow, as this type of flow has varying cross-sectional areas and flow velocities.

The continuity equation can be used to analyze flows through different channel sections. If the channel changes geometry, the velocity and the area of flow will vary between the different sections. However, the flow of water will remain the same. For divided channels, the concept functions just as if they were behaving as branch pipes (parallel pipes).

The sum of the flow at sections 1 and 2 in CERM Fig. 19.5 equals the total flow coming in before and after the channel division.

Figure 19.5 Divided Channel

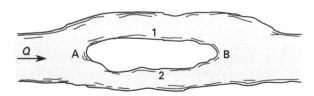

The continuity equation represents conservation of flow, meaning that even though velocity and cross-sectional area may change, total flow must remain the same in and out.

Exam tip: Use techniques for uniform and nonuniform flow measurement when determining the parameters necessary for calculating flow.

Energy Gradient

Handbook: Energy Line (Bernoulli Equation)

$$H = z_a + \frac{p_a}{\gamma} + \frac{v_a^2}{2g}$$

CERM: Sec. 19.10

CEWE: Sec. 2.3, Sec. 3.2

In open channel flow, if the flow is uniform, then the slope of the energy line will parallel the water surface and channel bottom. The energy gradient will equal the geometric slope.

CEWE Fig. 2.1 shows conservation of energy in a closed pipe system. The energy line equation calculates the total head, which is the sum of the elevation, z_a, pressure head, p_a/γ, and velocity head, $v_a^2/2g$. This is different from head loss, which is the change in height between points along a flow line.

Figure 2.1 Closed Conduit Flow Energy Equation Parameters

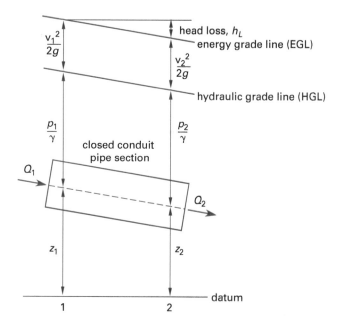

The variables in the energy line equation can change based on points along the flow path. Velocity can change based on cross-sectional area, pressure (depending on whether the flow is open to the atmosphere), and height above the datum. When considering the differences, the use of flow equations like the Manning equation may provide guidance if other variables along the flow path are known.

Exam tip: If using the energy line equation to compare two points along a flow, the difference in the equations should be the head loss from one point to the other.

Hydraulic Radius

Handbook: Hydraulic Radius; Geometric Elements of Channel Sections

$$R_H = \frac{\text{cross-sectional area}}{\text{wetted perimeter}}$$

CERM: Sec. 19.5, Sec. 19.6, Sec. 19.11, Sec. 19.12

$$R = \frac{A}{P} \qquad 19.2$$

Handbook section Hydraulic Radius and CERM Eq. 19.2 each give an equation for calculating hydraulic radius. The *hydraulic radius* is a ratio that represents the geometry of the channel with respect to the depth of the water flow. The hydraulic radius is used for estimating flow through a channel using the Manning equation and other flow measurement equations.

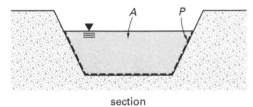

section

The illustration shows the area, *A*, and the wetted perimeter, *P*. It is important to consider the sides of the channel because they add a friction factor to the flow calculation. Note that the flow depth is not the same as the channel depth.

An open channel is not always at full flow, so the wetted perimeter (the sides of the channel touching the flow) and the cross-sectional area of the channel under water can be used to calculate the hydraulic radius. Equations related to the depth of flow are presented in CERM Table 19.2 and *Handbook* Table Geometric Elements of Channel Sections.

In constructed channels, the cross section parameters will be known; these may be less obvious, however, for natural channels or well-worn constructed channels. To calculate the hydraulic radius, select the geometric shape most like the channel and then estimate using the equation for the appropriate shape.

Exam tip: When calculating the hydraulic radius, the perimeter measured is the *wetted perimeter*, not the total, which means you should only count the surfaces that the water is touching as it flows through the channel.

Manning's Equation

Handbook: Manning's Equation; Approximate Values of Manning's Roughness Coefficient

$$Q = \left(\frac{1.486}{n}\right) A R_H^{2/3} S^{1/2}$$

$$v = \left(\frac{1.486}{n}\right) R_H^{2/3} S^{1/2}$$

CERM: Sec. 19.6, Sec. 19.7

$$v = \left(\frac{1.49}{n}\right) R^{2/3} \sqrt{S} \qquad \text{[U.S.]} \quad 19.12(b)$$

CEWE: Sec. 3.5

The first *Handbook* equation shows flow rate as the product of velocity and the cross-sectional area of the channel the water is traveling through. CERM Eq. 19.12 (b) is equivalent to the second *Handbook* equation but uses a rounded value for the correction factor.

The Manning equation is the primary formula for open channel flow and can be combined with the continuity equation to solve for flow rate. 1.486 is a correction factor used to convert the Manning roughness coefficient from SI to U.S. units. *Handbook* table Approximate Values of Manning's Roughness Coefficient provides Manning numbers that correspond to different materials.

The Manning roughness coefficient is an empirical value that simulates a friction factor. Used in combination with the hydraulic radius, it accounts for the reduction in the flow velocity based on the friction along the channel/water interface, which is the wetted perimeter. The Manning roughness coefficient varies with the flow depth and in most cases is assumed to be uniform.

Exam tip: If an exam problem does not give an exact roughness coefficient but presents a range on a table, use the average of the range.

Geometric Channel Sections

Handbook: Manning's Equation; Geometric Elements of Channel Sections

$$v = \left(\frac{1.486}{n}\right) R_H^{2/3} S^{1/2}$$

CERM: Sec. 19.6

$$v = \left(\frac{1.49}{n}\right) R^{2/3} \sqrt{S} \qquad \text{[U.S.]} \quad 19.12(b)$$

To use the Manning equation from the *Handbook* or the equivalent CERM Eq. 19.12(b), the channel geometry needs to be known. A common kind of exam problem asks for the flow depth. In such cases, it is necessary to determine R in terms of the depth of the flow. For example, the area of the flow is used to estimate spread calculations for gutters or roadways.

The depth and the bottom width of the channel are difficult factors to extract from the overall Manning equation. To solve for either the depth or the bottom width, an iterative process is needed. However, to simplify the mathematical exercise and reduce the possibility of computational errors, some alternatives are available. These include collecting multiple measurements along the channel and taking an average, and measuring flow and velocity and then estimating cross-sectional area based on the approximate geometric shape of the channel.

Exam tip: Average dimensions may be sufficient to calculate velocity in a channel.

Uniform Flow

Handbook: Manning's Equation

CERM: Sec. 19.6

CEWE: Sec. 3.4

Uniform flow has constant values for the depth and velocity of the flow and the depth and slope of the channel. The flow depth is also called *uniform flow depth* or *normal depth*. CEWE Fig. 3.3 shows uniform flow in a trapezoidal canal.

The Manning equation from the *Handbook* can be used to determine uniform flow.

Figure 3.3 Uniform Flow Through an Engineered Canal

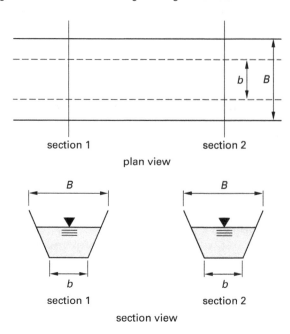

At a uniform flow rate through a channel, the variables and constants used in equations such as the Manning equation can be determined through observation. In controlled settings, weirs, flumes, and other constructed devices are used to help measure flow.

Exam tip: Engineered canals are designed for laminar flow and constant dimensions, although these may change if additional flow streams add flow.

Velocity Distribution

Handbook: Gradually Varied Flow Profile Diagrams

CERM: Sec. 19.4

Velocity through a channel varies based on depth. Velocity is at its lowest at the bottom of the channel and increases at higher levels. CERM Fig. 19.1 shows the velocity distribution in an open channel.

Figure 19.1 Velocity Distribution in an Open Channel

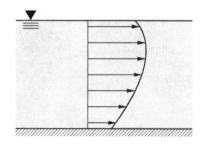

Flow is sometimes measured using several meters in different locations to estimate an average through a channel. This is because of how velocity is distributed through a channel based on proximity to the surrounding surfaces. In general, an average of the measured velocities is sufficient to produce an accurate flow rate, but the more locations measured, the more accurate the result.

Exam tip: In a symmetrical channel, the velocity profile will also be symmetrical.

Nonuniform Flow

CERM: Sec. 19.13

CEWE: Sec. 3.4

Nonuniform flow has changing values for the depth and velocity of the flow and the depth and slope of the channel. CEWE Fig. 3.4 shows nonuniform flow and the range of slopes in a river system.

Figure 3.4 Nonuniform Flow Through a Natural River System

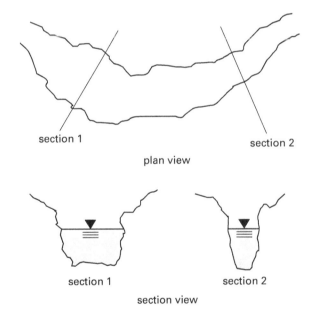

The change in the cross section of the channel and slope affects the depth and velocity of flow along the channel. For nonuniform flow, the conservation of energy equation is used to determine the flows and velocities at different segments along the channel.

Nonuniform flow channels are natural channels such as rivers or canals. Even a designed channel can deteriorate to nonuniform dimensions, requiring estimates for nonuniform flow.

Exam tip: When it is necessary to approximate flow through a natural or nonuniform channel, taking an average of the dimensions may be sufficient to estimate flow.

B. HYDRAULIC ENERGY DISSIPATION

Key concepts: These key concepts are important for answering exam questions in subtopic 11.B, Hydraulic Energy Dissipation.

- energy loss due to friction
- slope and erosion control

PE Civil Reference Manual (**CERM**): Study these sections in CERM that either relate directly to this subtopic or provide background information.

- Section 17.8: Energy Loss Due to Friction: Laminar Flow
- Section 17.9: Energy Loss Due to Friction: Turbulent Flow
- Section 17.20: Energy and Hydraulic Grade Lines with Friction
- Section 19.10: Energy and Friction Relationships
- Section 19.33: Hydraulic Jump
- Section 19.38: Determining Type of Culvert Flow
- Section 80.20: Slope and Erosion Control Features

Water Resources and Environmental Depth Reference Manual (**CEWE**): Study these sections in CEWE that either relate directly to this subtopic or provide background information.

- Section 3.8: Hydraulic Jump
- Section 3.9: Energy Dissipation

NCEES Handbook: To prepare for this subtopic, familiarize yourself with these sections in the *Handbook*.

- Depths and Flows
- Classification of Hydraulic Jumps
- Culvert Head Loss, Total
- Headwater and Tailwater Levels
- Entrance Head Loss
- Friction Head Loss
- Composite Culvert Roughness
- Grate (Bar Rack) Head Loss
- Inlet Control Design

- Estimation of Wave Heights Based on Surface Winds
- Ratio of Wind Speed of Any Duration U_t to the 1-hr Wind Speed $U_{3,600}$
- Duration of the Fastest-Mile Wind Speed U_f as a Function of Wind Speed (for Open Terrain Conditions)
- Revised Universal Soil Loss Equation

The following equations and figures are relevant for subtopic 11.B, Hydraulic Energy Dissipation.

Hydraulic Jump

Handbook: Depths and Flows

$$y_1 = -\tfrac{1}{2}y_2 + \sqrt{\frac{2v_2^2 y_2}{g} + \frac{y_2^2}{4}}$$

$$y_2 = -\tfrac{1}{2}y_1 + \sqrt{\frac{2v_1^2 y_1}{g} + \frac{y_1^2}{4}}$$

$$\Delta E = \left(y_1 + \frac{v_1^2}{2g}\right) - \left(y_2 + \frac{v_2^2}{2g}\right)$$

CERM: Sec. 19.33

$$d_1 = -\tfrac{1}{2}d_2 + \sqrt{\frac{2v_2^2 d_2}{g} + \frac{d_2^2}{4}} \quad \begin{bmatrix}\text{rectangular}\\ \text{channels}\end{bmatrix} \quad 19.91$$

$$d_2 = -\tfrac{1}{2}d_1 + \sqrt{\frac{2v_1^2 d_1}{g} + \frac{d_1^2}{4}} \quad \begin{bmatrix}\text{rectangular}\\ \text{channels}\end{bmatrix} \quad 19.92$$

$$\Delta E = \left(d_1 + \frac{v_1^2}{2g}\right) - \left(d_2 + \frac{v_2^2}{2g}\right) \approx \frac{(d_2 - d_1)^3}{4 d_1 d_2} \quad 19.95$$

CEWE: Sec. 3.8, Sec. 3.9

The equations in *Handbook* section Depths and Flows show how a hydraulic jump can reduce velocity and energy and, therefore, associated erosion. They are equivalent to CERM Eq. 19.91, Eq. 19.92, and Eq. 19.95. The only difference in the equations is the variable used to represent depth; CERM uses d while the *Handbook* uses y.

Flow remains constant through a channel based on the continuity equations. However, reducing velocity can help dissipate energy and reduce erosion. This can be done by increasing flow depth, which increases the cross-sectional area and decreases velocity, as shown in CERM Fig. 19.21. In a channel, providing a means to decrease flow will slow erosion and dissipate energy.

Figure 19.21 Conjugate Depths

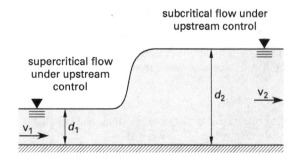

Exam tip: Knowing the parameters for one section of the channel can provide dimensions for the other portion.

Energy Dissipators

Handbook: Classification of Hydraulic Jumps

CERM: Sec. 80.20

Energy dissipators reduce flow velocity to protect downstream areas from erosion. Remember that the goal is to reduce the *velocity* of the flow; do not confuse this with reducing the flow itself.

There are two classifications for energy dissipators. They are either through-flow segments or located at the flow ends. For example, riprap is commonly used at the ends of culverts to prevent erosion.

Supercritical flows are high velocity, and subcritical flows are low velocity. Hydraulic jumps are often used as energy dissipators because they force water to "jump" into a subcritical flow regime that has slower velocities.

Exam tip: Combining energy dissipators along a flow, such as by using a geotextile fabric along the flow, making the slope less steep, or increasing the width of the outlet, can more effectively reduce flow velocity than using a single energy dissipator. When evaluating options, choose the most effective for the situation, but don't neglect using multiple options.

Energy Dissipation at Culvert Outlets

Handbook: Culvert Head Loss, Total; Headwater and Tailwater Levels; Entrance Head Loss; Friction Head Loss; Composite Culvert Roughness; Grate (Bar Rack) Head Loss; Inlet Control Design

$$H = H_c + H_f + H_0 + H_b + H_j + H_g$$

CERM: Sec. 19.38

The most common energy dissipation device used at the ends of culverts is riprap. However, riprap is often mistakenly designed. In design and construction, riprap should not be placed as an afterthought. Design procedures should be followed to select the appropriate material size. Riprap aprons have low energy dissipation through roughness and high energy dissipation by spreading flow.

The equation in *Handbook* section Culvert Head Loss, Total represents total culvert head loss at the entrance (subscript e) and exit (0), and due to friction (f), bends (b), junctions (j), and grates (g). The combination of head losses also represents the decrease in energy that can help reduce the velocity through a culvert.

Handbook sections Culvert Head Loss, Total, Headwater and Tailwater Levels, Entrance Head Loss, Friction Head Loss, Composite Culvert Roughness, Grate (Bar Rack) Head Loss, and Inlet Control Design show how energy dissipation is a function of head loss in the system.

Exam tip: Some head losses may be minor and can be neglected in the calculation. When reviewing the culvert properties, determine whether there are any significant joints, bends, or other characteristics that would impact head losses.

Erosion Prevention

Handbook: Estimation of Wave Heights Based on Surface Winds; Ratio of Wind Speed of Any Duration U_t to the 1-hr Wind Speed $U_{3,600}$; Duration of the Fastest-Mile Wind Speed U_f as a Function of Wind Speed (for Open Terrain Conditions); Revised Universal Soil Loss Equation

CERM: Sec. 80.20

Erosion is caused by water and air flows. Erosion creates sediment, pollutes water resources, and can undermine structural controls. Water-related erosion increases with flow velocity. Hydraulic energy dissipation is intended to prevent erosion.

To avoid erosion, barriers such as geotextiles, slope changes, and land cover can be implemented. However, note that most erosion occurs during large storm events with over one inch of precipitation.

Handbook sections Estimation of Wave Heights Based on Surface Winds, Ratio of Wind Speed of Any Duration U_t to the 1-hr Wind Speed $U_{3,600}$, Duration of the Fastest-Mile Wind Speed U_f as a Function of Wind Speed (for Open Terrain Conditions), and Revised Universal Soil Loss Equation present equations and illustrations on erosion based on the erosion type: erosion due to wind speed, wave height, or soil loss.

Exam tip: When designing for erosion prevention, multiple solutions may be necessary. Adding more ground cover, changing the slope, and broadening the width of the channel can all decrease erosion.

C. STORMWATER COLLECTION AND DRAINAGE

Key concepts: These key concepts are important for answering exam questions in subtopic 11.C, Stormwater Collection and Drainage.

- culvert flow types (USGS) and head losses
- full-flow energy and hydraulic grade lines

PE Civil Reference Manual **(CERM):** Study these sections in CERM that either relate directly to this subtopic or provide background information.

- Section 17.20: Energy and Hydraulic Grade Lines with Friction
- Section 19.10: Energy and Friction Relationships
- Section 19.37: Culverts
- Section 19.38: Determining Type of Culvert Flow
- Section 19.39: Culvert Design

Water Resources and Environmental Depth Reference Manual **(CEWE):** Study these sections in CEWE that either relate directly to this subtopic or provide background information.

- Section 3.11: Stormwater Collection
- Section 3.16: Culvert Design

NCEES Handbook: To prepare for this subtopic, familiarize yourself with these sections in the *Handbook*.

- Hydraulic Gradient (Grade Line)
- Approximate Values of Manning's Roughness Coefficient
- Culvert Flow Types (USGS)
- Culvert Flow Types (USGS): Type 1 Flow
- Culvert Flow Types (USGS): Type 2 Flow

- Culvert Flow Types (USGS): Type 3 Flow
- Culvert Flow Types (USGS): Type 4 Flow
- Culvert Flow Types (USGS): Type 5 Flow
- Culvert Flow Types (USGS): Type 6 Flow

Codes and standards: To prepare for this subtopic, familiarize yourself with these sections from codes and standards.

Hydraulic Design of Highway Culverts (FHWA HIF-12-026)

- Section 3.3: Culvert Design Using Nomographs

The following equations and figures are relevant for subtopic 11.C, Stormwater Collection and Drainage.

Hydraulic Grade Line

Handbook: Hydraulic Gradient (Grade Line)

CERM: Sec. 17.20, Sec. 19.10

The hydraulic grade line is the greater of the full-flow friction loss and the normal depth (depth of flow) using the Manning equation.

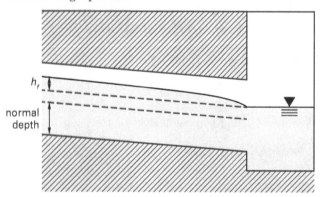

The illustration shows a specific case where the tailwater is above the normal depth. This leads to the hydraulic grade line being above the normal depth. However, the relationship can change based on pipe slopes, sizes, and tailwater elevations, and the normal depth may be above the full-flow friction loss. CERM Eq. 19.30 shows how to calculate the head loss along the grade line.

$$h_f = \frac{Ln^2 v^2}{R^{4/3}} \quad \text{[SI]} \quad 19.30(a)$$

$$h_f = \frac{Ln^2 v^2}{2.208 R^{4/3}} \quad \text{[U.S.]} \quad 19.30(b)$$

Handbook section Hydraulic Gradient (Grade Line) explains how the hydraulic grade line is the sum of the pressure and elevation heads at different points along a flow path.

> *Exam tip*: Hydraulic calculators used to solve these problems are considered slide rules and are not allowed in the PE exam.

> *Exam tip*: CERM Eq. 19.30 is not in the *Handbook*; it will be provided with any problem that requires it.

Storm Sewers

Handbook: Culvert Flow Types (USGS); Culvert Flow Types (USGS): Type 1 Flow; Culvert Flow Types (USGS): Type 2 Flow; Culvert Flow Types (USGS): Type 3 Flow; Culvert Flow Types (USGS): Type 4 Flow; Culvert Flow Types (USGS): Type 5 Flow; Culvert Flow Types (USGS): Type 6 Flow

CERM: Sec. 19.37

CEWE: Sec. 3.11

The methodology for the collection and drainage of stormwater includes various concepts. Hydrology provides the flow through the system. The energy equation provides the hydraulic gradient (the actual water elevation) for each pipe and structure. The use of stations is typical of transportation projects and other linear projects. A reference point, designated as 0+00, is selected where a distance measurement starts. Every station is 100 ft. For example, station 5+00 is equal to 500 ft from the outfall. Features do not always occur at 100 ft intervals. For example, a culvert located upstream might be indicated at station 12+57, which is equal to 1257 ft from the outfall.

> *Exam tip*: Storm sewers are designed for a 10-year storm event. To prevent clogging and plant growth, the minimum cleansing velocity is 2 ft/sec. To prevent damage, the minimum scouring velocity is 10 ft/sec.

Culverts

Handbook: Culvert Flow Types (USGS); Culvert Flow Types (USGS): Type 1 Flow; Culvert Flow Types (USGS): Type 2 Flow; Culvert Flow Types (USGS): Type 3 Flow; Culvert Flow Types (USGS): Type 4 Flow; Culvert Flow Types (USGS): Type 5 Flow; Culvert Flow Types (USGS): Type 6 Flow

CERM: Sec. 19.37, Sec. 19.39

CEWE: Sec. 3.16

FHWA HIF-12-026: Sec. 3.3

Culverts are drainage pipes that cross roads, railroads, or embankments. Although not provided for use by NCEES during the exam, Sec. 3.3 of FHWA HIF-12-026 is a great resource for nomographs that can be used to analyze different types of culverts and should be reviewed before taking the exam.

Handbook sections Culvert Flow Types (USGS): Type 1 Flow, Culvert Flow Types (USGS): Type 2 Flow, Culvert Flow Types (USGS): Type 3 Flow, Culvert Flow Types (USGS): Type 4 Flow, Culvert Flow Types (USGS): Type 5 Flow, and Culvert Flow Types (USGS): Type 6 Flow show how to calculate the culvert flow rates for different types of flow.

Exam tip: Culvert flows are classified by how flow passes through a culvert at the inlet and outlet. The entries in this chapter that address each culvert classification type have exam tips indicating the inlet and outlet traits that characterize the type.

Culvert Flow Classification: Type 1

Handbook: Approximate Values of Manning's Roughness Coefficient; Culvert Flow Types (USGS): Type 1 Flow

$$Q = A\sqrt{\frac{2g\Delta h}{\frac{2gn^2 L}{R^{1/3}} + k_e + 1}}$$

CERM: Sec. 19.38

$$Q = C_d A_c \sqrt{2g\left(h_1 - z + \frac{\alpha v_1^2}{2g} - d_c - h_{f,1\text{-}2}\right)} \quad 19.102$$

The equation in *Handbook* section Culvert Flow Types (USGS): Type 1 Flow is equivalent to CERM Eq. 19.102, as can be seen by substituting equivalent variables.

$$\Delta h = h_1 - z + \frac{\alpha v_1^2}{2g} - d_c - h_{f,1\text{-}2}$$

$$C_d = \left(\frac{2gn^2 L}{R^{4/3}} + k_e + 1\right)^{-1/2}$$

In type 1 flow, both the inlet and the outlet are submerged. Calculation of the critical depth, d_c, and the corresponding critical area, A_c, is not required. The hydraulic radius, R, must be calculated. The Manning roughness coefficient, n, and the entrance loss coefficient, k_e, must be looked up in *Handbook* table Approximate Values of Manning's Roughness Coefficient.

In flow types 1 through 3, the inlet is submerged, but flow is different at the outlet. The slope of the culvert is relatively flat.

Culvert Flow Classification: Type 2

Handbook: Approximate Values of Manning's Roughness Coefficient; Culvert Flow Types (USGS): Type 2 Flow

$$Q = A\sqrt{\frac{2g\Delta h}{\frac{fL}{4R} + k_e + 1}}$$

CERM: Sec. 19.38

$$Q = C_d A_c \sqrt{2g\left(h_1 + \frac{\alpha v_1^2}{2g} - d_c - h_{f,1\text{-}2} - h_{f,2\text{-}3}\right)} \quad 19.103$$

The equation in *Handbook* section Culvert Flow Types (USGS): Type 2 Flow is equivalent to CERM Eq. 19.103, as can be seen by substituting equivalent variables.

$$\Delta h = h_1 + \frac{\alpha v_1^2}{2g} - d_c - h_{f,1\text{-}2} - h_{f,2\text{-}3}$$

$$C_d = \left(\frac{fL}{4R} + k_e + 1\right)^{-1/2}$$

In type 2 flow, the inlet is submerged, there is full flow in the barrel of the culvert, and there is free outfall. Note that this equation is similar to the one discussed in the entry "Culvert Flow Classification: Type 1" in this chapter, but this equation uses the Darcy-Weisbach friction factor, f, instead of the Manning roughness coefficient, n. The entrance loss coefficient, k_e, must be looked up in *Handbook* table Approximate Values of Manning's Roughness Coefficient.

In flow types 1 through 3, the inlet is submerged, but the flow is different at the outlet. The slope of the culvert is relatively flat.

Culvert Flow Classification: Type 3

Handbook: Culvert Flow Types (USGS): Type 3 Flow

$$Q = C_d A \sqrt{2gh}$$

CERM: Sec. 19.38

$$Q = C_d A_3 \sqrt{2g\left(h_1 + \frac{\alpha v_1^2}{2g} - h_3 - h_{f,1\text{-}2} - h_{f,2\text{-}3}\right)} \quad 19.104$$

The equation in *Handbook* section Culvert Flow Types (USGS): Type 3 Flow is equivalent to CERM Eq. 19.104, as can be seen by substituting equivalent variables.

$$h = h_1 + \frac{\alpha v_1^2}{2g} - h_3 - h_{f,1\text{-}2} - h_{f,2\text{-}3}$$

In type 3 flow, the inlet is submerged but the outlet is not, and the normal depth is less than the culvert barrel height (entrance control). Note that the equation is identical to that of a fluid discharged from an orifice in a pressurized tank.

In flow types 1 through 3, the inlet is submerged, but flow is different at the outlet. The slope of the culvert is relatively flat.

Culvert Flow Classification: Type 4

Handbook: Culvert Flow Types (USGS): Type 4 Flow

$$Q = A_c \sqrt{2g\left(\Delta h + \frac{v_1^2}{2g} - h_i - h_f\right)}$$

CERM: Sec. 19.38

$$Q = C_d A_o \sqrt{2g\left(\frac{h_1 - h_4}{1 + \dfrac{29 C_d^2 n^2 L}{R^{4/3}}}\right)} \quad 19.105$$

$$Q = C_d A_o \sqrt{2g(h_1 - h_4)} \quad 19.106$$

The equation in *Handbook* section Culvert Flow Types (USGS): Type 4 Flow is equivalent to CERM Eq. 19.105, with the difference being the way in which the head is calculated. In addition, the CERM equation includes a discharge coefficient, C_d, which converts the units into cubic feet per second, as well as a term in the denominator to correct for friction. The difference between CERM Eq. 19.105 and Eq. 19.106 is that Eq. 19.106 is for culverts less than 50 ft wide. In such cases, the friction losses are negligible and the equation is simplified to ignore them.

In type 4 flow, the inlet is not submerged, the slope is mild, and there is low tailwater. The critical depth is at the outlet of the barrel of the culvert. Note that this equation uses the critical area at the exit, A_c, rather than the area of the culvert, A. The entrance loss, h_i, and the friction loss, h_f, must be calculated using the equations provided in the *Handbook*.

In flow types 4 through 6, the inlet is not submerged, and the slope and outlet conditions are different.

Culvert Flow Classification: Type 5

Handbook: Culvert Flow Types (USGS): Type 5 Flow

$$Q = A_c \sqrt{2g\left(\Delta h + \frac{v_1^2}{2g} - h_i\right)}$$

CERM: Sec. 19.38

$$Q = C_d A_o \sqrt{2g(h_1 - z)} \quad 19.107$$

The equation in *Handbook* section Culvert Flow Types (USGS): Type 5 Flow is equivalent to CERM Eq. 19.107, with the difference being the way in which the head is calculated. In addition, the CERM equation includes a discharge coefficient, C_d, which converts the units into cubic feet per second.

In type 5 flow, the inlet is not submerged, the slope is steep, and there is low tailwater. The critical depth is at the inlet of the barrel of the culvert. This equation is nearly identical to CERM Eq. 19.106 and the equation in *Handbook* section Culvert Flow Types (USGS): Type 4 Flow, but the friction loss, h_f, is not included.

In flow types 4 through 6, the inlet is not submerged, and the slope and outlet conditions are different.

Culvert Flow Classification: Type 6

Handbook: Culvert Flow Types (USGS): Type 6 Flow

$$Q = A \sqrt{2g\left(\Delta h + \frac{v_1^2}{2g} - h_i - h_f\right)}$$

CERM: Sec. 19.38

$$Q = C_d A_o \sqrt{2g(h_1 - h_3 - h_{f,2\text{-}3})} \quad 19.108$$

The equation in *Handbook* section Culvert Flow Types (USGS): Type 6 Flow is equivalent to CERM Eq. 19.108, with the difference being the way in which the head is calculated. In addition, the CERM equation includes a discharge coefficient, C_d, which converts the units into cubic feet per second.

In type 6 flow, the inlet is not submerged, the slope is mild, and the tailwater is above critical depth. This equation is nearly identical to that for type 5, but the area, A, is that of the culvert rather than the critical area, A_c.

In flow types 4 through 6, the inlet is not submerged, and the slope and outlet conditions are different.

D. SUB- AND SUPERCRITICAL FLOW

Key concepts: These key concepts are important for answering exam questions in subtopic 11.D, Sub- and Supercritical Flow.

- critical depth
- Froude number
- normal depth
- subcritical and supercritical flow

PE Civil Reference Manual (**CERM**): Study these sections in CERM that either relate directly to this subtopic or provide background information.

- Section 19.21: Specific Energy
- Section 19.25: Critical Flow and Critical Depth in Rectangular Channels
- Section 19.26: Critical Flow and Critical Depth in Nonrectangular Channels
- Section 19.27: Froude Number
- Section 19.29: Occurrences of Critical Flow
- Section 19.30: Controls on Flow

Water Resources and Environmental Depth Reference Manual (**CEWE**): Study these sections in CEWE that either relate directly to this subtopic or provide background information.

- Section 3.6: Subcritical and Supercritical Flow
- Section 3.7: Gradually Varied Flow

NCEES Handbook: To prepare for this subtopic, familiarize yourself with these sections in the *Handbook*.

- Froude Number
- Critical Depth
- Composite Slopes Channel Profiles

The following equations and figures are relevant for subtopic 11.D, Sub- and Supercritical Flow.

Froude Number

Handbook: Froude Number

$$\mathrm{Fr} = \frac{V}{\sqrt{gy_h}} = \sqrt{\frac{Q^2 T}{gA^3}}$$

CERM: Sec. 19.27

$$\mathrm{Fr} = \frac{\mathrm{v}}{\sqrt{gL}} \quad\quad 19.79$$

CEWE: Sec. 3.6, Sec. 3.7

$$\mathrm{Fr} = \frac{q}{\sqrt{gd_1^3}} \quad\quad 3.27$$

The *Handbook* equation or CERM Eq. 19.79 can be used to calculate the *Froude number*, a dimensionless value that is a convenient indication of the flow regime (i.e., subcritical or supercritical). When the Froude number is greater than 1.0, the flow is supercritical. When it is less than 1.0, the flow is subcritical. CEWE Eq. 3.27 can be used to estimate the Froude number in a rectangular channel. Refer to CEWE Sec. 3.6 for more information on subcritical and supercritical flow.

The Froude number characterizes flow and how flow may behave in a channel. Although equations from different sources use different variables, the calculations are similar, with a flow or velocity variable in the numerator and a dimensional variable, such as depth of flow or cross-sectional area, in the denominator.

Exam tip: Knowing the Froude number can help calculate flow rate in the channel and channel dimensions.

Critical Depth

Handbook: Critical Depth; Composite Slopes Channel Profiles

$$y_c = \left(\frac{q^2}{g}\right)^{1/3} \quad [\text{rectangular}]$$

$$Q_c = \left(\frac{A^3 g}{T}\right)^{1/2} \quad [\text{nonrectangular}]$$

CERM: Sec. 19.21, Sec. 19.25, Sec. 19.26

$$d_c^3 = \frac{Q^2}{gw^2} \quad \text{[rectangular]} \qquad 19.75$$

$$\frac{Q^2}{g} = \frac{A^3}{T} \quad \text{[nonrectangular]} \qquad 19.78$$

Critical depth results in critical flow, which is the most efficient flow; it has the maximum flow for the minimum amount of energy. Critical depth depends on the shape of the channel.

Handbook section **Composite Slopes Channel Profiles** illustrates a variety of channel profiles.

The equation for nonrectangular channels, given in different forms in *Handbook* section **Critical Depth** and CERM Eq. 19.78, uses the cross-sectional area, A, and surface width, T, to calculated critical flow. Since the channel is not rectangular, both variables can rapidly change based on the depth of flow. Note that this equation is incorrectly labeled as applying to rectangular channels in the *Handbook*.

In the equation for rectangular channels given in *Handbook* section **Critical Depth**, depth is referred to with the variable y (y_c is critical depth), while CERM Eq. 19.75 uses d (d_c is critical depth). The rectangular channel equation uses a single dimensional variable for critical depth, since the width of the channel remains constant. The nonrectangular equation uses critical flow, since the channel may be irregular, and critical depth may change as the channel depth and width change.

Exam tip: If calculating the dimensions for critical depth of a nonrectangular channel, establish a relationship between the surface width and area to simplify the equation.

Subcritical and Supercritical Flow

Handbook: Critical Depth; Composite Slopes Channel Profiles

$$\frac{\mathrm{v}^2}{2g} = \frac{y}{2}$$

CERM: Sec. 19.29, Sec. 19.30

CEWE: Sec. 3.6

Subcritical flow occurs when the critical depth is less than the normal depth. It has a slow velocity, ripples upstream, and is controlled through downstream control.

Supercritical flow occurs when the critical depth is greater than the normal depth. It has a fast velocity, no upstream ripples, and is controlled through upstream control.

CERM Fig. 19.15 shows critical depth flow over a short stretch of channel. *Handbook* section **Composite Slopes Channel Profiles** illustrates a variety of channel profiles.

Figure 19.15 Occurrence of Critical Depth

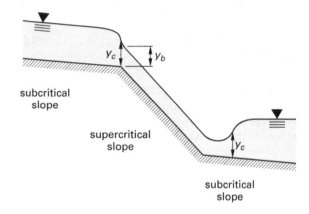

On the slope shown in CERM Fig. 19.15, supercritical flow at high velocity transitions to subcritical flow at low velocity.

Subcritical and supercritical flows describe how fluids can surge through a channel, and a change can cause a hydraulic jump.

Changes in channel depth can cause a change in flow characteristics as the channel is expanded or contracted. When determining whether flow is subcritical, supercritical, or critical, calculating the Froude number will indicate the nature of the flow.

The critical state of flow equation from *Handbook* section **Critical Depth** can be used to calculate critical depth at a given velocity. If the depth of flow is less than the depth calculated by this equation, the flow is supercritical. If above, it is subcritical.

Exam tip: Supercritical flow has a Froude number greater than 1.0, while subcritical flow has a Froude number less than 1.0.

48 Hydrology

Content in blue refers to the *NCEES Handbook*.

A. Storm Characteristics 48-1
B. Runoff Analysis 48-4
C. Hydrograph Development and Applications 48-6
D. Rainfall Intensity, Duration, and Frequency 48-9
E. Time of Concentration 48-11
F. Rainfall and Stream Gauging Stations.............48-13
G. Depletions 48-15
H. Stormwater Management 48-17

The knowledge area of Hydrology makes up between six and nine questions out of the 80 questions on the PE Civil Water Resources and Environmental exam. The organization of this chapter follows the order of subtopics given by NCEES for this knowledge area. Each subtopic is covered in the following sections.

A. STORM CHARACTERISTICS

Key concepts: These key concepts are important for answering exam questions in subtopic 12.A, Storm Characteristics.

- hydrologic cycle
- hyetograph
- probability of exceedance
- rainfall measurement
- return period
- storm distribution
- storm frequency
- water balance equation

NCEES Handbook: To prepare for this subtopic, familiarize yourself with these sections in the *Handbook*.

- Probability of Single Occurrence in a Given Storm Year
- Risk (Probability of Exceedance)
- Surface Water System Hydrologic Budget

PE Civil Reference Manual (**CERM**): Study these sections in CERM that either relate directly to this subtopic or provide background information.

- Section 20.1: Hydrologic Cycle
- Section 20.2: Storm Characteristics
- Section 20.3: Precipitation Data
- Section 20.7: Floods

Water Resources and Environmental Depth Reference Manual (**CEWE**): Study these sections in CEWE that either relate directly to this subtopic or provide background information.

- Section 1.1: Introduction
- Section 1.4: Storm Frequency
- Section 1.5: Characteristics of Precipitation

The following equations, figures, and tables are relevant for subtopic 12.A, Storm Characteristics.

Storm Frequency

Handbook: Probability of Single Occurrence in a Given Storm Year; Risk (Probability of Exceedance)

$$P(x \geq x_T) = p = \frac{1}{T}$$

$$P(x \geq x_T \text{ at least once in } n \text{ years}) = 1 - \left(1 - \frac{1}{T}\right)^n$$

CERM: Sec. 20.7

$$p\{F \text{ event in one year}\} = \frac{1}{F} \quad \quad 20.19$$

$$p\{F \text{ event in } n \text{ years}\} = 1 - \left(1 - \frac{1}{F}\right)^n \quad \quad 20.20$$

CEWE: Sec. 1.4

These equations are used to calculate the probability (risk) that a storm event in any given year will equal a design storm with a given recurrence interval frequency. The *Handbook* equation for the probability of a storm event in one year is the same as CERM Eq. 20.19, except that the *Handbook* uses the variable T for the storm frequency and CERM uses the variable F. The equation for the probability of a storm event in n years is the same as CERM Eq. 20.20, with the same difference in nomenclature.

The *storm frequency* (also known as *frequency of occurrence, recurrence interval,* or *return interval*) is the average number of years between storms of a given intensity. In general, the design storm may be specified by its recurrence interval (e.g., 100-year storm) or its annual probability of occurrence (e.g., 1% storm). A 100-year storm is one that would be exceeded in severity only once every 100 years on average.

The equations shown are valid for both storm and flood frequency. *Storm frequency* refers to the likelihood of occurrence of a weather event (a storm), whereas *flood frequency* refers to the likelihood of certain flows within a watershed. For more about flood frequency, see the entry "Floods" in this chapter.

CEWE Table 1.3 relates storm frequency (called the storm return interval) to probability.

Table 1.3 Storm Return Interval and Probability of Occurring Once in Any Given Year

return interval, t_r (yr)	probability, p
2	0.50
5	0.20
10	0.10
25	0.04
50	0.02
100	0.01
500	0.002

Exam tip: The *Handbook* equation for the probability of a storm event in n years can be used when considering the probability of exceedance of the design storm for the lifetime of a structure. For example, if a bridge has a proposed 50-year design life, the risk of a storm more intense than the design storm can be treated as the probability of such an event occurring in 50 years.

Water Balance Equation

Handbook: Surface Water System Hydrologic Budget

$$P + Q_{in} - Q_{out} + Q_g - E_s - T_s - I = \Delta S_s$$

CERM: Sec. 20.1

$$\begin{aligned}\text{total precipitation} &= \text{net change in surface water removed} \\ &+ \text{net change in ground water removed} \\ &+ \text{evapotranspiration} \\ &+ \text{interception evaporization} \\ &+ \text{net increase in surface water storage} \\ &+ \text{net increase in ground water storage} \end{aligned} \quad 20.1$$

$$P = Q + E + \Delta S$$

CEWE: Sec. 1.1

$$P = R + I + \text{ET} - \Delta S \quad 1.1$$

The water balance equation (or water budget equation) calculates the balance between water moving into and out of a watershed, lake, or other water system; it is, essentially, the application of the conservation of mass to the hydrologic cycle. The *Handbook* equation, CERM Eq. 20.1, and CEWE Eq. 1.1 are essentially the same, with differences in presentation. The *Handbook* equation is written with a focus on surface water systems (e.g., lakes, reservoirs) and assumes that groundwater infiltration is a loss, while CERM Eq. 20.1 makes no such assumption. Both equations are valid for groundwater and surface water systems.

CERM Eq. 20.1 can be restated and simplified as CERM Eq. 20.2.

$$\begin{aligned}\text{total precipitation} &= \text{initial abstraction} + \text{infiltration} \\ &+ \text{surface runoff}\end{aligned} \quad 20.2$$

Exam tip: Runoff is generally measured in units of volume over time or total volume of runoff, while factors such as precipitation are often measured in terms of depth of water over the area of the watershed. Runoff can be averaged over time and divided by the area of the watershed to convert the value to comparable units.

Hydrologic Cycle

CERM: Sec. 20.1

CEWE: Sec. 1.1

The hydrologic cycle (water budget) tracks how the precipitation that falls on the earth collects in a waterway, permeates into the ground (infiltration), is stored in the ground (the change in moisture storage), and then is dispersed back into the atmosphere through evaporation or plant transpiration. CEWE Fig. 1.1 illustrates this cycle.

Figure 1.1 The Hydrologic Cycle

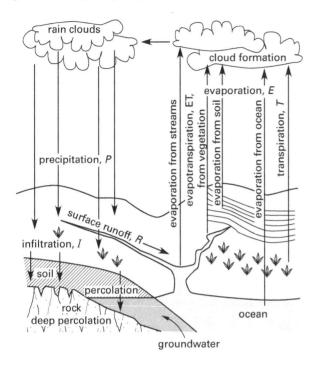

Adapted from NRCS. *Hydrology National Engineering Handbook.* Part 630. September 1997. USDA.

Exam tip: A good grasp of the hydrologic cycle provides a conceptual foundation for navigating hydrologic concepts and setting up equations.

Rainfall Measurement

CERM: Sec. 20.3

CEWE: Sec. 1.5

Rainfall data is measured in terms of the depth of precipitation that has accumulated within a rain gauge. Rainfall gauges vary in complexity and method but can be as simple as a bucket that collects rain.

The average rainfall over a specific area can be calculated in a number of ways, as outlined in CERM Sec. 20.3. For more information on the various rainfall measurement methods, see the entries "Thiessen Network Method" and "Inverse Square Method" in this chapter.

CEWE Fig. 1.6 shows a precipitation depth map for the middle and eastern parts of the United States. Maps such as this one are prepared using spatial statistical methods and are available for many different areas in the United States.

Figure 1.6 Typical 100 yr, 24 hr Precipitation Depth Curves

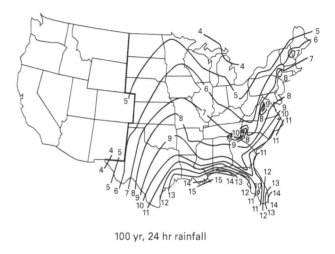

100 yr, 24 hr rainfall

Reprinted from *TR-55: Urban Hydrology for Small Watersheds.* NRCS. 1986. USDA.

Storm Distribution

CERM: Sec. 20.2

CEWE: Sec. 1.5

Storm distribution refers to the pattern of rainfall over the timeframe of a design storm for a region or location. CERM Fig. 20.1 shows examples of a storm hyetograph and cumulative rainfall curve, which are tools used to better describe these rainfall patterns.

Urban Hydrology for Small Watersheds (USDA TR-55) classifies rainfall distribution types in the United States according to the map shown in CEWE Fig. 1.7. Most of the United States has a Type II rainfall distribution, which means that over 90% of the rainfall is distributed within the middle 12 hours of a 24-hour storm. Maps also exist for more specific locations.

Figure 1.7 Rainfall Distribution in the United States

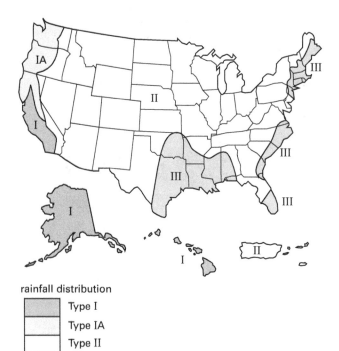

Reprinted from *TR-55: Urban Hydrology for Small Watersheds.* NRCS. 1986. USDA.

Exam tip: Hyetographs are often included with storm hydrographs as a way to better understand the rainfall and runoff correlations and the response of the watershed to rainfall events.

B. RUNOFF ANALYSIS

Key concepts: These key concepts are important for answering exam questions in subtopic 12.B, Runoff Analysis.

- NRCS curve number
- rainfall runoff analysis
- rational method
- SCS/NRCS method
- USGS regression method

NCEES Handbook: To prepare for this subtopic, familiarize yourself with these sections in the *Handbook*.

- Runoff Analysis: Rational Formula Method
- Runoff Coefficients for Rational Formula

- NRCS (SCS) Rainfall Runoff Method: Runoff
- NRCS (SCS) Rainfall Runoff Method: Peak Discharge Method
- SCS Hydrologic Soils Group Type Classifications
- Minimum Infiltration Rates for the Various Soil Groups
- Runoff Curve Numbers (Average Watershed Condition, $I_a = 0.2\ S_R$)
- Coefficients for SCS Peak Discharge Method
- USGS Regression Method—Peak Flow: Rural Equations
- USGS Regression Method—Peak Flow: Urban Equations

***PE Civil Reference Manual* (CERM):** Study these sections in CERM that either relate directly to this subtopic or provide background information.

- Section 20.15: Peak Runoff from the Rational Method
- Section 20.16: NRCS Curve Number
- Section 20.17: NRCS Graphical Peak Discharge Method

***Water Resources and Environmental Depth Reference Manual* (CEWE):** Study these sections in CEWE that either relate directly to this subtopic or provide background information.

- Section 1.11: Runoff Analysis

The following equations and tables are relevant for subtopic 12.B, Runoff Analysis.

Rational Method

Handbook: Runoff Analysis: Rational Formula Method; Runoff Coefficients for Rational Formula

$$Q = CIA$$

CERM: Sec. 20.15

CEWE: Sec. 1.11

The *Handbook* equation is identical to CERM Eq. 20.36, except for the subscripts for Q and A.

The instantaneous peak runoff is used to size culverts and storm drains. The rational method is applicable to small areas (less than 200 ac) but is seldom used for areas greater than 1–2 mi². The intensity depends on the time of concentration and the degree of protection desired (i.e., the recurrence interval).

The value of the runoff coefficient, C, depends on a variety of factors, such as surface cover, soil type, recurrence interval, antecedent moisture conditions, rainfall intensity, drainage area, slope, and percent of impervious land cover. Runoff coefficients can be found in *Handbook* table Runoff Coefficients for Rational Formula, and a weighted runoff coefficient, C_w, can be calculated for watersheds or drainage basins that include multiple types of land cover using the equation in *Handbook* section Runoff Analysis: Rational Formula Method.

$$C_w = \frac{A_1 C_1 + A_2 C_2 + \cdots + A_n C_n}{A_1 + A_2 + \cdots + A_n}$$

NRCS/SCS Method

Handbook: NRCS (SCS) Rainfall Runoff Method: Runoff; NRCS (SCS) Rainfall Runoff Method: Peak Discharge Method; SCS Hydrologic Soils Group Type Classifications; Minimum Infiltration Rates for the Various Soil Groups; Runoff Curve Numbers (Average Watershed Condition, $I_a = 0.2\ S_R$); Coefficients for SCS Peak Discharge Method

$$Q = \frac{(P - 0.2S)^2}{P + 0.8S}$$

$$S = \frac{1000}{\text{CN}} - 10$$

$$\text{CN} = \frac{1000}{S + 10}$$

$$q_p = q_u A_m Q$$

CERM: Sec. 20.16, Sec. 20.17

$$Q_{\text{in}} = \frac{(P_g - I_a)^2}{P_g - I_a + S} = \frac{(P_g - 0.2S)^2}{P_g + 0.8S} \quad\quad 20.44$$

$$Q_p = q_u A_{\text{mi}^2} Q_{\text{in}} F_p \quad\quad 20.45$$

CEWE: Sec. 1.11

The *Handbook* equation for runoff using the rainfall runoff method is equivalent to CERM Eq. 20.44, with Q and Q_{in} both representing the runoff and P and P_g both representing the precipitation. The *Handbook* equation for runoff using the peak discharge method is equivalent to CERM Eq. 20.45, the major difference being that CERM includes a factor, F_p, that takes into account the presence of ponds or swamps in the area being studied, while the *Handbook* equation assumes no ponds or swamps (which would make the factor equal to 1).

The Natural Resources Conservation Service (NRCS) method, formerly known as the Soil Conservation Service (SCS) method, classifies land use and soil type using a single parameter called the *curve number*, CN. Soil types are divided into Types A through D and are delineated based on the infiltration rate of the soils, as shown in *Handbook* section SCS Hydrologic Soils Group Type Classifications and *Handbook* table Minimum Infiltration Rates for the Various Soil Groups.

The NRCS/SCS method can be used for any homogeneous watershed with a known percentage of imperviousness. If the watershed varies in soil type or cover, it should generally be divided into regions to be analyzed separately.

Coefficients for use with the peak discharge method can be found in *Handbook* table Coefficients for SCS Peak Discharge Method, and curve numbers can also be found in *Handbook* table Runoff Curve Numbers (Average Watershed Condition, $I_a = 0.2\ S_R$). For the runoff method, a curve number can be found using the equation in section NRCS (SCS) Rainfall Runoff Method: Runoff. A composite curve number can be calculated by weighting the curve number for each region under study by the percentage of the total area made up by that region.

USGS Regression Method

Handbook: USGS Regression Method—Peak Flow: Rural Equations; USGS Regression Method—Peak Flow: Urban Equations

$$\text{RQ}_T = aA^b B^c C^d$$

$$\text{UQ2} = 2.35 A_s^{0.41} \text{SL}^{0.17} (\text{RI2}+3)^{2.04}$$
$$(\text{ST}+8)^{-0.65}(13-\text{BDF})^{-0.32}\text{IA}_s^{0.15}\text{RQ2}^{0.47}$$

$$\text{UQ5} = 2.70 A_s^{0.35} \text{SL}^{0.16} (\text{RI2}+3)^{1.86}$$
$$(\text{ST}+8)^{-0.59}(13-\text{BDF})^{-0.31}\text{IA}_s^{0.11}\text{RQ5}^{0.54}$$

$$\text{UQ10} = 2.99 A_s^{0.32} \text{SL}^{0.15} (\text{RI2}+3)^{1.75}$$
$$(\text{ST}+8)^{-0.57}(13-\text{BDF})^{-0.30}\text{IA}_s^{0.09}\text{RQ10}^{0.58}$$

$$\text{UQ25} = 2.78 A_s^{0.31} \text{SL}^{0.15} (\text{RI2}+3)^{1.76}$$
$$(\text{ST}+8)^{-0.55}(13-\text{BDF})^{-0.29}\text{IA}_s^{0.07}\text{RQ25}^{0.60}$$

$$\text{UQ50} = 2.67 A_s^{0.29} \text{SL}^{0.15} (\text{RI2}+3)^{1.74}$$
$$(\text{ST}+8)^{-0.53}(13-\text{BDF})^{-0.28}\text{IA}_s^{0.06}\text{RQ50}^{0.62}$$

$$\text{UQ100} = 2.50 A_s^{0.29} \text{SL}^{0.15} (\text{RI2}+3)^{1.76}$$
$$(\text{ST}+8)^{-0.52}(13-\text{BDF})^{-0.28}\text{IA}_s^{0.06}\text{RQ100}^{0.63}$$

$$\text{UQ500} = 2.27 A_s^{0.29} \text{SL}^{0.16} (\text{RI2}+3)^{1.86}$$
$$(\text{ST}+8)^{-0.54}(13-\text{BDF})^{-0.27}\text{IA}_s^{0.05}\text{RQ500}^{0.63}$$

Which of the United States Geological Survey (USGS) regression method equations is used for a given watershed is dependent on the level of urbanization of the basin and the return period of the design flow in question (i.e., the number given after the variable UQ).

Exam tip: Note that the urban equations are based on the characteristics for a two-hour storm with a two-year recurrence interval.

C. HYDROGRAPH DEVELOPMENT AND APPLICATIONS

Key concepts: These key concepts are important for answering exam questions in subtopic 12.C, Hydrograph Development and Applications.

- base flow
- Clark unit hydrograph
- development of hydrographs
- hydrograph analysis
- lag time
- lagging storm method
- NRCS synthetic unit hydrograph
- NRCS synthetic unit triangular hydrograph
- peak flow
- Snyder unit hydrograph
- synthetic hydrograph
- time of concentration
- time to peak

NCEES Handbook: To prepare for this subtopic, familiarize yourself with these sections in the *Handbook*.

- Unit Hydrograph
- Graphical Representation of a Unit Hydrograph
- Snyder Synthetic Unit Hydrograph
- SCS (NRCS) Unit Hydrograph
- Clark Unit Hydrograph
- Hydrograph Estimate of Stream Flow

PE Civil Reference Manual (**CERM**): Study these sections in CERM that either relate directly to this subtopic or provide background information.

- Section 20.8: Total Surface Runoff from Stream Hydrograph
- Section 20.9: Hydrograph Separation
- Section 20.10: Unit Hydrograph
- Section 20.11: NRCS Synthetic Unit Hydrograph
- Section 20.12: NRCS Synthetic Unit Triangular Hydrograph
- Section 20.14: Hydrograph Synthesis

Water Resources and Environmental Depth Reference Manual (**CEWE**): Study these sections in CEWE that either relate directly to this subtopic or provide background information.

- Section 1.6: Hydrographs

The following equations and figures are relevant for subtopic 12.C, Hydrograph Development and Applications.

Unit Hydrographs

Handbook: Unit Hydrograph; Graphical Representation of a Unit Hydrograph

CERM: Sec. 20.10

A *unit hydrograph* is a storm hydrograph that shows the distribution of runoff through a watershed for a theoretical storm event of a given duration. *Handbook* figure Graphical Representation of a Unit Hydrograph and CERM Fig. 20.13 illustrate the differences between the unit hydrograph and a regular storm hydrograph. For more about storm hydrographs, see the entry "Development of Hydrographs" in this chapter.

Figure 20.13 Unit Hydrograph

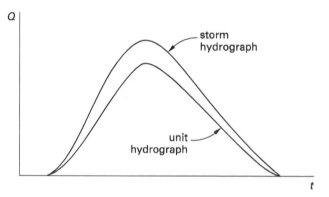

For more about the various methods of developing a unit hydrograph provided in the *Handbook*, see the entries "Snyder Synthetic Unit Hydrograph," "NRCS Synthetic Unit Hydrograph," "NRCS Synthetic Unit Triangular Hydrograph," and "Clark Unit Hydrograph" in this chapter.

Snyder Synthetic Unit Hydrograph

Handbook: Snyder Synthetic Unit Hydrograph

$$q_p = \frac{C_p A}{t_p}$$

$$t_p = C_t(LL_c)^{0.3}$$

$$T = 3 + \frac{t_p}{8}$$

If a watershed has no historical records available to produce a unit hydrograph from, a hydrograph can be reasonably approximated using a peak flow. Such a hydrograph is called a *synthetic unit hydrograph*. One of the earliest methods developed for synthetic unit hydrographs is the Snyder synthetic unit hydrograph, detailed in *Handbook* section Snyder Synthetic Unit Hydrograph.

The first *Handbook* equation shown calculates peak flow based on the storage coefficient, the watershed area, and the lag time. The second equation calculates lag time from the characteristics of the basin, based on the slope and length of the basin and stream. The time duration of the unit hydrograph is given by the third equation.

Exam tip: This method is not commonly in use, but be familiar with it nonetheless.

NRCS Synthetic Unit Hydrograph

Handbook: SCS (NRCS) Unit Hydrograph

$$t_L = \frac{L^{0.8}(S+1)^{0.7}}{1900\, Y^{0.5}}$$

$$q_p = \frac{K_p A_m Q_D}{t_p}$$

CERM: Sec. 20.11

$$t_{1,\text{hr}} = \frac{L_{o,\text{ft}}^{0.8}(S+1)^{0.7}}{1900\sqrt{S_{\text{percent}}}} \quad\quad 20.23$$

$$Q_p = \frac{484 A_{d,\text{mi}^2}}{t_p} \quad\quad 20.25$$

The first *Handbook* equation shown, which gives lag time as a function of the potential maximum retention, is the same as CERM Eq. 20.23, except that the variable for the watershed slope is Y in the *Handbook* and S_{percent} in CERM.

In order to draw the Natural Resource Conservation Service (NRCS) synthetic unit hydrograph, the peak flow, q_p or Q_p, and the time to peak, t_p, must both be calculated. The peak flow can be found either from the time to peak or from the time of concentration, t_c.

The second *Handbook* equation shown, which gives peak flow as a function of time to peak, is the same as CERM Eq. 20.25, but in the CERM equation, the values for the peaking constant ($K_p = 484$) and the volume of direct runoff ($Q_D = 1$, in units of square miles) are already given.

To calculate the peak flow from the time of concentration instead, the factor α' must be included; α' is equal to 1.5 when using customary U.S. units.

$$q_p = \frac{\alpha' K_p A_k Q_D}{t_c}$$

The equation for the time to peak is an approximation based on the time to concentration.

$$t_p = \tfrac{2}{3} t_c$$

The equation for the time of concentration as a function of the lag time is a different expression of the Soil Conservation Service (SCS) lag formula for time of concentration (discussed in the entry "NRCS Lag Method" in this chapter).

$$t_c = \tfrac{5}{3} t_L$$

Two other *Handbook* equations for the time of concentration are

$$t_c = \frac{L^{0.8}(S+1)^{0.7}}{1140\, Y^{0.5}}$$

$$t_c = \frac{L}{3600\text{v}}$$

Exam tip: The peak flow and time to peak contribute only one point to the construction of the unit hydrograph. To construct the remainder, a table such as CERM Table 20.3 must be used. If this method is needed on the exam, the necessary data should be provided in the problem.

NRCS Synthetic Unit Triangular Hydrograph

Handbook: SCS (NRCS) Unit Hydrograph

$$t_p = \tfrac{2}{3} t_c$$

CERM: Sec. 20.12

CEWE: Sec. 1.6

The Natural Resource Conservation Service (NRCS) synthetic unit triangular hydrograph is a simplified version of the NRCS synthetic unit hydrograph and can be used in unmonitored watersheds where only the peak flow is of interest. Unlike the hydrograph that it is based on, the triangular hydrograph captures only the peak flow, the time to peak, and the total duration of time for storm runoff. The peak runoff is calculated using the same equations as those used to calculate peak flow for the NRCS synthetic unit hydrograph, covered in the entry "NRCS Synthetic Unit Hydrograph" in this chapter. The peak flow used must be for a 1 in storm event.

CERM Fig. 20.14 illustrates the NRCS synthetic unit triangular hydrograph.

Figure 20.14 NRCS Synthetic Unit Triangular Hydrograph

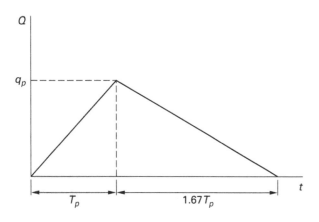

Clark Unit Hydrograph

Handbook: Clark Unit Hydrograph

$$\frac{A}{A_c} = 1.414 \left(\frac{t}{t_c}\right)^{1.5} \quad [\text{for } 0 \le t/t_c \le 0.5]$$

$$\frac{A}{A_c} = 1 - 1.414 \left(1 - \frac{t}{t_c}\right)^{1.5} \quad [\text{for } 0.5 \le t/t_c \le 1.0]$$

$$\frac{I_{t1} + I_{t2}}{2} - \frac{O_{t1} + O_{t2}}{2} = \frac{R(O_{t2} - O_{t1})}{\Delta t}$$

$$O_{t2} = CI_{t2} + (1-C)O_{t1}$$

$$C = \frac{2\Delta t}{2R + \Delta t}$$

$$S_t = RO_t$$

A *Clark unit hydrograph* models the watershed unit hydrograph as water moving through a reservoir. The area contributing to flows, A, as a percentage of the total watershed area, A_c, is a function of time, where the entire watershed contributes when the time at which the flow is measured is equal to the time of concentration.

The storage of water in the watershed is correlated with the outflow using a linear reservoir constant, R, which is based on the difference in inflows and outflows in the watershed over a given time step, Δt. Unlike other unit hydrograph methods, the Clark unit hydrograph method requires watershed data to generate a linear reservoir constant and develop the hydrograph.

Development of Hydrographs

Handbook: Hydrograph Estimate of Stream Flow

$$Q_t = Q_0 K^t$$

CERM: Sec. 20.8, Sec. 20.9

A *hydrograph* is a plot of the stream discharge versus time in a watershed. Hydrograph periods may be short (e.g., hours) or long (e.g., days, weeks, or months). A typical hydrograph is shown in CERM Fig. 20.7. The time base is the length of time that the stream flow exceeds the original base flow. The flow rate increases on the rising limb (concentration curve) and decreases on the falling limb (recession curve).

Figure 20.7 Stream Hydrograph

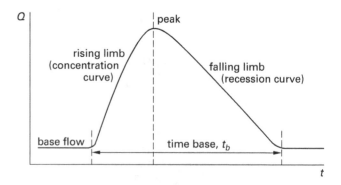

Hydrograph separation (hydrograph analysis) is used to separate runoff (surface flow, net flow, or overland flow) and base flow. There are several methods of separating base flow from runoff. CERM Fig. 20.8 through Fig. 20.11 are graphical representations of these methods, which are explained in more detail in CERM Sec 20.9.

Exam tip: Understanding the base flow impact on the hydrograph will allow you to determine the flows that are directly contributed by the storm of interest.

Lagging Storm Method

CERM: Sec. 20.14

The lagging storm method allows an engineer to develop a unit hydrograph for storms with durations that are whole multiples of the duration of the storm for which the original unit hydrograph was constructed. For example, this method can be used to construct a six-hour storm based on the information from a unit hydrograph for a two-hour storm. A visual representation of this method is shown in CERM Fig. 20.17.

Figure 20.17 Lagging Storm Method

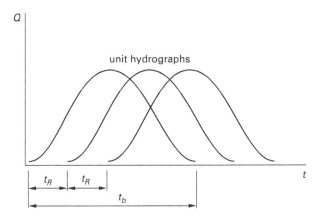

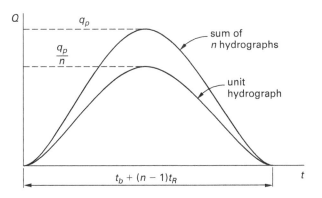

Exam tip: The *Handbook* does not discuss the lagging storm method, but it may prove useful when using unit hydrographs in the absence of direct gauge data.

D. RAINFALL INTENSITY, DURATION, AND FREQUENCY

Key concepts: These key concepts are important for answering exam questions in subtopic 12.D, Rainfall Intensity, Duration, and Frequency.

- floods
- intensity-duration-frequency (IDF) curves and equations
- rainfall intensity

NCEES Handbook: To prepare for this subtopic, familiarize yourself with these sections in the *Handbook*.

- Probability of Single Occurrence in a Given Storm Year
- Risk (Probability of Exceedance)
- Simplified Flood Frequency Equation
- Frequency Factor for Normal Distribution
- Point Precipitation
- National Weather Service IDF Curve Creation

PE Civil Reference Manual (**CERM**): Study these sections in CERM that either relate directly to this subtopic or provide background information.

- Section 20.6: Rainfall Intensity
- Section 20.7: Floods

Water Resources and Environmental Depth Reference Manual (**CEWE**): Study these sections in CEWE that either relate directly to this subtopic or provide background information.

- Section 1.4: Storm Frequency

The following equations and figures are relevant for subtopic 12.D, Rainfall Intensity, Duration, and Frequency.

Floods

Handbook: Probability of Single Occurrence in a Given Storm Year; Risk (Probability of Exceedance); Simplified Flood Frequency Equation; Frequency Factor for Normal Distribution

$$P(x \geq x_T) = p = \frac{1}{T}$$

$$P(x \geq x_T \text{ at least once in } n \text{ years}) = 1 - \left(1 - \frac{1}{T}\right)^n$$

$$x_T = \bar{x} + K_T s$$

CERM: Sec. 20.7

$$p\{F \text{ event in one year}\} = \frac{1}{F} \quad \text{20.19}$$

$$p\{F \text{ event in } n \text{ years}\} = 1 - \left(1 - \frac{1}{F}\right)^n \quad \text{20.20}$$

CEWE: Sec. 1.4

These equations are used to calculate the probability (risk) that a flooding event in any given year will cause a flood equal to a design flood with a given recurrence interval frequency. The *Handbook* equation for the probability of a flood event in one year is the same as CERM Eq. 20.19, except that the *Handbook* uses the variable T for the flood frequency and CERM uses the variable F. The *Handbook* equation for the probability of a flood event in n years is the same as CERM Eq. 20.20, with the same differences in nomenclature.

The *flood frequency* (*frequency of occurrence*, *recurrence interval*, or *return interval*) is the average number of years between floods of a given intensity. In general, the design flood may be specified by its recurrence interval (e.g., 100-year flood) or its annual probability of occurrence (e.g., 1% flood). A 100-year flood is one that would be exceeded in severity only once every 100 years on average.

The simplified flood frequency equation can be used to calculate the flood frequency for a normal distribution of frequency factors. The frequency factor can be found from *Handbook* table Frequency Factor for Normal Distribution.

The equations shown are valid for both storm and flood frequency. *Storm frequency* refers to the likelihood or occurrence of a weather event (a storm), whereas *flood frequency* refers to the likelihood of certain flows within a watershed. For more about storm frequency, see the entry "Storm Frequency" in this chapter.

Rainfall Intensity

Handbook: Point Precipitation

$$i = \frac{c}{T_d^e + f}$$

$$i = \frac{cT^m}{T_d + f}$$

$$i = \frac{cT^m}{T_d^e + f}$$

CERM: Sec. 20.6

$$I = \frac{K}{t_c + b} \quad \text{20.14}$$

The differences in nomenclature between the *Handbook* equations and CERM Eq. 20.14 are the use of i versus I for the rainfall intensity, T_d for duration versus t_c for time of concentration, and f versus b for the return period coefficient. In addition, the *Handbook* equations use region-specific climatological factors c and m, and the CERM equation uses the coefficient K, which is based on the rainfall region.

Rainfall intensity is the amount of precipitation per hour in a given watershed and is based on several regional factors, represented by the return period coefficients. Average rainfall intensity will be low for most storms.

> *Exam tip*: Be familiar with these equations, but also be prepared to find rainfall intensity information using maps or charts.

Intensity-Duration-Frequency Curves

Handbook: National Weather Service IDF Curve Creation

$$P = \frac{T_d + 20}{100}$$

$$i = \frac{60P}{T_d}$$

CERM: Sec. 20.6

Intensity-duration-frequency (IDF) *curves* are plots of peak rainfall intensity versus storm duration for several different return intervals. These curves are geographically specific and provide peak rainfall intensity values at a glance, which can be used to determine the correct rainfall intensity for hydrologic calculations. CERM Fig. 20.6 shows an example of an IDF curve.

The IDF curve equations from the *Handbook* are used to develop IDF curves, especially for areas where IDF curves are not readily available. These equations can be used to find the precipitation and rainfall intensity of a storm event from its duration. The constants used in the equations vary based on storm frequency and geographic location.

> *Exam tip*: Practice reading the rainfall intensity from IDF curves for a variety of durations and frequencies of storm events.

Figure 20.6 *Typical Intensity-Duration-Frequency Curves*

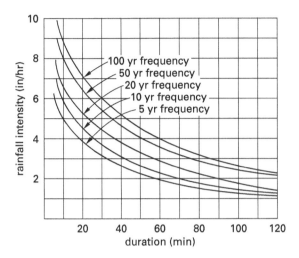

E. TIME OF CONCENTRATION

Key concepts: These key concepts are important for answering exam questions in subtopic 12.E, Time of Concentration.

- channelized flow
- shallow concentrated flow
- sheet flow
- time of concentration components

NCEES Handbook: To prepare for this subtopic, familiarize yourself with these sections in the *Handbook*.

- Kinematic-Wave Runoff Method
- Relations for Estimating α and m Based on Physical Characteristics of Channel Segments
- Time of Concentration: SCS Lag Formula
- Time of Concentration: Sheet Flow Formula
- Velocity Versus Slope for Shallow Concentrated Flow

PE Civil Reference Manual **(CERM):** Study these sections in CERM that either relate directly to this subtopic or provide background information.

- Section 20.5: Time of Concentration

Water Resources and Environmental Depth Reference Manual **(CEWE):** Study these sections in CEWE that either relate directly to this subtopic or provide background information.

- Section 1.11: Runoff Analysis

The following equations and figures are relevant for subtopic 12.E, Time of Concentration.

Channelized Flow

Handbook: Kinematic-Wave Runoff Method; Relations for Estimating α and m Based on Physical Characteristics of Channel Segments

$$Q = \alpha A^m$$

$$\alpha = \left(\frac{1.486}{N}\right)S_0^{1/2}$$

$$Q = \left(\frac{1.486}{n}\right)R^{2/3}S_0^{1/2}A$$

$$Q = \left(\frac{1.486}{N}\right)S_0^{1/2}h^{5/3}$$

The kinematic-wave runoff method is based on the Manning open channel equation and provides an estimated flow in a channelized section that requires limited parameter information to calculate. This estimated flow can be divided by an estimated cross-sectional channel area to estimate the velocity of flows in the channelized section.

Which equation to use to estimate channelized flow depends on the slope of the plane. If the slope of the plane is equal to the energy gradient (friction slope), the channelized flow can be found as a function of the cross-sectional area of the channel and the kinematic wave routing parameters, α and m (using the first and second equations shown).

If the slope of the plane and the energy gradient are not equal, the slope can be found in two ways. It can be found as a function of the slope of the plane, the cross-sectional area of the channel, the hydraulic radius, and the Manning roughness coefficient (the third equation shown). Alternatively, it can be found as a function of the slope of the plane, the depth of the channel, and the effective roughness parameter for overland flow (the fourth equation shown).

Handbook table Relations for Estimating α and m Based on Physical Characteristics of Channel Segments includes equations for the kinematic-wave routing parameters for different types of segments.

NRCS Lag Method

Handbook: Time of Concentration: SCS Lag Formula

$$t_L = 0.000526\, L^{0.8}\left(\frac{1000}{\text{CN}} - 9\right)^{0.7} S^{-0.5}$$

$$t_c = \frac{5}{3}t_L$$

CERM: Sec. 20.5

$$t_{c,\min} = 1.67 t_{\text{watershed lag time,min}}$$

$$= \frac{(1.67)\left(60\ \dfrac{\min}{\text{hr}}\right) L_{o,\text{ft}}^{0.8}(S_{\text{in}}+1)^{0.7}}{1900\sqrt{S_{\text{percent}}}} \quad 20.11$$

$$= \frac{(1.67)\left(60\ \dfrac{\min}{\text{hr}}\right) L_{o,\text{ft}}^{0.8}\left(\dfrac{1000}{\text{CN}}-9\right)^{0.7}}{1900\sqrt{S_{\text{percent}}}}$$

The Natural Resource Conservation Service (NRCS) lag formula, previously named the Soil Conservation Service (SCS) lag formula, finds the time of concentration as a function of the lag time, which is itself a function of the slope and length of the watershed and the curve number.

CERM Eq. 20.11 combines both *Handbook* equations into a single equation. The nomenclature differs only slightly.

Exam tip: This time of concentration found using this method can be used with the NRCS curve number method to calculate peak flow.

Sheet Flow

Handbook: Time of Concentration: Sheet Flow Formula

$$T_{ti} = \left(\frac{K_u}{I^{0.4}}\right)\left(\frac{nL}{\sqrt{S}}\right)^{0.6}$$

CERM: Sec. 20.5

$$t_{\text{sheet flow}} = \frac{0.007(nL_o)^{0.8}}{\sqrt{P_2}\, S_{\text{decimal}}^{0.4}} \quad 20.6$$

The *Handbook* equation uses the intensity of rainfall, I, with units of in/hr, while CERM Eq. 20.6 uses the precipitation, P_2, with units of in. CERM Eq. 20.6 also substitutes a value of 0.007 for the empirical coefficient, K_u.

Sheet flow is the overland flow of water prior to concentration; it contributes to the time of concentration, as shown in the sheet flow formula.

The sheet flow formula is valid for distances up to 300 ft. Over a distance greater than 300 ft, the flow usually becomes shallow concentrated flow. For irregularly shaped drainage areas, it may be necessary to evaluate several distances related to sheet flow.

Exam tip: For the sheet flow method, use the Manning roughness coefficient that best describes the land cover of the watershed.

Shallow Concentrated Flow

Handbook: Velocity Versus Slope for Shallow Concentrated Flow

CERM: Sec. 20.5

CEWE: Sec. 1.11

Shallow concentrated flow is a subset of overland flow where the flow becomes slightly concentrated but is not yet fully channelized. Both *Handbook* figure Velocity Versus Slope for Shallow Concentrated Flow and CERM Fig. 20.4 graph the relationship between the slope of a watercourse and the velocity of a shallow concentrated flow in that watercourse, with lines indicating the likely land cover category of a watershed with that watercourse slope and flow velocity. The *Handbook* figure uses a ratio of feet over feet for the slope, instead of a percentage value as in CERM Fig. 20.4. The *Handbook* figure also includes one additional land cover category (straight row crops) not included in the CERM figure.

Figure 20.4 NRCS Average Velocity Chart for Overland Flow Travel Time

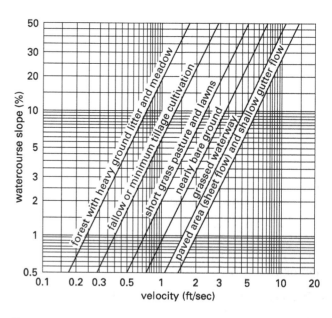

Reprinted from NRCS TR-55-1975. TR-55-1986 contains a similar graph for shallow-concentrated flow over paved and unpaved surfaces, but it does not contain this particular graph.

If any two parameters among the velocity of a shallow concentrated flow, the slope of the watercourse, and the land cover category of the watershed are known, the other parameter can be found using *Handbook* figure Velocity Versus Slope for Shallow Concentrated Flow or CERM Fig. 20.4.

Exam tip: Dividing the slope of the watercourse by the velocity of the shallow concentrated flow will give the travel time of that flow. *Handbook* figure Velocity Versus Slope for Shallow Concentrated Flow can be used to find values needed for this calculation.

Time of Concentration Components

CERM: Sec. 20.5

$$t_c = t_{\text{sheet}} + t_{\text{shallow}} + t_{\text{channel}} \qquad 20.5$$

Time of concentration is the time of travel from the hydraulically most remote point in a watershed to the watershed outlet or another design point.

For points along storm drains (e.g., manholes) being fed from a watershed, time of concentration is taken as the largest combination of

- overland flow time (sheet flow)
- swale or ditch flow (shallow concentrated flow)
- storm drain, culvert, or channel time

These three components are calculated separately and then added together, as shown in CERM Eq. 20.5. Time-of-concentration equations also exist that are based on the qualities of the watershed and thus combine the three components implicitly. For more about these alternate equations, see the entries "NRCS Lag Method" and "Sheet Flow" in this chapter.

F. RAINFALL AND STREAM GAUGING STATIONS

Key concepts: These key concepts are important for answering exam questions in subtopic 12.F, Rainfall and Stream Gauging Stations.

- estimating unknown precipitation
- inverse square/quadrant method
- rainfall gauging stations
- stream gauging stations
- Thiessen network method

NCEES Handbook: To prepare for this subtopic, familiarize yourself with these sections in the *Handbook*.

- Rainfall Gauging Stations: Thiessen Polygon Method
- Distance Weighting
- Rainfall Gauging Stations: Arithmetic/Station Average Method
- Rainfall Gauging Stations: Normal-Ratio Method
- Regression Method (Double-Mass Curve)
- Stream Gauging

PE Civil Reference Manual (**CERM**): Study these sections in CERM that either relate directly to this subtopic or provide background information.

- Section 20.3: Precipitation Data
- Section 20.4: Estimating Unknown Precipitation

Water Resources and Environmental Depth Reference Manual (**CEWE**): Study these sections in CEWE that either relate directly to this subtopic or provide background information.

- Section 1.5: Characteristics of Precipitation
- Section 1.12: Gauging Stations

The following equations and figures are relevant for subtopic 12.F, Rainfall and Stream Gauging Stations.

Thiessen Network Method

Handbook: Rainfall Gauging Stations: Thiessen Polygon Method

CERM: Sec. 20.3

CEWE: Sec. 1.5

Over large areas or watersheds, precipitation can vary depending on elevation, proximity to large waterways, and other factors. Precipitation can be measured with various gauges and weighted according to the method outlined in *Handbook* section Rainfall Gauging Stations: Thiessen Polygon Method.

The Thiessen method involves plotting stations on a map of the watershed area, drawing dotted lines between all stations, and bisecting these dotted lines with solid lines. The lines form polygons whose total area is the assumed watershed area. The weighted average precipitation depth can then be found using the process detailed in *Handbook* section Rainfall Gauging Stations: Thiessen Polygon Method. A Thiessen precipitation diagram is shown in CEWE Fig. 1.9.

Figure 1.9 Theissen Precipitation Diagram

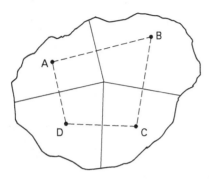

Exam tip: The Thiessen method is essentially a weighted average precipitation method where the resulting polygons represent the area of influence for each gauge.

Inverse Square Method

Handbook: Distance Weighting

$$P = \frac{\sum_{i=1}^{4} \frac{P_i}{L_i^2}}{\sum_{i=1}^{4} \frac{1}{L_i^2}}$$

CERM: Sec. 20.4

$$P_x = \frac{\dfrac{P_A}{d_{A\text{-}x}^2} + \dfrac{P_B}{d_{B\text{-}x}^2} + \dfrac{P_C}{d_{C\text{-}x}^2} + \dfrac{P_D}{d_{D\text{-}x}^2}}{\dfrac{1}{d_{A\text{-}x}^2} + \dfrac{1}{d_{B\text{-}x}^2} + \dfrac{1}{d_{C\text{-}x}^2} + \dfrac{1}{d_{D\text{-}x}^2}} \qquad 20.4$$

The *Handbook* equation is essentially the same as CERM Eq. 20.4, except that the *Handbook* uses the variable L for the distance between each station, and CERM uses the variable d.

The inverse square method is used to determine the unknown rainfall at a specific location, given the level of precipitation at other locations and the distances between those locations and the location for which the precipitation is unknown. The inverse square method can be used to interpolate missing data from a gauge or interpolate point precipitation for a specific location where no gauge exists.

Data Gaps in Precipitation Gauges

Handbook: Rainfall Gauging Stations: Arithmetic/Station Average Method; Rainfall Gauging Stations: Normal-Ratio Method; Regression Method (Double-Mass Curve)

CERM: Sec. 20.4

There are three commonly used methods for estimating unknown precipitation within a watershed based on the precipitation information provided by surrounding gauges.

The *arithmetic method* (or *station average method*) is used by choosing three stations close to and evenly spaced around the location with missing data. If the normal annual precipitations at the three sites do not vary more than 10% from the missing station's normal annual precipitation, the rainfall can be estimated as the arithmetic mean of the three neighboring stations' precipitations for the period in question, as shown in the equation in *Handbook* section Rainfall Gauging Stations: Arithmetic/Station Average Method.

$$P_x = \frac{1}{n}\sum_{i=1}^{n} P_i$$

If the normal annual precipitations do vary more than 10% from the missing station's, the *normal-ratio method* can be used. The equation for this method is found in Rainfall Gauging Stations: Normal-Ratio Method and CERM Eq. 20.4.

$$P_x = \sum_{i=1}^{n} W_i P_i$$

$$\frac{P_x}{A_x} = \frac{1}{n}\sum_{i=1}^{n}\frac{P_i}{A_i}$$

$$P_x = \frac{\dfrac{P_A}{d_{A\text{-}x}^2} + \dfrac{P_B}{d_{B\text{-}x}^2} + \dfrac{P_C}{d_{C\text{-}x}^2} + \dfrac{P_D}{d_{D\text{-}x}^2}}{\dfrac{1}{d_{A\text{-}x}^2} + \dfrac{1}{d_{B\text{-}x}^2} + \dfrac{1}{d_{C\text{-}x}^2} + \dfrac{1}{d_{D\text{-}x}^2}} \qquad 20.4$$

The *Handbook* equation allows for any number of gauges to be used, while the CERM equation assumes that the four closest gauges are used. These are essentially different terms for the same thing.

The *regression method* is used to find the rainfall at a specific gauge where there is existing data for that gauge and for other gauges in the watershed for earlier and later time steps. The equations for this method are found in *Handbook* section Regression Method (Double-Mass Curve).

$$S_1 = \frac{\Delta Y_i}{\Delta X_i}$$

$$y_1 = Y_i \frac{S_2}{S_1}$$

Exam tip: Be familiar with the main methods for addressing data gaps in gauge data to easily determine which method to use under different conditions.

Stream Gauging Stations

Handbook: Stream Gauging

Using the two-point method,

$$\overline{V} = \frac{V_{0.2} + V_{0.8}}{2}$$

Using the six-tenths-depth method,

$$\overline{V} = V_{0.6}$$

Using the three-point method,

$$\overline{V} = \frac{V_{0.2} + V_{0.6} + V_{0.8}}{4}$$

Using the five-point method,

$$\overline{V} = \frac{V_{\text{surf}} + 3V_{0.2} + 3V_{0.6} + 2V_{0.8} + V_{\text{bed}}}{10}$$

Using the six-point method,

$$\overline{V} = \frac{V_{\text{surf}} + 2V_{0.2} + 2V_{0.4} + 2V_{0.6} + 2V_{0.8} + V_{\text{bed}}}{10}$$

Using the two-tenths-depth method,

$$\overline{V} = V_{0.2}$$

Using the surface-velocity method,

$$\overline{V} = 0.87 V_{\text{surf}}$$

The stream gauging calculations are based on the use of velocity meters that can collect data at several different points along the vertical cross section of a stream or river. Once a cross section is identified, several points are chosen along the cross section from which to collect velocity data. At each point, a velocity profile is constructed by measuring the velocity at different depths based on a fraction of the total depth at that point.

Under the simplest conditions, the flow velocity is measured using the six-tenths-depth method, which consists of measuring the depth of water at a given point in the cross section, and then measuring the velocity at a depth that is 0.6 times the water depth at that point.

G. DEPLETIONS

Key concepts: These key concepts are important for answering exam questions in subtopic 12.G, Depletions.

- consumptive use
- detention
- diversions
- evapotranspiration
- Horton model (infiltration)
- infiltration
- runoff depletions

NCEES Handbook: To prepare for this subtopic, familiarize yourself with these sections in the *Handbook*.

- Minimum Infiltration Rates for the Various Soil Groups
- Surface Water System Hydrologic Budget
- Consumptive Use—Blaney-Criddle Method
- Evaporation: Pan Method
- Typical Values of Pan Coefficient, K_p
- Evaporation: Aerodynamic Method
- Infiltration: Richards Equation
- Infiltration: Horton Model
- Green-Ampt Equation

PE Civil Reference Manual **(CERM):** Study these sections in CERM that either relate directly to this subtopic or provide background information.

- Section 20.1: Hydrologic Cycle
- Section 20.16: NRCS Curve Number
- Section 21.17: Infiltration

Water Resources and Environmental Depth Reference Manual **(CEWE)**: Study these sections in CEWE that either relate directly to this subtopic or provide background information.

- Section 1.1: Introduction
- Section 1.9: Evapotranspiration
- Section 1.10: Infiltration
- Section 1.11: Runoff Analysis

The following equations, figures, tables, and concepts are relevant for subtopic 12.G, Depletions.

Infiltration

Handbook: Minimum Infiltration Rates for the Various Soil Groups; Infiltration: Richards Equation; Green-Ampt Equation

$$\frac{d\theta}{dt} = \frac{\partial y}{\partial z}\left(K \frac{dh}{d\theta} \frac{d\theta}{dz}\right) - \frac{dK}{dz}$$

$$f = K\left(\frac{Z + h_0 + S_z}{Z}\right)$$

For a time greater than the time when the water begins to pond, the infiltration rate is given by

$$f(t) = K + K\left(\frac{S_z \Delta \theta}{F}\right)$$

$$F_p = \frac{S_z K \Delta \theta}{i - K}$$

CERM: Sec. 20.16

CEWE: Sec. 1.10

The curve number method is the most common method used to account for infiltration losses in a watershed model. *Handbook* table Minimum Infiltration Rates for the Various Soil Groups provides infiltration rates for the various soil groups, for use with the curve number method.

The Richards equation is for use in situations with isotropic conditions and one-dimensional vertical flow. The Richards equation finds the water flux as a function of soil moisture content, depth of flow, hydraulic conductivity, and pressure head on the soil medium.

The Green-Ampt equation is for use with saturated soil. This equation finds the infiltration at a given time from the hydraulic conductivity of the saturated soil, the depth from the surface to the wetting front, the depth of the ponded water or pressure head, and the suction head (capillary head) at the wetting front.

A variant of the Green-Ampt equation can be used to find the infiltration at a time greater than the time when the water begins to pond, $f(t)$. For a time less than or equal to the time when the water begins to pond, the infiltration will be equal to the precipitation rate. Another variant of the equation can be used to find the amount of water that infiltrates before ponding, F_p.

Hydrologic Budget

Handbook: Surface Water System Hydrologic Budget

$$P + Q_{\text{in}} - Q_{\text{out}} + Q_g - E_s - T_s - I = \Delta S_s$$

CERM: Sec. 20.1

$$\begin{aligned}\text{total precipitation} &= \text{net change in surface water removed} \\ &+ \text{net change in ground water removed} \\ &+ \text{evapotranspiration} \\ &+ \text{interception evaporization} \\ &+ \text{net increase in surface water storage} \\ &+ \text{net increase in ground water storage}\end{aligned}$$

$$P = Q + E + \Delta S \qquad 20.1$$

CEWE: Sec. 1.1

$$P = R + I + \text{ET} - \Delta S \qquad 1.1$$

CERM Eq. 20.1, CEWE Eq. 1.1, and the *Handbook* equation are all mass balance equations that use different variables to capture the same essential concept.

Infiltration, evapotranspiration, and watershed storage are commonly referred to as *runoff depletions*; without them, runoff would equal precipitation. The equation in *Handbook* section Surface Water System Hydrologic Budget outlines a water budget and includes several depletions, including diversions (surface water outflows), infiltration, and transpiration.

When calculating runoff, these depletions are commonly accounted for through the use of the curve number or the runoff coefficient.

Exam tip: A hydrologic budget is essentially a conservation of mass equation. Be familiar with the many ways that this equation can be expressed so that you can easily express it as the inflows and outflows within any defined water system.

Diversions

Handbook: Consumptive Use—Blaney-Criddle Method

$$U = \sum K_t K_c t_m \left(\frac{p}{100\%}\right)$$

Flow diversions generally occur as a result of water use within a watershed. For instance, a certain amount of water may be diverted toward agriculture or drinking water uses. Diversions are generally site specific and need to be estimated based on information from local, state, or national water resource institutions. The Blaney-Criddle method equation can be used to estimate consumptive use within a watershed.

Evapotranspiration

Handbook: Evaporation: Pan Method; Typical Values of Pan Coefficient, K_p; Evaporation: Aerodynamic Method

$$E_L = K_p E_p$$

$$E_a = M(e_s - e)u_z$$

CEWE: Sec. 1.9, Sec. 1.11

Evaporation is a key factor in the design of water storage facilities. The *Handbook* provides equations for evaporation using either the pan method or the aerodynamic method.

The pan method requires a pan coefficient. The pan coefficient is based on the type of pan used to gather evaporation data, and average values and ranges of values for the coefficient can be found from *Handbook* table Typical Values of Pan Coefficient, K_p. The aerodynamic method requires the mass-transfer coefficient, the saturation vapor pressure, the actual pressure of the vapor at the given elevation, and the wind velocity.

CEWE Table 1.8 and Fig. 1.13 provide some general information related to average annual evaporation rates in various parts of the United States.

Equations for transpiration are not included in the *Handbook*.

Exam tip: Charts and tables similar to CEWE Table 1.8 and Fig. 1.13 may be present on the exam, so knowing how to read these is a useful skill.

Horton Model

Handbook: Infiltration: Horton Model

$$f = (f_0 - f_c)e^{-kt} + f_c$$

$$F = f_c t_p + \left(\frac{f_0 - f_c}{k}\right)(1 - e^{-kt_p})$$

CERM: Sec. 21.17

$$f_t = f_c + (f_0 - f_c)e^{-kt} \qquad 21.39$$

CERM Eq. 21.39 is the same as the *Handbook* equation for the infiltration rate except for the order of the variables.

The Horton model gives a lower bound on the infiltration capacity of an aquifer and can be used to find the cumulative infiltration over time.

Detention and Retention

Detention and retention systems detain flows within a stream to divert for other uses. For instance, retention systems may recharge groundwater aquifers, reservoir flows may be diverted toward agricultural or other uses, and rainwater catchment systems may be used for household water consumption. Detention and retention systems may also be used to slow flows within a watershed, meaning that these flows eventually do discharge through the watershed but do not contribute to the peak flow in the same manner. For information about the role of detention ponds in managing stormwater runoff, see the entry "Detention and Retention Ponds" in this chapter.

Exam tip: When considering depletions, think of detention in terms of how it impacts the overall water budget equation within a system.

H. STORMWATER MANAGEMENT

Key concepts: These key concepts are important for answering exam questions in subtopic 12.H, Stormwater Management.

- culverts
- erosion
- hydraulic gradient
- infiltration systems
- pond outlets
- pond routing
- retention/detention ponds
- swales
- weirs

NCEES Handbook: To prepare for this subtopic, familiarize yourself with these sections in the *Handbook*.

- Hydraulic Gradient (Grade Line)
- Culvert Flow Types (USGS)
- Velocity Versus Slope for Shallow Concentrated Flow
- Detention and Retention: Rational Method
- Detention and Retention: Routing Equation
- Detention and Retention: Modified Puls Routing Method
- Revised Universal Soil Loss Equation
- Computed K Values for Soils on Erosion Research Stations
- Values of the Topographic Factor, LS, for Specific Combinations of Slope Length and Steepness
- Mulch Factors and Length Limits for Construction Slopes
- Factor C for Permanent Pasture, Range, and Idle Land

***PE Civil Reference Manual* (CERM)**: Study these sections in CERM that either relate directly to this subtopic or provide background information.

- Section 16.15: Energy Grade Line
- Section 17.20: Energy and Hydraulic Grade Lines with Friction
- Section 17.26: Pressure Culverts
- Section 19.14: Flow Measurement with Weirs
- Section 19.15: Triangular Weirs
- Section 19.16: Trapezoidal Weirs
- Section 19.17: Broad-Crested Weirs and Spillways
- Section 19.37: Culverts
- Section 19.38: Determining Type of Culvert Flow
- Section 19.39: Culvert Design
- Section 20.5: Time of Concentration
- Section 20.20: Reservoir Sizing: Reservoir Routing

***Water Resources and Environmental Depth Reference Manual* (CEWE)**: Study these sections in CEWE that either relate directly to this subtopic or provide background information.

- Section 3.12: Spillway Capacity
- Section 3.14: Stormwater Detention/Retention Ponds
- Section 3.15: Pond and Reservoir Routing
- Section 4.9: Infiltration Trench Design

The following equations, figures, and tables are relevant for subtopic 12.H, Stormwater Management.

Hydraulic Gradient

Handbook: Hydraulic Gradient (Grade Line)

CERM: Sec. 16.15, Sec. 17.20

The *hydraulic grade line* (HGL) is a graph of the sum of the pressure and gravitational heads plotted as a position along the pipeline. CERM Fig. 16.4 illustrates an example of the HGL.

Figure 16.4 Energy and Hydraulic Grade Lines Without Friction

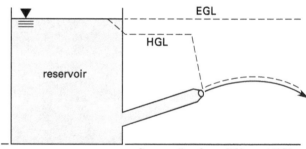

Exam tip: For stormwater management systems, the hydraulic grade line can be used to visually interpret how a given outlet structure may impact the hydraulics of the pond.

Pond Outlets—Culverts

Handbook: Culvert Flow Types (USGS)

For type 1 flow,

$$Q = A\sqrt{\frac{2g\Delta h}{\frac{2gn^2 L}{R^{4/3}} + k_e + 1}}$$

For type 2 flow,

$$Q = A\sqrt{\frac{2g\Delta h}{\frac{fL}{4R} + k_e + 1}}$$

For type 3 flow,

$$Q = C_d A\sqrt{2gh}$$

For type 4 flow,

$$Q = A_c \sqrt{2g\left(\Delta h + \frac{v_1^2}{2g} - h_i - h_f\right)}$$

For type 5 flow,

$$Q = A_c \sqrt{2g\left(\Delta h + \frac{v_1^2}{2g} - h_i\right)}$$

For type 6 flow,

$$Q = A \sqrt{2g\left(\Delta h + \frac{v_1^2}{2g} - h_i - h_f\right)}$$

CERM: Sec. 17.26, Sec. 19.37, Sec. 19.38, Sec. 19.39

A *culvert* is a water path (usually a large-diameter pipe) used to channel water around or through an obstructing feature. The *Handbook* classifies culvert flows into six different types, illustrated in the six parts of *Handbook* figure Culvert Flow Types (USGS). CERM Fig. 19.24 also illustrates flow profiles in culvert design.

Figure 19.24 Flow Profiles in Culvert Design

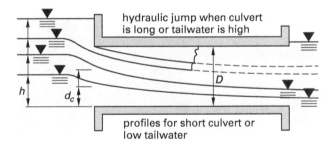

In the *Handbook*, Type 1 flow is for a culvert with a submerged outlet. Type 2 flow is for a culvert with an outlet where the normal depth is greater than the barrel height. Type 3 flow is for a culvert with entrance control and where the normal depth is less than the barrel height. Type 4 flow is for a culvert with a mild slope and low tailwater. Type 5 flow is for a culvert with a steep slope and low tailwater. Type 6 flow is for a culvert with a mild slope and tailwater of a given depth, shown as y_c in the *Handbook* figure.

Detention and retention pond outlets may also be modeled as culverts. This will often occur when the outflow pipe itself is located along the horizontal wall of the pond.

Exam tip: Other sources, such as CERM, use different type numbers to classify culvert flows. To prepare for the exam, become familiar with the culvert types as they are presented in the *Handbook*.

Swales and Infiltration Systems

Handbook: Velocity Versus Slope for Shallow Concentrated Flow

CERM: Sec. 20.5

$$v_{\text{shallow,ft/sec}} = 16.1345\sqrt{S_{\text{decimal}}} \quad [\text{unpaved}] \qquad 20.7$$

$$v_{\text{shallow,ft/sec}} = 20.3282\sqrt{S_{\text{decimal}}} \quad [\text{paved}] \qquad 20.8$$

CEWE: Sec. 4.9

Swales are shallow, broad ditches that convey shallow concentrated flow as it accumulates from sheet flow. These systems may be used along private properties or as drainage ditches along roads. Most are unlined. These systems are designed to ensure positive drainage away from structures and to move water toward more substantial stormwater management systems, such as channels or drainage inlets.

If the velocity is known, other parameters of the swale can be designed for optimal shallow concentrated flow. Velocity can be found from the Manning equation if the flow geometry is well defined, but if not, it must be determined from other correlations, such as those in CERM Eq. 20.7 and Eq. 20.8.

Handbook figure Velocity Versus Slope for Shallow Concentrated Flow illustrates the relationship of velocity and slope for a variety of shallow concentrated flows. This graph can be used for flow in swales.

Infiltration systems are used to encourage greater infiltration of water into the soil or groundwater table as a way to reduce the peak flows during a storm event. One example of this is an infiltration trench, which is illustrated in CEWE Fig. 4.9.

Sometimes these technologies are combined; for instance, a swale and an infiltration trench can be built together to create more conveyance for shallow concentrated flows.

Figure 4.9 Infiltration Trench

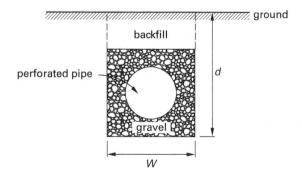

Detention and Retention Ponds

Handbook: Detention and Retention: Rational Method

$$V_{\text{in}} = i \sum A C t$$

$$V_{\text{out}} = Q_0 t$$

CEWE: Sec. 3.14

Detention ponds hold enough volume to store runoff and slowly release the flow to the waterway or storm sewer system, primarily through an outlet structure such as a pipe or weir (see CEWE Fig. 3.23). Retention ponds store the runoff volume and then release the runoff through infiltration and evaporation, not through an outlet structure.

Figure 3.23 Typical Stormwater Detention Pond Outlet

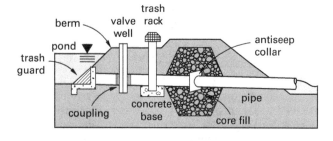

Reprinted from *Ponds—Planning, Design, Construction*, Agriculture Handbook 590, Natural Resources Conservation Services, 1997.

The *Handbook* includes the basic equations for using the rational method to calculate the storage required for a detention or retention pond, which can also be used to calculate the storage required for infiltration systems or swales. The storage required will be equal to the maximum possible value of the difference between the inflow and the outflow.

Exam tip: In a retention pond, V_{out} will usually consist of evaporation and groundwater infiltration unless pond overflow occurs, in which case the outflow can be modeled as flow over a weir. In a detention pond, the primary outflow mechanism will be surface flow through an outlet.

Pond Routing

Handbook: Detention and Retention: Routing Equation; Detention and Retention: Modified Puls Routing Method

$$\tfrac{1}{2}(I_1 + I_2)\Delta t + \left(S_1 - \tfrac{1}{2}O_1 \Delta t\right) = \left(S_2 + \tfrac{1}{2}O_2 \Delta t\right)$$

$$(I_1 + I_2) + \left(\frac{2S_1}{\Delta t} - O_1\right) = \frac{2S_2}{\Delta t} + O_2$$

The traditional equation used for routing flows through a detention or retention pond calculates the volume of water coming into the pond and leaving it over a given period of time. This requires that any instantaneous flow rates be converted into the average volume of water entering or exiting the system.

In contrast to the traditional routing equation, the modified Puls method calculates the inflow and outflow in terms of volume over time. Thus, if the available data for the system is in the form of instantaneous flow rates, and the desired storage time is amenable to the period of time of the flow rates, then routing can be performed in terms of the storage required over a specific period of time.

Revised Universal Soil Loss Equation

Handbook: Revised Universal Soil Loss Equation; Computed K Values for Soils on Erosion Research Stations; Values of the Topographic Factor, LS, for Specific Combinations of Slope Length and Steepness; Mulch Factors and Length Limits for Construction Slopes; Factor C for Permanent Pasture, Range, and Idle Land

$$A = RKLSCP$$

The revised universal soil loss equation (RUSLE) is used to estimate the erosion of soils in a watershed. *Handbook* table Computed K Values for Soils on Erosion Research Stations gives soil erodibility values for several types of soils. The value of the topographic factor can be found from *Handbook* table Values of the Topographic Factor, LS, for Specific Combinations of Slope Length and Steepness. The value of the crop and cover management factor can be found from either *Handbook* table Mulch

Factors and Length Limits for Construction Slopes* or *Handbook* table *Factor C for Permanent Pasture, Range, and Idle Land*, depending on the type of land.

Note that the RUSLE is not used to calculate instream erosive velocities.

The design of retention and detention ponds, use of suitable surface vegetation, and use of stormwater quality management methods are all common ways to reduce the intake of eroded soils into the stormwater system. See CEWE Sec. 1.14 for more information on soil erosion potential and rates.

Pond Outlets—Weirs

CERM: Sec. 19.14 through Sec. 19.17

CEWE: Sec. 3.12, Sec. 3.14

Flow through spillways from dams, reservoirs, and stormwater detention basins may be estimated using the weir flow equations. Generally, the three types of weirs used in water resources engineering are rectangular weirs, V-notch (triangular) weirs, and trapezoidal weirs.

See the entries "Sharp-Crested Weirs" and "Triangular (V-Notch) Weirs" in Chap. 5 in this book for more in-depth information on weir types and their respective equations.

Reservoir Routing

CERM: Sec. 20.20

$$V_{n+1} = V_n + (\text{inflow})_n - (\text{discharge})_n - (\text{seepage})_n - (\text{evaporation})_n \quad 20.47$$

CEWE: Sec. 3.15

$$\frac{S_{n2} - S_{n1}}{t} = \frac{I_1 + I_2}{2} - \frac{O_1 + O_2}{2} \quad 3.48$$

Reservoir routing is the process by which the outflow hydrograph of a reservoir is determined from the inflow hydrograph, the initial storage, and other characteristics of the reservoir.

The simplest method of reservoir routing is to keep track of increments in inflow, storage, and outflow period by period. This is the basis of the storage indication method. For this method, determine the starting storage volume for the first time step, and then use CERM Eq. 20.47 to calculate the volume at the next time step. This volume becomes the starting storage volume for the next iteration.

CEWE Eq. 3.48 is another reservoir routing equation based on the inflow and outflow within a reservoir and is a variation of the conservation of mass principle. As with other reservoir routing equations, it is intended to be used at time step intervals that allow the designer to emulate the storage capacity needed over time.

Exam tip: If the parts of the conservation of mass equation are broken down, exam problems on reservoir routing can be easily compartmentalized and solved. The key is to create tables that include columns for each component.

49 Groundwater and Wells

Content in blue refers to the *NCEES Handbook*.

A. Aquifers.. 49-1
B. Groundwater Flow 49-3
C. Well Analysis—Steady State......................... 49-5

The knowledge area of Groundwater and Wells makes up between three and five questions out of the 80 questions on the PE Civil Water Resources and Environmental exam. The organization of this chapter follows the order of subtopics given by NCEES for this knowledge area. Each subtopic is covered in the following sections.

A. AQUIFERS

Key concepts: These key concepts are important for answering exam questions in subtopic 13.A, Aquifers.

- aquifer characteristics
- types of aquifers

NCEES Handbook: To prepare for this subtopic, familiarize yourself with these sections in the *Handbook*.

- Pressure Field in a Static Liquid
- Porosity
- Soil Moisture Content
- Seepage Velocity
- Unconfined Aquifers
- Confined Aquifer

PE Civil Reference Manual (**CERM**): Study these sections in CERM that either relate directly to this subtopic or provide background information.

- Section 21.1: Aquifers
- Section 21.2: Aquifer Characteristics
- Section 21.7: Discharge Velocity and Seepage Velocity

Water Resources and Environmental Depth Reference Manual (**CEWE**): Study these sections in CEWE that either relate directly to this subtopic or provide background information.

- Section 4.1: Introduction
- Section 4.2: Hydraulic Properties of Aquifer Soils
- Section 11.14: Aquifers

The following equations and figures are relevant for subtopic 13.A, Aquifers.

Pressure Field

Handbook: Pressure Field in a Static Liquid

$$p_2 - p_1 = -\gamma(z_2 - z_1) = -\gamma h = -\rho g h$$

CERM: Sec. 21.1

$$H = \frac{p}{\rho g} \quad \text{[SI]} \quad 21.1(a)$$

$$H = \frac{p}{\gamma} \quad \text{[U.S.]} \quad 21.1(b)$$
$$= \frac{p}{\rho} \cdot \frac{g_c}{g}$$

The *Handbook* equation for the difference in pressure can be rearranged to calculate the rise of the water of a confined aquifer, as shown in CERM Eq. 21.1. This rise may occur when drilling a well. For example, groundwater may be at a depth of 100 ft below ground surface, and shortly thereafter, the groundwater level in the drill casing may rise to 90 ft below ground surface. This indicates that the drill has penetrated a confined aquifer; the groundwater rises to an equilibrium position inside the casing.

In an artesian well, a situation can occur in which the hydrostatic pressure head is greater than the aquifer thickness. Due to the higher-than-atmospheric pressure, water naturally escapes to the lower pressure space above the confining upper layer.

Exam tip: Knowing the type of aquifer will determine the equations that will subsequently be used. In most cases, the type of aquifer will be provided in the problem statement, but knowing the pressure field can identify this if it is unknown.

Porosity

Handbook: Porosity

$$\eta = \frac{\text{volume of voids}}{\text{total volume}}$$

CERM: Sec. 21.2

CEWE: Sec. 4.2

$$n = \frac{V_v}{V_t} \quad\quad\quad 21.3$$
$$= \frac{V_t - V_s}{V_t}$$

Porosity is the volume of open spaces divided by the total volume of the soil, represented as a percentage. This ratio is given by the *Handbook* equation for porosity, which is equivalent to CERM Eq. 21.3 and CEWE Eq. 4.1. CEWE Fig. 4.2 gives an illustrative example of soil porosity.

Figure 4.2 *Soil Porosity*

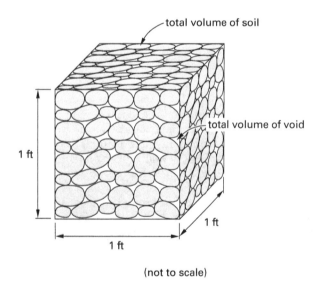

The total volume of a soil sample is the sum of the volume of solids and the volume of voids. The volume of voids is equal to the volume of water or the volume of air occupying the voids. The concept of porosity is similar to that of soil moisture content. One key difference is that porosity concerns volumes, while moisture content concerns mass or weight.

Exam tip: Porosity is the volume of space that is *not* filled by soil; this space could be filled with water or air. The greater the porosity, the easier air and water can pass through the soil.

Soil Moisture Content

Handbook: Soil Moisture Content

$$\theta = \frac{\text{volume of water}}{\text{total volume}}$$

CERM: Sec. 21.2

$$w = \frac{m_w}{m_s} \quad\quad\quad 21.2$$
$$= \frac{m_t - m_s}{m_s}$$

The *Handbook* equation for the moisture content of soil is a volume-based ratio. The equation assumes that the porosity of the water will occupy all voids and completely saturate the soil.

The more widely used soil moisture content equation is a mass-based ratio, where the ratio is the mass of the water to the mass of the solids, expressed as a percentage. CERM Eq. 21.2 shows the mass-based version of the equation. (Note, however, that mass and weight are often used interchangeably within civil engineering.) Do not confuse moisture content with *degree of saturation*, S, which is the ratio of the volume of water to the volume of voids.

Exam tip: Most equations will assume saturation of soil to make calculating flow easier. Aquifers are assumed to be saturated, so problems involving them require saturation to be accurate.

Hydraulic Gradient

Handbook: Seepage Velocity

$$V_v = \left(\frac{K}{\eta}\right)\left(\frac{\Delta h}{L}\right)$$

CERM: Sec. 21.2, Sec. 21.7

$$i = \frac{\Delta H}{L} \quad\quad\quad 21.6$$

$$v_{\text{pore}} = \frac{Q}{A_{\text{net}}} = \frac{Q}{nA_{\text{gross}}} = \frac{Q}{nbY} = \frac{Ki}{n} \quad\quad\quad 21.18$$

The *hydraulic gradient*, represented by i in CERM Eq. 21.6, is the measure of pressure within an aquifer. It is the change in pressure over the aquifer length, and it is unitless. The hydraulic head is determined through observation wells. The hydraulic gradient can then be used in the *Handbook* equation for seepage velocity, provided that the hydraulic conductivity, K, and effective porosity, η, are known. CERM Eq. 21.18 is equivalent to the *Handbook* equation.

The hydraulic gradient is typically calculated using a minimum of two groundwater monitoring wells, measuring the depth to groundwater, and calculating the difference in groundwater elevation. Finally, the distance between measuring points is determined. When determining the hydraulic gradient, the groundwater depth measurements among the monitoring points should be collected nearly simultaneously to avoid any temporal variations.

Exam tip: The hydraulic gradient can change based on season or due to well drawdown from pumps. When calculating the gradient, all pumps should be disabled to produce a natural result, since pumps can change the gradient artificially.

Types of Aquifers

Handbook: Unconfined Aquifers; Confined Aquifer

CERM: Sec. 21.1

CEWE: Sec. 4.1, 11.14

In an *unconfined aquifer*, the surface of the aquifer is at atmospheric pressure. *Handbook* figure Unconfined Aquifers shows an unconfined aquifer. In a *confined aquifer*, the surface of the aquifer is at hydrostatic pressure. *Handbook* figure Confined Aquifer shows a confined aquifer. An *artesian well* has a hydrostatic-pressured head greater than ground elevation.

CEWE Fig. 4.1 summarizes the different types of aquifers and the typical aquifer profile. The subsurface of a study area can have more than one aquifer, as shown. An aquifer can often be further subdivided into zones if there are differences in aquifer characteristics.

Figure 4.1 Typical Aquifer Profile

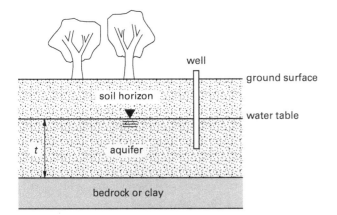

B. GROUNDWATER FLOW

Key concepts: These key concepts are important for answering exam questions in subtopic 13.B, Groundwater Flow.

- Darcy's law
- permeability/hydraulic conductivity
- seepage
- transmissivity

NCEES Handbook: To prepare for this subtopic, familiarize yourself with these sections in the *Handbook*.

- Darcy's Law
- Approximate Coefficient of Permeability for Various Sands
- Physical Properties of Water
- Properties of Water (SI Metric Units)
- Aquifers: Darcy's Law
- Hydraulic Conductivity
- Seepage Velocity
- Thiem Equation

PE Civil Reference Manual (**CERM**): Study these sections in CERM that either relate directly to this subtopic or provide background information.

- Section 21.3: Permeability
- Section 21.4: Darcy's Law
- Section 21.5: Transmissivity
- Section 21.7: Discharge Velocity and Seepage Velocity

Water Resources and Environmental Depth Reference Manual (**CEWE**): Study these sections in CEWE that either relate directly to this subtopic or provide background information.

- Section 4.4: Darcy's Law

The following equations, figures, and tables are relevant for subtopic 13.B, Groundwater Flow.

Darcy's Law

Handbook: Darcy's Law; Aquifers: Darcy's Law

$$Q = -KA\frac{dh}{dx}$$

$$Q = kiA$$

CERM: Sec. 21.4

$$Q = -KiA_{\text{gross}} \quad \quad 21.10$$
$$= -v_e A_{\text{gross}}$$

Darcy's law is a combination of the continuity equation and two key groundwater parameters, permeability and the hydraulic gradient. *Permeability* represents how easily groundwater can move through soil, and the *hydraulic gradient* represents the direction in which water flows (from high to low gradient).

The flow rate, Q, can be determined in cubic feet per day, gallons per day, or any other volume/time combination. Lowercase k and uppercase K mean the same thing here; the second *Handbook* equation is equivalent to CERM Eq. 21.10 and CEWE Eq. 4.1. The negative sign in the first *Handbook* equation accounts for the fact that flow is in the direction of decreasing head. That is, the hydraulic gradient is negative in the direction of flow.

Exam tip: Darcy's law functions similarly to the continuity equation, which states that flow equals the product of velocity and cross-sectional area. In this case, velocity is specific discharge or effective velocity.

Permeability

Handbook: Approximate Coefficient of Permeability for Various Sands; Physical Properties of Water; Properties of Water (SI Metric Units); Hydraulic Conductivity

$$K = \left(\frac{\gamma}{\mu}\right)k$$

CERM: Sec. 21.3

$$K = \frac{kg\rho}{\mu} \quad \quad \text{[SI]} \quad 21.7(a)$$

$$K = \frac{k\gamma}{\mu} \quad \quad \text{[U.S.]} \quad 21.7(b)$$

Permeability—also known as *hydraulic conductivity* or, less commonly, *the coefficient of permeability*—can be calculated with *Handbook* equation for hydraulic conductivity or CERM Eq. 21.7. *Intrinsic permeability*, k, is dependent only on the properties of the soil, not on those of the fluid. Other common units for permeability are gal/day-ft^2, ft/day, and cm/s.

Exam tip: Most of the variables used in these equations are for properties of water (μ, ρ, and γ) that can be determined quickly using *Handbook* table **Physical Properties of Water** or table **Properties of Water (SI Metric Units)**. Intrinsic permeability, k, can be estimated using *Handbook* table **Approximate Coefficient of Permeability for Various Sands**.

Seepage Velocity

Handbook: Seepage Velocity

$$V_v = \left(\frac{K}{\eta}\right)\left(\frac{\Delta h}{L}\right)$$

CERM: Sec. 21.7

$$v_{\text{pore}} = \frac{Q}{A_{\text{net}}} = \frac{Q}{nA_{\text{gross}}} = \frac{Q}{nbY} = \frac{Ki}{n} \quad 21.18$$

$$v_e = nv_{\text{pore}} = \frac{Q}{A_{\text{gross}}} = \frac{Q}{bY} = Ki \quad 21.19$$

Discharge velocity and seepage velocity have many different names. For the exam, be sure you are clear on the different names given in CERM Sec. 21.7, so that you know which type of velocity you need to find.

The *Handbook* equation or CERM Eq. 21.18 can be used to determine the *seepage velocity*, which is the velocity of water as it moves only through the voids (or the pores) in the soil.

CERM Eq. 21.19 can be used to determine the *discharge velocity*, which is the velocity of water as it moves through the entire cross section of the aquifer (both pores and solids).

To calculate the discharge velocity or the seepage velocity, the hydraulic gradient, i, and permeability, K, must be known. The difference between the two is the porosity, η, which represents how water seeps through the soil.

Exam tip: Seepage velocity represents water movement through the void spaces in soil. This should not be treated as though the flow is through an open channel.

Transmissivity

Handbook: Thiem Equation

$$T = Kb$$

CERM: Sec. 21.5

$$T = KY \qquad 21.13$$

Transmissivity is an index of the rate of groundwater movement, found by using the Thiem equation in the *Handbook* or CERM Eq. 21.13. For permeable soil, the thickness (b in the *Handbook* and Y in CERM) is the difference in elevations of the impermeable bottom and the water table.

Transmissivity can also be referenced as a coefficient. However, it has specific units that need to be adhered to. It is an index that compares the permeability to the thickness of the aquifer.

The permeability, also called the hydraulic conductivity, indicates how fast water can move through the aquifer. The aquifer thickness is the depth of the water in the aquifer. For short-term applications, the aquifer can be considered an endless supply of water. Multiply how fast the water can move through the aquifer by its thickness to find how much water can be extracted.

Exam tip: Transmissivity is used in a variety of aquifer calculations. Depending on whether the aquifer is confined or unconfined, transmissivity can be used in various ways to solve for other variables.

Flow Direction

CEWE: Sec. 4.4

The direction of groundwater flow is from the highest elevation to the lowest. This can be gauged using monitoring wells and measuring depth to water. Multiple wells must be placed to take measurements to determine where the hydraulic grade line is at its maximum. CEWE Fig. 4.5 gives an example of an aquifer cross section that shows the hydraulic grade line. The change in elevation of the water table at the upstream and downstream sections, z_1 and z_2, and the length of the section, L, are given in unit length.

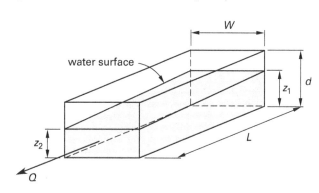

Figure 4.5 Aquifer Cross Section Defining Darcy's Law Parameters

Exam tip: To determine flow direction, calculate the slope between two points in an aquifer. Be aware that the slope is dependent on both the distance from the starting point and the depth to water, so it will not always be the lowest measured point.

C. WELL ANALYSIS—STEADY STATE

Key concepts: These key concepts are important for answering exam questions in subtopic 13.C, Well Analysis—Steady State.

- Thiem equation
- types of wells
- well components
- well hydraulics
- wellhead protection

NCEES Handbook: To prepare for this subtopic, familiarize yourself with these sections in the *Handbook*.

- Dupuit's Formula
- Thiem Equation
- Cooper Jacobs Equations
- Well Performance

PE Civil Reference Manual (CERM): Study these sections in CERM that either relate directly to this subtopic or provide background information.

- Section 21.6: Specific Yield, Retention, and Capacity
- Section 21.9: Wells
- Section 21.11: Well Drawdown in Aquifers
- Section 21.12: Unsteady Flow

Water Resources and Environmental Depth Reference Manual (**CEWE**): Study these sections in CEWE that either relate directly to this subtopic or provide background information.

- Section 4.5: Well Hydraulics

The following equations and figures are relevant for subtopic 13.C, Well Analysis—Steady State.

Dupuit's Formula

Handbook: Dupuit's Formula

$$Q = \frac{\pi K(h_2^2 - h_1^2)}{\ln \frac{r_2}{r_1}}$$

CERM: Sec. 21.11

$$Q = \frac{\pi K(y_1^2 - y_2^2)}{\ln \frac{r_1}{r_2}} \quad 21.25$$

CEWE: Sec. 4.5

$$Q = \frac{2\pi T(s_2 - s_1)}{\ln \frac{r_1}{r_2}} \quad 21.26$$

The assumptions for steady-state aquifers are the same as those for unconfined aquifers. The main assumption is that, since an unconfined aquifer is open to the atmosphere, it is not under any pressure. The water level in a well drilled in an unconfined aquifer will match the water level of the groundwater table.

The aquifer is assumed to be uniform; that is, it is made from the same soil material (homogeneous), and its physical properties remain constant (isotropic).

In CERM Eq. 21.25, the height of the water table above the impermeable layer is represented by y, whereas the *Handbook* equation uses h. In CERM Eq. 21.26 (and CEWE Eq. 4.7), the height of the water table is replaced by the drawdown, s. The *Handbook* and CERM also assign the subscripts differently to the larger and smaller radial distances and the larger and smaller aquifer depths. It doesn't matter which subscript is assigned to which, provided the aquifer depths also correspond correctly.

Exam tip: It is crucial that the water depth and distance are in the correct order. If calculations produce a negative result for flow, it usually means that the variables are not in the correct order.

Thiem Equation

Handbook: Thiem Equation

$$Q = \frac{2\pi T(h_2 - h_1)}{\ln \frac{r_2}{r_1}}$$

CERM: Sec. 21.11

$$Q = \frac{2\pi KY(y_1 - y_2)}{\ln \frac{r_1}{r_2}} \quad 21.27$$

CEWE: Sec. 4.5

Assumptions for steady-state aquifers apply to confined aquifers. Confined aquifers have an impermeable layer at both the top and the bottom, which means they are under pressure. When a well is drilled in a confined aquifer, the water level in the well will most likely exceed the water level of the groundwater table.

The aquifer is assumed to be uniform; that is, it is made from the same soil material (homogeneous), and its physical properties remain constant (isotropic).

In CERM Eq. 21.27, the height of the water table above the impermeable layer is represented by y, whereas it is h in the Thiem equation in the *Handbook*. Transmissivity, T, in the *Handbook* equation, is replaced by the product of hydraulic conductivity, K, and water table thickness, Y, in CERM. Also, the *Handbook* and CERM assign the subscripts differently to the larger and smaller radial distances and the larger and smaller aquifer depths; however, it doesn't matter which subscript is assigned to which, provided the aquifer depths also correspond correctly.

Exam tip: It is critical that the water depth and distance are in the correct order. In most cases, if calculations produce a negative result for flow, it means the variables are not in the correct order.

Cooper Jacobs Equations

Handbook: Cooper Jacobs Equations

CERM: Sec. 21.12

$$s = \frac{Q}{4\pi T} \ln \frac{2.25 Tt}{r_w^2 S}$$

$$T = \frac{2.3 Q}{4\pi} \ln \frac{\Delta \log t}{\Delta s}$$

The Cooper and Jacob's equations (called the Cooper Jacobs equations in the *Handbook*) are valid for confined aquifers. In the first *Handbook* equation, which is equivalent to that given by CERM Eq. 21.32, the drawdown, s, is a function of time, t; the well radius, r_w, the pumping rate, Q, and the aquifer transmissivity, T, are typically known. The coefficient of storage, S, has a common range of 5×10^{-5} through 5×10^{-3}.

If the transmissivity is to be determined, the second *Handbook* equation uses two observations (to determine $\Delta \log t$ and Δs) and the well pumping rate. If any other variables are unknown, the transmissivity can still be calculated as the product of water table height, Y, and hydraulic conductivity, K.

Exam tip: The Cooper and Jacob's equations can be used in conjunction with other confined aquifer equations, such as the Thiem equation, to determine drawdown at different points in the aquifer.

Well Performance

Handbook: Well Performance

CERM: Sec. 21.6, Sec. 21.11

$$s_t = s_a + s_w$$

$$q = \frac{Q}{s_t}$$

$$E(\%) = \frac{s_a}{s_t} \cdot 100\%$$

$$E(\%) = \left(1 - \frac{s_w}{s_t}\right) \cdot 100\%$$

Specific capacity is equal to the constant pumping discharge rate over drawdown at the well. It is a measure of the quantity of water that a well can produce per unit of drawdown. The opposite of specific capacity is *specific drawdown*.

The equation for well drawdown, s_t, includes both a linear term—aquifer loss, s_a—and a nonlinear term—well loss, s_w. The linear term implies that the drawdown will double if the pumping rate is doubled. The nonlinear term includes an exponent, n, and a constant, C, in its calculation.

$$s_w = CQ^n$$

Exam tip: These equations are used in determining well performance based on well drawdown, s_t, aquifer loss, s_a, and well loss, s_w. If the well is pumped faster than it can recharge, the well may need to be redesigned so as not to negatively impact the aquifer or disrupt any purpose it serves, such as for drinking water or groundwater treatment.

Well Components

CERM: Sec. 21.9

CERM Fig. 21.1 illustrates a typical gravity well. The grout at the top layer is intended to prevent the infiltration of water from the superficial aquifer, which has a lower water quality than the confined aquifer. The clay layers for this well show that there are two confined aquifers, since the clay is an impermeable layer. The screen and gravel are intended to reduce the infiltration of fine sediments into the water. The blank casing means it is not perforated. The sediment trap is where fine sediments can settle and be dredged as part of routine maintenance.

Figure 21.1 Typical Gravity Well

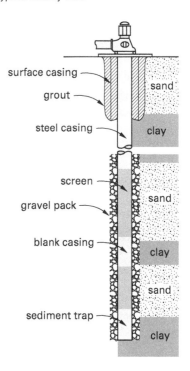

A well screen consists of a screen and a filter, both of which serve as barriers. The screen limits larger particles in the aquifer soil from clogging the well, while the filter prevents finer particles from contaminating the well water. Both barriers are intended to prevent sediment from entering the well and to allow water to pass into the well without accumulating excessive solids.

Exam tip: The type of aquifer will determine where the screen will be installed. In an unconfined aquifer, the screen is installed in the lower 30% to 40% of the well, while in a confined aquifer, it is in the middle 70% to 80%.

50 Wastewater Collection and Treatment

Content in blue refers to the *NCEES Handbook*.

A. Wastewater Collection Systems......................... 50-1
B. Wastewater Treatment Processes 50-3
C. Wastewater Flow Rates................................. 50-4
D. Preliminary Treatment.................................. 50-7
E. Primary Treatment 50-9
F. Secondary Treatment 50-11
G. Nitrification/Denitrification 50-13
H. Phosphorus Removal.....................................50-17
I. Solids Treatment, Handling, and Disposal 50-19
J. Digestion 50-21
K. Disinfection 50-23
L. Advanced Treatment.....................................50-24

The knowledge area of Wastewater Collection and Treatment makes up between five and eight questions out of the 80 questions on the PE Civil Water Resources and Environmental exam. The organization of this chapter follows the order of subtopics given by NCEES for this knowledge area. Each subtopic is covered in the following sections.

A. WASTEWATER COLLECTION SYSTEMS

Key concepts: These key concepts are important for answering exam questions in subtopic 14.A, Wastewater Collection Systems.

- force wastewater flow systems
- gravity wastewater flow systems
- wet wells

NCEES Handbook: To prepare for this subtopic, familiarize yourself with these sections in the *Handbook*.

- Principles of One-Dimensional Fluid Flow: Energy Equation
- Lift Station Pumping and Wet Wells
- Single Pump System Cycle Time
- Ideal Minimum Wet Well Volume
- Manning's Equation
- Approximate Values of Manning's Roughness Coefficient
- Wastewater Collection Systems: Wet Wells

PE Civil Reference Manual (**CERM**): Study these sections in CERM that either relate directly to this subtopic or provide background information.

- Section 17.19: Extended Bernoulli Equation
- Section 18.7: Sewage Pumps
- Section 19.6: Governing Equations for Uniform Flow
- Section 28.6: Sewer Pipe Materials
- Section 28.7: Gravity and Force Collection Systems
- Section 28.8: Sewer Velocities
- Section 28.9: Sewer Sizing
- Section 29.5: Wastewater Treatment Plants
- Appendix 19.B: Manning Equation Nomograph

Water Resources and Environmental Depth Reference Manual (**CEWE**): Study these sections in CEWE that either relate directly to this subtopic or provide background information.

- Section 7.9: Sewer Systems
- Section 7.10: Sewer Pipe Materials

Codes and standards: To prepare for this subtopic, familiarize yourself with these sections from codes and standards.

Recommended Standards for Wastewater Facilities (TSS)

- Chapter 40: Wastewater Pumping Stations

The following equations, figures, tables, and concepts are relevant for subtopic 14.A, Wastewater Collection Systems.

Force Wastewater Flow System

Handbook: Principles of One-Dimensional Fluid Flow: Energy Equation

$$\frac{p_1}{\gamma} + z_1 + \frac{v_1^2}{2g} = \frac{p_2}{\gamma} + z_2 + \frac{v_2^2}{2g} + h_f$$

CERM: Sec. 17.19

The energy equation accounts for loss of energy along a given length of a pipe network. This can be from head losses, friction, or other sources, such as additional wastewater streams.

Conservation of mass and energy apply, so both sides of the equation must be equal. In many cases, a sewer line will be open to the atmosphere at multiple locations due to manholes, and pressure will equalize with the atmosphere. In this case, change in velocity and sewer depth become the major factors. Velocity may change, but flow should not change unless additional wastewater streams are added. Remember that flow is equal to the product of cross-sectional area and velocity. As flow increases, if the pipe does not also increase proportionally, velocity will increase. If the flow remains the same, velocity should too, unless the physical properties of the pipe change.

Exam tip: For pressurized flow, the Bernoulli or energy equation should be used to solve for the pipe size. Friction (major) and minor losses will also need to be calculated.

Wet Wells

Handbook: Lift Station Pumping and Wet Wells; Single Pump System Cycle Time; Ideal Minimum Wet Well Volume; Wastewater Collection Systems: Wet Wells

CERM: Sec. 18.7

TSS: Chap. 40

Wet wells accumulate flow within a lift station. This flow can be from the wastewater collection system or other environmental sources (e.g., groundwater infiltration and rain). See *Handbook* figure Lift Station Pumping and Wet Wells for an illustration of a lift station. CERM Fig. 18.5 also illustrates a lift station.

Figure 18.5 *Typical Wastewater Pump Installation (greatly simplified)*

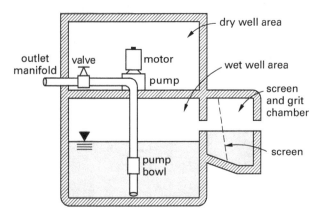

A lift station pump must be able to maintain the water level such that the station does not flood. To do so, calculating the pump running time, filling time, total cycle time, and minimum storage volume are critical. *Handbook* sections Wastewater Collection Systems: Wet Wells, Single Pump System Cycle Time, and Ideal Minimum Wet Well Volume provide useful equations to calculate the cycle times, inflows, outflows, and required volumes for pump systems.

$$t_r = \frac{\overline{V}}{D - Q}$$

$$t_f = \frac{V}{Q}$$

$$t = t_r + t_f = \frac{V}{D - Q} + \frac{V}{Q}$$

$$V_{\min} = \frac{T_{\min} Q_{\min}}{4}$$

TSS Chap. 40 contains recommendations for the design and application of wastewater pumping stations. Design guidance for wet wells is given in TSS Sec. 42.6.

Exam tip: In sizing a wet well, the well should have enough capacity that it does not flood. This is the storage volume of the wet well. If the storage volume is too small, the pump will run more often, which wears down the pump and leads to more frequent maintenance and replacement. A reasonable filling time must be determined between the pump running times.

Gravity Wastewater Flow System

Handbook: Manning's Equation; Approximate Values of Manning's Roughness Coefficient

CERM: Sec. 19.6, Sec. 28.7, Sec. 28.8, Sec. 28.9, App. 19.B

CEWE: Sec. 7.9, Sec. 7.10

A minimum slope is needed for gravity in sewer pipes to ensure cleansing velocities. Even with a lift station, the flow of wastewater is fed to the plant by gravity, so the pipe slope is crucial. Gravity flow tends to behave as an open channel. The equation in *Handbook* section Manning's Equation and *Handbook* table Approximate Values of Manning's Roughness Coefficient can be used to solve problems involving wastewater flow systems.

$$v = \left(\frac{1.486}{n}\right) R_H^{2/3} S^{1/2}$$

CEWE Table 7.1 shows the minimum pipe slope for various sewer pipe lengths.

Table 7.1 Minimum Pipe Slope for Various Sewer Pipe Lengths

diameter		slope (ft/100 ft)	
(in)	(mm)	pipe length ≤ 5 ft	pipe length > 5 ft
8	200	0.47	0.42
10	250	0.34	0.31
12	310	0.26	0.24
14	360	0.23	0.22
24	610	0.08	0.088
30	760	0.07	0.07

Adapted from *Collection Systems Technology Fact Sheet—Sewers, Conventional Gravity*, USEPA, 2002.

Sewer pipes should be designed with a maximum flow to accommodate future additions. In addition, remember that a pipe cannot be sized smaller than needed. If calculating a given flow results in a pipe diameter that falls between two standard diameters, always select the larger standard size.

Exam tip: Exam problems may require you to find the pipe size for a single network segment. Note that the Manning equation works only in U.S. units, and solving it requires the use of a Manning equation nomograph, which should be provided with the problem on the exam. An example of a Manning equation nomograph is shown in CERM App. 19. B.

B. WASTEWATER TREATMENT PROCESSES

Key concepts: These key concepts are important for answering exam questions in subtopic 14.B, Wastewater Treatment Processes.

- industrial wastewater
- strong and weak sewage
- wastewater processing

NCEES Handbook: To prepare for this subtopic, familiarize yourself with these sections in the *Handbook*.

- Common Chemicals in Water and Wastewater Processing

PE Civil Reference Manual **(CERM):** Study these sections in CERM that either relate directly to this subtopic or provide background information.

- Section 28.14: Solids
- Section 29.1: Industrial Wastewater Treatment
- Section 29.4: Disposal of Septage
- Section 29.5: Wastewater Treatment Plants

Water Resources and Environmental Depth Reference Manual **(CEWE):** Study these sections in CEWE that either relate directly to this subtopic or provide background information.

- Section 7.1: Domestic Wastewater
- Section 7.2: Industrial Wastewater
- Section 7.3: Municipal Wastewater
- Section 7.4: Municipal Wastewater Composition
- Section 7.5: Wastewater Characteristics
- Section 7.18: Wastewater Treatment Plant Loading
- Section 8.3: Wastewater Treatment

Codes and standards: To prepare for this subtopic, familiarize yourself with these sections from codes and standards.

Recommended Standards for Wastewater Facilities (TSS)

- Section 56: Essential Facilities
- Section 72: Design Considerations
- Section 92: Activated Sludge
- Section 93: Wastewater Treatment Ponds
- Section 111: Phosphorus Removal by Chemical Treatment
- Section 112: High Rate Effluent Filtration

The following tables and concepts are relevant for subtopic 14.B, Wastewater Treatment Processes.

Wastewater Processing

Handbook: Common Chemicals in Water and Wastewater Processing

CERM: Sec. 29.5

TSS: Sec. 56, Sec. 72, Sec. 92, Sec. 93, Sec. 111, Sec. 112

Wastewater treatment is a simple process of removing contaminants from sewage through physical, chemical, or biological means. *Handbook* table Common Chemicals in Water and Wastewater Processing shows various chemicals used to treat wastewater.

The treatment creates two products: a liquid effluent and sludge. Depending on the contaminants still present, the liquid effluent can be reused as graywater for irrigation, or it can be discharged into a receiving water body. The sludge can be used as compost in agricultural uses if it meets biosolids requirements, or it can be disposed of as waste in a landfill. The treatment of wastewater varies depending on the composition of the sewage, which varies based on the characteristics of the locale and on the effluent limitations found in the National Pollutant Discharge Elimination System (NPDES) permit issued to the facility.

Sewage varies based on local conditions and source, and it may include industrial, chemical, or residential wastes. *Recommended Standards for Wastewater Facilities* (TSS, also known as the Ten States Standards) provides advice on how to design for expected wastewater influent from residential and light commercial industry sources in Sec. 56, Sec. 72, Sec. 92, Sec. 93, Sec. 111, and Sec. 112. Expected loadings from heavy industry, the environment, and other sources must also be considered.

Exam tip: When designing wastewater treatment, examine the typical contaminants and expected flow to the facility, as well as any limitations to the facility's permits. Treatment steps are designed to handle the expected loading as well as low-flow and high-flow events. Wastewater discharge varies throughout the day and year, so treatment should vary accordingly.

Strong and Weak Sewage

Handbook: Common Chemicals in Water and Wastewater Processing

CERM: Sec. 28.14, Sec. 29.4

CEWE: Sec. 7.1, Sec. 7.4, Sec. 7.5, Sec. 7.6, Sec. 8.3

Municipal wastewater and sewage can be treated to varying degrees depending on the characteristics of the water body into which the treated water is to be discharged (or any other destination where it is approved for reuse). For instance, if sewage is to be discharged into the ocean, it may require only primary treatment, while if it is to be discharged into a small river, tertiary treatment may be required. (Wastewater treatment plants are typically described as *primary*, *secondary*, or *tertiary* to designate the level of treatment the plant employs.)

Handbook table Common Chemicals in Water and Wastewater Processing shows various chemicals used to treat wastewater. CEWE Table 8.2 lists expected pollution characteristics in municipal wastewater. In a community that is residential and light commercial, the wastewater may be weak, while heavy industry can produce wastewater with a high volume of contaminants, depending on what is being manufactured. In addition, some contaminants may be present only during certain seasons, such as chlorides in the winter and spring due to road salt.

Exam tip: Typical pollution characteristics can be used if sample data is not provided on an exam problem. In general, knowing the demographics of various communities will be the best guidance on how to utilize this information. Remember that all wastewater (residential, light commercial, and heavy commercial) is combined, so if there is a significant amount of one, it may dilute the others.

Industrial Wastewater

Handbook: Common Chemicals in Water and Wastewater Processing

CERM: Sec. 29.1

CEWE: Sec. 7.2

Industrial wastewater has different characteristics than domestic wastewater. In general, industrial wastewater is produced as a by-product of wet manufacturing processes (i.e., where water comes in contact with what is being produced). Depending on the process, different concentrations of pollutants may be weak while others may be strong. CERM Table 29.1 summarizes typical industrial wastewater effluent limitations. *Handbook* table Common Chemicals in Water and Wastewater Processing lists various chemicals used to treat wastewater.

Most industries that may disrupt wastewater treatment are required to treat their own wastewater before discharge. This *industrial pretreatment* is done so that the collection system and treatment facility are not overwhelmed by excessive pollution.

Exam tip: Exam questions on designing wastewater treatment with industry will typically provide the pollution characteristics in the problem statement. Industries vary such that average pollution characteristics are difficult to estimate, even if the industries are known.

C. WASTEWATER FLOW RATES

Key concepts: These key concepts are important for answering exam questions in subtopic 14.C, Wastewater Flow Rates.

- inflow and infiltration
- peak factors
- sewer design requirements
- wastewater flow

NCEES Handbook: To prepare for this subtopic, familiarize yourself with these sections in the *Handbook*.

- Runoff Analysis: Rational Formula Method
- Runoff Coefficients for Rational Formula
- SCS Hydrologic Soils Group Type Classifications
- Minimum Infiltration Rates for the Various Soil Groups
- Sewage Flow Ratio Curves

PE Civil Reference Manual (**CERM**): Study these sections in CERM that either relate directly to this subtopic or provide background information.

- Section 28.1: Domestic Wastewater
- Section 28.4: Wastewater Quantity
- Section 28.9: Sewer Sizing

Water Resources and Environmental Depth Reference Manual (**CEWE**): Study these sections in CEWE that either relate directly to this subtopic or provide background information.

- Section 7.11: Sanitary Sewers
- Section 7.12: Population Estimates
- Section 7.13: Peak Factors
- Section 7.14: Infiltration and Inflow
- Section 7.15: Extraneous Flows
- Section 7.16: Sewer Design Requirements

Codes and standards: To prepare for this subtopic, familiarize yourself with these sections from codes and standards.

Recommended Standards for Wastewater Facilities (TSS)

- Section 11.242: Hydraulic Capacity for Wastewater Facilities to Serve Existing Collection Systems
- Section 11.243: Hydraulic Capacity for Wastewater Facilities to Serve New Collection Systems
- Section 11.25: Organic Capacity
- Section 32: Design Capacity and Design Flow
- Section 33: Details of Design and Construction

The following tables and concepts are relevant for subtopic 14.C, Wastewater Flow Rates.

Inflow and Infiltration

Handbook: Runoff Analysis: Rational Formula Method; Runoff Coefficients for Rational Formula; SCS Hydrologic Soils Group Type Classifications; Minimum Infiltration Rates for the Various Soil Groups

CERM: Sec. 28.4

CEWE: Sec. 7.14, Sec. 7.15

TSS: Sec. 11.242, Sec. 11.243, Sec. 11.25

Inflow is stormwater that enters the wastewater system. *Infiltration* is groundwater that enters the wastewater system. Inflow and infiltration are seasonal variables as they are related to wet and dry seasons. For this reason, it is common for wastewater plants to consider these flows in addition to the base sewage flow.

Inflow and infiltration can be reduced with well-maintained sewers and separating storm and combined sewers. When performing treatment plant flow design, the overall condition of the collection system must be considered. Inflow and infiltration can be estimated by calculating the flow observed reaching the facility at its lowest rate during the year. The potential inflow to a wastewater system can be determined using the equation for peak discharge in *Handbook* section Runoff Analysis: Rational Formula Method and *Handbook* table Runoff Coefficients for Rational Formula, which contains runoff coefficients for several types of drainage areas. *Handbook* table Minimum Infiltration Rates for the Various Soil Groups provides minimum infiltration rates based on soil group types defined in *Handbook* section SCS Hydrologic Soils Group Type Classifications. The value of 100 gal/capita/day cited in TSS Sec. 11.242 and Sec. 11.243 should be used in conjunction with a peaking factor to cover normal infiltration.

Exam tip: Exam problems on sewage flow will typically provide any required infiltration and inflow information in the problem statement.

Wastewater Flow

Handbook: Sewage Flow Ratio Curves

CERM: Sec. 28.1

CEWE: Sec. 7.11, Sec. 7.12

TSS: Sec. 11.242, Sec. 11.243

Wastewater flows are closely related to potable water demand. These vary per jurisdiction, may be codified into regulations, and can be measured based on area, on number of units, or per person (gallons per capita per day). Estimates of wastewater flows vary from jurisdiction to jurisdiction; there are no national standards.

Per capita flows vary from 50 gpd to 140 gpd and may be as high as 160 gpd where industrial flows are included. Typically, a value of 125 gpd is used to convert population to average sewage flow, including commercial and industrial flow. Average design flows for different types of developments are reported in CEWE Table 7.2. Curve flows and equations can be found in *Handbook* figure Sewage Flow Ratio Curves.

Table 7.2 Average Sewer Design Flows for Various Types of Developments

type of development	design flow (gpd)
residential	
general	100/capita
single family	370/capita
townhouse unit	300/unit
apartment unit	300/unit
commercial	
general	2000/ac
motel	130/unit
office	20/employee
industrial (varies with type of industry)	
general	10,000/ac
warehouse	600/ac
school site (general)	16/student

(Multiply ac by 4046.87 to obtain m².)

TSS Sec. 11.242 and Sec. 11.243 recommend sizing wastewater facilities receiving flows from a new wastewater collection system to be based on an average daily flow of 100 gal per capita plus wastewater flow from industrial plants and major institutional and commercial facilities.

Exam tip: Different regions will have different design flows. TSS Sec. 11.243 gives an estimated 100 gal per capita per day as guidance on water demand, but that is based on use in the states around Lake Michigan. Arid parts of the country will use less, while areas with plentiful water resources may use more. If an exam problem provides a way to calculate the average water demand, use the value given.

Peak Factors

Handbook: Sewage Flow Ratio Curves

CERM: Sec. 28.4

CEWE: Sec. 7.13

TSS: Sec. 11.242, Sec. 11.243

Peak factors convert average annual flow to design flow. These factors take into consideration the dry and wet weather peaks.

Design flows are calculated by using peak factors. These peak factors are multiplied by the average annual flows to determine a design peak flow. Some jurisdictions will require the use of a given peak flow, and the peak factors vary from jurisdiction to jurisdiction.

Peak flows represent the maximum stress that a treatment facility is expected to handle in the event of environmental conditions such as intense rainfall or spring snowmelt. Flow in excess of the peak flow should result in emergency plans being enacted. As a rule of thumb, peak factors will range between 1.3 and 3.5 times the average flow. See *Handbook* figure Sewage Flow Ratio Curves for ratios of minimum and peak flows for a given population value. TSS Fig. 1 correlates the peaking factor to the average daily flows of 100 gal/capita/day cited in TSS Sec. 11.242 and Sec. 11.243.

Exam tip: Exam problems on peak flows will typically provide peak factors in the problem statement. If such data is not provided, estimate the approximate value (low, average, or high) based on sewer conditions and geographic region.

Sewer Design Requirements

Handbook: Sewage Flow Ratio Curves

CERM: Sec. 28.9

CEWE: Sec. 7.16

TSS: Sec. 32, Sec. 33

Sewers must be sized for minimum, maximum, and average flow. In general, a sewer line should maintain a flow velocity of 2 ft/sec so as not to accumulate solids or other debris in the pipe, and a maximum of 20 ft/sec to avoid damaging the pipe or creating excessive turbulence. When making calculations to determine pipe diameter, the pipe should not be full, since a full pipe has no space to take in excess flows. A pipe is best designed with the flow accounting for 70% of the pipe diameter. CERM Table 28.2 lists minimum flow velocities for various types of fluids. See *Handbook* figure Sewage Flow Ratio Curves for ratios of minimum and peak flows for a given population value.

Table 28.2 Minimum Flow Velocities

fluid	minimum velocity to keep particles in suspension (ft/sec)	(m/s)	minimum resuspension velocity (ft/sec)	(m/s)
raw sewage	2.5	(0.75)	3.5	(1.1)
grit tank effluent	2	(0.6)	2.5	(0.75)
primary settling tank effluent	1.5	(0.45)	2	(0.6)
mixed liquor	1.5	(0.45)	2	(0.6)
trickling filter effluent	1.5	(0.45)	2	(0.6)
secondary settling tank effluent	0.5	(0.15)		(0.3)

(Multiply ft/sec by 0.3 to obtain m/s.)

TSS Sec. 32 covers guidelines for design capacity and design flow for sewers. TSS Sec. 33 covers details of sewer design and construction.

Exam tip: On exam problems concerning sewer pipe design, you must account for current flows and anticipated future flows. A sewer pipe designed when a city is growing will need to be replaced before its expected lifespan ends if the population outgrows the area. In addition, monitoring flows contributed from industrial sources is critical to prevent excessive flow from nonresidential sources.

D. PRELIMINARY TREATMENT

Key concepts: These key concepts are important for answering exam questions in subtopic 14.D, Preliminary Treatment.

- bar screens
- fine screening
- grit chambers
- preliminary treatment process

NCEES Handbook: To prepare for this subtopic, familiarize yourself with these sections in the *Handbook*.

- Coarse Screening/Bar Rack Head Loss (Bernoulli Method)
- Bar Rack Head Loss—Kirschmer Method (Clean Racks)
- Bar Shape Factors
- Fine Screening Head Loss
- Clarifiers: Scour Velocity

PE Civil Reference Manual (**CERM**): Study these sections in CERM that either relate directly to this subtopic or provide background information.

- Section 29.12: Racks and Screens
- Section 29.13: Grit Chambers
- Section 29.14: Aerated Grit Chambers
- Section 29.15: Skimming Tanks
- Section 29.16: Shredders

Water Resources and Environmental Depth Reference Manual (**CEWE**): Study these sections in CEWE that either relate directly to this subtopic or provide background information.

- Section 8.6: Pretreatment Processes

Codes and standards: To prepare for this subtopic, familiarize yourself with these sections from codes and standards.

Recommended Standards for Wastewater Facilities (TSS)

- Section 61: Screening Devices
- Section 63: Grit Removal Facilities

The following equations, tables, and concepts are relevant for subtopic 14.D, Preliminary Treatment.

Bar Screens

Handbook: Coarse Screening/Bar Rack Head Loss (Bernoulli Method); Bar Rack Head Loss—Kirschmer Method (Clean Racks); Bar Shape Factors

CERM: Sec. 29.12

CEWE: Sec. 8.6

TSS: Sec. 61

Bar screens remove large objects such as litter or vegetation debris. The debris caught in the screens can be

- cleared by automatic scraping arms
- collected in bins
- disposed of in a landfill
- ground up and carried through the sewage flow

Bar screens have large openings, as they are intended to catch only large debris. CEWE Table 8.5 shows some common parameters to provide an idea of the different types of bar screens used. Head loss through a bar rack can be determined using the equation in *Handbook* section Coarse Screening/Bar Rack Head Loss (Bernoulli Method) or the equation in *Handbook* section Bar Rack Head Loss—Kirschmer Method (Clean Racks). Bar shape factors, β, can be found in the *Handbook* table Bar Shape Factors.

$$h_L = \left(\frac{1}{C}\right)\left(\frac{V^2 - v^2}{2g}\right)$$

$$h_L = \beta\left(\frac{w}{b}\right)^{1.33} h_v \sin\theta$$

The first equation is for coarse screens, and the second is for clean bar racks. When designing screens, the process will result in head losses that must be addressed, either by anticipating a reduction in flow rate or adding pumps to move the wastewater to the next treatment step. The *Handbook* equations provide an estimate of what those losses may be. A clean screen or rack will have a lower head loss than one that is partially clogged and needs to be cleaned. TSS Sec. 61 contains recommendations on the use of screening devices.

Exam tip: On the PE Civil Water Resources and Environmental exam, students will not be expected to look up standard values for screens. The problem will provide any specific details needed.

Fine Screening

Handbook: Fine Screening Head Loss

$$h_L = \frac{1}{C \cdot 2g}\left(\frac{Q}{A}\right)^2$$

CERM: Sec. 29.12

TSS: Sec. 61

Fine screens are used to remove small particles while relieving the load on grit tanks. A fine screen will generally have openings 1.6 mm to 3 mm wide. The head loss from the use of fine screens can be calculated using the *Handbook* equation. See CERM Sec. 29.12 for more information on using fine screens.

Fine screens are rare except when used with selected industrial waste processing plants. TSS Sec. 61 contains recommendations on the use of screening devices.

Exam tip: Exam problems should indicate whether fine screens are required. If fine screens are not explicitly required, they do not need to be considered when doing calculations.

Grit Chambers

Handbook: Clarifiers: Scour Velocity

$$v_c = \left(\frac{8\beta(s-1)gd}{f}\right)^{0.5}$$

CERM: Sec. 29.13

CEWE: Sec. 8.6

$$v = \sqrt{8k\left(\frac{gd_p}{f}\right)(\text{SG}_p - 1)} \qquad 8.1$$

TSS: Sec. 63

The *Handbook* equation, CERM Eq. 29.3, and CEWE Eq. 8.1 are equivalent, with some variables referenced differently depending on the source.

Handbook	CERM and CEWE	description
d	d_p	particle diameter
f	f	Darcy friction factor
β	k	dimensionless parameter for stickiness of the particle
s	SG_p	specific gravity of the particle

Grit chambers are facilities designed to capture sand and grit to protect pumps and other equipment within the wastewater treatment plant. They are mainly used to settle grit, which can be done by slowing the flow velocity, increasing the path taken by the grit, or pushing the grit to the sides of the chamber through centrifugal force. An additional benefit to using grit chambers is the removal of inorganic and nondigestible material that may inhibit downstream digestion. The most common grit chamber design is a horizontal flow design. Some designs incorporate oil and grease skimming. CEWE Table 8.6 and Table 8.7 give common values for design variables of different types of grit chambers. TSS Sec. 63 contains design considerations for grit removal facilities.

Exam tip: While flow through a wastewater treatment facility will remain consistent, the velocity must be kept below a certain level. Reducing velocity requires the flow to be applied to a larger cross-sectional area.

Preliminary Treatment Processes

CERM: Sec. 29.12, Sec. 29.13, Sec. 29.14, Sec. 29.15, Sec. 29.16

Preliminary treatment processes are the first steps in processing raw sewage. Large bar screens remove large solids, and grit chambers remove sand and grit. Preliminary treatment screens are considered part of the primary treatment system of a wastewater treatment plant.

Some preliminary treatments include removal of oil and grease that has passed through the screens, bars, or grit chambers. Removal of oil and grease is important to avoid foaming or scum growth later in the digesters. Shredding of waste solids is also sometimes necessary, allowing some solids that would otherwise have been caught earlier in preliminary treatment to pass through and settle later in the facility with other solids.

Exam tip: The need for additional preliminary treatment depends on the characteristics of the wastewater. In exam problems on the topic, the presence of oil, grease, or large solids typically means that preliminary treatment measures will be required.

E. PRIMARY TREATMENT

Key concepts: These key concepts are important for answering exam questions in subtopic 14.E, Primary Treatment.

- clarifier design equations
- clarifiers
- effective settling area in a clarifier
- phosphorus removal

NCEES Handbook: To prepare for this subtopic, familiarize yourself with these sections in the *Handbook*.

- Clarifiers: Settling
- Chemical Phosphorus Removal
- Chemical Phosphorous Precipitation
- Coagulation
- Stokes's Law
- Clarifiers
- Typical Primary Clarifier Efficiency Removal
- Water Treatment Weir Loadings
- Water Treatment Horizontal Velocities
- Water Treatment Clarifier Dimensions
- Design Criteria for Sedimentation Basins/Clarifiers

PE Civil Reference Manual (**CERM**): Study these sections in CERM that either relate directly to this subtopic or provide background information.

- Section 26.11: Sedimentation Physics
- Section 26.12: Sedimentation Tanks
- Section 26.15: Coagulants
- Section 26.16: Flocculation Additives
- Section 26.17: Doses of Coagulants and Other Compounds
- Section 26.21: Flocculation
- Section 29.28: Phosphorus Removal: Precipitation
- Section 30.1: Sludge
- Section 30.2: Activated Sludge Process
- Section 30.5: Sludge Parameters
- Section 30.6: Soluble BOD Escaping Treatment
- Section 30.7: Process Efficiency

Water Resources and Environmental Depth Reference Manual (**CEWE**): Study these sections in CEWE that either relate directly to this subtopic or provide background information.

- Section 8.4: Wastewater Treatment Plant
- Section 8.7: Primary and Secondary Treatment Processes

Codes and standards: To prepare for this subtopic, familiarize yourself with these sections from codes and standards.

Recommended Standards for Wastewater Facilities (**TSS**)

- Chapter 70: Settling
- Section 92.62: Biological Phosphorus Removal

The following equations, figures, tables, and concepts are relevant for subtopic 14.E, Primary Treatment.

Effective Settling Area in a Clarifier

Handbook: Clarifiers: Settling

CEWE: Sec. 8.7

TSS: Chap. 70

Solids removal depends upon an effective settling basin surface area. The effective settling area is reduced in two ways. The first is by the accumulation of sludge at the bottom of the tank. When a clarifier is first used, there is no sediment at the bottom. As time passes, the bottom quickly gets filled with sediment, or sludge. The sludge is removed at a constant rate, so the bottom depth of the sludge becomes constant. That depth of sludge reduces the working height of the clarifier because particles have less distance to travel and settle.

The second way the effective settling area is reduced is by the influence zones for the inlet and outlet. Since the liquid within the clarifier is moving, the sediment falls at an angle, which reduces the settling area space at those locations. CEWE Fig. 8.8 shows a sedimentation basin and an example of effective settling area.

Figure 8.8 Processes and Operational Scheme in a Sedimentation Basin

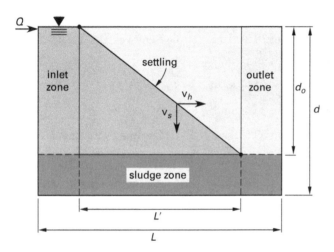

TSS Chap. 70 covers requirements and guidelines for settling units. TSS Sec. 72.21, Sec. 72.22, and Sec. 72.23 covers specifications for primary, intermediate, and final settling tanks, respectively. TSS Sec. 73.2 discusses sludge removal. TSS Sec. 74.1 and Sec. 74.3 discuss safety requirements for operator protection around settling tanks.

Exam tip: Exam problems on clarifiers may require you to find the sediment settling area, among other calculations.

Phosphorus Removal

Handbook: Chemical Phosphorus Removal; Chemical Phosphorous Precipitation; Coagulation

CERM: Sec. 26.15, Sec. 26.16, Sec. 26.17, Sec. 29.28

TSS: Sec. 92.62

Coagulation is the inclusion of a chemical to remove additional contaminants as they form larger particles. In wastewater treatment, the most common coagulants are aluminum sulfate (alum) and ferric sulfate or ferric chloride (ferric). These can be used to remove excess phosphorus if biological removal is not economical or feasible. *Handbook* section Coagulation provides stoichiometric equations for coagulation processes.

Phosphorus concentrations of 5 mg/L to 15 mg/L (as P) occur in untreated wastewater, most of which originates from synthetic detergents and human waste. Approximately 10% of the total phosphorus is insoluble and can be removed in primary settling. The amount removed by absorption in conventional biological processes is small. The remaining phosphorus is soluble and must be removed by converting it into an insoluble precipitate. *Handbook* sections Chemical Phosphorus Removal and Chemical Phosphorous Precipitation provide stoichiometric equations for phosphorus removal.

TSS Sec. 92.62 contains design considerations for biological phosphorus removal.

Exam tip: Exam problems on chemical removal treatment processes may require you to calculate the maximum amount of chemical to be removed based on a given clarifier and flow stream. Such problems will typically provide a site-specific safety factor in the problem statement. It is also important to note that chemical phosphorus removal will increase sludge production.

In the *Handbook*, *phosphorus* is misspelled in the section title Chemical Phosphorous Precipitation. Searching on "Chemical Phosphorus" will still get you to the right page.

Clarifiers

Handbook: Water Treatment Weir Loadings; Water Treatment Horizontal Velocities; Water Treatment Clarifier Dimensions; Design Criteria for Sedimentation Basins/Clarifiers; Typical Primary Clarifier Efficiency Removal

CERM: Sec. 26.11

CEWE: Sec. 8.4, Sec. 8.7

In general, there are three types of clarifiers: physical, chemical, and biological. Physical methods simply slow down the flow to allow dense particles to settle to the bottoms of basins or tanks. Chemical methods use coagulants that lump particles together so that they can easily settle. Coagulation is typically followed by flocculation, where slow mixing encourages larger particles (floc) to form and finally settle. Clarification basins can be circular, as shown in CEWE Fig. 8.3, or rectangular, as shown in CEWE Fig. 8.7. Note how weirs are used to control the velocity of the fluid moving through the clarifier. The CEWE figures show good examples of clarifiers, including one that uses multiple weirs. CEWE Fig. 8.2 shows a general schematic of a wastewater treatment plant.

Figure 8.3 Typical Circular Clarification Basin for Wastewater Treatment

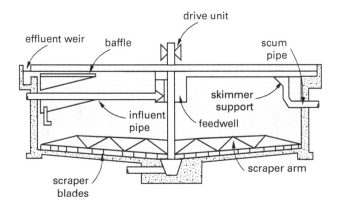

Clarifier efficiency values are given in *Handbook* table Typical Primary Clarifier Efficiency Removal. *Handbook* sections Water Treatment Weir Loadings, Water Treatment Horizontal Velocities, and Water Treatment Clarifier Dimensions give general design guidelines for clarifiers. *Handbook* table Design Criteria for Sedimentation Basins/Clarifiers provides design criteria for different types of basins.

Exam tip: Some exam problems may involve clarifiers with more than one weir, and some may require you to find the number of weirs needed based on given design criteria.

Clarifier Design Equations

Handbook: Stokes's Law; Clarifiers

$$\text{hydraulic loading rate} = \frac{Q}{A}$$

$$\text{hydraulic residence time} = \frac{V}{Q}$$

$$\text{WOR} = \frac{Q}{\text{weir length}}$$

CEWE: Sec. 8.4, Sec. 8.7

$$q_o = \frac{Q}{A_s} \qquad 8.2$$

$$t_d = \frac{V}{Q} \qquad 8.3$$

$$w_r = \frac{Q}{L} \qquad 8.4$$

TSS: Chap. 70

These are the main formulas for designing clarifiers, with the CEWE equations using only variables and the *Handbook* equations using words and variables. See CEWE Fig. 8.2 for a general schematic of a wastewater treatment plant.

Clarifiers are designed to slow the velocity of flow through a tank to allow physical, biological, and/or chemical processes to occur. If wastewater passes too quickly out of a clarifier, there will be insufficient time to treat it, and contaminants may remain that will pass through the treatment plant, appear in sludge, and/or upset other treatment steps. To avoid clarifiers being over- or undersized, facilities will have multiple smaller clarifiers that can be brought online or diverted from as flow increases and decreases.

TSS Chap. 70 covers requirements and guidelines for settling units. TSS Sec. 72.21, Sec. 72.22, and Sec. 72.23 cover recommended design dimensions for primary, intermediate, and final settling tanks, respectively. These guidelines may not apply if site-specific conditions do not require them.

Exam tip: All the units used in the formulas relate to each other. Conversions are required from milligrams to pounds of mass, from liters to cubic feet, and from hours to seconds.

F. SECONDARY TREATMENT

Key concepts: These key concepts are important for answering exam questions in subtopic 14.F, Secondary Treatment.

- clarifiers
- effective settling area in a clarifier
- food-to-microorganism ratio

NCEES Handbook: To prepare for this subtopic, familiarize yourself with these sections in the *Handbook*.

- Clarifiers: Typical Primary Clarifier Efficiency Percent Removal for Overflow Rates
- Clarifiers: Weir Loadings

- Activated Sludge Treatment: Organic Loading Rate
- Stokes's Law
- Clarifiers

PE Civil Reference Manual **(CERM):** Study these sections in CERM that either relate directly to this subtopic or provide background information.

- Section 26.12: Sedimentation Tanks
- Section 29.18: Chemical Sedimentation Basins/Clarifiers
- Section 30.1: Sludge
- Section 30.5: Sludge Parameters

Water Resources and Environmental Depth Reference Manual **(CEWE):** Study these sections in CEWE that either relate directly to this subtopic or provide background information.

- Section 8.4: Wastewater Treatment Plant
- Section 8.7: Primary and Secondary Treatment Processes
- Section 9.10: Food to Microorganism Ratio
- Section 9.13: BOD Removal Efficiency
- Section 9.16: Sludge Mass and Volume

Codes and standards: To prepare for this subtopic, familiarize yourself with these sections from codes and standards.

Recommended Standards for Wastewater Facilities (TSS)

- Chapter 70: Settling

The following equations, figures, and concepts are relevant for subtopic 14.F, Secondary Treatment.

Clarifiers

Handbook: Clarifiers: Typical Primary Clarifier Efficiency Percent Removal for Overflow Rates; Clarifiers: Weir Loadings; Stokes's Law; Clarifiers

$$\text{hydraulic loading rate} = \frac{Q}{A}$$

$$\text{hydraulic residence time} = \frac{V}{Q}$$

$$\text{WOR} = \frac{Q}{\text{weir length}}$$

CEWE: Sec. 8.4, Sec. 8.7

$$q_o = \frac{Q}{A_s} \quad \quad 8.2$$

$$t_d = \frac{V}{Q} \quad \quad 8.3$$

$$w_r = \frac{Q}{L} \quad \quad 8.4$$

TSS: Chap. 70

In general, there are three types of clarifiers: physical, chemical, and biological. Physical methods simply slow down the flow to allow dense particles to settle to the bottoms of basins or tanks. Chemical methods use coagulants that lump particles together so that they can easily settle. Coagulation is typically followed by flocculation, where slow mixing encourages larger particles (floc) to form and finally settle. *Handbook* table Clarifiers: Typical Primary Clarifier Efficiency Percent Removal for Overflow Rates provides overflow rates.

Clarifiers can be circular, as shown in CEWE Fig. 8.3, or rectangular, as shown in CEWE Fig. 8.7. Note how weirs are used to control the velocity of the fluid moving through the clarifier.

CEWE Fig. 8.2 shows a general schematic of a wastewater treatment plant. Clarifiers are designed to give enough time for solids to be removed from wastewater. Hydraulic residence time is therefore critical to allow for solids to either settle or coagulate and settle.

TSS Chap. 70 covers requirements and guidelines for settling units. TSS Sec. 72.21, Sec. 72.22, and Sec. 72.23 cover specifications for primary, intermediate, and final settling tanks, respectively. Recommendations for weir overflow rates are included in TSS Sec. 72.232. TSS Sec. 72.43 gives weir design rates.

Exam tip: Some exam problems may involve clarifiers with more than one weir, and some may require students to find the number of weirs needed. All the units used in the formulas relate to each other. Conversions are required from milligrams to pounds of mass, from liters to cubic feet, and from hours to seconds.

While CERM does not specifically discuss biological treatment technologies for nitrogen, CERM Chap. 30 does discuss the activated sludge process in detail. TSS Sec. 92.63 and Sec. 92.64 summarize biological nitrogen removal processes. TSS Sec. 72.232 covers design criteria for sludge settling tanks. TSS Section 73.2 summarizes the process of sludge removal. TSS Sec. 85.31 includes information on the required volume for digestion tanks.

CEWE Fig. 8.13 shows a typical scheme for nitrification and denitrification in wastewater treatment systems.

Exam tip: Nitrification is a form of aerobic digestion, and it requires oxygen to be completed. In contrast, denitrification is an anoxic process that does not require oxygen and is driven by denitrifying bacteria, which require an environment with low oxygen levels. As a result, these processes occur in separate steps within the overall wastewater treatment system.

Overall Nitrification Reaction

Handbook: Overall Nitrification Reaction

$$NH_4^+ + 1.863O_2 + 0.098CO_2$$
$$\rightarrow 0.0196C_5H_7NO_2 + 0.98NO_3^-$$
$$+ 0.941H_2O + 1.98H^+$$

CERM: Sec. 28.28

$$NH_4^+ + 2O_2 \rightarrow NO_3^- + H_2O + 2H^+ \quad 28.45$$

CEWE: Sec. 8.8

$$2NH_4^+ + 3O_2 \rightarrow 2NO_2^- + 4H^+ + 2H_2O \quad 8.8$$

$$2NO_2^- + O_2 \rightarrow 2NO_3^- \quad 8.9$$

TSS: Sec. 92.31

Handbook section Overall Nitrification Reaction shows the nitrification process as a single step and includes the microbial consumption of carbon dioxide, which produces additional cellular biomass.

CERM Eq. 28.45 and CEWE Eq. 8.8 and Eq. 8.9 remove this component because CO_2 is taken up from the atmosphere, the increase in biomass is relatively small, and the increase in biomass does not impact the stoichiometry of the remaining compounds in the equation.

CERM Sec. 28.28 briefly discusses nitrification and includes the chemical formula for this process in CERM Eq. 28.45. This formula does not show the intermediate step of ammonia conversion into nitrite. CEWE Sec. 8.8 provides a more in-depth discussion of the nitrification process and breaks it down into two chemical formulas: CEWE Eq. 8.8 shows the conversion of ammonia into nitrite and CEWE Eq. 8.9 shows the conversion of nitrite to nitrate.

CEWE Fig. 8.11 shows the stages of the nitrification and denitrification processes in typical wastewater treatment plants.

TSS Sec. 92.31 covers capacities and permissible loadings for aeration tanks in relation to other factors, including treatment plant size and degree of treatment required.

Exam tip: In the *Handbook*, this reaction contains an error, giving $0.0941H_2O$ instead of the correct $0.941H_2O$. Using the incorrect value on an exam problem may cause the calculated answer to be off by one order of magnitude.

Denitrification Reaction

Handbook: Biological Denitrification Reaction

$$NO_3^- \rightarrow NO_2^- \rightarrow NO \rightarrow N_2O \rightarrow N_2$$

$$C_{10}H_{19}O_3N + 10NO_3^-$$
$$\rightarrow 5N_2 + 10CO_2 + 3H_2O + NH_3 + 10OH^-$$
$$\text{[wastewater]}$$

$$5CH_3OH + 6NO_3^-$$
$$\rightarrow 3N_2 + 5CO_2 + 7H_2O + 6OH^- \quad \text{[methanol]}$$

$$5CH_3OOH + 8NO_3^-$$
$$\rightarrow 4N_2 + 10CO_2 + 6H_2O + 8OH^- \quad \text{[acetate]}$$

CERM: Sec. 28.28

$$2NO_3^- + \text{organic matter} \rightarrow N_2 + CO_2 + H_2O \quad 28.46$$

CEWE: Sec. 8.8

$$5CH_3OH + 6NO_3^- \quad \quad \quad 8.10$$
$$\rightarrow 3N_2 + 5CO_2 + 7H_2O + 6OH^-$$

$$5CH_3COOH + 8NO_3^- \quad \quad \quad 8.11$$
$$\rightarrow 4N_2 + 10CO_2 + 6H_2O + 8OH^-$$

Handbook section **Biological Denitrification Reaction** shows the pathway of nitrate to nitrogen gas. It also includes chemical formulas for the denitrification process in the presence of wastewater, methanol, and acetate. CERM Eq. 28.46 is a general denitrification formula with a generic source of organic carbon. In contrast, the *Handbook* and CEWE show equations with specific organic carbon sources, so the stoichiometric balance of these equations is different from that in CERM. CEWE Eq. 8.10 and Eq. 8.11 show denitrification using methanol and acetate as a carbon donor, respectively.

As shown in CERM Eq. 28.46, denitrification is the microbial decomposition of nitrate, NO_3, to nitrogen gas in the absence of oxygen and in the presence of organic carbon by facultative heterotrophic bacteria. By-products of this process include carbon dioxide and water. CERM Sec. 27.4 explains the denitrification process in wastewater treatment, which occurs separately from nitrification in wastewater treatment plants. CEWE Sec. 8.8 also includes a helpful explanation of the denitrification process. CEWE Fig. 8.13 shows a schematic of a typical nitrification and denitrification process in wastewater treatment plants.

Exam tip: Denitrification occurs in several wastewater treatment processes, including continuous flow stirred reactors, activated sludge systems, fixed film reactors, and fluidized bed biofilm reactors.

Denitrification Rate

Handbook: Rate of Denitrification

$$U'_{DN} = U_{DN} \cdot 1.09^{(T-20)}(1 - DO)$$

Handbook section **Rate of Denitrification** includes an equation used to calculate the overall denitrification rate based on the specific nitrification rate, wastewater temperature, and concentrations of dissolved oxygen in the wastewater.

Exam tip: The specific denitrification rate is based on the weight ratio of nitrogen, as nitrate, to the volatile material in the mixed liquor, otherwise known as the *mixed liquor volatile suspended solids* (MLVSS).

Ammonia Removal: Air Stripping

Handbook: Air Stripping: Mass Balance; Henry's Law; Air Stripper Packing Height; Minimum Air to Water Ratio; Stripping Factor; Stripping Factor: Gas Phase Equilibrium

CERM: Sec. 29.29

CEWE: Sec. 8.8

TSS: Sec. 92.33, Sec. 92.331

Air stripping is a chemical process for the removal of ammonia. In this method, lime is added to water to increase its pH to about 10. This causes the ammonium ions to change to dissolved ammonia gas. The water is passed through a packed tower into which air is blown at high rates. The air strips the ammonia gas out of the water. CEWE Sec. 8.8 provides more explanation of this process. CEWE Eq. 8.7 shows the basic chemical formula for the process.

$$NH_4^+ + OH^- \rightleftharpoons H_2O + NH_3 \quad \quad \quad 8.7$$

The *Handbook* shows several equations related to the mass balance, partial pressure, and design parameters required for air stripping.

- Air Stripping: Mass Balance
- Henry's Law
- Air Stripper Packing Height
- Minimum Air to Water Ratio
- Stripping Factor
- Stripping Factor: Gas Phase Equilibrium

Review TSS Sec. 92.33 and Sec. 92.331 for aeration equipment and nitrification process requirements.

Exam tip: While the preceding *Handbook* sections are not dedicated specifically to ammonia removal, the equations can be applied to ammonia stripping processes.

H. PHOSPHORUS REMOVAL

Key concepts: These key concepts are important for answering exam questions in subtopic 14.H, Phosphorus Removal.

- biological phosphorus removal
- chemical phosphorus precipitation and removal
- phosphorus concentrations
- phosphorus removal

NCEES Handbook: To prepare for this subtopic, familiarize yourself with these sections in the *Handbook*.

- Data on Selected Elements, Radicals, and Compounds
- Activated Sludge Treatment: Phoredox, VIP
- Activated Sludge Treatment: A_2O, UCT
- Activated Sludge Treatment: Bardenpho
- Biological Phosphorus Removal (BPR) Process
- Chemical Phosphorus Removal
- Chemical Phosphorous Precipitation

PE Civil Reference Manual **(CERM):** Study these sections in CERM that either relate directly to this subtopic or provide background information.

- Section 25.12: Phosphorus
- Section 28.13: Wastewater Characteristics
- Section 29.28: Phosphorus Removal: Precipitation
- Section 30.1: Sludge
- Section 30.2: Activated Sludge Process

Water Resources and Environmental Depth Reference Manual **(CEWE):** Study these sections in CEWE that either relate directly to this subtopic or provide background information.

- Section 8.8: Tertiary Treatment Process

Codes and standards: To prepare for this subtopic, familiarize yourself with these sections from codes and standards.

Recommended Standards for Wastewater Facilities (TSS)

- Section 72.232: Final Settling Tanks - Activated Sludge
- Section 92.62: Biological Phosphorus Removal
- Section 92.64: Combined Biological Nitrogen and Phosphorus Removal

- Section 92.7: Sequencing Batch Reactors
- Section 111: Phosphorus Removal by Chemical Treatment
- Section 112: High Rate Effluent Filtration
- Appendix: Handling and Treatment of Septage at a Wastewater Treatment Plant

The following equations, figures, tables, and concepts are relevant for subtopic 14.H, Phosphorus Removal.

Phosphorus Concentration

Handbook: Data on Selected Elements, Radicals, and Compounds

CERM: Sec. 25.12, Sec. 28.13, Sec. 29.28

CEWE: Sec. 8.8

High phosphorus concentrations lead to increases in algal blooms, which consume oxygen in waterways and lead to eutrophication. Phosphorus loading is an important consideration for successful wastewater treatment. CERM Table 28.4 shows the range of organic, inorganic, and total phosphorus concentrations common in strong and weak domestic wastewater. Most phosphorus in untreated wastewater originates from synthetic detergents and human waste. CERM Eq. 25.9 calculates the concentration of phosphorus in mg/L.

$$[X] = \frac{[P](MW_X)}{30.97} \qquad 25.9$$

CEWE Table 8.11 indicates that, once removed through biological and chemical treatment processes, wastewater effluent contains less than 1 mg/L of phosphorus.

Table 8.11 Typical Concentrations of Nitrogen and Phosphorus in Various Stages of Treatment

treatment	total nitrogen (mg/L)	total phosphorus (mg/L)
no treatment (raw wastewater)	15–100 (typically 50)	4–15 (typically 10)
primary treatment	40	7
secondary treatment	25–30	6
biological N removal	5–8	–
biological P removal	–	< 1

Exam tip: Total phosphorus consists of organic and inorganic phosphorus in the system. CERM Eq. 25.9 calculates the concentration of phosphorus in water as a product of its molar concentration and the molecular weight of the compound divided by the atomic weight of phosphorus. The molecular or atomic weights of several elements, including phosphorus, can be found in the *Handbook* table Data on Selected Elements, Radicals, and Compounds.

Biological Phosphorus Removal

Handbook: Activated Sludge Treatment: Phoredox, VIP; Activated Sludge Treatment: A_2O, UCT; Activated Sludge Treatment: Bardenpho; Biological Phosphorus Removal (BPR) Process

CERM: Sec. 30.1, Sec. 30.2

CEWE: Sec. 8.8

TSS: Sec. 72.232

Biological phosphorus removal occurs as part of an activated sludge treatment process, where wastewater is mixed with sludge and aerated. This mixture also includes a large concentration of aerobic bacteria that consume organic matter within the wastewater-sludge mixture. This process is the primary vehicle for biological phosphorus removal.

The *Handbook* provides more information about specific technologies used for biological phosphorus removal. *Handbook* table Biological Phosphorus Removal (BPR) Process provides design ratios of phosphorus to biochemical oxygen demand (BOD) and chemical oxygen demand (COD) for several biological phosphorus removal technologies, including Phoredox, anaerobic/anoxic/oxic, and Bardenpho systems. It also includes a recommended solids retention time for the design of each system. *Handbook* figures Activated Sludge Treatment: Phoredox, VIP, Activated Sludge Treatment: A_2O, UCT, and Activated Sludge Treatment: Bardenpho show the schematics of the activated sludge treatment processes cited in *Handbook* table Biological Phosphorus Removal (BPR) Process.

TSS Sec. 72.232 contains design guidelines on meeting thickening and solids separation requirements for activated sludge settling tanks.

Exam tip: Be comfortable sizing systems for phosphorus removal using each of the technologies listed in *Handbook* table Biological Phosphorus Removal (BPR) Process and be familiar with the activated sludge treatment process described in the *Handbook*.

Phosphorus Removal

Handbook: Biological Phosphorus Removal (BPR) Process; Chemical Phosphorus Removal; Chemical Phosphorous Precipitation

CERM: Sec. 25.12, Sec. 29.28

CEWE: Sec. 8.8

TSS: Sec. 92.62, Sec. 92.64, Sec. 111

Phosphorus is a nutrient found throughout the biosphere and in soil. It has various commercial and industrial uses and is a common contaminant known to drive eutrophication in lakes and other natural waterways. Phosphorus and phosphate are more common contaminants of concern in wastewater treatment than in drinking water. Phosphorus can be removed through settling, biological treatment (activated sludge), and chemically induced precipitation in a clarifier or settling basin using alum, lime, ferrous chloride, or ferric chloride.

CERM Sec. 25.12 and Sec. 29.28 discuss phosphorus and its removal in wastewater in more detail. CEWE Sec. 8.8 provides additional information. TSS Sec. 92.62, Sec. 92.64, and Sec. 111 provide information on the process requirements, feed systems, storage facilities, and hazardous chemical handling for phosphorus removal by chemical treatment. Finally, *Handbook* sections Chemical Phosphorus Removal and Chemical Phosphorous Precipitation provide an overview of key equations related to phosphorus removal. See the entry "Chemical Phosphorus Precipitation and Removal" in this chapter for a discussion of those equations.

Exam tip: Many wastewater treatment plants remove phosphorus in two steps: biological uptake occurs through activated sludge, and then the remaining phosphorus is chemically precipitated and removed.

Chemical Phosphorus Precipitation and Removal

Handbook: Chemical Phosphorus Removal; Chemical Phosphorous Precipitation

For phosphorus removal,

$$FeCl_3 + PO_4^{3-} \rightarrow FePO_4(\downarrow) + 3Cl^- \quad \text{[ferric chloride]}$$

$$3FeCl_2 + 2PO_4^{3-} \rightarrow Fe_3(PO_4)_2(\downarrow) + 6Cl^- \quad \text{[ferrous chloride]}$$

$$Al_2(SO_4)_3 \cdot 14H_2O + 2PO_4^{3-} \rightarrow 2AlPO_4(\downarrow) + 3SO_4^{2-} + 14H_2O \quad \text{[alum]}$$

For phosphate precipitation,

$$Al^3 + H_nPO_4^{3-n} \leftrightarrow AlPO_4 + nH^+ \quad \text{[with aluminum]}$$

$$Fe^{3+} + H_nPO_4^{3-n} \leftrightarrow FePO_4 + nH^+ \quad \text{[with iron]}$$

$$10Ca^{2+} + 6PO_4^{3-} + 2OH^-$$
$$\leftrightarrow Ca_{10}(PO_4)_6(OH)_2 \quad \text{[with calcium]}$$

CERM: Sec. 29.28

$$Al_2(SO_4)_3 + 2PO_4 \rightleftharpoons 2AlPO_4 + 3SO_4 \quad 29.21$$

$$FeCl_3 + PO_4 \rightleftharpoons FePO_4 + 3Cl \quad 29.22$$

CEWE: Sec. 8.8

Soluble phosphorus is removed by precipitation and settling. Aluminum sulfate (alum), ferric chloride ($FeCl_3$), and lime may be used depending on the nature of the phosphorus radical.

Chemical phosphorus precipitation and removal typically occur after activated sludge treatment via the use of precipitating chemicals during clarification.

Handbook sections Chemical Phosphorus Removal and Chemical Phosphorous Precipitation show the chemical reactions of ferric chloride, ferrous chloride, alum, and calcium (as lime) with phosphate. The formulas for ferric chloride and alum are essentially the same as CERM Eq. 29.21 and Eq. 29.22. CERM Eq. 29.21 and Eq. 29.22 are two of the common chemical reactions used to precipitate and remove phosphorus. CERM Eq. 29.22 does not include water molecules in the reaction of phosphate with alum, whereas the *Handbook* formula does. See CEWE Sec. 8.8 for more information on phosphorus removal.

Exam tip: Due to the many other possible reactions the phosphorus removal compounds can participate in, the dosage in wastewater should be determined from testing. The stoichiometric chemical reactions describe how the phosphorus is removed, but they do not accurately predict the quantities of coagulants needed.

In the *Handbook*, *phosphorus* is misspelled in the section title Chemical Phosphorous Precipitation. Searching on "Chemical Phosphorus" will still get you to the right page.

I. SOLIDS TREATMENT, HANDLING, AND DISPOSAL

Key concepts: These key concepts are important for answering exam questions in subtopic 14.I, Solids Treatment, Handling, and Disposal.

- sludge dewatering
- sludge disposal
- sludge quantity
- sludge treatment

NCEES Handbook: To prepare for this subtopic, familiarize yourself with these sections in the *Handbook*.

- Design and Operational Parameters for Activated-Sludge Treatment of Municipal Wastewater
- Solids Treatment, Handling, and Disposal

PE Civil Reference Manual **(CERM):** Study these sections in CERM that either relate directly to this subtopic or provide background information.

- Section 30.15: Quantities of Sludge
- Section 30.16: Sludge Thickening
- Section 30.17: Sludge Stabilization
- Section 30.18: Aerobic Digestion
- Section 30.19: Anaerobic Digestion
- Section 30.22: Sludge Dewatering
- Section 30.23: Sludge Disposal

Water Resources and Environmental Depth Reference Manual **(CEWE):** Study these sections in CEWE that either relate directly to this subtopic or provide background information.

- Section 9.16: Sludge Mass and Volume
- Section 9.17: Sludge Processing
- Section 9.18: Sludge Stabilization
- Section 9.20: Sludge Thickening
- Section 9.21: Sludge Dewatering
- Section 9.24: Sludge Disposal

Codes and standards: To prepare for this subtopic, familiarize yourself with these sections from codes and standards.

Recommended Standards for Wastewater Facilities (TSS)

- Section 88: Sludge Dewatering
- Section 89.3: Disposal

The following equations, figures, tables, and concepts are relevant for subtopic 14.I, Solids Treatment, Handling, and Disposal.

Sludge Treatment

Handbook: Design and Operational Parameters for Activated-Sludge Treatment of Municipal Wastewater

CERM: Sec. 30.16, Sec. 30.17, Sec. 30.18, Sec. 30.19, Sec. 30.22, Sec. 30.23

CEWE: Sec. 9.17, Sec. 9.18, Sec. 9.20, Sec. 9.21

There are four steps in the treatment, handling, and disposal of solids: thickening, digestion, dewatering/drying, and disposing of solids. In *thickening*, the water content of the sludge is reduced through either physical or chemical applications. Typical values for thickening are given in CEWE Table 9.4.

If an improvement to the quality of the sludge is needed, then digestion follows. Digestion is the principal process, as this is the only step in which the quality of the sludge is improved. After digestion (if needed), dewatering/drying is the third step in sludge treatment. *Dewatering* or *drying* dries the sludge to make it easier and cheaper to handle for disposal. *Handbook* table Design and Operational Parameters for Activated-Sludge Treatment of Municipal Wastewater provides parameters for various sludge treatment processes.

Disposal of solids can take several forms. Depending on the level of pollutant contamination, dried sludge may be landfilled as a waste product, reused as biosolids, or otherwise have its energy recovered. For sludge to be reused as biosolids, it must meet pollutant-specific limits and drying requirements outlined in EPA regulations and state rules. The reuse of sludge as biosolids, when feasible, reduces landfill costs and is a preferred alternative for most treatment facilities.

Exam tip: Sludge production and treatment must follow the same conservation of mass rules as any other material passing through treatment. When calculating sludge mass, check to make sure that all materials, both solids and water, are accounted for at each step.

Sludge Quantity

Handbook: Solids Treatment, Handling, and Disposal

$$\frac{W_s}{S_s \rho_w} = \frac{W_f}{S_f \rho_w} + \frac{W_v}{S_v \rho_w}$$

CERM: Sec. 30.15

$$V_{\text{sludge,wet}} = \frac{m_{\text{dried}}}{s\rho_{\text{sludge}}} \approx \frac{m_{\text{dried}}}{s\rho_{\text{water}}} \quad 30.48$$

CEWE: Sec. 9.16

$$V_{\text{sludge,wet}} = \frac{m_{\text{dry}}}{s\rho_{\text{sludge}}} \quad 9.38$$

Sludge volume is best determined from its dried mass. Sludge density is approximately equal to that of water at the reported operating temperature because the sludge is mostly water containing suspended and dissolved solids.

The equations presented represent the same mass conservation principles, but they are approached in different ways. CERM Eq. 30.48 and CEWE Eq. 9.38 are the same, with the CERM equation also stating that the wet sludge volume can be calculated with either the density of the sludge or water. This is because the density of sludge is approximately the same as water, as it is approximately 97% water prior to drying.

The *Handbook* equation in section Solids Treatment, Handling, and Disposal details the principle of conservation of mass (i.e., that the weight of solids is made up of both fixed solids and volatile solids). This is important when determining how much of the solids (specifically fixed solids) will remain after treatment if volatile solids are being consumed by microorganisms.

Exam tip: The majority of sludge is water, so while it includes solids, it will have many of the same characteristics as water (i.e., density). Even after drying is completed, sludge will still be approximately 80% water.

Sludge Dewatering

CERM: Sec. 30.22

CEWE: Sec. 9.21

TSS: Sec. 88

The purpose of sludge dewatering is to reduce the volume of sludge so that it is more cost-effective to handle and dispose of. By increasing the solids content from 2% to 20%, the volume of the sludge is reduced by 90%, requiring a tenth of the disposal cost. Once the solids volume reaches 25%, the sludge is dry enough to be handled with a shovel and is referred to as *sludge cake*. To design a sludge dewatering system, four key operational controls must be known.

- type of sludge
- dry weight

- time it needs to be held for dewatering
- pre- and post-dewatering requirements of the sludge

Sludge dewatering methods include vacuum filtration using a vacuum drum filter, pressure filtration using a belt filter press, and centrifugation using a solid bow centrifuge or sand and gravel drying bed. CEWE Fig. 9.7 shows an example of a sand and gravel drying bed.

Figure 9.7 General Design of a Sludge Drying Bed

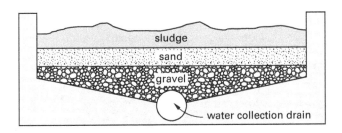

TSS Sec. 88 provides recommendations for sludge dewatering facilities, including sludge drying beds (TSS Sec. 88.2), mechanical dewatering facilities (TSS Sec. 88.3), and drainage and filtrate disposal (TSS Sec. 88.4).

Exam tip: If sludge drying includes allowing it to dry in the sun, remember that drying times depend on other environmental conditions, such as precipitation and outdoor temperature. In a temperate climate, sludge will not dry quickly during the rainy seasons or the winter. This limits when sludge can dry. However, this method is typically still more cost-effective than mechanical drying since the latter requires electricity and active maintenance.

Sludge Disposal

CERM: Sec. 30.23

CEWE: Sec. 9.24

TSS: Sec. 89.3

Sludge can take many forms: liquid sludge, sludge cake, screenings, and scum. Sludge disposal methods depend on the sludge composition and include landfilling, land applications, and incineration.

Landfilling is the least expensive disposal method. Sludge materials with high levels of pathogens are disposed of in municipal landfills so as to not spread disease, as they might in a land application. Sludge with monitored contaminants is disposed of in hazardous waste landfills. Sludge with low levels of pathogens and other contaminants can be disposed of through various land applications.

Sludge incineration is the most expensive disposal method since it requires high energy use, must meet strict air quality control requirements, and requires the disposal of ash by-products.

TSS Sec. 89.3 provides recommendations for sludge disposal.

Exam tip: Be familiar with the different disposal methods because a question based on these concepts could appear on the exam. Alternatively, a problem could provide sludge by-products with different compositions and ask you to select percentages or allocations to different disposal methods.

J. DIGESTION

Key concepts: These key concepts are important for answering exam questions in subtopic 14.J, Digestion.

- aerobic digestion
- anaerobic, aerobic, and anoxic treatments
- anaerobic digestion

NCEES Handbook: To prepare for this subtopic, familiarize yourself with these sections in the *Handbook*.

- Activated Sludge Treatment: Phoredox, VIP
- Activated Sludge Treatment: A_2O, UCT
- Activated Sludge Treatment: Bardenpho
- Activated Sludge Treatment: Bio-Chemical Sidestream
- Aerobic Design: Tank Volume
- Design Parameters for Anaerobic Digesters
- Anaerobic Digestion

PE Civil Reference Manual (**CERM**): Study these sections in CERM that either relate directly to this subtopic or provide background information.

- Section 30.18: Aerobic Digestion
- Section 30.19: Anaerobic Digestion

Water Resources and Environmental Depth Reference Manual (**CEWE**): Study these sections in CEWE that either relate directly to this subtopic or provide background information.

- Section 9.18: Sludge Stabilization

Codes and standards: To prepare for this subtopic, familiarize yourself with these sections from codes and standards.

Recommended Standards for Wastewater Facilities (TSS)

- Section 84: Anaerobic Sludge Digestion
- Section 85: Aerobic Sludge Digestion
- Section 92.3: Aeration

The following equations, figures, tables, and concepts are relevant for subtopic 14.J, Digestion.

Anaerobic, Aerobic, and Anoxic Treatment Steps

Handbook: Activated Sludge Treatment: Phoredox, VIP; Activated Sludge Treatment: A$_2$O, UCT; Activated Sludge Treatment: Bardenpho; Activated Sludge Treatment: Bio-Chemical Sidestream

CERM: Sec. 30.18, Sec. 30.19

TSS: Sec. 84, Sec. 85

In more advanced treatment processes, a mixture of aerobic, anaerobic, and anoxic treatment steps is used to target the removal of specific pollutants. *Handbook* figures Activated Sludge Treatment: Phoredox, VIP, Activated Sludge Treatment: A$_2$O, UCT, Activated Sludge Treatment: Bardenpho, and Activated Sludge Treatment: Bio-Chemical Sidestream show several different processes that can enhance treatment. Depending on the pollutant being removed, some of these processes will be more effective than others. Note that the *Handbook* diagrams are not to scale; an anaerobic step may last only 15 min and require only a small tank, while a much larger tank may be needed for an aerobic step with a hydraulic residence time of an hour or more.

TSS Sec. 84 provides information on anaerobic sludge digestion, and TSS Sec. 85 discusses aerobic sludge digestion.

Exam tip: Aerobic treatment is very common, as it provides oxygen to encourage biological treatment. Enhanced biological treatment is used to target specific nutrients that occur in excessive quantities. In some cases, the additional treatment steps are used to encourage microorganisms to accumulate these nutrients before entering the oxygen-rich environment.

Aerobic Digestion

Handbook: Aerobic Design: Tank Volume; Design Parameters for Anaerobic Digesters

$$V = \frac{Q_i(X_i + FS_i)}{X_d\left(k_d P_v + \dfrac{1}{\theta_c}\right)}$$

CERM: Sec. 30.18

CEWE: Sec. 9.18

TSS: Sec. 85, Sec. 92.3

Bacteria that require oxygen perform aerobic digestion. This requires an open holding tank and can remove up to 70% of volatile solids. CERM Table 30.9 provides typical characteristics for different parameters in aerobic digestion. Aerobic digestion is more commonly used than anaerobic digestion, as it is easier to manage. Aerobic digestion has a higher operating cost but can function at a wide range of wastewater parameters. See CERM Sec. 30.18 and CEWE Sec. 9.18 for more information on aerobic digestion. *Handbook* table Design Parameters for Anaerobic Digesters contains standard-rate and high-rate values for various digester parameters.

Aerobic digestion occurs when sufficient oxygen is present. For most treatment facilities, this requires pumping air into the treatment tank or channel to keep the treatment from becoming anaerobic or anoxic. TSS Sec. 92.3 contains guidelines on capacities and permissible loadings for aeration tanks, as well as required dimensions. TSS Sec. 85 contains information on aerobic sludge digestion, including tank capacity (TSS Sec. 85.3), air requirements (TSS Sec. 85.5), and digested sludge storage volume (TSS Sec. 85.9).

Exam tip: Aerobic digesters must be designed to be thoroughly mixed and sized to allow for thorough mixing. In exam problems on the topic, unless otherwise specified, you can assume that there are no significant dead zones that would inhibit the aerobic process.

Anaerobic Digestion

Handbook: Anaerobic Digestion; Design Parameters for Anaerobic Digesters

CERM: Sec. 30.19

CEWE: Sec. 9.18

TSS: Sec. 84

Bacteria that do not require oxygen perform anaerobic digestion. Anaerobic digestion is a three-step process: convert organic acids to fatty or amino acids, convert amino acids to organic acids such as acetic acid, and convert organic acids to methane and carbon dioxide. See CERM Sec. 30.19 and CEWE Sec. 9.18 for more information on anaerobic digestion. *Handbook* table Design Parameters for Anaerobic Digesters provides standard-rate and high-rate values for various digester parameters. *Handbook* section Anaerobic Digestion includes relevant equations for calculating standard and high rates.

Thermophilic digestion occurs when sludge is fermented at 131°F, while mesophilic digestion occurs when sludge is fermented between 86°F and 104°F.

Note the temperature ranges when the sludge is warmed. Heating up the sludge requires energy, but the anaerobic process produces methane that can be used in the process. Anaerobic bacteria take longer to digest sludge than aerobic bacteria. For this reason, larger holding tanks are needed, which leads to higher capital costs than for aerobic digestion processes. CERM Fig. 30.7 shows a simple anaerobic digester.

Figure 30.7 Simple Anaerobic Digester

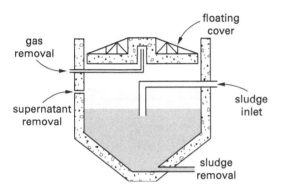

TSS Sec. 84 contains information on anaerobic sludge digestion, including tank capacity (TSS Sec. 84.3), gas collection and piping (TSS Sec. 84.4), and digested sludge production (TSS Sec. 84.7).

Exam tip: Anaerobic digestion can cause foaming if not carefully monitored for nutrient parameters. Although nutrients and oxygen are still necessary, the levels must be carefully monitored to maintain the anaerobic conditions for treatment.

K. DISINFECTION

Key concepts: These key concepts are important for answering exam questions in subtopic 14.K, Disinfection.

- chlorination
- dechlorination
- ultraviolet light

NCEES Handbook: To prepare for this subtopic, familiarize yourself with these sections in the *Handbook*.

- Chlorination
- Dechlorination
- Ultraviolet Light (UV)
- UV Dose Table for *Cryptosporidium*, *Giardia lamblia*, and Virus Inactivation Credit

PE Civil Reference Manual (**CERM**): Study these sections in CERM that either relate directly to this subtopic or provide background information.

- Section 26.37: Chlorination Chemistry
- Section 26.39: Advanced Oxidation Processes
- Section 26.41: Dechlorination
- Section 29.31: Chlorination
- Section 29.32: Dechlorination

Water Resources and Environmental Depth Reference Manual (**CEWE**): Study these sections in CEWE that either relate directly to this subtopic or provide background information.

- Section 5.9: Disinfection
- Section 8.8: Tertiary Treatment Processes

Codes and standards: To prepare for this subtopic, familiarize yourself with these sections from codes and standards.

Recommended Standards for Wastewater Facilities (TSS)

- Section 102: Chlorine Disinfection
- Section 103: Dechlorination
- Section 104: Ultraviolet Disinfection

The following equations, tables, and concepts are relevant for subtopic 14.K, Disinfection.

Chlorination

Handbook: Chlorination

CERM: Sec. 26.37, Sec. 29.31

CEWE: Sec. 5.9, Sec. 8.8

TSS: Sec. 102

Before wastewater can be discharged, it needs to be cleared of harmful pathogens. Disinfection is a process that removes pathogens prior to discharge. It destroys or deactivates pathogens by physical means, such as fine membranes or ultraviolet light, or by chemical means, such as chlorination. Physical processes, however, are often cost-prohibitive in wastewater applications because they may require fine and specialized membranes to remove pathogens.

The most common disinfection method is chlorination. Its efficiency depends on various parameters, such as turbidity, temperature, and nonorganic compounds reacting with chlorine. In water disinfection, a residual disinfectant must remain active in the water following the disinfection process.

CEWE Table 8.13 provides characteristics that affect the performance of the chlorination process. *Handbook* section Chlorination provides the chemical equations for chlorination. TSS Sec. 102 gives recommendations for chlorine disinfection in municipal wastewater systems. Topics include dosage (TSS Sec. 102.2), equipment (TSS Sec. 102.4), and sampling and control (TSS Sec. 102.6).

Exam tip: Similar to biological processes, chlorination requires a balance of chemicals for pollutants to be properly removed. Excessive chlorination can lead to the passing through of chlorine residuals, which is a public health risk at higher levels.

Dechlorination

Handbook: Dechlorination

CERM: Sec. 26.41, Sec. 29.32

CEWE: Sec. 5.9, Sec. 8.8

TSS: Sec. 103

The process of dechlorination removes residual chlorine, as excess chlorine is harmful to aquatic ecosystems. Dechlorination can be accomplished by chemical additives such as sulfur dioxide, sodium bisulfite, sodium metabisulfite, and hydrogen peroxide, or by carbon adsorption. Carbon adsorption is expensive, however, and is only used when all residual chlorine is to be removed.

Handbook section Dechlorination provides the chemical equations for dechlorination. TSS Sec. 103 gives recommendations for dechlorination of wastewater effluent. Topics include dosage (TSS Sec. 103.2), feed equipment (TSS Sec. 103.4), and sampling and control (TSS Sec. 103.6).

Exam tip: Different dechlorination processes can produce different by-products. In addition, the amount of dechlorination chemicals used must be balanced against the chlorination by-products being removed.

Ultraviolet Light

Handbook: Ultraviolet Light (UV); UV Dose Table for *Cryptosporidium*, *Giardia lamblia*, and Virus Inactivation Credit

$$D = It$$

CERM: Sec. 26.39

CEWE: Sec. 5.9

TSS: Sec. 104

Ultraviolet (UV) light treatment is used to inactivate organic contaminants in wastewater. In general, UV light is only effective if the wastewater is close to the light source (approximately 7 cm). Beyond that distance, the UV light is not as effective, and microorganisms may pass through to sludge or a local waterway to cause health risks.

UV radiation is very effective at inactivating *Cryptosporidium*. Its effectiveness at inactivating microorganisms is a function of both the intensity of the radiation and the exposure time. UV radiation damages the nucleic acid material in DNA and RNA and inhibits cell reproduction. The primary disadvantage is the absence of any residual disinfection for downstream protection. *Handbook* table UV Dose Table for *Cryptosporidium*, *Giardia lamblia*, and Virus Inactivation Credit contains log credits for various dosages.

TSS Sec. 104 gives recommendations for ultraviolet disinfection, including lamp type, channel design, transmittance, and dosage.

Exam tip: Ultraviolet light treatment is only effective with sufficient exposure. Note the distance at which water passes by the light source and how long it is exposed to light. In most cases, the light source is not a single point but a lamp that can be used to expose the water for a long duration.

L. ADVANCED TREATMENT

Key concepts: These key concepts are important for answering exam questions in subtopic 14.L, Advanced Treatment.

- ammonia removal
- filtration
- final clarifiers

NCEES Handbook: To prepare for this subtopic, familiarize yourself with these sections in the *Handbook*.

- Overall Nitrification Reaction
- Biological Denitrification Reaction
- Rate of Denitrification
- Typical Removal Credits and Inactivation Requirements for Various Treatment Technologies

***PE Civil Reference Manual* (CERM):** Study these sections in CERM that either relate directly to this subtopic or provide background information.

- Section 29.19: Trickling Filters
- Section 29.20: Two-Stage Trickling Filters
- Section 29.27: Final Clarifiers

- Section 29.29: Ammonia Removal: Air Stripping
- Section 30.4: Final Clarifiers

Water Resources and Environmental Depth Reference Manual (**CEWE**): Study these sections in CEWE that either relate directly to this subtopic or provide background information.

- Section 8.4: Wastewater Treatment Plant
- Section 8.8: Tertiary Treatment Processes

Codes and standards: To prepare for this subtopic, familiarize yourself with these sections from codes and standards.

Recommended Standards for Wastewater Facilities (**TSS**)

- Section 92.63: Biological Nitrogen Removal
- Section 112: High Rate Effluent Filtration

The following equations, figures, tables, and concepts are relevant for subtopic 14.L, Advanced Treatment.

Ammonia Removal

Handbook: Overall Nitrification Reaction; Biological Denitrification Reaction; Rate of Denitrification

CERM: Sec. 29.29

CEWE: Sec. 8.8

TSS: Sec. 92.63

Ammonia is a contaminant that can contribute to nitrogen pollution in wastewater. As a result, it must be removed through either a biological or physical process. *Handbook* sections Overall Nitrification Reaction, Biological Denitrification Reaction, and Rate of Denitrification contain equations applicable to nitrification and denitrification.

Ammonia can be removed by either biological processes or air stripping. In biological processes, ammonia is converted by bacteria first to nitrite and then to nitrate, which is then converted to nitrogen gas by denitrification. In air stripping, lime is added to increase pH, which causes the ammonia ions to become dissolved ammonia gas. This is then passed through a packed tower with air blown through at high rates, stripping the water of the dissolved nitrogen gas. Excess lime is then removed in a recarbonation process. See TSS Sec. 92.63 for general information on biological nitrification to remove ammonia.

Exam tip: When designing ammonia removal processes, both biological processes and air stripping can be used, first to remove a large portion and then to finish with a secondary process to remove any additional ammonia that could not be removed by the alternate treatment step.

Filtration

Handbook: Typical Removal Credits and Inactivation Requirements for Various Treatment Technologies

CERM: Sec. 29.19, Sec. 29.20

CEWE: Sec. 8.8

TSS: Sec. 112

Filtration is a process that removes flocculated materials that have escaped secondary treatment processes. Uniform filters use the same material throughout, usually poorly graded sand. Multi-media filters use different material gradation. *Handbook* table Typical Removal Credits and Inactivation Requirements for Various Treatment Technologies gives typical removal credits and inactivation requirements for various treatment processes.

CEWE Fig. 8.9 illustrates the concept of filter stratification, showing that denser materials settle faster than lighter ones.

TSS Sec. 112 contains information on filtration, including design considerations (TSS Sec. 112.12), filter types (TSS Sec. 112.2), backwash (TSS Sec. 112.4), and filter media selection (TSS Sec. 112.5).

Exam tip: In filtration, heavier and larger solids will be removed first, with smaller and lighter solids being removed at each step. When designing a filter, the design should take this into account so as to avoid clogging and the need for more regular maintenance.

Final Clarifiers

CERM: Sec. 29.27, Sec. 30.4

CEWE: Sec. 8.4

Final clarifiers are commonly used as the first step in a tertiary treatment process. They capture sludge and return it to secondary treatment trains. They can also be used as settling basins within different tertiary treatment processes. CEWE Fig. 8.2 shows a general schematic of a wastewater treatment plant.

Exam tip: The purpose of final clarifiers is to remove solids, not biological pollutants. Final clarifiers can be designed in a similar way to intermediate clarifiers but can have lower loading rates.

51 Water Quality

Content in blue refers to the *NCEES Handbook*.

A. Stream Degradation.. 51-1
B. Oxygen Dynamics.. 51-2
C. Total Maximum Daily Load............................. 51-7
D. Biological Contaminants 51-8
E. Chemical Contaminants 51-11

The knowledge area of Water Quality makes up between three and five questions out of the 80 questions on the PE Civil Water Resources and Environmental exam. The organization of this chapter follows the order of subtopics given by NCEES for this knowledge area. Each subtopic is covered in the following sections.

A. STREAM DEGRADATION

Key concepts: These key concepts are important for answering exam questions in subtopic 15.A, Stream Degradation.

- minimum base flow
- minimum environmental flow
- National Pollutant Discharge Elimination System
- physical parameters
- solids
- thermal, biological, and chemical requirements

NCEES Handbook: To prepare for this subtopic, familiarize yourself with these sections in the *Handbook*.

- Revised Universal Soil Loss Equation

PE Civil Reference Manual **(CERM):** Study these sections in CERM that either relate directly to this subtopic or provide background information.

- Section 22.22: pH and pOH
- Section 25.15: Turbidity
- Section 25.16: Solids

Water Resources and Environmental Depth Reference Manual **(CEWE):** Study these sections in CEWE that either relate directly to this subtopic or provide background information.

- Section 1.13: Sedimentation
- Section 1.14: Erosion

- Section 6.24: Stream Degradation
- Section 8.1: National Pollutant Discharge Elimination System

The following tables and concepts are relevant for subtopic 15.A, Stream Degradation.

Sediment

Handbook: Revised Universal Soil Loss Equation

CERM: Sec. 25.15, Sec. 25.16

CEWE: Sec. 1.13, Sec. 1.14

Erosion and sediment transport are two ways watershed characteristics directly impact instream water quality and stream degradation. Sediment transport within a stream system influences the biological and chemical quality of the system and directly impacts system requirements related to drinking water influents. CERM Sec. 25.15 and Sec. 25.16 discuss turbidity and solids within water supplies, and CEWE Sec. 1.13 and Sec. 1.14 discuss solids loading and erosion in the context of hydrology. In natural water systems, healthy solids loading will be watershed specific.

The *Handbook* does not include much information about total solids within stream systems, but *Handbook* section Revised Universal Soil Loss Equation includes equations related to erosion in watersheds, which directly contributes to solids within a stream system. The revised universal soil loss equation is discussed in more detail in Chap. 45 and Chap. 48 of this book.

Exam tip: Because the total solids and total suspended solids within a natural water system are related to mass balance within the entire watershed, water quality parameters traditionally used for water and wastewater systems will not apply.

Stream Degradation

CEWE: Sec. 6.24

Stream and other source water degradation occur when watersheds are exposed to an overabundance of pollution, when diversions reduce water quantity to a level that impairs the ecosystem, or when dams and other structures change fundamental physical characteristics of the stream. Stream degradation can manifest as changes in physical parameters such as turbidity (clarity), temperature, pH, solids contents, and biological or chemical oxygen demand. Degradation can also be evident in changes to the ecosystem, such as the increase or reduction of indicator organisms or the presence of algal blooms.

See subtopic 15.D, Biological Contaminants, and subtopic 15.E, Chemical Contaminants, in this chapter for an in-depth review on how biochemical and chemical oxygen demand can impact water.

Exam tip: Other entries in this chapter cover stream degradation parameters that are also used in drinking water and wastewater quality assessment. For this entry, understand the overall parameters that lead to waterway degradation.

Minimum Base Flow

CEWE: Sec. 6.24

The *minimum base flow* is the minimum flow required to sustain the ecosystems within a stream and to maintain water rights for downstream users. The minimum base flow will depend greatly on the characteristics of the watershed, the ecosystem's requirements, and the historic water rights.

According to water rights analysis, the minimum flow of water in a stream must be at least the amount required to ensure downstream water uses are preserved, which yields an arithmetic calculation. For environmental uses, the minimum flow depends on seasonal aquatic life and biological cycles such as spawning periods being supported, which may yield a more complicated analysis that also varies by season.

Exam tip: The base flow is related to the actual base flows within a stream prior to a precipitation event, which may be higher or lower than the minimum base flows required to meet the environmental and water rights needs within the watershed.

Physical Requirements

CERM: Sec. 22.22

CEWE: Sec. 1.13, Sec. 8.1

Physical properties in streams include temperature, turbidity, total solids, total suspended solids (TSS), and pH. The specific requirements to maintain aquatic life will depend greatly on the ecosystem and the indicator organisms used to assess ideal water quality parameters.

CEWE Table 8.1 includes wastewater effluent standards as specified by the National Pollutant Discharge Elimination System (NPDES), which manages the impact of point source pollution on instream flows within the United States. These requirements provide some indication of what healthy instream flows may require in most temperate environments; however, local watersheds may be more sensitive to some parameters than others.

Table 8.1 NPDES Wastewater Effluent Standards

parameter	maximum permitted
BOD_5 (mg/L)	30 *
oil and grease or TPH (mg/L)	15–55
total suspended solids (mg/L)	30–45
pH	6.0–9.0
temperature (°C)	< 40
color (color units)	2.0
NH_3/NO_3 (mg/L)	1.0–10
phosphates (mg/L)	0.2
heavy metals (mg/L)	0.1–5.0
surfactants (mg/L)	0.5–1.0 total
sulfides (mg/L)	0.01–0.1
phenol (mg/L)	0.1–1.0
toxic organics (mg/L)	1.0 total
cyanide (mg/L)	0.1

*30 day average value; the arithmetic mean of the BOD_5 values (by concentration) for effluent samples collected over a period of thirty consecutive days shall not exceed 15% of the arithmetic mean (by concentration) for influent samples collected at approximately the same time during the same period (Clean Water Act).

Exam tip: It is likely that any questions on the test that focus on instream water quality parameters will include any standards required to answer the question. Neither the *Handbook* nor the codes supplied in the exam include any standards for instream water quality. However, it may help to be familiar with the basic standards for TSS, pH, and temperature listed in CEWE Table 8.1.

B. OXYGEN DYNAMICS

Key concepts: These key concepts are important for answering exam questions in subtopic 15.B, Oxygen Dynamics.

- biochemical oxygen demand (BOD)
- chemical oxygen demand (COD)
- critical oxygen deficit
- deoxygenation
- initial oxygen deficit
- oxygen sag curves

- oxygen saturation
- reaction rate coefficient
- reoxygenation
- Streeter-Phelps equation
- theoretical oxygen demand (ThOD)

NCEES Handbook: To prepare for this subtopic, familiarize yourself with these sections in the *Handbook*.

- BOD Testing/Sampling
- BOD Exertion
- Typical Values for the BOD Rate Constant
- Stream Modeling—Streeter Phelps
- Oxygen Saturation
- Oxygen Deficit After Mixing
- Initial Deficit
- Deoxygenation Rate
- Stream Reaeration Rate
- Stream Oxygen Deficit

PE Civil Reference Manual (**CERM**): Study these sections in CERM that either relate directly to this subtopic or provide background information.

- Section 22.18: Solutions of Gases in Liquids
- Section 27.4: Microorganisms
- Section 28.16: Dissolved Oxygen in Wastewater
- Section 28.17: Reoxygenation
- Section 28.18: Deoxygenation
- Section 28.19: Tests of Wastewater Characteristics
- Section 28.20: Biochemical Oxygen Demand
- Section 28.21: Seeded BOD
- Section 28.22: Dilution Purification
- Section 28.23: Response to Dilution Purification
- Section 28.24: Chemical Oxygen Demand

Water Resources and Environmental Depth Reference Manual (**CEWE**): Study these sections in CEWE that either relate directly to this subtopic or provide background information.

- Section 6.16: Dissolved Oxygen
- Section 6.17: Oxygen Demand
- Section 6.19: Reoxygenation
- Section 6.20: Deoxygenation
- Section 6.21: Oxygen Sag Curve
- Section 11.17: Volatilization/VOC Removal

The following equations, figures, tables, and concepts are relevant for subtopic 15.B, Oxygen Dynamics.

Biochemical Oxygen Demand (BOD)

Handbook: BOD Testing/Sampling; BOD Exertion

$$\text{BOD} = \frac{D_1 - D_2}{P}$$

$$\text{BOD} = \frac{(D_1 - D_2) - (B_1 - B_2)f}{P}$$

$$\text{BOD}_t = L_0(1 - e^{-kt})$$

CERM: Sec. 27.4, Sec. 28.19, Sec. 28.20, Sec. 28.21

$$\text{BOD}_5 = \frac{\text{DO}_i - \text{DO}_f}{\dfrac{V_{\text{sample}}}{V_{\text{sample}} + V_{\text{dilution}}}} \quad 28.30$$

$$\begin{aligned}\text{BOD}_t &= \text{BOD}_u(1 - 10^{-K_d t}) \\ &= \text{BOD}_u(1 - e^{-K_d' t})\end{aligned} \quad 28.31$$

$$\text{BOD} = \frac{\text{DO}_i - \text{DO}_f - x(\text{DO}_i^* - \text{DO}_f^*)}{\dfrac{V_{\text{sample}}}{V_{\text{sample}} + V_{\text{dilution}}}} \quad 28.34$$

CERM Eq. 28.30 can be used to determine the BOD after a five-day incubation period. The test parameters include adding the sample to a volume of relatively clean water.

CERM Eq. 28.30 and the first *Handbook* equation are equivalent. CERM Eq. 28.30 includes all the volumetric parameters to calculate the ratio of wastewater to dilution water, whereas the *Handbook* equation uses a single variable, P. The *Handbook* equation for seeded BOD included in the same section is the same as CERM Eq. 28.34, with B_1, B_2, and f being used in place of DO_i, DO_f, and x, respectively. Additionally, *Handbook* section BOD Exertion shows an equation equivalent to CERM Eq. 28.31, with the variable defined as L_0 instead of BOD_u. This same section of the *Handbook* includes an equation for the oxygen demand in the stream over time.

CERM Eq. 28.31 can be used to calculate the BOD exertion or the oxygen demand in the stream over time. The BOD_u times $e^{-K't}$ term is the oxygen demand in the stream over time, so this equation subtracts the BOD at time t from the ultimate BOD (BOD_u) to determine the

BOD depleted over time. BOD_u is the total oxygen used by carbonaceous bacteria if the test is run for an extended time and is indicated by CERM Eq. 28.32.

$$BOD_u \approx 1.463 BOD_5 \qquad 28.32$$

For seeded BOD rates, CERM Eq. 28.34 applies.

> *Exam tip*: Practice calculating BOD_t using all the preceding equations as an efficient way to master BOD for any test question. Note that if a sample is not diluted, then the volumetric ratio in CERM Eq. 28.30 and Eq. 28.34 is 1.

Reaction Rate

Handbook: BOD Exertion; Typical Values for the BOD Rate Constant

$$k_T = k_{20}\theta^{T-20}$$

CERM: Sec. 28.18

$$K'_{d,T_1} = K'_{d,T_2}\theta_d^{T_1-T_2} \qquad 28.29$$

CEWE: Sec. 6.17

$$K'_T = K'_r \Phi^{T_1-T_2} \qquad 6.11$$

The reaction rate, k, for BOD exertion varies based on the temperature and amount of pollution in the water, as expressed in *Handbook* table **Typical Values for the BOD Rate Constant**. Reaction rates can be corrected for temperature using CERM Eq. 28.29, which is equivalent to CEWE Eq. 6.11. Often, reaction rates for a particular type of water body will be provided at room temperature, as is the case for the reaction rates provided in *Handbook* section **BOD Exertion**. The *Handbook* equation used to correct these reaction rates, therefore, uses 20°C in place of T_2, which is used in the CERM and CEWE equations.

> *Exam tip*: The *Handbook* equation can be used whenever the provided reaction rate is for room temperature. CERM Eq. 28.29 can be used for everything else.

Theoretical Oxygen Demand (ThOD)

Handbook: Stream Modeling—Streeter Phelps

$$D = \left(\frac{k_d L_a}{k_r - k_d}\right)(e^{-k_d t} - e^{-k_r t}) + D_a e^{-k_r t}$$

CERM: Sec. 28.23

$$D_t = \left(\frac{K_d BOD_u}{K_r - K_d}\right)(10^{-K_d t} - 10^{-K_r t}) + D_0(10^{-K_r t}) \qquad 28.36$$

CEWE: Sec. 6.17, Sec. 6.21

$$D_t = \frac{K'_d BOD_u (e^{-K'_d t} - e^{-K'_r t})}{K'_r - K'_d} + D_0 e^{-K'_r t} \qquad 6.17$$

Theoretical oxygen demand (ThOD) is the theoretical analysis of balanced chemical equations to determine the amount of dissolved oxygen needed to oxidize specific compounds.

The Streeter-Phelps equation, which accounts for both deoxygenation and reoxygenation processes, may be used to model the change in the dissolved oxygen deficit in a flowing water body after wastewater is discharged into it.

CERM Fig. 28.5 shows an oxygen sag curve, both with and without reoxygenation. The difference between the saturated dissolved oxygen, DO_{sat}, and the dissolved oxygen, DO, at any time, t, is the oxygen deficit, D_t. D_c is the critical deficit, or the lowest point in the sag.

Figure 28.5 Oxygen Sag Curve

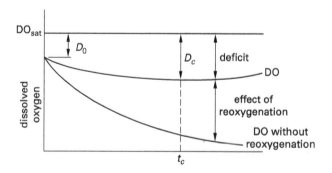

The Streeter-Phelps equation shown in *Handbook* section **Stream Modeling—Streeter Phelps** is the same as CEWE Eq. 6.17. CERM Eq. 28.36 is different in that it uses base-10 instead of base-e exponents. The correct equation will depend on the specific coefficients (K_d and K_r) being used.

Exam tip: Another way to approach the differences in base-e and base-10 equations, such as in the Streeter-Phelps equation, is to become adept at converting reaction rate constants using CERM Eq. 28.27.

Oxygen Saturation

Handbook: Oxygen Saturation

$$DO_{sat} = K_H P_{O_2}$$

CERM: Sec. 22.18

CEWE: Sec. 6.16, Sec. 11.17

Given enough time, at equilibrium, the concentration of a gas will reach a maximum known as the *saturation concentration*. The saturation concentration can be calculated using several versions of Henry's law (see CERM Sec. 22.18). There are several equations expressing Henry's law with different constants based on the units. CERM Eq. 22.7 shows the law in terms of the partial pressure of a gas as the product of the gas mole fraction and the Henry's law constant.

$$p_i = H_i x_i \qquad 22.7$$

CEWE Eq. 11.8 expresses the Henry's law constant in terms of the vapor pressure divided by the solubility of the gas in water.

$$H_m = \frac{p_{vap}}{S_w} \qquad 11.8$$

Handbook section Oxygen Saturation shows this equation in terms of dissolved oxygen concentration. It also provides two tables that show the dissolved oxygen levels in terms of mg/L for freshwater conditions at standard atmospheric oxygen and pressure conditions and for a variety of water salinity rates and temperatures. It also includes graphs of typical oxygen consumption over time and oxygen deficit rates over distance from a wastewater discharge point.

CEWE Eq. 6.4 is used to calculate dissolved oxygen saturation concentrations at nonstandard atmospheric pressure.

$$C_p = C^* p \left(\frac{\left(1 - \frac{p_{wv}}{p}\right)(1 - \theta_T p)}{(1 - p_{wv})(1 - \theta_T)} \right) \qquad 6.4$$

Exam tip: While Henry's law constants vary depending on the variable being solved for, the *Handbook* includes tables that can be used to directly determine the saturated dissolved oxygen content of water under a wide array of conditions.

Reoxygenation

Handbook: Oxygen Deficit After Mixing; Stream Reaeration Rate

$$D = DO_{stream} - DO$$

$$K_r = \frac{13.0 v^{0.5}}{H^{1.5}}$$

$$K_r = \frac{11.57 v^{0.969}}{H^{1.673}}$$

$$K_r = \frac{7.63 v}{H^{1.33}}$$

CERM: Sec. 28.16, Sec. 28.17

$$D = DO_{sat} - DO \qquad 28.17$$

$$r_r = K_r(DO_{sat} - DO) \qquad 28.18$$

CEWE: Sec. 6.19

$$r_r = K_r'(DO_{sat} - DO) = K_r' D \qquad 6.14$$

The difference between the saturated and actual dissolved oxygen concentrations is the oxygen deficit, D, which is expressed in CERM Eq. 28.17 and the *Handbook* equation, with DO_{sat} and DO_{stream} representing the same value. When an oxygen deficit exists in a stream, the water is reaerated through mixing and absorption of atmospheric oxygen.

CERM Eq. 28.18 (and the equivalent CEWE Eq. 6.14) indicates the rate of reaeration based on a base-10 reaeration constant. CERM Sec. 28.17 provides example rate constants based on the water body type. In some instances, a base-e constant is used. CERM Eq. 28.21 can be used to convert between base-10 and base-e constants. The reaeration constant can be calculated in base-e terms using CERM Eq. 28.22 and Eq. 28.23, which show the O'Connor and Dobbins formula and the Churchill formula, respectively.

Handbook section Stream Reaeration Rate provides equations to calculate the reaeration rate using the O'Connor and Dobbins method, the Churchill method, and the Langbein and Durum method. The Langbein and Durum method is not included in CERM. The other two equations vary slightly from CERM Eq. 28.22 and Eq. 28.23.

$$K'_{r,20°C} \approx \frac{3.93\sqrt{v_{m/s}}}{d_m^{1.5}} \quad \text{[SI]} \quad 28.22(a)$$

$$\begin{bmatrix} 0.3 \text{ m} < d < 9.14 \text{ m} \\ 0.15 \text{ m/s} < v < 0.49 \text{ m/s} \end{bmatrix}$$

$$K'_{r,68°F} \approx \frac{12.9\sqrt{v_{ft/sec}}}{d_{ft}^{1.5}} \quad \text{[U.S.]} \quad 28.22(b)$$

$$\begin{bmatrix} 1 \text{ ft} < d < 30 \text{ ft} \\ 0.5 \text{ ft/sec} < v < 1.6 \text{ ft} \end{bmatrix}$$

$$K'_{r,20°C} \approx \frac{5.049 v_{m/s}^{0.969}}{d_m^{1.67}} \quad \text{[SI]} \quad 28.23(a)$$

$$\begin{bmatrix} 0.61 < d < 3.35 \text{ m} \\ 0.55 \text{ m/s} < v < 1.52 \text{ m/s} \end{bmatrix}$$

$$K'_{r,68°F} \approx \frac{11.61 v_{ft/sec}^{0.969}}{d_{ft}^{1.67}} \quad \text{[U.S.]} \quad 28.23(b)$$

$$\begin{bmatrix} 2 \text{ ft} < d < 11 \text{ ft} \\ 1.8 \text{ ft/sec} < v < 5 \text{ ft/sec} \end{bmatrix}$$

Exam tip: CERM Sec. 28.17 includes common base-10 rate constants for different water body types. Become familiar with these values and use them as checks when calculating base-10 rate constants on the exam.

Deoxygenation and Oxygen Deficit

Handbook: Initial Deficit; Deoxygenation Rate; Stream Oxygen Deficit

$$k_d = 2.303 K_d$$

CERM: Sec. 28.18, Sec. 28.22

$$K'_d = 2.303 K_d \qquad 28.27$$

CEWE: Sec. 6.20

$$r_{d,t} = -K'_d \text{DO} \qquad 6.15$$

$$K'_{d,T_1} = K'_{d,T_2}\theta_d^{T_1-T_2} \qquad 6.16$$

CEWE Eq. 6.15 can be used to calculate the rate of deoxygenation based on the current dissolved oxygen content of the water source and the deoxygenation rate constant. The rate constant is based on the water body type and temperature. The temperature-corrected deoxygenation rate constant, $K'_{d,T}$, can be determined from CEWE Eq. 6.16. CERM Eq. 28.28 allows the conversion of base-e rate constants between different temperatures, given that the initial rate constant is provided at room temperature.

$$K'_{d,T} = K'_{d,20°C}\theta_d^{T-20°C} \qquad 28.28$$

Oxygen deficits within a stream include those that occur immediately after a waste stream is diluted by a freshwater system and those that accrue over time. CERM Eq. 28.35 can be used to calculate the final temperature, dissolved oxygen, biochemical oxygen demand (BOD), or suspended solids content immediately after two flows are mixed. This equation can also be used to calculate the dissolved oxygen in blended flows.

$$C_f = \frac{C_1 Q_1 + C_2 Q_2}{Q_1 + Q_2} \qquad 28.35$$

Handbook section Initial Deficit includes an equation for the initial deficit of dissolved oxygen, based on the premise of CERM Eq. 28.35. This number is then subtracted from the dissolved oxygen concentration in the stream prior to the effluent discharge to arrive at an initial deficit.

Handbook section Deoxygenation Rate includes an equation similar to CERM Eq. 28.27 that can be used to find the base-e version, k_d (K'_d in CERM), of the deoxygenation rate constant, K_d.

Exam tip: For highly polluted shallow streams, the deoxygenation constant can be assumed to be 0.25 per day.

Chemical Oxygen Demand (COD)

CERM: Sec. 28.24

CEWE: Sec. 6.17

Chemical oxygen demand (COD) is a measure of the oxygen requirements for the chemical oxidation of the waste in a water sample. The COD of wastewater is most often higher than biochemical oxygen demand (BOD) because more compounds can be chemically oxidized than can be biologically oxidized. COD is measured through oxidation of a sample by strong chemical oxidants.

COD does not differentiate between biological and nonbiological oxygen demand, but the test requires less time, usually 2 to 3 hours, making it less specific but more expedient than the tests described in the entry "Biochemical Oxygen Demand (BOD)" in this chapter.

For some industrial wastewater, a BOD test cannot be performed due to toxins in the sample that inhibit bacterial growth; under these circumstances, a COD test is used to indicate the oxygen demand of the wastewater.

Exam tip: Note that COD includes biological oxygen demand and oxygen demand for nonbiological processes.

C. TOTAL MAXIMUM DAILY LOAD

Key concepts: These key concepts are important for answering exam questions in subtopic 15.C, Total Maximum Daily Load.

- eutrophication
- impaired waters
- nutrients
- Section 303(d) of the Clean Water Act

NCEES Handbook: To prepare for this subtopic, familiarize yourself with these sections in the *Handbook*.

- Total Maximum Daily Load (TMDL)

PE Civil Reference Manual (**CERM**): Study these sections in CERM that either relate directly to this subtopic or provide background information.

- Section 25.12: Phosphorus
- Section 25.13: Nitrogen

Water Resources and Environmental Depth Reference Manual (**CEWE**): Study these sections in CEWE that either relate directly to this subtopic or provide background information.

- Section 6.14: Eutrophication
- Section 6.24: Stream Degradation
- Section 7.21: Nutrients in Wastewater

The following equations and concepts are relevant for subtopic 15.C, Total Maximum Daily Load.

Total Maximum Daily Load

Handbook: Total Maximum Daily Load (TMDL)

$$\text{TMDL} = \sum \text{WLA} + \sum \text{LA} + \sum \text{MOS}$$

Total maximum daily load (TMDL) is a process developed by the Environmental Protection Agency (EPA) as part of the enforcement of the Clean Water Act, which establishes the maximum pollutant loading without negative impact for a specific pollutant. Each pollutant has a TMDL established by the individual state based on the needs of the individual ecosystem. Calculations for TMDLs are developed for the specific waterway in question, and the process and level of analysis vary depending on the pollutant, the source of pollution, and the level of detail required by the study. You can learn more about TMDLs at the EPA website.

The *Handbook* equation can be used to calculate TMDLs based on the sum of point and nonpoint source pollution plus a margin of safety. Existing and future point sources are referred to as *waste load allocations* (WLAs), existing and future nonpoint sources and natural background concentrations of a pollutant are referred to as *load allocations* (LAs), and the margin of safety is abbreviated MOS.

Exam tip: Be familiar with the *Handbook* equation for this topic and use the information from the EPA to develop a qualitative understanding of how these concepts are applied.

Nutrients

CERM: Sec. 25.12, Sec. 25.13

CEWE: Sec. 6.14, Sec. 6.24, Sec. 7.21

Nutrients are chemical and organic compounds that are found in fertilizers, manures, and organic wastes in agriculture and land management. They are generally found in watershed runoff and wastewater. Left unchecked, an overabundance of nutrients in water bodies can lead to eutrophication. The two principal nutrients are nitrogen and phosphorus. Nitrogen can be in the form of nitrate, nitrite, or ammonium. Nitrogen and phosphorus can be present in fertilizers, manures, and organic wastes.

Nutrients are essential for a healthy aquatic ecosystem, but excess concentrations can degrade water quality (e.g., by causing algal blooms, which cause oxygen depletion).

Exam tip: In questions related to total maximum daily load (TMDL), nutrient loading will likely occur in the form of wastewater effluent or concentrations in runoff from agricultural watersheds or those with large areas dedicated to fertilizer use, such as golf courses.

Eutrophication

CEWE: Sec. 6.14

Eutrophication is a process in which water bodies receive excess nutrients that stimulate excessive growth of plants such as algae, periphyton-attached algae, nuisance plants, and weeds.

Eutrophication is an example of how a water body can become impaired through point and nonpoint source pollution. Total maximum daily loads (TMDLs) for nutrients are commonly required under such circumstances.

Exam tip: While the *Handbook* does not provide equations related to eutrophication, the presence of eutrophication is one way to identify that a water body is impaired, particularly by the nutrients discussed in the entry "Nutrients" in this section.

Impaired Waters

Total maximum daily loads (TMDLs) are regulated by states and tribes, as administered by the Environmental Protection Agency (EPA) under Section 303(d) of the Clean Water Act. The state identifies water bodies considered impaired waters. Essentially, this means that a water body does not meet the applicable water quality standards, which are dependent on state and local regulations. When specific pollutants can be identified as contributing to a water body's impairment, TMDLs can be developed that limit the mass per day of those pollutants discharged into the water body. The most common pollutants identified for impaired waters are pathogens.

Exam tip: While a basic equation for TMDLs is included in the *Handbook*, not much background information is provided. The laws and processes related to TMDL development are complex, and it isn't necessary to understand them in detail for the exam. Do understand, however, that TMDLs can be applied to any contaminant that may impair a watershed.

D. BIOLOGICAL CONTAMINANTS

Key concepts: These key concepts are important for answering exam questions in subtopic 15.D, Biological Contaminants.

- aerobic, anoxic, and anaerobic decomposition
- algae
- decay reactions
- decomposition of waste
- eutrophication
- growth rate constant
- indicator organisms
- limiting substrate
- microorganisms
- Monod kinetics
- pathogens
- steady-state retention times

NCEES Handbook: To prepare for this subtopic, familiarize yourself with these sections in the *Handbook*.

- BOD Testing/Sampling
- BOD Exertion
- Monod Kinetics—Substrate Limited Growth
- Multiple Limiting Substrates
- Nonsteady State Continuous Flow
- Steady State Continuous Flow
- Product Production at Steady State, Single Substrate Limiting
- Monod Growth Rate Constant as a Function of Limiting Concentration

- Comparison of Steady-State Retention Times (θ) for Decay Reactions of Different Order
- Comparison of Steady-State Performance for Decay Reactions of Different Order
- Removal and Inactivation Requirements
- Typical Removal Credits and Inactivation Requirements for Various Treatment Technologies

PE Civil Reference Manual (CERM): Study these sections in CERM that either relate directly to this subtopic or provide background information.

- Section 27.4: Microorganisms
- Section 28.15: Microbial Growth
- Appendix 25.A: National Primary Drinking Water Regulations

Water Resources and Environmental Depth Reference Manual (CEWE): Study these sections in CEWE that either relate directly to this subtopic or provide background information.

- Section 6.1: Microorganisms
- Section 6.2: Pathogens
- Section 6.6: Algae
- Section 6.8: Indicator Organisms
- Section 6.10: Decomposition of Waste
- Section 9.7: Bacterial Growth Kinetics

The following equations and figures are relevant for subtopic 15.D, Biological Contaminants.

Biochemical Oxygen Demand and Decomposition of Waste

Handbook: BOD Testing/Sampling; BOD Exertion

$$\text{BOD} = \frac{D_1 - D_2}{P}$$

$$\text{BOD} = \frac{(D_1 - D_2) - (B_1 - B_2)f}{P}$$

$$\text{BOD}_t = L_0(1 - e^{-kt})$$

CERM: Sec. 27.4

CEWE: Sec. 6.10

The *Handbook* equations only address decomposition from the standpoint of treatment processes. Microbial decomposition of waste involves oxidation/reduction reactions and is classified as aerobic or anaerobic. Molecular oxygen, O_2, must be present for decomposition to proceed by aerobic oxidation; its chemical end products are primarily carbon dioxide, water, and new cell material. A laboratory analysis of organic matter in water often includes a biochemical oxygen demand (BOD) test. Aerobic decomposition is the preferred method for large quantities of dilute (BOD5 < 500 mg/L) wastewater because decomposition is rapid and efficient and has a low odor potential. Some microorganisms can use nitrates in the absence of oxygen to oxidize carbon, in a process called *denitrification*. The end products from denitrification are nitrogen gas, carbon dioxide, water, and new cell material. Anaerobic decomposition occurs when molecular oxygen and nitrate are not present. The anaerobic decomposition of organic matter, also known as *fermentation*, produces carbon dioxide, methane, and water as the major end products.

CERM Table 27.4 provides a summary of various biological waste decomposition end products under aerobic, anoxic, and anaerobic conditions.

Exam tip: Decomposition of waste is a common method for the removal of many pathogens in treatment technologies. It is also the underlying biological basis for biological oxygen demand, as discussed in previous entries in this chapter.

Monod Kinetics

Handbook: Monod Kinetics—Substrate Limited Growth; Multiple Limiting Substrates; Nonsteady State Continuous Flow; Steady State Continuous Flow; Product Production at Steady State, Single Substrate Limiting; Monod Growth Rate Constant as a Function of Limiting Concentration; Comparison of Steady-State Retention Times (θ) for Decay Reactions of Different Order; Comparison of Steady-State Performance for Decay Reactions of Different Order

$$\mu = \frac{Yk_m S}{K_s + S} - k_d = \mu_{\max}\left(\frac{S}{K_s + S}\right) - k_d$$

CERM: Sec. 28.15

$$\mu = \mu_m\left(\frac{S}{K_s + S}\right) \qquad 28.9$$

$$\mu' = \mu_m\left(\frac{S}{K_s + S}\right) - k_d \qquad 28.16$$

CEWE: Sec. 9.7

$$\mu = \mu_{\max}\left(\frac{S}{K_s + S}\right) \qquad 9.5$$

For the large numbers and mixed cultures of microorganisms found in waste treatment systems, it is more convenient to measure biomass than numbers of organisms. This is accomplished by measuring suspended or volatile suspended solids. When one essential nutrient (referred to as a *substrate*) is present in a limited amount, the specific growth rate will increase up to a maximum value. The *Handbook* equation given in section Monod Kinetics—Substrate Limited Growth describes growth that is limited by a single substrate. This is essentially the same as CERM Eq. 28.16 and the Monod equation, given by CERM Eq. 28.9 and CEWE Eq. 9.5. CERM Eq. 28.16 is for the Monod equation net specific growth rate, which includes a decay coefficient. CERM Fig. 28.3 illustrates this concept.

Figure 28.3 Bacterial Growth Rate with Limited Nutrient

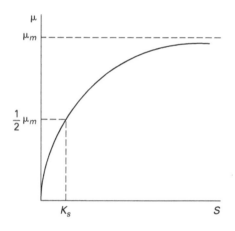

Additional equations for multiple limiting substrates, non-steady-state continuous flow, steady-state continuous flow, and product production at steady state are given in the following *Handbook* sections.

- Multiple Limiting Substrates
- Nonsteady State Continuous Flow
- Steady State Continuous Flow
- Product Production at Steady State, Single Substrate Limiting

Handbook figure Monod Growth Rate Constant as a Function of Limiting Concentration demonstrates Monod growth rates. *Handbook* sections Comparison of Steady-State Retention Times (θ) for Decay Reactions of Different Order and Comparison of Steady-State Performance for Decay Reactions of Different Order provide several equations related to the calculation of design variables for batch reactors.

Exam tip: *Handbook* section Monod Growth Rate Constant as a Function of Limiting Concentration and CERM Fig. 28.3 can be used to determine the maximum specific growth rate and the saturation constant. Be familiar with the *Handbook* equations used to calculate the specific growth rate under different conditions.

Indicator Organisms

Handbook: Removal and Inactivation Requirements; Typical Removal Credits and Inactivation Requirements for Various Treatment Technologies

CERM: Sec. 27.4

CEWE: Sec. 6.8

Isolating and identifying specific pathogenic microorganisms is a difficult task. Because of this, and because the number of pathogens relative to other microorganisms in water can be small, indicator organisms are used as a measure of the quality of the water. Total coliforms and fecal coliforms are the most common indicators. Total coliforms are mostly of intestinal origin, including *E. coli*, but can also be found in soils and plants. Fecal coliforms can be used to identify false positives from total coliform tests.

Handbook tables Removal and Inactivation Requirements and Typical Removal Credits and Inactivation Requirements for Various Treatment Technologies show the inactivation requirements for several indicator organisms, including *Giardia*, viruses, and *Cryptosporidium*. Table Removal and Inactivation Requirements indicates the general removal requirements, and Typical Removal Credits and Inactivation Requirements for Various Treatment Technologies lists these requirements for specific treatment technologies under a variety of water temperature, pH, and dosing conditions.

Exam tip: Traditional water and wastewater treatment design is primarily focused on the removal of indicator organisms. Where the *Handbook* information is more focused on *Giardia*, viruses, and *Cryptosporidium*, note that for drinking water, the goal is to completely remove fecal coliform.

Microorganisms

CERM: Sec. 27.4, App. 25.A

CEWE: Sec. 6.1

Microorganisms include viruses, bacteria, fungi, algae, protozoa, worms, rotifers, and crustaceans. Most biological contaminants are categorized as microorganisms that are either pathogens or environmental contaminants that deplete dissolved oxygen and drive eutrophication. Microorganisms are a key part of the overall ecosystem; it is only when they exist in overabundance or when parasitic microorganisms are present in drinking water that they can disrupt the ecological cycle and impact human health.

CERM Table 27.1 names some of the pathogenic organisms that may be found in water systems.

Exam tip: Microbial contaminants are a major focus for water and wastewater treatment systems. However, not all microbes are considered contaminants, particularly in natural waterways.

Pathogens

CERM: Sec. 27.4

CEWE: Sec. 6.2

A pathogen is any biological agent that causes disease or illness to its host. The term is most often used for agents that disrupt the normal physiology of a multicellular animal or plant. However, pathogens can also infect unicellular organisms. Pathogens may be introduced into water streams through animal and human fecal contamination. Pathogenic organisms include bacteria, protozoa, viruses, and worms.

CERM Table 27.1 names some of the pathogenic organisms that may be found in water systems.

Exam tip: Pathogenic microorganisms are the primary focus of biological contamination when considering water and wastewater treatment.

Algae

CERM: Sec. 27.4

CEWE: Sec. 6.6

Algae are autotrophic, photosynthetic organisms (photoautotrophs) and may be either unicellular or multicellular. Algae derive carbon from carbon dioxide and bicarbonates in water and derive energy through photosynthesis. Excessive algal growth (algal blooms) can result in supersaturated oxygen conditions in the daytime and anaerobic conditions at night. Some algae create tastes and odors in natural water. While they are not generally considered pathogenic to humans, algae do cause turbidity, and turbidity favors microorganisms that are pathogenic. Additionally, algal blooms are sometimes an indication of impaired watershed conditions.

Exam tip: While algae are not discussed in the *Handbook*, they can be an indicator of other contaminants in an impaired waterway when considering total maximum daily loads (TMDLs).

E. CHEMICAL CONTAMINANTS

Key concepts: These key concepts are important for answering exam questions in subtopic 15.E, Chemical Contaminants.

- bioaccumulation
- bioconcentration
- biomagnification
- chronic daily intake
- drinking water equivalent level
- human health risk
- maximum contaminant levels
- partition coefficient
- pollutant transport
- retardation factor
- Safe Drinking Water Act
- water quality mass calculation

***NCEES Handbook*:** To prepare for this subtopic, familiarize yourself with these sections in the *Handbook*.

- Water Quality: Mass Calculations
- Bioconcentration, Bioaccumulation, and Biomagnification Factors
- Bioaccumulation Factor
- Biomagnification Factor
- Octanol-Water Partition Coefficient
- Soil-Water Partition Coefficient
- Organic Carbon Partition Coefficient
- Retardation Factor R
- No-Effect Level (Reference Dose)
- Drinking Water Equivalent Level (DWEL)
- Toxic Unit Acute (TU_a)
- Toxic Unit Chronic (TU_c)
- Probability of Risk from Carcinogenic Substances: Linearized Multistage Model (EPA)
- Chronic Daily Intake for Ingested Contaminated Water

PE Civil Reference Manual (**CERM**): Study these sections in CERM that either relate directly to this subtopic or provide background information.

- Section 25.7: National Primary Drinking Water Standards
- Section 27.2: Cell Transport
- Appendix 25.A: National Primary Drinking Water Regulations

Water Resources and Environmental Depth Reference Manual (**CEWE**): Study these sections in CEWE that either relate directly to this subtopic or provide background information.

- Section 4.6: Water Quality Contamination and Prevention
- Section 6.13: Bioaccumulation
- Appendix 5.B: National Primary Drinking Water Regulations

The following equations and concepts are relevant for subtopic 15.E, Chemical Contaminants.

Water Quality Mass Calculation

Handbook: Water Quality: Mass Calculations

$$\frac{dM}{dt} = \frac{dM_{\text{in}}}{dt} + \frac{dM_{\text{out}}}{dt} \pm r$$

$$M = CQ = CV$$

The *Handbook* includes equations to determine the mass of a contaminant going into and out of a system. The first equation includes a reaction rate, which may either increase or decrease the mass of a given contaminant as it moves through the system.

Just as conservation of mass can be applied to water quantity, it can also be used to estimate the uptake of chemicals for water quality. In the second *Handbook* equation, the concentration of a chemical contaminant can be multiplied by the volume or flow of interest to determine the total mass of the contaminant being consumed or discharged on a larger time scale.

> *Exam tip*: Mass calculations such as that outlined in the *Handbook* can be used in conjunction with many of the concepts in this chapter to determine total masses of pollutants or, conversely, to calculate the concentration of a pollutant and its pathway based on the total that is discharged into the environment.

Bioconcentration

Handbook: Bioconcentration, Bioaccumulation, and Biomagnification Factors

$$\text{BCF} = \frac{C_{\text{org}}}{C}$$

CERM: Sec. 27.2

$$K_a = \frac{\text{concentration in lipid}}{\text{concentration in water}} \quad 27.1$$

Bioconcentration is the amount of chemical that accumulates in an aquatic organism with respect to its environment. This is expressed in CERM Eq. 27.1 as the concentration of the chemical in question in a lipid versus the concentration in water.

CERM Eq. 27.1 presents the topic in terms of partitioning at the cellular level, and the *Handbook* equation uses language that considers the entire organism. Additionally, the variables are represented differently. However, the two equations are essentially the same.

> *Exam tip*: As an example, the bioconcentration factor can be used when an exam question asks to make a correlation between the concentration of a contaminant in a waterway and the portion of that concentration being taken up by aquatic life in a stream system.

Bioaccumulation

Handbook: Bioaccumulation Factor

$$\text{BAF} = \frac{C_{\text{org}}}{C_s}$$

CEWE: Sec. 6.13

Bioaccumulation, described in detail in CEWE Sec. 6.13, is the accumulation of contaminants in an organism because of exposure to or ingestion of the contaminant. The level of bioaccumulation depends on the rate and mode of uptake, the rate at which the contaminant is removed from the organism, transformation of the substance by metabolic processes, the fat content of the organism, the hydrophobicity of the substance, and other biological environmental and physical factors. In general, more hydrophobic compounds are more likely to bioaccumulate in organisms such as fish.

The *Handbook* equation is for the partition of a chemical between an organism and a solid medium, such as soil.

Exam tip: While bioaccumulation occurs over time, the factor provided by the *Handbook* equation can be used to understand how a chemical will be taken up from an organism from exposure to a solid.

Biomagnification

Handbook: Biomagnification Factor

$$\text{BCF} = \frac{C_{\text{pred}}}{C_{\text{prey}}}$$

Biomagnification is the cumulative increase in the concentration of a persistent substance in successively higher levels of the food chain. As organisms consume food, the concentration of a chemical within a prey organism is absorbed by a predator at a rate that is dependent on several factors. Often, concentrations of the chemical will be absorbed by primary producers, such as microorganisms, plants, and algae, which are consumed by fish and herbivores. Herbivores are then consumed by carnivores, and a certain concentration of chemicals is passed on at each trophic level.

The *Handbook* equation is used to determine the percentage of a chemical contaminant that will be passed from prey to predator through consumption.

Exam tip: Be prepared to use the biomagnification, bioaccumulation, and bioconcentration equations in conjunction with each other to determine the partition pathway for various chemical contaminants.

Octanol-Water Partition Coefficient

Handbook: Octanol-Water Partition Coefficient

$$K_{\text{ow}} = \frac{C_O}{C_W}$$

CEWE: Sec. 6.13

The octanol-water partition coefficient indicates the concentration of a given chemical that will accumulate in water versus in organic materials. It is an indication of the chemical's hydrophobicity.

The *Handbook* equation is not included in CERM or CEWE but is briefly discussed in CEWE Sec. 6.13.

Exam tip: The octanol-water partition coefficient can be used to directly calculate the concentration of a chemical that will partition into water and into organic materials, such as humic substances or aquatic life. It can also be used in a qualitative manner to assess a chemical's hydrophobicity.

Soil-Water Partition Coefficient

Handbook: Soil-Water Partition Coefficient

$$K_{\text{sw}} = K_p$$

$$K_{\text{sw}} = \frac{X}{C}$$

The soil-water partition coefficient indicates the concentration of a given chemical that will accumulate in the water versus the soil. The *Handbook* equations are not included in CERM or CEWE.

Exam tip: The soil-water partition coefficient can be calculated as the concentration of a chemical in the soil divided by its concentration in the water. It can also be calculated as the organic carbon partition coefficient times the fraction of organic carbon in the soil.

Organic Carbon Partition Coefficient

Handbook: Organic Carbon Partition Coefficient

$$K_{\text{oc}} = \frac{C_{\text{soil}}}{C_{\text{water}}}$$

The organic carbon partition coefficient indicates the concentration of a given chemical that will accumulate in the water versus the organic carbon portion of the soil. The *Handbook* equation is not included in CERM or CEWE.

Exam tip: Note that all these environmental and biological partition coefficients can be used together to estimate the pathway of any chemical through the environment.

Retardation Factor R

Handbook: Retardation Factor R

$$R = 1 + \left(\frac{\rho}{\eta}\right) K_d$$

The retardation factor indicates the transport of a chemical by water or other solvent through a porous solid, such as soil.

The *Handbook* equation shows the retardation factor as a function of bulk density, porosity, and a distribution constant. This equation is not included in CERM or CEWE.

Exam tip: The retardation factor can provide an indication of how easily an environmental pollutant will move through soil or an aquifer when considering water flow patterns within the environment.

Risk to Human Health

Handbook: No-Effect Level (Reference Dose); Drinking Water Equivalent Level (DWEL); Toxic Unit Acute (TU_c); Toxic Unit Chronic (TU_c)

$$\text{RfD} = \frac{\text{NOAEL or LOAEL}}{\text{uncertainty factor}}$$

$$\text{DWEL} = \frac{\text{RfD} \cdot \text{body weight}}{\text{drinking water consumption}}$$

$$\text{TU}_a = \frac{100}{\text{LC}_{50}}$$

$$\text{TU}_c = \frac{100}{\text{NOEC}}$$

Human health impacts are often reviewed from the perspective of both acute and chronic levels of exposure. (*Acute* refers to the concentration of a given chemical that will result in death, and *chronic* refers to concentrations that will result in health problems related to limited or long-term exposure.) The *Handbook* includes equations that calculate several factors related to health risk evaluations for specific contaminants.

RfD is the *reference dose*, which is the amount of a chemical that the body can be exposed to with no observable effect, divided by an uncertainty factor that provides some room for error. For contamination in water, RfD is used to calculate the drinking water equivalent level, DWEL, which can be used as a maximum concentration of contaminant allowed in drinking water.

The acute and chronic toxic units—TU_a and TU_c, respectively—are the concentrations at which the human body is exposed to acute or chronic toxicity.

Exam tip: For exam questions on carcinogenic contaminants in water, RfD and DWEL can be used to calculate the concentration of the contaminant in water, which can then be used in combination with information from the problem statement to calculate the chronic daily intake (CDI) and probability of carcinogenic risk over time.

Risk from Carcinogenic Substances

Handbook: Probability of Risk from Carcinogenic Substances: Linearized Multistage Model (EPA); Chronic Daily Intake for Ingested Contaminated Water

$$p = \frac{d(\text{SF})}{mt} = (\text{CDI})(\text{SF})$$

$$\text{CDI} = C\left(\frac{(\text{IR})(\text{EF})(\text{ED})}{(\text{BW})(\text{AT})}\right)$$

The *Handbook* shows equations based on EPA models to determine the probability of exposure to carcinogenic substances based on exposure rates and the slope factor of the contaminant.

Safe Drinking Water Act

CERM: Sec. 25.7, App. 25.A

CEWE: Sec. 4.6, App. 5.B

The Safe Drinking Water Act (SDWA) provides maximum contaminant levels that are enforceable and others that are desirable (goals) to protect drinking water and its sources, such as rivers, lakes, reservoirs, springs, and groundwater wells. The primary standards establish maximum contaminant levels (MCLs) and maximum contaminant level goals (MCLGs) for materials that are known or suspected health hazards. The MCL is the enforceable level that the water supplier must not exceed, while the MCLG is an unenforceable health goal equal to the maximum level of a contaminant that is not expected to cause any adverse health effects over a lifetime of exposure.

Exam tip: The exam will focus on drinking water guidelines for chemical contaminants. The EPA values are not used by NCEES as a national standard. If there is a problem about chemical contaminants on the exam, you will be provided with the acceptable concentration levels.

52 Drinking Water Distribution and Treatment

Content in blue refers to the *NCEES Handbook*.

A. Drinking Water Distribution Systems 52-1
B. Drinking Water Treatment Processes 52-2
C. Demands 52-3
D. Storage 52-5
E. Sedimentation 52-6
F. Taste and Odor Control 52-7
G. Rapid Mixing 52-9
H. Flocculation 52-11
I. Filtration 52-13
J. Disinfection 52-14
K. Hardness and Softening 52-16

The knowledge area of Drinking Water Distribution and Treatment makes up between five and eight questions out of the 80 questions on the PE Civil Water Resources and Environmental exam. The organization of this chapter follows the order of subtopics given by NCEES for this knowledge area. Each subtopic is covered in the following sections.

A. DRINKING WATER DISTRIBUTION SYSTEMS

Key concepts: These key concepts are important for answering exam questions in subtopic 16.A, Drinking Water Distribution Systems.

- drinking water standards
- Ten States Standards

NCEES Handbook: The *Handbook* does not contain any material relevant to this subtopic. Be sure to study the listed sections in CERM and CEWE.

PE Civil Reference Manual (**CERM**): Study these sections in CERM that either relate directly to this subtopic or provide background information.

- Section 25.7: National Primary Drinking Water Standards
- Section 25.8: National Secondary Drinking Water Standards
- Appendix 25.A: National Primary Drinking Water Regulations
- Appendix 29.A: Selected Ten States Standards

Water Resources and Environmental Depth Reference Manual (**CEWE**): Study these sections in CEWE that either relate directly to this subtopic or provide background information.

- Section 5.2: Applicable Standards
- Appendix 5.A: Selected Ten States Standards

The following concepts are relevant for subtopic 16.A, Drinking Water Distribution Systems.

Drinking Water Standards

CERM: Sec. 25.7, Sec. 25.8, App. 25.A

CEWE: Sec. 5.2

The Safe Drinking Water Act is a set of federal requirements that must be met for potable drinking water throughout the United States. However, state and local jurisdictions may opt to have stricter requirements, so those regulations must be verified for each jurisdiction.

The Safe Drinking Water Act has two sets of standards. The first set consists of the mandatory federal regulations, which were established to protect the health of the public. The second set is encouraged but not mandatory; it addresses the aesthetics of the potable water supply. Water that meets the first set of standards is safe to drink, but if it does not meet the second set of standards, it may not be pleasant to drink, as it may have residual taste, odor, or color.

Exam tip: The variation between state and local jurisdictions is the reason that NCEES does not specify a set of specific standards on the exam.

Ten States Standards

CERM: App. 29.A

CEWE: App. 5.A

The Ten States Standards were developed in 1950 as a joint effort by mid-northern states. They have been updated a number of times since then, the latest revision for waterworks being from 2018. Today they are used by other states as a baseline for their individual standards.

The standards are organized in three parts, covering policy statements, interim standards, and recommended standards. *Policy statements* are recommendations for design and future processes. *Interim standards* are limited applications to new systems. *Recommended standards* are proven technologies applicable to all systems.

The Ten State Standards serve as guidelines, rather than an all-inclusive reference, for how to design water treatment and distribution systems.

B. DRINKING WATER TREATMENT PROCESSES

Key concepts: These key concepts are important for answering exam questions in subtopic 16.B, Drinking Water Treatment Processes.

- drinking water distribution
- water treatment plants
- water treatment processes

NCEES Handbook: The *Handbook* does not contain any material relevant to this subtopic. Be sure to study the listed sections in CERM and CEWE.

PE Civil Reference Manual **(CERM):** Study these sections in CERM that either relate directly to this subtopic or provide background information.

- Section 26.2: Process Integration
- Section 26.48: Storage and Distribution
- Section 29.5: Wastewater Treatment Plants

Water Resources and Environmental Depth Reference Manual **(CEWE):** Study these sections in CEWE that either relate directly to this subtopic or provide background information.

- Section 5.1: Introduction
- Section 5.6: Flocculation
- Section 5.7: Sedimentation
- Section 5.8: Filtration
- Section 5.12: Water Distribution Systems

The following concepts are relevant for subtopic 16.B, Drinking Water Treatment Processes.

Water Treatment Plants

CERM: Sec. 29.5

CEWE: Sec. 5.1

The location of a water treatment plant depends on a variety of factors; the availability of local water is most important, but power, sewerage, land, taxes, and traffic are other factors to consider. Water treatment plants need to be located above the floodplain to improve water distribution processes. They are best located at elevations between 15 ft to 20 ft above local grades.

Water treatment plants need pressurized systems to function and to eliminate the need to pump water between processes. An estimate of the amount of pressure used is about 15 ft of head for traditional systems and about 20 ft of head for specialized plants, such as reverse osmosis systems. The operational life cycle of equipment within treatment plants is commonly 25 years to 30 years (although the ideal goal is 50 years).

Exam tip: A treatment plant will pump the water that is to be treated to the elevation needed to feed it by gravity through all parts of the treatment. On the exam, unless the problem statement specifies otherwise, it can be assumed that the water is sufficiently pressurized to flow through the entire system.

Water Treatment Processes

CERM: Sec. 26.2

CEWE: Sec. 5.6, Sec. 5.7, Sec. 5.8

A water treatment plant process depends on the characteristics of the water supply. There is no standard water treatment plant. They are all custom-made facilities used to meet the specific criteria of the water supply. The traditional sequence of water treatment involves coagulation, sedimentation, and then filtration, which removes solids at specific sizes.

Direct filtration uses a modern sequence coagulant. It reduces the amount of sludge to be treated, but requires that the water supply be of high initial quality. Inline filtration uses coagulants applied at the filter intake. Mixing occurs when entering the filter.

CERM Fig. 26.1 shows an example of a water filtration process. Note that since there is no standard water treatment process, no schematic should be treated as a universal system.

Some of the chemicals typically added throughout the process to encourage coagulation are aluminum sulfate ($Al_2(SO_4)_3$), ferric chloride ($FeCl_3$), ferric sulfate ($Fe_2(SO_4)_3$), and soda ash (Na_2CO_3).

For equations related to specific treatment processes, see Sec. F through Sec. K in this chapter.

Drinking Water Distribution

CERM: Sec. 26.48

CEWE: Sec. 5.12

Water is commonly stored in both surface tanks and elevated tanks. The elevation of the water surface in the tank directly determines the distribution pressure.

Several methods are used to distribute water, depending on terrain, economics, and other local conditions. Gravity distribution is used when a lake or reservoir is located significantly higher in elevation than the population.

Distribution from pumped storage is the most common option when gravity distribution cannot be used. In this system, excess water is pumped during periods of low hydraulic and electrical demands into elevated storage. During periods of high consumption, water is drawn from the storage.

Using pumps without storage to force water directly into the mains is the least desirable option.

Exam tip: In a drinking water system, the design should not be exclusive to drinking water use, but any potable use. This includes commercial and industrial uses (which can be represented by population equivalents), fire systems, and any potential leaking pipes if the distribution system is preexisting.

C. DEMANDS

Key concepts: These key concepts are important for answering exam questions in subtopic 16.C, Demands.

- firefighting demand for a population
- firefighting demand for an individual building
- potable water demand
- typical water usage

NCEES Handbook: To prepare for this subtopic, familiarize yourself with these sections in the *Handbook*.

- Fire System Demands

PE Civil Reference Manual (**CERM**): Study these sections in CERM that either relate directly to this subtopic or provide background information.

- Section 20.19: Reservoir Sizing: Nonsequential Drought Method
- Section 26.43: Water Demand
- Section 26.44: Fire Fighting Demand
- Section 26.45: Other Formulas for Fire Fighting Needs

Water Resources and Environmental Depth Reference Manual (**CEWE**): Study these sections in CEWE that either relate directly to this subtopic or provide background information.

- Section 5.3: Water Demand and Supply

The following equations, figures, tables, and concepts are relevant for subtopic 16.C, Demands.

Firefighting Demand for a Population

Handbook: Fire System Demands

CERM: Sec. 26.45

$$Q = 1020\sqrt{P}(1 - 0.01\sqrt{P})$$

The combination of fire flow plus supply flow will vary from jurisdiction to jurisdiction as established by the fire department or fire marshal. A common requirement is to meet the average supply demand while providing firefighting flow for a short time duration, such as 2 hr.

The *Handbook* equation estimates that the fire flow demand, Q, is proportional to the population, P.

Exam tip: The equation for needed fire flow uses a population served value. While the current population can be applied to this value, the calculation should provide excess, as populations typically increase over time, and the system should be prepared in advance for excess need without requiring immediate additional construction.

Firefighting Demand for an Individual Building

Handbook: Fire System Demands

$$\text{NFF}_i = C_i O_i \big(1.0 + (X + P)_i\big)$$
$$C_i = 18 F A_i^{0.5}$$

CERM: Sec. 26.44

The combination of fire flow and supply flow will vary from jurisdiction to jurisdiction as established by the fire department or fire marshal. A common requirement is to meet the average supply demand while providing firefighting flow for a short duration, such as 2 hr.

Necessary flow to extinguish a fire in a nonsprinklered building is estimated by using the equation for needed fire flow from *Handbook* section Fire System Demands. The equation uses variables for type of construction, C_i, type of occupancy, O_i, a factor related to the exposure buildings, X, effective area of the building, A_i, and a coefficient for the type of construction, F. On the exam, values for these variables should be provided in the problem statement, since estimates are not given in the *Handbook*.

Exam tip: The equation for needed fire flow uses a population served value. While the current population can be applied to this value, the calculation should provide excess, as populations typically increase over time, and the system should be prepared in advance for excess need without requiring immediate additional construction.

Potable Water Demand

CERM: Sec. 20.19

CEWE: Sec. 5.3

Potable water demand estimates vary by jurisdiction. Each state or local agency typically establishes its own usage rates for different types of applications. Peak daily water demand, PDD, can be estimated using CEWE Eq. 5.1.

$$\text{PDD} = N_{du}\text{PWU}_{res} + A_f\text{PWU}_{com} + A_f\text{PWU}_{ofc} + A_f\text{PWU}_{ind} \quad 5.1$$

The process to estimate demand is similar for all water uses. CEWE Table 5.1 assumes 35 psi for all uses. This is the minimum recommended service pressure according to the Ten State Standards.

Table 5.1 Daily Water Demand Criteria

	PWU$_{res}$	PWU$_{com}$	PWU$_{ofc}$	PWU$_{ind}$
fire fighting flows (gpm) over 5 hr	1000	1000	1000	1500
minimum service pressure (psi)	35	35	35	35
lots ≤ 1 ac (gpd)	400 per detached du, 250 per attached du	0.5 per ft^2	0.3 per ft^2	0.5 per ft^2 or actual, whichever is more
lots ≥ 1 ac (gpd)	500 per du			

(Multiply ft^2 by 0.0929 to obtain m^2.)
(Multiply gal by 0.00379 to obtain m^3.)
(Multiply ac by 0.405 to obtain ha.)

Adapted from online forms created by University of Delaware, Water Resources Agency.

Demand can also be estimated using a Rippl diagram, as shown in CERM Fig. 20.23. A Rippl diagram allows an estimate based on demand from a preexisting reservoir. It shows how much storage is necessary throughout an average 24 hr period and can be used to estimate peak, average, and low water demands.

Figure 20.23 Reservoir Mass Diagram (Rippl Diagram)

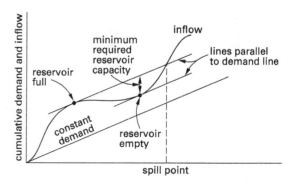

Exam tip: Exam problems on this topic will likely provide a table or chart in the problem statement for determining the combined potable water demand.

Typical Water Usage

CERM: Sec. 26.43

Water requirements can be estimated based on general usage. Residential (domestic) usage is roughly 100 gal per person per day, whereas usage in public locations is usually significantly less because water is used for fewer hours per day and for activities that are less water intensive. CERM Table 26.6 provides some rough estimates across various usage types. These can be applied based on total population in the service area or to a specific building being reviewed. These estimates may not be accurate for every part of the country. Locations with more accessible water reserves will use more water, while those in arid regions will use less.

Table 26.6 Annual Average Water Requirements[a]

	demand	
use	gpcd	Lpcd[b]
residential	75–130	284–490
commercial and industrial	70–100	265–380
public	10–20	38–80
loss and waste	10–20	38–80
totals	165–270	625–1030

(Multiply gpcd by 3.79 to obtain Lpcd.)
[a]exclusive of fire fighting requirements
[b]liters per capita-day

Exam tip: Values for local water usage will generally be provided on the PE Civil Water Resources and Environmental exam, but if none are supplied, an estimate of 100 gal per capita per day can be used for residences. Note that these all are ranges of usage; a reasonable estimate based on the community and facilities present should be taken into consideration.

D. STORAGE

Key concepts: These key concepts are important for answering exam questions in subtopic 16.D, Storage.

- impoundment storage capacity
- potable water storage
- treated water storage

NCEES Handbook: To prepare for this subtopic, familiarize yourself with these sections in the *Handbook*.

- Cross-Section Methods
- Principles of One-Dimensional Fluid Flow: Energy Equation

PE Civil Reference Manual (**CERM**): Study these sections in CERM that either relate directly to this subtopic or provide background information.

- Section 26.48: Storage and Distribution
- Section 80.14: Average End Area Method

Water Resources and Environmental Depth Reference Manual (**CEWE**): Study these sections in CEWE that either relate directly to this subtopic or provide background information.

- Section 3.15: Pond and Reservoir Routing
- Section 5.3: Water Demand and Supply
- Section 5.12: Water Distribution Systems

The following equations and concepts are relevant for subtopic 16.D, Storage.

Impoundment Storage Capacity

Handbook: Cross-Section Methods

CERM: Sec. 80.14

$$V = L\left(\frac{A_1 + A_2}{2}\right)$$

CEWE: Sec. 3.15

$$V = \Delta d_H\left(\frac{A_1 + A_2}{2}\right) \quad 3.49$$

The volume of a reservoir can be estimated with the average end area method using CEWE Eq. 3.49 or CERM Eq. 80.8, which is similar to the equation presented in *Handbook* section Cross-Section Methods. The only difference between the equations is that the *Handbook* and CERM use L for the distance between cross-sectional areas, while CEWE uses Δd_H.

A stage-area relationship can be used to determine a stage-volume relationship. *Stage-area* is an engineering term used to describe the relationship between depth and surface area; *stage-volume* similarly describes the relationship between depth and volume. Once the volume is determined, the time to drain the reservoir can be estimated by using the average discharge flow.

Exam tip: The cross-section method is meant to be an estimate. It treats the water storage volume as a prism, without considering slope or orientation.

Treated Water Storage

Handbook: Principles of One-Dimensional Fluid Flow: Energy Equation

CEWE: Sec. 5.3

$$P_1 - P_2 = \gamma h_f = \rho g h_f$$

Treated water storage helps supplement the capacity of a water treatment plant during demand peaks and supply lows. Tanks for treated water storage are typically elevated to supply pressure to the distribution system. The pressure-to-elevation relationship is shown in *Handbook* section Principles of One-Dimensional Fluid Flow: Energy Equation and CEWE Eq. 5.3.

This equation describes how much pressure water at a specific temperature and height can exert. The height of water storage is dependent on the desired water pressure.

Exam tip: Water pressure can be boosted or reduced external to storage, but gravity is the simplest method to meet the pressure requirements for water demand. Higher pressure is generally better than lower.

Potable Water Storage

CERM: Sec. 26.48

CEWE: Sec. 5.3, Sec. 5.12

$$\text{PDD} = N_{du}\text{PWU}_{res} + A_f\text{PWU}_{com} \\ + A_f\text{PWU}_{ofc} + A_f\text{PWU}_{ind} \qquad 5.1$$

Peak daily water demand can be calculated using CEWE Eq. 5.1. This equation calculates peak demand for peak water use, PWU, for residential uses based on the number of dwellings, N_{du}, and commercial, office, and industrial uses based on area, A_f. This peak demand advises how much storage is required to provide a given community with water.

There are two main methods for storing water for potable uses. The first is to store water before treatment, and the second is to store water after treatment. The two choices meet varying demand and supply levels.

Potable water is stored after treatment in order to have it ready at any given time to supplement the water distribution network during peak times. Just as demand varies, the supply can vary as well. To supplement periods when the supply is low, such as in the case of low lake levels or low river flows, water should be stored before treatment.

In some circumstances, water may need to be stored both before and after treatment to supplement demand peaks and supply lows.

Exam tip: If the treatment system depends on a low flow or inconsistent source, water storage before treatment is required to meet demand. This can be determined by evaluating the average flow from demand and the source, and then determining if supply will meet demand.

E. SEDIMENTATION

Key concepts: These key concepts are important for answering exam questions in subtopic 16.E, Sedimentation.

- minimum settling time
- settling velocity

NCEES Handbook: To prepare for this subtopic, familiarize yourself with these sections in the *Handbook*.

- Type 1 Settling - Discrete Particle Settling

PE Civil Reference Manual (**CERM**): Study these sections in CERM that either relate directly to this subtopic or provide background information.

- Section 26.11: Sedimentation Physics
- Section 26.12: Sedimentation Tanks

Water Resources and Environmental Depth Reference Manual (**CEWE**): Study these sections in CEWE that either relate directly to this subtopic or provide background information.

- Section 5.7: Sedimentation

The following equations, figures, and concepts are relevant for subtopic 16.E, Sedimentation.

Sedimentation Basics

Handbook: Type 1 Settling - Discrete Particle Settling

CERM: Sec. 26.11, Sec. 26.12

CEWE: Sec. 5.7

Sedimentation is the process of settling the floc so that it can be disposed of. Sedimentation depends on the properties of the water influent as well as the floc within the water. Sedimentation can be combined with flocculation in the same basin or tank. Flocculation changes the qualities of the sediment and can be a critical step before the sedimentation process.

Tank size, influent flow, and size of the particles that must be settled are all critical to the design. The figure in *Handbook* section **Type 1 Settling - Discrete Particle Settling** shows how these variables influence settling. CERM Fig. 26.3 presents a similar illustration.

Figure 26.3 *Sedimentation Basin*

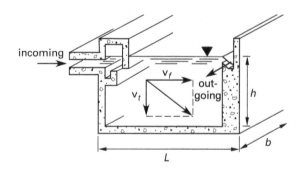

Exam tip: Sedimentation requires water to stay in treatment long enough for solids to settle. Calculating this duration depends on flow rate, particle size, and tank size. In most cases, treatment facilities are designed with multiple settling basins as redundancies and to accommodate low- and high-flow conditions.

Minimum Settling Time

Handbook: Type 1 Settling - Discrete Particle Settling

$$t_0 = \frac{H}{v_t}$$

CERM: Sec. 26.12

$$t_{\text{settling}} = \frac{h}{v_s} \quad \text{26.5}$$

CEWE: Sec. 5.7

$$t_s = \frac{d}{v_s} \quad \text{5.9}$$

Minimum settling time is the time it takes for the target particles to settle. It depends on the settling velocity and the tank depth. Given the size of a sediment particle, it is possible to determine the amount of time it will take that particle to settle to the bottom of the tank. The settling velocity is a function of the size and density of the particle as well as the viscosity of the water.

The distance traveled by the particle is represented as H in the *Handbook*, h in CERM, and d in CEWE. This is the minimum tank length required to allow for particles to fully settle.

Exam tip: Tank length can be longer than the minimum distance required for the target particles to settle but cannot be shorter. If an exam problem provides seemingly correct answers that are either shorter or longer than the solution, select the longer answer.

Settling Velocity

Handbook: Type 1 Settling - Discrete Particle Settling

$$v_0 = \frac{Q}{A} = \frac{Q}{WL} = \frac{g(\rho_s - \rho)d^2}{18\mu}$$

CERM: Sec. 26.11, Sec. 26.12

$$v_{s,\text{m/s}} = \frac{(\rho_{\text{particle}} - \rho_{\text{water}})D_m^2 g}{18\mu} \quad \text{[SI]} \quad \text{26.3(a)}$$

$$= \frac{(\text{SG}_{\text{particle}} - 1)D_m^2 g}{18\nu}$$

$$v^* = \frac{Q_{\text{filter}}}{A_{\text{surface}}} = \frac{Q_{\text{filter}}}{bL} \quad \text{26.7}$$

CEWE: Sec. 5.7

Settling time can be calculated from the settling velocity and the depth of the tank. Settling velocity and settling time depend on the water temperature, particle size, and particle specific gravity. Bacteria and colloidal particles are generally considered to be nonsettleable during detention periods available in water treatment facilities.

The equations found in the *Handbook* and CERM use the same variables with slightly different presentations. In all cases, the variables are represented by the following.

A	surface area	ft^2 (m^2)
d, D	particle diameter	ft (m)
L	tank length	ft (m)
Q	flow rate	ft^3/sec (m^3/s)
v	settling velocity	ft/sec (m/s)
ρ	fluid density	lbm/ft^3 (kg/m^3)

CEWE Fig. 5.3 and CERM Fig. 26.2 show the settling velocity for various particles.

Exam tip: Calculating settling velocity is important in determining the length of a settling tank. Particles will settle with gravity, but the tank must be long enough for particles to settle before discharge.

F. TASTE AND ODOR CONTROL

Key concepts: These key concepts are important for answering exam questions in subtopic 16.F, Taste and Odor Control.

- activated carbon adsorption
- air stripping
- odor and taste removal through dilution
- reverse osmosis

NCEES Handbook: To prepare for this subtopic, familiarize yourself with these sections in the *Handbook*.

- Freundlich Isotherm
- Air Stripping: Mass Balance
- Threshold Odor Number (TON)
- Reverse Osmosis: Osmotic Pressure of Solutions of Electrolytes

PE Civil Reference Manual (**CERM**): Study these sections in CERM that either relate directly to this subtopic or provide background information.

- Section 26.25: Filter Backwashing
- Section 26.26: Other Filtration Methods
- Section 26.30: Taste and Odor Control
- Section 29.30: Carbon Adsorption
- Section 34.47: Stripping, Air

Water Resources and Environmental Depth Reference Manual (**CEWE**): Study these sections in CEWE that either relate directly to this subtopic or provide background information.

- Section 6.15: Taste and Odor Issues
- Section 11.17: Volatilization/VOC Removal

The following equations, figures, and concepts are relevant for subtopic 16.F, Taste and Odor Control.

Activated Carbon Adsorption

Handbook: Freundlich Isotherm

$$q_e = K_f C_e^{1/n} = \frac{(C_0 - C_e)V}{w}$$

CERM: Sec. 29.30

Adsorption uses high-surface-area activated carbon to remove organic contaminants. Activated carbon is relatively nonspecific, and it will remove a wide variety of refractory organics as well as some inorganic contaminants. It should generally be considered for organic contaminants that are nonpolar, have low solubility, or have high molecular weights. CERM Sec. 29.30 gives definitions related to carbon adsorption.

For activated carbon adsorption to be an effective treatment process, media must be packed loosely enough for water to flow through. Flow can be reduced if contaminants plug the pores or the media becomes too closely packed, which will increase head losses and lower removal efficiency.

The equation in *Handbook* section Freundlich Isotherm calculates the loading on activated carbon of mass of contaminant to mass of activated carbon to maintain equilibrium. The calculation requires knowing the characteristics of the activated carbon and the contaminant concentrations of water entering the tank and desired at the effluent point.

Exam tip: The equilibrium loading on the activated carbon is limited by the carbon; it is not infinite. In some cases, calculating the equilibrium requires prior knowledge of what loading a treatment step can take.

Air Stripping

Handbook: Air Stripping: Mass Balance

$$QC_b(z) + Q_a y_0 = QC_e + Q_a y_b(z)$$

CERM: Sec. 34.47

CEWE: Sec. 11.17

In air stripping, volatile organic compounds are removed from aqueous solutions by greatly increasing the surface area of the contaminated water that is exposed to air. The equation in *Handbook* section Air Stripping: Mass Balance calculates the mass of a contaminant at both the air and water influent points and again at the air and water effluent points. After the process of air stripping, water concentration discharge will be less than at its entrance, and the opposite will be true for air. In addition, the flow rate of water and air should be the same at each influent and discharge point. See CEWE Sec. 11.17 for more details on air stripping.

CERM Fig. 34.12 shows how air stripping towers work, with clean air pumped in from the bottom and contaminated water spraying down from the top. Any pollutants that may volatilize then exit through the top of the tower (which may require air permits or additional treatment).

Exam tip: Air stripping functions with the same mass conservation principles as other water treatment processes.

Figure 34.12 Schematic of Air Stripping Operation

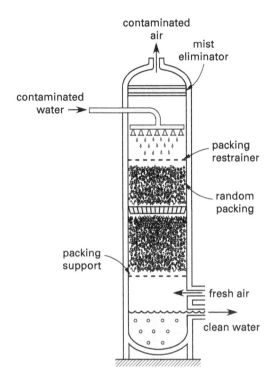

Odor and Taste Removal Through Dilution

Handbook: Threshold Odor Number (TON)

CEWE: Sec. 6.15

$$\text{TON} = \frac{A + B}{A}$$

CERM: Sec. 26.30

$$\text{TON} = \frac{V_{\text{raw sample}} + V_{\text{dilution water}}}{V_{\text{raw sample}}} \quad 26.39$$

Odor and taste in water can be reduced through dilution. One method to quantitatively determine the amount of odor and taste in water is to use the equation found in *Handbook* section Threshold Odor Number (TON), CEWE Eq. 6.3, or CERM Eq. 26.39. The TON quantifies the amount of dilution water needed to produce an odorless finished sample (of 200 mL). A larger TON value requires more dilution water to produce an odorless sample.

A TON of 3 or less is ideal, as the necessary dilution to an odorless sample is better with a lower value. For drinking water, a TON of 5 or greater may be detectable by customers.

Exam tip: The dilution water is meant to be odorless, so it should not contribute to any odor in treated water. If treated water has an excessive odor, an issue may exist with the source water, or specific treatment steps may need to be redesigned.

Reverse Osmosis

Handbook: Reverse Osmosis: Osmotic Pressure of Solutions of Electrolytes

$$\Pi = \phi \nu \left(\frac{n}{V} \right) RT$$

CERM: Sec. 26.25, Sec. 26.26

Osmotic pressure (in pascals) is calculated using the equation in *Handbook* section Reverse Osmosis: Osmotic Pressure of Solutions of Electrolytes. Reverse osmosis is a membrane-demineralization process. A thin membrane separates two solutions of different concentrations. Pore size is smaller (0.0001 μm to 0.001 μm) with reverse osmosis than with ultrafilter membranes, as salt ions are not permitted to pass through. (See CERM Sec. 26.26 for more details.) Typical large-scale osmosis units operate at 150 psi to 500 psi (1.0 MPa to 5.2 MPa).

Water filtered through a reverse osmosis system can collect a significant amount of minerals. The system must be backwashed routinely to keep the pores from clogging. (See CERM Sec. 26.25 for more details).

Exam tip: Reverse osmosis treats water by removing the ions of a contaminant. In calculating the pressure necessary for treatment, the specific quantity of contaminants being removed will determine pressure. Over time, removal efficiency will decrease as the minerals clog the filter, requiring backwashing. Monitoring the pressure across the membrane to see when this decrease occurs will guide the decision on how often backwashing will need to be performed.

G. RAPID MIXING

Key concepts: These key concepts are important for answering exam questions in subtopic 16.G, Rapid Mixing.

- rapid mixer impellers
- rapid mixing tanks
- reel and paddle or radial-flow mixers

NCEES Handbook: To prepare for this subtopic, familiarize yourself with these sections in the *Handbook*.

- Rapid Mix and Flocculator Design
- Reel and Paddle
- Turbulent Flow Impeller Mixer Power
- Values of the Impeller Constant K_T

PE Civil Reference Manual (CERM): Study these sections in CERM that either relate directly to this subtopic or provide background information.

- Section 26.19: Mixing Physics
- Section 26.20: Impeller Characteristics

Water Resources and Environmental Depth Reference Manual (CEWE): Study these sections in CEWE that either relate directly to this subtopic or provide background information.

- Section 5.5: Rapid Mixing

The following equations, figures, and concepts are relevant for subtopic 16.G, Rapid Mixing.

Rapid Mixing Tanks

Handbook: Rapid Mix and Flocculator Design

$$G = \sqrt{\frac{P}{\mu V}} = \sqrt{\frac{\gamma H_L}{t\mu}}$$

CEWE: Sec. 5.5

$$Gt_o = \frac{1}{Q}\sqrt{\frac{PV}{\mu}} \qquad 5.8$$

The *Handbook* equation and CEWE Eq. 5.8 represent the same calculation with different approaches; CEWE Eq. 5.8 uses flow while the *Handbook* equation does not. The CEWE equation can be rearranged into the *Handbook* equation as follows.

$$G = \frac{1}{Qt_0}\sqrt{\frac{PV}{\mu}} = \frac{1}{V}\sqrt{\frac{PV}{\mu}} = \sqrt{\frac{PV}{\mu V^2}}$$
$$= \sqrt{\frac{P}{\mu V}}$$

To encourage flocculation, raw water is treated using chemicals immediately following intake unless pretreatment is required. This is done for a short duration, and for this reason it is called *rapid mixing*. Mixing can be done using a plug flow method, where water flows through the basin, or a complete mixed tank method, where a chemical coagulant can thoroughly mix with the water prior to the sedimentation steps.

The equations from the *Handbook* and CEWE represent the velocity gradient or mixing intensity needed for rapid mixing. Calculating the velocity gradient uses power to the fluid, volume of the tank, dynamic viscosity, head loss, specific weight, time, and/or flow rate. Slow-moving paddle mixers operate with a velocity gradient of 20 sec^{-1} to 75 sec^{-1}, while rapid mixers operate at a higher gradient, approximately 500 sec^{-1} to 5000 sec^{-1}, depending on the mixing period.

Exam tip: In exam problems on the topic, the velocity gradient should be provided in the problem statement, requiring the calculation of one of the other variables. If a range is given, note that the range for some variables may be different depending on where they are in the equation. This may be important because, in such cases, the possible answers may not exactly match the calculated solution.

Reel and Paddle or Radial Flow Mixers

Handbook: Reel and Paddle

$$P = \frac{C_D A_P \rho_f v_r^3}{2}$$

$$F = 0.5 C_D \rho_f A v_p^2$$

CERM: Sec. 26.19

$$F_D = \frac{C_D A \rho v_{\text{mixing}}^2}{2} \qquad \text{[SI]} \qquad 26.20(a)$$

The *Handbook* equation and CERM Eq. 26.20 use the same variables, with the *Handbook* equation calculating the mixer's power and drag force and the CERM equation calculating only the drag force.

Reel and paddle mixers, also called radial flow mixers, are used in the same applications as axial flow mixers but are not as effective at keeping materials in suspension.

Exam tip: Radial flow mixers are less efficient in keeping materials in suspension than axial flow mixers, so it is less likely that the exam will present a problem that uses a radial flow mixer. However, some treatment facilities do use this type of mixer, so being familiar with its use may help on general knowledge questions regarding types of mixers and their applications.

Rapid Mixer Impellers

Handbook: Turbulent Flow Impeller Mixer Power; Values of the Impeller Constant K_T

$$P = K_T n^3 D_i^5 \rho_l$$

CERM: Sec. 26.20

$$P = N_P n^3 D^5 \rho \quad \text{[SI]} \quad 26.28(a)$$

Rapid mixers typically use axial flow impellers because these impellers are good at maintaining materials in suspension, which helps mix the coagulants. Note that rapid mixing uses the complete mixing model, not the plug flow model. The plug flow model assumes that water mixes as though it were being pushed with a piston, with chemicals mixing through the linear movement of water. Complete mixing assumes that the material is being distributed through the entire volume of water, such as by an impeller.

CERM Fig. 26.5 shows axial flow mixing impellers.

Figure 26.5 Typical Axial Flow Mixing Impellers

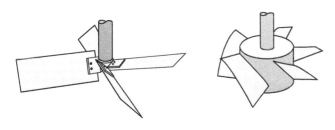

Vibration near the critical speed can be a major problem with modern, high-efficiency, high-speed impellers. Mixing speed should be well below (less than 80% of) the first critical speed of the shaft. Other important design factors include tip speed and shaft bending moment.

The equations presented in the *Handbook* and CERM are used to calculate the power of the impeller. These equations use the same variables, except that the impeller constant is K_T in the *Handbook* and N_P in CERM. Typical values for this constant can be found in *Handbook* table Values of the Impeller Constant K_T.

Exam tip: Impeller power is typically in the range of 0.5 hp/MGD to 1.5 hp/MGD for rapid mixers. On the exam, however, either the exact power will be provided or you will be able to calculate it from other values.

H. FLOCCULATION

Key concepts: These key concepts are important for answering exam questions in subtopic 16.H, Flocculation.

- flocculation basins
- flocculation process

NCEES Handbook: To prepare for this subtopic, familiarize yourself with these sections in the *Handbook*.

- Flocculation
- Typical Diffuser Wall Guidelines for Flocculation Basins

PE Civil Reference Manual (CERM): Study these sections in CERM that either relate directly to this subtopic or provide background information.

- Section 26.21: Flocculation
- Section 26.22: Flocculator-Clarifiers

Water Resources and Environmental Depth Reference Manual (CEWE): Study these sections in CEWE that either relate directly to this subtopic or provide background information.

- Section 5.6: Flocculation

The following equations, figures, tables, and concepts are relevant for subtopic 16.H, Flocculation.

Flocculation Process

Handbook: Flocculation

CERM: Sec. 26.21, Sec. 26.22

CEWE: Sec. 5.6

The flocculation process starts with the influent well. Baffles slow down the flow in the basin, and coagulants are applied, which cause the floc to form and then settle down. Floc forms during a period of gentle mixing, as continued rapid mixing would break up the forming floc. Fragile, cold-water floc mixes at 0.5 ft/sec, while warm-water floc mixes at 3 ft/sec.

The equation from *Handbook* section **Flocculation** is used to determine the mean velocity gradient. Calculating the velocity gradient uses power to the fluid, volume of the tank, dynamic viscosity, head loss, specific weight, time, and/or flow rate.

$$G = \sqrt{\frac{Q\gamma H}{\mu V}} = \sqrt{\frac{62.4H}{\mu t}} = \sqrt{\frac{P}{\mu V}}$$

Slow-moving paddle mixers operate with a velocity gradient between 20 sec^{-1} and 75 sec^{-1}, while rapid mixers operate at a higher gradient, between approximately 500 sec^{-1} and 5000 sec^{-1}, depending on the mixing period.

CEWE Fig. 5.2 shows an example of the flocculation process.

Figure 5.2 Circular Flocculation/Sedimentation Basin

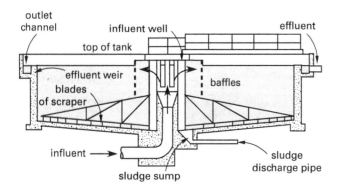

Adapted from *Unified Facilities Criteria, Water Supply: Water Treatment.* UFC 3-230-80A. January 2004. United States Department of Defense.

CERM Table 26.2 lists general characteristics of a flocculator-clarifier.

Table 26.2 Characteristics of Flocculator-Clarifiers

typical flocculation and mixing time	20–60 min
minimum detention time	1.5–2.0 hr
maximum weir loading	10 gpm/ft (2 L/s·m)
upflow rate	0.8–1.7 gpm/ft^2; 1.0 gpm/ft^2 typical (0.54–1.2 L/s·m^2; 0.68 L/s·m^2 typical)
maximum sludge formation rate	5% of water flow

(Multiply gpm/ft by 0.207 to obtain L/s·m.)
(Multiply gpm/ft^2 by 0.679 to obtain L/s·m^2.)

Exam tip: A flocculation basin that does not meet a minimum residence time will not remove the expected amounts of floc and solids. In addition, sludge will form and accumulate at the bottom of the basin, effectively reducing the size of the basin to less than its design volume. If this occurs, it will become necessary to remove the excess sludge through regular maintenance.

Flocculation Basins

Handbook: Typical Diffuser Wall Guidelines for Flocculation Basins

CERM: Sec. 26.22

CEWE: Sec. 5.6

A flocculation basin is designed to allow floc to settle after coagulation has occurred. The by-products from this step are sludge and clean water. Flocculation basins are typically circular but may be rectangular depending on specific site conditions or design requirements.

CEWE Table 5.2 provides typical dimensions of flocculation tanks, both rectangular and circular. *Handbook* table **Typical Diffuser Wall Guidelines for Flocculation Basins** provides typical diffuser wall guidelines for flocculation basins.

Table 5.2 Water Treatment Sedimentation/Flocculation Tank Dimensions

	parameter	value
rectangular tanks	tank depth	10–15 ft
	tank width	30 ft
	tank length	100–200 ft
	tank aspect ratio	3:1, length:width
	tank material	concrete
circular tanks	tank depth	9–14 ft
	diameter	25–200 ft

(Multiply ft by 0.3048 to obtain m.)

Exam tip: Unless otherwise stated in the problem statement, a circular basin can be assumed for exam problems on the topic. In addition, the basin must be large enough to allow for the floc to settle. When calculating the size, consider that the basin should be able to meet expected flow demands on the treatment facility.

I. FILTRATION

Key concepts: These key concepts are important for answering exam questions in subtopic 16.I, Filtration.

- filtration backwashing
- loading rate
- media filtration

NCEES Handbook: To prepare for this subtopic, familiarize yourself with these sections in the *Handbook*.

- Media Filtration: Loading Rate
- Media Filtration: Bed Expansion

PE Civil Reference Manual **(CERM):** Study these sections in CERM that either relate directly to this subtopic or provide background information.

- Section 26.24: Filtration
- Section 26.25: Filter Backwashing

Water Resources and Environmental Depth Reference Manual **(CEWE):** Study these sections in CEWE that either relate directly to this subtopic or provide background information.

- Section 5.1: Introduction
- Section 5.8: Filtration

The following equations, figures, and concepts are relevant for subtopic 16.I, Filtration.

Loading Rate

Handbook: Media Filtration: Loading Rate

CERM: Sec. 26.24

CEWE: Sec. 5.8

$$\text{loading rate} = \frac{Q}{A}$$

The loading rate is measured as the flow rate of water through the cross-sectional area of the filter that is perpendicular to the flow through the filter. The depth of the filter (the distance from the top of the filter where water enters and the bottom where it exits) is not used for this calculation.

For media filters, flow generally enters at the top of the filter and drains through the filter by gravity. Since filters are being used to remove solids, redundant filters are necessary.

Exam tip: Filter loading rates are critical to efficient treatment. A filter with too little water will not be effective, as pressure is necessary to provide the force through the media. A filter with too much water may result in overflowing the filter or creating too much pressure, pushing contaminants to pass through without treatment. For these reasons, keep in mind that redundant filters should be in place to both scale back and scale up treatment when necessary.

Filter Backwashing

Handbook: Media Filtration: Bed Expansion

CERM: Sec. 26.25

The filtration process requires periodic backwashing to remove contaminants that have accumulated in the packed media. If left alone, the filtered water may become more turbid, and the head loss through the filter will increase until the treatment processes require more pressure than the design can accommodate.

Backwashing typically occurs once head loss reaches between 6 ft and 8 ft (over approximately 1 to 3 days). The timing of backwashing is based on head loss rather than turbidity because head loss is more easily monitored and increases steadily over time, while turbidity may increase suddenly.

When backwashing is needed, the filter must be taken offline and cleaned with water so as to push contaminants out of the porous media. As a result, a water treatment facility needs redundant filters. Generally, this means that if a facility needs two filters, one filter is to allow for backwashing.

During an air prewash, air is pumped through the filter media, causing it to expand. Then clean water is passed through the media to remove contaminants. The amount of water needed is generally 1% to 5% of the total processed water. More wash water is necessary for larger filters.

The equations in *Handbook* section Media Filtration: Bed Expansion and CERM Eq. 26.37 can be used to calculate the necessary flow rates, volume, and time of backwash. The surface area of the filter is necessary for these calculations.

$$v_B = \frac{Q_B}{A_{\text{plan}}}$$

In the *Handbook* equation, v_B is the backwash velocity, and Q_B is the backwash flow rate.

$$V = A_{\text{filter}}(\text{rate of rise})t_{\text{backwash}} \qquad 26.37$$

Exam tip: When solving for how much wash water or time is necessary to backwash a filter, the exact measurements for the filter and the typical flow being treated by the filter are both needed. Remember that the number of filters is important, as each filter is expected to treat only some of the water needed to meet demand.

Media Filtration

CERM: 26.24

CEWE: Sec. 5.1

Filtration is a process that removes fine particles. It is a physical process, but it can be chemically aided. Filtration occurs before and potentially after sedimentation in the treatment process. Water treatment plants physically remove particles in a tiered system. CEWE Fig. 5.1 shows a typical water treatment process, in which large solids and debris are first filtered through screening, then flocculation and sedimentation take place to remove medium-size particles, and finally, fine particle filtration is performed. This allows particles to be removed in the order of their size to reduce clogging and other operational issues. This tiered system maximizes the efficiency of removal, and it protects more specialized systems downstream of the treatment plant.

Because filtration uses a tiered approach, chemicals that have been added for specific purposes (such as salt for water softening) may be removed through subsequent filtration. Any needed chemical processes should be performed only at stages where filtration will not detract from the benefits of their addition.

Exam tip: When designing filtration steps, do so in order of largest particles to smallest, keeping in mind that the sedimentation process will create larger particles, called *floc*, for the sedimentation step. If the filter steps are reversed or out of order, larger particles of floc can clog the fine filters, and fine particles can pass through the larger particle filtration steps later.

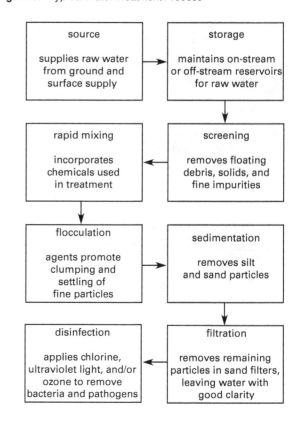

Figure 5.1 Typical Water Treatment Process

J. DISINFECTION

Key concepts: These key concepts are important for answering exam questions in subtopic 16.J, Disinfection.

- chlorination
- chlorination breakpoint
- dechlorination
- disinfection methods

NCEES Handbook: To prepare for this subtopic, familiarize yourself with these sections in the *Handbook*.

- Chlorination
- Dechlorination
- Chlorine Reactions in Water
- Chlorination Chart
- Hypochlorite Reactions in Water
- Baffling Factors

PE Civil Reference Manual (CERM): Study these sections in CERM that either relate directly to this subtopic or provide background information.

- Section 26.36: Disinfection
- Section 26.37: Chlorination Chemistry
- Section 26.38: Chlorine Dose
- Section 26.41: Dechlorination

Water Resources and Environmental Depth Reference Manual (CEWE): Study these sections in CEWE that either relate directly to this subtopic or provide background information.

- Section 5.9: Disinfection
- Section 6.23: Disinfection

The following equations, figures, tables, and concepts are relevant for subtopic 16.J, Disinfection.

Chlorination

Handbook: Chlorination; Chlorine Reactions in Water; Hypochlorite Reactions in Water; Baffling Factors

$$Cl_2 + H_2O \leftrightarrow H^+ + Cl^- + HOCl$$

$$HOCl \leftrightarrow H^+ + OCl^-$$

CERM: Sec. 26.37

CEWE: Sec. 5.9

A list of stoichiometric equations is provided in *Handbook* sections Chlorination, Chlorine Reactions in Water, and Hypochlorite Reactions in Water. Similar stoichiometric equations are provided in CERM Sec. 26.37 and CEWE Sec. 5.9. Chlorine can be added to the water as a gas, Cl_2, to form a liquid solution or as a solid (calcium hypochlorite). Residual chlorine must remain after the disinfection application to sustain the removal of pathogens throughout the distribution system. Note the common range of 0.2 mg/L to 0.5 mg/L provided in CEWE Sec. 5.9. The chemical equations show how chlorine reacts when applied as a gas or as a solid.

To be effective, chlorine must thoroughly mix with water to encounter as much of the biological contamination as possible. The different mixing methods are shown in *Handbook* table Baffling Factors.

Exam tip: The more thorough the mixing, the more efficient the use of chlorine will be in treatment.

Dechlorination

Handbook: Dechlorination

CERM: Sec. 26.41

A list of stoichiometric equations for dechlorination is provided in *Handbook* section Dechlorination. Excess chlorine can react with organic matter to produce trihalomethanes (THMs), including chloroform, which are detrimental to the health of people and aquatic life. For this reason, dechlorination is crucial to reducing public health hazards. The EPA limits the amount of total THMs to 80 parts per billion (ppb).

Dechlorination can be accomplished with a number of chemical additives. CERM Table 26.5 lists some of these and the dosages to be used. Monitoring of these chemicals is needed; excessive amounts may deplete the dissolved oxygen concentration and alter the pH.

Table 26.5 Dechlorinating Chemicals Used in Water Supply and Wastewater Facilities

name	formula	stoichiometric dose* (mg/mg Cl_2)
ascorbic acid	$C_6H_8O_6$	2.48
calcium thiosulfate	CaS_2O_3	1.19
hydrogen peroxide	H_2O_2	0.488
sodium ascorbate	$C_6H_7NaO_6$	2.78
sodium bisulfite	$NaHSO_3$	1.46
sodium metabisulfite	$Na_2S_2O_5$	1.34
sodium sulfite	Na_2SO_3	1.78
sodium thiosulfate	$Na_2S_2O_3$	0.556
sulfur dioxide	SO_2	0.903

*Doses are approximate and depend on pH and assumed reaction chemistry. Theoretical values may be used for initial approximations and equipment sizing. Under the best conditions, 10% excess is required.

Exam tip: Some dechlorination chemicals, such as sodium dioxide, are hazardous; this should be anticipated before design decisions are made. When calculating the amount of a dechlorinating chemical needed, anticipate that at least 10% more than the stoichiometric dose will be needed to completely remove all chlorine by-products, even with thorough mixing.

Chlorine Breakpoint

Handbook: Chlorination Chart

CERM: Sec. 26.38

Handbook figure Chlorination Chart shows a breakpoint chlorination curve. The purpose of breakpoint chlorination is to continue adding chlorine until the desired quantity of free residuals is reached. This cannot occur until the demand for combined residuals has been satisfied. CERM Fig. 26.7 shows a similar chlorination curve.

Figure 26.7 Breakpoint Chlorination Curve

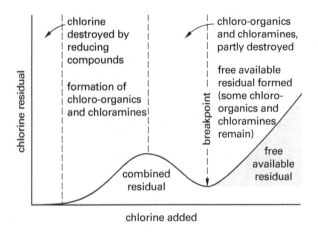

The amount of chlorine needed for disinfection varies with the organic and inorganic material present in the water, pH, temperature, and contact time. 30 min of chlorine contact time is generally sufficient to deactivate *Giardia* cysts. Satisfactory results can usually be obtained if a free chlorine residual of 0.2 mg/L to 0.5 mg/L can be maintained throughout the distribution system.

Exam tip: The chlorination breakpoint is reached when the amount of chlorine that has been added is just sufficient to treat the water. The goal is to be as close as possible to the breakpoint, fully treating biological contaminants without producing the by-products of excessive chlorination. Although an approximate concentration for chlorination is calculated, this will fluctuate throughout the year due to seasons and conditions of the source water.

Disinfection Methods

CERM: Sec. 26.36

CEWE: Sec. 6.23

Disinfection eliminates biological contaminants, such as bacteria and pathogens, from treated water. By the disinfection stage, particle removal is completed in the water treatment through sedimentation and/or filtration. It is also the last step in the health treatment of potable water. Disinfection is completed by means of two main methods: *physical*, which uses energy to inactivate pathogens, and *chemical*, which uses chemicals to inactivate pathogens.

Disinfection can come in the form of radiation, such as ultraviolet (UV) light or heat, or as a chemical, like chlorine or ozone. Chlorine disinfection is the method addressed in most exam problems on disinfection. UV treatment is addressed in Chap. 50 of this book. Other treatment processes are not addressed in the *Handbook*.

Exam tip: Chlorine disinfection produces by-products that may be a public health hazard, whereas other disinfection processes may not. The primary reason for chlorine use is cost and familiarity. UV light is expensive, requiring a lot of lamps and energy, while ozone often must be produced on site due to a short lifespan, impacting storage.

K. HARDNESS AND SOFTENING

Key concepts: These key concepts are important for answering exam questions in subtopic 16.K, Hardness and Softening.

- formula weight
- hardness
- neutral solutions
- softening

NCEES Handbook: To prepare for this subtopic, familiarize yourself with these sections in the *Handbook*.

- Periodic Table of Elements
- Lime-Soda Softening Equations
- Common Water Softening Compounds and Molecular Properties

***PE Civil Reference Manual* (CERM):** Study these sections in CERM that either relate directly to this subtopic or provide background information.

- Section 22.8: Formula and Molecular Weights
- Section 22.9: Equivalent Weight
- Section 25.1: Cations and Anions in Neutral Solutions
- Section 25.3: Alkalinity
- Section 25.5: Hardness
- Section 26.31: Precipitation Softening
- Appendix 22.A: Atomic Numbers and Weights of the Elements
- Appendix 22.C: Water Chemistry $CaCO_3$ Equivalents

Water Resources and Environmental Depth Reference Manual (**CEWE**): Study these sections in CEWE that either relate directly to this subtopic or provide background information.

- Section 5.10: Water Softening

The following equations, figures, tables, and concepts are relevant for subtopic 16.K, Hardness and Softening.

Formula Weight

Handbook: Periodic Table of Elements; Common Water Softening Compounds and Molecular Properties

CERM: Sec. 22.8, Sec. 22.9, Sec. 25.1, App. 22.A

The *formula weight* is the sum of atomic weights of all atoms in a molecule's empirical formula. The formula weight is not always equal to the molecular weight.

The molecular formula indicates the type and number of atoms in a molecule. The empirical formula is also known as the *simplest formula*. It is used to indicate the molecular ratio of elements present in a compound.

To determine the formula weight of molecules, consult *Handbook* figure Periodic Table of Elements. If a contaminant's mass is given, it will be given in grams per mole; this value is useful in determining treatment processes, especially chemical treatment, for the contaminant. *Handbook* table Common Water Softening Compounds and Molecular Properties lists several compounds and their properties.

Exam tip: An exam problem may ask for a given contaminant to be calculated equivalent to a primary contaminant. In the case of hardness or alkalinity, it might be asked for, or it might be presented as, for example, "mg/L as $CaCO_3$." This ratio is used to relate the contaminant to the treatment process or to establish a relationship between different contaminants.

Softening

Handbook: Lime-Soda Softening Equations

CERM: Sec. 25.5, Sec. 26.31

CEWE: Sec. 5.10

For calcium carbonate hardness removal,

$$Ca(HCO_3)_2 + Ca(OH)_2 \rightarrow 2CaCO_3(s) + 2H_2O$$

For calcium noncarbonate hardness removal,

$$CaSO_4 + Na_2CO_3 \rightarrow CaCO_3(s) + 2Na^+ + SO_4^{2-}$$

For magnesium carbonate hardness removal,

$$Mg(HCO_3)_2 + 2Ca(OH)_2$$
$$\rightarrow 2CaCO_3(s) + Mg(OH)_2(s) + 2H_2O$$

For magnesium noncarbonate hardness removal,

$$MgSO_4 + Ca(OH)_2 + Na_2CO_3$$
$$\rightarrow CaCO_3(s) + Mg(OH)_2(s) + 2Na^+ + SO_4^{2-}$$

Water softening removes the hardness in treated water caused by high levels of calcium and magnesium ions. Water hardness can be reduced using a lime-soda process, where lime ($Ca(OH)_2$) and soda ash (Na_2CO_3) are added to hard water to cause precipitation of excess sediments. Stoichiometric equations from the *Handbook*, CERM, and CEWE show the reactions that occur between these chemicals and hard water. Only a partial list of equations is shown here. The full list of equations can be found in *Handbook* section Lime-Soda Softening Equations.

Treated water for potable use should not exceed 500 mg/L as $CaCO_3$ of hardness, but it is more desirable for it to be at 100 mg/L to 150 mg/L as $CaCO_3$ of hardness. Excessively hard water can cause minerals to accumulate on surfaces (including the interiors of pipes) and may be considered a nuisance contaminant, exceeding allowable secondary drinking water standards.

Exam tip: Excessively hard water can clog pipes and contribute to other problems due to the excess minerals. When calculating hardness to determine whether it is within an allowable range or eligible for treatment, 100 mg/L as $CaCO_3$ is a good rule of thumb. It can be lower than this, but it should not be much higher because it can result in problems and user complaints.

Hardness

Handbook: Common Water Softening Compounds and Molecular Properties

CERM: Sec. 25.3, Sec. 25.5, App. 22.C

Hardness is caused by multipositive anions (also known as *polyvalent cations*). CERM Table 25.2 shows the most common cations that relate to hardness, and it highlights how they all have multipositive charges. Since the cations combine with anions, the table also gives a list of common anions related to hardness.

Table 25.2 Principal Cations and Anions Indicating Hardness

cations	anions
Ca^{++}	HCO_3^-
Mg^{++}	SO_4^{--}
Sr^{++}	Cl^-
Fe^{++}	NO_3^-
Mn^{++}	SiO_3^{--}

Depending on how common certain ions are, not all these cations and anions need to be quantified. Generally, calcium and magnesium are the most common cations, and bicarbonate, sulfate, and chloride are the most common anions. *Handbook* table Common Water Softening Compounds and Molecular Properties provides a summary of several water-softening compounds.

Exam tip: Anions are used as a surrogate measurement because the cations they bind with are responsible for water hardness. The measurement of anions is called *alkalinity*, which serves a similar role to hardness in water treatment considerations. Both alkalinity and hardness are measured in units as calcium carbonate ($CaCO_3$).

Neutral Solutions

CERM: Sec. 25.1

In neutral solutions, all positive charges (cations) must equal negative charges (anions). For the solution to be in a neutral state, the sum of the equivalent weights of the cations must equal the sum of the equivalent weights of the anions. To find the equivalent weight of elements that are already dissolved, divide the atomic weight of the elements by the number of their charges.

The table from CERM Ex. 25.1 shows a charge balance in solution. Many cations and anions may be in solution, but because solutions tend to be neutrally charged, including shifting pH, the equivalent concentration of ions will be neutral.

compound	concentration (mg/L)	equivalent weight (g/mol)	equivalent (meq/L)
cations			
Ca^{2+}	29.0	20.0	1.45
Mg^{2+}	16.4	12.2	1.34
Na^{2+}	23.0	23.0	1.00
K^+	17.5	39.1	0.44
		total	4.23
anions			
HCO_3^-	171.0	61.0	2.80
SO_4^{2-}	36.0	48.0	0.75
Cl^-	24.0	35.5	0.68
		total	4.23

On either side of the reaction, the positive and negative ions balance. This is related to other conservation principles, where molecular charge must be consistent throughout the equation. An unbalanced charge implies that electrons were lost or added in a reaction step that is not being shown but may still be present in solution.

Exam tip: Treatment of inorganic contaminants passing through a treatment facility is achieved by using the atomic charges and equivalently charged cations and anions to determine appropriate treatment chemicals. Knowing these charges helps in the dosing or ion exchange processes.

53 Engineering Economics Analysis

Content in blue refers to the *NCEES Handbook*.

Economics Analysis... 53-1

ECONOMICS ANALYSIS

The knowledge area of Engineering Economics Analysis makes up between one and three questions out of the 80 questions on the PE Civil Water Resources and Environmental exam. See Chap. 9, Engineering Economics, in this book for entries relevant to this knowledge area.

Index

A

AASHTO Soil Classification, 17-9
AASHTO Soil Classification System, 7-2, 10-12
Absolute Volume Method, 7-8, 14-7
Acceleration and Deceleration Length, 38-3
Accident Analysis, 35-11
Activated Carbon Adsorption, 52-8
Activated Sludge, 45-7
Active Earth Pressure, 24-2
Activity Identification and Sequencing, 1-4
Activity Time Analysis, 13-3
Activity-on-Arrow Diagram, 13-3
Activity-on-Node Diagram, 13-3
Additional Erosion and Sediment Control Measures, 8-6
Advanced Treatment, 50-24
Aerobic Digestion, 50-22
Aggregate Delay, 35-7
Aggregates, 21-14
Air Stripping, 52-8
Algae, 51-11
Allowable Compressive Load, 15-9, 34-6
Allowable Pile Load, 12-11
Allowable Strength Design Method, 2-1, 25-3
Allowable Stress Design (ASD), 15-1
Allowable Stress Design Versus Allowable Strength Design, 4-2
American Concrete Institute (ACI 318, 530), 33-3
American Welding Society (AWS D1.1, D1.2, and D1.4), 33-22
Americans with Disabilities Act (ADA) Design Considerations, 39-5
Ammonia Removal, 50-25
Ammonia Removal: Air Stripping, 50-16
Anaerobic Digestion, 50-22
Anaerobic, Aerobic, and Anoxic Treatment Steps, 50-22
Anchor Bolt Connections in Concrete, 29-9
Anchorage, 15-11, 34-9
Anchored Retaining Structures, 24-10
Aquifers, 12-6, 49-1
Areas of a Work Zone, 16-5
ASD Required Strength, 27-6
Asphalt Concrete, 42-6
Asphalt Concrete Design Methods, 42-6
Asphalt Service Life, 21-14
Asphalt-Concrete, 14-2
ASTM Material Testing, 7-8
At-Grade Intersection Layout, 38-4
At-Rest Coefficients, 3-2, 27-13
At-Rest Earth Pressure, 24-2
Atterberg Limits, 7-2
Auger Pile Installation, 12-11
Average Degree of Consolidation Versus Time Factor, 3-8
Average Depth of Pit Excavation, 10-6
Average End Area Method, 2-2, 8-2, 10-3, 11-2
Average Shear Stress, 28-10
Average Speed, 35-6
Average Speed (Mean Speed), 6-7
Axial, 4-7, 28-6
Axial Compression, 26-4
Axial Deformation, 28-7
Axial Load in Truss Members, 28-9
Axial Members, 4-8
Axial, Concentric, and Eccentric Loading, 28-6

B

Balance Line Between Two Points, 10-10
Bar Screens, 50-7
Barrier Design, 39-2
Barrier Design Criteria, 39-3
Barrier Flare Rate, 39-3
Barrier Installation Criteria, 39-3
Barrier Runout Length, 39-4
Basic Circular Curve Elements, 6-1
Basic Curve Elements, 36-1
Basic Vertical Curve Elements, 6-3
Basic Wind Speed, 27-5
Batch Reactors, Variable Volume, 45-3
Beam Deflection: Conjugate Beam Method, 28-15
Beam Deflection: Cracking Moment, 28-14
Beam Deflection: Double Integration Method, 28-14
Beam Deflection: Moment Area Method, 28-15
Beam Deflection: Strain Energy Method, 28-15
Beam Deflection: Superposition, 28-15
Beam Deflection: Table Lookup Method, 28-14
Beam Deflections, 4-9
Beam Deflections: Moment of Inertia, 4-10
Beam Lateral Bracing, 34-9
Beam Shape, 4-13
Beams, 4-10, 15-6, 28-5
Bearing, 28-17
Bearing Capacity, 3-10, 25-1
Bearing Capacity Equation: Concentrically Loaded Square or Rectangular Footings, 3-11
Bearing Capacity Equation: Concentrically Loaded Strip Footings, 3-10
Bearing Capacity Failure, 24-8
Bearing Capacity for Concentrically Loaded Square or Rectangular Footings, 25-2
Bearing Capacity for Concentrically Loaded Strip Footings, 25-1
Bearing Capacity Theory, 3-10
Bearing, Angles, and Azimuth, 36-2
Bearings, 15-7
Bending, 4-4
Bending Moment Diagrams, 4-5
Bending Moments, 4-6
Bending Stresses, 4-5, 4-9
Bending Stresses in Beams, 4-11
Benefit-Cost Analysis, 9-3
Benefit-Cost Ratio, 11-4
Bernoulli Equation, 5-16, 46-3
Bicycle Lanes, 39-5
Bills of Quantity (BOQs), 1-2
Bioaccumulation, 51-12
Biochemical Oxygen Demand (BOD), 51-3
Biochemical Oxygen Demand and Decomposition of Waste, 51-9
Bioconcentration, 51-12
Biological Contaminants, 51-8
Biological Nitrogen Removal, 50-14
Biological Phosphorus Removal, 50-18
Biomagnification, 51-13
Block Failure of Pile Groups, 26-7
Block Shear, 30-11
Bolt Testing, 14-4
Bolts, 30-12
Book Value, 9-3
Boring, 17-6
Boring Log, 7-1
Borrow Pit Grid Method, 2-2, 8-2, 10-4, 10-6, 11-2
Borrow Pit Volumes, 10-6
Boussinesq Stress Contour Chart, 25-4
Braced Cuts, 8-3
Bracing, 15-10, 34-8
Bracing and Anchorage for Stability, 15-9
Braking Distance, 36-4
Buckling, 28-17
Buffer Widths, 16-6
Buoyancy Resistance, 24-14

C

Cable Barriers, 39-2
California Bearing Ratio, 18-5
California Bearing Ratio Test, 42-2
Capacities of Earth-Handling Equipment, 10-11
Capacity (Signalized Intersection), 40-2
Capital Recovery, 9-2
Cast-in-Place Concrete Piers, 24-12
Cavitation, 46-10
Channelized Flow, 48-11
Check Dam, 8-6
Chemical Contaminants, 51-11
Chemical Oxygen Demand (COD), 51-7
Chemical Phosphorus Precipitation, 50-18
Chemical Phosphorus Removal, 50-18
Chezy Equation, 5-3, 45-13
Chlorination, 50-23, 52-15
Chlorine Breakpoint, 52-15
Chord Length, 6-3
Clarifier Design Equations, 50-11
Clarifiers, 50-10, 50-12
Clark Unit Hydrograph, 48-8
Clear Runout Areas, 39-2
Clear Zones, 39-1
Clearance on Sag Vertical Curves, 37-6
Clearances on Horizontal Curves, 36-5
Coefficient of Consolidation, 18-8
Coefficient of Curvature, 7-4
Cofferdam Components, 24-11
Cofferdam Design, 24-12
Cofferdam Overview, 24-11
Cofferdams, 15-12, 24-11
Collapsible Soil, 23-1
Column Base Plates, 30-11
Columns, 4-13, 15-6
Combined Stresses, 4-8
Common Stake Marking Abbreviations, 10-8
Compact, Noncompact, and Slender Elements, 30-2
Compaction, 7-8
Compaction and Settlement, 42-3
Compaction Equipment, 12-9
Compliance Inspections, 16-2
Composite Steel Beams, 30-5
Compound Horizontal Curves, 36-8
Compressible and Incompressible Fluids, 45-3
Concentrated and Uniformly Distributed Live Loads, 27-3
Concentric Loading, 4-9
Concrete, 7-6, 29-1

Concrete and Masonry Construction, 34-4
Concrete Axial and Combined Axial and
  Flexural Strength, 29-7
Concrete Beams, 29-3
Concrete Braced Frames, 34-8
Concrete Columns, 29-7
Concrete Compressive Strength, 7-6, 29-1,
  34-7
Concrete Construction, 34-1
Concrete Corbels, 29-9
Concrete Design, 42-7
Concrete Diaphragms, 29-4
Concrete Durability, 34-7
Concrete Exposure and Mix Requirements,
  29-2
Concrete Maturity, 34-7
Concrete Maturity and Early Strength
  Evaluation, 14-8, 34-6
Concrete Maturity: Time-Temperature
  Factor Method, 14-8
Concrete Mix Design, 14-6
Concrete Modulus of Elasticity, 29-2
Concrete One-Way Shear, 29-5
Concrete Overview, 7-6
Concrete Proportioning and Placement,
  14-6
Concrete Slabs, 29-4
Concrete Strength Gain Factors, 14-8
Concrete Strength Testing, 34-6
Concrete Strength-Gain Curve, 14-9
Concrete Testing, 14-1, 29-2
Concrete Torsional Strength, 29-6
Concrete Two-Way Shear, 29-6
Concrete Walls, 29-7
Cone Penetrometer Test (CPT), 17-8
Conical Spoil Banks, 10-7
Connections, 29-8, 30-9, 31-8, 32-8
Conservation of Mass, 45-3
Consolidation, 3-5
Consolidation and Normally Consolidated
  Soils, 18-7
Consolidation Curves for Normally
  Consolidated Soils, 3-6
Consolidation of a Layer of Clay Between
  Two Pervious Layers, 3-7
Constant- and Falling-Head Tests, 18-10
Constructing Bending Moment Diagrams,
  28-2
Constructing Shear Diagrams, 28-1
Construction, 18-11
Construction Activity Sequencing, 13-6
Construction Bracing, 34-9
Construction Documents, 34-3
Construction Loads, 2-1, 15-2, 27-3
Construction Loads, Codes, and Standards,
  15-1
Construction Methods, 2-2
Construction Sequencing, 13-1
Construction Site Layout and Control, 8-4
Construction Stakes for Storm Drains, 10-8
Continuity Equation, 5-16, 45-2, 46-3, 47-2
Conveyance, 5-3
Cooper Jacobs Equations, 49-6
Coordinate Method, 2-3, 8-3, 10-4, 11-2
Coring, 26-14
Corner Ramps, 39-6
Corrosion, 23-4
Corrosion Protection for Steel, 23-4
Cost Analysis for Resource Selection, 11-4
Cost Estimate Classification, 1-2
Cost Estimating, 1-2, 11-3
Cost Index, 1-2, 11-3
Cost per Unit of Time Reduction, 13-7
Cost Variance, 11-4
Coulomb Coefficients for Sloping Wall, 24-4
Crane Loading, 8-9
Crane Selection Process, 12-5
Crane Specifications, 12-5
Crane Stability, 12-3
Crane Use and Safety, 34-12
Cranes, 8-8
Crash Analysis, 35-11
Crash Cushions, 39-4

Crash Factors, 35-11
Crash Rates, 35-11
Crashing the Schedule, 13-7
Crew Distribution Charts, 13-6
Crew Hours, 11-6
Critical Depth, 47-11
Critical Path, 13-4
Critical Path Method Network Analysis,
  13-4
Critical Path Method Precedence
  Relationships, 1-3
Critical Volume-Capacity Ratio, 40-2
Cross Section Elements, 35-16, 39-4
Cross Sections, 42-4
Cross Slopes, 39-4
Crosshole Seismic Tomography, 17-5
Crosshole Sonic Logging (CSL) Test, 26-13
Culvert Flow Classification: Type 1, 47-9
Culvert Flow Classification: Type 2, 47-9
Culvert Flow Classification: Type 3, 47-10
Culvert Flow Classification: Type 4, 47-10
Culvert Flow Classification: Type 5, 47-10
Culvert Flow Classification: Type 6, 47-10
Culverts, 5-5, 47-8
Culverts and Culvert Design, 43-8
Curb Ramps, 39-6
Curve Elevation at Any Point on a Vertical
  Curve, 6-4, 37-2
Curve Radius, 6-2
Curve Radius and Superelevation, 35-15
Curve Widening, 35-16
Cut and Fill, 8-3
Cutoffs, 22-8
Cyclic Resistance Ratio (CRR), 20-2
Cyclic Stress Ratio (CSR), 20-2

D

Darcy-Weisbach Equation, 46-7
Darcy's Law, 7-6, 49-3
Data Gaps in Precipitation Gauges, 48-14
Dead and Live Loads, 4-1
Dead Loads, 4-1, 27-1
Dead Loads in Structural Analysis, 27-2
Dechlorination, 50-24, 52-15
Decision Sight Distance, 36-5
Deep Foundation Installation, 12-10
Deflection, 4-9, 28-13
Deformation Monitoring, 19-5
Degree of Curve, 6-1, 36-2
Degree of Saturation, 7-4
Demands, 52-3
Denitrification Rate, 50-16
Denitrification Reaction, 50-15
Density, 35-9
Density, Specific Weight, and Specific
  Gravity, 45-1
Deoxygenation and Oxygen Deficit, 51-6
Depletions, 48-15
Description and Classification of Soils, 17-9
Design Acceleration Response, 27-7
Design Considerations, 4-17, 27-4
Design for Small Eccentricity, 4-15
Design of Axial Load in Concrete Members,
  28-8
Design of Axial Load in Masonry Members,
  28-9
Design of Axial Load in Steel Members,
  28-9
Design of Axial Load in Wood Members,
  28-9
Design of Concrete Beams for Flexure,
  28-12
Design of Concrete Formwork, 15-5
Design of Masonry for Flexure, 28-12
Design of Masonry for Shear, 28-10
Design of Reinforced Concrete for Shear,
  28-10
Design of Retaining Walls, 4-18
Design of Wood for Flexure, 28-12
Design of Wood for Shear, 28-10
Design Sequencing Phases, 1-6, 13-2
Design Strength and Criteria, 4-13

Design Traffic Analysis and Pavement
  Design Procedures, 42-5
Design Traffic and Trucks, 42-6
Design Vehicles, 6-8
Detention and Retention, 48-17
Detention and Retention Ponds, 48-20
Detention and Retention: Modified Puls
  Routing Method, 5-12
Detention and Retention: Rational Method,
  5-11
Detention and Retention: Routing, 5-12
Detention/Retention Ponds, 5-11
Determining Crest Curve Length, 37-4
Determining Stake Location, 10-8
Detour Signs, 41-3
Development Length of Steel, 29-8
Development of Hydrographs, 48-8
Dewatering and Pumping, 2-3, 12-6
Dewatering Design, Methods, and Impact
  on Nearby Structures, 22-3
Diagrams, 28-1
Digestion, 50-21
Direct Shear Test, 18-4
Disinfection, 50-23, 52-14
Disinfection Methods, 52-16
Distributed Loads, 27-1
Distribution of Vertical Stress by the 2:1
  (60°) Method, 25-4
Distribution of Vertical Stress by the 2:1
  Method, 3-12
Disturbed Samples, 17-7
Diversions, 48-17
Downdrag, 26-5
Downdrag Designed as Settlement, 26-6
Downdrag Mitigation, 26-6
Drag Loads (Downdrag), 26-6
Drainage, 21-6
Drainage Design/Infiltration, 22-5
Drained Shear Strength of Clays, 3-9
Drilling Techniques, 17-5
Drinking Water Distribution, 52-2
Drinking Water Distribution Systems, 52-1
Drinking Water Standards, 52-1
Drinking Water Treatment Processes, 52-2
Drivability, 26-9
Driven Foundations, 26-8
Dry and Wet Soil Mixing, 21-5
Dupuit's Formula, 49-6
Dynamic Deep Compaction, 21-4
Dynamic Load Allowance, 27-11
Dynamic Testing, 26-10

E

Earned Value Method, 11-4
Earned-Value Analysis: Forecasting, 1-6
Earned-Value Analysis: Indices, 1-5
Earned-Value Analysis: Variances, 1-5
Earth Dams, Levees, and Embankments,
  21-9
Earth Pressure and Surcharge Loads, 27-11
Earth Pressure Coefficient, 24-1
Earthquake Principles, 20-3
Earthquake-Induced Ground Displacement,
  20-5
Earthwork Mass Diagrams and Haul
  Distance, 10-9
Earthwork Volumes, 42-3
Eccentric Bolt Group, 30-10
Eccentric Loading, 4-9
Eccentrically Loaded Welds, 30-12
Economic Factor Conversions, 9-1
Effective and Total Stresses, 3-8, 18-11
Effective Green Interval, 40-3
Effective Length, 4-14
Effective Length Factor, 28-7
Effective Length Factor, K, 30-7
Effective Settling Area in a Clarifier, 50-9,
  50-13
Effective Shear Strength, 3-9
Effective Stress, 18-12
Effective Stresses, 18-3
Effective Walkway Width, 35-13
Elastic Compression, 26-11

Elastic Method, 25-5
Elastic Method of Settlement, 3-13
Electrical Resistivity Tomography (ERT), 17-4
Electrical Safety, 8-9
Elements of an Intersection, 35-7
Elements of Horizontal Curves, 36-1
Embankment Construction, 19-2
Energy and Hydraulic Grade Lines, 46-4
Energy and/or Continuity Equation, 5-15, 46-1
Energy Dissipation at Culvert Outlets, 47-7
Energy Dissipators, 47-6
Energy Equation, 45-2
Energy Gradient, 47-2
Energy Loss Due to Friction: Laminar Flow, 46-6
Energy Loss Due to Friction: Turbulent Flow, 46-6
Engineering Economics, 9-1, 11-5
Environmental Loads, 15-3
Equilibrium, 4-3
Equipment Operations, 12-8
Equipment Productivity and Selection, 11-6
Equipment Selection, 26-9
Equivalent Lateral Forces, 27-7
Erosion and Sediment Control Permits, 8-6
Erosion Control Methods, 42-5
Erosion Prevention, 47-7
Establishing Slope Stake Markings, 10-9
Estimated Cycle Length, 40-3
Estimating Precipitation, 5-7
Euler Buckling, 28-8
Euler's Column Buckling Theory, 4-14
Eutrophication, 51-8
Evapotranspiration, 48-17
Excavation and Embankment, 8-1, 10-1, 42-4
Excavation and Embankment, Borrow Source Studies, Laboratory and Field Compaction, 19-1
Excavation and Loading Equipment, 12-9
Excavation by Soil Type, 2-4
Excavation Supports, 15-3
Excavations by Soil Type, 16-3
Excavations in Layered Soils, 16-3
Expansive Soil, 23-2
Experience Modification Rate, 8-8, 16-4
External Distance, 6-2

F

Factor of Safety: Liquefaction, 20-1
Failure by Breaching, 21-11
Failure by Piping, 21-10
Failure by Toe Heave, 21-11
Failure Modes, 21-10
Failure Surfaces, Pressure Distribution, and Forces, 3-2, 27-12
Fall Protection Anchor Point, 34-10
Falsework, 15-8, 34-5
Falsework and Scaffolding, 15-7, 34-5
Fast Tracking, 1-7, 13-2
Fatigue, 28-17
Field Compaction Methods, 7-9
Fillet Welds, 14-4
Filter Backwashing, 52-13
Filters, 22-6
Filtration, 21-6, 50-25, 52-13
Final Clarifiers, 50-25
Fine Screening, 50-8
Finishing and Grading Equipment, 12-10
Firefighting Demand for a Population, 52-3
Firefighting Demand for an Individual Building, 52-3
Flat Roof Snow Load, 27-9
Flexible Retaining Wall Site Considerations, 24-10
Flexible Retaining Wall Stability Analysis, 24-9
Flexural Strength of Singly Reinforced Concrete Sections, 29-4
Flexure, 28-10
Flocculation, 52-11

Flocculation Basins, 52-12
Flocculation Process, 52-11
Floods, 48-9
Flow Direction, 49-5
Flow Nets, 17-11, 22-2
Flow Rate, 22-2
Flow Reactors, Steady State, 45-4
Flow-Density Model, 6-6, 35-5
Fluid Power Equations, 46-9
Food-to-Microorganism Ratio, 50-13
Footings, 4-17
Force Wastewater Flow System, 50-1
Forces Behind a Gravity Wall, 24-6
Forces for Ordinary Method of Slices, 3-15
Forgiving Roadside Concepts, 39-1
Formula Weight, 52-17
Formwork, 2-5, 15-3, 34-3
Formwork Pressure, 2-6
Foundation Elements, 24-13
Foundation Settlement, 3-12
Foundation Subgrade Improvement Principles, 25-5
Foundation Uplift Resistance in Rock, 24-13
Foundations, 29-9
Four-Leg Intersections, 38-4
Frame Deflection: Unit Load Method, 28-14
Frames, 28-5
Free Float, 13-4
Free-Body Diagrams, 4-2
Free-Flow Speed, 35-3
Freehaul and Overhaul, 10-10
Friction Factor, 5-13
Friction Loss Coefficient, 46-7
Friction Losses, 5-3
Frost Depth, 3-11
Frost Heave Parameters, 23-5
Frost Susceptibility, 23-5
Frost Susceptibility Mitigation, 23-5
Frost-Susceptible Soils, 23-5
Froude Number, 47-11

G

Gantt Charts, 1-4, 13-2
General Types of Intersections, 38-4
Geologic Profiles, 18-12
Geometric Channel Sections, 47-4
Geometric Characteristics, 4-15
Geophysics, 17-4
Geosynthetic Applications, 21-5
Geotechnical Instrumentation, 19-4
Geotechnical OSHA Regulations, 19-3
Gradient and Uplift, 22-1
Grading Activities, 19-2
Gravity Model, 35-15
Gravity Wastewater Flow System, 50-2
Greenshields Maximum Flow Rate Relationship, 6-6, 35-3
Grit Chambers, 50-8
Ground Anchors, Tie-Backs, Soil Nails, and Rock Anchors for Foundations and Slopes, 24-13
Ground Improvement, 21-3
Ground-Penetrating Radar (GPR), 17-5
Groundwater Exploration, Sampling, and Characterization, 17-10
Groundwater Flow, 49-3
Groundwater Through an Aquifer, 17-11
Group Effects, 26-6
Group Efficiency, 26-7
Grouting, 22-7
Grouting and Other Methods of Reducing Seepage, 22-7
Guardrail End Treatment, 39-4
Guide Signs, 41-2

H

Hardness, 52-17
Hardness and Softening, 52-16
Hardy Cross, 46-14
Hauling and Placing Equipment, 12-10
Hazen Uniformity Coefficient, 7-3
Hazen-Williams Equation, 45-12, 46-7
Hazen-Williams Velocity, 5-3

Hazen's Equation for Permeability, 7-5
Headway, 6-5, 35-2
Heel Load Distribution, 24-6
Highway Safety Analysis, 35-15
Horizontal Broad-Crested Weirs, 45-11
Horizontal Circular Curves, 6-1
Horizontal Members, 29-3, 30-3, 31-2, 32-3
Horizontal Pressure on Walls from Compaction Effort, 24-5
Horton Model, 48-17
Hourly Volume Method, 35-9
Hydraulic Energy Dissipation, 47-5
Hydraulic Flow Measurement, 45-8
Hydraulic Grade Line, 47-8
Hydraulic Gradient, 22-3, 48-18, 49-2
Hydraulic Jump, 47-6
Hydraulic Loading, 45-3
Hydraulic Loading Rate, 45-4
Hydraulic Radius, 5-1, 47-3
Hydraulic Retention Time, 45-5
Hydraulics, 43-5
Hydrograph Development and Applications, 48-6
Hydrograph Separation, 5-11
Hydrograph Synthesis, 43-4
Hydrographs, 43-3
Hydrologic Budget, 48-16
Hydrologic Cycle, 5-6, 48-2
Hydrology, 43-1

I

Ice Loads, 27-10
Identification and Characterization of Problematic Soil and Rock, 17-2
IDF Curves, 48-10
Impact Loads, 27-10
Impact of Construction on Adjacent Facilities, 8-7
Impaired Waters, 51-8
Impoundment Storage Capacity, 52-5
Impulse-Momentum Principle, 5-16, 45-6, 46-2
In Situ Testing, 17-8
Incidence Rate, 16-4
Index Properties, 18-2
Index Properties and Testing, 18-1
Indicator Organisms, 51-10
Indirect In Situ Testing, 17-4
Industrial Wastewater, 50-4
Infiltration, 17-12, 22-6, 48-16
Infinite Slope Failure in Dry Sand, 3-15
Inflation, 9-3
Inflow and Infiltration, 50-5
Influence Area, 27-18
Influence Chart, 3-14
Influence Lines, 4-1, 27-8
Inspection, 14-5
Installation Methods/Hammer Selection, 26-8
Integrity Testing, 26-13
Integrity Testing Methods, 26-12
Intensity-Duration-Frequency Curves, 48-10
Interchanges, 38-2
Internal Forces and Stresses of a Truss, 4-8
International Building Code (IBC), 33-1
Interpretation of Available Existing Site Data and Proposed Site Development Data, 17-1
Interrupted Flow Pedestrian Facilities, 35-14
Intersection Capacity, 35-6, 35-8
Intersection Conflict Points, 38-5
Intersection Control Cases, 38-1
Intersection Delay, 35-7
Intersection Levels of Service, 35-7
Intersection Sight Distance, 36-6, 38-1
Intersection Sight Triangles, 38-1
Inverse Square Method, 48-14

K

K-Value Method for Crest Vertical Curves, 37-5

K-Value Method for Sag Vertical Curves, 37-5
Karst Mitigation, 23-3
Karst Topography, 23-3
Karst; Collapsible, Expansive, and Sensitive Soils, 23-1

L

Labor and Crew Hour Rates, 11-5
Lagging Storm Method, 48-9
Laminar Flow, 5-15
Landfill Design, 21-12
Landfill Liners and Caps, 21-12
Landfill Stability and Performance, 21-12
Landfills and Caps, 21-15
Lane Occupancy Used in Freeway Surveillance, 35-3
Lane Tapers, 16-6
Lane Widths, 39-5
Lateral Earth Pressure, 3-1, 24-1
Lateral Load and Deformation Analysis, 26-3
Lateral Load Capacity, 26-3, 26-7
Lateral Load Design Process, 26-3
Lateral Loads, 2-1
Lateral Pressure Due to Line Load Surcharges, 3-4, 27-14
Lateral Pressure Due to Point Load Surcharges, 3-3, 27-13
Lateral Pressure Due to Strip Load Surcharges, 3-4, 27-14
Lateral Pressure Due to Uniform Load Surcharges, 3-4, 27-14
Lateral Pressure on Formwork, 34-4
Leachate, 21-12
Lead and Lag Relationships, 1-5, 13-6
Length of Curve, 6-2
Level of Service, 6-8
Levels of Service, 35-4
Levels of Service for Freeways, 35-4
Levels of Service for Multilane Highways, 35-4
Levels of Service for Two-Lane Highways, 35-4
Lift Station and Wet Wells, 46-12
Lifting and Rigging, 12-1
Lifting Tensile and Compressive Forces, 12-1
Line of Sight for Trucks on Sag Vertical Curves, 37-7
Line of Sight on Sag Vertical Curves, 37-6
Linear Scheduling, 13-6
Liquefaction, 20-1
Liquefaction Analysis and Mitigation Techniques, 20-1
Lithology, 17-10
Live Load Reduction, 27-2, 27-18
Live Loads, 4-1, 27-2
Load and Resistance Factor Design (LRFD), 15-2
Load and Resistance Factor Design Method, 2-2, 25-3
Load Charts, 12-5
Load Combination Considerations, 27-17
Load Combinations, 27-16
Load Combinations for Concrete, Steel, and Masonry, 27-16
Load Combinations for Timber, 27-17
Load Distribution, 24-5
Load Duration Factors for Wood Structures, 27-6
Load Path Overview, 27-15
Load Paths, 27-15
Load Transfer Settlement, 26-5
Loading Rate, 52-13
Long-Term Deflections, 4-10
Loose and Compacted Volume, 8-2
Loss of Thickness for Buried Steel, 23-4
LRFD Bridge Design Specifications (AASHTO), 33-18

M

Manning Equation, 5-2, 43-6
Manning's Equation, 45-12, 47-3
Masonry, 32-1
Masonry Anchor Bolts, 32-8
Masonry Beams, 32-3
Masonry Columns, 32-5
Masonry Compressive Strength, 32-2
Masonry Construction, 34-2
Masonry Construction Components, 32-1
Masonry Design Methods, 32-2
Masonry Modulus of Elasticity, 32-3
Masonry Shear Walls, 32-5
Masonry Walls, 32-4
Mass Balance, 45-1
Mass Diagrams, 42-5
Material Properties and Testing, 14-1
Material Properties of Structural Steel, 30-1
Material Sampling, 17-4
Material Test Methods, 7-7
Material Testing and Observation, 19-2
Material Testing Tools, 19-2
Material Volume Change Characteristics, 10-1
Material Volume Change During Earthmoving, 10-3
Materials Strength Testing, 18-2
Media Filtration, 52-14
Method of Joints, 4-3
Method of Sections, 4-3
Microorganisms, 51-10
Middle Ordinate, 6-2
Middle Ordinate Distance, 37-3
Minimum Base Flow, 51-2
Minimum Column Anchorage, 34-10
Minimum Design Loads for Buildings and Other Structures (ASCE/SEI 7), 33-20
Minimum Runoff Length, 36-7
Minimum Settling Time, 52-7
Minimum Tangent Runout Length, 36-8
Minor Losses, 5-14
Mix Proportions by Weight, 14-7
Mobile Cranes and Their Components, 12-3
Modified Accelerated Cost Recovery System (MACRS), 9-3
Modulus of Subgrade Reaction, 18-7
Mohr-Coulomb Method, 26-1
Moisture Content, 7-4
Moment Effect on Bearing Capacity, 25-2
Moment, Shear, and Deflection Diagrams, 28-3
Momentum Equation, 5-17, 46-3
Monod Kinetics, 51-9
Mononobe-Okabe Method, 20-4
Moody Friction Chart, 46-5
Moving Loads, 27-8
Moving Loads on Beams, 27-8
Multiloop Systems, 46-15
Multiple Lift Rigging, 12-2
Multiple Pumps in Parallel, 46-11
Multiple Pumps in Series, 46-11

N

National Design Specification for Wood Construction (NDS), 33-15
National Design Specification Supplement, 33-17
Net Positive Suction Head, 46-9
Network Diagrams, 13-4
Neutral Solutions, 52-18
NIOSH Lifting Equation, 8-10
Nitrification/Denitrification, 50-13
Nitrogen Concentration, 50-14
Nomenclature in Moment, Shear, and Deflection Diagrams, 28-2
Non-Annual Compounding, 9-3
Nondestructive Weld Inspections, 14-4
Nonmotorized Facilities, 35-12
Nonrecoverable Slopes, 39-1
Nonuniform Flow, 47-5
Normal Stress, 4-8
Normally Consolidated Soils, 3-6
Notched Beams, 31-4
NRCS Curve Number, 43-5
NRCS Curve Number Method, 5-11
NRCS Lag Method, 48-11
NRCS Method, 48-5
NRCS Peak Discharge Method, 5-11
NRCS Synthetic Unit Hydrograph, 48-7
NRCS Synthetic Unit Triangular Hydrograph, 48-8
Nuclear Gauge Method, 7-8
Nutrients, 51-7

O

Octanol-Water Partition Coefficient, 51-13
Odor and Taste Removal Through Dilution, 52-9
One-Way Slab, 4-15
Open Channel Flow, 43-6
Open Channel Flow in a Rectangular or Trapezoidal Channel, 43-6
Open-Channel Flow, 5-1, 47-1
Organic Carbon Partition Coefficient, 51-13
OSHA 1910, 16-2
OSHA 1910 General Industry, 33-23
OSHA 1926, 16-2
OSHA 1926 Construction Safety Standards, 33-23
OSHA Construction Standards, 16-5
OSHA Regulations, 34-10
OSHA Regulations and Hazard Identification/Abatement, 16-1
OSHA Regulations for Crane Use, 8-9
OSHA Regulations for Hoisting and Rigging, 12-2
OSHA Standards Overview, 16-1
Osterberg Cell Compression Test, 26-12
Other Quality Measures, 14-6
Other Temporary Traffic Control Scenarios, 41-3
Overall Nitrification Reaction, 50-15
Overconsolidated Soils, 3-7, 18-8
Overconsolidation Ratio, 18-8
Overturning Failure, 24-9
Overview of Snow Loads, 27-9
Oxygen Dynamics, 51-2
Oxygen Saturation, 51-5

P

Parallel Pipe Systems, 5-13
Parshall Flume, 45-10
Particle Filtration, 22-8
Particle Size Distribution, 7-3, 10-11
Partition Loads, 27-3
Passing Sight Distance, 36-5, 37-4
Passive Earth Pressure, 24-3
Pathogens, 51-11
Pavement Damage from Frost and Freezing, 42-9
Pavement Design, 18-11
Pavement Design Categories, 21-13
Pavement Evaluation and Maintenance Measures, 42-8
Pavement Markings, 41-2
Pavement Preservation, 42-7
Pavement Problems and Defects, 42-8
Pavement Structural Numbers, 42-7
Pavement Structures, 21-13
Pavement Thickness, 21-14
Paving Equipment, 42-7
Peak Factors, 50-6
Peak Flow, 5-8
Peak Hour Factor, 35-9
Pedestrian Crossing Factors, 40-3
Pedestrian Crossing Time, 40-2
Pedestrian Detectors, 39-6
Pedestrian Islands and Medians, 39-6
Pedestrian Level of Service, 35-14
Pedestrian Platoons, 35-14
Pedestrian Unit Flow Rate, 35-13
Pedestrian Walking Space Requirements, 35-12
Pedestrian Walking Speed, 35-12
Permeability, 49-4
Permeability of Soils, 7-5
Permeability Testing, 18-3

Permeability Testing Properties of Soil and Rock, 18-9
Permissible Noise Exposure, 8-9, 16-3
Personal Protective and Life Saving Equipment, 34-11
Phase Diagram, 7-4
Phosphorus Concentration, 50-17
Phosphorus Removal, 50-10, 50-17, 50-18
Physical Requirements, 51-2
Piezometers, 22-3
Pile and Drilled-Shaft Load Testing, 26-11
Pile Driving Induced Settlement, 26-5
Pile Dynamics, 26-9
Pile Load Tests, 12-11
Pile-Driving Hammers, 12-11
Pipe Network Analysis, 46-12
Pipes in Parallel, 46-13
Pipes in Series, 46-13
Pitot Tube, 45-9
Pitot-Static Gauge, 5-14
Plastic Section Modulus and Moment, 28-12
Plasticity Index, 7-3, 10-12
Plate Bearing Value Test, 42-2
Pond Outlets—Culverts, 48-18
Pond Outlets—Weirs, 48-21
Pond Routing, 48-20
Porosity, 7-4, 49-2
Post-Construction Stabilization, 21-3
Potable Water Demand, 52-4
Potable Water Storage, 52-6
Precast/Prestressed Concrete Institute (PCI Design Handbook), 33-8
Precedence Diagrams, 13-1
Preliminary Treatment, 50-7
Preliminary Treatment Processes, 50-9
Pressure Conduit, 5-12, 46-4
Pressure Field, 49-1
Pressure Relationships in a Static Liquid, 46-1
Primary Consolidation, 21-15
Primary Treatment, 50-9
Prismoidal Formula, 2-2, 10-4, 11-2
Prismoidal Formula Method, 8-2
Probability of Flooding, 5-9
Profile Mass Diagrams, 10-10
Profiling, 18-12
Project Plans, 17-2
Project Schedules, 1-3
Properties of Pavement Quality, 42-8
Properties of Structural Steel, 7-7
Pseudostatic Analysis and Earthquake Loads, 20-4, 24-7
Pump Application and Analysis, 46-8
Pump Brake Horsepower, 46-10
Pump Design, 12-8

Q

Quality and Quality Control, 14-5
Quality Control Process, 14-5
Quality Control Stages, 14-5
Quantity and Cost Estimates, 11-1
Quantity Surveys, 1-1
Quantity Takeoff Methods, 1-1, 11-1
Quick Load Compression Test, 26-12

R

Radial Flow Mixers, 52-10
Raft Foundations, 2-5
Railroad Crossing Safety, 35-16
Rain Loads, 27-10
Rainfall and Stream Gauging Stations, 48-13
Rainfall Gauging, 43-2
Rainfall Intensity, 5-9, 48-10
Rainfall Intensity, Duration, and Frequency, 48-9
Rainfall Measurement, 48-3
Ramps and Intersections, 35-16
Range Diagrams, 12-5
Rankine Active and Passive Coefficients for Friction and Cohesion, 24-3
Rankine Active and Passive Coefficients for Friction Only, 24-3

Rankine Active and Passive Failure Zones, 3-2
Rapid Mixer Impellers, 52-11
Rapid Mixing, 52-9
Rapid Mixing Tanks, 52-10
Rate and Period Changes, 11-3
Rate of Grade Change per Station, 6-3, 37-2
Rational Formula Method, 12-7
Rational Method, 5-10, 43-3, 48-4
Reaction Rate, 51-4
Reactive/Corrosive Soils, 23-3
Recordable Injury or Illness, 16-4
Recoverable Slopes, 39-1
Red Clearance Interval, 40-3
Reel and Paddle Mixers, 52-10
Regulatory Signs, 41-2
Reinforced Earth Retention, 24-9
Reinforced Masonry Allowable Axial Compressive Stress: Allowable Stress Design, 32-6
Reinforced Masonry Allowable Combined Axial Compressive and Flexural Stresses: Allowable Stress Design, 32-7
Reinforced Masonry Allowable Flexural Stress: Allowable Stress Design, 32-4
Reinforced Masonry Allowable Shear Stress: Allowable Stress Design, 32-7
Reinforced Masonry Axial Compressive Strength: Strength Design, 32-6
Reinforced Masonry Combined Axial Compressive and Flexure Strength: Strength Design, 32-7
Reinforced Masonry Flexural Strength: Strength Design, 32-5
Reinforced Masonry Shear Strength: Strength Design, 32-7
Reinforced Slopes with Geotextiles, 21-9
Reinforcement Development Length, 32-9
Reinforcement Splices, 29-8
Reinforcing Steel, 29-3
Relative Compaction, 7-9
Relative Soil Compaction, 10-2
Relative Stiffness Factor, 26-4
Reoxygenation, 51-5
Reservoir Routing, 48-21
Reshoring, 15-9, 34-6
Resilient Modulus, 18-6
Resource Leveling, 1-4, 13-5
Resource Scheduling, 13-5
Resource Scheduling and Leveling, 13-5
Retaining Walls, 4-17, 29-10
Retardation Factor R, 51-14
Retention Structures, 21-9, 21-10
Retention Systems, 24-10
Reverse Osmosis, 52-9
Revised Universal Soil Loss Equation, 45-6, 48-20
Reynolds Number, 5-13, 45-9, 46-5
Rigid and Semirigid Barriers, 39-2
Rigid Pavements, 42-7
Rigid Retaining Wall Stability Analysis, 24-7
Risk from Carcinogenic Substances, 51-14
Risk to Human Health, 51-14
Road Designation, 35-3
Roadway Elements for Safe Design, 35-10
Rock Classification and Characterization, 17-9
Rock Classification Principles, 17-10
Rock Cores, 17-6
Rock Quality Designation (RQD), 17-10, 18-6
Rock Slope Failure, 3-15
Roundabouts, 38-4
Running Speed, 35-9
Running Time, 35-6
Runoff Analysis, 5-10, 48-4
Runoff Detention and Retention Basins, 43-5

S

Safe Drinking Water Act, 51-14
Safeguards During Construction, 34-12

Safety, 8-7
Safety Incidence Rate, 8-8
Safety Management, 34-11
Safety Management and Statistics, 16-3
Safety Recordkeeping, 16-4
Sag and Crest Curves, 6-4
Sag Vertical Curve Length for Comfort, 37-5
Sample Handling and Storage, 17-8
Sampling and Testing, 42-1
Sampling Considerations, 17-7
Sampling Techniques, 17-6
Sand Cone Method, 7-9
Saturation Flow Rate, 35-8, 40-2
Scaffolding, 2-6, 15-8, 34-5
Scaffolds and Temporary Structures, 8-10
Schedule Variance, 13-7
Schmertmann's Method, 3-13, 25-5
Screw Anchors and Micropiles, 24-13
SCS (NRCS) Unit Hydrograph, 5-8
SCS Method, 48-5
Secondary Consolidation, 21-15
Secondary Treatment, 50-11
Section Factor, 5-2
Sediment, 51-1
Sediment Load, 45-8
Sediment Structure, 8-6
Sedimentation, 8-5, 52-6
Sedimentation Parameters, 45-8
Seepage Analysis/Groundwater Flow, 22-1
Seepage Berms, 22-8
Seepage Control, 21-4
Seepage from Flow Nets, 17-12
Seepage Velocity, 49-4
Seismic Coefficients and Friction Angle, 24-5
Seismic Loads, 27-6
Seismic Resistance, 34-2
Seismic Site Characterization, 20-2
Seismic Site Class Categories, 20-3
Seismic Site Classification, 20-3
Sensitive Soils, 23-2
Separation, 21-5
Series Pipe Systems, 5-15
Settlement, 21-14, 21-15
Settlement for Normally Consolidated Soils, 3-6
Settlement Monitoring, 19-5
Settlement Ratio for Overconsolidation, 3-8
Settlement, Dust, and Runoff, 8-7
Settlement, Including Vertical Stress Distribution, 25-4
Settling Velocity, 45-7, 52-7
Sewer Design Requirements, 50-6
Shallow Concentrated Flow, 48-12
Shallow Foundation Settlement, 25-5
Shallow Foundations, 29-10
Shape and Rigidity Factors, 25-5
Sharp-Crested Weirs, 5-4, 45-10
Sharp-Crested Weirs with End Contractions (Francis Equation), 45-11
Shear, 4-6, 28-9
Shear and Moment, 28-5
Shear and Moment Diagrams, 28-10
Shear at a Point on a Beam, 4-6
Shear Force Diagrams, 4-7
Shear Modulus, 28-10
Shear Strength: Total Stress, 3-9
Shear Stress in Beams, 4-7
Shear Stress-Strain Relationship, 28-10
Shear Stresses in Beams, 4-11
Sheet Flow, 5-7, 48-12
Shop Drawings, 34-3
Shored Construction, 27-3
Shoring, 15-9, 34-6
Shoring and Reshoring, 15-8, 34-5
Shoulders, 39-5
Shrinkage and Swell, 42-4
Shrinkage Factor, 10-2
Shy Lines, 39-3
Side Friction, 36-7
Sieve Analysis, 7-3
Sight Distance, 35-16

Sight Distance Considerations, 36-3
Signal Matching, 26-10
Signal Timing, 40-1
Signal Warrants, 40-3
Signs and Pavement Markings, 41-1
Silt Fence, 8-6
Simple Shear Connections, 30-13
Simpson's Rule, 2-3, 8-3, 10-5, 11-3
Single Payment Compound Amount, 9-1
Single Payment Present Worth, 9-1
Single-Element Axial Capacity, 26-1
Single-Element Settlement, 26-4
Site and Subsurface Investigations, 10-11
Site Characterization, 17-2
Site Characterization Tasks, 17-2
Site Layout and Control, 10-8
Site Reconnaissance Overview, 17-3
Site Reconnaissance Primary Site Features, 17-3
Site Reconnaissance Secondary Site Features, 17-4
Site-Specific Seismic Loads, 27-7
Skin Friction Capacity: Mohr-Coulomb Method, 26-2
Skin Friction Capacity: α-Method, 26-2
Skin Friction Capacity: β-Method, 26-3
Slab on Grade, 21-1
Slabs, 4-15, 15-6
Slender and Nonslender Elements, 30-3
Slenderness Ratio, 4-14
Slope, 5-2
Slope Configuration: Type A, 19-4
Slope Configuration: Type B, 19-4
Slope Configuration: Type C, 19-4
Slope Drain, 8-6
Slope Stability, 20-5
Slope Stability and Slope Stabilization, 21-7
Slope Stability Factor of Safety, 21-7
Slope Stability Failure, 21-8
Slope Stability Guidelines for Design, 3-14
Slope Stabilization with Piles, 24-14
Slope Stabilization with Soil Nails and Screw Anchors, 24-14
Slope Stakes, 8-5
Slope Stakes Along a Highway, 10-8
Sloped Roof Snow Load, 27-10
Sludge Dewatering, 50-20
Sludge Disposal, 50-21
Sludge Quantity, 50-20
Sludge Treatment, 50-20
Snow Drift, 27-10
Snow, Rain, Ice, 27-9
Snyder Synthetic Unit Hydrograph, 48-7
Softening, 52-17
Soil and Rock Slopes, 21-8
Soil Classification and Boring Log Interpretation, 7-1
Soil Classification and Soil Excavations, 8-9
Soil Classification Systems, 42-3
Soil Consolidation, 3-5
Soil Dead Loads, 27-2
Soil Indexing Formulas, 10-5
Soil Moisture Content, 49-2
Soil Phases, 10-5
Soil Properties, 7-4
Soil Size, 21-3
Soil Stabilization Techniques, 42-2
Soil Testing, 14-1
Soil-Water Partition Coefficient, 51-13
Soldier Piles and Lagging, 15-12
Solids Loading, 45-5
Solids Loading Rate, 45-6
Solids Treatment, Handling, and Disposal, 50-19
Space Mean Speed, 6-6, 35-2
Special Design Provisions for Wind and Seismic, 33-18
Special Formwork, 34-4
Special Horizontal Curves, 36-8
Special Inspection Requirements in the International Building Code, 34-2
Special Inspections, 34-1

Special Machinery Impact Loads: Elevators, 27-11
Special Machinery Impact Loads: Machinery, 27-11
Special Reinforced Shear Wall Reinforcement Requirements, 27-6
Special Topics, 28-15
Specification Conformance, 7-7
Spectral Response Parameter, 27-7
Speed-Density Model, 6-6, 35-5
Speed-Flow Model, 6-7, 35-6
Spiral Curves, 36-8
Stability Analysis of Transitional Failure, 3-15
Stability of Rock Slope, 3-15
Stability of Walls and Slopes, 21-6
Stabilization Methods, 21-2
Stake Location, 8-4
Stake Markings, 8-5
Staking, 8-4
Standard Penetration Test (SPT), 17-8
Standardized Soil Testing Procedures, 42-1
Static Load Compression Testing, 26-11
Statically Determinate Trusses, 4-4
Stationing on a Circular Curve, 6-3
Stationing on a Horizontal Curve, 36-3
Stay-in-Place Formwork, 27-4, 34-4
Steady-State Slope Stability, 18-10
Steel, 30-1
Steel Anchor Breakout Strength in Concrete, 34-10
Steel Anchor Strength in Concrete, 34-10
Steel Beam Shear Capacity, 30-5
Steel Beams, 30-4
Steel Columns, 30-7
Steel Construction, 34-1
Steel Construction Manual (AISC), 33-11
Steel Erection: Site Layout, Site-Specific Erection Plan, and Construction Sequence, 34-11
Steel Erection: Structural Steel Assembly, 34-11
Steel Flexural Capacity, 30-4
Steel Members with Combined Axial and Flexural Loading, 30-8
Stiffness and Rigidity, 4-10
Stokes's Law, 45-7
Stopping and Passing Sight Distance, 37-3
Stopping Sight Distance, 36-3, 37-3
Storage, 52-5
Storm Characteristics, 5-5, 48-1
Storm Distribution, 48-3
Storm Frequency, 48-1
Storm Sewers, 47-8
Storm/Flood Frequency Probabilities, 43-2
Stormwater Collection and Drainage, 5-5, 47-7
Stormwater Management, 48-17
Stormwater Systems, 43-7
Straight-Line Depreciation, 9-3
Strain Gauges, 19-5
Stream Degradation, 51-1
Stream Gauging Stations, 48-15
Stream Hydrograph, 5-10
Street Gutter Flow, 5-5
Street Inlets, 5-5
Street Segment Interrupted Flow, 35-5
Strength Reduction Factor, 29-2
Strength Testing of Soil and Rock, 18-4
Stress States on a Soil Element, 3-1
Stress-Strain Curves for Steel, 7-7
Stress-Strain Testing, 18-2
Stress-Strain Testing of Soil and Rock, 18-6
Stress, Pressure, and Viscosity, 45-6
Stresses in Beams, 28-11
Strong and Weak Sewage, 50-4
Structural Integrity, 2-1
Structural Shape Designations, 7-7
Structural Steel, 7-7
Structural Steel Section Properties, 30-2
Sub- and Supercritical Flow, 47-11
Subcritical and Supercritical Flow, 47-12
Subgrade Compaction, 21-2

Subgrade Preparation, 21-2
Submittal Requirements, 34-2
Submittals, 34-2
Subsidence, 22-4
Substructure Construction, 34-2
Subsurface Drainage, 22-5
Subsurface Exploration Planning, 17-3
Suggested Advance Warning Sign Spacing, 16-6
Sulfates in Soils, 23-4
Sumps and Pumps, 22-4
Superelevation, 36-6
Superelevation Design, 36-7
Surcharge Loading, 27-15
Surface Drainage, 22-6
Surface Roughness Categories, 27-5
Survey Leveling, 10-9
Swales and Infiltration Systems, 48-19
Swell and Shrinkage, 8-2
Swell and Swell Factor, 10-3
Systems, 29-8, 30-9, 31-8, 32-8

T

Takeoff Report, 1-1
Taper Length Criteria, 41-3
Taste and Odor Control, 52-7
Taylor's Chart for Slope Angles or Friction Angles, 3-14
Temporary and Permanent Soil Erosion and Sediment Control, 8-5
Temporary Berms, 8-6
Temporary Seeding and Mulching, 8-6
Temporary Structures, 34-5
Temporary Structures and Facilities, 2-5
Temporary Support of Excavation, 15-11
Temporary Traffic Barriers, 41-3
Temporary Traffic Control, 41-2
Temporary Traffic Control Provisions, 39-7
Temporary Traffic Control Tapers, 41-3
Ten States Standards, 52-1
Tensile Capacity of Steel Members, 30-7
Test Pits, 17-6
Testing for Quality Assurance, 14-5
Theoretical Oxygen Demand (ThOD), 51-4
Theoretical Point-Bearing Capacity, 12-12
Thermal Deformation, 28-17
Thiem Equation, 49-6
Thiessen Network Method, 48-13
Three-Leg Intersections, 38-4
Timber, 31-1
Timber Design Method, 31-2
Time Mean Speed, 6-7, 35-6
Time of Concentration, 5-7, 43-2, 48-11
Time of Concentration Components, 48-13
Time of Concentration for Inlet Flow, 43-3
Time of Concentration for Sheet Flow, 43-2
Time Rate of Settlement, 3-8
Time to Peak, 5-8
Time-Cost Trade-Off, 1-4, 13-7
Tipping Load for a Horizontal Boom Tower Crane, 12-4
Tipping Load for a Lifting Boom Tower Crane, 12-4
Tipping Load for a Mobile Crane, 12-4
Toe Load Distribution, 24-6
Torsional Stress and Strain, 28-16
Total Dynamic Head, 12-7
Total Dynamic Pumping Head, 46-8
Total Float, 13-4
Total Maximum Daily Load, 51-7
Total Stopping Sight Distance, 36-4
Traffic Analysis, 35-8
Traffic and Noise, 8-7
Traffic Estimation, 6-7
Traffic Flow Rate, 6-5
Traffic Flow, Density, Headway, and Speed Relationships, 35-1
Traffic Forecast, 35-14
Traffic Sign Coloration, 41-2
Traffic Sign Placement, 41-1
Traffic Sign Shapes, 41-1
Traffic Volume, 6-5
Traffic Volume Abbreviations, 6-7

Transformed Section Properties, 28-12
Translational Failure, 3-15
Transmissivity, 49-5
Trapezoidal Method, 2-3, 8-3, 10-4, 11-2
Treated Water Storage, 52-5
Trench and Construction Safety, 19-3
Trench Excavations, 10-7
Trenching and Excavation, 2-4, 8-7, 16-3
Triangular (V-Notch) Weirs, 5-4, 45-11
Triangular Spoil Banks, 10-7
Triaxial Stress Test, 18-5
Tributary Area, 27-18
Tributary Areas, 27-17
Trip Generation and Traffic Impact Studies, 35-10
Trip Generation Curve, 35-10
Trip Generation Models, 35-10
Truss Basics, 4-3
Truss Deflection: Unit Load Method, 28-13
Trusses, 4-2
Turbulent Flow, 5-15
Turning Point, 37-2
Two-Way Slabs, 4-16
Types of Aquifers, 49-3
Types of Construction Equipment, 12-9
Types of Determinate Beams, 4-12
Types of Formwork, 34-3
Types of Indeterminate Beams, 4-12
Types of Interchanges, 38-3
Types of Retaining Wall Structures, 3-5, 4-18
Types of Wells, 12-8
Typical Consolidation Curve for Overconsolidated Soil, 3-7
Typical Soil Characteristics, 10-12
Typical Water Usage, 52-4

U

Ultimate Bearing Capacity, 12-12
Ultimate Pile Capacity, 12-11
Ultimate Strength, 27-2
Ultraviolet Light, 50-24
Unconfined Compressive Strength Test, 18-4
Underpinning, 24-12
Underpinning Overview, 24-12
Undisturbed Samples, 17-7
Undrained Shear Strength of Clays, 3-9
Unified Soil Classification, 17-9
Unified Soil Classification System, 10-12
Uniform Flow, 47-4
Uniform Gradient Future Worth, 9-2
Uniform Gradient Present Worth, 9-2
Uniform Gradient Uniform Series, 9-2
Uniform Series Compound Amount, 9-2
Uniform Series Present Worth, 9-2
Uniform Series Sinking Fund, 9-2
Uninterrupted Flow, 35-1
Uninterrupted Flow Pedestrian Facilities, 35-14
Unit Hydrograph, 5-8
Unit Hydrographs, 43-4, 48-6
Unit Weight, 7-5
USCS Soil Classification, 7-2
USDA Textural Classification, 17-9
USGS Regression Method, 48-5

V

Value Engineering, 11-5
Vehicle Loading, 27-8
Vehicular Live Loading, 27-9
Velocity Distribution, 47-4
Velocity Distribution in Open Channels, 5-4
Velocity Pressure, 27-5
Venturi Meters, 45-10
Vertical Clearance, 37-5
Vertical Curve Geometry, 37-1
Vertical Curves, 6-4
Vertical Curves Through Points, 6-4
Vertical Curves, Sag Curves, and Crest Curves, 37-1
Vertical Members, 29-6, 30-6, 31-5, 32-3

Vertical Strain Influence Factor Distributions, 3-13
Vertical Stress Contours for Continuous and Square Footings, 3-12
Vibrating-Wire Piezometers, 17-8
Vibration Monitoring, 19-6
Vibrocompaction, 21-4
Visual Impairments, 39-6
Void Ratio, 7-4
Volume-Capacity Ratio, 40-1

W

Wage Rates, 11-6
Walkway Slopes, 39-5
Wall Friction on Soil Wedges, 24-4
Walls, 15-4
Warning Signs, 41-2
Warrant 1: Eight-Hour Vehicular Volume, 40-4
Warrant 2: Four-Hour Vehicular Volume, 40-4
Warrant 3: Peak Hour, 40-5
Warrant 4: Pedestrian Volume, 40-5
Warrant 5: School Crossing, 40-5
Warrant 6: Coordinated Signal System, 40-5
Warrant 7: Crash Experience, 40-6
Warrant 8: Roadway Network, 40-6
Warrant 9: Intersection Near a Grade Crossing, 40-6
Wastewater Collection Systems, 50-1
Wastewater Flow, 50-5
Wastewater Flow Rates, 50-4
Wastewater Processing, 50-3
Wastewater Treatment Processes, 50-3
Water Balance Equation, 5-6, 48-2
Water Quality Mass Calculation, 51-12
Water Treatment Plants, 52-2
Water Treatment Processes, 52-2
Wave Equation Analysis, 26-10
Weaving and Maneuvering Sections, 35-16
Weaving Segments, 38-3
Weights of Fixed Service Equipment, 27-2
Weights of Materials and Constructions, 27-2
Weld and Bolt Installation, 14-2
Weld Designations, 14-3
Weld Symbols, 14-3
Welds, 30-10
Welds and Joints, 14-3
Well Analysis—Steady State, 49-5
Well Components, 49-7
Well Discharge, 12-7
Well Performance, 49-7
Wells, 22-4
Wet Wells, 50-2
Wind Analysis, 27-5
Wind and Seismic Loads, 27-17
Wind Exposure Categories, 27-5
Wind Force, 15-10
Wind Loads, 27-4
Wind Loads in Load Combinations, 27-4
Wind Resistance, 34-2
Wood Adjustment Factors, 31-2
Wood Beams, 31-3
Wood Combined Axial and Flexural Capacity, 31-7
Wood Compressive Capacity, 31-7
Wood Construction, 34-2
Wood Diaphragms, 31-3
Wood Fasteners, 31-8
Wood Flexural Strength, 31-4
Wood Posts, 31-5
Wood Shape and Species Properties, 31-1
Wood Shear Strength, 31-4
Wood Shear Walls, 31-6
Wood Stud Walls, 31-6
Wood Tensile Capacity, 31-6
Work Measurement and Productivity, 11-5
Work Zone, 16-5
Work Zone and Public Safety, 16-5

Y

Yellow Change Interval, 40-3

Z

Zero-Force Members, 4-4
Zone of Influence, 3-12